Cambridge Planetary Science Series
Editors: W. I. Axford, G. E. Hunt, R. Greeley

PHYSICS OF THE JOVIAN MAGNETOSPHERE

PHYSICS OF
THE JOVIAN MAGNETOSPHERE

Edited by A. J. DESSLER
Department of Space Physics and Astronomy
Rice University

CAMBRIDGE UNIVERSITY PRESS

Cambridge
London   New York   New Rochelle
Melbourne   Sydney

Published by the Press Syndicate of the University of Cambridge
The Pitt Building, Trumpington Street, Cambridge CB2 1RP
32 East 57th Street, New York, NY 10022, USA
296 Beaconsfield Parade, Middle Park, Melbourne 3206, Australia

First published 1983

Printed in the United States of America

Library of Congress Cataloging in Publication Data
Main entry under title:
Physics of the Jovian magnetosphere.
(Cambridge planetary science series)
Includes index.
1. Jupiter (Planet)–Atmosphere.
2. Magnetosphere. I. *Dessler, A. J.*
QB661.P47   538'.766'099925   81–21752
ISBN 0 521 24558 3   AACR2

CONTENTS

Contents   vii

TABLES

CONTRIBUTORS

Mario H. Acuña
Code 695
Goddard Space Flight Center
Greenbelt, MD 20771

J. K. Alexander
Code 695
Goddard Space Flight Center
Greenbelt, MD 20771

Sushil K. Atreya
Department of Atmospheric and
    Oceanic Science
Space Research Building
University of Michigan
Ann Arbor, MI 48109

Kenneth W. Behannon
Code 692
Goddard Space Flight Center
Greenbelt, MD 20771

J. W. Belcher
Room 37-695
Massachusetts Institute of Technology
Cambridge, MA 02139

Robert A. Brown*
Lunar and Planetary Laboratory
University of Arizona
Tucson, AZ 85721

T. D. Carr
Department of Astronomy
University of Florida
Gainesville, FL 32611

J. E. P. Connerney
Code 695
Goddard Space Flight Center
Greenbelt, MD 20771

M. D. Desch
Code 695
Goddard Space Flight Center
Greenbelt, MD 20771

A. J. Dessler†
Department of Space Physics and
    Astronomy
Rice University
Houston, TX 77251

C. K. Goertz
Department of Physics and Astronomy
University of Iowa
Iowa City, IA 52242

Melvyn L. Goldstein
Code 692
Goddard Space Flight Center
Greenbelt, MD 20771

D. A. Gurnett
Department of Physics and Astronomy
University of Iowa
Iowa City, IA 52242

T. W. Hill
Department of Space Physics and
    Astronomy
Rice University
Houston, TX 77251

S. M. Krimigis
Applied Physics Laboratory
Johns Hopkins University
Laurel, MD 20707

Carl B. Pilcher
Institute for Astronomy
2680 Woodlawn Dr.
Honolulu, HI 96822

---

* Present address: Space Telescope Science Institute, Homewood Campus, Johns Hopkins University, Baltimore, MD 21218

† Present address: Code ES01, Space Science Laboratory, NASA Marshall Space Flight Center, Huntsville, AL 35812

ix

E. C. Roelof
Applied Physics Laboratory
Johns Hopkins University
Laurel, MD 20707

F. L. Scarf
TRW Systems, Bldg. R1, Rm. 1176
One Space Park
Redondo Beach, CA 90278

A. W. Schardt
Code 660
Goddard Space Flight Center
Greenbelt, MD 20771

Darrell F. Strobel
Code 4780
Naval Research Laboratory
Washington, D.C. 20375

Richard Mansergh Thorne
Department of Atmospheric Sciences
University of California at Los Angeles
Los Angeles, CA 90024

James A. Van Allen
Department of Physics and Astronomy
University of Iowa
Iowa City, IA 52242

Vytenis M. Vasyliunas
Max-Planck-Institut für Aeronomie
D-3411 Katlenburg-Lindau 3
Federal Republic of Germany

FOREWORD

During the early 1960s the dominant emphasis of the space program of the United States was on manned space flight, looking toward landings on the moon and the detailed investigation thereof. Parallel with these activities, but at a much lower level of emphasis, was the development of a national program of planetary exploration. The nearby terrestrial planets Venus and Mars were the most readily accessible. Also, interest in search for extraterrestrial life on Mars provided a strong motivation for landing on its surface an elaborate device called an automated biological laboratory. The mission for accomplishing this was called Voyager, a name that was later changed to Viking. Still later, the name Voyager was adopted for an altogether different planetary mission.

The development of a national program of planetary exploration had many sources and many aspects. But to a very considerable extent, all of these aspects came into focus most clearly within the Space Science Board (SSB) of the National Academy of Sciences and more specifically within the National Aeronautics and Space Administration's Lunar and Planetary Missions Board (LPMB), created in early 1967 under the chairmanship of John W. Findlay of the National Radio Astronomy Observatory, with Homer E. Newell, John E. Naugle, Donald P. Hearth, Oran Nicks, and Robert Kraemer as the principal NASA participants. The minutes of the LPMB over the period 1967-70 reflect intensive and comprehensive consideration of every subsequently conducted lunar and planetary mission, as well as several that have not yet been conducted.

I was deeply involved in these considerations as a member of both the SSB and the LPMB and adopted as my special function the advocacy of missions to the outer gaseous planets – Jupiter, Saturn, Uranus, and Neptune. In response to this advocacy, Findlay appointed a Jupiter Panel, later expanded to the Panel on the Outer Planets, consisting of Von R. Eshleman, Thomas Gold, Donald Hunten, Guido Münch, James Warwick, Rupert Wildt, and myself, as chairman. One of the early documents of this panel reads in part as follows:

> From the standpoint of basic physical phenomena, Jupiter is perhaps the most interesting of the planets. On the basis of radio evidence, it has an immense radiation belt of relativistic electrons and a magnetic moment perhaps as great as 10^5 times that of the Earth. As a planetary object it is in a decidedly different class than are the Earth, Venus, Mars, and Mercury. Consequently in situ measurements in the vicinity of Jupiter should be made an early objective of the national planetary program.

To our great pleasure, NASA issued an invitation to the scientific community on June 10, 1968, for proposals for scientific investigations on planned Asteroid/Jupiter missions. The stated areas of interest were as follows:

1. Interplanetary magnetic and electric fields and interplanetary particles of solar and galactic origin out to large heliocentric radial distances.
2. Particulate matter in and beyond the asteroid belt.

xi

3. Particle and electromagnetic environment of the planet Jupiter.
4. Chemical and physical nature of the atmosphere of Jupiter and the dynamics thereof.
5. Thermal balance, composition, internal structure, and evolutionary history of Jupiter and its satellites.

This invitation elicited some 75 proposals by the December 2, 1968, deadline. Of these, 25 were ranked category 1 by the review committee. In early 1969, 11 instruments and 2 other investigations were selected, the missions were approved, and work got underway under the management of Charles F. Hall of the Ames Research Center of NASA. The two resulting missions – Pioneer 10, launched on March 3, 1972, and Pioneer 11, launched on April 6, 1973 – have been successful far beyond the expectations of their planners or even the participants. They have truly pioneered in making physical measurements in the outer solar system. Both spacecraft have successfully flown by Jupiter and have provided a greatly expanded knowledge of this planet and its satellites; Pioneer 11 continued onward to make the first exploration of Saturn in September 1979; and Pioneer 10, as of June 1982, is over 27 AU from the Sun. Both spacecraft continue to make unique observations of interplanetary phenomena at enormous heliocentric distances.

Parallel with the Pioneer 10/11 program, plans for follow-on missions to the outer planets were being formulated within the LPMB and elsewhere. One plan, called the Grand Tour of the Outer Planets, contemplated taking advantage of the uncommon configuration of Jupiter, Saturn, Uranus, and Neptune in the late 1970s and the 1980s by having a single spacecraft fly by all four of these planets. Although the Grand Tour as such was never approved, the plan was revived as a Jet Propulsion Laboratory program under the more modest title Mariner/Jupiter Saturn (MJS). I had the privilege of serving as chairman of the Science Working Group that developed the scientific rationale and general mission plan for MJS. The two spacecraft that emerged from these plans were, in due course, renamed Voyager 1 and Voyager 2, launched on September 5 and August 20, 1977, respectively.

By late 1981, each of the two Voyagers had flown through both the Jovian and Saturnian systems and had added a wealth of new knowledge of the physical properties of the planets themselves and their many satellites and rings. Voyager 2 is now targeted toward flybys of Uranus in January 1986 and Neptune in August 1989, thereby prospectively fulfilling the objectives of the Grand Tour, even though that term long ago became unmentionable in official circles.

The contents of this volume give ample testimony to the immense scientific achievements of these four outer planet missions. Each one of the hundreds, or perhaps thousands, of participants is entitled to an everlasting glow of pride in having had some part in their success.

James A. Van Allen

PREFACE

Why Jupiter? Is a book devoted solely to the magnetosphere of Jupiter too narrow, too specialized? With the present emphasis on solar-terrestrial relationships, why should we be studying other magnetospheres, and why Jupiter's? The primary reason is that Jupiter's magnetosphere is so unlike the Earth's in its fundamental workings. We study the Jovian magnetosphere because it is different. The difference challenges our understanding of magnetospheric physics. It leads us to a broader and more basic insight regarding both magnetospheric physics and the behavior of matter on a cosmic scale.

Jupiter is not an ordinary planet, nor does it have an ordinary magnetosphere. Although Jupiter's magnetosphere does most of the things Earth's does, it does them differently. For example, the Earth's magnetosphere extracts essentially all of its energy and some significant fraction of its plasma from the solar wind. In contrast, Jupiter's magnetosphere is powered by the slowing of Jupiter's spin, and nearly all of the magnetospheric plasma comes from internal sources – the satellite Io and the Jovian ionosphere. Jupiter also exhibits weak but genuine pulsar behavior. If we did not have the Earth's magnetosphere as a model, most theoretical work on the Jovian magnetosphere would probably be directed toward pulsar-type models.

The brief encounters of the two Pioneer and the two Voyager spacecraft with Jupiter have opened new frontiers of research in magnetospheric physics. Jupiter offers more than just another magnetosphere; it functions in a different mode and allows us to stretch our conceptions and develop better theories of the Earth's magnetosphere. The exciting promise of Jupiter's magnetosphere, lying within our solar system and accessible to us for direct measurement, is that it is also a link to distant astrophysical objects. The magnetosphere of Jupiter is wondrously complex; it seems to make room for nearly every idea that is proposed. Through in situ and ground-based measurements, we are developing a solid basis for extrapolation of space plasma physics to astrophysical objects. Such objects are so distant that all our information about them comes from various forms of remote sensing. Although some of our pre-Pioneer and pre-Voyager interpretations of ground-based (i.e., remote-sensing) data from Jupiter were correct (for example, synchrotron emission from a planetary magnetosphere as a source of decimetric radio emissions), the interpretations of other data were far off-track. This has shaken the confidence of many in our ability to understand the basic workings of a large, complex system by relying solely on information obtained by remote-sensing techniques. For Jupiter, the problem is alleviated because our interpretation of the remote-sensing data can now be guided by the results of the four planetary flybys.

The primary goal of this book is to provide a concise, authoritative distillation of the body of literature on Jupiter published thus far. Most of the original literature is easily accessible, but, as one would expect, there is a lot of it. One unusual feature is its distribution. Although many relevant papers are widely scattered, much of the work is concentrated in a few special issues of the leading journals. Specifically, the results of the four flybys and some attendant theory are described in special issues of *Science* (Vol. 183, No. 4122, 1974; Vol. 188, No. 4187, 1975; Vol. 204, No. 4396, 1979; Vol. 206,

No. 4421, 1979) and the *Journal of Geophysical Research* (Vol. 79, No. 25, 1974; Vol. 81, No. 19, 1976; Vol. 86, No. A10, 1981). There are also special issues of *Nature* (Vol. 280, No. 5725, 1979), *Icarus* (Vol. 27, No. 3, 1976; Vol. 44, No. 2, 1980), and *Geophysical Research Letters* (Vol. 7, No. 1, 1980). A book, *Jupiter* (edited by T. Gehrels, 1976), has been published that contains extensive discussions of pre-Voyager observations and theories. Finally, *The Satellites of Jupiter* (edited by David Morrison and published by the University of Arizona Press, 1982) is scheduled to appear contemporaneously with this book. Morrison's book should be considered a companion volume. It contains chapters that cover topics of direct interest to Jovian magnetospheric physics, such as Io's surface, volcanic emissions, atmosphere, and plasma torus. These topics are treated either in more detail or from a different viewpoint than the coverage in this book.

The task of providing an authoritative distillation of the literature on the Jovian magnetosphere is met by having a relatively small number of chapters written by authors who are, for the most part, willing to represent – or at least acknowledge the existence of – the work of distant colleagues. This book is an outgrowth of a meeting on Jovian magnetospheric physics held at Rice University in February 1980. The authors were invited, largely from among those who attended the meeting, on the basis of their expertise, breadth of interest, and writing ability. There were frequent communications among the authors to eliminate extensive duplication and to ensure that no significant topic would be inadequately covered. The result is a tightly interwoven book. The first eight chapters are largely descriptions of the experimental results and interpretations and conclusions derived therefrom. The final four chapters are devoted to the associated theoretical developments. Each chapter is reasonably self-contained, and they need not be read in any particular order. We suggest, however, that experimental results (first eight chapters) be read in conjunction with the appropriate theory (final four chapters). For example, Chapter 3, "The Low-Energy Plasma in the Jovian Magnetosphere," is closely tied to the theoretical developments in Chapter 11, "Plasma Distribution and Flow"; and Chapter 7, "Phenomenology of Magnetospheric Radio Emissions," should be read in conjunction with Chapter 9, "Theories of Radio Emissions and Plasma Waves." In some cases the reader may find it easier to read the theoretical chapter first. We definitely recommend some initial browsing.

Another goal is to make this book convenient for the reader and useful as a reference handbook. This is accomplished, in part, by the agreement in advance by the authors on a uniform system of symbol usage. Although there are stylistic differences from chapter to chapter, a degree of uniformity has been achieved. The principal symbols and acronyms used in this book, along with some brief definitions, are given in Appendix A; the coordinate systems used in the book are described in Appendix B, and a table of parameters for Jupiter and Io are in Appendix C. A list of tables follows the table of contents.

One of the delights in reading a book to which nearly two dozen authors have contributed is a revealing diversity of views and emphases. Jupiter's magnetosphere is complicated, and our understanding of its workings is primitive. Thus it should not be surprising, and certainly not alarming, to find some disagreement among chapters. Rather, such divergence is characteristic of an exciting field of research. There is much left to be done. If we are to understand Jupiter's magnetosphere and be able to generalize this understanding so that it can be applied to magnetized-plasma systems elsewhere in the universe, we must first argue among ourselves. Science knows no other way. The alert reader will spot such disagreement and might well use it as a basis for choosing a future research topic.

We are indebted to many for the production of this book: in particular, at Rice University, Dianne Drda †, editorial assistant, who organized and coordinated much of the diverse activity required to produce this work; Lorraine Dessler, who took over the task of editorial assistant during the final phases of coordinating and assembling this book; Georgette Burgess, who compiled the symbol and acronym list; and Jerry Mays, the best word processor operator I know. Tom Hill was a source of much good advice. The (relatively) low price of this book was made possible because of financial support from NASA Headquarters, specifically Robert Murphy and Henry Brinton of the Planetary Astronomy and Atmospheres Branch, and Erwin Schmerling and Michael Wiskerchen of the Upper Atmosphere and Magnetospheres Branch, who saw the value of this book and came to its aid while it was still in the conceptual stage. And finally, we wish to thank the staff of Composition Resources and of Cambridge University Press whose smooth professionalism solved many problems that might have otherwise become serious difficulties. The assistance of these people, of the anonymous referees, and of many others is implicit in this volume.

A. J. Dessler

† Dianne D. Drda (1943–1982)

During the past 17 years, most of the space physics community has had occasion, either directly or indirectly, to come in contact with Dianne Drda and her work. Starting late in 1965, she was copy editor for the Space Physics section of the *Journal of Geophysical Research*. She served as editorial assistant for *Reviews of Geophysics and Space Physics* from 1970 through 1974. At the time of her sudden and unexpected death from a cerebral aneurysm, she was working long hours on the production of this book. She worked with efficiency and with her typical cheerful demeanor both on this activity and as administrative secretary for the Department of Space Physics and Astronomy. The completion of this book, which she regarded with enthusiasm as a major reference work on Jovian magnetospheric physics, was important to her; she came within weeks of seeing it "in press."

As copy editor and as editorial assistant, she helped the community of space physicists in many ways. Her hallmarks were efficient professionalism in her craft and an infectious cheerfulness that brightened the day of those who talked with her. We shall miss her.

I

JUPITER'S MAGNETIC FIELD AND MAGNETOSPHERE

Mario H. Acuña, Kenneth W. Behannon, and J. E. P. Connerney

Jupiter has a magnetic moment second only to that of the Sun's. Estimates of dipole magnitude range from 4.208–4.28 G·R_J^3. The dipole is found to be tilted ~9.6° to the rotation axis toward ~202° System III (1965) longitude in the northern hemisphere. Jupiter also possesses substantial quadrupole and octupole moments, which implies that the interior sources of the field lie nearer the surface than in the case of Earth. The internally-produced field dominates the inner magnetosphere out to a distance of ~6 R_J, the orbital distance of Io, where measurement of perturbation magnetic fields yields an estimated 2.8 × 10^6 A for the current flowing in the Alfvén current tube linking Io with the Jovian ionosphere. The middle magnetosphere of Jupiter, extending from ~6 R_J to 30–50 R_J, is dominated by an equatorial azimuthal current disc. When the disc is modeled as a 5 R_J thick annular current sheet extending from 5 to 50 R_J, a fit to spacecraft data requires a current density of ~5 × 10^6 A/R_J^2 at the inner edge of the annulus. An asymmetric longitudinal variation in the radial component of the magnetic field in this region provides evidence for a thickening of the magnetodisc on the dayside. Between the outer edge of the disc and the sunward magnetopause, which can vary in distance from ~45 R_J to ~100 R_J there is a "buffer" zone which is very sensitive to variations to external pressure. The field there is southward on average, but highly variable in both magnitude and direction on timescales less than one hour, especially near the magnetopause. On the nightside of the planet, a magnetic tail of radius ~150–200 R_J and length of probably a few AU is formed in the interaction of the solar wind with the Jovian magnetosphere. The ~5 R_J thick current sheet separating the tail lobes is a night side extension of the magnetodisc, is on average parallel to the ecliptic plane and oscillates about the longitudinal axis of the tail as Jupiter rotates. In the dawn and tailward magnetosheath, the magnetic field has a predominant north–south orientation and varies between the two extremes with a quasiperiod of ~10 hr. There is strong correlation with the north–south plasma velocity components in those regions, suggesting convected wave production by the wobble of the flattened Jovian magnetosphere and current sheet system.

1.1. Introduction

Among the planets in the solar system, Jupiter is unique not only because of its immense size and mass and the variety of phenomena taking place in its environment, but also because of its large magnetic moment, second only to the Sun's. The Jovian magnetic field was first detected indirectly by radio astronomers who postulated its existence to explain observations of nonthermal radio emissions from Jupiter at decimetric and decametric wavelengths [see Chapter reviews and Carr and Gulkis, 1969; Carr and Desch, 1976; Berge and Gulkis, 1976; also de Pater, 1980b]. Some of the basic characteristics of the field were derived from these observations, such as its southward polarity, opposite Earth's, and the approximate value of the tilt angle between the magnetic dipole axis and the axis of rotation of the planet, determined to be ~10°. Coarse estimates were established for the surface field intensity and dipole moment based on trapped radiation models and the observed cutoff of decametric emissions at 40 MHz [Komesaroff, Morris, and Roberts, 1970; Carr, 1972a].

Since these early radio astronomical studies of the Jovian magnetosphere, four spacecraft, Pioneer 10 (1973), Pioneer 11 (1974), Voyager 1, and Voyager 2 (both

1979), have flown by the planet at close distances and have provided in situ information about the field geometry and its strength. The direct measurements confirmed the
zeroth order models derived from the radio data and added a vast amount of new and
detailed information about the global characteristics of the field, its dynamics and
interaction with the solar wind. For detailed reviews of early results derived from
ground based and the Pioneer 10 and 11 spacecraft observations the reader is directed
to the book *Jupiter*, edited by Gehrels [1976], and to Smith and Gulkis [1979], Davis
and Smith [1976], Acuña and Ness [1976a,b,c], and Carr and Gulkis [1969] and
references therein.

  As in the case of Earth, the magnetic field stands off the solar wind at a considerable
distance from the planet, creating a giant magnetospheric cavity. If it were visible, the
Jovian magnetosphere would appear from Earth to be the largest object in the sky.
Viewed head on, its width would be about four times as large as the Moon or Sun. The
boundary of this cavity comprises a detached bow shock wave, generated by the super-
Alfvénic flow of the solar wind past the magnetized obstacle, a magnetosheath layer in
which the deflected solar wind creates a turbulent flow regime, and a magnetopause
"surface" at which internal magnetospheric pressure balances that associated with the
impinging solar wind.

  It has become customary to describe the Jovian magnetosphere in terms of three
principal regions. The *inner magnetosphere* is the region where the magnetic field created by sources internal to the planet dominates, and contributions from current systems external to the planet are not significant. This region was initially defined as
extending from the planetary surface to a distance of $10-15 \, R_J$ (Jovian radii), but
recent results from the Voyager flybys indicate that the outer radius should be reduced
to approximately $6 \, R_J$, the orbit of Io. Outside of this region the effects of an azimuthal
current sheet in the equatorial plane produce a significant perturbation, leading to the
stretching of the magnetic field lines in the radial direction. The region in which the
equatorial currents flow will be denoted as the *middle magnetosphere*, and in our
definition, would extend from $\sim 6 \, R_J$ to approximately $30-50 \, R_J$ where the asymmetry due to the magnetopause and tail current systems becomes important. In the
outer magnetosphere, the field has a large southward component and exhibits large
temporal and/or spatial variations in magnitude and direction in response to changes in
solar wind pressure. This region extends from the magnetopause boundary to
approximately $30-50 \, R_J$, and includes as well the extensive Jovian magnetic tail.

  In this chapter we shall present a description of the magnetic field in these three
regions from the phenomenological point of view derived from spacecraft observations
(predominantly the recent Voyager results) and the quantitative models which have
been developed to date on the basis of these observations.

1.2. The inner magnetosphere

General description and mathematical models

As previously defined, the inner magnetosphere is that region where the magnetic
field generated by currents flowing in the interior of the planet dominates. The timescale for field variations, both in magnitude and direction, is long, and the field is
smoothly varying when observed from a spacecraft in a flyby trajectory. It is generally
accepted that the magnetic-field-generation mechanism is a thermal-convection-driven
dynamo operating in the electrically conducting regions of Jupiter's interior [Hide and
Stannard, 1976; Gubbins, 1974; Busse, 1979]. Thus knowledge about the magnetic
field provides means for studying Jupiter's interior [Smoluchowsky, 1975; Stevenson,

1974]. Beyond $6\,R_J$ the effects of external currents become too significant to be neglected; hence, we have chosen this radius as the outer boundary.

Because the region of interest contains no significant sources (i.e., $\mathbf{J} = \nabla \times \mathbf{B} \simeq 0$), the magnetic field $\mathbf{B}$ can, to good approximation, be derived from a scalar potential function V which represents the contribution of sources internal and external to the region. Thus, we have

$$\mathbf{B} = -\nabla V = -\nabla(V^e + V^i) \tag{1.1}$$

The potential function V is the sum of an external potential V^e and an internal potential V^i and is generally expressed in terms of spherical harmonic functions, that is, solutions to Laplace's equation of the form [Chapman and Bartels, 1940]

$$V = V^e + V^i = a \sum_{n=1}^{\infty} \left(\frac{r}{a}\right)^n T_n^e + \left(\frac{a}{r}\right)^{n+1} T_n^i \tag{1.2}$$

This is the classical formulation for the potential originated by Gauss in 1830 and used in studies of the Earth's magnetic field [Gauss, 1877]. In our case, r denotes the distance from Jupiter's center and a is Jupiter's equatorial radius, (1 R_J = 71,372 km). (A general discussion of the representation of magnetic fields in space has been given by Stern [1976].)

The functions T_n^e and T_n^i are given with respect to the coordinate system defined below, as:

$$T_n^e = \sum_{m=0}^{n} P_n^m(\cos\theta)\,[G_n^m \cos m\phi + H_n^m \sin m\phi] \tag{1.3}$$

for the contribution due to exterior sources and

$$T_n^i = \sum_{m=0}^{n} P_n^m(\cos\theta)\,[g_n^m \cos m\phi + h_n^m \sin m\phi] \tag{1.4}$$

for the internal sources.

The angles ϕ and θ are the conventional right-handed polar coordinates and therefore denote Jovigraphic System III (1965.0) *east* longitude and colatitude respectively, $P_n^m(\theta)$ are associated Legendre functions with Schmidt normalization and g_n^m, h_n^m, G_n^m, H_n^m are the Schmidt coefficients. In some studies the external potential V^e has been expressed in an inertial (nonrotating) coordinate system [see Smith et al., 1975].

If we assume a magnetic field representation of the form (1.1), the development of a magnetic field model from a set of M observations along a spacecraft trajectory requires the determination of the coefficients g_n^m, h_n^m, G_n^m, and H_n^m up to order $n = n_{max}$ ($n_{max} < \infty$) that minimize the sum of the vector residuals squared

$$X^2 = \sum_{k=1}^{M} |\mathbf{B}_{model}^k - \mathbf{B}_{obs}^k|^2 = \sum_{k=1}^{M} |\epsilon_k|^2 \tag{1.5}$$

The order of the model, $n = n_{max}$, represents the physical complexity associated with the mathematical formulation. From potential theory it is known that a unique

Table 1.1. *Spherical harmonic coefficients for models of the Jovian magnetic field (System III, 1965)*

Terms		O_4 (Acuña and Ness, 1976)	P 11 (3,2) A, Davis et al. (1975)
Internal:	g_1^0	4.218	4.144
	g_1^1	−0.664	−0.692
	h_1^1	0.264	0.235
	g_2^0	−0.203	0.036
	g_2^1	−0.735	−0.581
	h_2^1	−0.469	−0.427
	g_2^2	0.513	0.442
	h_2^2	0.088	0.134
	g_3^0	−0.233	−0.047
	g_3^1	−0.076	−0.502
	h_3^1	−0.580	−0.342
	g_3^2	0.168	0.352
	h_3^2	0.487	0.296
	g_3^3	−0.231	−0.136
	h_3^3	−0.294	−0.041
External:	G_1^0		−194.6 (nT)
	G_1^1		68.7
	H_1^1		80.3
	G_2^0		5.6
	G_2^1		9.1
	H_2^1		− 9.4
	G_2^2		− 16.2
	H_2^2		− 8.3

representation of the magnetic field in a source-free region is derivable from vector measurements over a simple surface that completely encloses the internal sources. Because a spacecraft trajectory only constitutes a single curve in space, the solutions obtained are not unique, and cross coupling or mutual dependence among the Schmidt coefficients occurs.

From the above, it is clear that trajectories that maximize the range of latitudes and longitudes covered within the radial distances considered lead to more complete representations of the planetary field. The characteristics of a given trajectory can be used to determine the order of the model n_{max} that can be used to model the observations with acceptable physical credibility. A model of the planetary field incorporating

Table 1.2. *Characteristics of dipole terms for models of the Jovian magnetic field*[a]

Model	$|M|$	Tilt	λ_{III} (1965)
O$_4$	4.28	9.6°	201.7°
P 11(3,2)A	4.208	10°	198.8°

[a] For the characteristics of dipole terms derived from other models of the field the reader is referred to *Smith and Gulkis* [1979].

terms of order higher than n_{max} may fit the data exceedingly well along the spacecraft trajectory, but not elsewhere. Hence, the predictive properties and basic purpose of the model are rendered useless. Detailed discussions on the selection of n_{max} for given spacecraft trajectories and the physical validity of higher order models have been given by Acuña and Ness [1976b], Davis and Smith [1976], and Connerney [1981a]. The latter author in particular has applied generalized inverse techniques to the least-squares problem represented by Equation (1.5), which yield direct estimates of confidence levels associated with the Schmidt coefficients for a given spacecraft trajectory.

In the case of Jupiter, the best trajectory for planetary magnetic field studies was that provided by Pioneer 11. This was a high latitude retrograde flyby with a closest Jovicentric approach distance of 1.6 R_J. Although Pioneer 10 and Voyager 1 approached the planet to within 2.85 and 4.9 R_J respectively, their prograde trajectories remained close to the equatorial plane; Voyager 2 only came to within 10 R_J of Jupiter and thus did not contribute significant data for studies of the planetary field. However, these latter three spacecraft provided a wealth of information about the middle magnetosphere and hence, indirectly, about the radial distance range over which the spherical harmonic representation is valid [Davis and Smith, 1976; Ness et al., 1979a,b,c].

On the basis of the Pioneer 10 and 11 measurements [Smith et al., 1975; Acuña and Ness, 1976a,b,c], several models of the Jovian magnetic field were developed. Each of these reflects not only the many approaches that are possible in obtaining estimates for the spherical harmonic coefficients, but they also point to slight measurement differences in the respective observations [Acuña and Ness, 1976c; Davis and Smith, 1976; Connerney, 1981a]. Two principal models have emerged as "best" representing the planetary field in terms of spherical harmonics: the O$_4$ model of Acuña and Ness [1976a,c] and the P 11(3,2)A model of Davis, Jones, and Smith [1975] (see also Smith, Davis, and Jones [1976]). They have been tested against independent evidence from ground based radio observations and in situ energetic-charged-particle measurements. The latter provide information about the global characteristics of the field not obtainable from the local magnetic field measurements and hence lend physical credibility to the models [Acuña and Ness, 1976a,c; Smith and Gulkis, 1979]. Table 1.1 lists the Schmidt coefficients for both models. The O$_4$ model includes terms up to the octupole ($n_{max} = 3$) but no external terms ($G_n^m = H_n^m \equiv 0$). The P 11(3,2)A model also includes internal terms up to the octupole but adds external terms which include quadrupole coefficients ($n_{max}^{ext} = 2$). These coefficients have been transformed to the System III (1965) longitude system values. The characteristics of the dipole terms are given in Table 1.2.

The global characteristics of the field predicted by these models are illustrated in Figures 1.1 through 1.3 where the total field intensity, dip, and declination angles are shown in the form of isocontour maps at the surface of the planet and at 2 Jovian radii.

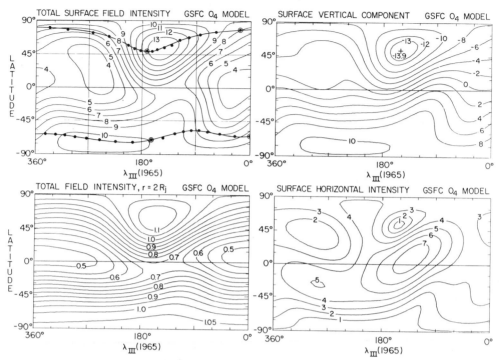

Fig. 1.1. Isocontour maps of total field intensity as predicted by the O_4 model at the surface and at 2 Jovian radii. The vertical and horizontal field intensities at the surface are also shown. Note that the dynamic flattening of 1/15.4 has been incorporated into the surface maps. Also shown in upper left panel are traces of the feet of the field line (flux tube) through Io (filled circles on solid curves).

The dip and declination angles are defined in the usual sense for Earth's case: dip = $\tan^{-1} Z/H$, where Z and H denote the local vertical (downward) and horizontal field components, respectively, and declination = $\tan^{-1} Y/X$, where X is the local field component directed towards the north and Y the local field component directed towards the east. Note that the surface maps take into account Jupiter's dynamic flattening, $f = 1/15.4$; those computed at 2 R_J assume a spherical surface. From the coefficients given in Table 1.1 and the maps shown in Figures 1.1 through 1.3 it is clear that Jupiter possesses substantial quadrupole and octupole moments which lead to a relatively complex field morphology extending to significant distances from the planet although the field is largely dipolar for distances as close as 2 R_J. On the surface the maximum field strength at the poles is asymmetrical, ~ 14 and 10.4 G in the north and south polar regions, respectively. The relatively large high-order terms also imply that the sources of the magnetic field, presumably in Jupiter's core, lie closer to the surface than in the case of the Earth.

A new estimate of Jupiter's planetary magnetic field has been obtained from the Voyager 1 observations. Connerney, Acuña and Ness [1982] combined an explicit model of the magnetodisc current system with a spherical harmonic model of the planetary field and obtained most of the parameters of an octupole internal field model. The resulting model fits the observations extremely well (Fig. 1.4) throughout the analysis interval ($r < 20 R_J$) and is very similar to the octupole Pioneer 11 models. Comparing dipole parameters of the epoch 1979.2 Voyager 1 model with the epoch 1974.9 O_4 model, Connerney, Acuña, and Ness find no statistically significant change in Jupiter's dipole moment, tilt angle or longitude.

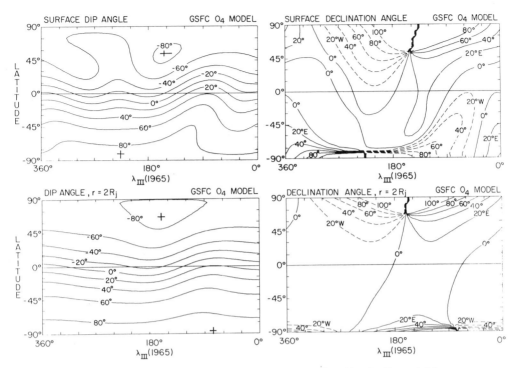

Fig. 1.2. Isocontour maps of declination and dip angles as predicted by the O_4 model for the surface and 2 Jovian radii. See text for the definition of these angles. A dynamic flattening of 1/15.4 has been incorporated into the surface maps.

Offset tilted dipole

A simplified model for the field can be obtained by fitting the data to a noncentered dipole. Alternatively, an offset, tilted dipole (OTD) representation can be derived from the quadrupole terms in the form given by Bartels [1936]. These offset, tilted dipole models of the Jovian field find application in problems that require a simple magnetic field model for computational ease and are not concerned with the detailed morphology of the field near the surface. A typical widely used offset, tilted dipole model is the D_4 model of Smith, Davis, and Jones [1976]; its characteristics are given in Table 1.3.

Table 1.3. *Characteristics of offset tilted dipole (OTD) models of the Jovian magnetic field*

Model	$|M|$	Tilt	λ_{III} (1965)	Offset (R_J)	δ^b	μ(1965)
D_4	4.225	10.8°	200.8°	0.101	5.1	155.6°
OTD[a]	4.35	9.5°	208.8°	0.068	−12.6°	174.2°
O_4	4.28	9.6°	201.7°	0.131	− 8.0°	148.57°
P 11(3,2)A	4.208	10°	198.8°	0.108	4.8°	143.07°

[a] Acuña and Ness [1976c].
[b] δ and μ are the latitude and longitude of the offset vector in System III (1965).

Fig. 1.3. Surface total field intensity, dip, and declination angles as predicted by the P 11(3.2)*A* model. See text for definition of the angles.

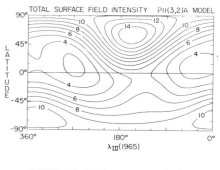

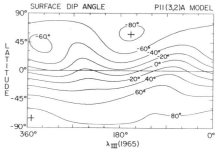

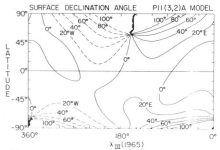

Also in the table are given the equivalent OTD models derived from the O_4 model and the P 11(3,2)*A*.

It can be argued that the OTD models represent the simplest morphological structure (dipole) associated with the planetary field. The degree of complexity is given by a measure of how far the real field deviates from this zeroth order representation. This has been illustrated in Figure 1.5 where we have plotted isointensity contours on the surface of the planet for the difference field between the O_4 and the OTD models of Acuña and Ness [1976c]. A strong quadrupolar component remains evident in this figure, indicative of a field complexity at the surface which cannot be adequately represented by OTD models.

An important point that must be kept in mind when using analytical models of planetary magnetic fields is their nonuniqueness. Magnetic field observations acquired along a flyby trajectory, for example, are insensitive to certain linear combination of parameters. This has been illustrated by the "invisible planet" constructed by Connerney [1981a] in connection with the Pioneer 11 flyby trajectory at Jupiter. This figure is reproduced in Figure 1.6 showing surface isocontour maps of Jovian model magnetic fields that would not have been detected by the magnetometers on Pioneer 11, assuming (Fig. 1.6*a*) a uniform noise distribution and (Fig. 1.6*b*) observation noise proportional to the local field magnitude.

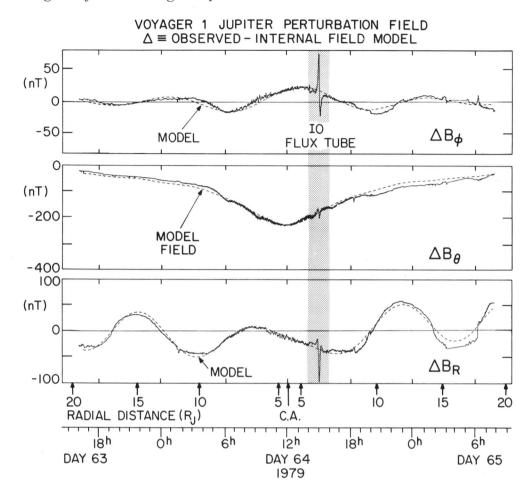

Fig. 1.4. Comparison of modeled magnetic field (dashed) with that observed for Voyager 1 (spherical coordinates are used). In this presentation the model internal field has been subtracted from the observations; the remaining model field (dashed) is due to Jupiter's magnetodisc currents. The total field at closest approach (C.A.) is ~ 3330 nT (from Connerney, Acuña, and Ness, 1982).

Magnetic field geometry relevant to energetic particle trapping

Energetic particles trapped in the Jovian magnetosphere serve as excellent tracers of the magnetic field morphology, since in their bounce and drift motions around the planet they trace out the magnetic field lines over a large region of space. Thus, energetic particle measurements aboard spacecraft play an important role in establishing the global characteristics of the field (see for example Van Allen et al. [1974a]), and in providing independent evidence against which the validity of the magnetic field models can be tested. The satellites in the Jovian system act to remove trapped energetic particles by absorption into their surfaces [Mead and Hess, 1973]. The observed absorption signatures and their relative locations with respect to the planet can be used effectively to infer characteristics of the magnetic field [Acuña and Ness, 1976c; Van Allen, 1976;

Fig. 1.5. Isocontour map of the differential in total surface field intensity computed by subtracting the OTD model of Acuña and Ness from the O_4 model. The large remaining quadrupolar and higher order contributions to the field cannot be adequately modeled by an offset, tilted dipole.

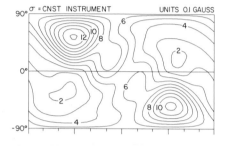

Fig. 1.6. "Invisible planet": Surface isointensity contour map of a Jovian model magnetic field that would not have been detected by a magnetometer on Pioneer 11: (a) (upper panel) assuming observations with a .005 G random noise component; (b) (lower panel) with a noise component proportional to the local field magnitude (see text). A dynamical flattening of 1/15.4 is assumed in the determination of the surface equipotential (from Connerney [1981a]).

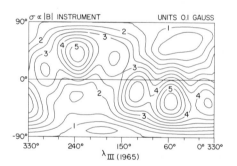

Simpson and McKibben, 1976; Fillius, 1976]. Conversely, the magnetic field models in conjunction with observed energetic charged particle absorption effects can be used to infer the presence and properties of undetected absorbers such as unknown rings and satellites [Van Allen et al., 1980; Acuña and Ness, 1976c; Fillius, 1976].

It is important therefore to determine the adiabatic particle parameters for particle detectors aboard a given spacecraft and trajectory. For example, L-shells and equatorial pitch angles for charged particles incident from a given direction constitute extremely useful quantities. (For a detailed discussion of adiabatic particle motion and parameters in a planetary magnetic field the reader is directed to Roederer [1970]). An example of this type of calculation using the O_4 model is given in Figure 1.7 for the Pioneer 11 spacecraft encounter trajectory and for particles with pitch angles equal to 90° at the spacecraft. Also shown in the figure are the range of L-shells swept by Io and Amalthea, as predicted by several models, where absorption effects should be observed [Van Allen et al, 1974a; Fillius, Mogro-Campero, and McIlwain, 1975]. Note that the range of L-shells swept by these two satellites as they move around in their orbits is

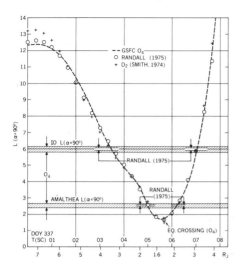

Fig. 1.7. McIlwain "L" parameter values computed for charged particles with 90° pitch angle at the spacecraft. The trajectory is that of Pioneer 11 at Jupiter. The shaded regions indicate the range of L-shells swept by the Jovian satellites Io and Amalthea, where particle absorption effects should be observable. See Van Allen [1976] for a description of the Randall (1975) model.

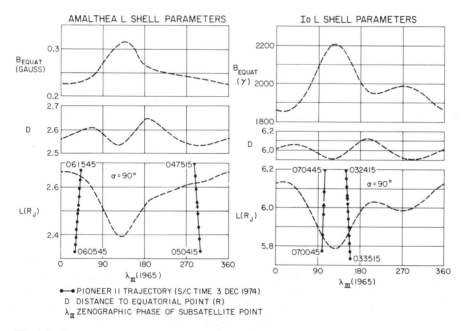

Fig. 1.8. L-parameter values computed for the "trajectories" (orbits) of Io and Amalthea around Jupiter. Note the large range of L-values which the satellites traverse around their orbits. The indicated longitude is System III (1965).

much broader for the O_4 and P 11(3,2)A models than predicted by dipole models [Acuña and Ness, 1976c], as illustrated in Figure 1.8. This effect becomes pronounced for satellites in the middle magnetosphere, such as Ganymede, where the field lines are stretched out radially by external currents. These effects are discussed in the next section.

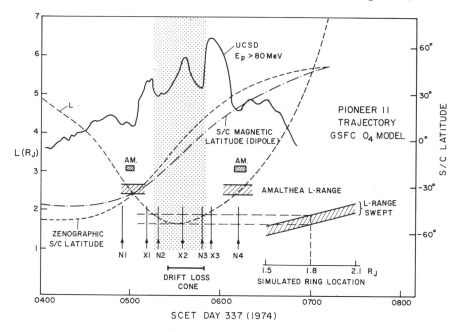

Fig. 1.9. Pioneer 11 observations of the Jovian ring. The charged particle data is that of Fillius [1976]. The *L*-values along the trajectory of Pioneer 11 and Amalthea were calculated using the O₄ model. A narrow ring located at 1.8 *R_J* provides an excellent fit to the observed absorption features N2 and N3.

One of the most interesting applications of adiabatic particle motion principles and satellite sweeping effects was the anticipation of the existence of a ring of particles around Jupiter at 1.8 *R_J* [Acuña and Ness, 1976c; Fillius, 1976; McLaughlin, 1980]. This is illustrated in Figure 1.9 where the Pioneer 11 spacecraft trajectory near closest approach in *L*-space and magnetic latitude is shown. Also shown are the energetic particle data of Fillius, Mogro-Campero, and McIlwain [1975] illustrating the existence of a multiply peaked structure in the count rate near closest approach. The two outer minima in the curve are due to absorption of energetic particles by Amalthea and correspond very closely to Amalthea's *L*-shell. The two inner minima are adequately explained by assuming the existence of a narrow absorber at 1.8 *R_J*, as indicated in the figure. This narrow ring was later observed by the Voyager 1 imaging experiment [Smith et al., 1979b]. The interpretation of the two minima being due to the presence of an absorber (ring or satellite) was considered remote since planetary rings had only been observed at Saturn [McLaughlin, 1980]. An analysis of the idealized response of energetic charged particle detectors aboard P 11 in an ambient magnetic field given by the O₄ model by Roederer, Acuña, and Ness [1977] indicated that the absorption signature could also be associated with geometric effects due to the intersection of the drift loss cone [Roederer, 1970] with the angular response of the detectors. This effect is indeed operative close to the planet but absorption by the ring material is the dominant loss mechanism.

It is also instructive to compute the configuration of the intersections of particle drift shells with the planetary ionosphere in order to locate areas where the mirror points for the trapped particles are low enough in altitude to interact strongly with the atmosphere–ionosphere. Constant *L* contours computed for the O₄ model are shown in

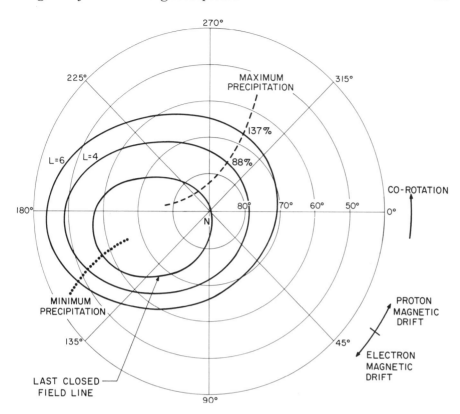

Fig. 1.10. L = constant drift shell intersections with the planetary ionosphere in the northern hemisphere of Jupiter. Dashed and dotted lines represent zones of maximum and minimum precipitation flux. The percentage values represent the relative variation of the precipitation function between minimum and maximum along a given L = const. contour for each hemisphere. [Roederer, Acuña, and Ness, 1977; Connerney, Acuña and Ness, 1981].

Figure 1.10 and 1.11 in Jovigraphic coordinates for the northern and southern hemisphere, respectively [Roederer, Acuña, and Ness, 1977]. Also included in this figure is the contour of last closed field lines as predicted by the model of Connerney, Acuña and Ness [1981] which is discussed in the next section. What is important in Figures 1.10 and 1.11 is the expected location of maximum and minimum particle precipitation flux along a given contour because these regions, in principle, could play an important role in the generation of decametric emissions from Jupiter and their control by Io [Dessler and Hill, 1979 and references therein]. The discovery and observations of the Io plasma torus [Kupo, Mekler, and Evitar, 1976; Broadfoot et al., 1979] have changed significantly this point of view, and although the precipitating particles are probably still responsible for the Io-independent emissions, their role is not completely understood. An attempt to correlate the precipitation regions with the decametric source regions as a function of central meridian longitude has been carried out by Roederer, Acuña, and Ness [1977] (see also Chap. 7).

Although our understanding of the emission mechanisms and source geometry is incomplete, the magnetic field models provide further insight into the decametric and decimetric emission phenomena. Synchrotron emission by relativistic electrons mirror-

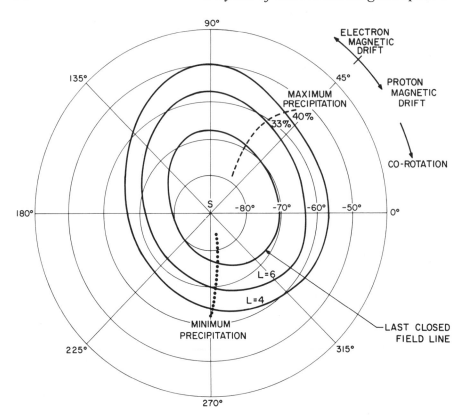

Fig. 1.11. Same as Figure 1.10, but for the southern hemisphere of Jupiter.

ing near the magnetic equatorial plane is the source of decimetric radiation. Acuña and Ness [1976a,b] first used the O_4 model to compute the charged particle equator at different radial distances from the planet, and the results of this calculation are shown in Figure 1.12 for $r = 1, 2, 4,$ and $6\ R_J$. The curves clearly illustrate the nonsinusoidal character of the magnetic equator or, equivalently, the "warping" of the decimetric emission region away from a simple flat disc. This explains the observed polarization characteristics at Earth [Carr and Gulkis, 1969; Berge and Gulkis, 1976; see also Chap. 3]. These preliminary studies have been extended by Smith and Gulkis [1979] and particularly by de Pater [1980b] to include the P 11(3,2)A model and realistic pitch angle distribution functions for the electrons. These studies confirm the good agreement between the decimetric observations and the analytical models of the magnetic field.

The Io flux tube

Theoretical explanations of the enigmatic control of Jupiter's decametric emissions by the satellite Io are based on a strong electrodynamic interaction with the corotating Jovian magnetosphere, leading to field-aligned (Birkeland) currents connecting Io with the Jovian ionosphere [Piddington and Drake, 1968; Goldreich and Lynden-Bell, 1969; Carr and Desch, 1976]. The interaction model is illustrated in Figure 1.13. The corotational electric field developed by the relative motion between Io and the magnetic field lines swept past the satellite (~ 57

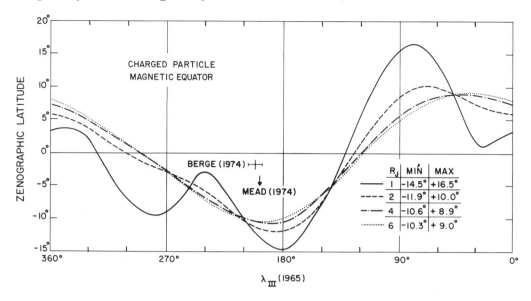

Fig. 1.12. Energetic charged particle magnetic equator ($|B|$ = min. along field line) computed from the O_4 model at several radial distances. Notice the deviation from a simple sinusoid predicted by the model, which is in good agreement with decimetric radio observations [Smith and Gulkis, 1979; de Pater, 1980b].

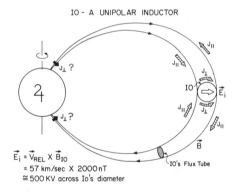

Fig. 1.13. Io as a unipolar inductor. Schematic representation of the induced current system flowing approximately parallel to the field lines in the flux tube linking Io to the Jovian ionosphere/atmosphere. The circuit is closed at the extremes by currents flowing in the Jovian ionosphere and in the ionosphere and/or interior of Io. A power of approximately 10^{12} W is dissipated in the system.

km/s) induces a potential difference of approximately 500 kV between the outer and inner faces of this satellite. This potential causes currents to flow from Io toward the Jovian ionosphere, both northward and southward, along the outer portion of the magnetic flux tube linking the satellite to the ionosphere. Return currents flow along the inner portions of the flux tube towards Io. In this simple model, the circuit is closed through the Jovian ionosphere, Ionian ionosphere and/or the interior of the satellite. The term "unipolar inductor" is applied to this type of interaction between a satellite and a planetary magnetosphere. The presence of the Io torus (i.e., a dense plasma) introduces additional plasma-inertial effects, and hence the simple model described above has to be expanded to include these effects leading to the concept of an "Alfvén current tube" as explained further below. Direct measurements of the perturbation magnetic fields due to the Io induced current system were obtained by the magnetic field experiment on Voyager 1 on 5 March 1979 when it passed within 20,500 km south of Io [Ness et al., 1979a; Acuña, Neubauer, and Ness, 1981].

Fig. 1.14. Perturbation magnetic field associated with the Io current system as detected by Voyager 1. In addition to the three orthogonal components in spherical coordinates, the angle between the background field and perturbation has been computed and is shown in the upper panel. The abscissa is the decimal day (SCET).

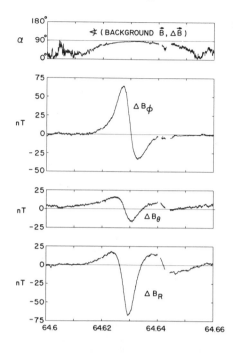

The Voyager 1 spacecraft encounter trajectory had been designed specifically to provide a passage at close range to the satellite through the south magnetic flux tube depicted in Figure 1.13. The geometrical position of the flux tube had been calculated on the basis of analytical models of the field without taking into account possible distortions introduced by the system of currents. During the time interval associated with the anticipated Io flux tube passage, the magnetic field experiment detected significant perturbations superimposed on the much larger background planetary field. The observations are shown in Figure 1.14 where the differences between a local model background field and the measurements are illustrated in Jupiter centered, spherical polar coordinates. The top panel in the figure shows that the angle between the perturbation field and the background field approached 90° during this interval, indicating the transverse nature (i.e., due to field aligned currents) of the perturbation. A preliminary report and analysis of quick look data was given by Ness et al. [1979a], which did not include the plasma inertial effects introduced by the presence of the Io plasma torus [Broadfoot et al., 1979; Bridge et al., 1979a]. The presence of the plasma leads to a "slowing down" of the magnetic field lines being swept past the satellite or, equivalently, to the generation of an Alfvén wave that propagates away from Io approximately parallel to the ambient field. The plasma density in the torus [Warwick et al., 1979a; Bridge et al., 1979a; Bagenal, Sullivan, and Siscoe, 1980] is such that plasma flow past Io is sub-Alfvénic, and hence no shock forms in front of the object. The interaction geometry is illustrated in Figure 1.15 for an Alfvén Mach number of 0.25 and a sonic Mach number of 1.

The general description of this type of interaction in terms of Alfvén waves propagating away from a satellite was first given by Drell, Foley, and Ruderman [1965]. Neubauer [1980] extended Drell's analysis to the nonlinear case and described Io's interaction with the Jovian magnetosphere as a nonlinear standing Alfvén current system (see also Goertz [1980a]). A detailed analysis of the Voyager 1 observations in terms of this Alfvén current system has been given by Acuña, Neubauer, and Ness

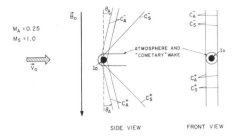

Fig. 1.15. Example of linear Alfvén and slow wave characteristics for $M_A = 0.25$ and sonic Mach number $M_s = 1$. Strong Alfvénic disturbances are expected to issue from Io and its vicinity along the appropriate tube of Alfvén characteristics [from Neubauer, 1980].

[1981] and leads to an estimate of 2.8×10^6 A for the currents flowing in the Alfvén current tube. This value for the current is strongly dependent upon the geometry assumed for the current tube due to the nonuniqueness of the model assumed to represent the observations. The smallest total current consistent with the Voyager 1 observations is obtained for a current sheet located just inside the flyby trajectory as seen from Io. Its value is given by $I_{min} = 6 \times 10^5$ A [Acuña, Neubauer, and Ness, 1981]. The total power dissipated in the system amounts to 1.8×10^{12} W. For additional discussions concerning alternative Alfvén current tube models the reader is referred to Southwood et al. [1980]. In addition to the magnetic field perturbations detected by the magnetometer experiment on Voyager 1, the plasma experiment detected plasma velocity perturbations consistent with the Alfvén standing wave system described above [Belcher et al., 1981; Goertz, 1980a, and see Chap. 3]. An important development associated with the Alfvén current system and the Io torus has been the work of Gurnett and Goertz [1981] concerning the generation of discrete decametric arcs [Warwick et al., 1979a,b] and the control of decametric emissions by Io. In this model the discrete arc structure is explained as due to multiple reflections of the Alfvén waves at the torus boundaries and Jovian ionosphere. The position of Io within the torus determines the degree of interaction with the atmosphere/ionosphere system (see Chap. 3).

1.3. The middle magnetosphere

We are concerned in this section with a description of the magnetic field in that region of space where the Jovian internal field is adequately represented by a tilted dipole and where the magnetic field of the magnetopause currents and tail currents is small. In this region the field morphology is dominated by the field of external currents in the Jovian magnetosphere, predominantly by the equatorial azimuthal currents of the magnetodisc. Various authors have defined the radial extent of this region to suit a particular application. Van Allen et al. [1974a] and Smith, Davis, and Jones [1976] chose an inner radius of 20 R_J, while Goertz [1976a, 1979] prefers $10R_J$. The outer radius is often assumed to be 30 R_J [Goertz, 1976a] or variable, that is, extending to within $15R_J$ of the magnetopause [Smith, Davis, and Jones, 1976]. Our definition of the extent of the middle magnetosphere is motivated by the observed magnetic field morphology and the extent of the equatorial azimuthal current system established by Voyager observations [Ness et al., 1979a,b; Connerney, Acuña and Ness, 1981]. These observations demonstrate that the equatorial currents extend inward towards Jupiter at least to the orbit of Io, at ~ 6 R_J. Earlier estimates of the inner edge of the current sheet range from ~ 30 R_J [Smith, Davis, and Jones, 1976] to 17 R_J [Engle and Beard, 1980; Gleeson and Axford, 1976]. Our choice of ~ 6 R_J as the inner edge of the middle magnetosphere thus leads conveniently to a current-free (curl-free) inner magnetosphere. Although the sheet currents contribute only a fraction of the total field at 6 R_J, the presence of a near equatorial radial component has significant consequences for charged particle motion and field line morphology. Thus, an adequate description of the field for $r > 6 \ R_J$ requires consideration of the magnetodisc current system.

We confine our discussion to the in situ magnetic field observations and models of the Jovian magnetosphere based upon these observations, excluding theoretical models of a rotating magnetodisc [e.g., Gledhill, 1967] and self-consistent magnetic field/plasma models [e.g., Goldstein, 1977]. We do not discuss the magnetic anomaly model [Dessler and Hill, 1975; Dessler and Vasyliunas, 1979; Vasyliunas and Dessler, 1981 and Chap. 10], where correlated particle intensity maxima and magnetic field minima are related to a surface magnetic field anomaly rather than the existence of an annular current sheet. Voyager observations of two minima in the magnetic field every 10 hr [Ness et al., 1979a,b] and charged particle observations [Bridge et al., 1979a,b] require an interpretation of these data based on local currents (i.e., magnetodisc currents). Therefore, we will consider only the disc models. See Chapter 10 for a discussion of the magnetic anomaly model.

Earlier studies

One of the most distinguishing characteristics of the Jovian magnetosphere revealed by the Pioneer 10 and 11 (P 10 and P 11) encounters in 1973 and 1974 was the marked distortion of the equatorial magnetic field morphology and associated energetic particle confinement. The pervasive 10-hr periodicity in the fluxes of energetic particles [Van Allen et al., 1974a] and magnetic field perturbations [Smith et al., 1975] particularly evident in the Pioneer 10 observations was identified with the repeated encounter of the spacecraft with an immense and relatively thin annular current sheet. The misalignment of the axis of the annular current sheet and the Jovian axis of rotation results in the periodic motion of the sheet past an observer fixed near the equatorial plane. Azimuthal (eastward) currents flowing in the disc contribute a radial field component to the ambient (nearly dipolar) field of internal origin, a contribution that reverses sign through the sheet, stretching field lines radially outward along the plane of the current sheet. Thus, the middle magnetosphere can aptly be regarded as the transition from a dipolar magnetospheric configuration to the disclike geometry of the "magnetodisc." Early models placed the current disc in the magnetic equatorial plane and suggested azimuthal symmetry about the magnetic dipole axis [Van Allen et al., 1974a] tilted ~10° with respect to the rotation axis. Hill, Dessler, and Michel [1974], Smith et al. [1975], and Smith, Davis, and Jones [1976] advocated a warping of the current sheet toward the rotational equator for $R \gtrsim 20 \ R_J$, reflecting the action of centrifugal forces on the corotating plasma [Hill, Dessler, and Michel, 1974], but Goertz et al. [1976] and Goertz [1979] argued that the disc deviated only slightly from the magnetic equator to at least 90 R_J. The magnetic anomaly advocates [Vasyliunas and Dessler, 1981] propose a current sheet that remains in the equatorial plane but exhibits longitudinal asymmetries related to magnetic anomaly effects.

Voyagers 1 and 2 (V 1 and V 2) traversed the Jovian magnetosphere outbound at 0420 and 0300 local time, respectively, supplementing the P 10 and P 11 outbound observations at 0500 and 1200 local time. Voyager investigations provided additional direct observations of the magnetic field [Ness et al., 1979a,b], as well as measurements of particles and plasmas not obtainable by P 10 and P 11 [Bridge et al., 1979a,b; Krimigis et al., 1979a,b; and Warwick et al., 1979a,b]. In addition, the V 1 and V 2 spacecraft trajectories were more favorable for observations of the azimuthal current sheet (in or near the magnetic equator), because the Voyager spacecraft approached

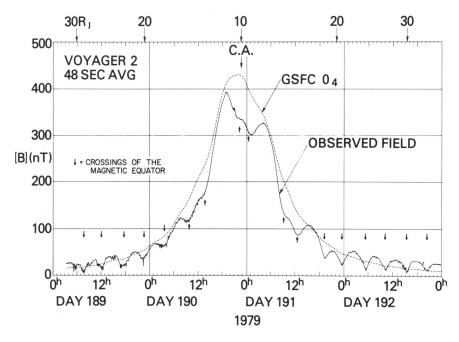

Fig. 1.16. Magnetic field magnitude observed by Voyager 2 during the encounter. The dashed curve is the model intrinsic magnetic field of Jupiter assuming the O_4 internal field model of Acuña and Ness [1976c], obtained from Pioneer 11 observations. Field magnitude minima occur at crossings of the magnetic equator, where the field of the azimuthal current sheet is antiparallel to Jupiter's (dipolar) intrinsic magnetic field. At larger radial distances, just above the disc currents (where the radial field component of the disc currents reaches a maximum), the (vector summed) total field magnitude exceeds the intrinsic field.

(and exited) closer to the Jovigraphic equator than the Pioneers. The Voyagers passed through the current sheet twice each planetary rotation, at times completely traversing the current disc. In contrast, the Pioneer inbound trajectories, in particular the P 10 approach at ~ 5° southerly Jovigraphic latitudes, brought these spacecraft close to the center of the current sheet (near the magnetic equator) and produced only one encounter with the sheet for each planetary rotation. P 10 outbound was similar, and P 11 exited at middle or high northerly latitudes, missing the sheet entirely. Thus, the Voyager encounters were instrumental in clarifying the observations of the magnetodisc obtained along the rather singular P 10 trajectory. Although these observations enabled a more complete description of the distant Jovian magnetotail [Behannon, Burlaga, and Ness, 1981; see also next section], it is important to realized that a major portion of the voluminous Jovian magnetosphere, for example, from 1200 LT to 0300 LT, remains unexplored.

Observations

The periodic magnetic field perturbations typical of each of the spacecraft traversals of the Jovian magnetosphere are illustrated in Figure 1.16 for the V 2 encounter. In this figure the observed magnetic field magnitude is compared with that predicted by the GSFC O_4 internal Jovian field model, a 15 coefficient, spherical harmonic expansion derived from P 11 observations [Acuña and Ness, 1976c]. The now familiar field magnitude depressions, coincident with crossings of the magnetic equator, are evident, as well as systematic enhancement of the field magnitude at larger radial distances whenever the spacecraft emerges from the current sheet. The P 10 and P 11 observations demonstrate a similar decrease in the field magnitude as the spacecraft nears the center of the azimuthal current sheet, although these field depressions occur only once each planetary rotation because of the trajectory constraints discussed earlier [Smith et al., 1975; Smith, Davis, and Jones, 1976].

A very useful and revealing presentation of the magnetospheric magnetic field data for studies of magnetic fields of external origin is that of a perturbation magnetic field plot. The perturbation field $\Delta\mathbf{B}$ is the difference between the observed magnetic field at any position and the predicted field of internal origin. In the present example, the GSFC O_4 model of the internal Jovian magnetic field is used. The perturbation computed field is increasingly sensitive to the model internal field at radial distances $< 6\ R_J$ where inaccuracies in predicting the rapidly increasing internal field become important. Consequently, we will concentrate at present on the observations at radial distances $> 6\ R_J$.

Examination of the perturbation field in spherical (1965 System III) coordinates for the Voyager 1 encounter (Fig. 1.16) reveals:

1. a slowly varying theta component ΔB_θ increasing in magnitude approaching Jupiter and antiparallel with the internal (nearly dipolar) field;
2. a small azimuthal (phi) component ΔB_ϕ;
3. a radial field component ΔB_r exhibiting a definite 10-hr periodicity, changing sign twice every 10 hours, and increasing in magnitude (more slowly than the theta component) with decreasing radial distance, and
4. a perturbation field direction, as determined by the angle between $\Delta\mathbf{B}$ and $\mathbf{B}_0$ which periodically varies, as the planet rotates, from being almost perpendicular to being almost antiparallel to the internal field direction.

This perturbation field due to the current sheet possesses the solenoidal field geometry appropriate to a system of azimuthal equatorial currents. At small radial distances the perturbation field is largely vertical and antiparallel to the internal field. At the magnetic equator, the radial field is zero, and the total field magnitude (e.g., Fig. 1.16) reaches a minimum. Above the magnetic equator, the radial field is positive outward and below the equator, inward.

Comparison of the perturbation magnetic field (Fig. 1.17) and the spacecraft trajectory in magnetic equatorial coordinates (Fig. 1.18) suggests that the plane of symmetry of the current system lies in or near the magnetic equator. The important features in the perturbation field plot (Fig. 1.17) can be understood by application of Ampere's Law to a narrow circuit centered about the plane of symmetry of the sheet (Fig. 1.19). The contribution of $\int \mathbf{B} \cdot \mathbf{dL}$ along the vertical segments of the circuit is negligible because the vertical field component varies little over the width Δr (Fig. 1.17). Only the radial field component, B_r, which changes sign above and below the plane of symmetry of the sheet, contributes to the integral. By Ampere's Law,

$$\int \mathbf{B} \cdot \mathbf{dL} = \mu_o I_{encl}$$

The radial component of the field is a local measure of the azimuthal current density. Thus, the current density at $r = 10\ R_J$, assuming a total sheet thickness of $5R_J$, is easily estimated from the magnitude of the radial field component above the sheet ($\sim 50\ \mathrm{nT}$) to be $\sim 0.224\ \mathrm{mA/km^2}$ ($\sim 1.13 \times 10^6\ A/R_J^2$). In Figure 1.17, we note that the

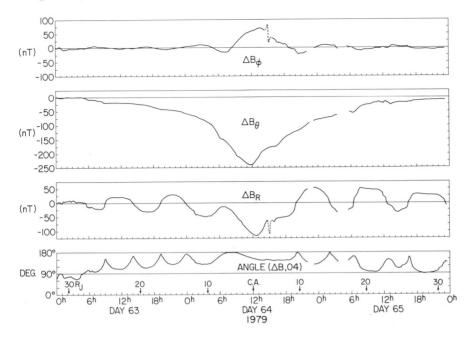

Fig. 1.17. Perturbation magnetic field $\Delta \mathbf{B}$ observed by Voyager 1 during passage through the Jovian magnetosphere for $R < 30\ R_J$. The reference field used is the GSFC O_4 model of Acuña and Ness [1976c]. Spherical coordinates are used. The angle between the perturbation field and the reference field is shown in the bottom panel.

radial field increases linearly with increasing z from zero at $z = 0$ to a maximum B_r just above the sheet (for example, V 1 outbound at approximately 0730 on day 65). Above the sheet, exterior to the current-carrying region, the radial field varies slowly with z, as is appropriate for a two-dimensional current geometry. The spacecraft reenters the current sheet at approximately hour 1120 at greater radial distance where the radial field above the sheet is slightly diminished, and again approaches zero at the magnetic equator. Note that the transition from the current sheet to the current-free region is both well-defined and unambiguous. The decrease in B_r just above the sheet from 0730 on day 65 to 2200 on day 65 is clear evidence for a decrease in the current density with increasing r, very nearly approaching a $1/r$ dependence in this region of the magnetodisc. The theta component of the perturbation field is likewise consistent with the assumed solenoidal field geometry, and the phi component is understood as a consequence of the misalignment of the symmetry plane of the annular current sheet and the System III coordinate system.

Magnetodisc magnetic field models

Magnetic field observations are obtained only along spacecraft trajectories through the Jovian magnetosphere, but we would like to have a description of the magnetic field throughout the entire magnetosphere. The difficult task of reconstructing the Jovian magnetosphere from the data obtained during the Pioneer and Voyager passes through it can and has been approached in several different ways. Engle and Beard [1980] and Beard and Jackson [1976], concerned primarily with the overall configuration of the Jovian magnetosphere and in particular the magnetopause surface shape, assumed an

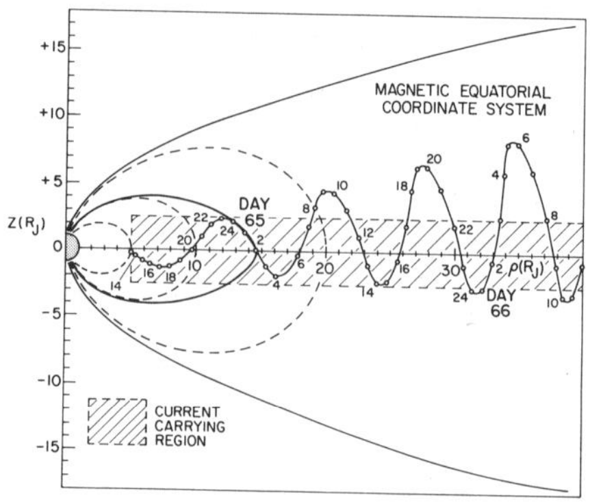

Fig. 1.18. Trajectory of Voyager 1
spacecraft in 1979 in magnetic latitude
and radial distance. The location of the
Connerney, Acuña, and Ness [1981]
model azimuthal current sheet is shown
in cross section. The near-equatorial
trajectory of Voyager 2 is similar.

Fig. 1.19. Model of the azimuthal
current sheet and appropriate ampere's
circuit. The radial magnetic field
component is a local measure of
azimuthal current density.

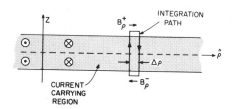

infinitesimally thin equatorial current sheet and an internal dipole field, and computed
this surface by equating the solar wind plasma and magnetospheric magnetic pressures. These models are not intended as detailed fits to the observations, but rather are
constructed to be qualitatively consistent with the observed P 10 magnetopause crossings. They use assumed models of the equatorial current sheet. Barish and Smith [1975]
confined their distended disclike model field to a spherical surface in order to approximate the blunt dayside magnetopause observed by P 10 [Smith et al., 1975]. Each of
these models utilized, in addition to the available magnetic field observations, a physical constraint to shape the distant field morphology.

The more quantitative models of Goertz et al. [1976] and Connerney, Acuña and
Ness [1981] are examples of two dissimilar approaches to magnetospheric modeling.
The versatile Euler potential method (e.g., Stern [1970, 1976]), exemplified by the
model of Goertz et al. [1976], is a mathematical representation of the magnetic field
based on a description of the field morphology. Functions appropriate to the observed
or assumed field line geometry are selected, and adjustable parameters are chosen to
minimize the differences between the computed and observed field along the trajectory. The magnetic field at points not on the trajectory is obtained by extrapolation. In
contrast, the magnetospheric magnetic field model of Connerney, Acuña, and Ness
[1981] is derived from a mathematically tractable current distribution inferred from
the observations. Adjustable parameters directly related to currents in the model
Jovian magnetosphere are selected to minimize the differences between the modeled
and observed fields along the trajectories. The magnetic field elsewhere is computed
directly from the model current distribution. The distinction between these two
approaches is essentially the following: in the Euler potential method, one chooses simple functions to fit the field line geometry, and accepts whatever current distribution
is implied by the representation chosen. In the source modeling method, one chooses
a physically reasonable current distribution, and accepts the limitations imposed
by mathematical tractability and the probability of a mathematically more complex
field-line description.

Euler potential models

In the Euler potential representation, the divergence-free vector magnetic field **B** is expressed as

$$\mathbf{B} = \nabla f \times \nabla g$$

where the functions f and g are scalar functions of position known as Euler potentials. Surfaces of constant f and g are everywhere tangent to **B**, because

$$\mathbf{B} \cdot \nabla f = 0$$

and

$$\mathbf{B} \cdot \nabla g = 0$$

Lines along which the surfaces $f = $ constant and $g = $ constant intersect are thus field lines. Stern [1976] describes the properties of Euler potentials and presents illustrative examples of their use in magnetospheric modeling.

The approximate magnetospheric model of Barish and Smith [1975] was designed to confine field lines to a sphere of radius $r = 100\ R_J$ and yield an equatorial field magnitude proportional to $1/r^2$ for $r \geq 20R_J$. They used the Euler potentials

$$f = M\alpha$$

$$g = \phi$$

where M is a scale factor and α is given by

$$\alpha^6 - \frac{\alpha^5 \sin^2 \theta}{r} + \frac{\ln (.01\ r) \sin^2 \theta}{7.7 \times 10^{11} (\cos^2 \theta + .008)} = 0$$

in a spherical coordinate system with θ denoting colatitude, ϕ the usual azimuth angle, and lengths in units of R_J. The field lines in this model do not deviate from meridional planes. The adjustable parameters were chosen to produce a field magnitude along the P 10 inbound trajectory similar to that reported by Smith et al. [1975]. The model is intended for use in the dayside magnetosphere at low latitudes.

Goertz et al. [1976] chose the functions

$$f = M \frac{\rho^2}{(\rho^2 + z^2)^{3/2}} + \frac{b_0}{(\rho^2 + z^2)^{0.7}} [\log \cosh (z) + 10]$$

and

$$g = \phi + 175 (e^{\rho/500} - 1)$$

in a cylindrical coordinate system with z aligned with the magnetic dipole axis and ρ in the magnetic equatorial plane, with $M = 4GR_J^3$, $b_0 = 9 \times 10^{-2}$ G, and units of length are expressed in R_J. In this model the field lines lie in azimuthally warped surfaces given by $g = $ constant, reproducing the spiral configuration inferred by Smith et al. [1975] from P 10 outbound observations. The agreement between the observed and computed field magnitudes along the outbound P 10 trajectory is quite good for $r \geq 20R_J$, where the model is applicable. Jones, Melville, and Blake [1980] have offered two additional Euler potential models that are very similar to the Goertz et al. [1976] model.

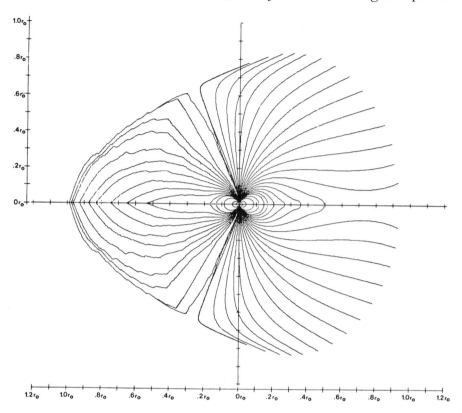

Fig. 1.20. The model Jovian magnetosphere of Engle and Beard [1980].

Source models

The only models to include the Jovian magnetopause surface currents and shape are the models of Beard and Jackson [1976] and Engle and Beard [1980]. The latter is a more complete version of the former and we will confine our discussion to it. Engle and Beard [1980] assumed an infinitesimally thin equatorial current sheet carrying an azimuthal current proportional to $1/r^{1.7}$ between $r = 18 R_J$ and $r = 100\ R_J$. The choice of current sheet is intended only to be qualitatively consistent with the P 10 observations and is not the result of a fit to the observations. The resulting field morphology, consisting of contributions from the internal dipole, equatorial current sheet, and magnetopause currents, is illustrated in the noon–midnight meridian plane in Figure 1.20. The magnetopause surface is considerably flattened due to the geometry of the current sheet as compared with the magnetopause formed about a dipolar internal field. The computation of the magnetic field of the current sheet in this model is unwieldy, but the paper by Engle and Beard contains ample graphical presentations of the variations of their modeled field across the magnetosphere. They also have expressed the modeled field due to magnetopause currents as a 65 coefficient spherical harmonic expansion, which they argue is relatively insensitive to the details of the equatorial current sheet. The reader is referred to the paper by Engle and Beard [1980] for details of the computations and lists of model coefficients.

Connerney, Acuña, and Ness [1981] model the Jovian magnetosphere for $r \lesssim 30\ R_J$, where the field due to magnetopause currents is negligible. The model consists of a finite thickness ($5\ R_J$) annular current sheet with an azimuthal current density

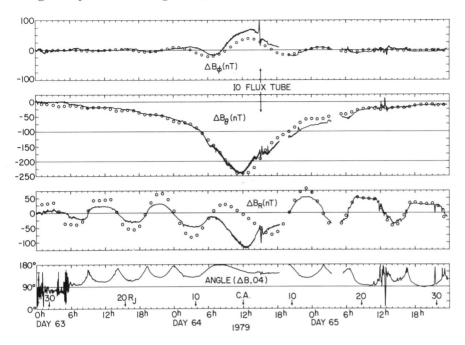

Fig. 1.21. Comparison of modeled perturbation magnetic field with that observed for Voyager 1. Open circles are (hourly) computed values using the Voyager 1 and Pioneer 10 model parameters of Connerney, Acuña, and Ness [1981].

inversely proportional to radial distance, extending from 5 to 50 R_J in the magnetic equatorial plane. Computation of the field quantities is accomplished by numerical integration, but excellent analytical approximations (for all but a small region near the inner sheet edge) can also be applied. These are given in Appendix A. A model fit to the V 1 and P 10 vector observations required a current density at the inner edge of the annulus of $\sim 5 \times 10^6$ A/R_J^2 (~ 1 mA/km^2). The same model fit to the V 2 observations required $\sim 1/3$ less current. A comparison of the model fit to the V 1 perturbation field illustrated earlier appears in Figure 1.21. These simple azimuthally symmetric models are capable in each case of producing a generally good and self-consistent fit to all three field components. Details of the perturbation plots, such as the flattening of the B_r component that occurs when the spacecraft exits the current-carrying region (e.g., V 1 outbound at 0730 on Day 65), appear quite naturally in the modeled field as well. The magnetic field morphology of the annular current disc used in these models is illustrated in Figure 1.22. The magnetic field of the azimuthal sheet current has a vertical component that is antiparallel to the internal Jovian field. The total magnetospheric magnetic field, consisting of the internal Jovian dipole (4.2 GR_J^3) and the contribution due to the sheet currents, is illustrated in Figures 1.23 and 1.24. The field morphology is quite similar to that envisioned by the early P 10 investigators (e.g., Van Allen et al. [1974a]) with respect to the distention of field lines along the symmetry plane of the azimuthal current sheet.

A much improved fit to the V 1 observations (compare Figures 1.4 and 1.21) has been obtained (Connerney, Acuña, and Ness, 1982) by simultaneous inversion of both the internal field and the field due to the magnetodisc currents. The best fitting magnetodisc is similar to that described above, but lies in the centrifugal equator, 2/3 of the way between the rotational and magnetic equators.

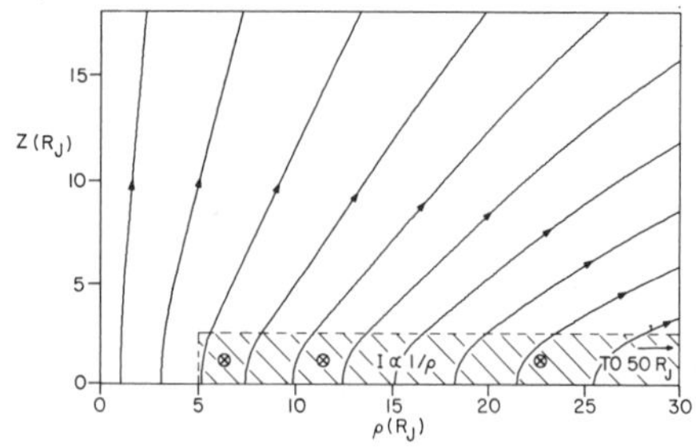

Fig. 1.22. Meridian plane plot of field
lines associated with the magnetodisc
current sheet model of Connerney,
Acuña, and Ness [1981].

Fig. 1.23. Meridian plane projection of
magnetosphere field lines (heavy) and
isointensity contours (light) for Voyager
1 (and Pioneer 10) model of Connerney,
Acuña, and Ness [1981]. Values on field
lines indicate colatitude of field line; field
magnitude contours are expressed in
nanoteslas.

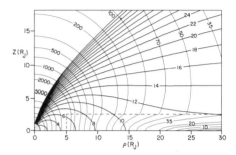

Fig. 1.24. Meridian plane projections of
magnetosphere field lines (heavy) and
isointensity contours (light) for Voyager
1 (and Pioneer 10) model of Connerney,
Acuña, and Ness [1981]. Values on field
lines indicate equatorial distance of field
line crossing in absence of magnetodisc
current sheet. Field magnitude contours
are given in nanoteslas.

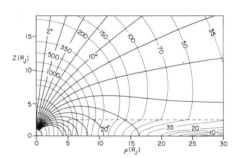

Discussion

All of the magnetic field models, with the exception of the Engle and Beard [1980]
model, have assumed azimuthal symmetry. Goertz et al. [1976] fit the observed
P 10 field magnitude outbound at ~ 0500 local time, whereas Barish and Smith [1975]
fit the observed P 10 inbound field magnitudes at ~ 1000 local time. These models
apply on either of the inbound or outbound passes only and are not used for both.
Connerney, Acuña, and Ness [1981] also noted a departure from azimuthal symmetry
evident in the V 1, V 2, and P 10 observations within 30 R_J. Because they used vector
observations instead of field magnitudes, they were able to determine that the radial
field component was primarily responsible for the inbound/outbound asymmetry, the
inbound radial component being ~ 1/3 less than the outbound radial component. The
observed vertical field component, however, did not show an inbound/outbound asymmetry. This, they suggested, could readily be explained by a thickening of the annular
current sheet in the dayside magnetosphere, that is, a displacement to higher latitudes
of ~ 1/3 of the azimuthal current. This possibility is supported by P 10 (inbound) observations of 1.79–2.15 MeV protons [Northrop, 1979], which indicate a thickening of
the dayside plasma sheet. Additionally, Voyager observations [McNutt et al., 1981] of

plasma fluxes toward the magnetic equator at local midnight and away from the magnetic equator at local noon provide direct and independent evidence of such a dayside thickening.

The Euler potential models are most useful in providing an uncomplicated analytical description of field line morphology facilitating studies of particle motion or field line motion (e.g., Stern [1970]). They are less useful in providing insight into the large-scale current systems responsible for the magnetodisc magnetic field. Goertz et al. [1976] and Jones, Melville, and Blake [1980] have noted that the Euler potential models imply (by taking the curl of the modeled field) unphysical currents in certain regions of space, for example, infinite currents at the origin or large currents at great distances. The model of Barish and Smith [1975] is a good example of how these fictitious currents arise. Their model uses a dipolar field at high latitudes plus an additional term effective only at low latitudes to stretch out the field lines along the equator. The model results in azimuthal currents in the equator to pull the field lines radially outward, but also azimuthal currents at higher latitudes of the opposite sign. The currents are essentially required to cancel the magnetic field of the equatorial currents so that the total high latitude field is dipolar. In this example, the current system obtained is largely a result of the unreasonable requirement that the magnetic field of the equatorial current sheet be localized to lower latitudes. Such features make the estimates of local current density (in the magnetodisc) untrustworthy, unless it can be shown that the contributions to the observed magnetic field of all other currents is small. Clearly, it is necessary to exercise caution in the interpretation of the quantity $\nabla \times \mathbf{B}$ computed from the Euler potential models. For the same reason, caution is also advised in extrapolation of the field away from the equatorial region using the Euler potential models.

Consequences of a magnetodisc field morphology

Each of the magnetic field models describes a magnetodisc field geometry in which field lines are radially stretched along the plane of symmetry of the azimuthal current disc (the magnetic equator in the models of Goertz et al. [1976] and Connerney, Acuña, and Ness [1981]). Phenomena that are sensitive to the field line morphology, such as the absorption of charged particles by satellites and the magnetopause surface geometry, are essentially consequences of the gross magnetodisc magnetic field geometry. Any of the models exhibiting the magnetodisc geometry would suffice to illustrate qualitatively such phenomena; for convenience we choose to utilize the model of Connerney, Acuña, and Ness [1981] in our discussion of charged particle absorption and auroral effects.

The magnetic field of the Jovian azimuthal current system is comparable in magnitude to the field of internal origin at a radial distance of $\sim 15\ R_J$. At greater distances the resultant vertical field is much smaller than the radial component of the sheet field; thus the field is largely radial above and below the sheet. At $30\ R_J$, the field magnitude is twice that expected of a 4.2 G-R_J^3 dipole, and it is due almost entirely to the radial component of the sheet field. Thus, in magnitude and configuration the distant field is dominated by the annular current sheet.

We envision a Jovian magnetopause surface that is everywhere extended along the magnetic equator, that is, flattened in the direction of the Jovian dipole axis. This is entirely consistent with the static model calculations of Engle and Beard [1980] who assumed a rather different equatorial current sheet in their determination of the magnetopause surface. Indeed, the observations of Lepping et al. [1981] of 10-hr periodicities in the antisolar magnetosheath are understood in terms of the dynamics of a

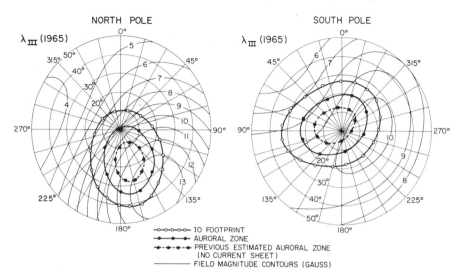

Fig. 1.25. Comparison of Io flux tube footprint for O_4 model with location of auroral zones computed with and without the magnetodisc current sheet. (The Io flux tube footprint is enlarged by a small amount [$< 2°$] when the magnetodisc currents are included. Only field lines at higher latitudes are significantly affected.)

flattened magnetopause surface. The five and ten hour periodicities in field magnitude, direction, and bow shock/magnetopause observations at greater distances all can be regarded as a consequence of the 10-hr "rocking" of the azimuthal current system due to the misalignment of the Jovian rotational and magnetic axes (see Sec. 1.4).

Another interesting consequence of the annular current sheet concerns the mapping of field lines from the (magnetic) equatorial plane to the planet's surface. Field lines at lower latitudes are unaffected by the presence of sheet currents, to $L \sim 6$. The radial stretching of higher latitude field lines results in a very narrow range of latitudes on the Jovian surface that map onto, or link to, the magnetodisc. In Figure 1.25 we show the intersection of the last closed field lines corresponding to the model magnetosphere illustrated in Figure 1.24 with the oblate surface of Jupiter. We have assumed the O_4 internal field model and simply traced field lines from just above the sheet at $\sim 30\ R_J$ down to the Jovian cloud tops. All of the field lines linking to the magnetodisc in this model originate between the Io footprint and the "auroral zone" footprint on the surface of Jupiter. Although we have tentatively identified the foot of the "last closed field lines" as the Jovian auroral zone, it is necessary to bear in mind that the asymmetries that have been neglected here, for example, the tail and magnetopause currents, are important in the determination of the auroral zone. In particular, we anticipate a dynamic situation, in which a variable Jovian auroral zone moves about within the indicated auroral zone in response to the 10-hr "rocking" of the azimuthal current sheet. A further variability in the size of the auroral oval may result if the azimuthal currents undergo a significant time variation, or as the tail and magnetopause currents change in response to solar wind variations.

The current sheet has important implications for charged particle motion in the Jovian magnetosphere, particularly in the absorption of trapped radiation by satellites. The stretching of the magnetic field lines in the magnetic equatorial plane leads to a significant distortion of the charged particle drift shells away from the simple quasidipolar geometry derived from the internal field model calculations for radial distances greater than $\sim 6\ R_J$. A complete description of charged particle motion requires knowledge of the magnetic field line geometry and the field gradient and curvature drifts over the particle's bounce path and longitudinal drift motion. Such a drift-shell tracing computation including the effects of the current sheet given in this chapter is impractical due to the numerical labor and computer time required. For the purposes of this discussion, it will suffice to address the "radial distance to the equatorial point" associated with each magnetic field line. This identifies the equatorial crossing point of a field line, determined by numerically tracing the field line to the equator. Although useful for discussing some aspects of satellite absorption of charged particles (e.g., possible absorption event times), such a field-geometric approach does not benefit the study of energy, species and time dependencies inherent in satellite absorption signatures. In studying the geometry of trapped particle drift shells, it is customary to identify the field lines by their "*L*-value." Such terminology is only valid for strictly dipolar or quasidipolar fields and is not applicable to the highly distorted equatorial fields associated with the current sheet. The reader is referred to Roederer [1970] and Stern [1976] for extended discussions on the subject.

Qualitatively, the stretching of field lines in the magnetic equatorial plane implies that the charged particle absorption features created by the sweeping effect of the Jovian satellites can occur at radial distances far removed from those predicted by the internal field models. The timing and duration of the absorption event(s) depends on the positions of the satellite and observing spacecraft relative to the magnetic equator. The inclusion of the current sheet field must result in significantly broader particle absorption regions (in radial extent) where the radial magnetic field component is large. Figure 1.26 clearly illustrates these effects for the case of the V 2 encounter trajectory. For comparison, we have shown the radial distance to the equatorial point of magnetic field lines encountered by the spacecraft for both the O_4 model and a model including the current sheet [Connerney, Acuña, and Ness, 1981] as a function of time. Also shown is the range of radial distances (to the equatorial point) of field lines encountered by Ganymede in one Jovian day for both models of the magnetic field. Several interesting results can be derived from the figure. The inclusion of the current sheet field significantly enhances the "wobbling effect" of the magnetic field past the spacecraft. This increases the radial range of magnetic field lines that are morphologically connected to the spacecraft over that predicted by the internal field model alone. In the case of the Ganymede absorption, this effect is dramatic, owing to Ganymede's motion relative to the current sheet (Ganymede traverses the entire sheet width twice each rotation) and the near radial geometry of the field at ~ 15 R_j. The time interval during which V 2 is located within the Ganymede sweeping region is extended from a few tens of minutes predicted by the internal field model to over *5 hours* for the model including the current sheet. Note that V 2 entered (briefly) the expanded absorption region at ~ 0400 SCET, and again approximately 5 hr later; in the absence of a current sheet the absorption region is encountered only once, briefly, near 0900 SCET. In this case, the sweeping region is defined as the range of radial distances to the equatorial point of magnetic field lines which are connected to Ganymede around its orbit. This effect is, we expect, largely responsible for the location and extent of the plasma voids observed near Ganymede by V 2 on the inbound leg [Burlaga, Belcher, and Ness, 1980] from 0400–1200 SCET.

Although the model used here does not include the effects of the magnetopause or tail currents, it is clear from the figure that at times the radial distance to the equatorial point of the field line connected to the spacecraft can become comparable to the magnetopause radial distance or the "hinge" point in the Jovian tail. This exercise is similar to the identification of "open field lines" by Goertz et al. [1976] along the P 10 trajectory outbound.

Fig. 1.26: The radial distance to the equatorial point of magnetic field lines encountered by Voyager 2 as computed from the O_4 model and the model including the current sheet field.

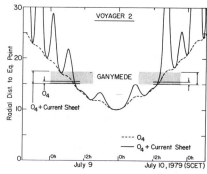

The results given here are necessarily approximate owing to the variability to the ring current and the external effects associated with the magnetopause and solar wind compression of the magnetosphere. In addition, the assumption of an axisymmetric ring current is only partially satisfied in the Jovian magnetosphere because the models given here point to a significant day–night asymmetry. This, of course, leads to significant drift-shell splitting, drift-loss cones and quasistable trapping regions. The radially stretched magnetic field geometry in the vicinity of the equatorial plane leads to an enhancement of the relative maxima observed in charged particle fluxes when traversing the magnetic equator. These effects, as observed by P 10 [Van Allen et al., 1974a], led to the initial concept of the magnetodisc (see also Sec. 1.4).

Nonmeridional field

In addition to, and quite distinct from, the radial extension of field lines in meridional planes due to azimuthal currents, Smith et al. [1975] observed a (westward) warping of the magnetic field lines out of meridional planes, most evident on the P 10 outbound trajectory near the dawn meridian. The P 10 inbound observations, as well as the P 11 (inbound and outbound) observations in the late morning sector of Jupiter's magnetosphere, demonstrated considerably less warping and less regular behavior. The P 10 outbound observations taken alone seem to suggest a spiral magnetic field configuration [Smith et al., 1975] reminescent of the Parker spiral of the solar wind. The assumed spiral field configuration was interpreted as evidence for the centrifugally driven planetary wind of Hill, Dessler, and Michel [1974] beyond the critical Alfvén radius of 30 R_J (for a review of outflow models see Kennel and Coroniti [1977a]). More recently, Hill [1980] has revised his estimate of the critical Alfvén radius to 20 R_J, based on V 1 plasma observations [Bridge et al., 1979a]. Empirical models of the azimuthal magnetic field, appropriate to the P 10 outbound pass only, are given by Goertz et al. [1976] and Jones, Melville, and Blake [1980]. A spiral configuration of the magnetic field was inferred also from the P 10 energetic particle flux modulation observed along the outbound trajectory [Northrop, Goertz, and Thomsen, 1974].

Assuming azimuthal symmetry, Parish, Goertz, and Thomsen [1980] found a radial current of $\sim 6 \times 10^7$ A in Jupiter's equatorial plane necessary to fit the P 10 outbound azimuthal field component. Connerney [1981b] proposed a physically more intuitive model of a polar inflow, equatorial radial outflow current system, requiring $\sim 10^8$ A, but questioned the assumption of azimuthal symmetry essential to either determination.

Observations by V 1 and V 2 [Ness et al., 1979a,b,c] outbound at 0400 and 0240 local time, respectively (midtail-transit positions), demonstrated unambiguously the development of an extensive Jovian magnetic tail (see Sec. 1.4 for a complete discussion) and not a spiral magnetic field configuration. Thus the deviation of the magnetic field lines from meridional planes appears to be driven by the solar wind interaction rather than Jupiter's rotation. Magnetic field lines are swept back in the antisolar direction and approach the direction of the magnetopause near that boundary. The so-called spiral field configuration observed by P 10 outbound was largely an artifact of the sample of observations in the dawn meridian. None of the aforementioned models reflect the appropriate symmetry (about the noon–midnight meridian plane), and therefore none are applicable in general to the Jovian magnetosphere. The model of Engle and Beard [1981] possesses the appropriate meridional symmetry but lacks a dusk to dawn tail current required by the observations. In a recent reanalysis of all Pioneer passes Jones, Thomas, and Melville [1981] proposed a model current system consisting of an equatorial azimuthal disc current and an equatorial dusk to dawn current extending well into the dayside magnetosphere. A more complete discussion of the Jovian magnetotail is found in Section 1.4.

1.4. The outer magnetosphere

Dayside outer magnetosphere

In the outer magnetospheric regions the cylindrical symmetry prevalent in the inner regions disappears due to the effects of the solar wind interaction with the magnetosphere. We thus have to differentiate between a "sunward side" outer magnetosphere and a "magnetospheric tail" extending away from the planet in the antisolar direction. The description and discussion of the outer magnetosphere on the *sunward* side of Jupiter will be restricted to the magnetospheric region denoted as "outer zone" in Figure 1.27 (also see Fig. 1.37). This zone just inside the magnetopause (MP) was described by Smith et al. [1975] on the basis of Pioneer observations as a region of irregular magnetic field with, however, a persistent southward component indicative of a closed magnetosphere. According to Davis and Smith [1976], during the Pioneer encounters it was a region approximately 15 R_J thick, at least near the equator, with a field strength of ~ 5 nT. The average field magnitude observed by V 1 in that region was ~ 16 nT for a final MP crossing at 47 R_J, and at the time of the V 2 encounter the field strength there was ~ 7 nT on average for a final MP traversal distance of ~ 62 R_J. Smith, Fillius, and Wolfe [1978] have concluded that when the Jovian magnetosphere is at its greatest extent, most of the field just inside the MP is not derived from the planetary (dipole) field, but from currents within the magnetosphere, the major contributor probably being the equatorial current sheet. When the magnetosphere is compressed, as during the passage of a high speed stream in the solar wind, the relative contribution of the planetary dipole field to the field near the magnetopause becomes increasingly signficant, increasing, for example, from a contribution of $\sim 20\%$ at 100 R_J to $>40\%$ at 50 R_J.

The Voyager spacecraft also observed the outer zone to be approximately 15 R_J thick and to consist of predominantly southward fields, as illustrated in Figure 1.27b, which shows V 2 inbound 1-hr average magnetic field vectors projected on the plane containing the planet–Sun line and the magnetic dipole axis. At ~ 45 R_J, the field is seen to become more nearly parallel to the equatorial plane and then to increase in steepness relative to that plane outside the equatorial current sheet as the planet was approached. The field geometry observed by V 1 and V 2 inbound in the outer Jovian magnetosphere is thus in qualitative agreement with the closed magnetosphere model discussed by Smith, Davis, and Jones [1976] and shown schematically in Figure 1.27a.

a.)

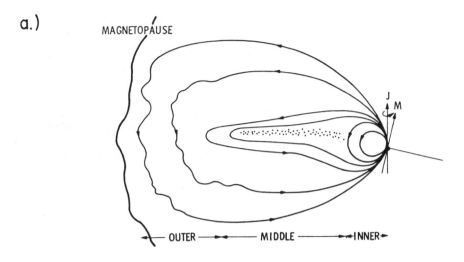

b.)

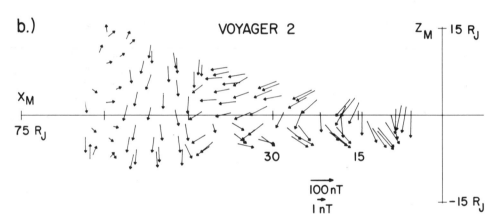

Fig. 1.27. (a) Closed Jovian magnetosphere model based on Pioneer observations [Smith, Davis, and Jones, 1976]. For Pioneer the inner region extended to ~ 20 R_J, the middle to ~ 60 R_J, and the outer to ~ 90 R_J. For Voyager the dimensions were ~ 15 R_J, ~ 45 R_J, and ~ 60R_J, respectively (Voyager 1's data were similar to Voyager 2's).
(b) Projection of hourly average vector magnetic field on the dipole coordinate x–z plane along the Voyager 2 trajectory inbound at Jupiter. In the dipole system, z_M is parallel to the magnetic dipole axis of Jupiter, positive northward, and the x_M and y_M axes lie in the magnetic equatorial plane such that the x_M–z_M plane contains the planet–Sun line, x_M positive toward the Sun.

In this view, the high latitude field lines are displaced outward by the presence of the equatorial current sheet and combine with current sheet and MP current fields to form the relatively "soft" zone of predominantly southward fields between the outer edge of the current sheet and the MP. It is the trapped plasma associated with the current sheet which ultimately must balance the pressure of the external thermalized sheath plasma. The "buffer" zone between the edge of the current sheet and the MP is very sensitive to variations in external pressure, functioning as a shock-absorber for fluctuations of small to moderate amplitude but allowing large-scale compression of the dayside magnetosphere in the presence of major perturbations, as illustrated by differences in the scale of the dayside magnetosphere observed by the Pioneer and the Voyager spacecraft.

Although we have defined here the thickness of the outer zone to be approximately 15 R_J as suggested by the Pioneer and Voyager magnetic field observations, this only applies to the inner part where relatively steady, southward pointing fields are seen. Although southward-directed on average, the fields nearest the MP are highly variable and disorganized on shorter timescales (the left-most, nonsouthward-directed hourly data in Figure 1.27b are a mix, however, of magnetosheath, magnetosphere and a small fraction of interplanetary data). This outer region in particular can vary significantly in radial extent, resulting in a sunward magnetopause distance at least as great as ~ 100R_J [Smith et al., 1975; Smith, Fillius, and Wolfe, 1978; Trainor et al., 1974; Van Allen et al., 1974a; Wolfe et al., 1974]. Kivelson [1976] analyzed the fluctuations observed in the buffer zone by P 10 and P 11 and described the region as a "layer of turbulence." The variance of the fluctuating field was found to decrease with increasing frequency, approximately as f^{-2}. V 1 and V 2 also observed high levels of field variability there, but it is not clear without further analysis that "turbulence" is the most accurately descriptive term for the observed fluctuations. Various models for the plasma motions in this outer region have been discussed by both Smith, Davis, and Jones [1976] and Kivelson [1976].

The Jovian magnetic tail

A major effort in the analysis and interpretation of the data obtained from the Pioneer spacecraft in the Jovian magnetosphere was directed towards a description of the characteristics of the magnetodisc current sheet. The magnetosphere was viewed as being dominated by this assumed axially symmetric distension of the near equatorial magnetic field containing enhanced plasma and energetic particle populations. On the P 10 outbound pass near the dawn terminator, the distortion of the magnetic field out of the magnetic meridian plane was viewed by some as a spiraling of the magnetic field due to a radial outflow of plasma or to waves in the magnetodisc itself [see Sec. 1.3 and Northrop, Goertz, and Thomsen, 1974; Kennel and Coroniti, 1975; Goertz, 1976b; Kivelson et al., 1978]. Smith et al. [1975] suggested, however, that the bending of field lines out of meridional planes was probably due to drag forces between the rapidly rotating magnetosphere and the slower magnetosheath flows.

The existence of a Jovian magnetotail of a few AU in length had been inferred by means of cosmic ray electron propagation studies [Krimigis et al., 1975; Mewaldt, Stone, and Vogt, 1976; Pesses and Goertz, 1976]. It was not until the V 1 observations at a local time of approximately 0400 that the existence and characteristics of the Jovian magnetotail were first identified by direct measurement [Ness et al., 1979a,b]. These results showed that the magnetic field at increasingly larger radial distances from the planet tended to lie parallel to the equatorial plane of Jupiter and also to the ecliptic plane, because the latter is inclined only 3° with respect to the former. More importantly, the azimuthal direction of the field showed an alignment paralleling the magnetopause surface, according to a reasonable model of the MP shape (see Sec. 1.4). These data were combined retrospectively with the P 10 results and compared to the geometry of the magnetic field in the dawn to midnight sector of Earth's magnetosphere. This comparison suggested that the transition of the field in the far magnetosphere near dawn to a magnetotail configuration was being observed. The V 2 results added substantially to a validation of this interpretation, showing that at a local time

Fig. 1.28. Projection of hourly average
magnetic field components on the solar
magnetospheric (*SM*) *x–y* plane along the
Voyager 1 and 2 outbound trajectories.
Only the field vectors corresponding to
alternate hours have been plotted for
clarity. The data show that the field is
distorted from magnetic meridian planes
near the dawn terminator and bent back
to parallel the magnetopause (MP)
boundary. The MP shown is the Voyager
2 model boundary.

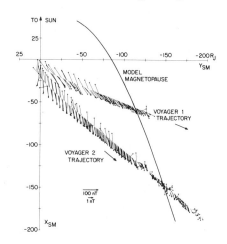

of ~0240 the magnetic field direction clearly indicated the sweeping back of magnetic fields lines into the tail region. Standard plots of 16 min average field magnitude,
direction, and rms deviation for the complete tail passes of both V 1 and V 2 have been
published by Behannon, Burlaga, and Ness [1981].

A clear visualization of the geometry of the field in the Jovian magnetotail region
may be obtained by means of vector data plots. Figure 1.28 shows the projection on
the *xy* plane of subsets of hourly averaged data obtained by the V 1 and V 2 spacecraft
during their entire transits through the Jovian magnetosphere outbound from periapsis
in March and July, 1979, respectively. The model MP shown corresponds to the surface determined by V 2 (see Sec. 1.4). A logarithmic scale has been used to represent
field vector intensity. The coordinate system used is *solar magnetospheric*, in which
the *x* axis points from the planet to the sun and the *z* axis lies in a plane defined by the *x*
axis and the instantaneous position of the magnetic dipole axis. Close to the planet the
field is seen to lie in magnetic meridian planes, although farther out the direction is best
described as a sweeping back of of the field lines so as to parallel the magnetopause
surface.

The companion views for these data are presented in Figure 1.29 and show the
orientation of the magnetic field observed by V 2 in both the *xz* and *yz* projections.
This figure also illustrates the differences between the tail and the magnetosheath (MS)
fields. Following the initial entry of V 2 outbound into the MS, it found itself back
inside the tail again for an extended period of time, out to a radial distance of $> 250\ R_J$.
There is essentially no difficulty in differentiating between MS and magnetotail fields
on the basis of the high inclination of the sheath fields relative to the *xy* plane as
illustrated by Lepping et al. [1981] (see last topic of this section).

The dominant variations observed by the Voyager spacecraft outbound at Jupiter to
a distance of at least 140 R_J were produced by the recurrent passages through the tail
current sheet, the motions of which are controlled by the ~10-hr planetary rotation
period. Figures 1.28 and 1.29 illustrate these periodic traversals of the current sheet
and the alternate location of the spacecraft in either the north or south tail lobe; this is
identified more readily in Figure 1.28 by the changing polarity of the magnetic field. In
Figure 1.29 the apparent excursion of the spacecraft in the *xz* and *yz* planes is due to
the precessional motion of the magnetic dipole axis, which results from planetary rotation and tilt of the dipole axis to the rotation axis, and the subsequent rocking of the
coordinate axes about the *x* axis. An alternative explanation for the observed magnetic
field periodicities is that they are caused by a magnetic anomaly effect, which intro-

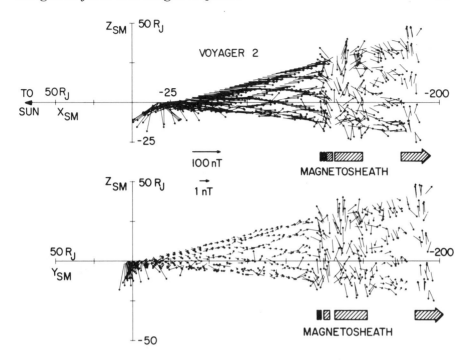

Fig. 1.29. Voyager 2 hourly average *SM x–z* (above) and *y–z* (below) magnetic field components. All hours have been plotted for more complete coverage in these views. Note in the upper panel the predominance of north lobe (outward-directed) fields as a result of the increasingly more northern location of the Voyager 2 outbound trajectory relative to the mean current sheet position. Also note in both panels the marked contrast between magnetosheath and tail field orientation, although during the more distant period in the tail more variation of the field was observed than in the near-planet region of the tail.

duces asymmetries in the magnetic field and asserts an enhancement of plasma and energetic particle intensities over the longitudes of an active sector (region of anomalously weak magnetic field) in the northern hemisphere of Jupiter [Dessler and Vasyliunas, 1979; Vasyliunas and Dessler, 1981; and Chapts. 10 and 11].

Field magnitude dependence on radial distance along the magnetotail

Detailed studies of the physical conditions leading to the development of a planetary magnetotail, its geometry and eventual disappearance or merging with the interplanetary medium require the knowledge of how the pressure balance is maintained or varies along the cavity-bounding surface, that is, the magnetopause. Of fundamental importance is the knowledge of the field magnitude dependence as a function of radial distance from the planet down the magnetotail. Analyses of this type have been performed for the Earth's magnetotail [Behannon, 1968; Mihalov et al., 1968] and yield a fall-off rate of the field intensity as a function of radial distance proportional to a power law, where the exponent was estimated to lie between -0.3 and -0.7. Similar studies have been conducted for the Jovian magnetotail using Pioneer and Voyager data.

The maxiumum hourly average field magnitude during each 10-hr period, B_{max}, was used to determine the radial variation of the total magnetic field as the V 1 and V 2 spacecraft traveled outbound from periapsis. A similar analysis was performed for

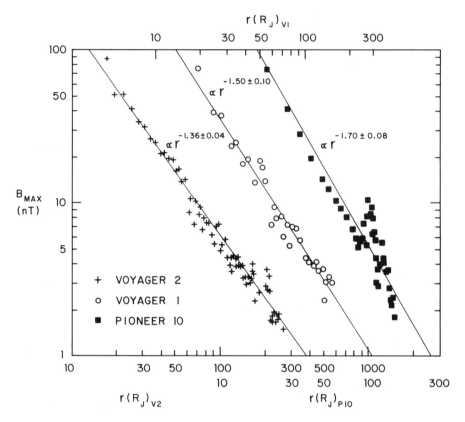

Fig. 1.30. Variation of average magnetotail lobe field with Jovicentric distance for Voyagers 1 and 2 and Pioneer 10. The observed variations are coded as indicated for the respective spacecraft. The least squares best fit inverse power law decreases are also shown in each case as solid curves. Best fit power law exponents are also given.

comparison purposes using P 10 data obtained from the NSSDC. The results are shown in Figure 1.30. A detailed description of the least-squares fitting procedure has been given in Behannon [1976], Appendix A.

For clarity of presentation, each of the three data sets have been separated along the abscissa and, as indicated, are to be referenced to radial distances scales that have been similarly shifted. The differing trends of the respective data sets reflect the different outbound trajectories, which had Sun–planet–spacecraft angles at the magnetopause of 99°, 115°, and 135° for P 10, V 1, and V 2, respectively. An indication of the degree of difference among individual gradients is given by the best-fit power law exponents, which were -1.70 for P 10, -1.50 for V 1, and -1.36 for V 2. There is no overlap in the standard error ranges on the three exponent values, suggesting that the difference between them is significant. These results indicate a departure from axial symmetry and require a tail-like current geometry. We note that the fall-off rates determined from these spacecraft observations are all steeper than those of the Earth's magnetotail.

There are systematic deviations from the simple power law dependencies. These effects are most noticeable in Figure 1.30 at and beyond $r = 100 \, R_J$ in the V 2 and P 10 data. V 2 observed another, somewhat broader increase in the neighborhood of $r = 50 \, R_J$ (see data in Behannon, Burlaga, and Ness [1981]), as did also the V 1 spacecraft. Although these features may well signify temporal variations, we cannot

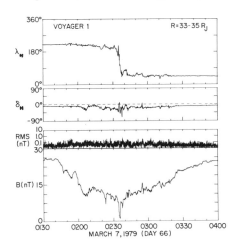

Fig. 1.31. Average magnitude, B, and direction, λ_H (longitude) and δ_H (latitude) of the magnetic field in heliographic coordinates (see text for definition) during a traversal of the current sheet by Voyager 2 at a radial distance of $\sim 34\,R_J$. Also given is the rms deviation of the total field over the 9.6 s averaging period.

uniquely separate temporal and spatial variations with single spacecraft traversals of the tail.

Tail current sheet observations

As in the case of the Earth, the Jovian magnetotail includes a large-scale current sheet separating the northern and southern lobes and across which the field reverses direction. Illustrating some of the general characteristics of this magnetotail current sheet, Figure 1.31 shows 2.5 hr of V 1 9.6 s average magnetic field data taken on March 7, 1979, during a crossing of the tail sheet at a Jovicentric distance of $34\,R_J$ (and a distance of $22\,R_J$ tailward of the dawn–dusk meridian plane). At the bottom is displayed the average field magnitude B, with the rms deviation of the field over the 9.6 s averaging period shown immediately above B. Note that the rms deviation is given on a logarithmic scale running from 0.01 to 10 nT. In the top half of the figure are shown the heliographic (HG) coordinate longitude and latitude angles λ_H, δ_H of the average vector field. By definition of the heliographic coordinate system, the angles λ_H, δ_H are given by

$$\lambda_H = \tan^{-1}(X_T/X_R)$$

and

$$\delta_H = \sin^{-1}(X_N/X)$$

where the spacecraft centered orthogonal unit vectors are defined with $\hat{R}$ along the Sun–spacecraft line, positive away from the Sun; $\hat{T}$ perpendicular to $\hat{R}$ and parallel to the Sun's equatorial plane, positive in the direction of planetary motion; $\hat{N} = \hat{R} \times \hat{T}$; and where

$$X = (X_R^2 + X_T^2 + X_N^2)^{1/2}$$

Figure 1.31 illustrates a classic traversal of a plasma sheet and embedded "neutral" sheet. In the field magnitude variation we see a broad and somewhat unsteady depression produced by the plasma sheet population of protons and ions of heavier nuclei as discussed by Lanzerotti et al. [1980]. At the minimum point of the depression is seen a relatively narrow (~ 2 min.), deeper depression simultaneous with the major variation in field azimuth, λ_H, and the largest negative deflections of the δ_H angle. Within this narrower depression a minimum average field of ~ 1 nT was observed. The field was unsteady in direction as well as magnitude for most of the extent of the plasma sheet and tended to display an enhanced southward component throughout the plasma sheet region. The rms shows the increase in fluctuation energy within this region, although in this case the increase does not show up as dramatically as it would on a more expanded linear vertical scale.

Only approximately 25% of all Voyager sheet crossings resemble the example shown in Figure 1.31. Distant tail sheet crossings are more typically complex, consisting often of a series of partial or full traversals presumably caused by unsteady sheet motions. The fluctuations observed in the current sheet can at best be qualified as quasisinusoidal with amplitudes of ~ 1 nT and characteristic times ranging from 5 to 30 s.

Spectral analysis of the magnetic field in the *lobes* of the magnetotail, that is, the regions of the tail outside the plasma sheet, shows those regions to be magnetically quiet in general. Spectra computed over frequency ranges from 0.2 to 5 Hz had power spectral densities that are essentially at the Voyager magnetometer noise level ($\sim 1 \times 10^{-5} f^{-1}$ nT/Hz). These results are consistent with the extremely low plasma cutoff frequencies, corresponding to the electron concentrations of 10^{-5} cm^{-3} or less, found by the plasma wave experiment in the tail lobes [Gurnett, Kurth, and Scarf, 1980].

Voyager 2 magnetic field data obtained during crossings of the tail current sheet were analyzed by means of the minimum variance method of Sonnerup and Cahill [1967]. The objective was to determine the orientation, relative to the planet and the Sun, of the tail current sheet. The results, as discussed by Behannon, Burlaga, and Ness [1981], show that the normals are closely distributed in heliographic longitude about 90° and 270°, indicating that these normals oscillate primarily in the dawn–dusk meridian plane, with little tendency for pitching or oscillation in the forward (sunward) or tailward directions, and that they also tend to lie preferably in a fan centered on the normal to the heliographic equatorial plane ($\delta_N = 90°$). This result suggests that the tail current sheet tends to have a mean orientation approximately parallel to the heliographic equatorial plane (or the Jovigraphic equatorial plane, because the two planes were nearly equal at the time of the Voyager encounters). Motions of the sheet are limited primarily to oscillations about the planet–Sun line, with little evidence for rigid north–south motion or bending about the east–west axis.

Shape and motions of the Jovian current sheet

The literature contains several models of the current sheet treated as a mathematical surface and compared with P 10 observations. Four distinct approaches have taken the current sheet as (a) a rigid magnetodisc [Van Allen et al., 1974a; Goertz, 1976a], (b) a magnetodisc that bends over at distances beyond $r = r_0$ to become parallel to the planet's equatorial plane [Hill, Dessler, and Michel, 1974; Smith et al., 1975], (c) a rigid disc for $r < r_0$ with a propagating wave for $r \geq r_0$ [Northrop, Goertz, and Thomsen, 1974; Kivelson et al., 1978], (d) the same configuration as in (c) but with constant wave amplitude beyond $r = r_0$ [Eviatar and Ershkovich, 1976] and (e) a rigid disc but with longitudinally varying spiralling [Vasyliunas and Dessler, 1981]. Models (a) through (d) have been illustrated schematically by Carbary [1980]. All of these models except (e) have axial symmetry with respect to Jupiter's rotation axis, that is, they imply that all observers at a given distance from Jupiter would see the same sequence of current sheet crossings, except for a phase lag, regardless of their longitude. Observations of

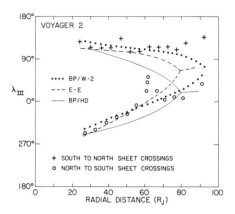

Fig. 1.32. System III longitude of Voyager 2 at the times of current sheet crossings as a function of radial distance from Jupiter, together with the predictions of three axial models (see text). The bent plane and the hinged-disc models predict a symmetry between the $S \rightarrow N$ and $N \rightarrow S$ crossings which is not observed.

current sheet crossings by V 1 [Ness et al., 1979a,b], which moved farther tailward than P 10, indicated that the axial models have limited applicability, and it was suggested that the current sheet on the night side is influenced by the formation of a tail as a result of the interaction of the solar wind with Jupiter's magnetosphere. V 2, which moved much farther tailward than V 1, provided further evidence against the early models and for the effects of a tail [Ness et al., 1979c; Bridge et al., 1979b]. Carbary [1980] reviewed the axial models referenced above and confirmed that they do not satisfactorily describe the times of the tail current sheet crossings observed by V 1 and V 2 (or, equivalently, the System III longitude at the time of a crossing as a function of the distance of the spacecraft from Jupiter); he showed that the model of Eviatar and Ershkovich [1976] gave a $\lambda_{III}(r)$ profile that qualitatively resembles the particle observations of current sheet crossings, but he did not show a fit of the model to the data.

Behannon, Burlaga, and Ness [1981] have used the measured times of the night side current sheet crossings determined by the magnetometers on V 1 and V 2 in an analysis that gives the best fit of those crossings to an assumed surface in order to evaluate and compare various models of the current sheet configuration, including both the published axial models (and some modifications of them) and some new (nonaxial) models which describe the effect of a tail configuration more explicitly. The axial model fitting results generally supported the conclusions of Carbary [1980]. Only the model of Eviatar and Ershkovich [1976] has been found to do a reasonably good job in modeling the observations, providing a fair fit to the V 1 crossings and a good fit to those of V 2. A modified model which allowed the wave to propagate from $r = 0$ rather than $r = r_0$ gave good fits to both data sets. The quality assessments were based on best-fit rms deviation values. Comparative results are illustrated in Figure 1.32 which shows the V 2 sheet crossing longitudes (System III) compared with curves predicted by the bent plane or hinged disc model (BP-HD), the limited-amplitude wave model of Eviatar–Ershkovich (E-E) and a model similar to the latter with wave propagating from the origin (BP/WP), which clearly gives a better fit (see also Chap. 5).

If the shape and position of the current sheet on the night side of the planet is influenced by the formation of a magnetotail, then there should be at least one nonaxial model which provides a still better global description of the current sheet surface, particularly at large distances in the tail. The simplest nonaxial model for a tail sheet is a plane surface that is approximately parallel to the Jovian equatorial plane and which rocks about the longitudinal axis of the tail as Jupiter's magnetic dipole axis cones about the rotation axis. More globally accurate would be a composite model that reduces to a rotating disc close to Jupiter and a rocking plane at large distances in the tail.

Fig. 1.33. Current sheet elevation (z) and azimuth of the dipole axis (ϕ_D) at the time of current sheet crossings are shown as functions of distance from Jupiter along $\hat{X}$, together with the predictions for these parameters based on a best fit to the rocking plane/rotating disc (RP/RD) model. This one-parameter model gives a good fit to the data over the range of distances that was considered.

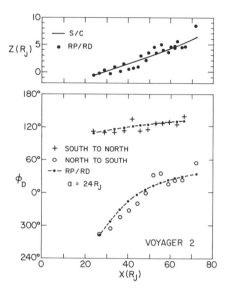

The nonaxial data-fitting study considered both the simple rocking plane and the composite rocking plane/rotating disk (RP/RD), as well as variants of the latter with both longitudinal crosstail bending and wave propagation [Behannon, Burlaga, and Ness, 1981]. The coordinate system used in the analysis was a fixed Cartesian system with origin at Jupiter's center, $\hat{Z}$ northward along Jupiter's rotation axis, $\hat{X}$ radially away from the Sun, and $\hat{Y}$ completing the right-handed triad. Within a certain radial distance ($\sim 30\ R_J$), a rotating disc approximates the near-planet current sheet. The 1-parameter RP/RD model, which reduces to a rotating disc close to Jupiter and to the rocking plane at large distances, is a ruled surface given by

$$z/\tan \alpha = x \operatorname{sech} (x/a) \cos \phi_D + y \sin \phi_D$$

where α is the "hinge point" distance, and ϕ_D denotes the longitude of Jupiter's dipole axis relative to the tail axis at the time of the current sheet crossings.

Fits of the nonaxial models to the V 1 and V 2 current sheets crossings yielded results that ranged from fair for V 1 for both the rocking plane and the RP/RD models to good and very good, respectively, for the rocking plane and the RP/RD modeling of V 2 data. Fits with modified models also have shown that wave propagation is relatively unimportant for nonaxial models. Detailed descriptions of the models and tabulations of fitting results are given by Behannon, Burlaga, and Ness [1981].

The validity of the RP/RD model can be demonstrated by comparing the observed ϕ_D with the predictions of the RP/RD model. These results are shown in Figure 1.33 for the V 2 tail current sheet crossings. The model gives an excellent fit to both the south-to-north and the north-to-south crossings over the entire range of distances, $24\ R_J \lesssim x \lesssim 72\ R_J$. More distant crossings were not used in the analysis because of considerable temporal variation. Figure 1.33 also shows the observed z positions of the spacecraft at the times of the crossings as a function of x together with the predictions of the RP/RD model. The model gives a very good fit to $z(x)$, the scatter being only $\sim \pm 1\ R_J$. Although the results of applying statistical tests to the models suggest that the "best-fitting" axial and nonaxial models are equivalent in terms of "goodness-of-fit" [Behannon, Burlaga, and Ness, 1981], it should be noted that the simple nonaxial model has fewer parameters.

Because the data sets are rather restricted, it cannot be claimed that any one of the models is better than another. The foregoing exercise merely illustrates that a one-parameter nonaxial model fits the data, particularly that of V 2, with less deviation than a particular two-parameter axial model. The essential feature of these models seems to be that in the tail, the current sheet is extended parallel to the x axis with a limited elevation along z at distances $\geq 30\ R_J$. It is interesting to note that none of the models tested gave as good a fit for V 1 as for V 2, suggesting that at the position of V 1 there were spatial distortions of the current sheet geometry that were not evident along the V 2 trajectory (see Behannon, Burlaga, and Ness [1981] for a comparison of the results of fitting the various models to the V 1 and V 2 data).

With the axial and nonaxial model it is assumed that the current sheet moves periodically, with a period $2\pi/\Omega$ equal to Jupiter's rotation period. For each model one can then predict the speed in the z direction as a function of position. The Voyager results suggest a maximum speed of the current sheet along z of $\leq 5\ R_J/\mathrm{hr}$, depending on the maximum elevation of the sheet and on the spacecraft position. Using the RP/RD model, the thickness of the Jovian tail sheet was estimated by Behannon, Burlaga, and Ness [1981] under the assumptions that (a) the sheet motion was to a good approximation in the z direction, (b) the component of spacecraft velocity in the z direction was negligibly small, (3) the sheet motion during each traversal was steady. The latter is probably the weakest assumption. The average estimated thickness was $\langle \tau \rangle = 4.8 \pm 0.3\ R_J$ for 10 cases where the RP/RD model accurately predicted the z coordinate and ϕ_D. Using the average field strength observed during each of these crossings gives a sheet thickness range of 6–50 gyroradii (R_L) for 30 keV O^+ ions and greater by a factor of four for protons of the same energy. Such a sheet is relatively thin in comparison with the scale of the magnetotail, in contrast with the thickness of the terrestrial plasma sheet, which can be 40% of the radius of the tail. This result is consistent with the current sheet thickness estimate of $\sim 4.2\ R_J$ from plasma wave measurements [Barbosa et al., 1979] and with that the inner magnetosphere sheet model of [Connerney, Acuña, and Ness, 1981]; it is somewhat larger than the estimates of $\sim 1\ R_J$ by Smith et al. [1975] and 2.2–3.4 R_J by Kivelson et al. [1978] based on current sheet observations in other regions of the magnetosphere.

Bow shock and magnetopause shapes

The first detection of the Jovian bow shock and measurements of magnetosheath fields were performed by the P 10 and P 11 spacecraft [Smith et al., 1975; Davis and Smith, 1976]. The V 1 and V 2 encounters with Jupiter [Ness et al., 1979a,b] provided additional opportunities to investigate the physical characteristics of the planet's bow shock (BS), magnetosheath (MS), magnetopause (MP), and the outermost regions of the magnetosphere. The closest approach trajectories of V 1 and V 2 in Jupiter's orbital plane are shown in Figure 1.34. The times at which the two spacecraft crossed the BS and MP boundaries, as identified in the magnetic field measurements, were tabulated by Lepping, Burlaga, and Klein [1981]. Figure 1.35 shows magnetic field magnitude and root-mean-square (rms) deviation as functions of time for both V 1 and V 2 inbound at Jupiter from the preshock interplanetary magnetic field (IMF) through the BS and MP crossings to closest approach (CA). Both the number of crossings and the range of crossing positions observed in each case indicate significant variability in boundary locations. A marked variability in the BS and MP positions was also observed by the P 10 and P 11 spacecraft [Smith, Fillius, and Wolfe, 1978].

In the case of the Pioneers, published model BS and MP shapes were based on the gas dynamic analog and scaled from the Earth's boundaries to the actual Pioneer

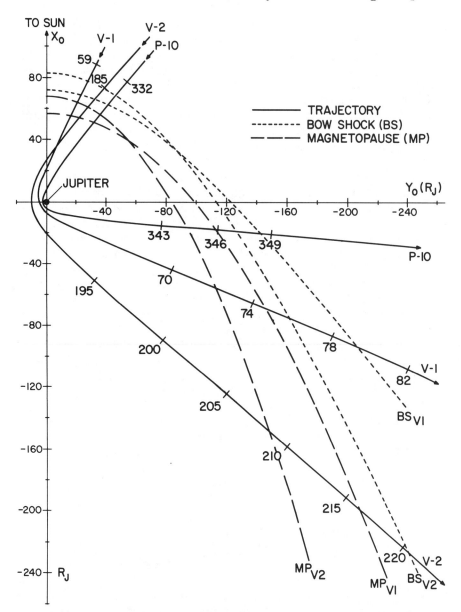

Fig. 1.34. The trajectories of Pioneer 10 and Voyagers 1 and 2 in the plane of Jupiter's orbit. Also shown are model magnetopause (MP) and bow shock (BS) boundaries (see text).

observations [Wolfe et al., 1974; Mihalov et al., 1975]. For each Voyager spacecraft encounter, model hyperbolic BS and parabolic MP curves were fitted to the average observed boundary positions [Lepping, Burlaga, and Klein, 1981]. These curves are also shown in Figure 1.34. The hyperbola and the parabola have been used success-fully to describe mathematically the terrestrial BS and MP shapes, respectively [e.g., Fairfield, 1971]. Assuming that the BS is an axially symmetric hyperbola in the orbital

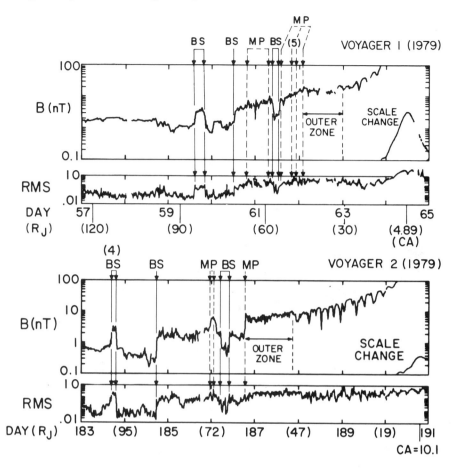

Fig. 1.35. Magnetic fields measured by Voyager 1 (above) and Voyager 2 (below) inbound at Jupiter to closest approach (CA). Given are the 16-min. average magnetic field magnitude B in nanoteslas (nT), field direction in heliographic longitude λ_H and latitude δ_H, and rms deviation in nT. λ is measured in a plane parallel to the sun's equator with $\lambda_H = 180°$ in the sunward direction. $\delta_H = 90°$ is northward with respect to that plane. Bow shock (BS) and magnetopause (MP) crossings are indicated (see text).

plane and that the MP is an axially symmetric parabola, and that these curves pass through the *average* Voyager boundary positions, the following curves resulted:

Voyager 1 MP: $y_0 = [168\,(57 - x_0)]^{1/2}$

Voyager 2 MP: $y_0 = [102\,(68 - x_0)]^{1/2}$

Voyager 1 BS: $y_0 = 0.712\,[(242 - x_0)^2 - 28{,}900]^{1/2}$

and

Voyager 2 BS: $y_0 = 0.380\,[(569 - x_0)^2 - 236{,}000]^{1/2}$

where x_0 and y_0 are coordinate positions relative to the planet-centered Jupiter Orbital Coordinate System in units of Jovian radii, and where $\hat{X}_0$ is positive sunward, $\hat{Z}_0$ is per-

Table 1.4. *Summary of Voyager boundary crossings*

	Bow shock			Magnetopause						
		Distance (R_J)				Distance (R_J)		Average	Normal	
									Est	Model
Spacecraft	No	First	Last	No.	(No.)*	First	Last	δ_H	λ_H	λ_H
Voyager-1										
Inbound	5	86	56	9	(8)	67	47	3°	165°	168°
Outbound	7	199	258	3	(3)	158	165	7°	124°	120°
Voyager-2										
Inbound	7	99	66	3	(2)	72	62	−1°	155°	152°
Outbound	9	282	>380	17	(2)	170	279	?	?	109°

(No.)* refers to the number of well-estimated cases; see text.

pendicular to the planet's orbital plane, positive northward, and $\hat{Y}_0 = \hat{Z}_0 \times \hat{X}_0$. $\hat{X}_0$ defines the axis of symmetry. Because neither Voyager spacecraft traversed the dusk-side boundaries (which was also true for P 10 and P 11), little can be said about the duskside portions of the models. Aberration due to planetary motion with respect to the solar wind is negligible, but an asymmetry may still exist, at least at lower latitudes near Jupiter, because of the rapid rotation of the planet and plasma loading at lower latitudes.

Table 1.4 summarizes the number of BS and MP boundary crossings, both the total number in each case and for the MP crossings the number for which normals to the surface could be acceptably well estimated [Lepping and Behannon, 1980] using the Sonnerup and Cahill [1967] minimum variance technique; the latitude (δ_H) and longitude (λ_H) of the average of the "well-estimated" MP normals, and the longitude of the normal to the model MP are also given in HG coordinates. Table 1.4 shows that, except for the outbound V 2 MP crossings, the agreement between the average estimated and the model normals is quite good; it has been assumed that $\delta_H\{\text{Model}\} \equiv 0°$. Identification of the V 2 outbound MP traversals was based on rapid changes in field direction together with changes in 10–140 eV plasma electron flux and plasma ion density (J. Belcher and H. Bridge, private communication, 1980). Only two of the 15 crossings provided results consistent with the model normal. In almost all ($\geq 87\%$) of the "well-estimated" MP cases, the boundary surface discontinuities were consistent with being tangential discontinuities according to the criteria of Lepping and Behannon [1980].

Magnetosheath field structure

The magnetic field in Jupiter's magnetosheath, particularly along the dawn magnetospheric flank, exhibits quasiperiodic variations in orientation that appear to be correlated with planetary rotation, and it is appropriate to include a brief description of the MS observations in this outer magnetosphere survey. Specifically, the fields observed by V 1 and V 2 outbound in the MS were oriented predominantly in a north or south direction and were found to vary between the two extremes of orientation with a quasiperiod of ≈ 10 hr, the Jovian rotation period [Lepping, Burlaga, and Klein, 1981; Lepping et al., 1981]. Examination of the P 10 MS outbound data has revealed a similar large-scale structuring of the MS in those observations, also. Similar directional changes in the field were observed by the Voyager spacecraft inbound through the MS near the MP, but the occurrence was much less frequent, no quasiperiodicity was apparent, and the scale lengths were on average much shorter. Highly inclined fields have also been observed occasionally in the Earth's MS, although to date there have been no reports of sheath field variations that are controlled by planetary rotation. Thus, the phenomenon detected at Jupiter may be unique to and characteristic of the Jovian system.

A broad view of the phenomenon is given in Figure 1.36, which shows the latitude (δ_H) of the magnetic field as observed by V 2 outbound for a period of approximately ten days. The data are plotted with respect to the System III (1965) longitude of the subspacecraft position, with two 360° cycles (plus 45° overlap) per panel. The calendar day (and fraction of a day) at the beginning of each panel is shown at the left. Periods during which the spacecraft reentered the magnetosphere or entered the solar wind are denoted by shading.

It can be seen in the figure that a significant number of field variation "events" occurred near $\lambda_{III} \approx 360°$ (indicated by vertical arrows at bottom). Although "events" occurred at other longitudes as well, there was no tendency for the events to cluster

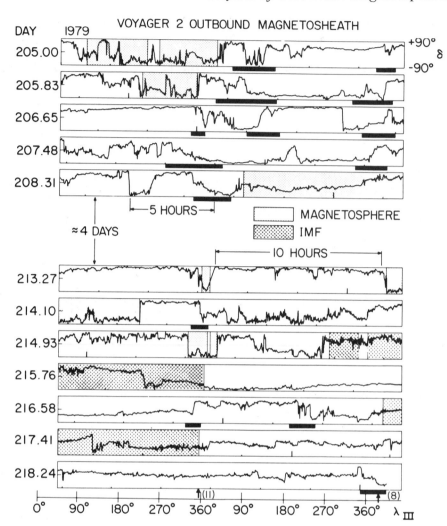

Fig. 1.36. Latitude (δ_H) of the magnetic field in the outbound MS along the Voyager 2 trajectory for a total of 12 Jupiter rotation periods; the central four days consisted primarily of magnetosphere data. Periods within the magnetosphere or in the interplanetary magnetic field (IMF) are denoted by shading. The data are plotted with respect to Jupiter's System III longitude of 1965.0 (λ_{III}) and day of the year, with major tick marks denoting starts of days.

about any longitude other than 360°. Here an "event" is defined as one of four occurrences: (1) a major north-to-south change of the latitude δ_H; (2) a major south-to-north change of δ_H; (3) a MP crossing; or (4) a BS crossing. Eleven such events occurred in association with the first arrow and eight with the second. This suggests a synchronization of field latitude changes and MP and BS motions jointly with the planetary rotation period of ≈ 10 hr (9.925 hours for System III, 1965). Some events also occur with a 5-hr period, but with a lower occurrence frequency. Note that there is much less variability in the later sheath measurements taken at greater distance from the MP.

A minimum variance analysis according to the method of Sonnerup and Cahill [1967] was carried out using 78 1-hr averages of V 2 sheath data and 64 V 1 1-hr aver-

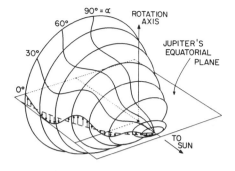

Fig. 1.37. A model of the dynamical magnetopause of Jupiter. The MP surface is shown here as it would be seen by an observer 30° away from the Jupiter–Sun line in the equatorial plane and 30° above that plane. It is assumed that near the vicinity of the nose of the magnetosphere the intersection of the MP surface with a plane perpendicular to the Jupiter-Sun line is an ellipse due to an internal disc-like current sheet. The ratio of the semimajor and semiminor axes was chosen equal to that given by Engle and Beard [1980] for the ellipse in the plane which contains the rotation axis. As the planet rotates, an ellipse near the nose oscillates about the Jupiter–Sun line. Information about the orientation of this ellipse propagates tailward at a speed which equals the sum of the Alfvén speed and the bulk speed in the magnetosheath near the MP (see text). The dashed line is the model MP derived from observed crossing positions [Lepping et al., 1980].

ages [Lepping et al., 1981]. This analysis showed the tendency for the undulatory variations in the MS field to be contained in a plane which was nearly parallel to the local model MP for each data set analyzed. Similar analyses were carried out for portions of most of the large sheath field features individually using 48 s average V 1 and V 2 field data. The specific analysis intervals for V 2 are indicated by the black horizontal bars in Figure 1.35. With only a few exceptions, the individual minimum variance-estimated normals were found to cluster closely around the model MP normals.

It was additionally found by Lepping et al. [1981] that the north–south component of the magnetic field observed outbound in the MS by V 1 and V 2 was strongly correlated with the north–south plasma velocity component. The components orthogonal to the north–south direction showed only weak correlations.

A plausible explanation for the observed phenomenon described briefly here can be formulated in terms of the draping of MS fields around a flattened (approximately in the $\hat{Z}_0$ direction) and dynamic Jovian magnetosphere, as suggested in the previous section. The motion and shape of such a magnetosphere depend on the rapid rotation of the current sheet within it. The flowing plasma in the MS is probably being deviated alternatively "northward" and "southward," as viewed in Jupiter's equatorial plane, primarily by the rocking motion of the magnetospheric current sheet about the Sun–Jupiter line. The field lines frozen into this deflected plasma are constrained to lie in a plane parallel to the MP, especially close to the MP, and to move with the plasma in a convected wave generated by the rocking motion of the magnetosphere (the geometry is illustrated in Fig. 1.37 and further described in the caption). Similarly, the quasi-periodic V 2 observation of the MP may be related to the motion and configuration of the magnetospheric current sheet, particularly that of the magnetotail current sheet. Pioneer observations suggested that the magnetospheric current sheet contributes a substantial disturbance field in Jupiter's outer magnetosphere [Smith et al., 1975]. The Voyager observations provide strong evidence that the effects of this current sheet probably extend beyond the magnetosphere into the magnetosheath.

Prior to the Voyager encounters with Jupiter, Dessler and Vasyliunas [1979] speculated that the preferential concentration of magnetospheric plasma within the active sector hypothesized in the magnetic anomaly model [Dessler and Hill, 1975] might have an observable influence on the location and motion of the MP. On this basis, a range, in λ_{III} (1965) coordinates, was predicted within which the Voyager MP crossings should occur. The longitude of greatest occurrence of V 2 outbound crossings, $\lambda_{III} \approx 360°$, lies just outside the range suggested by Dessler and Vasyliunas. A 2.5 hr delay ($\Delta\lambda_{III} \simeq 90°$) is required to force agreement. Following the Voyager encounters, Vasyliunas and Dessler [1981] have reiterated what had been stated as a qualifier in Dessler and Vasyliunas [1979]: that there is no clear theoretical basis for believing that the active sector can affect the MP, as for example, through the formation of an active sector-associated asymmetrical bulge in the MP surface. It is clear from the Voyager magnetometer results that a possible validation of the magnetic anomaly model of the Jovian magnetosphere will require further coordinated interdisciplinary studies and perhaps additional observations.

1.5. Summary

From the foregoing it is clear that significant progress has been made toward the understanding and modeling of the Jovian magnetosphere, both from an absolute knowledge point of view as well as from the standpoint of comparative studies of planetary magnetospheres in the solar system.

The largely complementary trajectories of the Pioneer and Voyager spacecraft have allowed the in situ exploration of a reasonably representative fraction of the Jovian magnetospheric cavity although the coverage of important regions is far from complete, as illustrated in Figure 1.34. The intrinsic planetary field outside ~ 1.5 R_J and the inner and middle magnetosphere, including the equatorial current sheet, are reasonably well understood on a global scale. The same applies to the large scale configuration of the magnetosphere, although this understanding is restricted to regions relatively close ($< 300 \ R_J$) to the planet and our understanding is as yet qualitative. Much of the Voyager observations and the associated initial interpretations and models have yet to be fully digested. The retrospective and correlative studies that will necessarily be undertaken in the next few years will probably eclipse our present understanding quickly, as the pieces of the puzzle laid out in this volume are assembled. From the magnetic field point of view there remains, however, a large number of problems which require investigation; among these are: (1) the small scale geometry of the field near the surface and the large scale configuration of Alfvén standing waves in the Io torus which undoubtedly play a significant role in determining decametric emission characteristics; (2) additional studies concerning the electrodynamic interaction of satellites and rings with the corotating plasma; (3) large scale current systems in the magnetosphere and on its boundaries; (4) the adiabatic motion of charged particles in the Jovian magnetodisc; (5) self-consistent magnetic field/plasma models; (6) long term secular variations of the intrinsic planetary field; and (7) possible intrinsic magnetic fields of the Galilean satellites.

The polar regions remain largely unexplored except for remote optical and UV observations from spacecraft and ground-based observatories. These regions are certain to play a crucial role in the dynamics of the magnetosphere-solar wind interaction, as in the case of the Earth, and their study is of fundamental importance.

Future spacecraft placed in orbit around Jupiter will contribute essential data to carry out these studies. Ground-based observations will augment and complement spacecraft observations and will provide the long term observational baseline required

Fig. 1.38. Contour map showing maximum errors in current sheet model magnetic field components, B_ρ or B_z, in three regions of ρ, z space for which simple analytic forms exist.

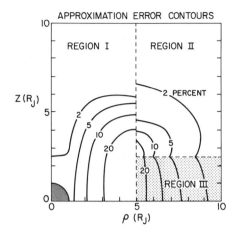

to monitor large-scale changes in the morphology of the Jovian system as well as discover potentially new phenomena in the magnetosphere. These studies will contribute significantly to the understanding of our own Earth's magnetosphere and the general problem of the interaction of rapidly rotating magnetized bodies with stellar winds.

APPENDIX

For the purposes of computing the field due to the model current sheet the numerical integration of the integral results quoted by Connerney, Acuña, and Ness [1981] may be cumbersome. The following analytical approximations, applicable in the regions defined in Figure 1.38 can be used with negligible error except in the immediate vicinity of the inner sheet edge:

$$0 \le \rho \le 30\ R_J;\ |z| \le 30\ R_J$$

$$B_\rho^{I} = (\mu_0 I_0/2)\,(\rho/2)\,(1/F_1 - 1/F_2)$$

$$B_z^{I} = (\mu_0 I_0/2)\,\{\,5(z^2 + 5^2)^{-1/2} - (\rho^2/4)\,[\,(\,(z-2.5)/F_1^3) - (\,(z+2.5)/F_2^3)\,]\,\} - B_z'$$

$$B_\rho^{II} = (\mu_0 I_0/2)\,\{\,(1/\rho)\,[F_1 - F_2 + 5] - (5^2\rho/4)\,[1/F_1^3 - 1/F_2^3]\,\}$$

$$B_z^{II} = (\mu_0 I_0/2)\,\{\,5[z^2 + \rho^2]^{-1/2} - (5^2/4)\,[\,(z - 2.5)/F_1^3 - (z + 2.5)/F_2^3]\,\} - B_z'$$

$$B_\rho^{III} = (\mu_0 I_0/2)\,\{\,(1/\rho)\,[F_1 - F_2 + 2z] - (5^2\rho/4)\,[1/F_1^3 - 1/F_2^3]\,\}$$

$$B_z^{III} = B_z^{II}$$

and

$$B_z' = (\mu_0 I_0/2)\,(1/10)$$

where in Region I:

$$F_1 = [(z - 2.5)^2 + 5^2]^{1/2}$$

$$F_2 = [(z + 2.5)^2 + 5^2]^{1/2}$$

Physics of the Jovian magnetosphere

and in Regions II and III:

$$F_1 = [(z - 2.5)^2 + \rho^2]^{1/2}$$

$$F_2 = [(z + 2.5)^2 + \rho^2]^{1/2}$$

With z and ρ in units of Jovian radii. A choice of $(\mu_0 I_0/2) = 225$ (150) corresponding to the V 1 (V 2) model of Connerney, Acuña, and Ness [1981] leads to a computed field in nanoteslas. The quantity contoured in Figure 1.38 is the percent error of the approximate solution for either B_ρ or B_z, whichever is greatest, determined by comparison of the computation using the approximate solutions given here and the appropriate numerical integration [Connerney, Acuña, and Ness, 1981].

ACKNOWLEDGMENTS

The authors are grateful to their colleagues N. F. Ness and A. W. Schardt, Editor A. J. Dessler, and two anonymous referees for their helpful comments and suggestions at various stages of the development of this review. We also thank wholeheartedly Barbara C. Holland for producing and correcting the typescript and Floyd H. Hunsaker and Larry A. White for producing most of the illustrations.

2

IONOSPHERE

Darrell F. Strobel and Sushil K. Atreya

Our understanding of Jupiter's ionosphere has been enhanced by the Voyager encounters. Vibrationally excited H_2 ($v \geq 4$) probably played an important role in providing a rapid loss mechanism for protons (the major topside ion) during the Voyager encounter. A straightforward calculation of the Voyager 1 entry electron concentration profile with chemistry and physics adequate to understand the Pioneer radio occultation profiles yields significant differences from the Voyager radio science measurements and suggests that substantial improvements in models may be necessary if the preliminary results of the observations remain unaltered after improved reduction and analysis of the data. It is argued that, although solar EUV radiation probably controlled the ionosphere during the Pioneer observations, particle precipitation as evidenced by strong H_2 airglow emissions on a planetwide basis appears to have played an essential role in both heating the thermosphere and as an ionization source during the Voyager encounters. The main contribution to the Pedersen conductivity in Jupiter's ionosphere occurs in the region where multilayer structure is dominant and accurate reductions of the Voyager radio occultation measurements are not yet available. Theoretical estimates of the integrated Pedersen conductivity are in the range of 0.02–10 mho; the former values are representative of an ionosphere produced by solar EUV radiation and the latter of the auroral ionosphere under intense particle precipitation.

2.1. Introduction

The original interest in an ionosphere on Jupiter was generated by the discovery of strong radio-frequency emissions at ~ 20 MHz that were thought to be plasma frequencies associated with Jupiter's ionosphere [Gardner and Shain, 1958]. A historical summary of developments in our understanding of Jupiter's ionosphere may be found in Strobel [1979]. The ionosphere of Jupiter provides a means to couple the magnetosphere to the atmosphere by virtue of its high conductivity and collisional interaction with the neutral atmosphere. The Pioneer and Voyager spacecraft have done much to stimulate interest in Jupiter's ionosphere by providing direct measurements of profiles of electron concentration at selected locations on Jupiter. Detailed measurements of ionospheric structure provide important information on the neutral atmosphere and physical processes that affect its structure in addition to data of purely ionospheric interest. Because a number of reviews on Jupiter's ionosphere have appeared previously [Hunten, 1969; McElroy, 1973; Huntress, 1974; Atreya and Donahue, 1976; Strobel, 1979], this chapter emphasizes the contributions from the Voyager program to our understanding of Jupiter's ionosphere.

2.2. Basic principles

The most fundamental quantities needed to understand the ionosphere are the electron and ion concentrations as a function of altitude, latitude, and, in the case of Jupiter, longitude. They are solutions of the electron continuity equation

$$\frac{\partial n_e}{\partial t} + \nabla \cdot \overline{\phi}_e = P_e - \alpha n_e^2 \tag{2.1}$$

and ion continuity equation

$$\frac{\partial n_i}{\partial t} + \nabla \cdot \overline{\phi}_i = P_i - L_i n_i \tag{2.2}$$

where n is the concentration, $\overline{\phi}$ the flux, P the production rate per unit volume, L the loss rate, and α the recombination rate. The subscripts e and i refer to electrons and ions, respectively. The constraint of electrical neutrality requires $n_e \equiv \Sigma_i\, n_i$. The only adequately understood sources of electrons and ions are solar extreme ultra-violet (EUV) photoionization of neutral atmospheric species and subsequent ioniza-tion by the ejected photoelectrons. Jupiter's ionosphere is peculiar when compared to other planetary ionospheres in that energetic-particle precipitation exhibits distinct longitudinal asymmetry and day–night asymmetry as evidenced by optical signatures obtained in the ultraviolet [Broadfoot et al., 1981]. Appropriate asymmetries in $P_{e,i}$ and hence $n_{e,i}$ are expected. An unfortunate aspect of particle precipitation is that direct measurements are not available and only indirect inferences from optical signatures can be used to estimate source strength and spatial variation of $P_{e,i}$. During the period of the Voyager encounter it appears that, in high magnetic latitudes, particle precipita-tion was the major source of ionization, atomic hydrogen, and heat for the thermo-sphere. In contrast, ion chemistry and plasma recombination processes for Jupiter's ionosphere are fairly well understood, as reviewed below.

Redistribution of plasma as represented by the flux divergence terms is not under-stood for Jupiter with the exception of ambipolar diffusion along magnetic flux tubes. Horizontal wind systems in Jupiter's thermosphere could transport plasma along the flux tubes by ion drag to raise or lower ionospheric layers. Intense heating of auroral regions as observed by Voyager would drive a wind system similar to that modeled by Roble et al. [1979] for the Earth. At midlatitudes, upward plasma drifts of as much as 50 ms^{-1} could be generated by this wind system. Dynamo electric fields and $\mathbf{E} \times \mathbf{B}$ drifts could be as important for vertical transport across the horizontal, equatorial magnetic fields as in the Earth's equatorial ionosphere. Unfortunately the magnitude of these electric fields in Jupiter's dynamo region is not known.

2.3. Ionization sources

From our terrestrial experience, we would expect solar EUV radiation to dominate the production of ionization in the main ionospheric layers of Jupiter in spite of the reduc-tion in solar flux by a factor of 27 relative to Earth. Solar EUV ionization is most rele-vant for the equatorial ionosphere outside of the Lyman-α bulge region, whereas particle ionization plays a significant role at high latitudes. The major ionizable constit-uents H_2, He, and H require solar photons shortward of 804, 504, and 912 A (1 A $= 10^{-10}$ m), respectively, to eject an electron. Direct photoionization of H_2 is the major ionization channel. At the time of the Pioneer observations, direct pho-toionization of H was a negligible source of protons for the ionosphere. The most potent source of H^+ was dissociative photoionization of H_2 [McElroy, 1973] as illus-trated in Figure 2.1. The principal reason is that for average solar activity, atomic hydrogen remains a minor constituent ($\sim 10^{-3}$ [H_2]) in the main ionospheric region. However, during Voyager observations the increased Lyman-α albedo implies much greater atomic hydrogen abundances [Yung and Strobel, 1980] and thus enhanced direct photoionization of H [Atreya, Donahue, and Waite, 1979]. Helium, with a greater abundance than H (during the Pioneer encounter), has a large ionization production rate in spite of its high ionization potential (see Fig. 2.1). Photoelectrons

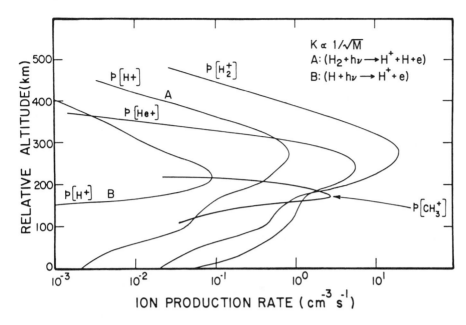

Fig. 2.1. Ion production rates $p(X^+)$ by solar EUV radiation with solar zenith angle of 60° for a model atmosphere with the eddy diffusion coefficient $K \propto [M]^{-1/2}$, where M is the neutral concentration, and equal to 3×10^7 cm^2 s^{-1} at the turbopause appropriate for Pioneer observations. The height scale refers to the altitude above the level at which the atmospheric density is 10^{16} cm^{-3}. Note that the CH$_3^+$ production rate is based on revised CH$_3$ densities from Yung and Strobel [1980] (after Atreya and Donahue [1976]).

produced in the ionization process are not an important source of protons according to the analysis of Cravens, Victor, and Dalgarno [1975]. In Jupiter's ionosphere, the CH$_3$ radical may play a comparable role to that of NO in the Earth's lower ionosphere. With a low ionization potential, it is ionized by absorption of solar Lyman-α and initiates the formation of a low-lying ionization layer [Prasad and Tan, 1974, Fig. 1].

The importance of galactic cosmic rays as an ionizization source is enhanced relative to the Earth by the drop in solar radiation intensity with increasing distance from the Sun [Capone et al., 1976, 1977]. Calculation of ion production rates by cosmic ray bombardment of Jupiter is complicated. The most recent computation was by Capone et al. [1979], who obtained a peak ionization rate of ~10 cm^{-3} s^{-1} at a magnetic latitude of 60° and neutral concentration of 10^{19} cm^{-3}.

Another potentially important ionization source on Jupiter is neutral atoms and molecules ejected from the Galilean satellites under the intense radiation environment of the extensive Jovian magnetosphere. Under bombardment by keV or MeV particles from the Jovian magnetosphere, material can be sputtered off the surfaces of the Galilean satellites [Matson, Johnson, and Fanale, 1974]. In the case of Io, this process probably results in the formation of an extensive, but incomplete, torus of neutral sodium [R. A. Brown, 1974, Chapter 6]. From the Voyager encounters with Jupiter we know that Io is the site of intense volcanic activity [Morabito et al., 1979], has an SO$_2$ atmosphere [Pearl et al., 1979], and is the source of S and O ions observed in the Io plasma torus [Bridge et al., 1979a Broadfoot et al., 1979]. Although the details of how and where neutral particles are injected in the torus are uncertain, the ionization of neutrals occurs primarily by electron impact and at a rate of $(2–20) \times 10^{28}$ ion s^{-1} (see

Chap. 3 and 6). The newly created ions are accelerated and swept away by the Jovian magnetic field that rotates more rapidly than the satellites.

Siscoe and Chen [1977] have argued that the major loss of this ionization is radial diffusion by flux tube interchange, which is driven by electric fields associated with winds in Jupiter's ionosphere [Brice and McDonough, 1973; Coroniti, 1974]. Siscoe [1978a] has estimated probabilities for inward radial diffusion from individual satellites. In the case of Io, he estimated 0.52 for low concentrations of torus plasma. From analysis of Voyager data, Richardson et al. [1980] suggest that 0.1 is more appropriate for high concentrations of torus plasma. These inward-diffusing ions are eventually captured by the parent planet, whereas the outward-diffusing ions are lost by interaction with the solar wind. Based on the supply rate of ions to the plasma torus, we estimate that the flux of oxygen and sulfur ions and associated electrons into Jupiter's atmosphere may be as large as 3×10^7 cm^{-2} s^{-1}, which corresponds to a total flux of about 10^{28} s^{-1}. The energy distribution of these entering ions and associated electrons is unknown. The loss of particles into the atmosphere may be asymmetric as discussed below.

In spite of these uncertainties there is considerable evidence for precipitating particles (probably electrons) into Jupiter's equatorial and polar atmospheres from observations of intense H_2 band emission by the Voyager UVS [Sandel et al., 1979, 1980; McConnell, Sandel, and Broadfoot, 1980]. The nightside Lyman-α observed by the Voyager UVS also requires particle precipitation to explain the measured intensity [McConnell, Sandel, and Broadfoot, 1980]. A distinct peculiarity of the equatorial H_2 emissions is that their intensity seems to depend on whether the Sun is shining (Sandel et al., 1979; McConnell, Sandel, and Broadfoot, 1980; Broadfoot et al., 1981). During the day the H_2 Lyman and Werner band intensity was observed to be 2.8 ± 1.0 kR, whereas at night an upper limit of 500 R was obtained [Broadfoot et al., 1981; McConnell, Sandel, and Broadfoot, 1980]. In addition to this day–night asymmetry, there is a pronounced longitudinal asymmetry in the Lyman-α brightness of Jupiter with peak intensity at $\lambda_{III} \sim 80$–$120°$ W, and latitude ~ 8–$12°$ N [Sandel, Broadfoot, and Strobel, 1980; Clarke et al., 1980; Dessler, Sandel, and Atreya, 1981]. Although the Lyman-α bulge is a manifestation of resonance scattering of solar Lyman-α from an enhanced H column density region above the absorbing CH_4 layer [Sandel, Broadfoot, and Strobel, 1980], the atomic hydrogen bulge is created by convecting, corotating magnetospheric electrons (~ 100 keV and ~ 1 erg cm^{-2} s^{-1}) precipitating preferentially in the longitude and latitude region of the Lyman-α bulge. This magnetospheric convection pattern is driven by a longitudinal mass asymmetry in the Io plasma torus [Dessler, Sandel, and Atreya, 1981]. Substantial dissociation and ionization of H_2 as well as heating occur in this preferred region of precipitation, but magnetospheric electrons penetrate too deeply for H_2 band emission to escape absorption by overlying CH_4 [Strobel, 1979].

Another ionization source that involves satellites is the electromagnetic interaction of Io with Jupiter. Goldreich and Lynden-Bell [1969] argued that Io is a unipolar generator that develops a potential drop of 400 kV across its radial diameter and drives a current system along its magnetic flux tube. Each foot of the flux tube imbedded in the Jovian ionosphere carries a current of $\sim 10^6$ A principally by keV electrons with power dissipation that can approach 4×10^{10} W [Goldreich and Lynden-Bell, 1969; see also Dessler and Chamberlain, 1979]. The mechanisms for charged-particle acceleration are still uncertain. Ionization and UV radiation constitute part of the power dissipation by energetic charged particles. Atreya et al. [1977] measured localized Lyman-α emission of ~ 300 kR (assumed from a spot with diameter 1000 km) that almost certainly originated from the foot of Io's flux tube. (A revision by a factor of about 2.6 in the geocoronal Lyman-α calibration factor changes the hot spot Lyman-α intensity reported by Atreya et al. [1977] from 120 to 300 kR.) Based on H_2 cross sections, approximately 10 ion pairs are produced for each Lyman-photon emitted [Cravens, Victor, and Dalgarno, 1975]. Thus at each flux tube foot approximately 3×10^{12} ionization events cm^{-2} s^{-1} accompany the observed Lyman-α emission. An alternate explanation of the observed hot spots is a transient manifestation of the interaction between the active sector in the magnetic-anomaly model of the Jovian magnetosphere and the Io plasma torus as a consequence of longitudinal density gradients and current driven plasma instability [Dessler, Sandel, and Atreya, 1981].

During the Voyager encounters with Jupiter, when the Io plasma torus densities were high, the auroral activity occurred over an extensive area of Jupiter's polar region (width ~ 6000 km), which corresponded to magnetic field lines that map the Io plasma torus into Jupiter's atmosphere, rather than being confined to the Io flux tubes [Broadfoot et al., 1979; Sandel et al., 1979]. The measured intensities were approximately 40 kR of Lyman-α and 80 kR of H_2 Lyman and Werner bands [Broadfoot et al., 1981] and corresponded to an energy deposition rate of ~ 10–30 ergs cm^{-2} s^{-1} or 2–6×10^{13} W of 3–30 keV electrons and a column ionization rate of $\sim 1 \times 10^{12}$ cm^{-2} s^{-1} [Yung et al., 1982]. These large ionization rates are considerably greater than elsewhere on the planet and thus generate locally enhanced electron-number densities and conductivities in the polar regions as compared to the equatorial and midlatitude ionosphere.

2.4. Ion recombination

The principal atomic ion, H^+, recombines radiatively at a rather slow rate of $\sim 3.5 \times 10^{-12}$ cm^3 s^{-1} at low plasma concentrations and room temperature [Bates and Dalgarno, 1962]. Radiative recombination typically exhibits an electron temperature dependence of the form $T_e^{-0.7}$ [Bates and Dalgarno, 1962]. A similar recombination rate is also appropriate for He^+, but is unimportant due to the rapid removal of He^+ by reaction with H_2.

By virtue of the rapid ion–molecule reaction, $H_2^+ + H_2 \rightarrow H_3^+ + H$, H_3^+ recombination acquires a central role in plasma recombination in Jupiter's ionosphere. Leu, Biondi, and Johnsen [1973] have measured an H_3^+ dissociative recombination rate of $(2.3 \pm 0.3) \times 10^{-7}$ $(T_e/300)^{-1/2}$ cm^3 s^{-1}. At partial pressures of $\sim 10^{-2}$ Torr (1 Torr = 133 Pascal = 1 mm Hg) and low temperatures (~ 205 K) they obtained a recombination coefficient for the H_5^+ dimer ($H_3^+ \cdot H_2$) of $(3.6 \pm 1.0) \times 10^{-6}$ cm^3 s^{-1}. The available evidence for H_3^+ strongly suggests that it can only recombine slowly as a result of unfavorable potential curve crossings in the ground vibrational level [Bates, 1962; see also references in Gross and Rasool, 1964]. A conservative upper limit of 10^{-8} cm^3 s^{-1} can be put on this recombination rate.

Hydrocarbon ions play an important role in the lower ionosphere, principally around and below the homopause. Only limited laboratory studies have been performed on these ions. In many instances the identity of the major recombining ion and the products of recombination are not accurately known. The parent gas serves to identify the recombining plasma. For CH_4 a plasma recombination rate of 3.9×10^{-6} cm^3 s^{-1} has been measured by Maier and Fessenden [1975] at room temperature, somewhat larger than the Rebbert, Lias, and Ausloos [1973] value of 1.9×10^{-6}. The dominant recombining ions are probably CH_5^+ and $C_2H_5^+$. As the complexity of the heavy hydrocarbon ions increases, the recombination rate increases initially

Fig. 2.2. Schematic of the important
ionospheric chemical reactions [Atreya
and Donahue, 1981]. The hatched circles
represent terminal ions that are removed
by electron recombination.

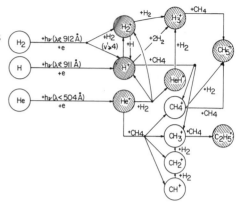

(e.g., 5.8×10^{-6} cm^3 s^{-1} for C$_3$H$_8$) but eventually it decreases with greater complexity
(e.g., 4.2×10^{-6} cm^3 s^{-1} for C$_5$H$_{12}$ [Maier and Fessenden, 1975]). Unlike hydrocarbon
plasmas, recombination in an NH$_3$ plasma is pressure dependent with a rate varying
from 4.6×10^{-6} at 1 Torr to 11.2×10^{-6} cm^3 s^{-1} at 10 Torr [Maier and Fessenden,
1975].

2.5. Ion chemistry

In this section, a brief review of the most important considerations relevant to understanding the ion chemistry of Jupiter is given. A detailed discussion of Jovian ion
chemistry is given by Huntress [1974] and Atreya and Donahue [1976], and a recent
summary of relevant laboratory data may be found in Huntress [1977]. Important
chemical reactions in the Jovian ionosphere are listed in Table 2.1.

The atmosphere of Jupiter consists predominantly of H$_2$ with $\sim 10\%$ He by volume
and trace amounts of CH$_4$, NH$_3$, H, and other hydrocarbons. The dominant ion produced at high altitudes by photon and particle impact is H$_2^+$ (see Fig. 2.1), which
quickly reacts with H$_2$ to form H$_3^+$ (cf. Table 2.1 and Fig. 2.2). The charge transfer reaction H$_2^+$ + H is also fast and is important in the topside ionosphere as a H$^+$ source
where [H] $\gtrsim$ [H$_2$]. From the previous discussion, the most important source of H$^+$ is
photodissociative ionization of H$_2$. Similarly, energetic particles precipitating into the
Jovian atmosphere generate secondary electrons that produce considerable H$^+$ by electron impact dissociative ionization of H$_2$. An additional source of H$^+$ is the reaction of
He$^+$ + H$_2$, which is particularly important if He is uniformly mixed with H$_2$ to very
high altitudes. There is appreciable production of He$^+$ (see Fig. 2.1) and it primarily
reacts with H$_2$ to form H$^+$ approximately 80% of the time and H$_2^+$ approximately 20%
of the time [Johnsen and Biondi, 1974]. At high altitudes, H$^+$ does not react with He
and H$_2(v)$ unless $v \geq 4$ and thus must either diffuse downward to chemical sinks at lower
altitudes or slowly recombine radiatively. These chemical sinks are the reactions

$$H^+ + CH_4 \rightarrow CH_3^+ + H_2 \qquad (2.3)$$

$$\rightarrow CH_4^+ + H$$

$$H^+ + H_2 + H_2 \rightarrow H_3^+ + H_2$$

[Dalgarno, 1971; McElroy, 1973]. During the Voyager encounters with Jupiter, when
high thermospheric temperatures were produced by particle precipitation and elevated
photoelectron fluxes by high solar activity; photoelectrons with less than 10 eV lose

Table 2.1. *Important reactions in the ionosphere of Jupiter*
[Atreya and Donahue, 1976]

Reaction number	Reaction	Rate constant[a]
	Ion production	
$p1$	$H_2 + h\nu \rightarrow H_2^+ + e$	
$p2$	$\rightarrow H^+ + H + e$	
$p3$	$H_2 + e \rightarrow H_2^+ + 2e$	
$p4$	$\rightarrow H^+ + H + 2e$	
$p5$	$H + h\nu \rightarrow H^+ + e$	
$p6$	$H + e \rightarrow H^+ + 2e$	
$p7$	$He + h\nu \rightarrow He^+ + e$	
$p8$	$He + e \rightarrow He^+ + 2e$	
	Ion exchange	
$e1$	$H_2^+ + H_2 \rightarrow H_3^+ + H$	2.1×10^{-9}
$e2$	$H_2^+ + H \rightarrow H^+ + H_2$	$\sim 6.4 \times 10^{-10}$
$e3$	$He^+ + H_2 \rightarrow H_2^+ + He$	$\lesssim 20\%$
$e4$	$\rightarrow HeH^+ + H \quad$ sum	1.0×10^{-13}
$e5$	$\rightarrow H^+ + H + He$	$\gtrsim 80\%$
$e6$	$He^+ + CH_4 \rightarrow CH^+ + H_2 + H + He$	2.4×10^{-10}
$e7$	$\rightarrow CH_2^+ + H_2 + He$	9.3×10^{-10}
$e8$	$\rightarrow CH_3^+ + H + He$	6.0×10^{-11}
$e9$	$\rightarrow CH_4^+ + He$	4.0×10^{-11}
$e10$	$H^+ + H_2 + H_2 \rightarrow H_3^+ + H_2$	3.2×10^{-29}
$e11$	$H^+ + CH_4 \rightarrow CH_3^+ + H_2$	2.3×10^{-9}
$e12$	$\rightarrow CH_4^+ + H$	1.5×10^{-9}
$e13$	$HeH^+ + H_2 \rightarrow H_3^+ + He$	1.85×10^{-9}
$e14$	$H_3^+ + CH_4 \rightarrow CH_5^+ + H_2$	2.4×10^{-9}
$e15$	$CH^+ + H_2 \rightarrow CH_2^+ + H$	1.0×10^{-9}
$e16$	$CH_2^+ + H_2 \rightarrow CH_3^+ + H$	7.2×10^{-10}
$e17$	$CH_3^+ + CH_4 \rightarrow C_2H_5^+ + H_2$	9.6×10^{-10}
$e18$	$CH_4^+ + CH_4 \rightarrow CH_5^+ + CH_3$	1.15×10^{-9}
$e19$	$CH_4^+ + H_2 \rightarrow CH_5^+ + H$	4.1×10^{-10}
	Ion removal/electron–ion recombination	
$r1$	$H_3^+ + e \rightarrow H_2 + H$	see text
$r2$	$H_2^+ + e \rightarrow H + H$	$< 1.0 \times 10^{-8}$
$r3$	$HeH^+ + e \rightarrow He + H$	$\sim 1.0 \times 10^{-8}$
$r4$	$H^+ + e \rightarrow H + h\nu$	see text
$r5$	$He^+ + e \rightarrow He + h\nu$	see text
$r6$	$CH_5^+ + e \rightarrow$ neutral	1.9×10^{-6}
$r7$	$C_2H_5^+ + e \rightarrow$ products	1.9×10^{-6}

[a] The rate constants are in units of $cm^3 s^{-1}$ for two-body reactions, and $cm^6 s^{-1}$ for three-body reactions (see Atreya and Donahue [1976] for references).

their energy primarily by vibrational excitation of H_2 [Henry and McElroy, 1969]. Significant departures from a Boltzmann distribution occur at low densities to enhance high vibrational levels. Thus, at high altitudes and low H_2 densities, the reaction

$$H^+ + H_2 \, (v \geq 4) \rightarrow H_2^+ + H \tag{2.4}$$

is an important loss path for protons [McElroy, 1973; Atreya, Donahue, and Waite, 1979].

Below the homopause or turbopause, the level where an atmosphere ceases to be mixed and where the CH_4 mixing ratio is close to 10^{-3}, He^+ reacts preferentially with CH_4 to yield CH_2^+ and CH^+. Both of these ions react rapidly with H_2 to produce CH_3^+ and CH_2^+, respectively. The ions CH_3^+ and CH_2^+ react quickly with CH_4 to form $C_2H_5^+$ and CH_5^+, respectively. Also CH_4^+ reacts fairly rapidly with H_2 to generate CH_5^+. If CH_4 were the only trace gas in the atmosphere, CH_5^+ and $C_2H_5^+$ would be the terminal ions and would dissociatively recombine to yield CH_4, C_2H_2, and C_2H_4.

In addition to CH_4, there are trace amounts of photochemically produced C_2H_2, C_2H_4, and C_2H_6 in Jupiter's atmosphere [Strobel, 1975; Hanel et al., 1979; Atreya, Donahue, and Festou, 1981]. The ions CH_5^+ and $C_2H_5^+$ are no longer terminal but form other hydrocarbon ions by the following reactions

$$CH_5^+ + C_2H_2 \rightarrow C_3H_5^+ + H_2 \tag{2.5}$$

$$CH_5^+ + C_2H_4 \rightarrow C_2H_5^+ + CH_4 \tag{2.6}$$

[Munson and Field, 1969] and

$$C_2H_5^+ + C_2H_2 \rightarrow C_3H_3^+ + CH_4 \tag{2.7}$$

$$\rightarrow C_4H_5^+ + H$$

$$C_2H_5^+ + C_2H_4 \rightarrow C_3H_5^+ + CH_4 \tag{2.8}$$

$$C_2H_5^+ + C_2H_6 \rightarrow C_4H_9^+ + H_2 \tag{2.9}$$

[Huntress, 1977]. As can be seen, the complexity of ions and possible reactions increase rapidly as additional trace gases are considered. Further insight into hydrocarbon chemistry must await additional laboratory studies and accurate measurements of the hydrocarbon concentrations in the atmosphere.

2.6. Observations of Jupiter's ionospheres

The Pioneer 10 and 11 spacecraft obtained a total of four radio occultation measurements of the Jovian ionosphere. Measurements were made during immersion (latitude 26° N, solar zenith angle, $\chi = 81°$, late afternoon and at latitude 79° S, $\lambda_{III} = 239°$, $\chi = 93°$, early morning), where λ_{III} is System III (1965) longitude, and emersion (latitude 58° N, $\chi = 95°$, early morning and at latitude 20° N, $\lambda_{III} = 91°$, on the evening side) with Pioneer 10 and 11, respectively [Fjeldbo et al., 1975, 1976]. The electron-number density profiles obtained from inversion of the data are shown in Figure 2.3. There are two distinctive features of these profiles. All have large plasma scale heights on the topside (~ 400–800 km) and exhibit pronounced layered structure on the bottomside. The profiles also reveal a strong possibility of spatial and temporal variations. The assumptions of a topside ionosphere comprised of H^+ and electrons in diffusive equilibrium and thermal equilibrium among the electrons, ions, and neutrals [Henry and McElroy, 1969; Nagy et al., 1976] implies temperatures of 600–900 K.

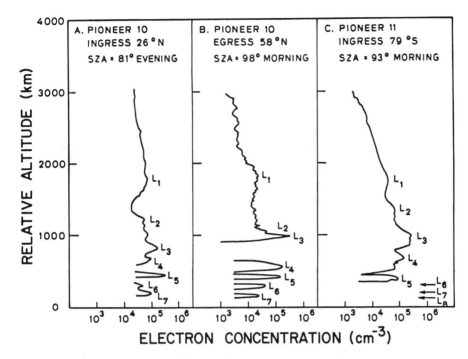

Fig. 2.3. Pioneer 10 and 11 radio occultation measurements of the electron-number density profiles [Fjeldbo et al., 1975]. The Fjeldbo et al. figures have been redrawn on a common height scale by Chen [1981]; the altitude reference is at a density of 10^{19} cm^{-3}, which lies approximately 50 km above the ammonia cloud tops.

The calculations of Nagy et al. [1976] suggest that thermal equilibrium breaks down at an altitude slightly above the peak in electron concentration for high exospheric temperatures. The earlier photoelectron heating calculations of Henry and McElroy [1969] neglected transport effects, although the calculations of photoelectron fluxes and heating rates by Swartz, Reed, and McDonough [1975], Kutcher, Heaps, and Green [1975], and Nagy et al. [1976] took transport effects into account. The two-stream calculations of Nagy et al. [1976] yield a maximum electron heating rate of ~ 50 eV cm^{-3} s^{-1} at an altitude of ~ 450 km above the ammonia cloud tops ($P \simeq 0.6$ bar, $T \approx 150$ K); and a maximum photoelectron escape flux of $\sim 3 \times 10^6$ cm^{-2} s^{-1} at 10 eV.

Although ionospheric scintillations and multipath propagation effects were encountered and in some regions were severe, some bottomside layers are real structure in the ionosphere, e.g., in Figure 2.3a layers L_3 and L_5 [Fjeldbo et al., 1975]. The origin of these layers was attributed to the presence of metallic ions (such as Na$^+$ from the Io plasma torus) similar to the Earth's sporadic E layers [Atreya, Donahue, and McElroy, 1974] and ionization of CH$_3$ radicals [Prasad and Tan, 1974]. Chen [1981] has calculated the flux of Na$^+$ and S$^+$ to be 3×10^4 cm^{-2} s^{-1} and 4×10^3 cm^{-2} s^{-1}, respectively, in order to produce layer L_6.

The Voyager 1 occultation measurements were made at $12°$ S, $\lambda_{\text{III}} = 63°$, $\chi = 82°$, dip angle $I = 45°$, late afternoon during entry and $1°$ N, $\lambda_{\text{III}} = 314°$, $\chi = 98°$, $I = 16°$, predawn during exit [Eshleman et al., 1979a]. In Figure 2.4 the entry profile has a peak electron-number density of 2.2×10^5 cm^{-3} at 1600 km above the 1 mb level and topside scale heights of 960 and 590 km above and below the 3500 km level,

Fig. 2.4. Voyager 1 and 2 measurements of the electron-number density profiles. The geometry of the Voyager 1 measurements is: entry: 12° S, 63° W, solar zenith angle 82°, late afternoon: exit: 1° N, 314° W, solar zenith angle 98°, predawn. The latitude Jovigraphic and the longitude is System III (1965). The Voyager 2 occultations occurred at: entry: 66.7° S, solar zenith angle 87.9°, evening conditions where the ionosphere has been sunlit for hours; exit: 50.1° S, solar zenith angle 92.4°, morning conditions, the lower regions of the ionosphere have just emerged into sunlight after having been in darkness for hours. The zero of the altitude scale corresponds to the 1 mb pressure level that lies approximately 160 km above the ammonia cloud tops (from Eshleman et al. [1979a] and Voyager RSS team).

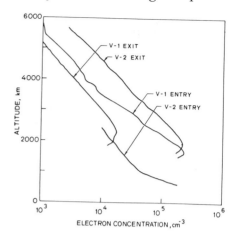

respectively. The exit profile, where the ionosphere was in darkness for ~5 hours, has a peak electron-number density of 1.8×10^4 cm^{-3} at 2300 km and constant topside scale height of ~960 km. The high altitude of the peak is indicative of an equatorial ionosphere. The poor assumption of protons and electrons in ambipolar diffusive equilibrium across magnetic field lines yields a plasma temperature of approximately 1250 K for the larger scale height [Eshleman et al., 1979a; Atreya, Donahue, and Waite, 1979]. Only if the electron-number density is chemically controlled can an accurate temperature be extracted from the plasma scale height in the equatorial ionosphere. A comparison of the two profiles indicates a net chemical loss rate of ~10^{-4} s^{-1} at 2300 km during the night. It should be noted that the entry profile was obtained in the hydrogen Lyman-α bulge region, which has higher H and presumably higher H$^+$ densities than nonbulge regions [Sandel et al., 1980]. Thus, the chemical loss rate could be smaller than inferred above. The entry profile, being a midlatitude profile, because of its 45° dip angle, implies that the observed change in the electron concentration could be simply a manifestation of an equator to midlatitude variation. An examination of the numerous Pioneer electron concentration profiles, however, reveals no such variation. On the other hand, there is a striking increase in the topside temperature from the Pioneer to the Voyager epoch, making the chemical loss of H$^+$ via vibrationally excited H$_2$ ($v' \geq 4$) a viable mechanism during the Voyager observations [Atreya, Donahue, and Waite, 1979]. One should further note that on Saturn where the topside temperatures are relatively low (700-800 K), the peak electron concentration is virtually the same between the equatorial [10° S, Kliore et al., 1980], and high latitude [73° S, Tyler et al., 1981] observations, and no diurnal change was detected. At the low exospheric temperatures prevalent on Saturn, the population of the vibrationally excited H$_2$ is ignorable, and unless the non-LTE effects are important, the loss of H$^+$ by charge exchange with H$_2$ is negligible [Atreya and Waite, 1981].

The Voyager 2 occultation measurements were made at 66.7° S, $\lambda_{III} = 254.8°$, $\chi = 87.9°$, $I = 78°$, evening during entry and 50.1° S, $\lambda_{III} = 148.1°$, $\chi = 92.4°$, $I = 76°$, sunrise during exit [Eshleman et al., 1979b]. The respective plasma scale heights and temperatures are 880 km, 1200 K and 1040 km, 1600 K under the reasonable assumption of ambipolar diffusive equilibrium at midlatitudes [Eshleman et al.,

1979b]. It should be noted that the Voyager 2 entry occurred at a latitude where the Io plasma torus maps into Jupiter's southern polar ionosphere. On exit, a peak electron concentration of $\sim 2 \times 10^5$ cm^{-3} at 2000 km was measured.

2.7. Structure of Jupiter's upper atmosphere

The large plasma scale heights inferred from radio occultation data imply a hot thermosphere because the electrons, ions, and neutrals are approximately in thermal equilibrium [Henry and McElroy, 1969; Nagy et al., 1976]. Similarly, the Voyager ultraviolet solar and stellar occultation data yield large neutral scale heights that indicate a hot thermosphere [Atreya et al., 1979, 1981; Festou et al., 1981]. McElroy [1973] raised the possibility by analogy with the solar corona that upward propagating wave energy could, in principle, create a hot thermosphere on Jupiter. Analysis of the β Scorpii occultation by Jupiter revealed temperature oscillations suggestive of wave propagation in Jupiter's mesosphere [Veverka et al., 1974]. French and Gierasch [1974] evaluated the propagating wave hypothesis and determined that inertia-gravity wave propagation was consistent with the available information. The observed thermal wave amplitudes imply an upward energy flux of ~ 3 erg cm^{-2} s^{-1} and heating rate of ~ 2 K day^{-1}, in considerable excess of the solar EUV flux absorbed above the same level [$\sim 1.3 \times 10^{-2}$ erg cm^{-2} s^{-1}, Strobel and Smith, 1973] and sufficient to account for a hot thermosphere if dissipation of wave energy occurs preferentially within five to ten scale heights above the observed homopause level ($\sim 10^{13}$ cm^{-3}) [Veverka et al., 1974; Atreya and Donahue, 1976; Atreya et al., 1979]. Although the altitude where waves dissipate is not known, the local heating caused by wave dissipation probably results in a steeper temperature gradient than expected for thermal conduction of heat from higher altitudes. This is in contradiction to the oberved temperature (cf., Fig. 2.5).

The same hot thermosphere could also be maintained with less heat input by steady precipitation of soft electrons and protons ($\lesssim 100$ eV) that deposit their energy higher in the atmosphere (~ 0.5 erg cm^{-2} s^{-1} was needed at 10^{11}-10^{12} cm^{-3} density level during the time of Pioneer encounter, [Hunten and Dessler, 1977]; up to ~ 3 erg cm^{-2} s^{-1} at the time of the Voyager encounters [Atreya et al., 1979]). Hunten and Dessler suggested that ionospheric plasma could undergo centrifugal acceleration along flux tubes and be heated by adiabatic compression as a result of flux tube interchange. Based on source strength estimates for ionospheric plasma, they indicated that the heating rate would be deficient by at least one order of magnitude, but noted that plasma associated with Io's torus might supply a sufficient energy flux.

The strong emission of H Lyman-α and H$_2$ Lyman and Werner bands in the polar and equatorial regions during the Voyager encounter is conclusive evidence for particle precipitation on almost a global basis [Broadfoot et al., 1979; Sandel et al., 1979; McConnell, Sandel, and Broadfoot, 1980]. The energy deposition rate in the auroral regions corresponds to a globally averaged deposition rate of ~ 1 erg cm^{-2} s^{-1}, approximately 80 times the solar heating rate [Yung et al., 1982]. In addition, the 2.8 kR of dayside equatorial H$_2$ Lyman and Werner band emission requires a particle deposition rate of 0.4 erg cm^{-2} s^{-1} or diurnally averaged rate of ~ 0.2 erg cm^{-2} s^{-1}. These deposition rates are lower than the 3 erg cm^{-2} s^{-1} heating rate inferred by Atreya et al. [1979] from the determination of the Jovian exospheric temperature with the Voyager UVS solar occultation data. However, it is probable that the particle heating is due to softer electrons [Hunten and Dessler, 1977] and occurs higher in the atmosphere than assumed by Atreya et al. [1979]; then only 0.3-0.8 erg cm^{-2} s^{-1} would be required. But this is still an order of magnitude larger than the soft electron deposition inferred from Jupiter's nightside airglow by McConnell, Sandel, and Broadfoot [1980]. It would

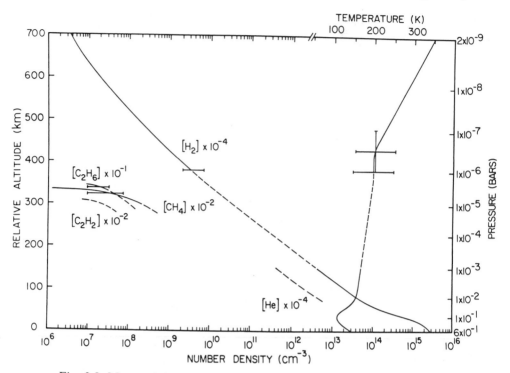

Fig. 2.5. Measured distribution of gases and temperature in the atmosphere of Jupiter [Atreya, Donahue, and Festou, 1981; Festou et al., 1981]. The upper atmospheric data ($z \geq 330$ km) are from Voyager UVS measurements, although the lower atmospheric data are from Voyager IRIS, and RSS experiments.

appear that energy deposition in the auroral regions may be the major source of heat for the thermosphere during periods characteristic of the Voyager encounters.

Of primary importance to ionospheric modeling studies are the distributions of the major gases and temperature at ionospheric heights. The structure of the neutral upper atmosphere has been determined predominantly as a result of the UVS solar and stellar occulation experiments; some significant complementary data have been provided by IRIS (infrared spectrometer) and RSS (radioscience) measurements as well. In a Voyager 1 solar UV occultation experiment, analysis of the continuum absorption by H_2 in the 600-800 Å region yielded a neutral temperature of 1450 ± 250 K at 1750 ± 250 km [Atreya et al., 1979]. An alternate analysis of the solar occultation data yields temperatures somewhat lower; however, with the statistical uncertainties of the two analyses considered, the range of deduced temperatures overlap. This temperature is nearly the same as the plasma temperature determined at the same equatorial latitude and at approximately the same time by the RSS experiment [Eshleman et al., 1979a]. The solar occultation experiment did not provide information on the lower thermosphere and mesosphere owing to the large angular diameter of the Sun (0.1°) and the range of the spacecraft (7 R_J). The ultraviolet occultation of a star Regulus (α Leo, HD 87901, B7V, V 1.35) as monitored by Voyager 2 provided temperature and composition profiles above ~330 km altitude measured from the ammonia cloud tops which are located at $P \simeq 0.6$ bar, $T \approx 150$ K. Analysis of these data yielded density profiles of CH_4, C_2H_2, C_2H_6, and H_2, and the temperature profile shown in Figure 2.5

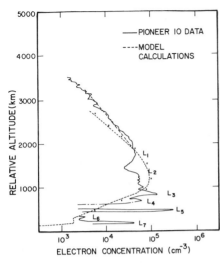

Fig. 2.6. Comparison of electron concentration profile measured by the Pioneer 10 radio occultation experiment (Fig. 2.3b) and model calculations assuming a 1000 K exospheric temperature. The short-dashed curve assumes a $T^{-0.5}$ variation for radiative recombination rate, α_r of H^+ and He^+. Joining of X's will yield the profile when $\alpha_r \propto T^{-0.75}$. The long-dashed curve shows the distribution under the assumptions of diffusive equilibrium, and $\alpha_r \propto T_e^{-0.5}$. The main ion is H^+, except that the secondary peak at the location of layer L_7 is where production of CH_3^+ peaks. The height scale is the same as in Figure 2.3b [Atreya and Donahue, 1976].

[Atreya, Donahue, and Festou, 1981; Festou et al., 1981]. The hydrocarbon mixing ratios were also measured in the stratosphere by IRIS [Hanel et al., 1979], and considerable variability, both temporal and latitudinal, were found in those values. Because the main ionosphere lies far above the region of the hydrocarbons, ionospheric models there are insensitive to the hydrocarbon densities.

2.8. Ionospheric modeling

In this section, the significant points concerning Jovian ionospheric calculations and including attempts at modeling one ionospheric profile measured by the Voyager RSS experiment are presented. Chemical reactions important in the ionosphere of Jupiter are listed in Table 2.1 taken from Atreya and Donahue [1976]. These reactions are illustrated in a flow diagram [Fig. 2.2, Atreya and Donahue, 1981]. In addition to these reactions, the reaction of protons with vibrationally excited H_2 with subsequent removal of the resulting molecular ion by reactions (e1) and (r1) in Table 2.1 becomes important under conditions of high exospheric temperature [McElroy, 1973; Atreya, Donahue, and Waite, 1979a].

$$(e20)\ H^+ + H_2\ (v \geq 4) \rightarrow H_2^+ + H \tag{2.10}$$

The Pioneer 10 and 11 observations were made near minimum solar activity. The assumption of solar EUV being the principal ionization source in the Jovian atmosphere appears to be adequate for reproducing most of the observed characteristics. Given absorption of solar EUV and associated ionization and the ion atom/molecule interchange and removal in Table 2.1, a set of nonlinear differential equations can be solved numerically to generate a model of the ionosphere. Atreya and Donahue [1976] attempted to calculate the ionospheric profile appropriate to the observation geometry of a Pioneer 10 radio occultation (Fig. 2.3b). They assumed an eddy diffusion coefficient of 2×10^7 cm^2 s^{-1} at the homopause, hydrocarbon density distribution as given by Strobel [1975], and an atmospheric density distribution corresponding to a high exospheric temperature of 1000 K. The results of their calculations for different assumed radiative recombination rates of the terminal ion H^+ and for diffusive equilibrium conditions are shown in Figure 2.6. The main ionospheric distribution is fairly well reproduced by these model calculations. The layers L_1-L_7 are probably due to sporadic E-type clustering of ions such as Na^+ [Atreya, Donahue, and McElroy, 1974], and protons [Atreya and Donahue, 1976]. The source of the former ions could be meteoric in-fall and/or Io. No measurements of the temperature or neutral species in the upper atmosphere of Jupiter were available at the time of the Pioneer observations; therefore, the model of the neutral atmosphere is constrained principally by ionospheric measurements. It should be remarked also that considerable variations in terms of the number and location of the layers, height of the peak, and the plasma scale height were observed in the three successful radio occultation experiments on Pioneers 10 and 11. In general, however, an average peak electron number density of $(1$–$3) \times 10^5$ cm^{-3} at 1000-1300 km with an average topside scale height of about 600 km is a good representation of the ionospheric conditions at Pioneer encounter time.

Considerable variations in the ionospheric characteristics were also seen as a function of solar zenith and angle and latitude in both the Pioneer and Voyager observations. The Voyager 1 data (Fig. 2.4) show a dramatic decrease in the peak electron-number density from dayside to nightside. A preliminary attempt at modeling one of the observations, the Voyager 1 entry occultation, is described here. Assuming the photochemical scheme (Table 2.1) applicable for the Pioneer observations, but including the reaction $e20$, which is important for elevated exospheric temperatures [Atreya et al., 1979], Atreya and Donahue [1981] have repeated the calculations with the solar flux (February, 1979 values, H. E. Hinteregger, personal communication) adjusted for the latitude and solar zenith angles of the Voyager 1 entry occultation. The thermal and neutral composition distributions in the atmosphere were taken from the actual observations as opposed to a model as in the Pioneer calculations. The eddy diffusion coefficient at the homopause, K_h, was assumed to be 1×10^6 cm^2 s^{-1} based on the analysis of the equatorial solar and stellar occultation data by Atreya, Donahue, and Festou [1981] and Festou et al. [1981]. The globally averaged value of K_h was also found to be $\sim 10^6$ cm^2 s^{-1} by McConnell et al. [1981], Yung and Strobel [1980], and Broadfoot et al. [1981].

We show in Figure 2.7 the calculated electron concentration density profile taken from Atreya and Donahue [1981], along with the Voyager 1 entry data. The calculation yields H^+ as the major ion in the upper ionosphere, while below 700 km, production of H_3^+ dominates and results in heavier molecules such as $C_nH_m^+$. A comparison between the model and the measurements reveals numerous apparent discrepancies. The following points are noteworthy:

(1) The peak of the measured electron concentration appears greater by about a factor of 3 than the calculated local concentration. We believe that the calculated value is well within the range of uncertainties of the measurements.

(2) The location of the peak in the calculated electron concentration profile appears to be much lower than the measurements. If one believes that the observed maximum at about 1600 km indeed represents the peak in the electron concentration, several possible scenarios for reconciling the calculations with the measurements are possible. First, one could invoke an upward drift of ~ 10 m s^{-1} to raise the peak to the desired height. Aurorally driven thermospheric winds are directed equatorward in a manner similar to the Earth's response to magnetic disturbances [Roble et al., 1979] and induce an upward drift by ion drag. Simple expressions given by Strobel [1970] indicate an equatorward wind of 20 m s^{-1} displaces the electron concentration peak upward by ~ 4 neutral scale heights when $I = 45°$, in reasonable agreement with the height differences in Figure 2.7. Second, if the atomic H number density is equal to or greater than the local H_2 number density in the 1000–2000 km range, a large production of H^+

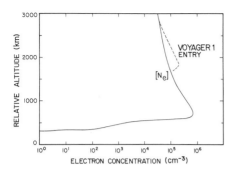

Fig. 2.7. Electron concentration calculated using photochemical scheme described in Table 2.1, and for the geometry of observation corresponding to the Voyager 1 entry occultation (Fig. 2.4). The distribution of gases and temperature in the atmosphere is taken from Figure 2.5 (after Atreya and Donahue [1981]).

by direct ionization of H would result in a completely altered ionospheric structure. This is a distinct possibility because the Voyager 1 entry profile was obtained in the Lyman-α bulge region where atomic hydrogen is probably enhanced [Sandel, Broadfoot, and Strobel, 1980; Clarke et al., 1980; McConnell, Sandel, and Broadfoot, 1981]. On the other hand, it is also likely that the measured maximum in the electron concentration profile does not represent the main ionospheric peak because there may be ionization layers at this altitude or just below giving the appearance of a peak in the electron concentration [Atreya and Donahue, 1981]. This is readily noticeable in the profiles measured by Pioneer 10 and 11 (Figs. 2.3a,b,c). Numerous ionization layers are expected to be present in the atmosphere (V. Eshleman, 1979, personal communication); their analysis, however, is yet incomplete.

(3) The calculated profile shows a slightly greater scale height than the measurements. This is a direct reflection of the assumed exospheric neutral temperture, which is somewhat greater than the plasma temperature. The error analysis of the Voyager RSS data has not yet been completed; it is believed that, within the range of statistical uncertainty, the two temperatures are equal and the scale heights are equivalent.

Although no attempt has been made here to reproduce the equatorial nightside ionosphere, the ideas presented earlier by Atreya, Donahue, and Waite [1979] for the reduction of nighttime electron-number density appear valid. In a relatively hot thermosphere, the conversion of protons to molecular ion H_3^+ takes place with subsequent faster removal of H_3^+ once the ion production source – presumably the Sun – has been turned off (reactions $e20, e1, r1$). A slow decay of the ionosphere throughout the night is expected. For the electron concentration to drop by a factor of 20 from dayside to nightside, one requires an overall rate constant for the reaction $e20$ to be 4.3×10^{-16} cm^3 s^{-1}. The rate constant for H_2 excited to the fourth vibrational level or higher would have to be $(1-4) \times 10^{-9}$ cm^3 s^{-1}. Although the rate of this reaction has not been measured in the laboratory, chemical kinetics considerations do not rule out such a value. In addition the vibrational distribution may depart significantly from local thermodynamic equilibrium [McElroy, 1973].

The present attempts to model the measured ionospheric profiles may require improvements, especially if the results of preliminary analysis of the Voyager data remain unaltered after further reduction and analysis of the data. It should be remarked, however, that improved analysis of the data has already begun to give the indication that the measured peak in electron concentration does not represent the actual peak and that the actual peak indeed appears to lie well below 1600 km (R. H. Chen, private communication, 1981) as has been surmised earlier by Atreya and Donahue [1981]. The lack of knowledge of the particle precipitation, including the possible low-energy electron drizzle in the equatorial region, atomic hydrogen distribution, and statistical uncertainties of the dynamics have hampered efforts to carry out realistic calculations.

2.9. Concluding remarks

It would appear on the basis of the Pioneer and Voyager obervations, which occurred at approximately solar minimum and maximum conditions, respectively, that Jupiter's ionosphere and thermosphere undergo significant solar cycle changes. However, the differences between the Pioneer and Voyager encounters may reflect differences in volcanic activity on Io, which ultimately supplies mass to the plasma torus. The Io plasma torus densities appear to be significantly higher during the Voyager encounter period than the Pioneer encounter period [Pilcher and Strobel, 1981; Intriligator and Miller, 1981]. During times of high plasma torus densities, pitch-angle scattering of energetic magnetospheric electrons is enhanced and occurs at lower energies than possible during low-density conditions. Precipitation of energetic electrons is preferentially into Jupiter's polar regions along magnetic field lines connected to the Io plasma torus. Thus, high densities in the plasma torus lead to intense auroral precipitation, which generates significant increases in dissociation and ionization of H_2 as well as intense localizing heating. As a consequence of aurorally driven thermospheric wind systems, atomic hydrogen and heat are redistributed on a global basis and result in an enhanced Lyman-α albedo and elevated temperatures in the thermosphere, in agreement with the differences between the Pioneer and Voyager encounters.

For Pioneer conditions, the ionosphere outside of the auroral zone is produced primarily by solar EUV radiation. The hot thermosphere is maintained either by wave dissipation or soft electron precipitation. Auroral activity appears to be confined to local hot spots that coincide with flux tubes of the Galilean satellites mapped into the atmosphere. Voyager conditions are characterized by an even hotter thermosphere ($\sim$ 400 K warmer), primarily owing to additional particle precipitation. The auroral zones are complete ovals, delineated by strong emissions of H_2 Lyman and Werner bands and H Lyman-α and have enhanced electron concentrations. In the equatorial regions, strong airglow emission from H_2 suggests that particle precipitation may also contribute to the ambient electron concentrations. Because the Voyager 1 entry profile was in the hydrogen bulge region, the initial modeling attempt illustrated in Figure 2.7 does not allow definite conclusions on the relative contributions of solar EUV and precipitating electrons as ionization sources.

Comparison of Pioneer and Voyager electron concentration profiles suggests that Jupiter's ionosphere is at least as complex and variable as the Earth's ionosphere. Of potential importance but still poorly understood is the multilayered structure of Jupiter's ionosphere. The Pedersen conductivity plays a fundamental role in the electrical coupling of the magnetosphere and ionosphere. This quantity, which when divided by the electron number density maximizes at $[H_2] \sim 3 \times 10^{13}$ cm^{-3}, precisely in the layered region of the ionosphere. Note that in Figure 2.6 for Pioneer conditions this is the L_6-L_8 region, whereas for Voyager conditions in Figure 2.7 the main contribution to the Pedersen conductivity occurs in the 300–700 km region where no RSS measurements have yet been reduced. Theoretical estimates of the Pedersen conductivity

$$\sigma_{Ped} = \frac{n_e e}{B} \frac{\nu_{in} f_{ci}}{\nu_{in}^2 + f_{ci}^2} \tag{2.11}$$

where B is the magnetic field strength ($\sim$ 8 G), e is charge, ν_{in} is the ion–neutral collision frequency, and f_{ci} is the ion cyclotron frequency, can be given on the basis of previous discussion and Figures 2.5-2.7. The integrated Pedersen conductivity is

$$\Sigma = \int \Sigma_{Ped} dz \simeq (1 \times 10^{-11} \, \bar{n}_e + 1 \times 10^{-2}) \text{ mho} \tag{2.12}$$

where $\bar{n}_e$ (in m^{-3}) is the average electron-number density over the neutral H$_2$ concentration region of $(1-300) \times 10^{12}$ cm^{-3}, and the second term represents the upper ionospheric contribution. When the ionosphere is produced solely by solar EUV radiation, $\bar{n}_e \sim 10^9$ m^{-3} and $\Sigma \sim 0.2$ mho. In the presence of sporadic E layers and/or moderate energetic electron precipitation $\bar{n}_e \sim 3 \times 10^{10}$ m^{-3} and $\Sigma \sim 0.3$ mho. In the auroral regions observed by Voyager where intense energetic particle precipitation is present, $\bar{n}_e$ may be as large as 1×10^{12} m^{-3} and $\Sigma = 10$ mho. Thus considerable variability in the Pedersen conductivity is expected in Jupiter's ionosphere, especially in the auroral regions. It should be noted that B varies from 3.3 to 14 G on Jupiter's surface, and B was assumed to be 8 G in these estimates.

The other important ionosphere in Jovian magnetospheric studies is that of Io. A recent excellent review of Io's atmosphere and ionosphere may be found in Kumar and Hunten [1981]. No additional measurements of Io's ionosphere were made by the Voyager spacecraft. The Pioneer 10 radio occultation experiment detected the presence of an ionosphere on Io comparable to the ionospheres observed on Mars and Venus [Kliore et al., 1973, 1975]. On the dayside ($\chi = 81°$) and downstream they inferred a peak electron concentration of $\sim 6 \times 10^4$ cm^{-3} at an altitude of ~ 100 km with a topside plasma scale height of ~ 200 km and total extent of 750 km. On the nightside and upstream the peak electron concentration was $\sim 10^4$ cm^{-3} but at 40 km with an abrupt termination of the ionosphere at ~ 200 km altitude and much smaller plasma scale height. The corresponding neutral SO$_2$ concentrations at the surface appropriate for these measurements is $\sim 10^{11}$ cm^{-3} [Kumar and Hunten, 1981].

ACKNOWLEDGMENTS

We acknowledge research support provided by grants from the NASA Planetary Atmospheres Program. We thank the Voyager RSS Team for Figure 2.4. Dr. M. Acuña kindly supplied the dip angles for the Voyager radio occultation points.

3

THE LOW-ENERGY PLASMA IN THE JOVIAN MAGNETOSPHERE

J. W. Belcher

The magnetosphere of Jupiter is unique in the solar system because of its large extent and rapid rotation, and because of the prodigious source of plasma provided by the satellite Io. Io and its associated neutral clouds inject $> 10^{29}$ amu/s of freshly ionized material into the magnetosphere, producing a plasma torus with a density maximum near the L-shell of Io and a total mass of $\sim 10^{36}$ amu. The innermost region of this torus contains a cool plasma corotating with the planet and dominated by S^+ ions with temperatures of a few eV. At greater distances, beyond 5.6 R_J, the plasma ions are warmer, consisting primarily of sulfur and oxygen ions with temperatures of ~ 40 eV. The plasma electrons here have mean energies of 10 to 40 eV, and exhibit distribution functions which are non-Maxwellian, with both a thermal and suprathermal component. Near Io itself, the Alfvén wave generated by Io gives rise to observed perturbations in the magnetospheric velocities as the ambient plasma flows around the Io flux tube. In the middle magnetosphere, between ~ 8 and ~ 40 R_J, the ions and electrons tend to be concentrated in a plasma disc or sheet that is routinely cooler than its higher latitude surroundings. This plasma tends to move azimuthally but does not rigidly corotate with the planet. The electron density enhancements at the plasma sheet are due primarily to an increase in the electron thermal population with little change in the suprathermal population. At larger radial distances, the suprathermal electrons constitute an increasing fraction of the total electron density and pressure. The mean ion and electron temperature in the plasma sheet increases with increasing distance. Although the mass densities in the middle magnetosphere are dominated by ions with energies of less than 6 keV, the thermal pressure there appears to be due primarily to an energetic ion component with characteristic energies of ~ 25 keV and above. Values of the Alfvénic Mach number range from < 0.1 in the cold inner torus to 2–3 in the plasma sheet.

3.1. Introduction

In 1954, Burke and Franklin [1955] discovered radio emissions from Jupiter at 22.2 MHz. Subsequent observations established the strong control of these decametric emissions by the satellite Io. Thus, a strong satellite–magnetosphere interaction at Jupiter was known well before the start of planetary exploration, and this provided a strong motivation for direct measurements at Jupiter, especially near Io. The first in situ observations of Jupiter were made in December of 1973 with the Pioneer 10 flyby and a year later with the Pioneer 11 flyby. Extensive data were obtained on the magnetic field of Jupiter and on the energy spectrum and spatial distributions of charged particles with energies above ~ 0.5 MeV/nucleon. The Pioneer plasma instrument also provided evidence concerning the plasma environment. Low-energy positive ions (100 to 4800 V), interpreted as protons, were detected in the inner magnetosphere with reported concentrations ranging from 10 to 100/cm^3 and temperatures of the order of 100 eV [Frank et al., 1976]. More recently, Intriligator and Miller [1981] have reexamined the Pioneer data in the Io torus in light of the Voyager results, and report an ion composition and topology similar to that seen by Voyager, as we discuss in the subsection entitled "Comparison with the Pioneer 10 plasma observations." Apparently, the low-energy electron measurements by the Pioneer plasma instruments were seriously compromised by trapped photoelectrons and secondary electrons [Grard, DeForest, and Whipple, 1977; Scudder, Sittler, and Bridge, 1981].

Just prior to the Pioneer 10 encounter, Brown [1973] reported ground-based optical observations of neutral sodium associated with Io. This observation was followed by the discovery of a torus of singly ionized sulfur [Kupo, Mekler, and Eviatar, 1976]. The S^+ optical radiation these authors observed came from a region inside the orbit of Io (the inner cold torus) and was interpreted by Brown [1976] as originating in a dense ring of cold plasma in that region ($n_e \simeq 3200/cm^3$ and $T_e \simeq 2$ eV). Although this conclusion provoked some discussion at the time, it was not fully appreciated by the space plasma community. However, it clearly marks the discovery of the cold plasma torus at Io. Optical observations of the torus and associated physical processes are discussed at length in Chapter 6.

The Voyager encounters with Jupiter in 1979 have resulted in a dramatic increase in our detailed knowledge of the plasma properties of the magnetosphere. The warm plasma torus was discovered from remote observations of strong emissions from S^{2+}, S^{3+}, and O^{2+} by the Voyager 1 ultraviolet spectrometer [Broadfoot et al., 1979; Sandel et al., 1979]. Measurements by the Planetary Radio Astronomy experiment [Warwick et al., 1979a] and the Plasma-Wave Science experiment [Scarf, Gurnett, and Kurth, 1979] allowed the determination of electron densities along the spacecraft trajectory from characteristic frequencies of various plasma wave emissions. The Low-Energy Charged Particle experiment [Krimigis et al., 1979a] measured the characteristics of the energetic ion population to much lower energies (≥ 25 keV) than those of the Pioneer experiments. And, finally, details of the distribution functions of electrons between 10 and 5950 V, as well as those of positive ions in the same energy range, were measured by the Plasma Science experiment [Bridge et al., 1979a]. Such experiments provide a wealth of critical information that is not available from inferences based on remote observations alone. In this chapter, we present a primarily phenomenological account of the in situ observations of low-energy plasma in the Jovian magnetosphere, beginning with the Io torus and proceeding outward. We concentrate primarily on results from the Voyager Plasma Science experiment [Bridge et al., 1977]. The physical mechanisms underlying this phenomenology are considered to a limited extent in Section 3.6, and in more detail in Chapters 10, 11, and 12.

3.2. The Io plasma torus

Bagenal and Sullivan [1981] have constructed comprehensive models of the positive-ion morphology of the Io plasma torus, using positive ion and electron measurements from the Plasma Science experiment on Voyager 1. These authors used the observed density and kinetic temperature of each ionic species and the electron temperatures of Scudder and Sittler (private communication, 1980), in conjunction with the theory appropriate for the ambipolar diffusion of plasma in a rotating, tilted dipole with mirror symmetry about the centrifugal equator. The resulting two-dimensional model of the elementary-charge concentration of positive ions in the Io torus is shown in Figure 3.1. The apparent asymmetry about the centrifugal equator is due to the geometry of the tilted dipole. The distribution of plasma along a given magnetic field line is determined by the balance of electrostatic, centrifugal, and thermal pressure gradient forces. It is apparent from this figure that the torus is divided into two parts: a cold inner region inside of 5.6 R_J where the ions are closely confined to the centrifugal equator, with a vertical scale height of only ~ 0.2 R_J; and a warm outer region with a much larger scale height of ~ 1 R_J. This warm/cold dichotomy of the Io plasma torus is most

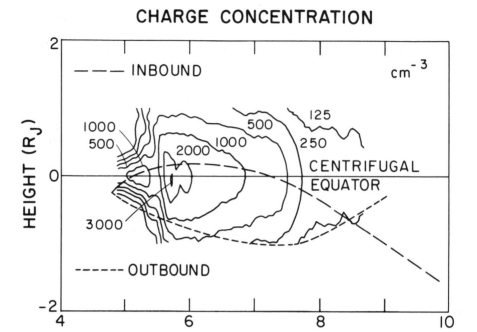

Fig. 3.1. A contour map of elementary-charge concentration in a cylindrical coordinate system based on the centrifugal equator. The map has been constructed from plasma measurements made along the inbound spacecraft trajectory of Voyager 1 by using a theoretical expression for the distribution of plasma along dipolar field lines. Both the inbound and outbound trajectories are indicated by dashed lines. The contours are in units of elementary-charge/cm^3.

dramatically shown in the data presentation of Figure 3.2, which displays the low-energy positive-ion spectra measured along the inbound trajectory of Voyager 1 between 4.9 and 7.0 R_J. To understand this and subsequent figures, we briefly describe the plasma experiment.

The Voyager Plasma Science experiment

The Plasma Science experiment consists of four modulated-grid Faraday cups, three of which (A, B, C) are symmetrically positioned about an axis that generally points toward the Earth, and a fourth (the side sensor, D) oriented at right angles to this direction. Positive-ion measurements are made in all four sensors, and electron measurements in the D sensor alone. The electron measurements are discussed in detail in the subsection entitled "Electron distribution functions in the torus." Throughout most of the inbound leg of the Voyager 1 trajectory (before 0500 UT on March 5, 1979, outside of ~ 10 R_J) the side sensor was pointed in the direction of azimuthal flow of plasma around Jupiter. As the spacecraft approached the planet, the viewing geometry changed rapidly so that after 0500 UT, the look direction of the main sensors was swept into the direction of azimuthal flow and then rapidly away after the closest approach at 1204 UT (Spacecraft Event Time) on March 5, 1979, near 4.9 R_J. Thus, positive-ion data are obtained primarily from Voyager 1 inbound observations; in the

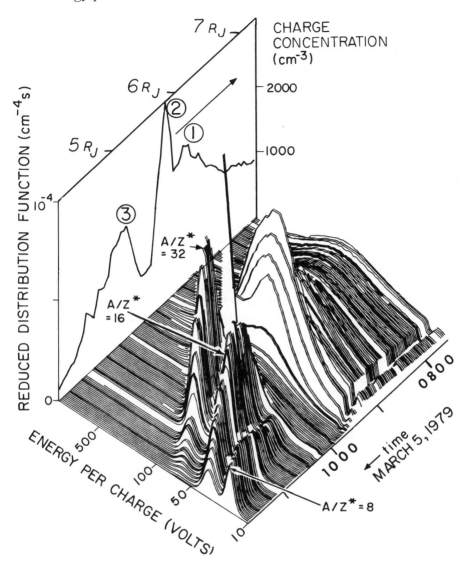

Fig. 3.2. A three-dimensional plot of reduced distribution function against energy per charge for spectral measurements made in the C cup of the main sensor between 0730 UT (7 R_J) and 1145 UT (4.9 R_J) on March 5, 1979. A total of 160 spectra are shown. Two spectra are omitted every 48 min. when the instrument was in a different measurement mode. Every tenth spectrum is emphasized with a darker line. The back panel shows the total positive-ion elementary-charge concentration as a function of time determined from fits to the corresponding spectra. The number features in the back panel are discussed in the text. The notch near 20 V in many of the positive ion spectra is due to interference from another instrument on the spacecraft.

torus proper, the data came from the three Faraday cups in the main cluster, whereas in the middle magnetosphere (outside of 10 R_J), they are obtained primarily from the side sensor. The positive ions are generally transonic to highly supersonic. Because of the unfavorable look directions, positive-ion data outbound are of much poorer quality than inbound, and they will not be discussed here. During the Voyager 2 encounter, the viewing geometry was similar to that of Voyager 1, but the closest approach distance of 10 R_J precluded observations in the torus.

Each of the four PLS sensors provides an energy-per-charge scan of the positive-ion plasma between 10 and 5950 V. The scan in velocity space is integral in the directions perpendicular to the sensor normal and differential in the direction in velocity space along the sensor normal. Thus, the four energy-per-charge scans provide reduced one-dimensional ion distribution functions for four different directions in velocity space, convolved with the response functions of the sensors (see Appendix A of McNutt, Belcher, and Bridge [1981]). The data in Figure 3.2 are reduced distribution functions from the C sensor of the main cluster, which is essentially aligned with corotational flow during this time interval. The analysis of these spectra is complicated by the presence of many different ionic species in the Jovian magnetosphere. Fortunately, however, for any energy-per-charge instrument, an ion with mass number A and charge number Z^* moving at the same velocity as H^+, appears at an energy-per-charge of A/Z^* times the energy-per-charge of H^+. In the cold region of the torus, after 1016 UT on March 5, the distinct peaks in the energy-per-charge spectra in Figure 3.2 correspond to values of A/Z^* of 8, 16, and 32, assuming a velocity component into the C sensor appropriate for rigid corotation. A different assumption concerning the velocity would, of course, give different values of A/Z^*, but the values of 8, 16, and 32 are plausible (O^{2+}, O^+ or S^{2+}, and S^+). The energy-per-charge of corotating H^+ at this time is below the 10 V threshold of the instrument.

In the cold torus, at least, the interpretation of the positive-ion spectra is thus straightforward: the location of a given ionic peak helps to identify its mass-to-charge ratio as well as providing a quantitative measure of the velocity component into the sensor; the width of the peak provides a measure of the thermal speed for that ionic species and the area under the peak provides a measure of its concentration. In other parts of the magnetosphere, the interpretation is more complex because the peaks overlap (e.g., the warm ion spectra before 1016 UT in Fig. 3.2), but the analysis proceeds in the same spirit – that is, we take advantage of the separation in energy-per-charge of the comoving ionic species due to their different mass-to-charge ratios and their transonic to highly supersonic Mach numbers. In some respects, the electron analysis is less complex because there is only one species of electron and because the electrons are always highly subsonic.

With this brief sketch of the operation of the instrument, we now discuss in turn the observed thermal structure of the torus (ions and electrons), the elementary-charge concentration in the torus, the positive-ion composition in the torus, and finally, plasma velocities in the torus.

Positive-ion temperatures in the torus

In the cold torus, the spectra in Figure 3.2 can be quantitatively analyzed for the density and temperature of each ionic species present, as well as for their common velocity component into the C sensor, by simultaneous fits to a sum of convected Maxwellians. An example of such a fit is given in Figure 3.3. For such cold spectra, the only serious ambiguity in the determination of these plasma parameters arises from the fact that some peaks in A/Z^* do not correspond to a unique ionic species. The most troublesome example in the Io torus and throughout the magnetosphere is the peak at A/Z^* values of 16, the common ratio for O^+ and S^{2+}. Composition and temperature estimates for these two ions are dependent on what fit assumptions are made (e.g., the various ions can be assumed to have either equal temperatures or equal thermal speeds).

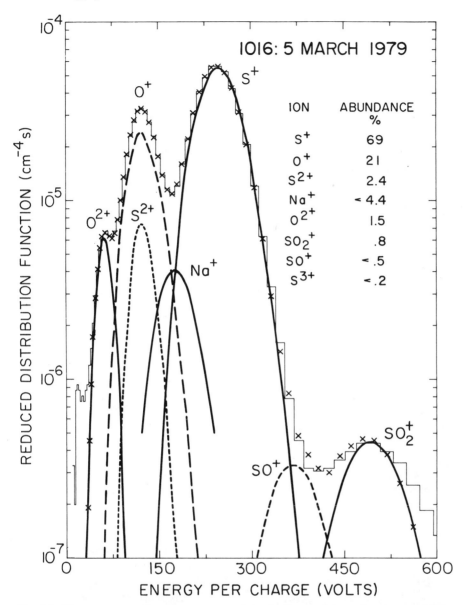

Fig. 3.3. The energy-per-charge spectrum made in the C cup of the main sensor at 1016 UT (5.3 R_J) March 5, 1979. The data are shown as a histogram and the fit to the current in each measurement channel is shown by ×'s. The individual Maxwellian distributions of each ion that make up the reduced distribution function of the fit are shown by the curved lines.

In the warm torus, the ambiguities in the determination of the fit of plasma parameters for each ionic species are more severe than in the cold torus and are more strongly dependent on a variety of assumptions. Whereas, in the cold torus, the Mach numbers of the various adjacent peaks are high enough (≥ 6) so that each A/Z^* peak is well separated from adjacent peaks, in the warm torus the Mach numbers are lower (≥ 3), and the various heavy-ion peaks overlap (see Appendix A of McNutt, Belcher,

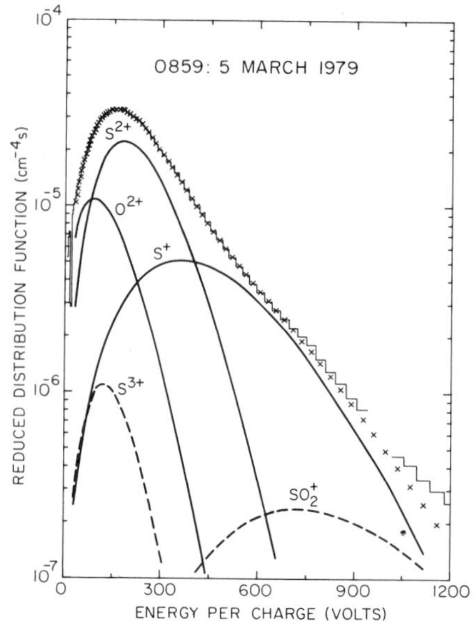

Fig. 3.4. The C cup energy-per-charge
spectrum for 0859 UT on March 5, 1979
(6.0 R_J), which has been fit under the
assumption that all the ions are
corotating with Jupiter and have the
same temperature. The data are shown
as a histogram and the fit to the current
in each measurement channel is shown
by ×'s. The individual Maxwellian
distributions of each ion that make up
the reduced distribution function of the
fit are shown by the curved lines.

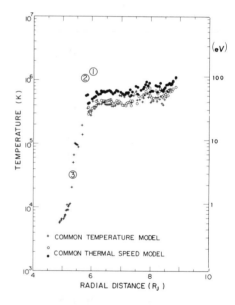

Fig. 3.5. The radial profile of average
ion temperature derived from fits to the
energy-per-charge spectra of positive
ions. The +'s are from the fits where the
ions are assumed to have the same
temperature. The circles are from the fits
where the ions are assumed to have the
same thermal speed. With the common
thermal speed model the average ion
temperature has been calculated
assuming the $A/Z^* = 16$ spectral peak to
be all S^{2+} (●) or all O^+ (○). The numbered
features refer to maxima in the
concentration profiles
shown in Figures 3.2 and 3.7.

and Bridge [1981]). The plasma parameters obtained depend critically on assumptions
concerning composition, temperature, and bulk motion. An example of a fit to a spectrum in the warm torus is shown in Figure 3.4. Bagenal and Sullivan [1981] discuss in
detail the results of the fitting procedures in the torus under various assumptions and
the subsequent range of plasma parameters that are consistent with the data obtained
there.

Figure 3.5 shows the radial temperature profile along the inbound trajectory
determined from fits to the ion spectra that are shown in Figure 3.2; uncertainty limits
associated with model dependencies are indicated. It is these ion temperatures along
the spacecraft trajectory, in conjunction with the electron temperatures discussed in
the subsection entitled "Electron distribution functions in the torus," that are used to

construct the model plasma torus in Figure 3.1. The ions beyond about 5.8 R_J have a fairly constant temperature of $(6.0 \pm 1.5) \times 10^5$ K (~ 50 eV), resulting in a vertical scale height of ~ 1 R_J in that region. This average ion temperature of 50 eV is an order of magnitude less than that expected if the cyclotron speed were equal to the full corotational velocity (see the subsection entitled "Sources of plasma"). As the spacecraft moved inward through the torus, the average temperature decreased sharply inside 5.6 R_J, dropping by a factor of ~ 50 to less than ~ 1 eV. This decrease in ion temperature is the cause of the decrease in vertical scale height shown in Figure 3.1.

Electron distribution functions in the torus

In addition to the measurements of positive-ion spectra of Figure 3.2, simultaneous measurements of the electron distribution functions are available. These measurements are made in the D sensor only. As with the ions, the measurements are differential in velocity space for that component of electron velocity along the D cup normal, and integral for velocity components transverse to that normal. Because only one sensor is used, the electron measurements do not contain complete information on the full three-dimensional electron distribution function. However, the electrons are highly subsonic, and with the assumption of isotropy, one can unfold an excellent representation of the electron distribution function. Unfortunately, the low-energy instrumental threshold of 10 V leaves a gap in the energy coverage, and this gap is particularly severe in the cold region of the torus, where the thermal electron temperatures are ~ 5 eV or less. An equally serious problem in the electron analysis is the effect of spacecraft charging in the high-density regions of the torus. In these regions the spacecraft did apparently achieve a negative potential on the order of a few tens of volts, which has left the final analysis of the electron spectra in the torus (especially the absolute concentrations) incomplete pending a more thorough treatment of the charging problem at high concentrations. In the lower density environment outside of the torus, the charging problem is reasonably well understood and the effects of spacecraft potential can be self-consistently taken into account [Scudder, Sittler, and Bridge, 1981]. In any case, the electron temperature can be determined independently of the concentration, and the temperature profile of the thermal electrons in the torus qualitatively resembles that of the ions. Scudder, Sittler, and Bridge [1981] find three relatively well-defined subregimes of the plasma torus: (a) the cold inner torus, with $T_e \lesssim 5$ eV; (b) the temperate middle torus, with $T_e \simeq 10$–40 eV; and (c) the hot outer torus just beyond 8 R_J with $T_e \gtrsim 100$ eV. This last regime is the inner edge of the radially extended plasma sheet (see Section 3.4).

For illustration, Figure 3.6 shows electron distribution functions as measured at 5.5, 7.8, and 8.9 R_J. All of these spectra have well developed thermal populations that are well fit by Maxwellian distributions (the "core" electrons, in analogy with solar wind nomenclature). They also exhibit distinct suprathermal tails (the "halo" electrons). In each panel, the subscript "e" refers to the parameters characterizing the electron distribution as a whole, the subscript "C" to parameters characterizing the Maxwellian fit to the thermal population, and the subscript "H" to the parameters of the suprathermal population. The coldest electron spectrum shown ($T_c = 5$ eV), at 5.5 R_J, occurs near the precipitous drop in ion temperature shown in Figure 3.5. In addition to illustrating the cooling of the thermal electron population with decreasing radius, Figure 3.6 also illustrates the increasing importance of the suprathermal electron population with increasing radius. For example, the suprathermal fraction by number varies from 0.02% at 5.5 R_J to 8% at 8.9 R_J. We will find that this trend continues as we move into the middle magnetosphere. The existence of these "hot" and "cold" populations of

Physics of the Jovian magnetosphere

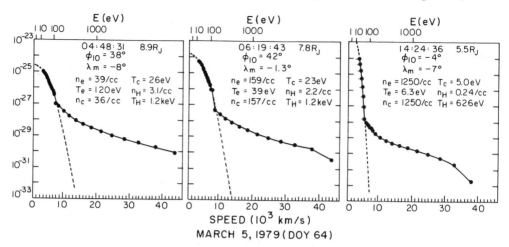

Fig. 3.6. Plots of the electron distribution function measured at three different times in the inner magnetosphere by the Voyager 1 Plasma Science experiment. The measurement times, radial distance of spacecraft from Jupiter, System III longitude of spacecraft relative to Io, and dipole magnetic latitude of the spacecraft are indicated. The connected points are data points, and the dashed lines are Maxwellian fits to the thermal population of electrons.

electrons has also been deduced from the characteristics of various plasma wave emissions. In particular, at the time of the center distribution function shown in Figure 3.6, Coroniti et al. [1980] concluded that the observations of plasma chorus by the Plasma Wave Science experiment required the presence of suprathermal electrons with a characteristic energy of 1 keV and a fractional density of 1.1%, in good agreement with the population actually measured. In addition, Birmingham et al. [1981] using the characteristics of emissions observed by the Planetary Radio Astronomy experiment, have inferred the existence of a hot electron population. These authors also predicted a variation of T_c/T_H that decreases as the spacecraft moves from the hot outer torus to the cool inner torus, a prediction qualitatively in agreement with the distributions of Figure 3.6. Finally, an analysis of the thermal emissions from the torus also calls for the presence of such electrons (see Chap. 12).

Elementary-charge concentrations in the torus

Even though we cannot at present derive absolute electron concentrations from the Plasma Science electron measurements in the high density torus, we can derive an estimate of that concentration from the positive-ion measurements (which are less affected by spacecraft charging) and the requirement of charge neutrality. Figure 3.7 show the elementary-charge concentration along the inbound trajectory for ions with values of $A/Z^* \geq 8$, as determined from fits to the ion spectra shown in Figure 3.2. The three local maxima labeled in Figure 3.7 are the same as those labeled in the back panel of Figure 3.2. We also show in Figure 3.7 the determinations of the total electron concentration by the Planetary Radio Astronomy experiment, which should match the elementary-charge concentration from the positive-ion fits, if contributions from ions with $A/Z^* < 8$ are small. Because the agreement between these two determinations is good, we conclude that lighter ions do not contribute a significant amount of charge concentration at these latitudes and distances. For an estimate of their contribution at higher latitudes at these distances, see Tokar et al. [1982].

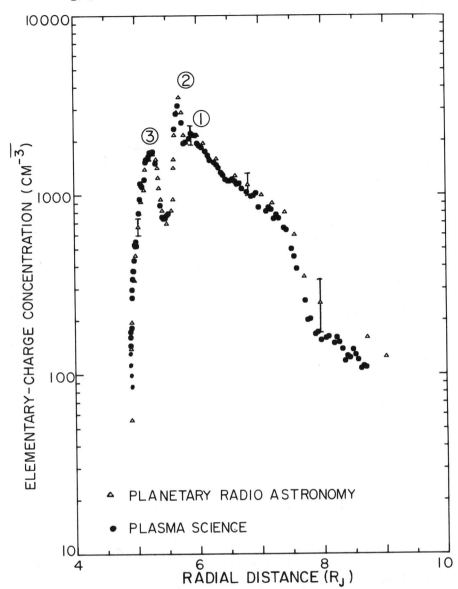

Fig. 3.7. The radial profile of charge concentration along the inbound trajectory of Voyager 1. The Plasma Science measurements (•) are of the elementary-charge concentration of positive ions derived from fits to positive-ion energy-per-charge spectra. •The Planetary Radio Astronomy data (Δ) from Birmingham et al. [1981] are electron concentrations determined from the cutoff frequency of plasma wave modes (the uncertainties in the Planetary Radio Astronomy determinations are shown by vertical bars). The numbered features are described in the text, and are also indicated in Figures 3.2 and 3.5.

The detailed features of the inbound elementary-charge concentration shown in Figure 3.7 are best understood in close comparison with the inbound temperature profile of Figure 3.5 and the two-dimensional contour model of Figure 3.1. The outer edge of the torus is indicated by the rapid increase in concentration in Figure 3.7 as the spacecraft moved inside of $\sim 7.5\ R_J$. After crossing the centrifugal equator at 7.1 R_J

(see Fig. 3.1), the spacecraft remained less than 0.15 R_J above that equator and traversed the core of the warm torus. The in situ concentration built up to a broad local maximum around the orbit of Io at 5.95 R_J (Peak 1 in Fig. 3.7). The spacecraft then passed over the small region of peak concentrations in the contours shown in Figure 3.1, giving rise to the sharp spike (Peak 2) in the in situ concentration profile of Figure 3.7 ($\sim$ 3100/cm^3 at $\sim$ 5.7 R_J). The Planetary Radio Astronomy experiment recorded a peak concentration of 3500/cm^3 around this time. However, there were few measurements of such large values and these were all measured near 5.75 R_J. Radially inward of 5.7 R_J, the elementary-charge concentration dropped rapidly by a factor of $\sim$ 5 to a local minimum of 740/cm^3 at $\sim$ 5.4 R_J. The spacecraft then crossed the centrifugal equator again, near the crest of the inner localized knoll in the concentration contours of Figure 3.1. This inner knoll is caused by the collapse of the torus plasma toward the centrifugal equator as it diffuses inward and cools (see Fig. 3.5 and the subsection entitled "Diffusive transport"). The crossing of this knoll resulted in a third local maximum in elementary-charge concentration of 1740/cm^3 at $\sim$ 5.2 R_J (Peak 3 of Fig. 3.7) before a final rapid decrease as the spacecraft made its closest approach to Jupiter at 4.89 R_J.

Bagenal and Sullivan [1981] found that the global structure on the outbound pass was nearly indistinguishable from that on the inbound pass, with similar features at similar L-shells, when the offset of the tilted dipole was taken into account. The main effect of including the offset is to change the apparent outbound trajectory shown in Figure 3.1. When this was done, the electron concentration profile predicted for the outbound structure from inbound data was in good agreement with that measured by Warwick et al. [1979a]. Consequently, the Plasma Science measurements show no clear evidence that the electron concentration varies in either longitude or local time, even though enhanced ultraviolet emission has been reported from the warm torus in the dusk quadrant [Sandel, 1980], and enhanced S$^+$ emission has been reported from the cold torus in the active sector [Trafton, 1980]. In contrast, radical changes in plasma properties are observed over radial spatial scales as short as $\sim$ 10^4 km, for example the temperature gradient at 5.5 R_J ($\sim$ 7 $\times$ 10^5 K R_J^{-1}).

Positive-ion composition in the torus

Major ionic species. In addition to these variations in overall concentration in the inner magnetosphere, there is considerable variation in the relative abundances of the different ionic species. Table 3.1 from Bagenal and Sullivan [1981] presents in situ concentrations determined from fits to energy-per-charge spectra at 4.96 and 5.3 R_J (in the inner cold torus); at 6.0 R_J (in the outer warm torus); at 8.6 R_J (just outside the torus); and at four of the many locations in the plasma sheet of the middle magnetosphere where there are well-resolved spectral peaks (11.8, 20, 28, and 42 R_J). The spectra from the middle magnetosphere are included here for completeness in the discussion of composition (the middle magnetosphere is discussed in detail in Sec. 3.4). At 6.0 and 8.6 R_J, where the various peaks are unresolved, the results of fitting the spectra assuming different thermal models are tabulated.

From the warm torus outward into the middle magnetosphere, the composition remains fairly constant. The major ionic species are some combination of O$^+$ and S^{2+} at an A/Z^* value of 16. Next in importance are O^{2+}, S^{3+}, and S$^+$, with the relative abundance of S^{3+} and O^{2+} varying considerably depending on assumptions about the thermal state of the plasma. In the cold inner torus the composition is significantly different. The dominant ion is S$^+$, with O$^+$ next in importance, and with few ions of higher ionization states (S^{2+}, S^{3+}, O^{2+}).

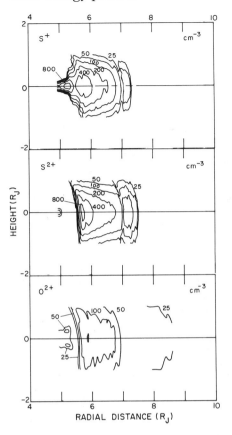

Fig. 3.8. Contour maps of the concentration of S^+, S^{2+}, and O^{2+} ions in a cylindrical coordinate system based on the centrifugal equator. The maps have been constructed from plasma measurements made along the inbound spacecraft trajectory using theoretical expressions for the distribution of plasma along dipolar magnetic field lines. The contours are in units of ions/cm³.

Figure 3.8 shows two-dimensional concentration maps for three individual ionic species: S^+, S^{2+}, and O^{2+}. These contour maps are constructed in the same manner as in Figure 3.1. The contrast between the S^+ and S^{2+} contours shows clearly that lower ionization states (S^+) dominate the inner torus, with very little contribution from the higher ionization states. The contour map for S^{3+} ions is similar to the S^{2+} map shown here in that there are insignificant concentrations of S^{3+} ions in the inner torus. The contour map for O^+ is similar to that of S^+ in shape but with lower concentration.

Bagenal and Sullivan [1981] point out an interesting aspect of the O^{2+} contour map in Figure 3.8. The minor quantity of O^{2+} in the cold inner torus has been drawn away from the centrifugal equator by the field-aligned polarization electric field that is set up by the electrons and the heavier sulfur ions. Near $5.3\ R_J$, sufficient O^{2+} ions have drawn off the equator so that double maxima form in the contours approximately $0.2\ R_J$ off the centrifugal equator, with a relative minimum at the equator itself.

In general, lighter ions of higher charge state are more easily pulled off the centrifugal equator, and thus their distribution along the field lines has a larger effective scale height. Eventually, at higher latitudes, protons must become the dominant ionic species because of their low mass relative to sulfur and oxygen (see, for example, Tokar et al. [1982]).

Present quantitative estimates of the relative abundance of the major ionic species from in situ observations should be taken as indicative, but not definitive, especially in the warm torus. In addition to the model dependencies introduced by our lack of knowledge of the thermal state of the warm torus, there are also uncertainties in the analysis due to a number of other effects. For example, all of the published analyses

Table 3.1. *Composition of the plasma in the dayside magnetosphere of Jupiter*

A/Z*		64 1120 4.96	64 1016 5.3	64 0859 6.0	64 0527 8.6		64 0150 11.7	63 1550 20	63 0505 28	62 1031 42
	UT:DOY R_J									cm^{-3}
1	H^+						2.2	0.21^a	0.06^a	0.06^a
8	O^2	48	26	160	28	20	1.5	0.10^b	0.03^b	0.03^b
				—	26^b	9^b				
10	S^{3+}	<5.6	<3.5	27	—	0.5	0.6	0.09^b	0.05^b	0.04^b
				170^b	—	11^b				
16	S^{2+}	14	39	430	—	19	2.2	0.39^d	0.17^d	0.10^d
				560^d	16^d	12^d				
16	O^+	250	350	130	34	—	2.9	0.78^c	0.35^c	0.20^c
				1100^c	32^c	24^c				
23	Na^+	<21	<72					0.08^b	0.05^b	0.02^b
32	S^+	91	1100	430	8	28	1.4	0.09^b	0.07^b	0.02^b
				470^b	11^b	23^b				
48	SO^+	<2	<8							

64	SO$_2^+$	3.5	13	73	7	7		
				8[b]	8[b]	8[b]		
V_c		62	66	75	108	148	251	351	527	kms^{-1}
V/V_c		1.0	1.0	1.0	0.9	0.77	0.76	0.49	0.43	

[a]Determined from low resolution spectra (McNutt, Belcher and Bridge 1981). All densities are derived from the isothermal model for the ions unless marked as follows: [b]From constant thermal speed model. [c]From constant thermal speed model when $A/Z^* = 16$ spectral peak is assumed to be all O$^+$. [d]From constant thermal speed model when $A/Z^* = 16$ spectral peak is assumed to be all S^{2+}.

assume that the spacecraft potential in the warm torus is zero, when in fact it is probably a few tens of volts negative. Negative potentials of this magnitude in the warm torus imply that the present analyses underestimate the concentration of the species with lower values of A/Z^*, for example, O^{2+} and S^{3+}, and overestimate the concentrations of the species with higher values of A/Z^*, for example, S^+ and S^{2+} [Bagenal, 1981]. There are similar problems associated with uncertainties in the plasma velocity in the outer parts of the warm torus (see the subsection entitled "Plasma velocities in the torus").

Finally, the assumption that the distribution functions of the ions are Maxwellian may be open to question in the warm torus, although it is demonstrably valid in the cold torus (compare Figs. 3.3 and 3.4). In particular, throughout the warm torus significant particle fluxes were detected at energies well above the bulk of the plasma. From this, Bagenal and Sullivan [1981] concluded that either very heavy molecular ions were present [Sullivan and Bagenal, 1979] or recently created sulfur ions were present that had not yet thermalized. This latter possibility is consistent with the recent analysis of ground-based observations of S^+ in terms of a hot and cold population [Brown and Ip, 1981]. There is no a priori reason to exclude such nonequilibrium distributions. A more sophisticated analysis of the in situ plasma data in the torus is in progress, and may produce quantitative differences in estimates of composition in the warm torus. However, the qualitative picture of low (higher) ionization states in the cold (warm) torus is well established, and in good agreement with other spacecraft and ground-based measurements [see Bagenal, 1981].

Minor ionic species. In addition to the major ionic species, there is at times clear evidence for various minor species, such as the peak at an A/Z^* value of 64 (probably SO_2^+) in Figure 3.3. When the presence of the minor species is indicated by a distinct peak or shoulder, we give a quantitative estimate of its concentration in Table 3.1. When there is no clear indication of the presence of a given minor ion, we place an upper limit on its concentration, as indicated in Table 3.1 and illustrated in Figure 3.3 for Na^+ and SO^+.

The well-resolved spectral peaks at A/Z^* values of 23 found in cold regions of the middle magnetosphere suggest sodium ions form about 10% of the ionic composition there. In the inner torus filling the gap between the peaks at A/Z^* values of 16 and 32 with appropriate amounts of ions with A/Z^* values of 23 suggests sodium is relatively less abundant closer in, forming less than $\sim 5\%$ of the ion composition. If the spectral feature at A/Z^* values of 64 is SO_2^+, then, in the cold torus, its concentration is $\sim 1\%$ of the total ion concentration. In the warm torus, the concentration of SO_2^+ near Io's orbit is $\sim 1\%$ of the total for the common thermal speed model and $\sim 5\%$ of the total for the isothermal model.

In the middle magnetosphere, protons comprise up to $\sim 30\%$ of the number density with their importance increasing with distance away from the plasma sheet, as discussed in the subsection entitled "The plasma sheet." In the inner magnetosphere the kinetic energy of H^+ is generally below the energy-per-charge threshold of the plasma instrument. However, there are a few spectra before closest approach with a feature in the lowest channels that might be the tail of a distribution function below the well-resolved spectral peak at an A/Z^* value of 8. If this feature corresponds to H^+ (He^{2+}) ions with the same temperature as the heavy ions, then a fit to the data gives concentration estimates of 3% (0.4%) of the total ion population. Although these percentages should be regarded with caution (because the observed distribution does not include the peak), their low values are consistent with the good agreement in the torus between

Table 3.2. *Vector velocities in the cold torus as seen in the corotating frame[a]*

Time UT on March 5, 1979	Distance R_J	Corotation speed (km/s)	$\dfrac{\mathbf{V}\cdot(\hat{\mathbf{r}}\times\hat{\mathbf{b}})}{\|\hat{\mathbf{r}}\times\hat{\mathbf{b}}\|}$ ~Azimuthal	$\dfrac{\mathbf{V}\cdot(\hat{\mathbf{b}}\times(\hat{\mathbf{r}}\times\hat{\mathbf{b}}))}{\|\hat{\mathbf{r}}\times\hat{\mathbf{b}}\|}$ ~Radial	$\mathbf{V}\cdot\hat{\mathbf{b}}$ ~Southward
1016	5.28	66.3	−0.1	−0.9	1.4
1120	4.95	62.1	−0.1	−0.7	1.7

[a] After Table 2 of Bagenal [1981]

the electron concentration from the Planetary Radio Astronomy experiment and the elementary-charge concentration of positive ions with $A/Z^* \geq 8$ (see the subsection entitled "Elementary-charge concentrations in the torus"). The fact that protons are a major ion by number in the middle magnetosphere but apparently a minor ion in the torus may be indicative of the source – for example, the H^+ ions might originate in the Jovian ionosphere, or perhaps from the icy Galilean satellites.

Plasma velocities in the torus

In the cold inner torus, corotating flow is essentially parallel to the symmetry axis of the main sensor cluster, and the various ionic species are cold enough so that the response of each of the Faraday sensors can be reliably taken to be a constant. In such a situation, the full vector velocity can be reconstructed with high accuracy from the simultaneous measurements in the A, B, and C sensor [Bridge et al., 1977]. For example, consider the C sensor data at 1016 UT on March 5, 1979 (Fig. 3.3). Data from the A and B sensors are similar, and each of the three sensors can be used to determine on independent component of the full vector velocity. Although the three velocity components so determined are not orthogonal, it is a simple matter to reconstruct the orthogonal components from them. In Table 3.2, we give two determinations of the vector velocity in the cold torus, from Bagenal [1981], based on simultaneous fits to all three sensors. These velocity components are quoted in the corotation frame of reference (i.e., after subtraction of rigid corotation from the inertial velocity). The unit vector $\hat{\mathbf{b}}$ is the local magnetic field direction as determined by the magnetometer experiment [Ness et al., 1979a], and the velocities are resolved into a coordinate system based on $\hat{\mathbf{b}}$ and the unit vector in the radial direction $\hat{\mathbf{r}}$. From Table 3.2, we see that the plasma in the cold torus moves at the corotational velocity to within a few kilometers per second. The probable error in the determination of this velocity is of the same order, so that there is no evidence for any systematic deviation from strict corotation in the cold torus.

In the warm torus, it is more difficult to determine full vector velocities, both because the plasma is warmer, and because the flow is more oblique to the A sensor normal. However, estimates of the velocity component into the C sensor are still reasonably straightforward. These estimates are consistent with strict corotation in the inner part of the warm torus, although there is some indication that the flow falls below corotation at the outer boundary of the warm torus [Bagenal and Sullivan, 1981]. This result is in keeping with the Planetary Radio Astronomy observations of Kaiser and Desch [1980], who find evidence for subcorotation of the plasma flow by 3% to 5% near 8 R_J, and it foreshadows the striking deviations from strict corotation found in the middle magnetosphere (see the subsection entitled "Plasma velocities in the middle magnetosphere").

Comparison with the Pioneer 10 plasma observations

Heavy-ion dominance of the Jovian magnetospheric plasma is a relatively new concept. In particular, all of the initial results of the Pioneer particle experiments were interpreted in terms of light ions. With the advantage of hindsight, it is worth reconsidering the results of Frank et al. [1976] in light of current views. In the Pioneer 10 data in the inner magnetosphere, positive-ion flux density vs. energy-per-charge spectra showed peaks well above the corotation energy for protons. With the assumption that the plasma was entirely composed of protons, Frank et al. [1976] concluded that the plasma was subsonic and analyzed the Pioneer 10 data accordingly. It now seems plausible that instead of hot, subsonic protons, the Pioneer 10 experiment was detecting transonic or supersonic heavy ions. For example, at 2.85 R_J, Frank et al. [1976] found a characteristic energy per charge of 105 V compared to corotational energies per charge of only 6.5 V for protons. However, heavy ions with mass-to-charge ratios around 16 would have corotational energies per charge of 100 V at this distance and thus could account for the observed energy-per-charge spectrum with no need for a hot proton component. This suggestion that Pioneer 10 detected heavy ions from Io is not a new one, and has been previously postulated by Hill and Michel [1976], Neugebauer and Eviatar [1976], and Goertz and Thomsen [1979b].

Intriligator and Miller [1981] have recently reexamined the Pioneer 10 positive-ion data between 5.4 and 6.9 R_J, and their results give additional weight to these earlier suggestions. Among other findings, these authors present evidence for the corotating ions S^{2+} and O^{2+}, with temperatures of tens of eV. The plasma current profile shows a relative variation with distance which is qualitatively similar to the density profile shown in Figure 3.7. There is a well-defined maximum at ~ 6 R_J (as contrasted to the 5.7 R_J maximum for Voyager), with a steep fall-off toward Jupiter and a gradual decrease away from Jupiter. Although Intriligator and Miller [1981] do not give estimates of ion concentrations, the concentrations quoted by Frank et al. [1976] are qualitatively consistent with the Voyager results, if allowance is made for the A/Z^* dependence of the estimates (the interpretation as protons underestimates the true concentration of heavy ions). These similarities between Pioneer 10 and Voyager 1 observations are evidence for the presence of an Io torus in December of 1973 with many of the properties of the torus as observed in March of 1979. Quantitative comparisons of the two data sets may provide insight into the long-time-scale variability of the torus.

3.3. The Io flux tube

Velocity perturbations near the flux tube

The nature of the direct interaction of Io with the magnetospheric plasma is an area of long-standing interest in space plasma physics. For this reason, Voyager 1 was targeted for passage through the Io flux tube at some 20500 kilometers south of Io. The subsequent Voyager 1 measurements of the perturbations in the magnetic field and in the fluxes of low-energy positive ions near the Io flux tube clearly established the presence of a large amplitude Alfvénic wave pattern propagating southward along the magnetic field lines [Ness et al., 1979a; Neubauer, 1980; Acuña, Neubauer, and Ness, 1981; Belcher et al., 1981]. The close correlation between the magnetic field and ion-flux-density perturbations near Io is shown in Figure 3.9. The ion flux density in this figure is from the B sensor of the main cluster. All sensors in the main cluster exhibited this same increase in flux density near the flux tube, closely correlated with the magnetic field perturbation.

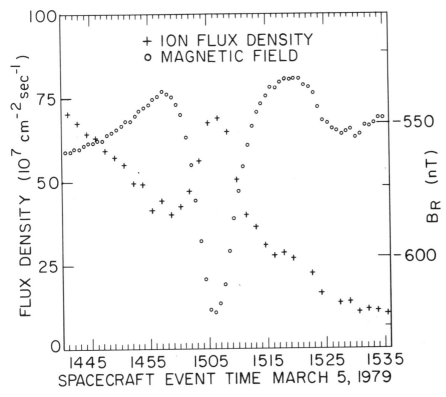

Fig. 3.9. Plots of the ion flux density from the B sensor and of the radial component of magnetic field as a function of time near the Io flux tube. The close anticorrelation between these two measurements indicates southward propagation of the Ionian Alfvén wave.

Figure 3.10 is a schematic illustration of the probable flow pattern sampled by the spacecraft. The location of the flux tube relative to the spacecraft trajectory is determined by fits to the magnetic field measurements [Acuña, Neubauer, and Ness, 1981]. If Io is a good conductor, the flow pattern of the ambient plasma external to the flux tube should be similar to incompressible flow around a cylinder [Scholer, 1970]. Streamlines for such a flow are illustrated in Figure 3.10. We indicate in this figure the cone on which the normals to the sensors in the main cluster lay, and also the time of the maximum in the flux density perturbation shown in Figure 3.9 (1506 UT on March 5, 1979).

The interpretation of the Plasma Science observations is straightforward in the context of this flow pattern. The perturbation in both field and plasma parameters occurred as the ambient magnetospheric plasma deviated from strict corotation to flow around the Io flux tube, so as to avoid the plasma "frozen" to the field lines threading Io. At the closest approach to the flux tube, this deviation from corotation was such as to bring the flow more nearly into the main cluster, thus increasing the ion flux density into the main cluster sensors. The anticorrelation between the radial component of the field and the flux density perturbations indicated southward propagation of the pattern [Belcher et al., 1981]. This qualitative interpretation can be made quantitative by an analysis of the multisensor data for oblique angles of incidence (see the subsection

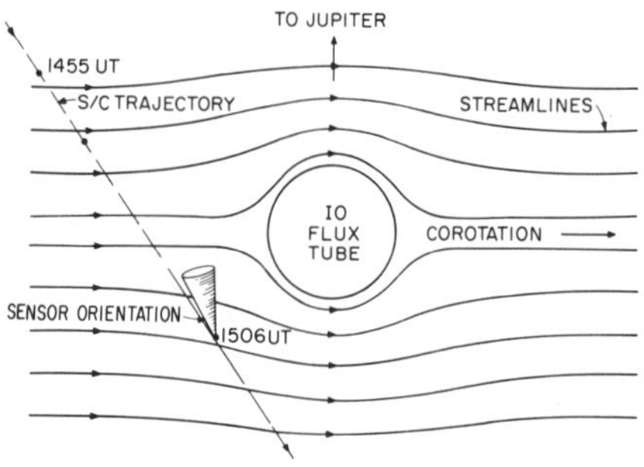

Fig. 3.10. The Voyager 1 trajectory with respect to the Io flux tube looking down from the north. The streamlines are for incompressible flow around a cylinder. The cone indicates the orientation of the main sensor normals.

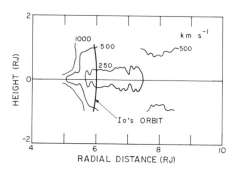

Fig. 3.11. A contour map of local Alfvén speed calculated from the O4 magnetic field model [Acuña and Ness, 1976c] and the total ion mass density, measured along the spacecraft trajectory and extrapolated along the magnetic field lines.

entitled "Plasma velocities in the middle magnetosphere"), and such a study is in progress (Barnett and Olbert, private communication, 1981).

Alfvén speeds in the torus

Bagenal and Sullivan [1981] have constructed contours of constant Alfvén speed for the Io torus, using their two-dimensional model for the total mass density of the plasma and the O4 model of Acuña and Ness [1976c] for the magnetic field configuration. This contour map is shown in Figure 3.11. Although the calculated local Alfvén speed everywhere exceeds the local corotation speed, it displays a region of uniformly low values in the outer torus with minimum speeds of ≤ 250 km s^{-1} occurring near the centrifugal equator. The position of Io with respect to the centrifugal equator also varies as shown in Figure 3.11. The Alfvén speed of the plasma in the vicinity of Io therefore varies by a factor of two with the System III longitude of the satellite. The large Alfvén speeds outside the torus mean that the time for Alfvén waves generated near Io to reach the ionosphere is largely determined by the length of the propagation path in the torus. This transit time is therefore modulated by the System III position of Io. Similarly, other properties of the propagating waves, such as geometry and damping, will also vary with longtitude. The subsequent changes in the field-aligned current associated with the Alfvén wave may explain the Io-modulation of the decametric radiation [Gurnett and Goertz, 1981].

3.4. The middle magnetosphere

General morphology

As we move into the middle magnetosphere (between ~ 10 and $\sim 40\ R_J$), the dominant plasma structure becomes the concentration enhancements associated with crossings of the plasma sheet. Because of the increasing plasma velocity with increas-

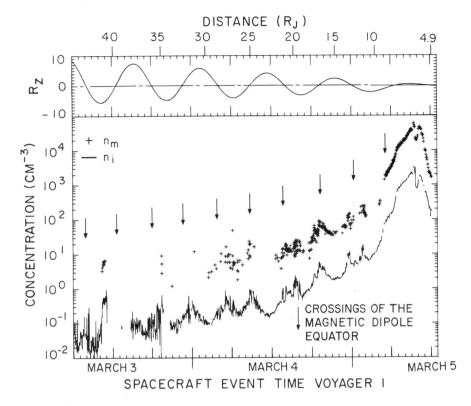

Fig. 3.12. Estimates of the total elementary-mass and elementary-charge concentrations of positive ions, n_m and n_i, in units of amu per cm^3 and elementary charges per cm^3, respectively, for the Voyager 1 encounter. The top panel shows the distance of the spacecraft from the magnetic dipole equatorial plane and arrows in the bottom panel indicate the crossings of this plane. Dates are given at 1200 UT.

ing distance, H^+ moves into the energy range of the Plasma Science instrument. Figure 3.12 shows values of the elementary-charge concentration n_i for positive ions with energy per charge between 10 and 5950 V, for the inbound Voyager 1 trajectory [McNutt, Belcher, and Bridge, 1981]. These elementary-charge concentrations are obtained by taking the total flux density of positive charge for the sensor that is least oblique to corotating flow and dividing by a model velocity component into that sensor (see the subsection entitled "Plasma velocities in the middle magnetosphere"). Such an estimate of n_i can be obtained for every positive ion spectrum, regardless of whether the various ionic species are resolved, and thus the n_i curve is essentially continuous except for data gaps. Simultaneous estimates of n_e, the total electron concentration, can be obtained from the electron measurements, and in the middle magnetosphere the estimates of n_i and n_e show good agreement [Scudder, Sittler, and Bridge, 1981]. Figure 3.12 also gives values of the positive-ion elementary-mass concentration n_m as obtained from fits of Maxwellians to the energy-per-charge spectra of positive ions. As discussed in detail by McNutt, Belcher, and Bridge [1981], in the middle magnetosphere we require for such fits a proton peak that is well resolved from the heavy ion peak in the positive-ion spectra. Frequently the ions are of sufficiently low Mach number that the protons and the heavy ions are not well resolved, especially at larger radial distances, so fit estimates of n_m and the ion temperature cannot be obtained for every

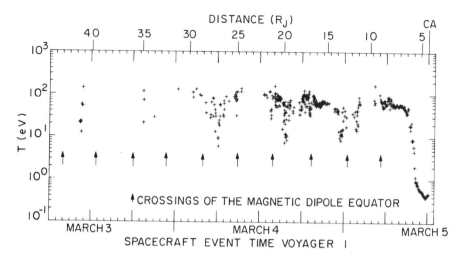

Fig. 3.13. Positive ion temperatures for Voyager 1 inbound from Bagenal and Sullivan [1981] and McNutt, Belcher, and Bridge [1981]. The time period covered is the same as that in Figure 3.12.

spectrum, as is obvious from the coverage in Figure 3.12. Resolved spectra tend to occur at local maxima in concentration, because the ions are cooler there.

The concentration profiles in Figure 3.12 are dominated by the Io plasma torus in the inner magnetosphere, with a rapid decrease in concentration as we move into the middle magnetosphere. Superimposed on this overall decrease are two local increases every Jovian rotation period (9 hr 55 min.), with the local maxima near crossings of the magnetic-dipole equatorial plane. The local increases occur as the plasma sheet sweeps across the spacecraft twice each planetary rotation period. These increases in concentration are accompanied by marked decreases in the temperatures of both positive ions and electrons. For example, Figure 3.13 shows the available positive ion temperatures from fits to ion spectra, over the same time period as covered in Figure 3.12. A close examination of the two figures shows that the local maxima in concentration are associated with local temperature minima on Voyager 1 inbound.

On Voyager 2 inbound, the plasma sheet crossings are not as clearly defined. Figure 3.14 shows the positive-ion concentrations from the middle magnetospheric passage of Voyager 2, in a format similar to Figure 3.12. The smaller number of estimates of n_m as compared to Voyager 1 reflect a smaller number of positive-ion spectra in which the protons are well resolved from the heavy ions. This lack of resolution is due to an overall decrease in the Mach number of the flow on Voyager 2 inbound, which we could attribute to an increase in temperature, but is more likely due to a decrease in flow velocity during the Voyager 2 encounter (see the subsection entitled "The breakdown of corotation"). Before approximately 1400 UT on July 8, 1979 ($\sim 25\ R_J$ inbound), concentration peaks near crossings of the magnetic equatorial plane are reasonably well defined. After this time on the inbound trajectory, there were no well-defined peaks near the crossings, except for the peak near closest approach, at ~ 2100 UT on July 9. In part, the Ganymede associated dropouts centered at 15 R_J [Burlaga, Belcher, and Ness, 1980; Connerney, Acuña, and Ness, 1981] may obscure some of the plasma sheet increases. However, the lack of pronounced crossings after ~ 1400 UT on July 8, probably reflects a thickening of the dayside plasma sheet due to the arrival of a solar wind pressure ridge on July 8. The increase in upstream solar wind pressure begins at

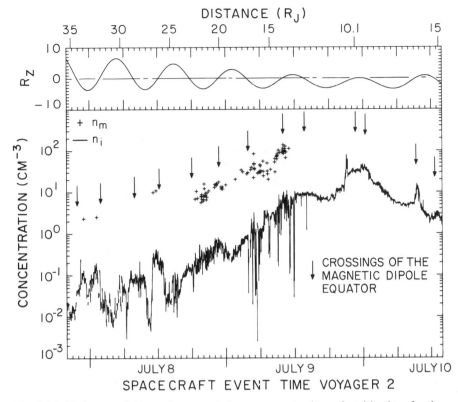

Fig. 3.14. Estimates of the total mass and charge concentrations of positive ions for the Voyager 2 encounter, similar in format to Figure 3.12.

approximately 1800 UT on July 7 rising from a value of $\sim 4 \times 10^{-11}$ N/m^2 to a peak of $\sim 27 \times 10^{-11}$ N/m^2 at ~ 1200 UT on July 8, with a slow decrease over the next three or four days. Using a scaling law inversely proportional to the cube root of the pressure, Goodrich, Sullivan, and Bridge [1980] find that the standoff distance to the Jovian bow shock should have decreased from ~ 107 to ~ 58 R_J over this period. This overall compression of the magnetosphere probably leads to a thickening of the plasma sheet on the dayside, as has been previously suggested on the basis of Pioneer energetic particle data [MacDonald, Schardt, and Trainor, 1979]. There are also suggestions in the data that the solar-wind compression may cause plasma flows away from the dayside current sheet, as discussed in the subsection entitled "Nonazimuthal flows."

Even when well-defined concentration maxima occur near crossings of the magnetic equatorial plane, we note that crossings of the plasma sheet (as defined by the plasma-concentration signature) do not necessarily coincide with crossings of the current sheet (as defined by the magnetic field signature). The most prominent example of this distinction between plasma sheet and current sheet is the bifurcated plasma sheet crossing near 0000 UT on March 5, at 13 R_J. Figure 3.15 shows the high resolution positive-ion spectra during this crossing, from 2200 UT on March 4 to 0215 UT on March 5, 1979. The peaks in relative flux density in this figure correspond to the peaks in concentration. The anticorrelation between concentration and temperature is readily apparent (see also Figs. 3.12 and 3.13). Of the two peaks in the concentration profile, one peak is north of the expected crossing of the magnetic dipole equatorial plane, at 4.0° in magnetic latitude (2330 UT on March 4), and one peak is distinctly south of the expected

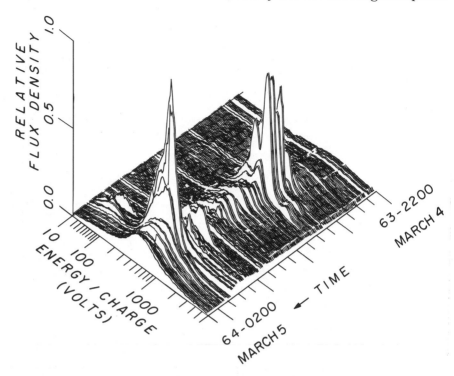

Fig. 3.15. The relative flux density of positive ion charge vs. energy per charge as measured in the D sensor between 2200 UT on March 4 and 0215 UT on March 5. The two main peaks in the prominent ion spectra near 2330 UT are due to ions with A/Z^* ratios of 10.67 (S^{3+}) and 16 (S^{2+} or O^+), with a much smaller peak above the 16 peak at an A/Z^* value of 32 (S^+). The cold peaks around 2330 UT on March 4 occur near a magnetic dipole latitude of $+4.0°$, whereas the cold peaks around 0139 UT on March 5 are at $-7.9°$ magnetic latitude. The current sheet crossing is at 0000 UT on March 5 near 1.3° magnetic dipole latitude.

crossing, at $-7.9°$ magnetic latitude (0139 UT on March 5). The current sheet crossing, as defined by the change in the poloidal field direction from away to toward the planet, occurred close to the dipole equatorial plane, at 1.3° magnetic dipole latitude (0000 UT on March 5 [Connerney, Acuña, and Ness, 1981]). The crossing near 21 R_J on Voyager 1 inbound is somewhat similar, in that there is a peak in concentration near the crossing of the magnetic dipole equatorial plane, close to but not exactly centered on the current sheet crossing at 1405 UT on March 4, near 0° magnetic dipole latitude. In addition, there is a more pronounced increase in concentration, 7.1° south of the expected crossing, at 1530 UT on March 4, far from the current sheet crossing (see Fig. 3.12). Figure 25 of McNutt, Belcher, and Bridge [1981] shows the high-resolution positive-ion spectra during this crossing. Again, the increases in ion concentration are accompanied by pronounced decreases in ion temperature by as much as a factor of 10 (from ~ 100 to ~ 10 eV, see Fig. 3.13).

The plasma sheet

With the caveat that the plasma sheet and the current sheet do not always coincide, we examine in detail both the electron and positive ion structure of a sheet crossing near 17 R_J on Voyager 1 inbound, for which the plasma and current sheet signatures do, in

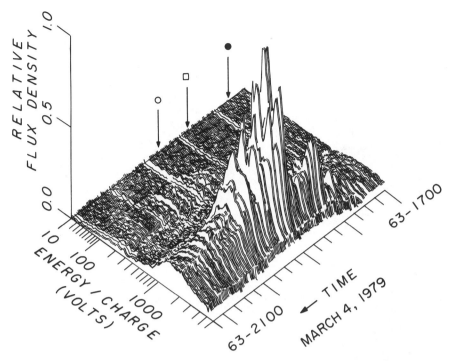

Fig. 3.16. The relative flux density of positive-ion charge as measured in the D sensor between 1700 UT and 2115 UT on March 4, 1979, similar in format to Figure 3.15. The arrows mark the times of the three electron spectra shown in Figure 3.17.

fact, coincide. Although the cooling effect in the positive ions is not as pronounced as during other crossings, the ion spectra show enough resolution throughout the crossing for detailed quantitative analysis. Figure 3.16 shows the high-resolution positive-ion spectra during this crossing, from 1700 UT to 2115 UT on March 4. The heavy-ion species occupy the energy-per-charge scan from ~600 to ~4000 V. The increase in concentration and decrease in temperature which defines the sheet crossing are apparent. The dominant cold peak near 1900 UT in this figure is due to S^{3+}. The proton signal is lost in the noise in the high-resolution ion spectra, but it is clearly present in the low resolution spectra. Electron distribution functions at three different locations in the sheet (at the times indicated by arrows in Fig. 3.16) are shown in Figure 3.17. The electron spectrum at 1723 UT (filled circles) is representative of spectra away from the sheet crossing; the spectrum at 1906 UT (open circles) occurs at the peak concentration in the crossing; the spectrum at 1824 UT (open boxes) occurs during a subsidiary density enhancement south of the main peak. As in the inner magnetosphere (see Fig. 3.6), the electron distribution functions exhibit a cold, thermal component as well as a higher-energy, suprathermal component. It is qualitatively clear from Figure 3.17 that the concentration increase in the current sheet is due primarily to an increase in the concentration of the core electrons, accompanied by a simultaneous decrease in their temperature.

Figures 3.18 and 3.19 display quantitative plasma parameters for both electrons and positive ions in this crossing, as well as magnetic field parameters from Ness et al. [1979a]. The bottom panel of Figure 3.18, as well as the second panel from the top of the figure, refer to positive-ion velocities, which will be discussed in the subsection entitled "Plasma velocities in the middle magnetosphere." The second panel from the bottom

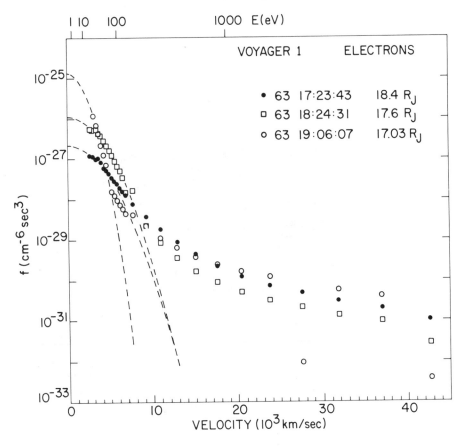

Fig. 3.17. Electron distribution functions vs. electron speed during the Voyager 1 inbound plasma sheet crossing near 17 R_J. For reference, the corresponding electron energy in eV is given at the top of the figure. The spectra are from the times indicated by arrows in Figures 3.16 and 3.19. Gaussian fits to the cold "core" electron component are indicated by the dashed lines.

of Figure 3.18 gives estimates of n_m (in amu/cm^3) obtained from fits to positive-ion spectra with resolved ion species, when available. The middle panel of the figure shows the ratio of total elementary-mass concentration to that of the protons, n_p. The top panel shows the temperature of the protons (triangular), which McNutt, Belcher, and Bridge [1981] found to be a good measure of the heavy-ion temperatures, and also the temperature of the core electrons (crosses), obtained from Maxwellian fits to the thermal electron component. In Figure 3.19, the top panel displays the total electron number density n_e obtained from integration of the full electron distribution function. Also shown is the halo number density n_H, obtained by direct integration over the full distribution function minus the Maxwellian fit to the core. In the second panel from the top, T_e is the mean energy of the electrons as obtained by direct integration over the observed distribution function, T_c is the temperature of the core electrons from the Maxwellian fit, and T_H is the characteristic energy of the halo electrons as determined from partial pressures. Also displayed in Figure 3.19 are the magnetic field strength, two direction angles of the field, and the Pythagorean variance as determined from 48 s averages.

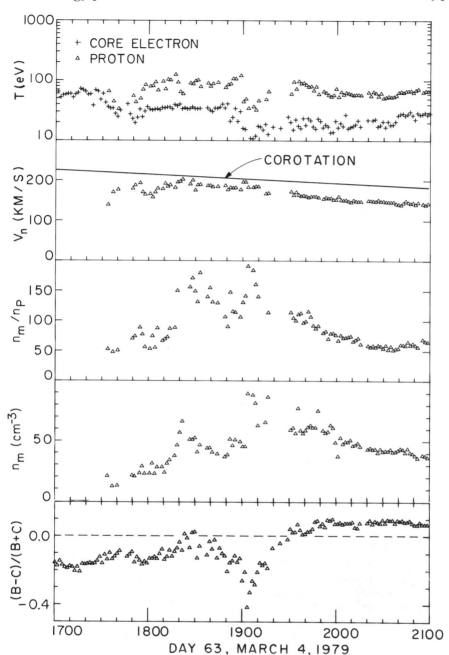

Fig. 3.18. A plot of plasma parameters for the plasma sheet crossing near 17 R_J on Voyager 1 inbound, corresponding to the time interval shown in Figure 3.16.

From the detailed analysis of this and other current sheet crossings in the middle magnetosphere, Scudder, Sittler, and Bridge [1981] and McNutt, Belcher, and Bridge [1981] conclude that:

1. The plasma sheet concentration maxima are routinely cooler than their surroundings, with respect to both electrons and positive ions. There is a tendency for the more abrupt crossings to exhibit lower ion temperatures.

Fig. 3.19. A plot of electron and magnetic field parameters for the plasma sheet crossing near 17 R_J on Voyager 1 inbound, corresponding to the time intervals shown in Figures 3.16 and 3.18.

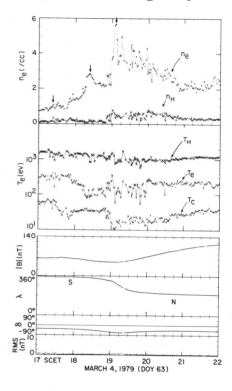

2. The heavy ions are greatly enhanced with respect to protons in the plasma sheet, presumably because of the larger scale height of the lighter protons. Both in and out of the sheet, the protons are essentially negligible in terms of total elementary-mass concentration ($\sim 2\%$ or less), at least over the limited range of magnetic latitudes sampled by the Voyager spacecraft.

3. At the center of the crossing of the plasma sheet by Voyager 1 at 17 R_J (see Figs. 3.18 and 3.19 near 1900 UT), $\rho V_n^2/2$ exceeds the magnetic energy density by about a factor of three – that is, the flow has become super-Alfvénic in the sheet owing to the high concentrations and low field strengths. Away from the maximum sheet density, where the mass concentration is lower and the magnetic field strength higher, the flow is sub-Alfvénic. As we move outward from this 17 R_J crossing on the dayside in the Voyager 1 encounter, every subsequent sheet crossing shows this transition from sub-Alfvénic to super-Alfvénic flow. As we move inward from this crossing, the flow is always sub-Alfvénic. Thus, on Voyager 1, the Alfvénic critical point in the plasma sheet must occur inside of 17 R_J. Note that the Alfvénic Mach number exceeds unity in the sheet maxima beyond 17 R_J on Voyager 1 without any drastic consequences, in accord with theoretical conclusions [Goldstein, 1977; Goertz, 1979]. On Voyager 2, however, the sheet crossings for which we have mass concentration estimates do not tend to show super-Alfvénic velocities. We attribute this difference to lower azimuthal velocities on the Voyager 2 pass (see the subsection entitled "Plasma velocities in the middle magnetosphere"), and to the lack of large concentration enhancements in the plasma sheet between ~ 12 and ~ 25 R_J, as discussed in the subsection entitled "General morphology."

4. The enhancement in the total electron concentration in the sheet is due to an increase in the concentration of the thermal population; the suprathermal density

enhancement is less pronounced, and is nearly symmetric about the magnetic equator.

5. The reduction in the mean electron temperature T_e is due predominantly to a decrease in the core temperature T_C, with the suprathermal energy T_H relatively unaffected. Scudder, Sittler, and Bridge [1981] suggest that the suprathermal population observed within the plasma sheet is a global population that is not especially confined to the vicinity of the sheet. By contrast, the thermal population is localized by the presence of the heavy ions, and by the consequent large polarization potential of the order of the mean electron energy they must overcome to leave the sheet.

6. The mean electron temperature T_e is asymmetric about the plasma sheet, being on average cooler on the centrifugal side of the sheet (the generally positive Jovigraphic latitude of the spacecraft implies that the centrifugal equator will be encountered south of the magnetic dipole equator). Steady-state arguments imply that cooler plasma should be found toward the centrifugal equator and hotter plasma toward the magnetic equator [see Appendix B, and Hill, Dessler, and Michel, 1974; Goertz, 1976; Cummings, Dessler, and Hill, 1980]. In some average sense, this is the pattern observed in the electron parameters by Scudder, Sittler, and Bridge [1981], both in the middle and outer magnetosphere.

7. The fraction of suprathermal electrons in the plasma sheet increases with increasing Jovicentric distance, from $\lesssim 1\%$ in the torus, to $\sim 10\%$ at $\sim 15\ R_J$, to $\sim 50\%$ at $\sim 40\ R_J$. As a consequence, the mean electron temperature in the sheet proper increases with increasing Jovicentric distance. Scudder, Sittler, and Bridge [1981] envisage this positive temperature gradient as due to the intermixing of a cool, dense plasma that is transported directly outward from the Io torus without diversion from the centrifugal equator and a hot suprathermal bath of electrons at midlatitudes, which is an older and possibly recirculated population (with possibly an ionospheric component). As the cool plasma become more dilute at greater distances, the suprathermal bath of electrons increasingly dominates the number density of electrons in the sheet proper.

8. The sonic Mach number of the positive ions decreases with increasing radial distance, presumably owing to an increasing temperature. There is some evidence for this in the ion temperature estimates of Figure 3.13, but a stronger argument involves the lack of resolved proton/heavy-ion peaks at larger distances. A Mach-one proton distribution with a bulk speed of 200 km/s has a temperature of ~ 200 eV, and it will not be resolved from the heavy-ion signature. Gaps in the temperature profile in Figure 3.13 are due to a selection effect related to temperature, because we have only analyzed spectra with resolved proton peaks for this figure. Thus, we have systematically eliminated ion spectra exhibiting temperatures greater than ~ 200 eV, and the large number of gaps at greater radial distances implies an increased incidence of hotter ion spectra. If we relax the requirement that a resolved proton peak be present, and fit all low-resolution ion spectra, we qualitatively find that within $40\ R_J$ the ion temperature is typically ~ 1 keV or less. This estimate reflects the fact that in almost all cases, the distribution functions of the measured spectra peak below 6 keV, a result that is consistent only with a plasma temperature of less than ~ 6 keV.

9. Even at $\sim 15\ R_J$ where the suprathermals are $\sim 10\%$ by number in the sheet proper, Scudder, Sittler, and Bridge [1981] find that the suprathermals account for the majority of the electron energy density ($\gtrsim 75\%$) in the sheet. That is, the thermal population dominates the number density, and the suprathermal population, the energy density.

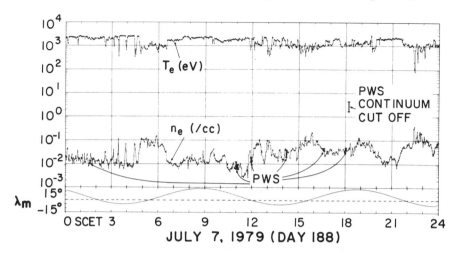

Fig. 3.20. Time series day plot of Voyager 2 inbound electron parameters on July 7, 1979, when the spacecraft is between 46 and 33 R_J from Jupiter. In the top panel, the Plasma Science determination of n_e and T_e are displayed. In addition, the Plasma Wave Science broadband continuum cutoff determinations of the electron concentration are indicated by the symbol I. In the next panel down, the tilted dipole magnetic latitude of the spacecraft λ_m in degrees has been added for reference, where the dashed horizontal line indicates when the spacecraft passed the magnetic dipole equator.

10. The plasma pressure at the sheet crossings due to both ions and electrons in the energy range of the Plasma Science experiment is insufficient to balance the decrease in magnetic pressure. If the plasma sheet is in quasistatic equilibrium, pressure balance must be caused by ions with energies above the energy per charge range of 6 kV of the Plasma Science experiment, consistent with the results of Krimigis et al. [1981] and Lanzerotti et al. [1981] (see Chapter 4). It is tempting to speculate that the ion distribution functions have both a thermal and suprathermal component, as do the electrons, perhaps for similar reasons. In particular, the suprathermal ion component may dominate the ion pressure even though it may not dominate the concentration as suggested by Belcher, Goertz, and Bridge [1979]. However, we also note that there may be severe problems with the quasistatic approximation, at least for some of the dayside current sheet crossings, particularly those on Voyager 1 inbound at 13 and at 21 R_J. It is hard to understand how any quasistatic model of the plasma sheet could produce mass concentration peaks north and south of the field reversal region, with a relative minimum at the reversal.

11. Even in the outer parts of the middle magnetosphere, the anticorrelation between n_e and T_e is a pronounced effect. This is illustrated in Figure 3.20, which displays electron parameters from 46.5 to 33.1 R_J (July 7) on Voyager 2 inbound. Note especially from 0200 UT to 0500 UT of July 7, the "spikey" appearance and disappearance of an electron plasma that is much cooler and denser than the surrounding regions. These cool electron concentration enhancements are well correlated with depressions in the magnetic field strength; they may represent phenomena associated with the proximity of the plasma sheet. The broader concentration increases in Figure 3.20 are, of course, associated with expected

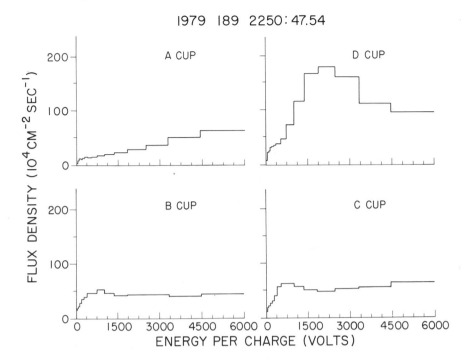

Fig. 3.21. The flux density of positive ion charge, in units of elementary charges per square centimeter per second, vs. energy per charge for a low resolution ion measurement from Voyager 2 inbound at 20.0 R_J.

current sheet crossings. Note, however, that even within these broad increases, there are local spikes in concentration containing cool electrons (e.g., the spikes in crossing near 2230 UT on July 7 in Fig. 3.20). These spikes appear to be uniformly distributed in magnetic latitude within the sheet, and lead to a sheet structure that is complex and highly time variable.

Plasma velocities in the middle magnetosphere

Azimuthal flows. Bridge et al. [1979a,b] report that the low-energy plasma in the Jovian magnetosphere tends to move azimuthally about Jupiter. This statement is based on the observation that the relative magnitudes of the positive ion fluxes in the different sensors during the encounters correspond qualitatively to those expected for a cold corotating beam (Figs. 1 and 2 of McNutt, Belcher, and Bridge [1981]). An example of this behavior is shown in Figure 3.21. This is a low-resolution positive-ion spectrum obtained by Voyager 2 on July 8 at 2250 UT, when the spacecraft was 20 R_J from Jupiter. Rigidly corotating flow at this distance would appear at a speed of 240 km/s, with the aberrated flow velocity at angles of 95°, 69°, 64°, and 19° to the A, B, C, and D cup axes, respectively (0° is directly into the cup). Thus, one would expect to see corotating H^+ (O^{2+}) at an energy per charge of ~ 300 V (~ 2400 V) in the D cup, and at an energy per charge of ~ 75 V (~ 600 V) in the B and C cups. The positive-ion spectra in the B, C, and D cups in Figure 3.21 are consistent with the unresolved presence of H^+

plus heavy ions at roughly these energies. The low-energy shoulder in the D cup is probably H^+, with the higher-energy peak due to heavy ions. In the B and C cups, the H^+ shoulder disappears because of projection effects, and the heavy-ion peak moves to lower energies, and also lower fluxes, because of the cup response. For a cold beam, the A cup should show no response at this time, and thus the fluxes in the A cup must be due to the finite thermal spread of the beam. Qualitatively, it is clear from these spectra that at this distance the positive ions are reasonably supersonic and moving azimuthally to first order (radial and vertical motions may also occur, as discussed in the subsection entitled "Nonazimuthal flows," although in general these are second-order effects). This pattern holds through both the Voyager 1 and 2 encounters. Note that the equal flux levels in the B and C sensors in Figure 3.21 suggest that the vector velocity is more or less in the Jovian equatorial plane at this time, because the B and C sensors are symmetrically oriented south and north with respect to that plane (see Belcher and McNutt [1981], or McNutt, Belcher, and Bridge [1981]).

The D cup spectrum in Figure 3.21 is representative of ~70% or more of the low-resolution spectra obtained during the Voyager 1 encounter. The remaining 30% of the low-resolution spectra from Voyager 1 display a distinct separation between H^+ and the heavy ions, and they can be analyzed for temperatures, concentrations, and velocity components. In contrast, the only quantitative parameter that can be easily derived from unresolved ion spectra such as in Figure 3.21 is the elementary-charge concentration of positive ions in the energy range of the Plasma Science experiment. For example, the total flux density of positive-ion charge in the D cup spectrum of Figure 3.21 is 1.1×10^7 proton charges per square centimeter per second. This charge flux density is the product of the total elementary-charge concentration in this energy range, n_i, times the (common) ion-velocity component into the cup, V_n, provided that the latter is reasonably supersonic. Thus, if V_n is 200 km/s at the time of the spectrum in Figure 3.21, n_i must be 0.54 proton charges per cubic centimeter. The estimates of n_i in Figures 3.12 and 3.14 were obtained in this way, using a model for V_n obtained from the resolved ion spectra, as we now discuss.

For the resolved spectra in the D sensor on the inbound trajectory, the response function of that sensor to the plasma can reasonably be approximated by a constant. It is then directly possible to obtain from the measured currents estimates of plasma parameters describing the positive-ion distribution functions – for example, composition, concentration, temperature, and (in the case of a single sensor) that component of velocity along the sensor look direction. To obtain all three components of the plasma velocity, however, requires information from at least three sensors. As we have previously noted in the subsection entitled "Plasma velocities in the torus," in the cold inner torus, inside the orbit of Io, corotating flow is essentially parallel to the symmetry axis of the main cluster, and the response of each of the Faraday cups in the main cluster can be taken to be constant [Bridge et al., 1977]. In such a situation, the full vector velocity can be reconstructed with high accuracy from measurements in the A, B, and C sensors (see Table 3.2). In the dayside middle magnetosphere, however, all sensors except the D sensor are usually significantly oblique to the flow. Thus, vector velocites can only be obtained using an analysis based on the instrument response for ions that arrive at large angles to the detector normals in the main cluster. The numerical methods required for a multisensor analysis using the full response are well understood in principle, but they are complex to implement in practice. A systematic program to analyze quantitatively multisensor data in the middle magnetosphere and at the Io flux tube is in process (Barnett and Olbert, private communication, 1981). However, at present, quantitative estimates of velocities in the middle magnetosphere are limited to that component of velocity along the D sensor axis.

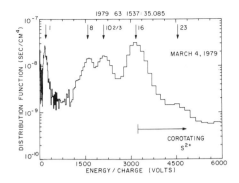

Fig. 3.22. A high resolution *M*-mode measurement from the D sensor of the positive-ion distribution function at 19.8 R_J. A least squares fit to the 8, 10.67, and 16 peaks yields a common velocity component of 195 km/s into the D sensor, and this velocity is used to draw the various arrows in the figure. The velocity component expected on the basis of rigid corotation is 238 km/s. The expected energy per charge of an S^{2+} ion moving at the rigid corotation speed is indicated.

The breakdown of corotation. In terms of single sensor analysis, the best determinations of the component of plasma flow into the D sensor in the middle magnetosphere are obtained from the cold ion spectra occurring in the plasma sheet crossings (see the subsection entitled "The plasma sheet"). Figure 3.22 is one of the best examples of a high-resolution ion spectrum with excellent species resolution. The common velocity component of the heavy ions in Figure 3.22 is 195 km/s, as compared to a component of 238 km/s expected for rigid corotation. The thermal width of the peak at an A/Z^* value of 16 in Figure 3.22 is only 11 km/s, so that this peak is 3.9 thermal widths below the expected corotation speed. One might argue that the lack of corotation displayed in this spectrum is due to various nonphysical effects – for example, spacecraft charging, or improper identification of ionic species. However, for this type of spectrum, the arguments for the observation of a true departure from rigid corotation are incontrovertible [McNutt et al., 1979; Belcher, Goertz, and Bridge, 1980; McNutt, Belcher, and Bridge, 1981]. The upper panel in Figure 3.23 shows values of V_n obtained from fits to such high-resolution spectra with resolved heavy ion peaks, as in Figure 3.22. There are not a large number of such spectra because the requirement of resolved heavy-ion peaks requires Mach numbers greater than about 6. However, these cold spectra occur systematically at the concentration maxima associated with the plasma sheet (see Figure 3.12). Thus, although they are few in number, they represent the speed of the plasma at the peak concentration in the middle magnetosphere. The departure from rigid corotation is systematic, growing with increasing Jovicentric distance.

The bottom panel of Figure 3.23 shows the values of V_n determined from fits to low-resolution ion spectra with a proton peak well separated from the heavy-ion peak. The value of V_n is determined mainly by the proton peak. There are many more of these spectra because the protons are more prominent in the wider energy windows of the low resolution mode and are well-resolved from the heavy ions down to Mach numbers as low as ~2. However, the uncertainties in the value of V_n so determined are much larger, because the energy resolution is coarser, and because the position of the light-ion peak is more strongly affected by possible spacecraft charging. Although these determinations agree with the high-resolution values when comparison is possible, it is the high-resolution values that argue convincingly for the breakdown of rigid corotation. Figure 3.24 shows values of V_n divided by the expected rigid component, using both high and low-resolution spectra, for both Voyager 1 and Voyager 2. There is some indication that the magnetospheric plasma was rotating more slowly through the dayside during the Voyager 2 passage.

Independent of and essentially simultaneous with these observations of a departure from strict corotation, Hill [1979, 1980] predicted on theoretical grounds the existence of such an effect owing to inertial loading (see the subsection entitled "Inertial loading"

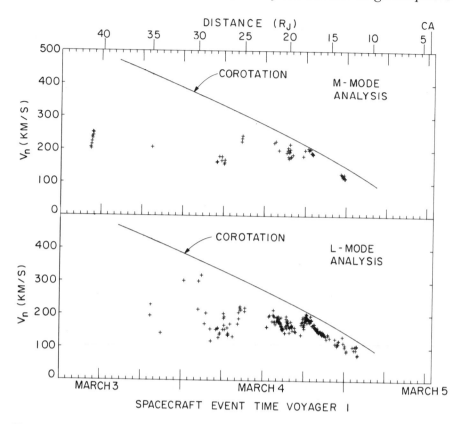

Fig. 3.23. Value of the component of velocity into the D sensor for Voyager 1 as determined from fits to low resolution (*L*-mode) and high resolution (*M*-mode) ion spectra. The component of velocity expected for rigid corotation is also shown.

and Chaps. 10 and 11). This theory is now the accepted explanation for the subcorotational velocities observed in the middle magnetosphere.

Nonazimuthal flows. In addition to the quantitative least-squares fit values of V_n from single-sensor analysis, more qualitative information about the full vector velocity can be obtained from multiple sensor data by taking advantage of the nearly symmetric orientation of the B and C cups with respect to the Jovian equatorial plane. Such an analysis suggests a general pattern of nonazimuthal flow that appears to be away from the equatorial current sheet on the dayside and toward it on the nightside [Belcher and McNutt, 1981; McNutt, Belcher, and Bridge, 1981]. Figure 3.25 illustrates this pattern qualitatively for Voyager 2; a similar phenomenon is seen in the Voyager 1 fluxes. This figure displays the difference of the total flux density of positive ions into the B cup and the total flux density into the C cup, divided by the sum of these two flux densities. Because the B cup opens southward and the C cup opens northward, this difference will be positive if there is flow to the north and negative if there is flow to the south. The horizontal axis is linear in spacecraft event time. Also indicated are radial distance from the planet and Universal Time at the top and local time at the bottom. The figure also shows the spacecraft position perpendicular to the magnetic equatorial plane.

Recall that passing through the magnetic equatorial plane essentially marks crossings of the Jovian current sheet, especially within 20 R_J of the planet. The main

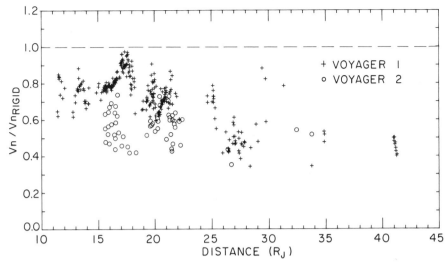

Fig. 3.24. The observed component of velocity into the D sensor divided by the component expected for rigid corotation for Voyagers 1 and 2.

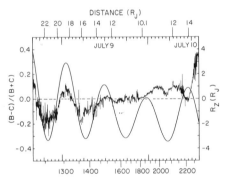

Fig. 3.25. A plot of the differences between the flux densities in the B and C cups divided by their sum for Voyager 2. Also shown is the vertical distance of the spacecraft from the magnetic dipole equatorial plane. Local time is indicated at the bottom and both distance from Jupiter and universal time at the top (the date is given at 1200 UT).

sensor is oriented such that there would be slightly more purely azimuthal flow into the C cup than the B cup on the inbound leg, and vice versa on the outbound leg. Even in the absence of any north/south flow, one would expect to see a negative signature in $(B - C)/(B + C)$ inbound and a positive signature outbound. In fact, the differences in Figure 3.25 oscillate about a negative average value inbound and a positive average value outbound, as expected. The oscillations about this mean imply that in the day-side magnetosphere near noon, there is a tendency for plasma flow to be northward when the spacecraft is north of the magnetic dipole plane, and southward when the spacecraft is south of that plane – that is, the flow is away from the current sheet. The amplitude of the difference decreases to zero with decreasing radius and increasing local time toward dusk. In the dusk to midnight sector, the difference again increases with increasing radius and increasing local time toward midnight. However, in this sec-tor the plasma flow tends to be northward when the spacecraft is south of the magnetic equator, and vice versa – that is, the flow is now toward the current sheet.

Belcher and McNutt [1980] have suggested that these apparent flows away from and toward the equatorial region are caused by dynamic expansion and contraction of the plasma sheet resulting from the compression of the dayside magnetosphere by the solar wind. As flux tubes rotate into the dayside magnetosphere, they move closer to Jupiter and the plasma flows away from the equator due to the compression and to the decreasing centrifugal force. On the nightside, the flux tubes move outward, and the plasma collapses back toward the equator due to the expansion and to the increasing centrifugal force. One obvious problem with this qualitative picture is the phase of the effect, because in the quasistatic approximation one would expect maximum compression and thus zero velocity near noon, not near dusk. On the other hand, the plasma sheet may not be in quasistatic equilibrium [see McNutt, Belcher, and Bridge, 1981], and thus there may be a phase lag due to dynamic overshoot and finite-propagation time effects. At present, there is no quantitative model, and such explanations remain conjectural.

3.5. The outer magnetosphere

In the dayside magnetosphere beyond about 50 R_J, the organized plasma sheet structure disappears. Even so, we still observe sporadic appearances of cold, dense concentrations of electrons and ions, correlated with decreases in magnetic field strength. Such regions appear similar in morphology to the spikey structure in the middle magnetosphere (see the subsection entitled "The plasma sheet"), except that they occur randomly, and not near expected crossings of the magnetic equator. Outside of the cool, dense regions, hot electrons with characteristic energies of a few keV appear to be the exclusive electron population. Positive-ion measurements in these hot regions are difficult to interpret both because of the low flux levels at these distances and because of the possibility of significant hot-electron feedthrough into the positive-ion measurements (see the discussion in McNutt, Belcher, and Bridge [1981]). Overall, the dayside outer magnetosphere appears to be a disordered and turbulent region in its plasma characteristics.

In contrast, the nightside magnetosphere is well organized by the plasma sheet structure out to great distances. On Voyager 1 outbound, regular enhancements in electron concentration were measured by the Plasma Science experiment, occurring twice per planetary rotation period out to 85 R_J, and once per rotation period out to 120 R_J. The location and time of these crossings of the predawn plasma sheet have been studied extensively [e.g., Goertz, 1981], and will not be discussed here. As we have noted above in the subsection entitled "The Voyager Plasma Science experiment," positive-ion observations outbound on the Voyagers were of poor quality for the most part because of the unfavorable viewing geometry.

3.6. Discussion

Sources of plasma

Given the observed properties of the low energy plasma at Jupiter, a number of authors have considered the nature of the sources, transport mechanisms, and energy and angular momentum balance in that plasma. We briefly review some of these considerations in light of the above phenomenology. The detection of an SO_2 atmosphere by Pearl et al. [1979] and SO_2 frost on the surface of Io [Fanale et al., 1979; Smythe, Nelson, and Nash, 1979] make SO_2 an obvious source material for the magnetosphere. However, Bagenal and Sullivan [1981] observe more sulfur in the torus than expected from the full dissociation and ionization of SO_2. Unfortunately, the characteristics of the dissociation and ionization processes for SO_2 are not well known. Shemansky [1980b] suggests that they probably involve O_2 and SO molecules and their ions. The existence of intermediate dissociation products of SO_2 as well as the fact that oxygen is less readily ionized than sulfur may explain why more sulfur was observed in the torus than expected. In addition, there must be some mechanism (such as sputtering) for the removal from Io of material that is not a major constituent of the atmosphere (e.g., consider neutral sodium, which has been observed in the magnetosphere for many years). Because there is probably a significant amount of elemental sulfur on the surface of Io [Masursky et al., 1979], additional sulfur may be supplied to the torus by such a process.

Although the ultimate source for the torus plasma is Io, it is not clear where the ionization takes place. The plasma could come either from the ionization of a diffuse cloud of neutral gases spread over a large portion of Io's orbit, or from the immediate vicinity of Io itself in the complex interaction of the magnetospheric plasma and the satellite. For a diffuse source, freshly ionized particles gain a cyclotron speed equal to the magnitude of the difference between the original velocity of the neutral atom and the local corotation velocity. At 6 R_J, sulfur and oxygen would gain cyclotron energies of 545 and 270 eV, respectively. Because the observed temperature of the bulk of the ions in the torus is ~ 40 eV, a diffuse source requires an efficient mechanism for the removal of most of this initial cyclotron energy. In contrast, the localized source can produce ions that initially pick up much smaller cyclotron energies, owing to the reduction of the local corotational electric field in the immediate vicinity of Io [Goertz, 1980a]. However, the Voyager ultraviolet measurements indicated that there was no enhancement of emission in the vicinity of Io [Shemansky, 1980b] as would be expected if the source were localized there. In addition, the existence of an extensive cloud of neutral sodium and the recent observation of neutral oxygen in the vicinity of Io's orbit [Brown, 1981a] suggest that the sulfur and oxygen ions may come from extended clouds of neutral atoms. The conclusion of Brown and Ip [1981], and Bagenal and Sullivan [1981], that there may be significant fraction of S^+ ions with energies far higher than the average energy of the torus, also supports the notion of diffuse pickup of new ions at large cyclotron energies followed by rapid dissipation of energy to the observed ion temperatures (see also Chap. 6).

Diffusive transport

In any case, new ions enter the torus from a source region localized either to the vicinity of Io or its orbit. They eventually diffuse away from the source under the controlling infuence of centrifugal force, which aids outward but inhibits inward transport [Richardson et al., 1980; Richardson and Siscoe, 1981; Siscoe and Summers, 1981]. To investigate the nature of this diffusive transport of plasma, Bagenal and Sullivan [1981] computed the total number of ions in a unit L-shell times L^2 (i.e., NL^2). This quantity is plotted in Figure 3.26 as a function of distance from Jupiter. The total tube content had a maximum at ~ 5.7 R_J and decreased monotonically radially inward and outward from the peak. Although the concentration along the spacecraft trajectory exhibited three local maxima at different radial distances, the flux tube content has only a single peak, which supports the model of diffusion of plasma from a single source near Io.

Siscoe et al. [1981] have made the most detailed effort to date to interpret these observations of flux tube content in terms of the mass transport theory appropriate for flux tube interchange diffusion. As illustrated in Figure 3.26 from Siscoe et al. [1981], the inner magnetosphere can be divided into four different diffusion regions according to the differing slopes of NL^2 vs. L in these regions. Large slopes indicate zones of weak diffusion and small slopes zones of strong diffusion. The innermost region contains the cool inner torus, and is referred to as the precipice to describe the sharp decrease in total mass and temperature observed there. The warm Io torus is referred to as the ledge, and is separated from the inner edge of the plasma disc or sheet by the ramp. The

Fig. 3.26. Radial profile of the magnetic flux shell density $Y = NL^2$, where N is the number of ions in a flux shell per unit L.

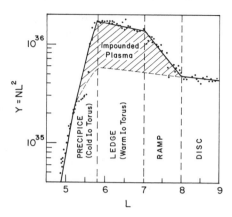

sharp break in the slope of NL^2 at $L = 5.8$ is the likely location of the inner edge of the source region, and therefore marks the separation between the domains of inwardly and outwardly diffusing ions. The steep slope in the precipice demonstrates that inward diffusion "uphill" against the centrifugal force is comparatively feeble. The relatively small amount of inwardly diffusing plasma is transported slowly, and so has time to cool by radiation. As the plasma cools, it collapses toward the centrifugal equator, and the resulting increase in the local density enhances the emission so that the plasma cools further. This runaway process leads to the sharp transition in temperature and ionization states between the regions of inward and outward diffusion.

In contrast, 90% of the injected plasma rapidly diffuses outward to replenish the ledge, and does not have time to cool by radiation. The shallow gradients distinguishing the plasma ledge and the plasma disc illustrate the situation expected when centrifugally driven interchange instability is restrained only by ohmic dissipation in Jupiter's ionosphere [Siscoe and Summers, 1981]. The much steeper decline delineating the plasma ramp signifies a substantially diminished level of diffusive activity. The origin of this feature is attributed by Siscoe et al. [1981] to pressure gradient inhibition of the interchange motion enforced by a prominent precipitation edge to the ring current that coincides with the ramp [Krimigis et al., 1979a]. The base of the hatched area in Figure 3.26 depicts the shape the Io plasma formation would have if there were no ring current. The hatched area itself indicates the amount of Io plasma supported against the outward centrifugal force by the inwardly directed pressure of the inner surface of the ring current. The quantity of matter impounded in this way is very nearly equal to the total bulk of the plasma ledge alone. Bagenal and Sullivan [1981] estimate that the total mass of the ledge (i.e., the warm torus) is 1×10^{36} amu.

Inertial loading

As diffusion continues outward through the ramp and into the extended plasma disc, the diffusing plasma must be accelerated to higher azimuthal velocities with increasing radial distance in order to maintain corotation with the planet. At larger distances, the energy and angular momentum requirements for this continued acceleration becomes an increasing strain on the frictional coupling between the Jovian ionosphere and the neutral atmosphere, and eventually at large enough distances the magnitude of this frictional coupling is no longer sufficient to maintain rigid corotation [Hill, 1979]. Hill [1980] has calculated a mass-loading rate of $\sim 10^{30}$ amu/s in order to account for the marked deviation from corotation observed by the Plasma Science experiment beyond

20 R_J [McNutt et al., 1979]. It is worth emphasizing that the Plasma Science instrument cannot determine velocity components beyond ~ 40 R_J because of the lack of well-resolved, cold ionic spectra in the outer magnetosphere. However, measurements by the Low-Energy Charged Particle experiment also provide estimates of the plasma velocity under certain assumptions, and on Voyager 1 inbound the two different velocity measurements agree when they can be compared [Carbary et al., 1981]. The estimates of velocities from the Low-Energy Charged Particle experiment in the outer magnetosphere show that ion velocities there again become reasonably close to those expected of full corotation (see Chapter 4). Such behavior of velocity with distance cannot be explained by the inertial loading model of Hill, and is not understood theoretically at the present time.

The middle and outer magnetosphere

In contrast to the large effort that has been directed at understanding the physics of the Io torus, there have been relatively few post-Voyager attempts to understand the physical mechanisms that control the thermal properties and dynamics of the plasma in the middle and outer magnetosphere. Most obviously, it is unclear what source of energy keeps the bulk of the low-energy ions from adiabatically cooling as they diffuse outward. The heating mechanism may be related to the compression of the dayside magnetosphere due to the solar wind interaction [i.e., Goertz, 1978; Belcher and McNutt, 1980]. This same mechanism may be responsible for the energic ion component discussed in Chapter 4. Just as puzzling is the thermal stratification in the sheet. It may be that the hotter, higher latitude ions have a different temporal history as compared to the cool ions at the concentration maxima, as has been suggested for the suprathermal electron population adjacent to the sheet. Scudder, Sittler, and Bridge [1981] suggest that the suprathermal electrons (i.e., the 2–3 keV "bath" of hot electrons at higher latitudes), are a natural consequence of the peculiar situation of closed field lines being simultaneously above two different exobases, one at the ionospheric foot of the flux tube, with the other at the centrifugal equator. In this situation, the small population of electrons that can escape the respective exobases will find a natural place in the midlatitude regions. Thus, there may be no need for an energization process for these suprathermal electrons.

In addition to the thermal properties, the existence of nonazimuthal flow velocities in the middle magnetosphere (e.g., flows away from and toward the plasma sheet) is of interest since it bears on the overall dynamics of the plasma sheet. Quantitative models of the plasma sheet dynamics in light of the Voyager observations require additional observational and theoretical input. Future progress in our understanding of this region depends on more thorough analysis of the data from the various experiments, as well as on detailed cooperative studies. The middle and outer magnetosphere is the region of greatest promise in terms of potentially dramatic increases in our understanding of the important physical processes in the Jovian magnetosphere.

ACKNOWLEDGMENT

The success of the Voyager Plasma Science experiment is due to the leadership of Professor H. S. Bridge, Principal Investigator. I am grateful to F. Bagenal, H. S. Bridge, R. McNutt, Jr., J. Scudder, and E. Sittler, Jr. for a careful reading of the manuscript and many constructive suggestions. This work was supported under JPL contract 953733 and NASA contract NGL22-009-015.

4

LOW-ENERGY PARTICLE POPULATION

S. M. Krimigis and E. C. Roelof

Voyager 1 and 2 performed the first unambiguous low-energy ($E \geq 30$ keV) ion measurements in and around the Jovian magnetosphere in 1979. The magnetosphere contains a hot ($kT \sim 30$ keV), multicomponent (H, He, O, S) ion population dominated by convective flows in the corotation direction out to the dayside magnetopause and on the nightside to ~ 130–$150\ R_J$ beyond which the ion flow direction changes to predominantly antisolar, but with a strong component radially outward from Jupiter. This tailward flow of hot plasma, the magnetospheric wind, accounts for the loss of $\sim 2 \times 10^{27}$ ions/s and $\sim 2 \times 10^{13}$ W from the magnetosphere. Comparison of energetic (≥ 30 keV) ion to magnetic field pressure reveals that particle and magnetic pressures are comparable from the magnetopause inward to at least $\sim 10\ R_J$, that is, magnetosphere dynamics is determined by pressure variations in a high-β plasma. This particle pressure is responsible for inflation of the magnetosphere and it (rather than the planetary magnetic field) determines the standoff distance with the solar wind. The ion spectrum can be described by a convected Maxwellian component at $E \leq 200$ keV, and a nonthermal tail at higher energies described by a power law of the form $E^{-\gamma}$. New theoretical techniques were developed in order to interpret the low-energy solid-state detector measurements of temperature, number densities, pressures, and flow velocities in this novel hot-plasma environment. Detailed analysis of composition measurements shows the presence of roughly equal numbers of protons and heavier ($A > 1$) ions, with likely charge states for the ions of He^+, O^{+2}, S^{+3}, and C^{+6}. In addition, molecular hydrogen (H_2 and H_3) has been identified, suggesting that the Jovian ionosphere is a source of magnetospheric plasma, in addition to the satellite sources of O and S. The ion spectra at $E \geq 200$ keV/nucleon are best organized in terms of energy/charge, suggesting an electric field acceleration mechanism. Finally, measurement of convection velocities shows that the plasma is more or less corotating at large ($\geq 30 R_J$) distances from Jupiter, but slows to well below corotation closer to the planet.

4.1. Introduction

Our knowledge of Jupiter's magnetosphere before the Pioneers and Voyagers actually flew by the planet was conveyed to us by relativistic electrons. Those trapped in the Jovian radiation belts produce the decimetric and decametric radio emissions (see Chap. 5) first detected at Earth in the 1950s, and those electrons escaping into interplanetary space fill the heliosphere and dominate the galactic component up to energies of ~ 40 MeV. However, our experience with the Earth's magnetosphere has shown that its dynamics are primarily determined by energetic ions in the range of tens to hundreds of keV. For example, the energy density in the ring current consists primarily of ions in the range of 20–200 keV [Williams, 1979], whereas the dynamics of the magnetotail are primarily determined by plasma flows in the range of a few keV to several tens of keV [Frank, Ackerson, and Lepping, 1976; Keath et al., 1976; Roelof et al., 1976; Coroniti et al., 1980b]. At Jupiter, the low-energy ions dominate magnetospheric processes even more dramatically than at Earth. Consequently, although we shall not ignore the low-energy electrons altogether, our main interest concerns the discovery by Voyager of the crucial roles played by ions ≤ 1 MeV/nucleon in Jupiter's huge and complex magnetosphere.

Magnetic field and energetic particle measurements obtained during the Pioneer 10 and 11 encounters with Jupiter in 1973 and 1974, respectively, revealed substantial information concerning the morphology and overall structure of the Jovian magnetosphere. Some of the most important findings included the sustained 10-hr periodicity in particle fluxes and magnetic field magnitude, and the overall inflation of the magnetosphere to distances that were twice as large as those predicted using the magnitude of the Jovian dipole moment measured by the spacecraft. An excellent review and synthesis of the Pioneer results published through 1977 is contained in a paper by Kennel and Coroniti [1979]. The Pioneer instrumentation, however, did not include detectors capable of obtaining comprehensive ion measurements in the energy range of a few keV to several hundred keV, so that the characteristics of the low-energy ($\lesssim 200$ keV) ion component of the magnetospheric particle population (which was presumably responsible for the inflation of the magnetosphere), and its composition remained largely unknown. As a consequence, questions of the dynamics and the overall balance between the magnetic field and plasma population in the magnetosphere of Jupiter could only be speculated upon [Walker, Kivelson, and Schardt, 1978; Goertz et al., 1979].

In what follows, we review and interpret measurements of the intensities, energy spectra, angular variations, and composition characteristics of the low-energy ion population obtained by both Voyager spacecraft in the Jovian magnetosphere. We also describe some novel analysis techniques that we have employed to generate density, pressure, composition, and plasma-flow profiles in the magnetosphere, and we compare these to results reported by other investigations on the spacecraft.

The energetic ion component of the Jovian, magnetosphere provides most of the plasma pressure and thus determines the dynamics of magnetospheric motions. In this sense, the magnetosphere of Jupiter is unique among those with intrinsic magnetic fields (i.e., Mercury, Earth, and Saturn), in that these particles rather than the magnetic field provide the primary pressure for deflecting the solar wind. In addition, these ions account for a substantial fraction of the plasma number density in the outermost portion of the magnetosphere and represent a significant part of the number density at distances as close as $\sim 10~R_J$ to the planet. The composition of the energetic ion population is dominated by hydrogen, helium, oxygen, and sulfur, the heavier ions being present in amounts comparable with the protons. Measurements of the ion flux anisotropies show that these are dominated by convective flows in the corotation direction on the dayside inside the magnetopause, and on the nightside to distances of $\sim 130\text{–}150~R_J$; beyond this point, to distances well past the dawn bow shock of the planet, the ion flows change predominantly to the antisolar direction, but with a strong component projected radially outward from the planet.

4.2. Observational overview

The Low-Energy Charged Particle (LECP) investigation on the Voyager spacecraft utilizes a variety of solid state detectors in the LEMPA (Low-Energy Magnetospheric Particle Analyzer) head to obtain measurements of energetic electrons (14 keV $\leq E_e \leq$ 20 MeV) and ions (30 keV $\leq E_i \leq$ 150 MeV) in several energy intervals with good energy, species, time, and spatial resolution. Above an energy of ~ 200 keV/ nucleon, individual ion species can be identified separately and their energy spectra and angular distributions determined using the LEPT (Low-Energy Particle Telescope) head. Sensor geometry factors range from $\sim 2 \times 10^{-3}$ cm^2 sr to ~ 2.3 cm^2 sr and are designed to handle fluxes ranging from quiet-time cosmic ray medium nuclei ($< 10^{-6}$ cm^2 s sr)$^{-1}$ to intense proton and electron populations ($> 10^8$ cm^2 s sr)$^{-1}$ in the inner

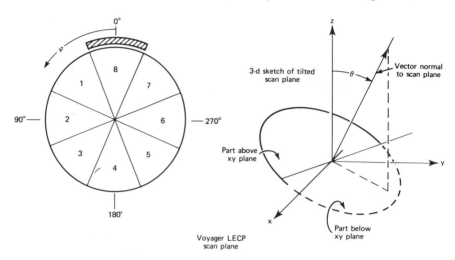

Fig. 4.1. Voyager LECP scan plane: (*a*) scan plane sectoring and angle conventions (angles are measured counterclockwise from the center of the shield sector (sector 8)); and (*b*) three-dimensional sketch indicating how scan plane is tilted from the Jupiter ecliptic *xy* plane. When Voyager is oriented on the star Canopus, the polar angle $\theta \sim 100°$. The *z* axis points toward the north ecliptic pole [Carbary et al., 1981].

Jovian magnetosphere. A full description of the instrument design characteristics and capabilities have been given elsewhere [Krimigis et al., 1977, 1981].

The LECP detector heads are mounted on a rotating platform on the spacecraft so that angular measurements in eight 45° intervals may be obtained. Figure 4.1 shows the rotation plane of the detector telescopes where the *x* and *y* axes lie roughly in the ecliptic plane. The sector scheme in the rotation plane of the instrument is shown on the left side and depicts the eight 45° sectors. The detector apertures can be rotated through 360° in 45° steps at rates ranging from 1 step every 6 s to 1 step every 48 s; it is also possible to park the detector in any one of the eight sectors. Sector 8 is permanently blocked by a shield so that an accurate estimate of background contributions to the count rate of every channel can be obtained once per scan. The viewing directions of each sector depend on the particular spacecraft orientation. These directions are shown in Figure 4.2 for two of the most common spacecraft references, that is, the stars Canopus and Arcturus. The lower part of the figure shows the viewing directions for most of the Voyager 2 encounter, indicating that the LECP instrument was viewing directions quite close to the ecliptic plane. The tick marks at the zero latitude line labeled by day number indicate the direction of corotation on that particular day. It is evident from the upper part of Figure 4.2 that the Arcturus reference is not optimum for measuring corotational flow in the Jovian magnetosphere. This orientation occurred on days 192–195 during the Voyager 2 encounter and for most of the outbound pass of the Voyager 1 encounter.

The LECP instruments on Voyager 1 and Voyager 2 detected the approach to Jupiter's magnetosphere through a number of upstream ion intensity enhancements several hundred R_J before encounter of the planetary bow shock. A trajectory plot indicating the time and location of each ion event observed inbound from both spacecraft and outbound from Voyager 1 is shown in Figure 4.3 [Zwickl et al., 1980; 1981]. The solid circles represent the location of measurements of short-lived events lasting from a few minutes to three hours in the low-energy (≥ 50 keV) ion channels. The solid

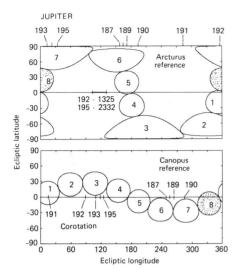

Fig. 4.2. Viewing of the LECP sectors in ecliptic coordinates. Shown are the viewing areas for the eight LECP sectors when the Voyager 2 spacecraft was in the Arcturus reference orientation (top panel) and in the Canopus reference orientation (lower panel). The corotation direction at 0000 UT of each day is given by the tick marks on the zero latitude line. The Jupiter direction for each day is given at the top of the figure. The viewing directions for the Voyager 1 instrument are effectively the same as shown here [Krimigis et al., 1981].

squares in Figure 4.3 represent long-lived flux increases that last at least 8 hr. Figure 4.3 also shows two mass histograms of ions in the range ~ 0.6 to 1 MeV/nucleon; the one on the left was obtained in the magnetosheath whereas the one on the right was measured during one of the upstream events. The upstream event contained substantial enhancements of oxygen and sulfur, similar to the composition in the magnetosheath and magnetosphere [Hamilton et al., 1981].

The measurements obtained by the LECP instrument in the magnetosphere and its vicinity have been synthesized into a phenomenological model [Krimigis et al., 1979b, 1981], which is shown in Figure 4.4. Here an attempt is made to present the three-dimensional view of the magnetosphere surface. The horizontal transparent plane represents the ecliptic plane, and the rotating magnetodisc at this particular orientation of the planetary magnetic-moment vector appears on the nightside below this plane. The arrows indicate the direction of the corotational plasma flow in the magnetodisc, and its wavy nature suggests a propagating Alfvén wave. The large arrows to the right indicate the direction of solar-wind flow. The bulge in the magnetodisc at about 0800 local time has been exaggerated to indicate the possibly irregular nature of the plasma boundary/magnetopause surface owing to variation in solar wind pressure. Beyond the nightside corotational boundary, the small arrows indicate plasma outflow in the magnetospheric wind, which flares out and expands beyond the nominal width of the nightside magnetodisc.

In the following overview, we present data that are to be viewed in the context of the concepts incorporated in the schematic view of the magnetosphere in Figure 4.4. Figure 4.5 presents the averages of selected energy channels from the Voyager 1 LECP instrument for a 40-day period, beginning at distances $> 200 \ R_J$ inbound and extending to $> 350 \ R_J$ outbound. In the top curve, the intensity of the low-energy ions ($Z \geq 1$), begins to increase inside $\sim 200 \ R_J$, long before the spacecraft encountered the planetary magnetopause. These discrete increases are emissions of magnetospheric ions and are discussed in detail by Zwickl et al. [1981]. The overall increase in rate is well over six orders of magnitude for the data shown, not including the particle intensities at closest approach, which will be presented separately.

Following periapsis, the basic asymmetry of the magnetosphere between day and night begins to manifest itself with the first encounter of the nightside magnetodisc occurring at $\sim 22 \ R_J$; these encounters continue in a periodic fashion with a 10-hr

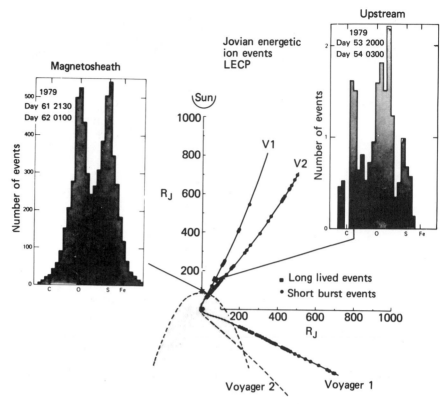

Fig. 4.3. Voyager 1 and 2 trajectories within 1000 R_J of Jupiter. Shaded region around Jupiter represents the area encompassed by the bow shock as measured by Voyager 1. The increased number of events measured by Voyager 2 inbound and by Voyager 2 outbound strongly suggests that interplanetary field-line connection to the bow shock is necessary for the observation of Jovian energetic ion events in interplanetary space. Voyager 2 outbound data are not included because of poor data quality during solar conjunction. The orientation of the eight LECP look directions are shown inbound and outbound with respect to the Sun, the nominal direction of the magnetic field and Jupiter [Zwickl et al., 1981].

period to ~ 130 R_J. Various discontinuities in the basic structure are apparent at ~ 55–60 R_J, at ~ 80–90 R_J, and finally at ~ 130 R_J where the basic periodicity breaks down prior to the encounter of the magnetopause at ~ 160 R_J. We note that activity in the intensity profile of low-energy ions continues to at least 350 R_J, long after magnetopause passage.

The second data trace in the top panel shows the intensity profile of uniquely identified protons ($Z = 1$). The proton profile is grossly similar to that of the ions ($Z \geq 1$) although there exist significant differences; in particular, the decrease in proton intensity to interplanetary values soon after the spacecraft entered the magnetosheath from the nightside of the planet is in contrast to the continued presence of low-energy ions. The third curve represents ions with $Z \geq 6$ which are judged to be a combination of sulfur and oxygen by extrapolation from high-energy channels [Krimigis et al., 1979a]. Again, the gross structure of the intensity profile is similar to that of lower energy ions and of the protons but with some notable exceptions. For example, the overall increase in the ion profile is well over seven orders of magnitude, that is, larger than that of

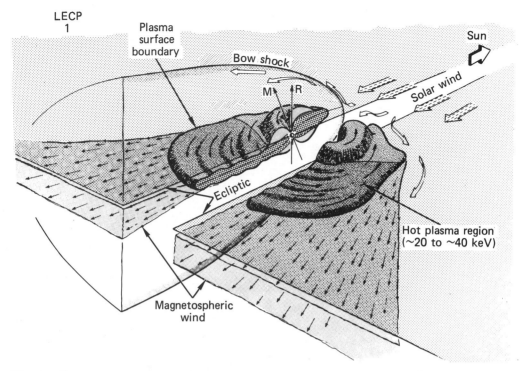

Fig. 4.4. The hot plasma model of the Jovian magnetosphere gives a three-dimensional view. The grey region denotes the hot plasma of the corotating magnetodisc. Solar-wind pressure (arrows), causes the disc to be blunt on the dayside and extended on the nightside. In the far magnetotail (thin arrows), the disc is disrupted and particles are expelled away from the planet in the magnetospheric wind. The temperature of the plasma in most of the outer magnetosphere is 20–40 keV.

protons; in addition, it is possible to deduce that the overall ion intensity increases began as early as day 51, long before the spacecraft encountered the Jovian bow shock for the first time. The relative increase in these ions towards the end of day 60 was greater than that of protons or lower-energy ions. Following the breakdown in periodicity at $\sim 130\ R_J$, the $Z \geq 6$ ions resemble more the profile of the low-energy ions ($Z \geq 1$) than of the protons.

The middle panel of Figure 4.5 shows the profile of low-energy electrons, which generally exhibit the same gross behavior as the ions, although the upstream increase on days 53 and 54 is not clearly evident in the electrons; also, the downstream intensity increases observed for the ions are not always present in the electrons. The electrons, however, continue to be detected far beyond the nominal magnetopause crossing, but terminate abruptly at the first bow-shock crossing on day 77 and later at the shock crossing on day 81.

Finally, the bottom panel on the figure presents the proton to alpha particle (p/α) ratio in the indicated energy range for the duration of the encounter, with the exception of the period around closest approach. The ratio has relatively large values (~ 50–80) even before encounter of the Jovian magnetopause with the largest values (~ 200) obtained at the end of day 53. Generally, the ratios within the magnetosphere, and especially outbound out to the end of day 70 are larger than typically observed in the solar wind or solar-energetic-particle events (~ 10–30).

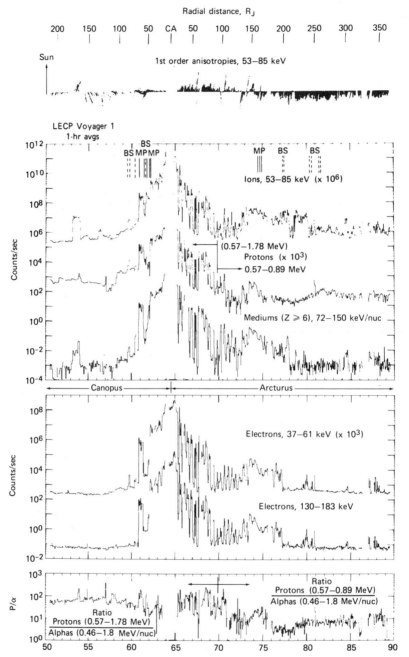

Fig. 4.5. Overview of Voyager 1 encounter. Plotted in the two center panels are 1-hr averages of several of the LECP ion and electron channels. In addition, the anisotropy of the low-energy ions is plotted at the top of the figure while the p/α ratio is at the bottom. Energetic particles from the Jovian magnetosphere were first detected at a distance of $> 600 \ R_J$ inbound and extended to $> 1500 \ R_J$ outbound [Zwickl et al., 1981]. The spacecraft entered the magnetosphere initially on day 60 and remained inside until ~ day 75. Closest approach to the planet occurred at ~ 1200 UT on day 64. The asymmetric nature of the sunward and tailward magnetosphere [Fillius, 1976] is clearly evident as is the corotational nature of the anisotropy.

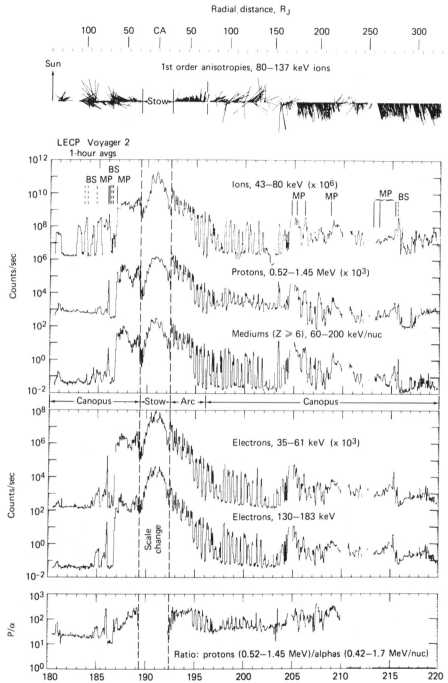

Fig. 4.6. Overview of the Voyager 2 encounter. The format here is the same as in the previous figure. Voyager 2 periapsis occurred at ~2300 UT on day 190. During the interval 189, 0540 UT through 192, 1345 UT, the LECP instrument was in a non-stepping "stow" mode to reduce the counting rate so that compositional measurements could be made through closest approach. As in Figure 4.5, the corotation nature of the anisotropy during intervals when the spacecraft was in the magnetosphere is clearly evident.

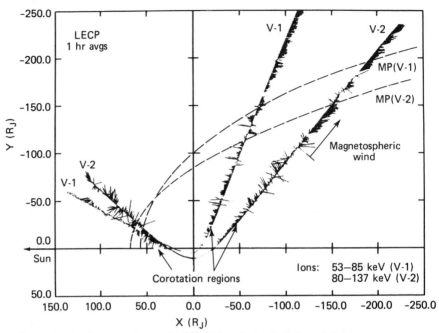

Fig. 4.7. Summary of the anisotropy [Krimigis et al., 1981]. Flows in the corotational directions were observed by both Voyager 1 and 2 throughout the inbound leg of magnetosphere passage and out to 130–160 R_J on the outbound legs. Beyond this distance, the flow changed to an antisunward/anti-Jupiter direction. The dashed line gives the model magnetopause locations for both encounters [Ness et al., 1979c].

The delineation of different regions of the Jovian magnetosphere environment is evident from the ion anisotropy data shown at the top of Figure 4.5 in the form of vectors. These vectors, obtained to first order by a harmonic fit to the sector data [Carbary et al., 1981], point in the direction of particle flow. Note the abrupt flow change to a direction away from the Sun coincident with the breakdown in periodicity at ~130 R_J, but prior to the first magnetopause crossing. We take this to be the spacecraft entry into the magnetospheric wind indicated in Figure 4.4.

Figure 4.6 presents data from the Voyager 2 encounter using the same format as that for Voyager 1. The top curve in the upper panel shows significantly more upstream ion activity during this encounter. This difference is probably due to the fact that the interplanetary field connection from the spacecraft to the Jovian bow shock is more likely to occur for the trajectory of Voyager 2, because the field tends to be nearly transverse to the Jupiter–Sun line (see Fig. 4.3). The gross intensity profile has many features similar to those seen by Voyager 1 including the breakdown in periodicity at ~150 R_J and continued low-energy ion intensity fluctuations downstream to at least 340 R_J. Note the large enhancement of $Z \geq 6$ ions relative to that of protons on day 203. This marks the first change in the plasma flow from the corotational to the tail-ward direction (vector trace on top). As was the case for Voyager 1, the p/α ratio is larger than typical solar wind or solar-interplanetary particle values. In both Figures 4.5 and 4.6, the transition to the magnetospheric wind shown schematically in Figure 4.4 is evident and coincident with the breakdown in periodicity inside the magneto-pause. Note that the p/α ratio in the magnetospheric wind region closely resembles that of the inner part of the magnetosphere and is at least an order of magnitude higher than values expected in the solar wind.

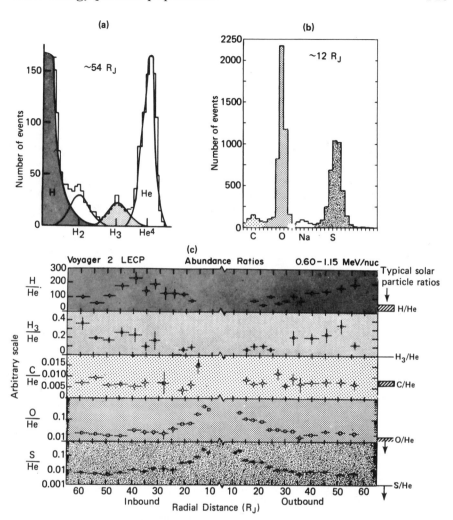

Fig. 4.8. Elemental composition measurements by the LECP instrument within the Jovian magnetosphere: (*a*) mass histogram showing the presence of molecular hydrogen, H_2 and H_3; (*b*) mass histogram representative of the composition throughout the magnetosphere; and (*c*) variation with distance to the planet of the abundances of the elements indicated relative to He (adapted from Krimigis et al. [1979a,b] and Hamilton et al. [1981]). Solar abundances are shown on the left-hand ordinate for comparison.

A most revealing view of the structure of the magnetosphere is provided by the first-order ion anisotropies, which can be taken as indicative of convective plasma flow direction. A summary of the overall anisotropy profile at low energies for both Voyagers 1 and 2 is presented in Figure 4.7 in a more convenient format. Here the projections of the first-order-anisotropy vectors onto the ecliptic plane are drawn on the spacecraft trajectories from $\sim 150\ R_J$ upstream to $\sim 325\ R_J$ downstream of the planet. The model magnetopause locations from both encounters [Ness et al., 1979c] are shown for reference. We note that there exist two general flow directions in the Jovian magnetosphere, one which points in the direction of planetary corotation and obtains throughout the dayside, and on the nightside to distances of ~ 130–$160\ R_J$; beyond that distance and before crossing the magnetopause (identified by the plasma instrument [Bridge et al., 1979a,b]), the flow changes to an antisunward/anti-Jupiter direction and continues to large distances away from the planet. These results are the basis for the schematic representation of corotational flow and antisunward flow in Figure 4.4.

The composition characteristic of most of the Jovian magnetosphere at higher energies is presented in Figure 4.8. Figure 4.8a shows a histogram of the light elements which reveals the presence of molecular hydrogen (H_2, H_3) in addition to the expected peaks of hydrogen and helium [Hamilton et al., 1980]. The presence of H_3 which is expected to be a constituent of the Jovian upper ionosphere is taken as evidence that the ionosphere is a significant plasma source, [as originally suggested by Ioannidis and Brice, 1971], at least for the outer part of the Jovian magnetosphere. The ionospheric source of hydrogen may well be the cause of the large values of the H/He ratios shown in Figure 4.6. Figure 4.8b shows a histogram of heavier ions [Krimigis et al., 1979b; Hamilton et al., 1981], which reveals the presence of peaks at the location of carbon, oxygen, sodium, and sulfur, and is characteristic of the composition throughout the Jovian magnetosphere. As pointed out previously O, S, and Na originate at Io (see also Chap. 3).

Figure 4.8c presents the ratios of five different species relative to helium, computed over equal energy/nucleon intervals near ~ 1 MeV/nucleon. We note three basic types of behavior. He and H_3 both decrease relative to helium with decreasing radial distance, except for a period between 60 and 40 R_J inbound. The C/He ratio remains essentially constant throughout the magnetosphere. Both O and S increase rather smoothly relative to He with decreasing distance within $\sim 30\text{--}40 R_J$. The magnitude of the ratio change from outer to inner magnetosphere is much larger for O and S than for H and H_3 (note the logarithmic scale for the oxygen and sulphur ratios).

As pointed out by Krimigis et al. [1979b], the C/He ratio observed by Voyager 2 in Jupiter's magnetosphere is quite similar to that found in solar particles. The O/He ratio in the outer magnetosphere is about a factor of 2–3 higher than for solar particles. On the other hand, S/He is more than a factor of 10 higher than solar values even in the outer magnetosphere. These facts, plus the fact that the O/He and S/He ratios increase as Voyager 2 approached Jupiter although the C/He ratio remains constant, led Krimigis et al. [1979b] to suggest that the energetic C and He are of solar origin (probably from the solar wind), that S is of local origin (probably from Io), whereas O could have both a local and solar component. Because H_3 has not been observed among solar particles, its local origin is certain.

The H/He ratio was quite large in the outer Jovian magnetosphere, typically > 100. This is certainly a much larger value than found in the solar wind (typically 10–20) although, as will be discussed later, the nature of the acceleration process can greatly alter the abundance ratio in the source material. Because H_3 is of ionospheric origin, Hamilton et al. [1980] have argued that H^+ ions should also participate in the processes that bring H_3^+ ions from the ionosphere. Helium, on the other hand, may not participate in these processes because it is found lower in the Jovian atmosphere. This would explain the large hydrogen to helium ratios.

To illustrate the ion spectral form at low energies, we show in Figure 4.9 a typical spectrum measured in the outer magnetosphere by Voyager 2. The shape of the spectrum does not conform with a power law over the entire energy range, but rather exhibits significant flattening at the low-energy end. Although ions are not specifically identified at lower energies (round points), the detector response at $90°$ to the flow direction is generally due to protons [Krimigis et al., 1981]; the pulse height analyzed proton spectrum (dashed line) and a square point that represents a specific proton rate

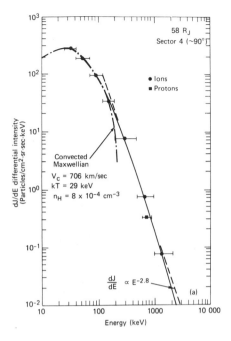

Fig. 4.9. Detailed spectral fit to the low energy ion channels [Krimigis et al., 1981]. Plotted (closed circles) are the intensities measured in sector 4 (~ 90° from convection direction) of the low energy ion channels. In this direction, the detector response is thought to be due to protons only. The dotted curve shows the thermal distribution obtained using parameters listed in the figure. The dashed curve indicates a power-law fit from the pulse-height-analyzed proton data with a spectral index of 2.8. The closed square is from the composition detector head channel, which is sensitive only to protons.

from the composition detector support this conclusion. Assuming that the detector is responding to protons at all energies, we proceed to obtain a fit to the three lowest energy points by assuming a convected Maxwellian distribution [Krimigis et al., 1979a]. The parameters providing the best fit are a velocity of ~ 700 km s^{-1}, temperature of ~ 29 keV and density of ~ 8×10^{-4} cm^{-3}. Similar temperatures have been obtained for most of the magnetosphere for both encounters and represent the basis for labeling the magnetodisc in Figure 4.4 with the temperature of 20–40 keV.

The importance of the information contained in the spectrum and composition of the low-energy charged particle populations inside and outside the Jovian magnetosphere should now be apparent. The process of deducing this information from the LECP measurements requires some understanding of how such an instrument responds to hot, flowing, multispecies plasma. The necessary techniques are described in the next section.

4.3. Measurement of hot multispecies convected plasmas using energetic particle detectors

The surprise for the LECP team when the data came back from the Voyager 1 Jupiter encounter was not that LECP had responded to thermal plasma in the outer magnetosphere, but rather that there was such a relatively dense (~ 10^{-3} cm^{-3}) plasma component with so high a temperature ($kT \simeq 20$ keV) that was so rich in heavy ions (like oxygen and sulfur). To see why we expected the LECP to respond to plasmas, even if the temperature had been considerably lower and the magnetosphere even less rich in heavy ions, we have to examine in some detail the response of a total energy detector to a moving plasma.

Well before the Voyager 1 encounter with Jupiter's magnetosphere, we were aware that solid state detectors with low electronic thresholds ($\lesssim 30$ keV) could respond to flowing hot plasmas of sufficiently high density [Roelof et al., 1976]. In the Earth's plasma sheet and magnetosheath, there are strong, sporadic proton plasma flows with bulk velocities (V) of several hundred km s^{-1}, densities in the range 10^{-3}–10^{-1} cm^{-3} and temperatures (T) of several keV (where the thermal energy kT may be expressed in eV using the relation $10^6\,K = 86$ eV). Direct plasma measurements had been made with electrostatic analyzers on the IMP-6, 7, and 8 spacecraft, for example, Frank, Ackerson, and Lepping [1976] and Hones [1976]. Also on the IMP-7 and 8 spacecraft were energetic particle detectors [Williams, 1977] that had an energy threshold for incident protons of ~50 keV: 30 keV electronic threshold for energy deposited in the live volume of the detector, with an additional ~20 keV undetected energy loss in the dead layer and the ~40 micrograms cm^{-2} aluminum contact. Unlike the electrostatic analyzers, which respond at considerably lower energies to the bulk of the thermal plasma, the solid state detectors usually respond to the high-energy tail of the plasma distribution. In the Earth's plasma sheet, we showed that this tail contained sufficient information to provide estimates of plasma temperature and bulk velocity from the angular and spectral response of the IMP detectors [Roelof et al., 1976]. Once we understood the conditions that exist in Jupiter's magnetosphere, we realized that the Voyager LEMPA detectors could yield estimates not only of temperature and bulk velocity, but also of lower bounds on density and pressure.

It is the motion of the plasma that dramatically enhances the response of an energetic particle detector. The energy of a thermal particle with relative velocity v, averaged over the distribution function for a plasma with temperature T flowing with a bulk velocity $\mathbf{V}$, is

$$<E> \;=\; \frac{m}{2}\, < (\mathbf{V} + \mathbf{v})^2 > \;=\; mV^2/2 \,+\, 3\,kT/2 \qquad (4.1)$$

because $\langle v^2 \rangle = 3\,kT/m$. A handy relationship to remember is that an ion with instantaneous velocity v (expressed in km s^{-1}) has an energy per nucleon given by $(v/439 \text{ km s}^{-1})^2$ in keV/nucleon. Equation (4.1) then immediately illustrates how bulk motion of a thermal plasma greatly enhances the detector's sensitivities to the heavier species. For a plasma with $kT = 20$ keV and a bulk velocity $V = 600$ km s^{-1}, protons have a translation energy $mV^2/2 = 1.9$ keV, which is considerably smaller than their mean thermal energy $3\,kT/2 = 30$ keV. By contrast, even though oxygen ions (^{16}O) at the same temperature also have 30 keV mean thermal energy, their energy of translation adds another 30 keV to their total mean energy.

Thinking in terms of Jupiter's magnetodisc, if the hot plasma component "corotates" with the planet's ionosphere, then the magnetic field lines are "rooted" in the conducting ionosphere in the sense that the polarization electric field $\mathbf{E} = -(\mathbf{V} \times \mathbf{B})/c$ maps out from the ionosphere into the magnetodisc along the magnetic field lines. Thus, at radius $\mathbf{r}$ where $\mathbf{V} = \Omega \times \mathbf{r}$ (Ω being the Jovian System III angular velocity of 1.76×10^{-4} s^{-1}), the $\mathbf{E} \times \mathbf{B}$ drift velocity of a charged particle would be $\mathbf{V} = (c/B^2)\,(E \times B) = \Omega \times \mathbf{r} - \hat{\mathbf{B}}\,\hat{\mathbf{B}} \cdot (\Omega \times \mathbf{r})$, the second term vanishing if the field lines lie in meridional planes. In the equatorial plane, $V = |\Omega \times \mathbf{r}| = \Omega r = (12.6 \text{ km s}^{-1})\,(r/R_J)$, so the radius at which the translation energy term in Equation (4.1) equals the mean thermal energy would be $(42.7\ R_J)\,(kT/A)^{1/2}$, where kT is expressed in keV and A is the atomic mass number of the ions. Thus, for $kT = 20$ keV, the corotation energy of hydrogen is negligible compared to its thermal energy within the magnetosphere, but the corotation and thermal energies are comparable for ^{16}O at 47.7 R_J, or for ^{32}S at 33.8 R_J.

An aside may be appropriate here: we have assumed that all ion species have the same temperature and the same bulk velocity, and indeed our LECP results for Jupiter appear to be approximately self-consistent within this assumption. However, in another space plasma (the solar wind near 1 AU), it is a better approximation to say that all ion species have about the same mean-square velocity. Because $\langle v_a^2 \rangle = 3\, kT_a/m_a$ for species "a" with atomic mass number A, equal thermal speeds would imply $T_a \simeq A\, T_H$, where T_H is the hydrogen temperature. Moreover, the bulk velocities $\mathbf{V}_a$ can differ from the hydrogen bulk velocity $\mathbf{V}_H$ by a velocity increment $\Delta\mathbf{V}_a$ which is magnetic-field aligned and seems to be bounded above by the local Alfvén velocity. These results were established for solar wind ^{4}He by Asbridge et al. [1976] and Marsch et al. [1982] and recently extended to solar wind ^{56}Fe by Mitchell and Roelof [1980] and Mitchell et al. [1981].

The discussion above of mean thermal and translational energies was intended only as an introduction and is hardly a complete discussion of what an energetic particle detector actually measures. What we must examine is the instrumental response to the incident unidirectional differential flux $j_a(E, \mathbf{p})$: the number of particles of species "a" with energies in the differential range $(E, E + dE)$ and momenta $\mathbf{p}$ in a differential solid angle $d^2\omega$ about the unit vector $\mathbf{p}$ crossing an area $d^2\sigma$ (normal to $\mathbf{p}$) per unit time is $j_a\, dE\, d^2\omega\, d^2\sigma$. From the measurements, we can estimate the temperature, number density, pressure, and convection velocity of the dominant hot ions.

To put things on a solid theoretical ground, we begin with the phase space density function $f^a(\mathbf{x}, \mathbf{p})\, d^3x\, d^3p$ which gives the number of particles of species "a" in a volume d^3x about position $\mathbf{x}$ and with momenta in the range $d^3p = p^2\, dp\, d^2\omega$ centered about $\mathbf{p}$. The unidirectional flux function $j^a(E, \mathbf{p})$, which we have been using is just the velocity v times the number of particles in the volume d^3x about $\mathbf{x}$ in the energy range $dE = v\, dp$ with momenta in d^3p about $\mathbf{p}$:

$$j^a(E,\hat{\mathbf{p}})\, d^3x\, dE\, d^2\omega = v f^a(\mathbf{x},\mathbf{p})\, d^3x\, p^2\, dp\, d^2\omega \qquad (4.2)$$

from which we obtain the useful relationship

$$j^a(E,\hat{\mathbf{p}}) = p^2 f^a(\mathbf{x},\mathbf{p}) \qquad (4.3)$$

Allow us a slight lapse from rigorous notation in not being explicit about the $\mathbf{x}$ dependence of j^a because we usually are talking about the measured flux at the spacecraft, and we always know where that is.

One consequence of the covariance of the laws of statistical mechanics is that the phase space density is invariant under the Lorentz transformation. Using the nonrelativistic (Galilean) limit for the velocity transformation $\mathbf{v} = \tilde{\mathbf{v}} + \mathbf{V}$, where $\mathbf{v}$ is the velocity of a particle in the rest frame (e.g., the spacecraft), and $\tilde{\mathbf{v}}$ is the velocity which the same particle would have in a frame moving with velocity $\mathbf{V}$ relative to the rest frame, the equality of the distribution function in the spacecraft frame $f^a(\mathbf{x},\mathbf{p})$ to the distribution function $\tilde{f}^a(\tilde{\mathbf{x}},\tilde{\mathbf{p}})$ in a frame moving with the bulk velocity of ion plasma species "a" implies, from (4.3), that at the same point in configuration space, the fluxes are related by

$$j^a(E,\hat{\mathbf{p}})/p^2 = \tilde{j}^a(\tilde{E}, \tilde{\hat{\mathbf{p}}})/\tilde{p}^2 \qquad (4.4)$$

where $\tilde{E} = \tilde{p}^2/2m_a$ and $\tilde{p}^2 = m_a^2\,(v^2 - 2\mathbf{v}\cdot\mathbf{V} + V^2)$ for a given ion species.

The transformation, (4.4), of fluxes (valid also when the relativistic momentum transformation is used), finds many applications in energetic particle space physics. In our discussion, its main utility lies in describing the response of a detector to a moving plasma when we have some theoretical ground for saying what the distribution function $\tilde{f}$ should look like in the plasma frame (e.g., $\tilde{f}$ might be approximately

Fig. 4.10. Detailed fit of sectored data using Equation (4.5) of the text. The data represent a 73-min. average for a quiet time on the Voyager 2 inbound trajectory. The top two panels represent fits to proton and helium data. The bottom panel represents a fit using a $Z \geq 6$ channel assuming a mixture of oxygen and sulfur (O/S = 1, 10) and a convection speed of 700 km/s [Carbary et al., 1981].

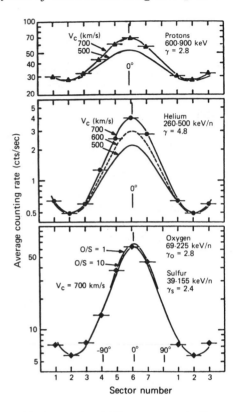

Maxwellian owing to thermalizing wave/particle interactions). However, we shall see later that (4.4) can also be used without any foreknowledge of $\tilde{f}$.

The most direct application of (4.4) can be made in the energy range where the LECP instrument can identify individual ion species and measure their spectra. This is accomplished using the pulse height analysis data from the LEPT for energies ≥ 0.5 MeV/nucleon. If we assume that $\tilde{j}^a \simeq j_0^a(E_a/\tilde{E})^{\gamma_a}$, that is, an isotropic power-law differential spectrum in the vicinity of the ion total energy E_a, then (4.4) becomes [Ipavich, 1974],

$$j^a(E,\phi) = \tilde{j}_0^a (E_a/E)^{(\gamma_a+1)}\left(1 - 2\,\frac{V}{v}\sin\theta\cos\phi + \frac{V^2}{v^2}\right)^{-(\gamma_a+1)} \tag{4.5}$$

where ϕ measured from the projection ($V\sin\theta$) of the convection velocity onto the scan plane of the instrument. The flux anisotropy produced by convection can be represented more simply when the convection velocity (V) is significantly less than the particle velocity. Then (4.5) can be written in the normalized form

$$\ln\frac{j^a(E,\phi)}{j^a(E,\pi/2)} \simeq 2(\gamma_a + 1)\,\frac{V}{v}\sin\theta\cos\phi + \mathrm{O}(V^2/v^2) \tag{4.6}$$

For the outer parts of the Jovian magnetosphere where the LECP scan plane is roughly perpendicular to the local magnetic field (so $\sin\theta \simeq 1$), one can therefore make several independent estimates of the convection velocity directly from the anisotropy amplitudes of several separate ion species and the dependence of the amplitude on velocity (energy) for each species. Figure 4.10 shows the angular distribution for protons (0.6–0.9 MeV), helium nuclei (1.0–2.0 MeV) and $Z \geq 6$ ions (1.1–3.6 MeV) during a 73-min. interval when the intensity remained relatively constant at 58 R_J on the inbound leg of the Voyager 2 trajectory [Carbary et al., 1981]. Sectors 6, 2, and 4 pointed the sensors into, away from, and perpendicular to the corotation direction, respectively. The differential energy spectra measured during this time are well represented by $j = KE^{-\gamma}$. In sector 4, approximately perpendicular to the corotation direction, $\gamma_p = 2.8$ and $\gamma_{He} = 4.8$ in the indicated energy intervals [Krimigis et al., 1981].

The expected directional dependence of the fluxes for each ion species can be computed from (4.5) using the observed power-law spectra, assuming negligible anisotropy in the plasma rest frame. The smooth curves in Figure 4.10 represent the fluxes computed from (4.5) for various values of $V \sin \theta$. An excellent fit for all sectors is obtained when $V \sin \theta = 700$ km/s for both protons and helium. The corotation velocity of $V_c = \Omega R$ is 730 km/s at 58 R_J. The remarkable agreement between the measured and computed fluxes in each of the seven sectors for particles having different velocities and spectral exponents provides strong evidence that the plasma indeed corotates rigidly at this distance from the planet. The bottom panel shows the results of a similar analysis where γ was independently determined from anisotropies of $Z \geq 6$ ions from a rate channel for which $V_c = 700$ km/s was assumed. Again, an excellent fit to the data was obtained.

Before proceeding, let us consider the assumption underlying (4.5) that $\tilde{j}^a$ was isotropic in the convected frame. It has been pointed out that spatial gradients in the distribution function and flows parallel to the local magnetic field may make significant contributions to the anisotropies measured in the Jovian magnetosphere [e.g., Northrop, Birmingham, and Schardt, 1979; Thomsen et al., 1980; Birmingham and Northrop, 1979]. For the particular example just given, Carbary et al. [1981] determined that spatial gradients were relatively small, and that the flows were essentially convective. However, the resolution of the question of gradient anisotropies is particularly important at lower energies where LEMPA provides less exact information on ion composition than does the LEPT at higher energies.

If the particle flux is not too anisotropic in some frame of reference, the first-order anisotropy produced by a spatial gradient in flux intensity is given by the well-known expression

$$\mathbf{A}_{grad} = \rho \, \hat{\mathbf{B}} \times \nabla \ln j \qquad (4.7)$$

where ρ is the particle's local gyroradius (pc/qB) and j is the flux averaged over all directions. This expression may be generalized to strong anisotropies without losing its essential character as long as the anisotropies remain unidirectional [Roelof, 1975]. The essential point to remember from (4.7) is that gradient anisotropies are large only when the spatial scale of the flux perpendicular to the magnetic field gets as small as a cyclotron radius.

For comparison, the anisotropy (in the spacecraft frame) due to convection alone can be written down directly from (4.6)

$$\mathbf{A}_{conv} = \frac{2}{v} (\gamma + 1) \mathbf{V} + O(V^2/v^2) \qquad (4.8)$$

The coefficient of $\mathbf{V}/v$ in (4.8) may be recognized as three times the "Compton-Getting Factor," well-known to cosmic-ray physicists, and A_{conv} is referred to as the "Compton-Getting Anisotropy" [see Gleeson and Axford, 1968].

It has been suggested to us that gradient anisotropies could mimic corotation anisotropies near the "neutral sheet" of Jupiter's magnetodisc. The reason given was that the cyclotron radius of partially ionized ions can become large ($> 1\ R_J$) because the residual north–south field is small ($\lesssim 1\ nT$), so that a positive radial flux gradient on the scale of several R_J could produce an azimuthal anisotropy in the corotation direction. However, the data from the LECP does not support this conjecture. Carbary, Krimigis, and Roelof [1982] show that the form of the energy spectrum does not vary strongly with radius, so the $\nabla \ln j$ term in (4.7) is approximately energy independent. Consequently, A_{grad} should increase with velocity, but the measured anisotropies show no significant change between 0.1 and 1.0 MeV, a factor of 3 change in velocity. The lack of energy dependence in this energy range is more consistent with that predicted by (4.8) because the observed spectra steepen noticeably between 0.1 and 1.0 MeV, tending to keep the fraction $(\gamma + 1)/v \simeq$ constant. Moreover, as we shall see in the next section, we find anisotropies whose amplitudes and directions both are consistent with convection throughout the magnetodisc, including the immediate vicinity of the neutral sheet. We have not yet seen any evidence in the data in support of gradient anisotropies dominating convective anisotropies.

Because we have been discussing convective anisotropies for both power-law and thermal spectra, it is worthwhile comparing their energy and species dependence. For a thermal spectrum, Roelof et al. [1976] showed that convected thermal distributions have the same general dependence as given by (4.6) if γ, the local power-law exponent, is calculated from the definition $\gamma = -(\partial \ln \tilde{j}/\partial \ln \tilde{E})$ using the thermal flux $\tilde{j} = \tilde{p}^2 (2\pi mkT)^{-3/2} \exp(-\tilde{E}/kT)$

$$\ln \frac{j(E,\phi)}{j(E,\pi/2)} \simeq \frac{mvV}{kT} \sin\theta \cos\phi + O(V^2/v^2) \tag{4.9}$$

Thus both power-law and thermal convected fluxes have anisotropies obeying an approximate log-cosine law when $V \ll v$:

$$\ln \frac{j(E,\phi)}{j(E,\pi/2)} \simeq \alpha \cos\phi + O(V^2/v^2) \tag{4.10}$$

where $\alpha = mvV \sin\theta/kT$ for a thermal spectrum, and $\alpha = 2(\gamma + 1)(V/v)\sin\theta$ for a local power-law spectrum.

Note that the anisotropy amplitude α varies inversely with velocity for a power law, but directly with momentum for a thermal spectrum. But if we now fix the particle's total energy for ion species with different atomic mass numbers A, then $v \propto A^{-1/2}$ and $mv \propto A^{1/2}$, so we have (rather remarkably) $\alpha \propto A^{1/2}$ for either a thermal or a power-law spectrum (assuming that either T or γ is the same for all species). An immediate consequence of either (4.6) or (4.10) is that if a heavy ion species has an integral flux in the convective frame comparable to that of the protons, the forward fluxes ($\phi = 0$) in the spacecraft frame will be dominated by the heavy ions. For example, consider 20 keV ions of atomic mass number A: $\alpha \simeq A^{1/2}\gamma\ (V/1000\ \mathrm{km\ s^{-1}})$ for $2 \leq \gamma \leq 4$, or for a thermal distribution, $\alpha \simeq A^{1/2}\ V/50\ kT$ (where V is in km s^{-1} and kT in keV). Then for $V = 1000\ \mathrm{km\ s^{-1}}$, protons would have anisotropies of $\alpha = \gamma$ or $\alpha = 20/kT$ either of which can be of order unity in the magnetodisc, while ^{16}O or ^{32}S nuclei of the same energy would have anisotropies 4 or 5.7 times stronger.

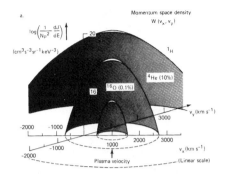

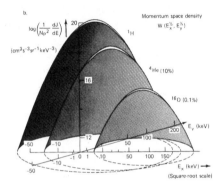

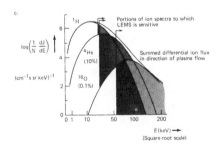

Fig. 4.11. Distribution functions and instrument response for a hot flowing plasma incident on a low energy magnetic spectrometer (LEMS), such as the ion channels of the LEMPA. Upper and middle panels show momentum-space densities of ^{1}H, ^{4}He (10%) and ^{16}O (0.1%) nuclei as a function of (a) particle velocity, or (b) total energy, for all directions relative to hot plasma flow ($kT = 5$ keV, $V = 1000$ km s^{-1}). Lower panel (c) shows the incident differential particle spectra for each species when the detectors look into the plasma flow; actual portions of spectra to which LEMS respond are shaded. The summed response of the LEMS is not only sensitive to hot-plasma flow, but it also displays the spectral signature (flattening) of the presence of heavy nuclei.

One means of visualizing the way an energetic particle detector responds to a con-vected hot plasma including several ion species is to represent the different components of the plasma in terms of the momentum space density $W(\mathbf{p})$, which is related directly to the plasma density $N(\text{cm}^{-3})$ and the differential unidirectional particle flux $j(E)$ through $W(\mathbf{p}) = j(E)/Np^2$. If the bulk plasma velocity is taken along the p_x direction, and if the distribution function is symmetric about this axis, we may chose p_y as the transverse momentum $p_\perp$ in an arbitrary transverse direction. Then the z axis may be used to represent the momentum space density itself $W(p_x, p_y)$, and from the momen-tum space density surface we can directly derive the directional dependence of the flux incident on a detector from the intersection of the right circular cylinder $p_x^2 + p_y^2 = p^2$ with $W(p_x, p_y)$. Of course, W must be multiplied by Np^2 to obtain the true flux $j(E)$.

In Figure 4.11, we present a set of Maxwellian momentum-space densities $W(\mathbf{p})$ for the hydrogen, helium (10% abundance relative to ^{1}H) and oxygen (0.1% abundance relative to ^{1}H) components of a $kT = 5$ keV plasma moving with a bulk velocity of $\mathbf{V} = 1000$ km s^{-1}. These are shown in two novel representations in terms of energy. First, let us define as the independent variable in our first representation $(E_x/A)^{1/2} = (E_\parallel/A)^{1/2}$ and $(E_y/A)^{1/2} = (E_\perp/A)^{1/2}$, which have the same mass-independent proportionality to the velocity components $(v_x, v_y) = (v_\perp, v_\parallel)$ for all species. Because $\tilde{v} = v - \mathbf{V}$ and $\hat{W} = (2\pi mkT)^{-3/2} \exp(-m\tilde{v}^2/2kT)$, the thermal density surfaces in a $\log W [(E_\parallel/A)^{1/2}, (E_\perp/A)^{1/2}]$, plot, as in Figure 4.11a are coaxial paraboloids of revolution centered on $(E_x/A)^{1/2} = V(m_1/2)^{1/2}$, where m_1 is the average mass/nucleon, since $m\tilde{v}^2/2kT = (A/kT)[(E_x/A)^{1/2} - V(m_1/2)^{1/2}]^2 + (A/kT)(E_y/A)$. The paraboloids are broader for the lighter elements, since the e^{-1} (thermal) width is $(kT/A)^{1/2}$. Although Figure 4.11a portrays the more familiar aspects of a multispecies thermal distribution (characterized by a single temperature), it does not provide the insight we need into the response of instruments like the LEMPA, which measure the total energy that is deposited in the detector, not the velocity. Consequently, in Figure 4.11b we plot $\log W$ as a function of $E_x^{1/2} = E_\parallel^{1/2}$ and $E_y^{1/2} = E_\perp^{1/2}$. The surfaces of constant W are still paraboloids of revolution, but now they all have identical shapes because $m\tilde{v}^2/2kT = [E_x^{1/2} - V(Am_1/2)^{1/2}]^2/kT + E_y/kT$ and the thermal (e^{-1}) width is simply $(kT)^{1/2}$. However, their axes are now displaced from each other along the $E_x^{1/2}$ axis at the values $V(Am_1/2)^{1/2} \propto A^{1/2}$. It is the distribution function as shown Figure 4.11b that is relevant to the LEMPA response, rather than more familiar representation as in Figure 4.11a.

Of course, what the instrument really "sees" is the incident differential flux $j = Np^2W$, and this function is plotted for the flux (due to each of the three species) in the convection direction ($\phi = 0$) as a function of $E_x^{1/2}$ in Figure 4.11c. In Appendix A we describe how the actual instrument response is obtained by integrating the incident flux over the detector solid angle and energy passband efficiencies $\epsilon_k^a(E)$, and then summing over all species; see Equation (4.19) and Figure (4.31). We have given a suggestion of this all important step in Figure 4.11c by indicating approximately the lowest energies of the ^{1}H, ^{4}He, and ^{16}O fluxes to which the LEMPA responds (e.g., the lower extent of the ϵ_k^a for the lowest detector energy channel). Immediately one sees from the Figure 4.11b the very dramatic enhancement of heavy ion fluxes in the convection direction, and Figure 4.11c raises the possibility that ions such as ^{16}O, if their abundance approaches that of ^{1}H, could dominate the LEMPA response in the corotation direction in the Jovian outer magnetosphere. Consequently, the problems of estimating the convection velocity and deducing the composition of the hot plasma from the LEMPA channel rates are intertwined.

The exact expression for the incident integral flux due to a convected thermal distribution was calculated by Roelof et al. [1976]

$$J(E,\phi) = \pi^{-3/2} Nv_0 e^{-V^2/v_0^2} e^{y^2-x^2} H(x,y) \tag{4.11}$$

where N is the plasma density, v is the velocity of a particle with the detector threshold energy E, v_0 is the particle velocity with energy $E_0 = kT$, $x = (v - V\cos\phi)/v_0$, $y = V\cos\phi/v_0$, V is the plasma bulk velocity, and

$$2H(x, y) = x^2 + 3xy + 3y^2 + 1 + (\pi/2)^{1/2}(3y + 2y^3) e^{x^2} erfc(x) \tag{4.12}$$

Selected calculations of integral fluxes incident in the forward direction, $J(E,0)/N$,

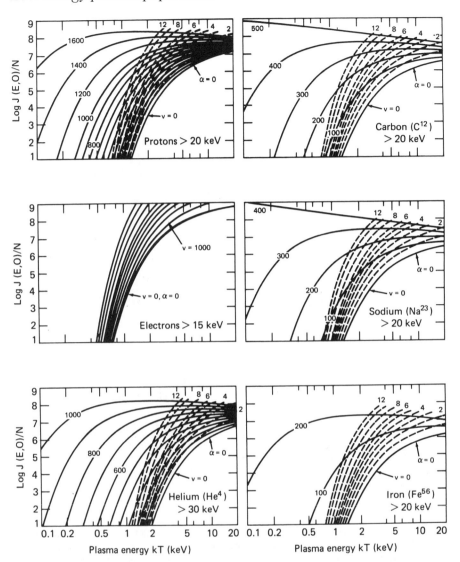

Fig. 4.12. Incident integral flux (energy > E) in the ion flow direction of thermal plasma $J(E,O)$, normalized by density of each species (N), as a function of plasma temperature (kT) for various values of plasma velocity. Dashed lines give contours of anisotropy parameter $\alpha = mvV/kT$; see Equation (4.10).

from (4.11) and (4.12) are presented in Figure 4.12 for plasma velocities above 100 km s^{-1} and temperatures from 100 to 20 keV for 20 keV protons, 15 keV electrons, 30 keV ^{4}He and 20 keV ^{12}C, ^{23}Na, and ^{56}Fe. Also included in Figure 4.12 as dashed lines are contours of the exponential anisotropy parameter $\alpha = mvV/kT$ from (4.9) which gives an approximate indication of the strong anisotropies of the fluxes. The strong dependence on temperature and bulk velocity is particularly dramatic for the heavier species.

To illustrate the sensitivity of the LEMPA detector to flowing thermal plasmas, we have plotted in Figure 4.13 the differential forward fluxes in the lowest channel of Voyager 2 versus kT for selected flow velocities of a hydrogen plasma at a density of

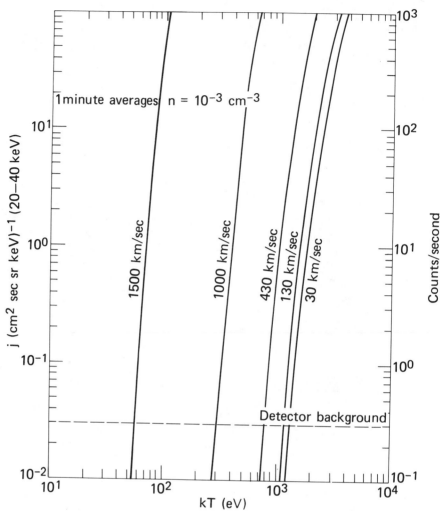

Fig. 4.13. A figure to demonstrate the ability of the LEMPA detector in responding to a very tenuous hot flowing plasma. The curves labeled by the plasma flow speed indicate the proton differential fluxes at 20–40 keV as a function of plasma temperature for a plasma with density 10^{-3} cm^{-3}. The detector background is shown to illustrate the sensitivity of the instrument. For example, for a plasma of density 10^{-3} cm^{-3} flowing at 30 km/s, the proton fluxes are above our detector background when the plasma temperature is above 1.3 keV.

10^{-3} cm^{-3}. At such low densities, most plasma instruments have difficulty in measuring plasma velocities and temperatures on relatively short timescales (~ 1 min). It is evident from the figure that LEMPA can readily measure small (~ 50–100 km s^{-1}) velocities at plasma temperatures ≥ 1 keV, and responds to higher temperature (≥ 10 keV) plasma without requiring large flow velocities, even at these low densities. The presence of heavier ions enhances this sensitivity.

Krimigis et al. [1979a] made use of these dramatic ion anisotropies to identify oxygen as a major component in the Jovian magnetosphere. Shortly after the first (inbound) magnetopause crossing, the lowest energy ion channel (PL01) of LEMPA on Voyager 1 measured the angular distribution shown in the inset to Figure

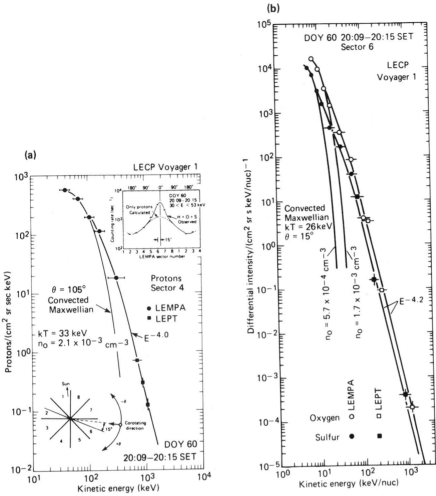

Fig. 4.14. Evidence from flux anisotropies for comparable densities of He and O: (a) Angular distribution in Voyager 1 LEMPA channel PL01 just inside dayside magnetopause, with LEMPA/LEPT spectrum in the sector (4) transverse to corotation fit by corotating ($V \sim 800$ km s^{-1}) thermal ($kT = 33$ keV) proton distribution with non-thermal, power-law tail ($\gamma = 4.0$); and (b) Spectrum from forward sector (6) with proton contribution subtracted, fit by similar distribution to (a), but for corotating hot O and S ions (O/S $\simeq 3$) [from Krimigis et al., 1979a].

4.14a. The sector rates in the directions away from corotation (2,3,4), as well as the spectrum in the transverse direction shown in Figure 4.14a, can be fit by a corotating ($V \sim 800$ km s^{-1}) thermal ($kT = 33$ keV) proton distribution that makes a transition to a nonthermal power-law tail ($\gamma = 4.0$) at an energy $E \simeq (\gamma + 1) kT$; we discuss this "gamma-thermal" distribution in detail in Appendix B. It is clear from the inset to the figure, however, that the low-energy sectors in the direction of corotation (5,6,7) must be responding to heavier ions, in addition to the protons (dashed lines). Subtracting this proton contribution from the rates in the forward sector (4.6), and assuming the "excess" ions are oxygen and sulfur with O/S ~ 3, as measured at ≥ 200 keV/nucleon by the LEPT, a satisfactory fit is obtained in Figure 4.14b for essentially the same distribution parameters used to fit the protons in Figure 4.14a. Therefore Krimigis et al.

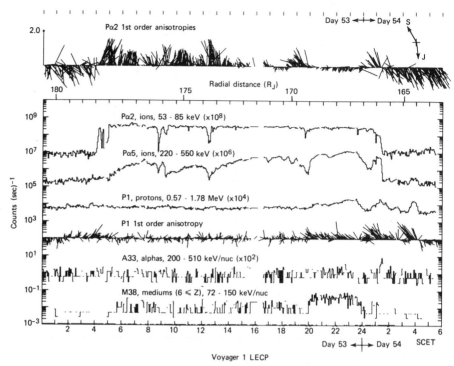

Fig. 4.15. Long-lived Jovian particle event observed on day 53 by Voyager 1 inbound. The count rate data are averaged over one scan, 192 seconds. The flat top PL02 profile is typical for long-lived events, as is the gradual rise in the PL05 count rate. A clear increase in the P1 count rate occurred after the main event on day 54 [Zwickl et al., 1981]. The anisotropy direction is noted at upper right, with arrows pointing towards the Sun (S) and Jupiter (J).

[1979a] were able to conclude that the concentrations of H and O were comparable in this region of the magnetosphere.

This then completes our presentation of techniques for deducing convection velocities and composition of flowing multicomponent plasmas. In Section 4.4, we estimate hot plasma pressures and number density. The theoretical basis of these estimates can be found in Appendix C.

4.4. Presentation of results

Upstream regions

As noted in connection with Figure 4.3, the first inbound observation of a "Jovian Ion Event" occurred at $\sim 600\ R_J$ on Voyager 1 and $\sim 860\ R_J$ on Voyager 2, although outbound observations by Voyager 1 show small ion events to beyond $1200 R_J$. Although not shown here, the maximum flux level of the short-lived events falls off slowly with increasing radial distance as does their frequency [Zwickl et al., 1981].

Figure 4.15 presents the details of one long-lived event observed by Voyager 1 on day 53, 1979 at a distance of $\sim 175\ R_J$ upstream from Jupiter [Zwickl et al., 1981]. The flux profile of the lowest energy ions in $P\alpha2$ has a rapid onset followed by a nearly constant flux level for ~ 21 hr, then displays an equally rapid decay. Several of the short-lived decreases observed in the flux of the $P\alpha2$ time profile occurred during the passage of what appear to be tangential discontinuities (TD's) in the interplanetary magnetic field, which was nearly radial throughout this event (N. F. Ness, private communication, 1981). Note that these decreases tend to occur at 5-hr intervals, suggesting Jovian magnetosphere control throughout the event. The connection, if any, between the TD's and the suggested periodicity is not understood. The ion anisotropy at 53 keV changes abruptly at event onset from a solar convective direction to a direction consistent with flow outward away from Jupiter as shown in the vector plot on top of the figure. The anisotropy magnitude is initially very large but decays with time. In this event, the flow direction of 53 keV ions actually reverses to an inward flow later in the event. The higher energy ions have a much slower onset whereas the more energetic protons in channel P1 do not increase until 2000 hr on day 53, that is, toward the end of the event. Note that the spectrum becomes harder at this time, and actually exhibits a peak at about 500 keV early on day 54 [Zwickl et al., 1981].

There are other aspects of this event, however, that are worth noting. First, the alpha particle intensity shown in channel A33 remains essentially at preevent background throughout this entire period while the $Z \geq 6$ ion population (channel M38) shows an increase at the beginning of the event and again at about 2000 UT, suggesting that the lower energy ions in $P\alpha 2$ and $P\alpha 5$ are primarily $Z \geq 6$. As noted in Figure 4.3, pulse-height analysis spectra at ~0.5 MeV/nucleon during the later part of this event show the higher energy ions to consist primarily of oxygen and sulfur. Because this composition is known to be characteristic of ions within the Jovian magnetosphere, we are led to suggest that upstream events are probably due to ions escaping directly from the magnetosphere. In view of the complexity of the structure, however, and the strange intensity-time profile, which are not unlike those observed upstream from Earth [Sarris et al., 1978; Scholer et al., 1980], there could well be additional mechanisms at work.

Outer magnetosphere

Magnetosheath-magnetopause. Scan-averaged (192 s) ion and electron count rates and first-order particle anisotropies are shown in Figure 4.16 for the interval around the first magnetopause crossing by Voyager 1 [Krimigis et al., 1979a]. The ion rates are enhanced immediately behind the bow shock and the anisotropy vectors initially point in the direction of the deflected solar wind. The ion count rate in the magnetosheath is at a level that is essentially identical to that observed during the long-lived upstream ion event on days 53 and 54. As the spacecraft approached the magnetopause, we note that the anisotropy becomes more radial in nature, that is, ions flow away from Jupiter. Upon crossing the magnetopause, the flows changed from the radial to the corotational direction. Details of the particle count rate in each sector at selected time intervals during the magnetopause crossing are shown as insets to Figure 4.16. Prior to magnetopause crossing, the flow was directed away from Jupiter; during the crossing, the flow began to change to a corotational direction; after the crossing, the flow was strongly corotational. This convective flow in the corotational direction is characteristic of most plasma flow in the Jovian magnetosphere both on the dayside and nightside, as previously summarized in Figure 4.7. The magnetopause crossing for the electrons is much better defined than for the ions.

Energy spectra. As explained in Section 4.3, it is possible to use the angular and spectral measurements of the LECP instrument to deduce the temperature at lower energies (i.e., ≤ 200 keV) and the spectral slope at higher energies (≥ 200 keV) throughout the magnetosphere, all in the rest frame of the plasma. Figure 4.17 pre-

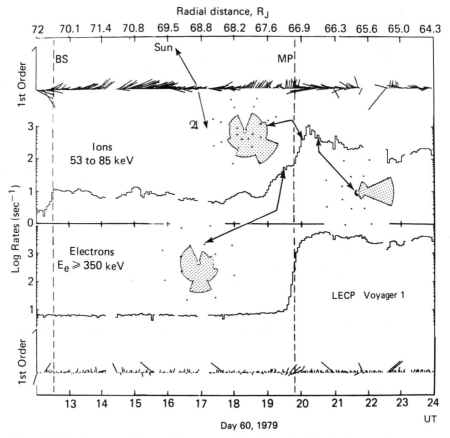

Fig. 4.16. Particle intensities and flows around the interval of the first magnetopause crossing. Pie diagrams at selected points show detector count rate as a function of view direction [Krimigis et al., 1979a]. Note that some radial outflow is observed, notably at ~2230 to 2300 UT.

sents the results of these computations for the inbound pass of Voyager 1 using 15-min. count rate averages. The top panel shows the count rate for 53–85 keV ions averaged over the detector scan and includes the nominal crossing times of the magnetic equator. The middle panel shows the spectral exponent γ; the spectra tend to be softer (larger γ) at magnetic equatorial crossings, especially in the outer magnetosphere, with systematic spectrum variations well over 1 unit of γ. The amplitude of the fluctuations is considerably reduced after the final magnetopause entry at the beginning of day 62.

The lower panel displays the calculated proton temperature kT; there appears to be a significant correlation between peaks in the temperature profiles and times of crossing of the magnetic equator. Specifically, on days 62–63 the four major peaks are ~10 hr apart and occur at the $\lambda_{\mathrm{III}} \sim 300°$ crossings (active sector, see Chap. 10); corresponding peaks at $\lambda_{\mathrm{III}} \sim 100°$ crossings are more difficult to discern. There are however exceptions to this general correlation, the most obvious of which is the temperature peak at 1700 UT on day 63. Undoubtedly there exist time variations that can distort the apparent association between equatorial crossings and kT, as evidenced by the Voyager 2 encounter [Krimigis et al., 1981]. In general it can be stated that the temperature increases at the equator and decreases at high latitude, and is thus anti-correlated with the spectral behavior of the power law at higher energies.

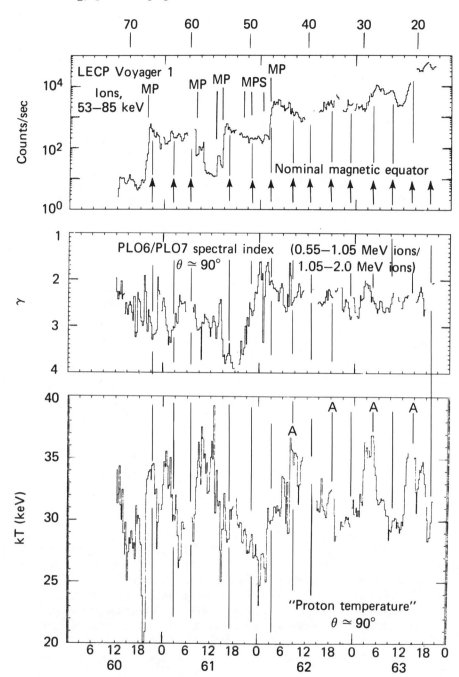

Fig. 4.17. Voyager 1 inbound spectral parameters [Krimigis et al., 1981]. The lower panel displays the calculated proton temperature derived from 15 min. averaged data from the sector perpendicular to the flow direction. The second panel shows the spectral index derived from the PL06 and PL07 channels in the same sector. The top panel shows the counting rate of the PL02 channel and includes the magnetopause (*MP*), bowshock (*S*), and nominal magnetic equator crossings (arrows). *A* indicates the active sector, $\lambda_{III} \sim 260°$ (see Chap. 10).

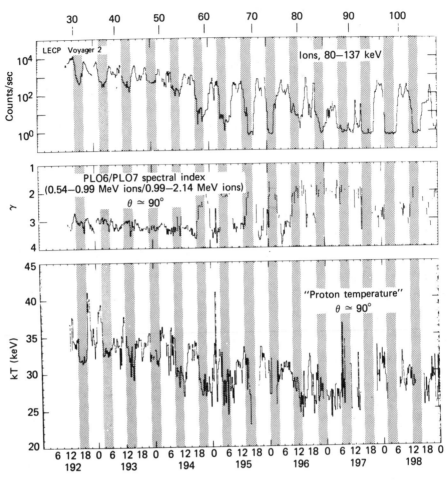

Fig. 4.18. Similar to Figure 4.17 but for Voyager 2 outbound [Krimigis et al., 1981].

Figure 4.18 shows the variation of spectral parameters for Voyager 2 for the outbound trajectory to ~ 105 R_J. The γ parameter displays little variation with distance from the magnetodisc through late on day 194, at which time the spectrum becomes much harder outside the magnetodisc, due to the presence of solar energetic protons on nightside field lines [Hamilton et al., 1981]. The temperature profile shown in the lower panel, however, closely follows the intensity profiles on the top panel; that is, temperature peaks occur at density maxima and temperature valleys occur at density minima. Although spectral data are not shown for the magnetospheric wind region (i.e, after day 203), the values for γ and kT are in the same range as those determined for the magnetodisc.

The presentation of the spectra would be incomplete without some indication of the behavior of energetic electrons. Figure 4.19 shows representative differential spectra (upper panel) at low energies for selected time intervals (lower panels) during the inbound pass of Voyager 1. The data points are plotted at the geometric mean for each of the five differential channels of the low-energy electron detector. The spectra can be described by $dj/dE = KE \exp(-E/E_0)$ with values of E_0 ranging from 25 to 35 keV. Note that if E_0 is interpreted as kT, the temperatures are quite similar to those obtained

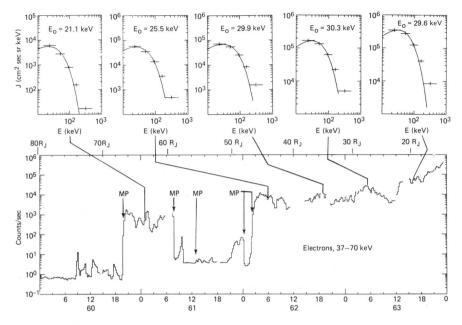

Fig. 4.19. Characteristic spectra of low energy electrons in outer magnetosphere. Lower panel shows detailed intensity profile. Upper panels show spectral plots at the indicated times. "Goodness of fit" to a γ-thermal spectrum is typically ~ 0.08 in fitting the thermal part to the lowest three points.

for the ions. Further, there is a tendency for E_0 to increase as the distance to the planet decreases. A more detailed investigation [Carbary and Krimigis, 1980] has shown that a pure exponential is an equally good fit and is characteristic of both the dayside and nightside magnetosphere.

Number-density and pressure profiles. A lower bound on the number density of energetic ions can be obtained from the measured differential intensities using relation (4.23) of Appendix C:

$$N > \sum_{n} \sum_{k} \frac{2\pi \Delta\mu_n}{Gv_k} R_k(\mathbf{n}) \tag{4.13}$$

$R_k(\mathbf{n})$ is the rate measured by the kth energy channel (geometric factor G) in the solid angle $2\pi \Delta\mu_n$ about the unit vector $(-\mathbf{n})$; see the definition of R_k in (4.17) of Appendix A. The lower bound is valid if the "scaling velocity" v_k for the kth channel satisfies the condition (4.22) of Appendix C. The density lower bound calculated in this manner is independent of any model, makes use of the observed differential intensity and angular distribution, and only depends on the selection of v_k, which represents the velocity of a specific ion in a particular energy channel. To obtain a lower limit on the number density, we assume that the detector response in all directions is entirely due to protons; in this case the velocity we use in Equation (4.13) is that appropriate for the proton energy passband of each particular channel. The likely upper bound on the densities is obtained by assuming that the detector responds entirely to oxygen in every direction; that is, the velocity used in the denominator of Equation (4.13) is that appropriate for oxygen ions in each channel. See Krimigis et al. [1981] and the derivation of Equation (4.23) in Appendix C for further discussion.

Physics of the Jovian magnetosphere

The computation of a lower bound on the average energetic (i.e., ≥ 30 keV) particle pressure $\bar{P}$ can be done in a similar manner to that employed in the computation of particle density, that is, by simple integration of the observed spectra, and is given by expression (4.29) of Appendix C:

$$\bar{P} > \sum_k \frac{4\pi p_k}{3G} R_k(\mathbf{n}_\perp) \tag{4.14}$$

where p_k is a scaling momentum, not necessarily related to the quantity v_k in (4.13), and the rate of the kth channel is now taken only in the direction $(\mathbf{n}_\perp)$ perpendicular to the convection velocity. The quantities in the summation are known or observable, and we proceed to obtain a lower and upper limit on the pressure in a manner similar to that of the density by using the appropriate p_k for a particular channel. This approximation is closely related to the stricter lower bound on pressure given by Equation (4.23) in Appendix C, which is valid if condition (4.28) is satisfied. The number density and pressure profiles for the encounter of Voyager 2 are presented in the next three figures.

Figure 4.20 shows density and pressure data for the inbound portion of the Voyager 2 trajectory through $\sim 30~R_J$. Estimated electron densities from the plasma wave (PWS) instrument [Gurnett et al., 1981b], and ion densities from the plasma (PLS) instrument are shown for comparison. It is evident that there appear to be significant differences between the LECP densities and the PWS electron densities at the magnetopause encounters at ~ 72 and $\sim 62~R_J$. However, from the second magnetopause crossing and to distances as close as $45~R_J$, there is reasonable agreement between the energetic particle ion concentrations and the PWS electron concentrations, given the limitations of this comparison [for a full discussion see Krimigis et al., 1981].

Inside $\sim 45~R_J$ there are significant increases in the electron concentrations that are not reflected in the energetic ion concentrations, especially at 43 and $39~R_J$ and most importantly $\sim 33~R_J$. We note that during this last electron concentration enhancement, the low-energy plasma densities from Belcher, Goertz, and Bridge [1980] appear to agree well with the PWS data, suggesting that at these times copious fluxes of ions were present with energies below the LECP energy threshold. The bottom panel of Figure 4.20 shows the particle pressure profile, including the comparison with magnetic field energy density. It is observed that energetic particle pressure can easily balance the magnetic field pressure and that the field energy density is generally depressed during energetic particle pressure peaks. This appears to be the case even during electron concentration peaks at ~ 43, ~ 38, and $\sim 33~R_J$, suggesting that even at those times when the cold plasma appears to dominate, the pressure is carried entirely by the energetic particles.

The presentation of the number-density and pressure profile is continued in Figure 4.21 although the LECP instrument was in its stow mode [for assumptions regarding the derivation of these densities see Krimigis et al., 1981]. It is evident that electron densities reported by the PWS near closest approach are generally higher than the ion densities derived from the LECP measurements. Gurnett et al. [1981b] find that there is good agreement between the energetic particle densities and those obtained by the plasma wave instrument, if the assumption is made that the LECP detector response is primarily due to doubly ionized oxygen.

The lower panel shows energetic particle pressure compared to the observed and expected magnetic field signatures, from the work of Ness et al. [1979c]. It is evident from the data that the large diamagnetic depressions observed at ~ 1100 and ~ 2200 UT on day 190 and ~ 1000 UT on day 191, are coincident with large peaks in the ener-

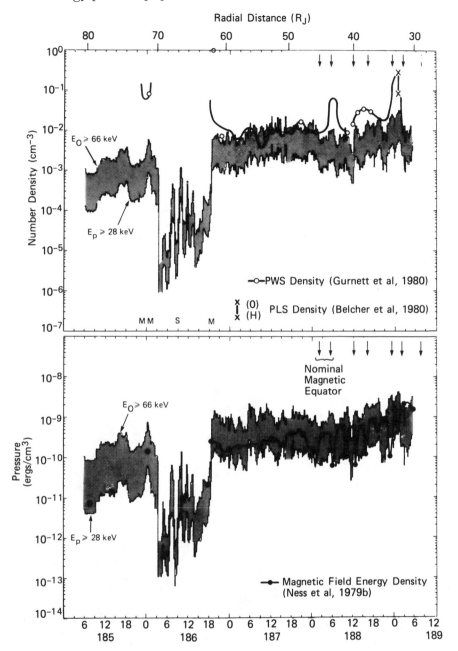

Fig. 4.20. Voyager 2 inbound ion density and pressure [Krimigis et al., 1981]. Shown are the computed density (top panel) and pressure (bottom panel) derived from an integration of the observed spectra. The upper curve corresponds to the assumption that all particles are singly charged oxygen and the lower curve to the assumption that all particles are protons. In the top panel, the bow shock (*S*) and the magnetopause (*M*) are indicated; crossings of the nominal magnetic equator are indicated by the short arrows. The solid line indicates the electron density derived from the plasma wave experiment with the open points representing exact density determinations. The crosses are plasma densities from the plasma experiment. The solid line in the lower panel indicates the magnetic field energy density ($B^2/8\pi$) obtained from the magnetic field experiment [Ness et al., 1979c].

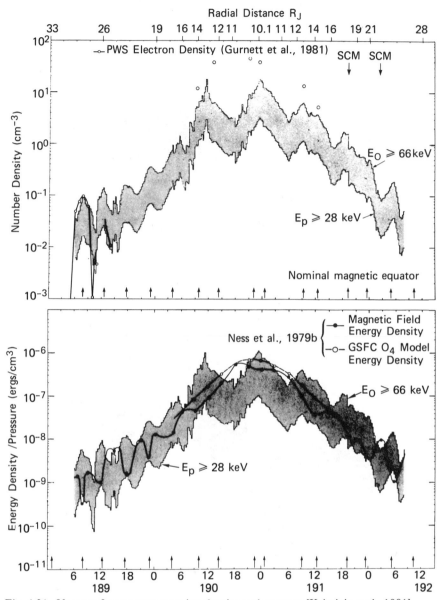

Fig. 4.21. Voyager 2 near encounter ion density and pressure [Krimigis et al., 1981]. SCM denotes a spacecraft maneuver.

getic particle pressure. In addition to the main peaks, good correlation between energetic-particle-pressure maxima and magnetic-field depressions are found both prior to and well after closest approach.

To summarize, it appears that for the duration of the entire Voyager 2 inbound pass the plasma pressure can be accounted for with ions of energies ≥ 30 keV, that is, a situation that also obtains in the Earth's ring current [Williams, 1979]. With regard to particle number density, it is evident that a large fraction of the density in the outer magnetosphere ($\geq 40\ R_J$) resides with ions of energy ≥ 30 keV. Inside $\sim 40\ R_J$ there appear to be substantial fluctuations in the number density, with a fraction accounted

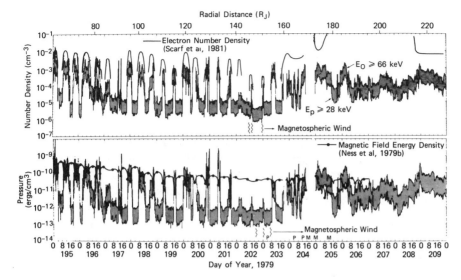

Fig. 4.22. Voyager 2 outbound densities and pressures [Krimigis et al., 1981]. As in the two previous figures, the electron density and magnetic field energy density is given by the solid lines in the upper and lower panels.

for by energetic particles ranging from a few percent to well over 50%, if the assumption is made that the detector response in this region of the magnetosphere is primarily due to oxygen and sulfur.

Figure 4.22 presents data in a format similar to that of Figures 4.18 and 4.19 during the outbound portion of the Voyager 2 trajectory from ~ 70 to $\sim 225\ R_J$. The solid lines are estimated electron concentration profiles from Gurnett et al. [1981b], representing upper limits of the electron concentration during magnetodisc crossings. We note that the measured energetic particle ion densities are typically a factor of 2–5 below the PWS electron densities, although in some cases (for example, early on day 195, late in day 196, middle of day 199, and especially on day 203), the energetic ion densities are in good agreement with the estimated electron density by PWS. The lower panel in the figure shows estimated particle pressures and the corresponding magnetic field energy density. We again observe that the energetic particle pressure in the magnetodisc and the magnetic field pressure in the lobes are similar, as shown by the detailed study of Lanzerotti et al. [1980]. It is important to note that the particle pressure is comparable to the field pressure well outside the boundary of corotational plasma flow (day 203) and into the magnetospheric wind region, up to at least day 206 (i.e., outside the magnetopause).

Anisotropies. The LECP instrument scan plane was optimized to obtain pitch-angle distributions and anisotropies in the outer Jovian magnetosphere as described in Figure 4.2. This was in contrast to the Pioneer spacecraft anisotropy studies, which were primarily obtained in a plane essentially perpendicular to the Jovian equatorial plane and thus were optimized for observing pitch-angle distributions in the inner, dipolar region of the magnetosphere. A standard harmonic analysis was performed using the full 8-sector set of data [Carbary et al., 1981]. The count rate is least-square fit to the function

$$C(\phi) = C_0 [1 + A_1 \cos(\phi - \phi_1) + A_2 \cos 2(\phi - \phi_2)] \tag{4.15}$$

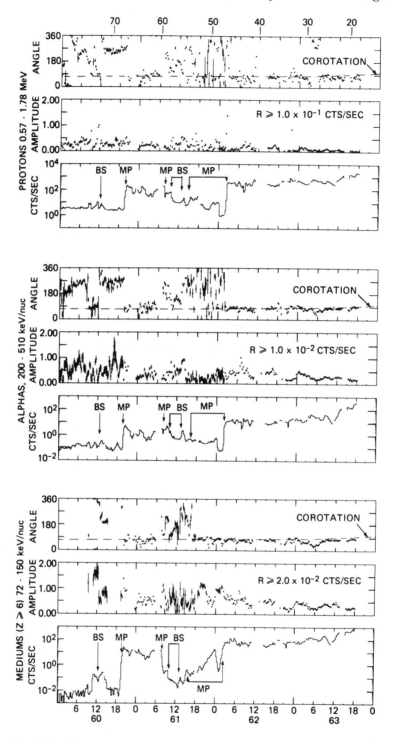

Fig. 4.23. Voyager 1 inbound, detailed anisotropy plots [Carbary et al., 1981]. First order anisotropies for three representative species are shown. Data shown here are 15-min. averages. Bow shock and magnetopause locations are taken from Bridge et al. [1979a] and Ness et al. [1979a].

where C_0 represents a scan average, A_1 is the anisotropy amplitude, and ϕ_1 is the anisotropy angle ($\phi_1 = 0$ represents a flux maximum in sector 8). This analysis will concentrate on the first order (or streaming) anisotropy given by A_1 and ϕ_1. The vector A_1, ϕ_1 defines a flow strength and direction. In a large majority of cases, LECP anisotropies in the Jovian magnetosphere are well described by such a unidirectional fit. To obtain a unique determination of the amplitude, only channels whose response is primarily due to one particle species are used. These are the 0.57–1.78 MeV protons, 0.2–0.5 MeV/nucleon alpha particles, and 0.07–0.15 MeV/nucleon medium nuclei ($Z \geq 6$).

An example of these fits is given in Figure 4.23 that displays anisotropies on the Voyager 1 inbound trajectory. These 15-min. averages reveal a great deal of structure in regions both external and internal to the magnetopause. Anisotropy angles at $\sim 110°$–$190°$ represent solar wind flow on day 60. Crossing into the magnetosheath on day 60 after ~ 1300 UT, LECP measured sheath flows at angles of $\sim 250°$. Entry into the magnetosphere on day 60 was signaled by a change in the anisotropy angle from $\sim 260°$ to $\sim 80°$, the latter angle being that of nominal corotation (see also Fig. 4.16). The last magnetopause crossing occurred during a time span of several hours beginning in the latter half of day 61 [Ness et al., 1979a]. Anisotropy angles of protons and alphas were variable during this time interval and so the angle is characteristic of both magnetosheath flow and corotation. However, the $Z \geq 6$ particles have strictly corotational anisotropy directions throughout the series of magnetopause crossings. Within the magnetosphere, anisotropy amplitudes can be variable on timescales of 15 min., although the anisotropy angles of all ion species remained essentially in the sense of corotation ($70°$–$80°$). Proton anisotropies are weaker than alpha particle anisotropies, which in turn seem weaker than the $Z \geq 6$ anisotropy; this is expected from the velocity dependence of the convective anisotropy. Except at the magnetopause, the inbound data do not indicate strong spatial gradients in the particle population (scale lengths of ~ 10 R_J suggested by the rate profiles seem to be appropriate for ions on the dayside). Finally, we note that the anisotropy angles tend to be somewhat less than the nominal corotation angle after the beginning of day 63. This fact is most apparent in the medium Z channel, and is suggestive of a small steady component of radial outflow.

Figure 4.24 displays the Voyager 2 anisotropies within the magnetodisc region from the beginning of day 195 to the end of day 199. The anisotropies reveal a striking correlation with position relative to the midplane of the magnetodisc plasma sheet. The LECP ion anisotropy amplitudes appear to be strongest near the midplane of the plasma sheet and weakest away from the midplane. This effect occurred for all ion species shown. The anisotropy angles do not show similar signs of dependence on distance from the midplane. Generally, the anisotropy angles remained close to the nominal corotation angle ($\sim 315°$).

The anisotropies of the $Z \geq 6$ particles show the most pronounced effect with radial distance in the region of the magnetodisc. Figure 4.25 shows 1-hr averages of the $Z \geq 6$ particle anisotropies for most of the magnetodisc crossings by Voyager 2. No selective data sorting with distance from the midplane has been done and some of the scatter in the data is probably due to changes in amplitude with Z_{DSC} (i.e., distance from midplane). In spite of the scatter, the anisotropy amplitudes show an easily recognizable linear increase with R. In a more detailed analysis, Carbary et al. [1981] show that the anisotropy amplitude is largest at the equator and decreases as one moves away from the magnetodisc at large values of Z_{DSC}.

Figure 4.26 displays the several days surrounding the Voyager 1 encounter with the wind region. Shown here are low-energy ions with energies 53–85 keV. The space-

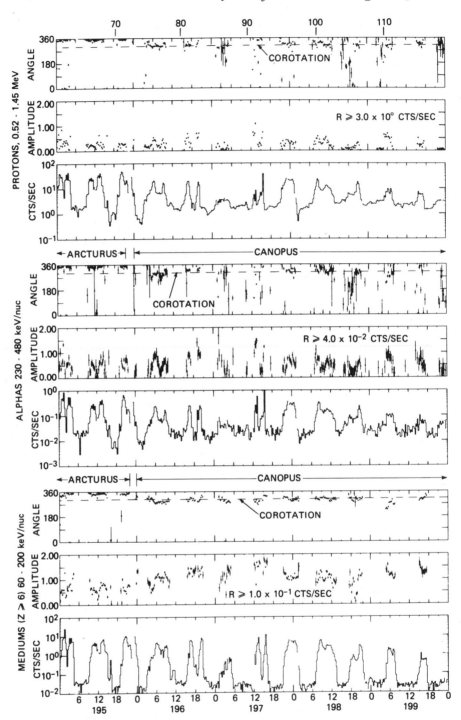

Fig. 4.24. Voyager 2 outbound magnetodisc, detailed anisotropy plots [Carbary et al., 1981]. First order anisotropies are shown in the same format as the previous figure. The data are 15-min. averages. Bow shock and magnetopause locations are taken from Bridge et al. [1979b] and Ness et al. [1979c].

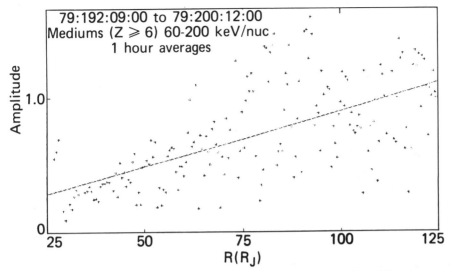

Fig. 4.25. First order anisotropy amplitudes for the medium Z ($Z \geq 6$) particles observed on Voyager 2 outbound [Carbary et al., 1981]. All data points are included, irrespective of spacecraft distance from the magnetodisc. The line represents a least-squares fit to the data.

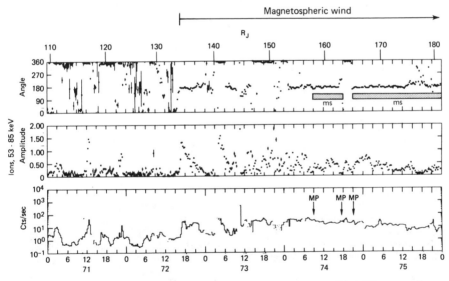

Fig. 4.26. Voyager 1 magnetospheric wind region [Carbary et al., 1981]. The transition from corotational magnetospheric flow (angles near 360° for this orientation) to outward or antisolar flow (angles near 180°) first occurs on day 72, well before the first magnetopause encounter. Magnetopause locations are taken from Bridge et al. [1979a] and Ness et al. [1979a]. The data represent 15-min. averages.

craft's first entry into the wind region occurred near 1200 UT on day 72, ~ 2 days prior to encounter of the first magnetopause crossing; this can be seen by the abrupt change in the anisotropy angle from ~ 350° (corotation) to a relatively steady 180° (magnetospheric wind). The 10-hr periodicities associated with the magnetodisc break down in this region (bottom panel), although the anisotropies continue to be very strong. Sudden ~ 180° changes in the anisotropy angle occurring on day 73 indicate that the

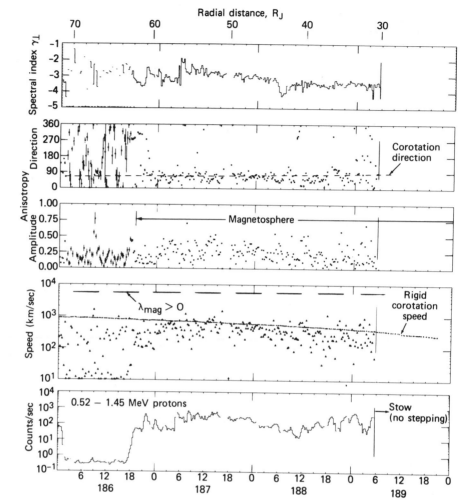

Fig. 4.27. Voyager 2 Compton-Getting speed for 0.57–1.78 MeV protons [Carbary et al., 1981]. The plasma speeds observed during the inbound phase of the encounter are shown in the fourth panel. The corotation speed for a rigid magnetosphere is given by the dotted line. The periods where the spacecraft was in the northern magnetic hemisphere are given by the bar along the top of the panel. The lines in the anisotropy direction panel show the corotation direction. The bar at the top of the anisotropy amplitude panel shows the periods when the spacecraft was inside the Jovian magnetosphere.

disc/wind boundary passed over the spacecraft several times. The spacecraft entered the magnetosheath near the middle of day 74 and reentered the magnetosphere for a short time and again encountered corotational flow. A similar observation was made during the Voyager 2 trajectory, again with the flow vector turning antisunward ~ 1.5 days prior to the first crossing of the magnetopause. The change from corotational to tailward flow was even more dramatic for Voyager 2 in that it was accompanied by a heavy ion beam described by a relatively cold (~ 5 keV) Maxwellian with a peak energy at ~ 265 keV [Krimigis et al., 1980; Krimigis, 1981].

Velocities. Because the convection velocities are much smaller than the velocities of ~ 1 MeV protons, we can relate the convective speed (V_c) to the particle anisotropy amplitude (A_1) by the simple rearrangement of Equation (4.8):

$$V_c = \frac{A_1 v}{2(\gamma + 1)} \tag{4.16}$$

We have used Equation (4.16) to compute a radial profile of the convection speeds in the dayside Jovian magnetosphere. A_1 is the measured first order Compton-Getting anisotropy amplitude as defined in Equation (4.8), where γ is the observed spectral index (for 0.5–4 MeV protons) perpendicular to corotation [Krimigis et al., 1981], and v is determined from the mean energy of the proton channel used. Figure 4.27 shows the speeds derived from Equation (4.16) during the inbound portion of the Voyager 2 encounter using the indicated proton channel. The time profile of the count rate and computed speed are shown in the lower two panels whereas the corresponding anisotropy amplitude, phase, and spectral index are shown in the upper three panels. The dotted curve in the fourth panel indicates the speed corresponding to a rigidly corotating magnetosphere, although the dashed line in the second panel indicates the expected corotation direction. It is evident from the speed profile that velocities are close to that of corotation in the outer part of the magnetosphere but exhibit considerable structure especially toward the inner magnetosphere. The corresponding Voyager 1 profile (Chap. 5) shows (~10 hr) variation in the speed inside ~50 R_J. In agreement with the speed profile of McNutt et al. [1979], the LECP speeds are consistently less than corotation within ~20 R_J of Jupiter. It should be emphasized that even though individual measurements of anisotropy angles often depart significantly from the corotational direction, the flow orientation is generally near the corotation direction.

Velocities in the nightside magnetodisc are not as readily obtainable, primarily due to unfavorable orientation of the LECP scan plane with respect to the corotation direction. In those cases where the orientation was favorable, it has been determined [Lanzerotti et al., 1980] that convective speeds are well below corotation for radial distances beyond ~80 R_J. Using similar techniques, it has been determined that velocities in the magnetospheric wind region are in the range of 300–900 km s^{-1}.

Composition and charge states. The most unusual aspect of the Voyager 1 and 2 encounters with Jupiter in terms of energetic particles was undoubtedly the composition. As shown in Section 4.2, the composition of magnetospheric particles is primarily made up of He, S, and O in addition to protons, with a small percentage of Na. The Pioneer encounters in 1973 and 1974 gave no hint that such composition might exist in the magnetosphere primarily because the instruments were not optimized to make measurements in the appropriate energy range. As noted in Figure 4.8c, there are important changes in the abundance ratios as functions of radial distance, that can be used to infer sources and/or acceleration mechanisms. Similar variations of the relative abundances are also observed in the magnetodisc and the magnetospheric wind regions.

Figure 4.28 shows selected abundance ratios during the Voyager 2 outbound pass out to ~200 R_J. The data from 10–60 R_J in Figure 4.8c are repeated, but in a slightly different format with the lower three panels showing the variation of various elements with respect to oxygen. Beyond ~60 R_J, all ratios show departures from their values in the middle magnetosphere to values that are more typical of solar particles. Another qualitative change in the ratios occurs at ~160 R_J. This is approximately the point at which the anisotropy of the lower energy ions switched from being predominantly in the corotation direction to being directed away from Jupiter, forming the magnetospheric wind. The transition occurred before the first magnetopause crossing that is

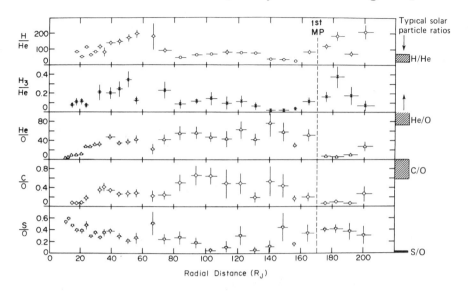

Fig. 4.28. Selected abundance ratios from 10–200 R_J during the outbound pass of Voyager 2 [Hamilton et al., 1981]. Between 43 and 160 R_J, the ratios were calculated only for periods near the current sheet. The first of several magnetopause crossings is indicated at ~ 170 R_J. Voyager 2 entered the magnetospheric wind region at between 160 and 170 R_J. Solar abundances of the ratio are noted on the right-hand ordinate.

indicated in Figure 4.28, and was apparently not affected by several other magnetopause crossings detected between ~ 170–185 R_J [Bridge et al., 1979b].

In the wind region, the abundance ratios at ~ 1 MeV/nucleon ions can be described as being more Jovian than solar in nature. S, O, H_3^+, and H abundances are all high relative to C and He. From Figure 4.6, we see that the energetic particle intensities were also higher in the wind region than in most of the magnetotail. Hamilton et al. [1981] note that even beyond 200 R_J there were instances when the composition of heavy ions was more similar to that in the inner magnetosphere than to that characteristic of solar/interplanetary particles. The overall impression from Figure 4.28 is that although the magnetodisc may be relatively poor in terms of heavy ions [see also Lanzerotti et al., 1980], the magnetospheric wind exhibits a distinctly Jovian composition suggesting that most of the flux comes from the inner part of the Jovian magnetosphere.

The changes in the relative abundance ratios are crucially dependent on the form of the energy spectrum; changes in the ratios with location or time could reflect either changes in the relative densities of the various species or only the relative changes in spectral slopes or both. The obvious question is whether there exists a representation where the spectra of all particle species are identical so that the relative abundances actually reflect the concentrations of the elements in the source plasma.

To investigate the spectral dependence noted above, a set of spectra for the various species in the outer magnetosphere is presented in Figure 4.29a. As is evident from the figure, all species have different slopes so that the relative abundances depend on the energy at which the ratios are evaluated. Hamilton et al. [1981] have shown that the most appropriate representation of the spectra in Jupiter's magnetosphere is on the basis of energy/charge. Their result is shown in Figure 4.29b which uses the identical data as those presented in Figure 4.29a, and also extends the energy range to lower

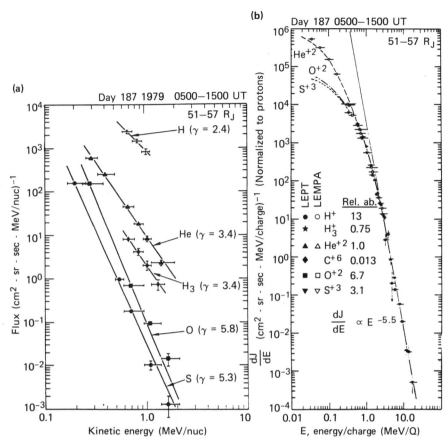

Fig. 4.29. (*a*) Energy spectrum of the indicated species plotted versus energy per nucleon (equal velocity); (*b*) differential intensities of the same species but plotted as a function of energy per charge [from Hamilton et al., 1981].

energies where the species are not identified uniquely. Here the intensities have been evaluated at a given energy/charge and the curves representing each element were displaced vertically by a factor indicated in the figure as "relative abundances." A charge state of $+2$ was assumed for oxygen and $+3$ for sulfur, primarily based on the charge states inferred from the UV experiment data [Broadfoot et al., 1979]. It is evident that in this representation, the energy spectra of all species fit on a common curve, and the relative displacements form the basis for estimating the relative number densities for all species. We note from the relative abundances that the number of $Z \geq 2$ ions is comparable to the number of protons. Thus, by inference from the energetic ions, the Jovian plasma is unique among magnetospheric plasmas in that it is dominated by ions heavier than hydrogen (see also Chap. 3).

Hamilton et al. [1981] have also demonstrated that the energy per charge representation is most appropriate for the entire magnetosphere and the magnetospheric wind region. In this representation, the large increases in the flux ratios of oxygen and sulfur relative to helium at equal energy/nucleon between the outer and inner magnetosphere (Figure 4.8*c*) are in fact consistent with flux ratios of ~1 at equal energy/charge, with no change in the value of this ratio with radial distance. The same procedure applied to the H/He ratio at equal energy/charge gives a roughly constant value of ~15. By

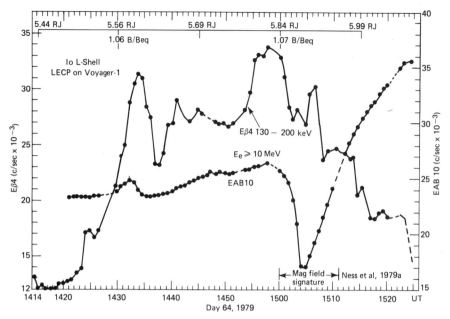

Fig. 4.30. Low and high energy electron profiles during close encounter of Io. Points are 48-s averages; missing data are indicated by dashed lines.

formalizing the procedure of expressing the energy spectra in terms of energy/charge, Hamilton et al. [1981] were able to parametrize the problem and deduce the charge state of heavier ions as follows: C^{+6}, $O^{+2,+3}$, and $S^{+3,+4}$. Thus, the energy/charge representation not only removes the radial dependence of the abundance ratios shown in Figure 4.8c, but enables the deduction of the relative charge states of all the ions and results in relative abundances that probably reflect those of the source plasma (see also Chap. 3).

Inner magnetosphere

The LECP instrument obtained a large number of measurements in the inner magnetosphere as well, some of which have been reported by Krimigis et al. [1979a], Armstrong et al. [1981], and Lanzerotti et al. [1981] (see also Chap. 12). The significance of these measurements lie primarily in the extended energy spectra and species coverage, and their relationship to the location of the Io plasma torus. These measurements complement and extend the measurements obtained by the Pioneer spacecraft, and their significance and implications are discussed in other chapters (Chaps. 5, 8, and especially 12). The original material may be found in the references cited above.

One particular aspect of the inner magnetosphere passage of Voyager 1 is worthy of note, namely the close approach to Io. During this passage only the energetic (> 10 MeV) electron fluxes show definite evidence of an interaction of the particles with field lines that may have passed through Io [Krimigis et al., 1979a]. A detailed plot of the intensity of low and high energy electrons in the vicinity of Io's L-shell is shown in Figure 4.30. Here, we note that the high energy electrons decreased by approximately 30% at the same time as a perturbation in the local magnetic field was observed [Ness et al., 1979a]. However, no significant change specifically associated with the magnetic field signature is clearly observed in the low energy electrons. These electrons display an overall enhancement in association with the Io flux tube, reminiscent of the electron "spike" observed by Fillius [1976] during the Pioneer 10 passage through the Io *L*-shell. Additional analysis is necessary before this increase in intensity can be ascribed to acceleration processes expected to take place in the vicinity of Io. It must be noted, however, that increases of this magnitude or larger in the intensities of low energy electrons have been seen throughout the inner magnetosphere passage [Krimigis et al., 1979a; also Fig. 12.3]. The possibility exists that after analysis of the full data set an upper limit to a possible magnetic moment associated with Io could be made [Krimigis et al., 1979a; Southwood et al., 1980; Ip, 1981; Kivelson and Southwood, 1981].

4.5. Recapitulation and open questions

We now review those properties of the magnetosphere environment that were established by the observations (Secs. 4.2 and 4.4) and identify areas where open questions still exist. We find it helpful to view the results in the context of the model presented in the summary Figure 4.4.

It has been established that most of the plasma in the outer magnetosphere is hot (~ 20 to 40 keV), and that it more-or-less corotates with the planet all the way to the magnetopause on the dayside; on the nightside, the hot plasma suddenly breaks into an antisunward flow inside the magnetopause at a distance from ~ 130 to $\sim 160\ R_J$, which continues to large distances past the downstream bow shock. Further, the plasma is hotter at the magnetic equator than off it.

The ion energy spectrum (≥ 30 keV) is characterized by a thermal part described well by a convected Maxwellian, and a nonthermal component that can be fit by a power law in energy, with the transition point between the two components at ~ 150 to 250 keV. The composition of the nonthermal component is primarily H, He, O, and S. This composition is also consistent with the angular distributions measured at lower (< 200 keV) energies. In fact, by expressing the spectra in terms of energy/charge all particle species display identical spectral forms, suggesting an electric field acceleration process. Using this procedure, it is possible to deduce the relative densities and charge states of the source plasma: O^{+2}, S^{+3}, C^{+6}, and relative densities of hydrogen to heavier ($A > 1$) ions of $\sim 1:1$. Finally, the electron spectrum below ~ 200 keV displays a thermal distribution not unlike that of the ions.

The energetic ions ($E \geq 30$ keV) appear to dominate the plasma pressure inward to at least $\sim 10\ R_J$ and they thus determine the statics and dynamics of the magnetosphere. These ions are responsible for the inflation of the dayside magnetosphere to about twice the distance expected on the basis of the observed dipole magnetic moment. This interaction with the solar wind is different from that at Earth or any of the planets investigated so far. Because of the high-beta plasma (thermal energy density exceeding magnetic energy density), we expect that the shape of the magnetopause is not likely to be smooth and could exhibit "bulges" owing to variations between solar wind and Jovian plasma pressure. There is evidence from low-energy ion streaming events that Jovian plasma escapes frequently into the upstream and downstream solar-wind regions.

It was theoretically expected even before the Pioneer flyby [Ioannidis and Brice, 1971] that the transport of ionospheric plasma in the Jovian magnetosphere might be predominantly outward, in contrast to the Earth where it is predominantly inward. Mestel [1968] pointed out that a cold plasma can break out into radially outflowing wind when the corotational energy density approximates the magnetic energy density at the critical point of the flow (Alfvén radius). Michel and Sturrock [1974] proposed that the Jovian internal plasma sources would centrifugally drive a "planetary wind" similar to the solar wind. This idea was expanded on by Coroniti and Kennel [1977] and Kennel and Coroniti [1977b] who retained the features of an axially symmetric outflow interacting with the solar wind. Hill, Dessler, and Michel [1974] took the view that the solar wind would block the outflow on the dayside and the wind would only flow out the tail.

The observations have brought out a number of serious contradictions to these early expectations. The basis for the models is that the opening distance occurs when the cold (ionospheric) plasma corotational energy density equals the magnetic field energy density, resulting in pre-Voyager predictions for the opening radius from ~ 30 to $\sim 70\ R_J$. Using the actually observed cold plasma densities and composition at Io (Chap. 3) in these models, reduces the opening distance to $\sim 10\ R_J$ [see, for example, discussion in Kennel and Coroniti, 1979]. The distance where plasma outflow begins was observed by Voyager to be in the range of ~ 130 to $\sim 160\ R_J$ on the nightside; some evidence exists that there is magnetospheric plasma escape upstream of Jupiter as well (see Fig. 4.3). In addition, all theories assumed the plasma to be cold at the opening radius, but the observations give temperatures of ~ 30 keV. Finally, the models assumed initially that the plasma source is the ionosphere, rather than the satellites. The observations also show (see Fig. 4.28) that the composition in the wind is similar to that of the magnetosphere inside $\sim 30\ R_J$, rather than the magnetodisc, as the models suggest. For these reasons, the plasma outflow discovered by Voyager was named "magnetospheric wind" [Krimigis et al., 1979b] rather than planetary wind. It is evident that the wind phenomenon modeled for the magnetosphere of Jupiter by a direct application of Mestel's [1968] stellar wind concept does not correspond to the Voyager findings. The early theoretical models, although valuable in expanding our perspective, need to be reviewed critically, in view of the observations (see Chaps. 10 and 11 for additional discussion).

The magnetospheric wind apparently represents the principal loss process of hot plasma from the magnetosphere. Krimigis et al. [1981] estimate the ion loss rate at $\sim 2 \times 10^{27}\ s^{-1}$ and the energy loss rate at $\sim 2 \times 10^{13}$ W. These compare with supply rates of $\sim 3 \times 10^{28}\ s^{-1}$ (dominated by Io and, in the case of protons, by Jupiter), and $\sim 10^{15}$ W (dominated by Jupiter) [Krimigis et al., 1981; Chap. 10].

Despite the overall consistency of the results and interpretations, we must keep in mind that our conclusions are of necessity based on only the all too brief Pioneer and Voyager flybys through a small part of a very large, heterogeneous and dynamic magnetosphere. Some of the results are probably more valid than others. For example, the hot plasma is probably characeristic of the magnetosphere at all local times, as is the rather exotic composition. The magnetospheric wind region, on the other hand, has only been sampled twice in the ~ 0300 to 0500 local time regions, and could be drastically different at other parts of the nightside magnetosphere, such as local midnight or dusk. It is unlikely that the plasma circulation pattern on the nightside is as simple as has been drawn in Figure 4.4.

There exist other areas of uncertainty, some of which will probably be addressed by more detailed studies of the combined Voyager data sets, whereas others must await the more definitive measurements to be obtained by the Galileo spacecraft such as the spectral region between the PLS and LECP measurements (~ 6 to ~ 28 keV for ions, and ~ 6 to ~ 14 keV for electrons), although no major surprises should be expected there. It is likely that the temperature and number density of the plasma are highly variable, and there may be substantial variations in the relative concentrations of the hot and cold components. Three-dimensional flux measurements should help illuminate the overall circulation pattern, which could only be inferred indirectly by the Voyager

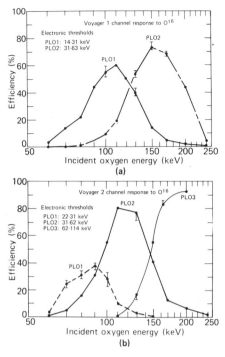

Fig. 4.31. Efficiency functions for the detection of ^{16}O nuclei in the Voyager 1 and 2 LEMPA detectors [Krimigis et al., 1981]. The reduced efficiencies for the PL01 and PL02 channels are a result of the relatively large FWHM of the deposited energy in the detector (~ 16 keV) compared to the width of the channels.

measurements. We certainly hope that, as the data analysis proceeds, additional features of essential significance to the physics of the Jovian magnetosphere will emerge that will allow the construction of more sophisticated models, the predictions of which can be checked by future spacecraft or ground-based observations.

APPENDIX A: INSTRUMENT RESPONSE

For all ion channels of LEMPA, a sufficiently accurate expression for the contribution from ion species "a" to the rate $R_k^a(\mathbf{n})$ of the kth energy channel when the detector looks in the direction of the unit vector $\mathbf{n}$ is

$$R_k^a(\mathbf{n}) = G \int_0^\infty dE \, \epsilon_k^a(E) \, j^a(E, -\mathbf{n}) \tag{4.17}$$

where $j^a(E, \hat{\mathbf{p}})$ is the incident undirectional differential particle flux for species "a", and $\epsilon_k^a(E)$ is the detector efficiency function for the kth energy passband, averaged over the detector area $\Delta\sigma$ and solid angle $\Delta\omega$. The "geometric factor" G is the integral of $|\hat{\mathbf{p}} \cdot \mathbf{n}|$ over the same area $\Delta\sigma$ and solid angle $\Delta\omega$. Efficiency curves $\epsilon_k^a(E)$ for the response to ^{16}O in the lowest ion channels on Voyagers 1 and 2 are shown in Figure 4.31 and additional response curves for ^{4}He, ^{32}S, and ^{40}Ar may be found in Krimigis et al. [1981]. For very steep or highly peaked spectra, the energy dependence of the channel efficiencies is an important consideration, and the use of a "band-passed" efficiency $\bar{\epsilon}_k^a$ in order to write

$$R_k^a \simeq G \, \bar{\epsilon}_k^a \, j_k^a \, \Delta E_k^a \tag{4.18}$$

where j_k^a is the differential flux centered within an appropriately chosen channel passband ΔE_k^a, must be considered a rough approximation. These approximate channel passbands for the LEMPA may be deduced from curves such as shown for ^{16}O in Figure 4.31.

Of course – and here is our main challenge – the rate of kth channel is due to the sum of the rates produced by all the species present, so the measured quantity is the rate

$$R_k(\mathbf{n}) = G \sum_a \int_0^\infty dE \, \epsilon_k^a(E) \, j^a(E, -\mathbf{n}) \tag{4.19}$$

APPENDIX B: IMPROVED SPECTRAL MODELS

When calculating the contribution of a given ion species to a particular energy channel, the most useful quantity is the integral flux $J(E,\phi)$ above an energy E in the scan direction ϕ (because the incident flux in a finite energy band with uniform efficiency can be deduced from this function). For a convected thermal spectrum, it is easily seen from the formulas (4.11) and (4.12) that $J(E,\phi) = Nv \, F(kT/E, V/v,\phi)$, where N is the density of thermal density and v is the particle speed corresponding to energy E. The function F is then a "universal" function of only the dimensionless ratios kT/E (thermal energy normalized by particle energy) and V/v (bulk velocity normalized by particle velocity). Consequently only a single computation of the universal function F over its three dimensionless variables is required for a theoretical estimate of the directional integral flux due to streaming ions with a Maxwellian distribution. For any particular species, energy, temperature, and bulk velocity, one simply evaluates the dimensionless variables kT/E and V/v to obtain the integral flux $J = NvF$.

The universal function $F(x,y,\phi)$ is presented in Figure 4.32a as a logarithmic contour plot for $\phi = 0$ and $\phi = \pi/2$ (integral fluxes parallel and transverse to the flow direction). The contours reveal the extreme "front-to-side" anisotropy ratios $J(E,O)/J(E,\pi/2)$ expected from a convected Maxwellian for $kT/E \leq 1$ and $V/v \leq 0.1$, that is, particle energies well above the thermal energy and particle velocities approaching the order of the bulk velocity. This suggests that the particle anisotropy would be strongly affected by any nonthermal tail on the energy spectrum, particularly in the directions $\phi \gtrsim 90°$ for $E > kT$ and $v \sim V$.

Nonthermal tails in plasma accelerated from substrate thermal distributions are apparently intrinsic to ion acceleration processes upstream of the Earth's bow shock [Lin, Meng, and Anderson, 1974], in the Earth's magnetotail [Sarris, Krimigis, and Armstrong, 1976], on the Sun [Zwickl et al., 1978], and in corotating shock structures in the solar wind [Gloeckler, Hovestadt, and Fisk, 1979; Gosling et al., 1981]. They are also evident in the Jovian ion spectra presented by Krimigis et al. [1979a,b], as can be seen from Figures 4.9 and 4.14. Consequently, when analyzing the LEMPA measurements of convective plasma in the magnetodisc, we face the compound problem of an unknown spectral form and an unknown ion composition.

Therefore we developed a model spectrum incorporating a very simple parametrization of a nonthermal tail called a "γ-thermal" distribution which has been applied successfully to flowing plasma in the earth's magnetosphere by Sarris et. al., 1981. In the plasma frame, we set the differential flux $\tilde{j} \propto \tilde{E} \exp(-\tilde{E}/kT)$ for $\tilde{E} \leq E_1$, and $\tilde{j} \propto (\tilde{E}/E_1)^{-\gamma}$ for $\tilde{E} \geq E_1$, where $\gamma + 1 = E_1/kT$. This spectrum makes the transition from a

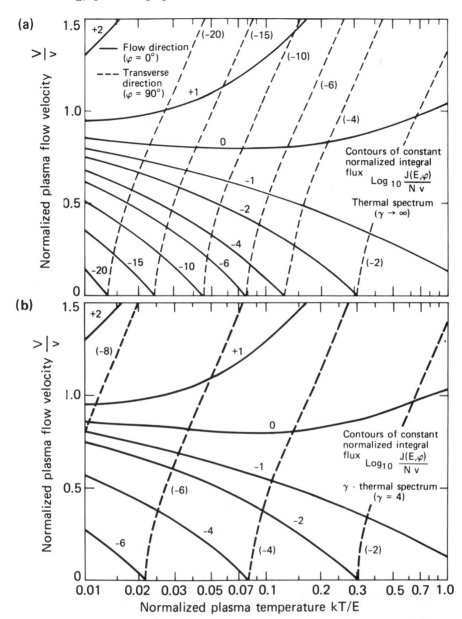

Fig. 4.32. (a) Logarithmic contours of integral particle flux of convected thermal particles J normalized by Nv (total density times velocity), as a function of temperature (kT) normalized by particle energy (E), and convection velocity (V) normalized by particle velocity (v). Solid line contours are for integral ion fluxes parallel to convection ($\phi = 0$), and dashed are for those transverse to convection ($\phi = \pi/2$). The front-to-side ratio $J(E,0)/J(E, \pi/2)$ may be read off at intersection of solid and dashed curves. (b) Same representation as (a), but for a "gamma-thermal" spectrum ($\gamma = 4$), such as shown in Figure 4.33.

Maxwellian to a power law in the plasma frame at energy $\tilde{E} = E_1 = (\gamma + 1)kT$ with no discontinuity in intensity or slope. Several sample spectra are shown in Figure 4.33 and should be compared with observed spectra which can be fit with the γ-thermal form, as

Fig. 4.33. The "γ-thermal" model differential flux spectrum, which makes a continuous transition from a thermal distribution with temperature T to a nonthermal tail described by a power law $E^{-\gamma}$ at energy $E = (\gamma + 1)\,kT$.

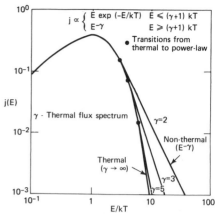

shown in Figures 4.9 and 4.14. This functional form has significant computational advantages over other functions (such as the "kappa" distribution), when comparison with energetic particle measurements is required. One can still define a "universal" function $F(x,y,\phi,\gamma) = J(E,\phi,\gamma)/Nv$ with $x = kT/E$ and $y = V/v$, but F now depends on γ as well. The effect of the nonthermal tail on an integral flux plot like that of Figure 4.32a (one of which can be done for each value of γ), is to shift drastically the positions of the contours in the lower left-hand corner of the plot ($kT/E < 0.1$, $V/v > 0.7$). For example, we show in Figure 4.32b a logarithmic plot for the universal function $F(x,y,\phi,\gamma)$ with $\gamma = 4$. Comparing with Figure 4.32a, the contour $J/Nv = 10^{-20}$ in the extreme corner of the thermal plot becomes $J/Nv = 10^{-7}$ for the γ-thermal distribution. This parametrization is quite useful for the LECP instrument, because the power-law indices are readily determined from the higher energy LEPT channels where protons, alphas, and medium nuclei are separately identified.

APPENDIX C: MEASUREMENT OF DENSITY AND PRESSURE IN THE PRESENCE OF CONVECTION

The composition of the hot Jovian plasma is one of the most interesting of the scientific questions raised by the LECP observations. However, it is precisely the composition that must be assumed for any mathematical modeling of the LEMPA response. Consequently, an alternative, but complementary approach was taken where lower bounds on density (N), pressure (P), and convection velocity (V) could be obtained from the general properties of distribution functions summarized by (4.3) and (4.4) without invoking parametrized models.

Density. By definition, the density of species "a" is

$$N^a = \int d^3 p\, f^a = \int_0^\infty \frac{dE}{v} \int_{4\pi} d^2\omega\, j^a(E,\hat{\mathbf{p}}) \tag{4.20}$$

This looks tantalizingly like (4.19), except for the inverse power of v that occurs in (4.20) but not in (4.19). This factor causes real difficulty, because, when we try to sum

(4.20) over species to get the total density, v is smaller (for a given energy) for the heavier ions. Also, (4.20) involves fluxes in all directions whereas our rates are only measured in the scan plane. In strong convective flows, the anisotropies may be quite strong (particularly, as we shall see, for the heavy-ion species). However, if the scan-plane measurements happen to contain the direction of convective flow, these usually provide sufficient information to reconstruct the omnidirectional average needed for (4.20). Let us therefore introduce into (4.19) a "scaling velocity" v_k for the kth channel to try to handle the troublesome factor of v. We then form the sum

$$\sum_k \sum_n \frac{2\pi\Delta\mu_n}{Gv_k} R_k(\mathbf{n}) = \sum_a \int_0^\infty \frac{dE}{v} \sum_n \frac{2\pi\Delta\mu_n}{G} \tag{4.21}$$

$$\left[\sum_k \frac{v}{v_k} \epsilon_k^a(E) \right] j^a(E, -\mathbf{n})$$

By $\sum_n$ we mean the sum over the sector orientations $\mathbf{n}$ of the detector scanning in azumith, and $2\pi\,\Delta\mu_n$ is the representative solid angle assigned to the detector in the nth position, $\theta = 0$ being assigned to the direction in the scan plane closest to the convection velocity (i.e., within the peak sector).

Comparison of (4.20), summed over all species, with (4.21) reveals that if we chose the v_k such that, for all energy ranges of the dominant species over which $j^a(E)$ makes a significant contribution to the energy integral,

$$\sum_k \epsilon_k^a(E) \frac{v}{v_k} < 1 \quad \text{(dominant ``a'')} \tag{4.22}$$

then we obtain a lower bound on the density from (4.21)

$$N > \sum_n \sum_k \frac{2\pi\Delta\mu_n}{Gv_k} R_k(\mathbf{n}) \tag{4.23}$$

The inequality (4.22) can be satisfied trivially by choosing a sufficiently large v_k, but the larger the value of v_k, the less information there is in the lower bound given by (4.23). Also, reference to Figure 4.31 reveals the importance of the channel efficiency functions ϵ_k^a in identifying the dominant species in a given channel. Krimigis et al. [1979] used a form close to (4.23) that replaced the weighted sum over $\mathbf{n}$ by 4π times the scan averaged rates for each channel. By using v_k as the velocity of protons at the center of each energy band, they argued that they obtained a lower bound on N because the protons had the highest velocity among the expected dominant species present in the outer magnetosphere (see Sec. 4.4 and Figs. 4.20, 4.21, and 4.22).

These same authors also argued that an upper bound on the density could similarly be obtained from the RHS of (4.23) by choosing the v_k to be the midchannel ^{16}O velocity, assuming that this low velocity would reverse the inequality. This procedure will yield a good upper bound only if there are no significant fluxes of ions at energies below the LEMPA ion channel passbands because these would contribute to N^a in (4.20) but would not appear in (4.23). However, it was the lower bound (4.23), as it was evaluated by Krimigis et al. [1979b] with v_k chosen for ^1H, which demonstrated that the bulk of the hot magnetodisc plasma is in the tens of keV temperature range (see Sec. 4.4).

Pressure. There is another approach that leads to an independent estimate of density, but more importantly, to an estimate of pressure which is minimally affected by the strong anisotropies produced by the presence of substantial numbers of heavy ions. Figure 4.11b suggests, intuitively, that j^a and $\tilde{j}^a$ should be roughly comparable when the detector direction $\mathbf{n}$ is transverse to the convection velocity. Examination of the figure shows that this approximation will underestimate the contributions of the heavier ions, but we are interested only in a lower bound on the pressure. If we write $\mathbf{v} \cdot \mathbf{V} = vV\cos\theta$, then for $\theta = \pi/2$, $\tilde{v}^2 = v^2 + V^2$ and $\tilde{\theta} = \tan^{-1} v/V$. Thus, a transverse measurement of j^a is related to $\tilde{j}^a$ evaluated at a higher energy $\tilde{E} = (m_a/2)(v^2 + V^2)$ and at an angle $\tilde{\theta}$ different from $\pi/2$. The ratio of the energies is $\tilde{E}/E = 1 + V^2/v^2$, and this is near unity if $V << v$, in which case $\tilde{\theta} \simeq \pi/2 + V/v$. If we have reason to believe that the flux $\tilde{j}^a$ is not strongly anisotropic in the plasma frame, this permits the use of the transverse flux to estimate density and pressure in that frame [Lanzerotti et al., 1980].

First, we write the expression for the average scalar pressure, $\tilde{P}^a$, which must be computed in the frame where the bulk velocity $\mathbf{V}^a = 0$ [Spitzer, 1962]:

$$\tilde{P}^a = \frac{m_a}{3} \int_0^\infty d\tilde{E}\, \tilde{v} \int_{4\pi} d^2\tilde{\omega}\, \tilde{j}^a(\tilde{E}, \tilde{\mathbf{p}}) \qquad (4.24)$$

Applying the transformation (4.4) for the transverse direction $\hat{\mathbf{p}} = (-\mathbf{n}_\perp)$ gives

$$\tilde{j}^a(\tilde{E}, \hat{\mathbf{p}})\tilde{v} = \left(1 + \frac{V_a^2}{v^2}\right)^{3/2} j^a(E, -\mathbf{n}_\perp)v \qquad (4.25)$$

By fixing the direction $\hat{\mathbf{p}} = -\mathbf{n}_\perp$, the vector $\tilde{\mathbf{p}}$ becomes a function of E through the transformation $\tilde{\mathbf{p}} = -m_a(\mathbf{n}_\perp v + \mathbf{V}_a)$. However, in the approximation that $\tilde{j}^a$ is nearly isotropic, this $\tilde{\mathbf{p}}$ dependence has little effect; otherwise it would greatly increase the complexity of the calculation. From the energy transformation $\tilde{E} = E + m_a V_a^2/2$ we have $d\tilde{E} = dE$, so the range of integration $0 < E < \infty$ corresponds to $m_a V_a^2/2 < \tilde{E} < \infty$ because particles with $\mathbf{v} = (-\mathbf{n}_\perp)v \to 0$ in the spacecraft frame still have $\tilde{\mathbf{v}} \to -\mathbf{V}_a \neq 0$ in the plasma frame. Consequently a detector looking in the transverse direction in the spacecraft frame is not responding to energies $\tilde{E} < E_a = m_a V_a^2/2$ in the plasma frame. Thus, under the weak-anisotropy approximation $\tilde{j}^a(\tilde{E}, \hat{\mathbf{p}}) \simeq \tilde{j}^a(\tilde{E})$.

$$\overline{P}^a = \frac{4\pi m_a}{3} \int_0^{E_a} d\tilde{E}\, \tilde{v}\, \tilde{j}^a(\tilde{E}) + \frac{4\pi m_a}{3} \int_0^\infty dE\, v \left(1 + \frac{V_a^2}{v^2}\right)^{3/2} j^a(E, -\mathbf{n}_\perp) \quad (4.26)$$

Using an approach similar to (4.21), we form the sum

$$\sum_k \frac{p_k}{3G} R_k(\mathbf{n}_\perp) = \sum_a \frac{m_a}{3} \int_0^\infty dE\, v \left[\sum_k \epsilon_k^a(E) \frac{p_k}{m_a v} \left(1 + \frac{V_a^2}{v^2}\right)^{-3/2} \right] \qquad (4.27)$$

$$\left(1 + \frac{V_a^2}{v^2}\right)^{3/2} j^a(E, -\mathbf{n}_\perp)$$

Thus, if a scaling momentum p_k can be found for each channel k such that

$$\sum_k \epsilon_k^a(E) \frac{p_k}{m_a v} \left(1 + \frac{V_a^2}{v^2}\right)^{-3/2} < 1 \quad \text{(dominant "}a\text{")} \tag{4.28}$$

for all energies in which there is significant transverse flux for the dominant ion species, then the LHS of (4.27) exceeds the second term on the RHS of (4.26), and we have a lower bound on the average scalar pressure

$$\bar{P} > \Delta P + \sum_k \frac{4\pi p_k}{3G} R_k(\mathbf{n}_\perp) \tag{4.29}$$

where ΔP represents the sum over all species of the first integral on the RHS of (4.26), that is, the average partial pressures for energies $\tilde{E}^a$ in the plasma frame below the translation energy $m_a V_a^2/2$.

The lower bound on $\bar{P}$ used by Krimigis et al. [1981] and Lanzerotti et al. [1980] was obtained with $\Delta P = 0$, because they did not estimate the integral contribution ΔP (for $E < E_a$) which appears explicitly in (4.29) (see Sec. 4.4 and Figs. 4.18 through 4.20). As with the number density, no upper bound is offered here for the pressure because that would require a priori knowledge of the contributions of ion fluxes at energies below the lowest LEMPA channel.

By an identical line of reasoning, we can obtain an analogous lower bound on the number density because $\tilde{N}^a = N^a$.

$$N > \Delta N + 4\pi \sum_k R_k(\mathbf{n}_\perp)/G v_k \tag{4.30}$$

where ΔN represents the summed partial densities ($\tilde{E}^a < m_a V_a^2/2$). The lower bound is valid if, for all energies in which there is significant transverse flux for the dominant ion species, v_k satisfies

$$\sum_k \epsilon_k^a(E) \frac{v}{v_k} \left(1 + \frac{V_a^2}{v^2}\right)^{-1/2} < 1 \quad \text{(dominant "}a\text{")} \tag{4.31}$$

Thus, (4.30) with condition (4.31) provides an estimate of the lower bound on the density, independently from our other estimate (4.23), and with a different scaling condition than (4.22).

If only the sums on the RHS of (4.23) or (4.30) and (4.29) are used, they yield conservative lower bounds for N and $\bar{P}$ because $\Delta N > 0$ and $\Delta P > 0$. All bounds always exist because the conditions (4.28) or (4.22) and (4.31) can always be satisfied for large enough v_k or small enough p_k. Note that v_k and p_k are just scaling parameters that should be chosen so as to maximize the lower bounds although still satisfying the parameter conditions involving $\epsilon_k^a(E)$ for the significant energy ranges of all dominant ion species. They need not be related to any particular physical set (v_a, $m_a v_a$) for a given ion species, although association of the scaling velocities and momenta with those of the fastest ions actually being measured often provides valid choices.

ACKNOWLEDGMENTS

We thank the many persons at APL/JHU, the Universities of Maryland and Kansas, and Bell Laboratories for their efforts in the design, fabrication, and implementation of the LECP experiment. We are especially grateful to the LECP coinvestigators, T. P. Armstrong, W. I. Axford, C. O. Bostrom, C. Y. Fan, G. Gloeckler, E. P. Keath, and L. J. Lanzerotti, for their contributions to the hardware and analysis efforts, and to our coauthors in previous publications, J. F. Carbary, D. C. Hamilton, and R. D. Zwickl, whose efforts were essential to the overall success of the data reduction and analysis. The efforts of Voyager project personnel at JPL and at NASA Headquarters were essential to the success of the LECP program. This analysis has been supported at APL/JHU by NASA under Task I of Contract N00024-78-C-5384 between The Johns Hopkins University and the Department of the Navy and by NASA Grant NAGW-154 to The Johns Hopkins University.

5

HIGH-ENERGY PARTICLES

A. W. Schardt and C. K. Goertz

In the Jovian magnetosphere, electrons, protons, and heavier ions are accelerated to energies well above 10 MeV. These energetic particles constitute a valuable diagnostic tool for studying magnetospheric processes and produce the Jovian radio emissions. In the inner magnetosphere, both the electron and proton fluxes with energies above 1 MeV build up to $\sim 10^8$ per cm^2 s and constitute a major radiation hazard to spacecraft passing through this region. Surprisingly, high fluxes of energetic oxygen and sulfur (>7 MeV/nuc) are also found in the inner magnetosphere. Of particular interest are the interactions of these particles with the inner Jovian moons and with the Io plasma torus. Throughout much of the middle magnetosphere and magnetospheric tail, highest fluxes are found in the plasma sheet, which coincides closely with the tilted dipole equator out to 45 R_J (Jupiter radii). This plasma sheet has not been identified beyond 45 R_J in the subsolar hemisphere; however, on the night side, it extends to 200 R_J. On the day side, fluxes near the equator are relatively independent of distance (15 to 45 R_J) and fall into the range 10^4 to 10^5 per cm^2 s each for protons and electrons above ~ 1 MeV. In the predawn direction, proton and electron fluxes decrease by three orders of magnitude from 20 to 90 R_J (10^5 to 10^2 per cm^2 s) and then remain relatively constant to the boundary layer near the magnetopause. In the middle and outer magnetosphere, particle fluxes change rapidly (2 to 10 min.), and it appears that the energetic particle population is subject to a number of different dynamic processes. Jupiter is a strong source of interplanetary electrons and $\sim 10^{14}$ W are required to energize these electrons. The electron flux above 2.5 MeV in interplanetary space and in large regions of the outer magnetosphere is modulated by Jupiter's rotation period. This modulation has been ascribed to the interaction of the solar wind with a rotating longitudinal asymmetry within the Jovian magnetosphere.

5.1. Introduction

Jupiter, as the biggest planet in the solar system, also has the largest magnetosphere. In many respects this magnetosphere is quite different from that of the Earth:

1. The energy required to drive the Jovian magnetosphere is apparently extracted from Jupiter's rotational energy rather than from the solar wind as is the case for the equivalent terrestrial system. In this sense, Jupiter, as a rapid rotator (Chap. 10), may serve as a laboratory for the study of pulsar physics.
2. The Jovian magnetosphere contains several satellites and a ring that not only absorb energetic particles, but are also a source of plasma that significantly affects the structure and dynamics of the Jovian radiation belts.
3. Jupiter is a strong source of energetic charged particles that can be detected as far away as the orbit of Mercury and that carry away energy at a rate of several million megawatts (much more than the present energy consumption of terrestrial civilization). Studying Jupiter may, hence, throw some light on astrophysical acceleration processes.

On the other hand, similarities also exist with the terrestrial magnetosphere. In fact, it is surprising how successfully concepts derived from the study of the Earth's magnetosphere have been applied.

Whereas fundamental magnetospheric properties are obtained from magnetic field and plasma data, energetic particles yield essential additional information. For the purpose of this chapter, energetic particles are defined as particles with an energy significantly higher than the "temperature" of the plasma. At Jupiter, this temperature may be as high as several tens of keV. Thus, particles with energies in excess of 0.5 MeV are mainly dealt with in this chapter. The lower-energy particles are described in Chapters 3 and 4. The energy density represented by the energetic particles is negligible compared to the energy densities of the plasma and magnetic field. Thus, they do not themselves affect the magnetic field, but are test particles that trace out field lines and, as such, yield information about the overall magnetic field topology. Energetic particles result from acceleration processes, and their characteristics help us to find the energization mechanisms that are active in the Jovian magnetosphere.

It has become traditional to divide the Jovian magnetosphere into distinct spatial regions, defined either by their special magnetic field properties, particle distributions or combinations thereof. The terms "inner," "middle," and "outer magnetosphere" are generally accepted (Fig. 5.1a). Usually, the boundaries between these are placed at 10–20 R_J and 30–45R_J, respectively.

There is little evidence for a well-organized current sheet throughout the front side of the outer magnetosphere, and no strong periodic modulation of the proton population has been observed, but an apparent 10-hr clocklike modulation of >6-MeV electrons was found. The azimuthal drift velocity is highly variable, being some of the time (but not always) lower than that expected from corotation.

The dawnside outer magnetosphere is characterized by a well-developed thin current sheet and open field lines relatively close to the current sheet. In many ways it resembles the Earth's magnetotail. Very close to the magnetopause it becomes similar in character to the front-side outer magnetosphere. Evidence for rapid local acceleration has been found in the taillike region. It is very likely that the energetic particles produced in the Jovian magnetosphere are released through the tail into interplanetary space.

The middle magnetosphere is characterized by the presence of a distinct azimuthal current sheet which becomes increasingly important at larger radial distances. The local-time asymmetry is less pronounced than in the outer region. The subsolar current sheet is much thicker than in the dawn region. On the day side, the orientation of the current sheet is predominantly rotationally controlled, whereas in the dawn magnetotail the current sheet evolves at larger distances into the more solar-wind-controlled configuration. In this middle region, field-aligned proton streaming has been observed. The proton phase space density changes little with distance, indicating that this region may be a source for these particles. Acceleration by magnetic pumping (Chap. 10) presumably occurs here. Evidence for the recirculation model (Chap. 10) is also strongest here. The azimuthal drift velocity is generally lower than that expected from corotation.

The inner magnetosphere ($<15\ R_J$) is the best understood region. Whereas the properties of the outer magnetosphere depend sensitively on local time, very little of such a dependence is discernible in the inner magnetosphere. Except for the presence of the Galilean satellites, this region resembles the radiation belts of the earth. Energetic particles are transported inward by radial diffusion, gaining energy through adiabatic compression effect. They are pitch-angle scattered by electromagnetic waves and absorbed by the satellites and ring. As the electrons diffuse closer to the planet and, consequently, become more energetic, they begin to radiate significant amounts of synchrotron radiation, the decimetric radio emission observed at Earth. Particles scattered into the atmosphere cause auroral activity.

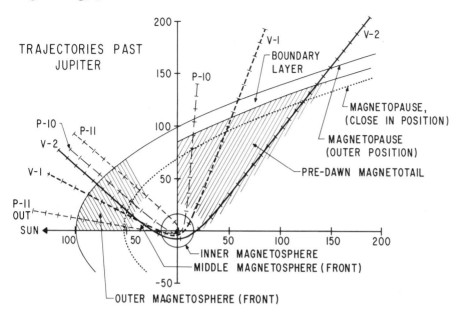

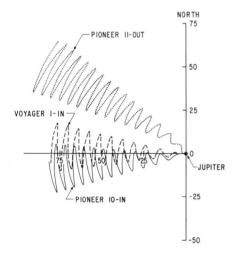

Fig. 5.1. Jupiter encounter trajectories: (a) Distance from Jupiter plotted at its local time or azimuth relative to the Sun–Jupiter line. Pioneer 10 and 11 trajectories are designated with P-10 and P-11; similarly, the Voyager 1 and 2 trajectories by V-1 and V-2. Tick marks indicate the start of a new day, and the two magnetopause positions shown correspond to high and low solar wind pressure; (b) Magnetic dipole coordinates of representative Pioneer and Voyager trajectories.

The data described here come from four flybys whose trajectories are shown in Figure 5.1a. The local-time coverage in the middle and outer magnetospheres is incomplete, extending only from 2:00 to 14:00 (Fig. 5.1a). For energetic particles, the magnetic latitude of the spacecraft is the other important parameter. Due to the offset between the spin axis and the magnetic dipole axis, the dipole latitude (Fig. 5.1b) oscillates around the geographic latitude by about ±10° every 9 hr 55 min. Because the actual magnetic field deviates from a dipole in the middle and outer magnetospheres, Figure 5.1b represents only a first approximation. Particle flux modulations with this period are primarily due to latitudinal effects; however, a given latitude is encountered only at two longitudes and only once at the extremes. Therefore, the available data do not permit a unique separation between latitudinal and longitudinal effects. These limitations are unavoidable with flybys and can only be removed by a Jupiter orbiter.

Of all the properties of the Jovian magnetosphere, the characteristics of the energetic particles have been measured most extensively. Table 5.1 lists the instru-

Table 5.1. *Energetic particle detectors*

Group	Designation	Particle sensitivity	Energy range (MeV)	Spacecraft
University of Chicago Simpson et al., 1974a	ECD	Electrons Protons	> 3 > 30	Pioneer 10,11
	Fission Cell	Protons Electrons α Heavy Nuclei	> 35	Pioneer 10,11
	LET	Protons He^{++}	0.54 – 8.8 0.30 – 5.0	Pioneer 10,11
University of Iowa Van Allen et al., 1974a	GNA	Electrons Protons	> 5 > 30	Pioneer 10,11
	GMB	Electrons Protons	> 0.55 > 6.6	Pioneer 10,11
	GMC	Electrons Protons	> 21 > 77.5	Pioneer 10,11
	GMD	Electrons Protons	> 31 > 77.5	Pioneer 10,11
	GMG G	Electrons Protons	> 0.06 0.61 – 3.4	Pioneer 10 Pioneer 11

Institution	Detector	Particle	Energy	Units	Spacecraft
University of California at San Diego (UCSD) Fillius and McIlwain, 1974a	C1	Electrons	> 6 > 5		Pioneer 10 Pioneer 11
	C2	Electrons	> 9 > 8		Pioneer 10 Pioneer 11
	C3	Electrons	> 13 > 12		Pioneer 10 Pioneer 11
	M1	Electrons	> 35		Pioneer 10,11
	M2	Background	> 0.85		Pioneer 10,11
	M3	Protons	> 80		Pioneer 10,11
GSFC/University of New Hampshire Trainor et al., 1974	HET	Electrons Protons and He^{++} Medium Nuclei	2.1 – 8.0 20 – 500 40 – 120	MeV MeV/nucl. MeV/nucl.	Pioneer 10,11
	LET I	Protons Protons and He^{++} Medium Nuclei	0.4 – 3 3 – 21 6 – 40	MeV MeV/Nucl. MeV/nucl.	Pioneer 10,11
	LET II	Electrons Protons plus Ions Protons	0.05 – 2.1 0.2 – 2.1 3.2 – 21	MeV MeV MeV	Pioneer 10,11

Table 5.1. *(cont.)*

Group	Designation	Particle sensitivity	Energy range (MeV)	Spacecraft
CALTECH/GSFC Cosmic Ray System (CRS) Stone et al., 1977	LET	Protons and He^{++} Medium Nuclei	0.43 – 8 MeV 7 MeV/nucl.	Voyager 1,2
	HET	Electrons	0.1 – 12 MeV	Voyaber 1,2
	TET	Electrons	> 0.6 MeV	Voyager 1,2
Applied Physics Lab (APL) Low-energy Charged Particle Krimigis et al., 1977	LECR	Electrons Ions (mainly concerned with low-energy particles)	0.014 – 10 MeV 0.030 – 150 MeV	Voyager 1,2

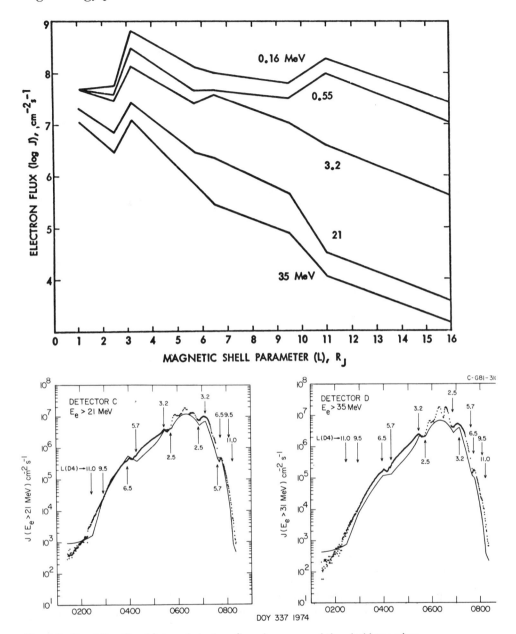

Fig. 5.2. Omnidirectional integral electron flux above several threshold energies: (*a*) Radial dependence according to the Divine (1976) model (reprinted with the permission of Divine [1976] and the Jet Propulsion Laboratory). (*b*) and (*c*) Comparison between the Divine (1976) model, solid line, and observations by the University of Iowa detector, dots [from Thomsen, 1979].

ments that have provided the data on which this chapter is based. For details of the instruments, for example, their ability to distinguish species and resolve pitch angle distribution and time variations, we refer the reader to the original literature.

In view of the accessibility of the original literature (see Preface), and in the interest of readability, we will not refer to all original papers but use their results freely.

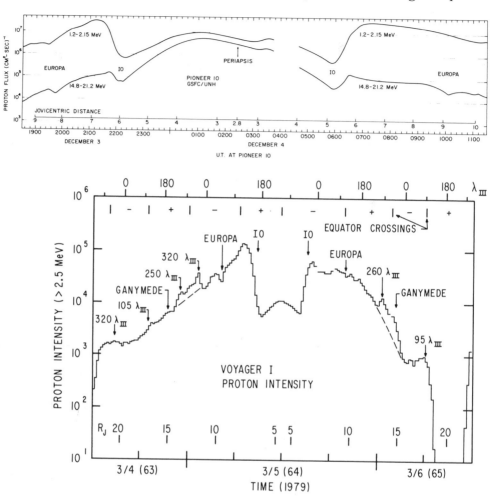

Fig. 5.3. (*a*) Omnidirectional, integral proton flux observed with the Goddard space Flight Center/University of New Hampshire (GSFC/UNH) experiment on Pioneer 10, showing the intensity variations at the orbits of Io and Europa [from Trainor et al., 1974]. (*b*) Relative proton intensities (16-min. averages) observed in the inner magnetosphere with the Voyager 1 experiment of the California Institute of Technology/Goddard Space Flight Center (CIT/GSFC). Owing to large deadtime corrections, the magnitude of the drop at Io is uncertain, and the detector threshold increased with counting rate. Equatorial crossings are indicated with tick marks and the + indicates when Voyager 1 was in the northern hemisphere.

5.2. Inner magnetosphere ($R < 15\ R_J$)

In discussing the structure and dynamics of the energetic particle distribution, there are four important questions to be answered:

1. How many particles are there?
2. How are they transported?
3. What are the losses they suffer as they are transported?
4. What, finally, is their source?

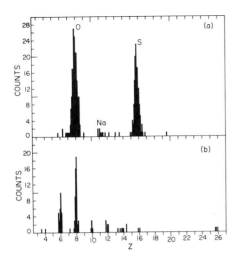

Fig. 5.4. Elemental (Z) distribution of heavy nuclei observed with Voyager 1 in the Jovian environment (CIT/GSFC experiment). The energy was ≥ 7 MeV per nucleon: (*a*) Within 5.8 R_J inside Io's orbit, (*b*) Outside 11 R_J [From Vogt et al., 1979a].

Observations

The first question has been fairly well answered by observations obtained during the four flybys. Although the list of instruments from which measurements are available is quite impressive (Table 5.1), few of these instruments provided unambiguous data in the very intense radiation levels of the inner magnetosphere. There are problems with penetrating background radiation, decreased efficiency and dead-time uncertainties. Particle species identification and absolute intensity levels are, thus, not always completely reliable; however, the experimenters have, to a large extent, corrected for these problems and, due to the energy overlap of different instruments, a fairly coherent picture has emerged.

A few years after the Pioneer 10 and 11 encounters, several authors [Divine, 1976, unpublished, 1978; Mihalov, 1977] combined the data from different instruments into time-stationary models of the distribution of energetic particles. Figure 5.2*a* shows the empirical model produced by Divine. Figures 5.2*b* and 5.2*c* show the comparison of the model with observed omnidirectional intensities of electrons with energies $E > 21$ MeV and $E > 31$ MeV. Clearly, for these high energies, the model fits the data quite well; however, the fit at lower energies is much less impressive. This discrepancy is probably due to larger time-variability of low-energy fluxes. It seems that the success of the time-stationary Divine and Mihalov models in fitting the energetic electron data indicates that there exists a steady long-term average electron distribution about which temporal fluctuations occur. The amplitude of the fluctuations decreases with increasing energy. Although no detailed comparisons of Voyager 1 and 2 data with the Divine or Mihalov models are available as yet, it appears that the general flux level of electrons above 1 MeV did not change significantly between the Pioneer and Voyager encounters. This is an agreement with the observed long-term stability of decimetric radiation.

Figures 5.3*a* and *b* show the radial variation of the intensity of energetic ions in the inner region obtained from different instruments on different flybys. Whereas the Pioneer data were always interpreted in terms of protons being the dominant ion, it has become apparent since the Voyager encounters that energetic sulfur and oxygen ions are also quite significant. Figure 5.4 shows the masses of heavy ions with energies ≥ 7 MeV per nucleon. They reflect the solar composition in the outer magnetosphere, but only oxygen, sulfur and, to a lesser degree, sodium are present in the inner region

[Vogt et al., 1979a]. (See Chap. 4 for a discussion of ion fluxes with energies below 7 MeV/nuc.) Krimigis [1979, private communication] has pointed out that the Pioneer data may be due in part to heavier ions and that the absolute intensity levels of "protons," as derived from Pioneer data, are difficult to obtain.

Figures 5.2 and 5.3 display similarities, but also significant differences. Both ions and electrons increase in intensity with decreasing distance. This has been interpreted as a signature of inward radial diffusion. For both electrons and ions the energy spectrum becomes harder with decreasing radial distance. This may also be a result of radial diffusion. The ions are much more affected by the satellite Io and its plasma torus than the energetic electrons. (This is not necessarily true for lower-energy particles.) This difference has been interpreted as being the result of reduced access of energetic electrons to Io and enhanced pitch angle scattering of ions in the plasma torus (see section on Satellite Interactions). Near Europa, the electron and ion signatures are more similar because of the absence of a plasma torus.

Our knowledge of the pitch angle distributions of the energetic particles in the inner region is less complete than our knowledge of their spatial and energy distributions, in part because the response of a number of the high-energy detectors is omnidirectional and the overlap in energy coverage is reduced. Conflicting results have been published regarding the nature of the proton pitch angle distribution outside Io's orbit. The University of Chicago group [Simpson et al., 1974a] report dumbbell proton distributions (peaked at $\sim 0°$ pitch angle) outside of $L = 6$, which change dramatically to pancake distributions (peaked at $\sim 90°$ pitch angles) at Io's orbit. Other instruments on Pioneers 10 and 11 did not see dumbbell distributions in the region outside of Io's orbit, but rather pancake distributions. A pancake distribution was also observed with Voyager [Lanzerotti et al., 1981]. However, the pancake distributions do become more pronounced inside of Io's orbit. Note that this is true only for the energetic particles discussed here. For lower-energy ions, the reverse is true. Lower-energy ions are depleted near 90° inside Io's orbit, which could be due to charge exchange [Lanzerotti et al., 1981].

The electron pitch angle distributions are pancake throughout the inner magnetosphere, with the pancake nature becoming more pronounced with decreasing radial distance. At the orbit of Io, the trend towards more sharply peaked distributions is halted. Inside of Io's orbit, the sharpness of the pancake distribution of $E_e > 9$-MeV electrons remains constant. There is even some indication for a decrease at the orbit of Io [Fillius, 1976]. This is not an unexpected signature for absorption of charged particles by a moon.

Transport and losses in the inner magnetosphere

Before the flybys of Pioneer 10 and 11, the theoretical discussion of the structure of Jupiter's (inner) magnetosphere centered on the question of diffusive transport and on the probable loss mechanisms that would operate on the trapped energetic particles. It was recognized that a solar wind-driven convection would not extend into the inner regions of the Jovian magnetosphere and that radial diffusion was proposed as the most likely transport mechanism [Brice and McDonough, 1973]. Regarding loss mechanisms, considerable attention was paid to the role of wave-particle interactions as a mechanism for causing pitch angle scattering of particles into the loss cone [Thorne and Coroniti, 1972].

The Pioneer flybys obtained experimental evidence from the energetic particle data to support the theoretical expectations that pitch angle diffusion significantly affects the energetic charged particles in the inner magnetosphere, and the intensity and frequency spectrum of waves responsible for the pitch angle diffusion were estimated [Sentman and Goertz, 1978]. These waves have now been observed in situ by plasma wave instruments on board Voyager 1 and 2 (Chaps. 8 and 12). It was also realized that the particles precipitating into the Jovian ionosphere would deposit a significant amount of energy and could give rise to a number of possible effects, including ionospheric heating, plasma loading of the magnetosphere due to the production of secondary electrons that can escape the gravitational potential barrier, and optical and x-ray emissions. Both optical (auroral type) and x-ray emission from the Jovian ionosphere have recently been found (Chaps. 2 and 6). Auroral emission has been observed to come preferentially from the foot of the magnetic field lines threading the Io plasma torus. It has been argued by several authors that there should be enhanced wave activity and, consequently, enhanced pitch-angle scattering into the atmospheric loss cone in the Io torus. Such an enhanced wave activity is, indeed, observed (Chaps. 8 and 12). Considering the potential complexity of the problem, the list of successful predictions is quite impressive.

The theoretical model, on the basis of which many of the predictions were made, had been developed to explain the structure of the earth's radiation belts. It is gratifying, indeed, to find that the model is also applicable to the Jovian magnetosphere. The starting point of the analysis is the equation for radial diffusion in the magnetic field of the planet with conservation of the first and second adiabatic invariants

$$\frac{\partial f}{\partial t} = L^2 \frac{\partial}{\partial L} \left(\frac{D_{LL}}{L^2} \frac{\partial f}{\partial L} \right) - \mathscr{L} + S \tag{5.1}$$

In Equation (5.1), f is the phase space density of particles with first and second invariant μ and J_2; it is related to the "measured" differential unidirectional particle flux per cm^2 sr s, by the expression

$$f(\mu, J_2, L) = \frac{dN}{d^3 \times d^3 p} = \frac{j(E, \delta_0, L)}{p^2} \tag{5.2}$$

where

$$E = E(\mu, J_2, L) = \text{particle kinetic energy} \tag{5.3a}$$

$$\delta_0 = \delta_0(\mu, J_2, L) = \text{equatorial pitch angle} \tag{5.3b}$$

and

$$p^2 c^2 = E^2 + 2E m_0 c^2 \tag{5.3c}$$

The variable L is the normal L-shell parameter which in the complex magnetic field of Jupiter depends on the particle's energy and pitch angle. D_{LL} is the diffusion coefficient. Usually one assumes a parametric form $D_{LL} = D_0 L^n$. The symbols $\mathscr{L}$ and S represent loss and source terms, and the assumption of steady state ($\partial f / \partial t = 0$) is made, which is probably valid for high-energy particles in the inner magnetosphere. The loss and source terms are often combined and written as

$$\mathscr{L} - S = f / \tau \tag{5.4}$$

where $\tau(\mu, J_2, L)$ is a characteristic lifetime of particles. This combined loss and source term (note that τ may be negative if sources exceed losses) is comprised of a number of possible processes: pitch-angle diffusion, satellite sweep up, satellite injection, interparticle collisions, synchrotron radiation, etc. It is not clear that each physical process can be represented by a form like Equation (5.4); however, the obvious mathematical advantages of Equation (5.4) have generally outweighed such doubts. Combining Equations (5.4) and (5.1) and assuming that the particle population does not change ($\partial f/\partial t = 0$), we arrive at what has become known as the "lossy radial diffusion model,"

$$L^2 \frac{\partial}{\partial L} \left(\frac{D_{LL}}{L^2} \frac{\partial f}{\partial L} \right) - f/\tau = 0 \tag{5.5}$$

The analysis of the particle data in the framework of this model has proceeded in two ways: One either assumes a loss rate and determines the diffusion coefficient or vice versa.

Satellite interactions

An example of the first method is the analysis of the particle flux profile as the spacecraft crosses the orbit of a Jovian moon [for details see Mogro-Campero, 1976, and Thomsen, Goertz, and Van Allen, 1977a,b]. The characteristic loss time for absorption by a moon sweeping the region between L_1 and L_2 is given by

$$\tau = \frac{3}{2} \frac{t_s(\Delta L/r_m)}{\alpha_L \, \alpha_D} \tag{5.6}$$

where ΔL is the range of L values covered by the moon's orbit and r_m is its effective radius (Chap. 11). The value of ΔL depends on the eccentricity of the satellite's orbit, the magnetic field topology and the particle's energy and pitch angle. A complete treatment must take into account factors such as L-shell splitting in order to evaluate the L values along the moon's orbit in a nondipolar magnetic field. The time t_s is the time between two successive encounters of a particle with the moon. If ω_C, ω_D, and ω_m are the angular velocities of corotation, of magnetic curvature and gradient drifts, and of the moon, respectively, we have

$$t_s = \frac{2\pi}{\omega_C - \omega_m \pm \omega_D} \tag{5.7}$$

The plus sign is for positive particles and the minus sign is for electrons. For the inner moons like Io, ω_C is equal to Jupiter's angular velocity. If a particle's mirror latitude is less than the moon's magnetic latitude, it can escape absorption. The factor α_L is the fraction of a moon's orbit where this is the case.

If the drift velocity and bounce period of the particles are such that in one-half bounce period they would drift more than one moon's diameter relative to the satellite, then some particles may escape absorption by "leapfrogging" over the satellite. This effect is represented by α_D, which is the fraction of particles in a given flux tube which are absorbed in one encounter of the moon with the flux tube [for details see Thomsen, Goertz, and Van Allen, 1977a,b].

The effective satellite radius r_m depends on electric and magnetic field perturbations in the vicinity of the satellite. τ may be poorly known because it is not easy to determine r_m a priori. Analysis of the energy dependence of τ for different models of the electric fields in Io's vicinity [Thomsen, 1979] shows that little can be said with certainty about τ and, hence, about D. Published estimates of τ and D are based on modeling the moons with spherical absorbers of infinite resistivity. In this model r_m is equal to the particle's cyclotron radius plus the moon's radius; however, the reader should keep in mind the drastic assumptions underlying the model. In that case, one finds for protons of $\mu = 1.7$ MeV/G at $L \simeq 6$, a value for $\tau = 1.1 \times 10^6$s. Outside the range from L_1 to L_2, the losses are usually assumed to be small, that is,

$$0 \quad L < L_1 \qquad \text{region I}$$

$$\frac{1}{\tau} = \frac{1}{\tau} \quad L_1 \leq L \leq L_2 \quad \text{region II}$$

$$0 \quad L > L_2 \qquad \text{region III}$$

Requiring continuity of f and its derivative at L_1 and L_2, one can solve Equation (5.1) and fit the solution to the observed phase space density profiles in the vicinity of a moon. The L dependence of D_{LL} near the moon's orbit may be neglected.

Figure 5.5a shows examples of such calculations at the moon Io ($L = 5.9$). We see that a reasonable fit can be obtained; however, the proton data require that $D = 2.6 \times 10^{-8} R_J^2 \, s^{-1}$, while the electron data are consistent with $D = 4 \times 10^{-7} R_J^2 \, s^{-1}$. The difference is unexpected on theoretical grounds. The rapid change in proton phase-space density extends over a larger radial range than specified as Region II in Figure 5.5a, and a better fit is obtained for a larger sweeping region as shown in Figure 5.5b. Because the sweeping range $5.6 \leq L \leq 6.4$ is much larger than the range covered by Io, the losses of protons may not be entirely due to absorption by Io but also due to pitch angle scattering in a toruslike region of enhanced plasma density (Chap. 3). Using a minimum scattering lifetime, Thomsen, Goertz, and Van Allen [1977a] estimated an upper limit for the diffusion coefficient in Io's torus of $D = 1.5 \times 10^{-7} R_J^2 \, s^{-1}$. For a combination of pitch angle scattering and Io absorption, the diffusion coefficient is $D = 6 \times 10^{-7} R_J^2 \, s^{-1}$ in good agreement with the value obtained from the electrons. Other estimates for D range from $10^{-9} R_J^2 \, s^{-1}$ for protons to $10^{-6} R_J^2 \, s^{-1}$ for electrons [Simpson et al., 1974a]. Mogro-Campero and Fillius [1976] found $D = 6 \times 10^{-8} R_J^2 \, s^{-1}$ or $D = 3 - 4 \times 10^{-7} R_J^2 \, s^{-1}$, depending on the analysis method they used.

To obtain a value for the exponent n in the variation of $D_{LL} = D_o L^n$, one can go through the same exercise at other moons, for example, Europa. This indicates that n is probably less than 6, most likely even less than 4. One can also use the shape of the phase space density profile in Regions I and III to determine n. From the proton data, one obtains $n = 2.3 \pm 0.5$, which is more consistent with diffusion coefficients theoretically derived by assuming that the radial diffusion is driven by ionospheric dynamo fields than by assuming fluctuating solar wind-driven convection electric and magnetic fields. The reader must, however, be cautioned against an uncritical acceptance of this widely stated conclusion. The theoretically expected values of $n = 3, 7-10$ and $\simeq 10$ for the respective processes are all based on ad hoc assumptions concerning the fluctuation spectrum and its radial dependence. A recent determination of n using Voyager 1 and 2 data of >0.5-MeV ions by Armstrong et al. [1981] does not yield a low value of n but indicates a best value of $n \sim 7.5$. However, only 15 data points were

Fig. 5.5. Phase space density profile of protons with $N = 1.7$ MeV/G and $K = 12.2\ G^{1/2}\ R_J$ for the Pioneer 11 inbound trajectory (University of Iowa experiment). The heavy line shows the phase space density derived from observations, and the light lines show solutions as a function of the parameter $D\tau$ (in units of R_J^2). (*a*) The loss region is taken to be $5.8 \geq L \geq 6.2$. (*b*) An extended loss region is used, $5.6 \geq L \geq 6.4$ [from Thomsen et al., 1977a].

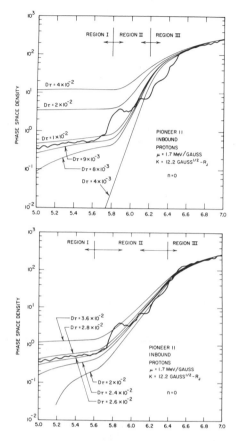

used; furthermore, the experimental data do not allow for an accurate determination of n because all loss processes are not properly known.

Using these estimates for the diffusion coefficient, one can fit phase space density profiles in regions remote from moons and obtain lifetimes τ, that is, loss rates. All authors have concluded that for energetic electrons the lifetime is not infinite, thus, local losses do occur. The total loss rate of energetic electrons ($E_e > 200$ keV) between $L = 10$ and $L = 3$ has been estimated by Thomsen and Sentman [1979] as 10^{24} particles/s. These particles, if precipitating into the Jovian atmosphere, would constitute a considerable amount of heating, namely a few tenths of a mW m^{-2}. The cause of these losses has been suggested to be pitch-angle scattering by resonant interaction with electromagnetic waves in the whistler mode. Indeed, such waves have been observed in the Jovian magnetosphere. For more detail about wave-particle interaction in the Jovian magnetosphere, we refer the reader to Chapter 9.

Electron synchrotron radiation

Inside $L \simeq 3$, loss due to synchrotron radiation becomes important. In fact, one can use the well known equations for synchrotron radiation to calculate τ in this region [see e.g., Birmingham et al., 1974]. The estimates for the diffusion coefficient obtained from balancing radial diffusion with synchrotron losses (observed decimetric, DIM, radiation from Jupiter) agree very well with the ones quoted above. Furthermore, the

DIM radiation intensities calculated from observed electron fluxes agree with observed values, confirming quite nicely the synchrotron radiation theory of DIM.

The observations of DIM are described in Chapter 7. DIM consists of two components: the thermal radiation from Jupiter's atmosphere and the nonthermal, or synchrotron, radiation. It is this second component which can be used to deduce the characteristics of the energetic particles and magnetic fields in the innermost region of the Jovian magnetosphere ($r \leq 3 R_J$).

Recently, de Pater [1981a,b] completed an elaborate model calculation for Jupiter's synchrotron radiation. The full multipole magnetic field configuration is used, as well as a realistic electron distribution in energy and pitch angle (see Chap. 7). Outside Amalthea's orbit, her particle distribution agrees well with Pioneer observations; however, inside this orbit, where no comprehensive data are available, the pitch angle distribution is considerably sharpened towards 90° pitch angles because only these particles can escape absorption by Amalthea. A similar effect occurs across the L shell of the newly discovered ring. These effects account for the observed degree of circular polarization and the sharp increase in the intensity of the radiation peaks near the ring.

Whereas the overall agreement between the calculated radiation parameters and the observed ones is good, certain observed features cannot be accounted for by the model. The conspicuous "hot spots" of the DIM radiation at longitudes $\lambda_{III} = 225° \pm 10°$ require a relative overabundance of energetic electrons at longitudes 240° to 360° and perhaps a dusk to dawn electric field in the inner magnetosphere. Conceivably, our knowledge of Jupiter's magnetic field is still inadequate to reproduce this detail; however, it should also be noted that longitudinal asymmetries are quite common in the Jovian magnetosphere and form the basis of what is commonly known as the "magnetic anomaly model" (see Chap. 10). It is quite likely that many diverse effects, including corotating convection driven by a surface magnetic anomaly, combine to cause specific asymmetries.

We summarize this section by stating that energetic particles in the inner magnetosphere are transported radially inward by a diffusive process. It seems that the diffusion is driven by ionospheric dynamo fields produced by winds in the atmosphere of Jupiter. The particles are subject to strong losses resulting from pitch angle scattering into the atmospheric loss cone by resonant cyclotron interaction with electromagnetic waves. The particles are also absorbed by the moons Europa and Io, although the efficiency of that process is poorly known. Inside $L \simeq 3$ synchrotron radiation (see Chap. 7) becomes an important loss mechanism for energetic electrons.

5.3. The subsolar hemisphere

The outer and middle magnetosphere show considerable variations with local time. The fact that there are differences, to be described below, in particle intensities, pitch-angle distributions, energy spectra, and modulation phase and amplitude cannot be stressed enough. Such differences indicate that the external influence of the solar wind is important for the outer and middle magnetosphere. Models that neglect azimuthal effects cannot claim to be realistic global models although they may work well over a limited range of local time. (Most quantitative models proposed to date apply only to a limited region of the magnetosphere. In light of our severely restricted local time coverage, global models must still be regarded as speculative.)

For energetic particles, the transition from the inner to the middle magnetosphere occurs at about 15 R_J because, beyond that distance, the magnetic field distortion affects the trapped energetic particles in a major way. The plasma sheet responsible for

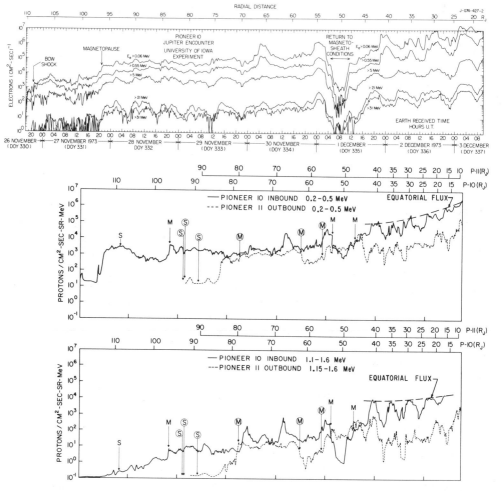

Fig. 5.6. Omnidirectional proton and electron fluxes observed with Pioneer: (*a*) Electron fluxes between 20 and 110 R_J observed on the inbound pass of Pioneer 10 with the University of Iowa Experiment [from Baker and Van Allen, 1976]. (*b*) Differential proton fluxes from the Pioneer 10-in and Pioneer 11-out passes. The bow shock and magnetopause crossings are indicated by arrows labeled S and M, respectively, with those for Pioneer 11 enclosed in a circle [from McDonald, Schardt, and Trainor, 1979; GSFC/UNH experiment].

the field distortion has been identified to only about 45 R_J in the subsolar hemisphere. Beyond 45 R_J, the plasma sheet is either absent or very thick in this hemisphere and cannot be identified by its magnetic field signature. The size of this region, called the outer magnetosphere, depends sensitively on solar wind pressure. The outer magnetosphere extends from 45 R_J to the magnetopause, which has been observed between the extremes of 46 and 97 R_J.

Particle fluxes in the subsolar magnetosphere

Electron and proton fluxes in the front magnetosphere are shown in Figures 5.6, 5.7, and 5.8. Generally, several magnetopause crossings are observed (Fig. 5.6*b*) as the size of the magnetosphere changes in response to variations in solar wind pressure. The

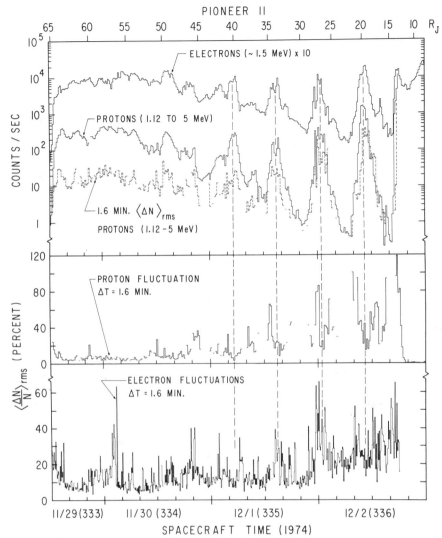

Fig. 5.7. Sixteen-minute averages of proton and electron intensities observed during the inbound pass of Pioneer 11 (GSFC/UNH experiment). Also given are the rms fluctuations, averaged over 16 min., of counting rates observed 1.6 min. apart. Fluctuations are shown only for periods during which the χ^2 test indicates a $>95\%$ probability that the fluctuations are not due to counting statistics.

average flux has little radial dependence; however, in the outer magnetosphere, fluxes of particles below 2 MeV have an irregular dependence on time and appear to respond primarily to changes in interplanetary conditions. In the middle magnetosphere, the wobble of the plasma sheet produces a strong periodic modulation, which is particularly well illustrated in Figure 5.7. As may be seen in Figure 5.1*b*, the Pioneer spacecraft entered the plasma sheet only once per Jovian period (~ 10 h) while the low-latitude Voyagers 1 and 2 crossed the plasma sheet twice. Accordingly, the Pioneer fluxes (Figs. 5.6 and 5.7) were modulated with a 10-hr period and the Voyager fluxes with a 5-hr period (Fig 5.8). Also, the Pioneer modulation is much deeper than observed with Voyager because of the larger latitude excursion (Fig. 5.1*b*). As may be

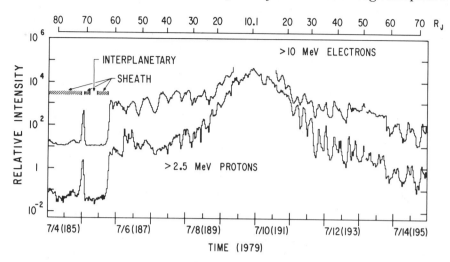

Fig. 5.8. Electron and proton intensities (32-min. averages) observed during the Voyager 2 encounter [from Vogt et al., 1979b; CIT/GSFC experiment].

seen in Figures 5.6*a* and 5.8, the 10-hr modulation of > 5-MeV electrons persists in the outer magnetosphere. This modulation is due to a different mechanism than the one in the middle magnetosphere and will be discussed further as "clock" modulation.

Superimposed on the average fluxes are large-intensity fluctuations on a minute timescale. These fluctuations are illustrated in Figure 5.7 in terms of their rms values. Except for special periods, the percentage fluctuations became quite small (< 3%) in the inner magnetosphere. Examples of larger fluctuations in the inner magnetosphere occurred near satellite orbits, for example, Ganymede [Burlaga, Belcher, and Ness, 1980; Connerney, Acuña, and Ness, 1981]. The existence of such rapid flux changes is indicative of dynamic processes in the middle and outer magnetospheres. Because little correlation exists between electron and proton fluctuations, different or spatially separate processes must affect the two types of populations.

Middle magnetosphere

The plasma sheet, which occurs near the magnetic dipole equator, has a dominant influence on the particle population in the middle magnetosphere. The approximately 10° tilt of the dipole relative to the Jovian spin axis caused the magnetic latitude of Pioneer 10 to oscillate between $+1°$ and $-19°$ and, hence, between regions of stronger and weaker particle fluxes. The Pioneer 11-out observations between 20 and 45 R_J extend the magnetic latitude range from 24° to 44°. As might be expected, the maxima in the Pioneer 11-out proton flux (Fig. 5.6*b*) are just slightly less than the minima in the Pioneer 10 flux; however, no such match was observed in the electron flux. The latitudes of Voyagers 1 and 2-inbound oscillate between $+7°$ and $-13°$.

The approximate value of flux vs. distance at the equator is indicated by the dashed line in Figure 5.6*b*. It should be remembered, however, that Pioneer 10 inbound sampled the equatorial flux at only one system III (1965) longitude near $\lambda_{III} = 200°$. A longitudinal asymmetry in the energetic particle flux is a possible consequence of the magnetic-anomaly model (see Chaps. 10 and 11). Both the Voyager 1 and Pioneer 10-in data are consistent with a maximum flux in the active hemisphere. Voyager 1 crossed the magnetic equator twice each Jovian period near $\lambda_{III} = 100°$ and 310°, and

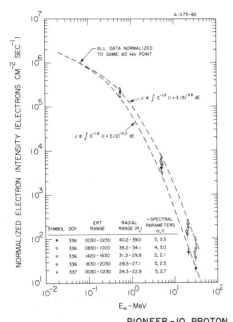

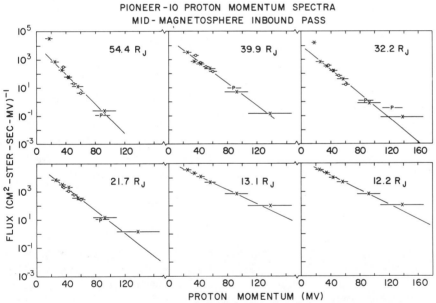

Fig. 5.9. (*a*) Electron spectra in the middle magnetosphere, 22–40 R_J, observed with the University of Iowa experiment on Pioneer 10-in. See Equation (5.8) for meaning of *H* and *n* [from Baker and Van Allen, 1976]. (*b*) Proton momentum spectra (1-hr averages) from the inbound pass of Pioneer 10 [from McDonald, Schardt, and Trainor, 1979; GSFC/UNH experiment].

a flux enhancement of 30% to 70% was observed at the 310° crossing [Vogt et al., 1979a]. The maxima in the Pioneer 10-in proton flux occurred one to two hours after the equatorial crossing, based on magnetometer data, also indicating a longitudinal component to the flux modulation. However, no longitudinal dependence was found in the Voyager 2 data (Fig. 5.8). A longitudinal effect may not always be found because

Fig. 5.10. Geometry of angular
distribution measurements. The circle
illustrates the path of the detector as the
spacecraft (Pioneer) or as the scan
platform (Voyager) turns about the y
axis. The directions shown for the
magnetic field and corotation velocity
are typical of the Pioneer inbound
passes.

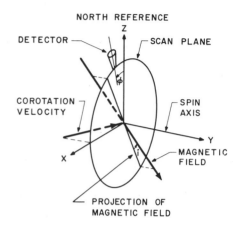

any longitudinal dependence in particle source strength is averaged out by curvature
and gradient drifts, provided particles remain trapped for several drift periods.

The differential spectra of energetic particles do not obey a simple power law in
energy, but become steeper at higher energies. Below 0.5 MeV, the electron spectrum is
an exponential in energy (Chap. 4), above 0.5 MeV it (Fig. 5.9a) can be represented by

$$j_e(E) = C E^{-1.5} \left(1 + \frac{E}{H}\right)^{-n} \tag{5.8}$$

where C, H, and n are free parameters [Baker and Van Allen, 1976; for a different
representation, see McIlwain and Fillius, 1975]. In the middle magnetosphere, $5 \leq H
\leq 35$ MeV and $2 \leq n \leq 5.5$. The off-equatorial spectra are somewhat softer than the
equatorial spectra. As discussed in Chapter 4, the differential energy distribution of
protons below 0.5 MeV is well described by a Maxwellian spectrum plus a high energy
tail that follow a power-law distribution. However, proton spectra above 0.5 MeV are
best expressed with an exponential dependence on momentum (Fig. 5.9b)

$$j(E) = \frac{C}{\sqrt{E}} e^{-\sqrt{E/E_o}} \tag{5.9}$$

where E_o corresponds to the e-folding momentum. This momentum falls into the range
from 7 to 14 MV, or E_o equals 25 to 100 keV [McDonald, Schardt, and Trainor, 1979].
As in the case of electrons, the hardest spectra occur at the magnetic equator and there
is little dependence of E_o on radial distance. Departures from the average shape of the
proton spectrum have been observed and will be discussed later. Ion spectra have been
observed primarily at energies below 0.5 MeV/nuc [Hamilton et al., 1981] and are
discussed in Chapters 3 and 4.

Information about the magnetosphere can be derived from the angular distribution
of the energetic particles. Angular distributions at Jupiter have been observed in a scan
plane. During the Pioneer passes, this plane was nearly perpendicular to the ecliptic
plane, it was generally parallel during the Voyager passes (Figs. 4.1 and 4.2). The orientation of the magnetic field, **B**, relative to the plane affects the type of information that
can be derived (Fig. 5.10). If **B** makes an angle i with the plane, then only pitch angles
in the range i to 180-i are sampled, and each pitch angle is sampled at two gyrophases.
The angular distributions have been subjected to Fourier analysis in the form

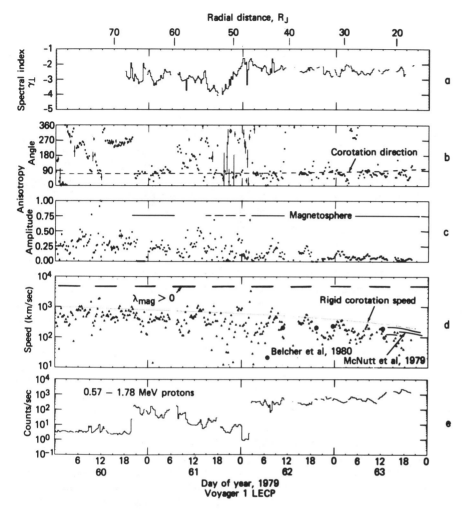

Fig. 5.11. Plasma speed derived from Compton–Getting effect of 0.57- to 1.78-MeV protons observed during the inbound pass of Voyager 1 with the Applied Physics Laboratory/University of Maryland Experiment: (*a*) Index of power law fit, $E^{-\gamma}$, to proton spectra; (*b*) Magnitude of first-order anisotropy; periods when the spacecraft was in the magnetosphere are identified with the solid line; (*c*) Direction of first-order anisotropy; the lines show the corotation and radial outflow directions; (*d*) Plasma speed with the speed for rigid corotation shown by the light line. Periods during which the spacecraft was in the northern magnetic hemisphere are given by bars along the top. Selected values of the plasma speed derived from the plasma experiment are shown by heavy dots [Belcher, Goertz, and Bridge, 1980; McNutt et al., 1979]; (*e*) Counting rate of 0.57- to 1.78-MeV protons [from Carbary et al., 1981].

$$j(v,\phi) = A_o(v) \left[1 + A_1/A_o \cos(\phi - \phi_1) + A_2/A_o \cos 2(\phi - \phi_2) + \ldots\right] \quad (5.10)$$

In this expression, $j(v,\phi)$ is the flux of particles with a given velocity entering the detector from the direction ϕ (Fig. 5.10). $A_o(v)$ is the spin averaged flux and A_n is the nth order anisotropy. The even terms ($n = 2, 4, \ldots$) reflect primarily the pitch angle distribution for trapped particles. The odd terms reflect the effects of intensity gradients, of electric fields or the equivalent motion of the reference frame, and of field-aligned flow of the particle population.

The first-order anisotropy (A_1, ϕ_1) has been used to determine whether the magnetosphere corotates with the planet and to observe field-aligned particle flow. The former quantity can be measured most easily if the scan plane is perpendicular to $\overline{B}$. In that case

$$A_1/A_0 \simeq \left[\frac{1}{A_0}\frac{\partial A_0}{\partial v} - \frac{2}{v} \right] \overline{\phi}_1 \cdot \overline{v}_c + \rho_g \overline{\phi}_1 \cdot \frac{\nabla A_0}{A_0} \times \frac{\overline{B}}{B} \qquad (5.11)$$

where v is the particle velocity, $\overline{v}_c$ is the corotation or $\overline{E} \times \overline{B}$ velocity, $\overline{\phi}_1$ is a unit vector in the direction of the first-order anisotropy, ρ_g is the gyroradius, and $\overline{B}$ the magnetic field. A number of terms have been dropped which are normally small in a dipole field [Birmingham and Northrop, 1979]. The first term in Equation (5.11) is the Compton–Getting effect, which permits a determination of the local plasma velocity, provided the gradient effect given by the second term can be evaluated.

In principle, a more accurate determination of the local plasma velocity is possible from the angular distribution of low energy particles, because the anisotropies are larger and the gyroradii (gradient effects) are smaller. These data have shown (Chap. 4) that the plasma velocity is generally parallel to the corotation direction. However, the magnitude of the velocity was derived mostly from high-energy data (Fig. 5.11) because the low energy ion composition, hence, ion velocity, could be determined accurately at only a few places (Chaps. 3 and 4). The middle magnetosphere moves in the general direction of corotation much of the time, but significant departures in magnitude (and sometimes direction) occur beyond 20 R_J (see also McNutt, Belcher, and Bridge [1981]). First-order anisotropies in energetic electron distributions are negligibly small because, at equal energies, electrons have a much higher speed and smaller gyroradii than protons. The question of corotation is further discussed in Chapters 3, 4, and 11.

Angular distributions of protons in the middle magnetosphere have also been observed when the magnetic field made a small angle relative to the scan plane. In that case, the first-order anisotropy contains an additional term that is proportional to particle flow along the magnetic field. This has been interpreted in terms of an actual transport of protons away from the equator [McDonald, Schardt, and Trainor, 1979; Northrop, Birmingham, and Schardt, 1979] and a short residence time [Northrop, 1979]. However, these observations were made during periods when the magnetic field was swept back and had a large component parallel to the $\Omega \times \overline{r}$ corotation velocity, and, therefore, at least a fraction of the field-aligned velocity does not represent an actual transport of particles. The surprising discovery of large fluxes of energetic oxygen and sulfur ions [Krimigis et al., 1979a] in the middle magnetosphere raises the possibility that the combined response to protons plus heavy ions may account for much of the first-order anisotropies and that any actual transport of particles is below previous estimates.

Proton pitch-angle distributions, as reflected by the second-order anisotropy, are pancake near the equator. Such a distribution would be a consequence of the magnetic-pumping model suggested by Goertz [1978]. The distribution is often sufficiently flat to be near or just beyond the onset of the mirror instability [Northrop and Schardt, 1980] which redistributes pitch angles, as is required by the magnetic-pumping model.

Energetic-electron pitch angle distributions near the equator are almost isotropic between 25 and 45 R_J. The A_2/A_0 term is only a few percent, giving a slightly pancake shape. Between 20 and 25 R_J, however, the electron distributions are strongly field aligned (that is, dumbbell) as would be expected from the recirculation model

(Chap. 10). In this model, electrons diffuse at a high latitude (small pitch angles) from the inner magnetosphere to field lines that extend into the middle magnetosphere.

A major effort has gone into defining the dynamic processes in the middle magnetosphere and the degree to which the three adiabatic invariants [Northrop, 1963] are conserved. One approach is to use Liouville's theorem and adiabatic theory to correlate particle distribution functions, and, hence, particle fluxes at different points along a field line. This permits calculating the intensity along a field line by using the equatorial distribution function and the magnetic field configuration. If stable trapping conditions exist in the middle magnetosphere, the particle energy is constant along a field line and pitch-angle scattering can be ignored. Under these conditions, the flux at a pitch angle δ is the same as the flux at a pitch angle δ_2 at a different point along the same field line, provided the pitch angles are related by

$$\sin \delta = \sqrt{B/B_2} \, \sin \delta_2 \tag{5.12}$$

Usually one relates the flux to the equatorial values; in this case, δ_2 and B_2 are equatorial pitch-angles and field magnitudes. Consequently, the intensity of an isotropic distribution is invariant along a field line, and a pancake distribution of the form $C(1 + \sin^2 \delta_0)$ at the equator will transform into $C(1 + B_0/B \sin^2 \delta)$ at a point where the field strength has increased from B_0 to B. Calculations using a local field model developed for this purpose have been performed for 0.5- and 1.3-MeV electrons, and for 0.5- and 1.9-MeV protons in the region from 20 to 30 R_J [Schardt and Birmingham, 1979]. The electron flux modulation observed during the Pioneer 10-inbound pass was closely predicted by this picture; however, the predicted proton flux modulation is much smaller than observed. Observed fluxes about 8 R_J below the magnetic equator are only 1/5 to 1/3 as large as predicted by this model.

Three different factors may cause or contribute to this discrepancy.

1. A flux change by a factor of 3 to 5 with System III longitude could produce the effect; however, there is no evidence to support such a large longitudinal dependence.
2. The proton intensity depends on local time (see section entitled Predawn magnetosphere). The magnetic field lines through an observation point 8 R_J below the magnetic equator cross the equator at an earlier local time. The field model used to fit the Pioneer 10 data in this region [Schardt and Birmingham, 1979] gives a difference of only 10° or 40 min. in local time; however, the available field data do not permit a unique solution. Based on the change in flux with local time (Pioneer 10 inbound vs. outbound), a difference of 2 to 3 hr would probably be required to account for the observations.
3. Some of the assumptions about stable trapping may be invalid. Extensive electromagnetic and plasma wave activity were observed in the middle magnetosphere (Chap. 8); however, specific waves that could account for the energetic particle observations have not been identified.

At times, proton spectra changed rapidly owing to preferential acceleration between 1 and 3 MeV. Figure 5.12 shows successive spectra taken 15 min. apart at about 37.5 R_J when the spacecraft was well below the magnetic equator. The 0345 spectrum is a typical exponential in momentum. Thirty minutes later, a distinct peak appeared near 2 MeV which then relaxed to a plateau between 1 and 3 MeV. This plateau was observable for about two hours. Essentially no change in the slope of the spectrum was

Fig. 5.12. Proton spectra (15-min. average) observed by Pioneer 10-in between 37.3 and 37.8 R_J on December 2, 1973. Note the enhanced flux between 1.5 and 3 MeV at 4:15 spacecraft time. The data were taken with the GSFC/UNH instrument. The solid circles and P's represent LET-I data and the crosses LET-II data (see Table 5.1).

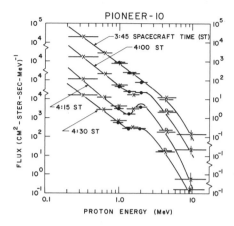

seen at lower or higher energies, and only a small change was observed in the total flux. A probable explanation is resonant proton acceleration over a limited energy range by a plasma wave. Unfortunately, the absence of simultaneous wave measurements prevented the identification of the specific mechanism.

Large, rapid changes in proton flux provide additional strong evidence for particle acceleration. These events were first described by Simpson et al. [1975; see also Simpson and McKibben, 1976]. An outstanding example of such a flux change was a tenfold increase in the 0.5- to 5-MeV proton flux that occurred in 10 min. at 32 R_J when the spacecraft was about 7 R_J below the magnetic equator [Schardt, McDonald, and Trainor, 1978]. Before and after the event, the flux was characteristic of the low level normally encountered at this distance from the magnetic equator. The relative proton fluxes and their pitch-angle distributions are shown in Figure 5.13. Owing to an inclination between the magnetic field and the scan plane, fluxes at small pitch angles could not be sampled. At 1255, the small proton flux was relatively isotropic with a small field-aligned component indicating a flow towards Jupiter. A considerably higher flux was observed at 1302 but with little change in angular distribution. At peak intensity (1305), a strong field-aligned injection was observed. Because of the rapid flux increase and a 5-minute half-bounce period, most of the protons had not yet mirrored. The magnetic field signature showed that a current sheet passed across the spacecraft at the same time. Alpha particles and/or heavier ions were accelerated as well as protons, but their flux peaked at a slightly different time.

Because the proton spectrum was unchanged by this event, it is quite possible that this type of process is responsible for the normally observed proton population in the middle magnetosphere. Numerous such acceleration events of various magnitudes may occur with the greatest number in the plasma sheet; however, the spacecraft would only see a time average because it is normally not at the exact site of the acceleration process. The superposition of numerous events could give rise to the rms fluctuations in the counting rate shown in Figure 5.7. These fluctuations have been defined in terms of the difference between counting rates N_i and N_{i+1} taken 1.6 min. apart:

$$< \Delta N >_{\text{rms}} = \left(\frac{1}{10} \sum_{i=1}^{10} (N_i - N_{i+1})^2 \right)^{1/2} \tag{5.13a}$$

$$< \frac{\Delta N}{N} >_{\text{rms}} = \left(\frac{1}{10} \sum_{i=1}^{10} \left(\frac{N_i - N_{i+1}}{1/2(N_i + N_{i+1})} \right)^2 \right)^{1/2} \tag{5.13b}$$

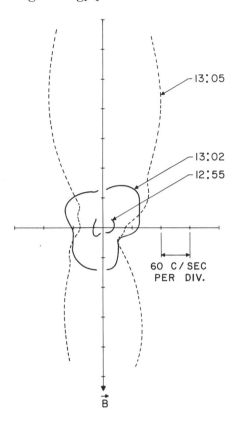

Fig. 5.13. Pitch angle distribution (in the corotating frame) of 1.15- to 2.15-Mev protons at 32 R_J in subsolar hemisphere. The data were taken with LET-II of the GSFC/UNH experiment on Pioneer 10.

13:05

13:02

12:55

60 C/SEC PER DIV.

$\vec{B}$

Because counting statistics also produce random fluctuations, the observed rms values are plotted only when the probability is greater than 95% (χ^2 test) that the effect is not statistical.

As can be seen from Figure 5.7, the largest fluctuations ($<\Delta N>_{rms}$) occur close to the plasma sheet of the middle magnetosphere, as would be expected if the equatorial plasma sheet is the region of greatest activity. In contrast, the percentage proton fluctuations are largest in regions where the flux drops most rapidly, perhaps because the proton flux along a field line does not follow simple trapping theory.

Fluctuations in proton fluxes at different energies are highly correlated with less than 6 min. time delay due to a differential drift. In contrast, cross correlation between fluctuations in electron and proton fluxes is small. A lack of cross correlation would be expected if the fluctuations are due to wave particle interactions because protons and electrons interact with an entirely different part of the spectrum. In contrast, radial gradients in the proton and electron fluxes are in the same direction, and strong cross-correlation would be expected if the fluctuations were caused by bulk motion, that is, expansion and contraction of the magnetosphere.

These observations lead to the following conclusions: (a) the observed fluctuations are primarily due to actual changes in the population of a flux tube rather than to motion of drift shells with different particle intensities; (b) any specific flux enhancement is dissipated in a few minutes; (c) most, if not all, the flux changes are produced by processes occurring near the equator; and (d) the same process affects a wide range of particle energies, but discriminates between protons and electrons.

Outer magnetosphere

Pioneer observations of electron and proton fluxes in the subsolar outer magnetosphere (45 R_J to the magnetopause) are shown in Figures 5.6, 5.7, and 5.8. Because the
magnetopause position responds sensitively to solar wind pressure [Smith, Fillius, and
Wolfe, 1978], the outer magnetosphere is repeatedly being compressed and then
allowed to expand. This continuous pumping should contribute substantially to the
irregular magnetic field configuration and changes in energetic particle fluxes. That
fluxes increased when the magnetosphere was compressed is suggested by the higher
fluxes observed when the magnetopause crossed the spacecraft in response to compression (change from magnetosphere to magnetosheath) than when the magnetosphere
expanded. This effect is particularly noticeable for the large compression observed by
Pioneer 10 inbound near 55 R_J (Fig. 5.6).

Particle spectra in the outer magnetosphere are similar to those in the middle
magnetosphere. Electron spectra are somewhat harder, as reflected in a value for H
(Equation 5.8) between 15 and 35, rather than 2 to 5, as in the middle magnetosphere.
Owing to a relative increase in < 1-MeV protons, the proton spectra become a power
law, $E^{-\gamma}$, with γ between 3 and 4. An acceleration event of protons near 2 MeV has
been observed at 73 R_J and resembled the one illustrated in Figure 5.12.

First-order anisotropies dominate angular distributions in the outer magnetosphere.
For electrons, field-aligned streaming was observed (Sentman and Van Allen, 1976)
with Pioneer 10 at 64 and 52 R_J. The latter coincided with a known compression of the
magnetosphere. First-order anisotropies of protons are generally in agreement with
corotation out to about 65 R_J, although they fall frequently below the rigid-corotation
value (Fig. 5.11). This was particularly evident during the Pioneer 10 inbound pass.
The first-order anisotropies did not reflect corotation in either magnitude or direction
beyond 75 R_J [McDonald, Schardt, and Trainor, 1979]. This variability is not surprising, because changes in magnetopause position could produce high local plasma velocities throughout much of the outer magnetosphere. Second-order anisotropies in the
outer magnetosphere are variable and may be either pancake or dumbbell.

A regular 9 hr 55 min. modulation of the high-energy electron flux (> 5 MeV) was
found in the outer magnetosphere (Figs. 5.6 and 5.8). It was originally suggested that
the equatorial plasma sheet may extend to the magnetopause and produce the modulation by the same mechanism as in the middle magnetosphere. Such an interpretation is
not borne out by the magnetic field configuration [Smith et al., 1976]. Throughout the
outer magnetosphere, the field is quite irregular; it generally has a southward component, and clear-cut plasma sheet crossings are not observed. A study of the phase of this
modulation is also inconsistent with a plasma sheet modulation and will be discussed in
the section on Modulation [McKibben and Simpson, 1974; Fillius and Knickerbocker,
1979].

Although lower energy ions (< 1 MeV/nuc., Chap. 4) already show an enhancement
of oxygen and sulfur, energetic ions ($\sim$ 7 MeV/nuc) in the outer magnetosphere (Fig.
5.4) have the composition of solar cosmic rays rather than the enhanced oxygen and
sulfur abundance found in the inner magnetosphere. Their intensity was, however, significantly higher than in interplanetary space; thus, acceleration has to be invoked to
raise their energy above the instrumental threshold. It may be concluded that energetic
particles in the outer Jovian magnetosphere are subject to dynamic processes in this
region of high-β plasma. The primary driving mechanism is presumably due to changes
in solar wind pressure and plasma diffusing out from the inner magnetosphere. Magnetospheric asymmetries may provide a modulation mechanism for high-energy electrons. It should be noted that Kivelson [1976] has identified a turbulent boundary layer

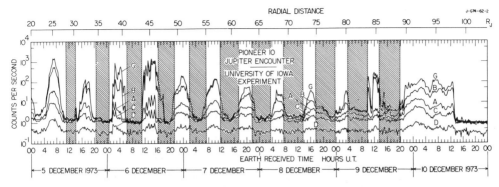

Fig. 5.14. Count rates of the University of Iowa experiment observed during Pioneer 10-out. The shaded areas correspond to times when Pioneer 10 was on open field lines [from Goertz et al., 1976b].

that may be identical with the outer subsolar magnetosphere. Such a boundary layer may be formed by outflowing magnetospheric plasma [Chap. 11; Dessler, 1979]. As yet, this layer has not been theoretically discussed in any detail for the Jovian magnetosphere but should prove a very challenging concept for theoreticians.

5.4. Predawn magnetosphere

Observations

In the antisolar hemisphere, the magnetospheric plasma is no longer compressed by the solar wind and expands into a relatively thin plasma sheet which has been identified by its magnetic signature to distances beyond 200 R_J [Behannon, Burlaga and Ness, 1981]. The particle fluxes are heavily modulated (Figs. 5.14 and 5.15), and the modulation of electrons and protons is in phase. Intercomparison with magnetic field data shows that flux maxima occur at crossings or approaches to the plasma sheet. A close anticorrelation is always found between the magnetic field strength and both the electron [Van Allen, 1979] and proton fluxes [Walker, Kivelson, and Schardt, 1978; Goertz, Schardt, and Van Allen, 1979]. Because particles above 200 keV are responsible for only a minor fraction of the field decrease, we believe that energetic particles constitute the high-energy tail of the thermal plasma in the sheet.

Whereas Pioneer 10-out encountered only one particle maximum per rotation (Fig. 5.14), the lower-latitude Voyager spacecraft saw two maxima at least out to 80 R_J and Voyager 2 frequently to beyond 150 R_J (Fig. 5.15). The following picture emerges: the predawn magnetosphere contains a current sheet in which energetic particles are trapped. The sheet is inclined with respect to the rotational axis and thus moves up and down relative to the equatorial plane, passing a low latitude spacecraft twice per rotation and approaching a high latitude spacecraft once per rotation. Near the central plane of the current sheet the magnetic field lines are closed, but field lines at higher latitudes are presumably open and cannot trap particles (Chap. 1) [Goertz et al., 1976b; Connerney, Acuña, and Ness, 1981]. In the region of open field lines, the flux drops to interplanetary levels and the plasma density (Chaps. 3 and 4) reaches extremely low values. This effect is illustrated in Figure 5.14, where shading indicates when Pioneer 10 was on open field lines. The large difference between the subsolar and antisolar magnetosphere beyond 20 R_J is quite obvious if one compares Figures 5.6, 5.7, and the left side of Figure 5.8 with Figures 5.14 and 5.15. The plasma sheet modulation of

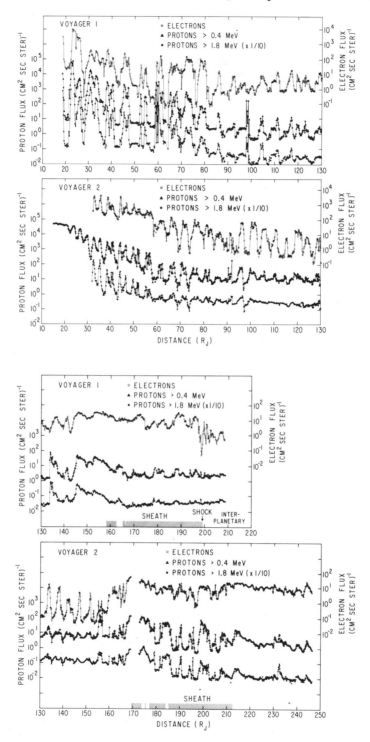

Fig. 5.15. Electron (2.6–5.1 MeV) and proton fluxes (16-min. average) observed by the CIT/GSFC experiment on Voyagers 1 and 2 towards −115° and −137°, respectively, from the Jupiter–Sun line. Electron and >1.8-MeV proton fluxes above 10^3 cm^{-2} s^{-1} st^{-1} are uncertain because of large correction and are plotted only to show relative trends [from Schardt, McDonald, and Trainor, 1981].

particle fluxes is generally less pronounced at local times 0800 to 1200 and extends to only 45 R_J. Plasma sheet configurations that are consistent with the observed modulation will be discussed in the next section.

An intercomparison of absolute flux levels between inbound and outbound passes shows that energetic-particle fluxes in the predawn plasma sheet are generally $\leq 10\%$ of equivalent fluxes at the same distance in the subsolar hemisphere. Unfortunately, time variability of the trapped energetic-particle flux and problems associated with intercomparing data from different instruments preclude a determination of the local time at which minimum flux is reached. However, indications are that the flux does not change much between local times of 0200 and 0600 but increases rapidly between 0600 and 1100. This observation confirms predictions of the magnetic pumping process (Chap. 10) provided the major azimuthal change in the field configuration occurs between 0600 and 1200.

Strong flux gradients make it difficult to determine the plasma convection velocity from the first-order anisotropy of energetic protons (see discussion under middle magnetosphere). The low energy data (Chap. 4) show that the convection velocity is in the direction of corotation and the higher energies indicate [Carbary et al., 1981; Schardt et al., 1981] that the magnitude is often consistent with rigid corotation, but more detailed analysis is required to confirm the preliminary results. Based on observations in the subsolar hemisphere, it is to be expected that the plasma convection velocity is substantially lower than corotation, but no free expansion into the tail takes place (Chaps. 3 and 4).

First-order anisotropies from Pioneer 10-out showed, starting at $\sim 15\ R_J$, a consistent radial outflow of energetic protons [Sentman, Van Allen, and Goertz, 1975]. This indicates that the trapping lifetime should be relatively short. In contrast, the second-order anisotropy in the plasma-sheet region is either zero (isotropic) or slightly pancake, which is an indication of a long trapping lifetime. No such radial outflow was observed during the Voyager passes [Carbary et al., 1981], thus indicating that trapping lifetimes are normally long.

Near flux maxima, the proton and electron spectra were nominally the same as in the subsolar hemisphere. A typical proton spectrum at $64.9\ R_J$ under quiescent conditions is shown in Figure 5.16c. Consistent with the prediction of open field lines at flux minima, the proton spectra and fluxes in the tail lobe resemble those of interplanetary protons (Fig. 5.16a).

As in the terrestrial magnetotail, disturbed conditions occur also in the Jovian predawn magnetosphere. Periods of enhanced and irregular proton flux can be seen in Figure 5.15a for Voyager 1 near 60 and 100 R_J. Under these conditions, the spectral shape depends on pitch angle (Figs. 5.16b and d). Protons with small pitch angles (LET-B) have a softer spectrum than locally mirroring protons. During the 59 R_J event, a large local intensity gradient must have existed (2.6 times per R_J toward center of plasma sheet) to account for the intensity ratio (5–10) between two detectors that pointed perpendicular to the magnetic field. The event near 100 R_J is an example of protons streaming away from Jupiter along magnetic field lines. Several short bursts, lasting about 10 min. each, were observed during which the flux of protons coming from Jupiter was greatly enhanced (up to 40:1) over the flux at $\sim 90°$ pitch angles. As can be seen from Figure 5.16d, the two spectra were quite different.

Three different types of activity have been observed near the plasma sheet [Schardt, McDonald, and Trainor, 1981]:

1. General flux increase with a hardening of the spectrum and moderate anisotropies were observed at 45 and 59 R_J.

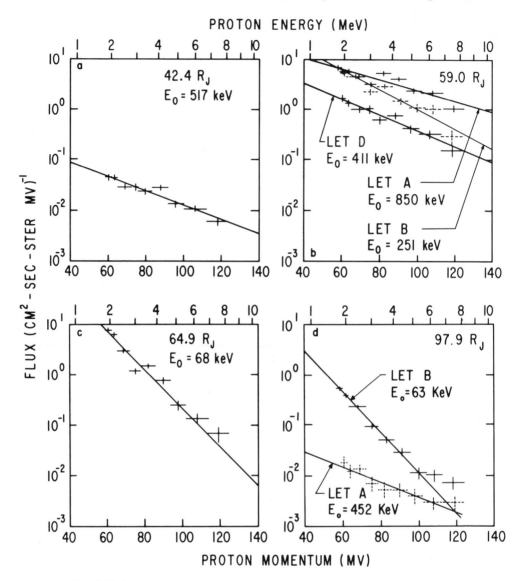

Fig. 5.16. Proton momentum spectra (64-min. average) observed during the outbound pass of Voyager 1 with the CIT/GSFC LET detector. The spectrum at 42.4 R_J was taken in the tail lobe, the 59.0 R_J spectra represents an acceleration event, the 64.9 R_J spectrum gives a typical quiescent plasma sheet population and the 97.9 R_J spectra were observed during field-aligned streaming. LET-B pointed towards Jupiter within about 20° of the magnetic field direction, and LETs A and D pointed approximately perpendicular to the field direction [from Schardt, McDonald, and Trainor, 1981].

2. Outward flow along an ordered field configuration was observed at 98 and 133 R_J.

3. Outward flow in an irregular field configuration was observed in the boundary layer inside the magnetopause at 155 R_J with Voyager 2. This was also observed at lower energies [Krimigis et al., 1980] and can be explained in terms of a rapid convective motion of the thermal plasma.

At a distance of 10 to 20 R_J inside the magnetopause, the strong periodic modulation of the energetic-particle flux ceases; the electron flux increases to its prior value in the plasma sheet and remains that high even in the sheath region outside the magnetosphere (Figs. 5.14 and 5.15). Intensity changes are smaller and less regular. This region appears to be a boundary layer, which has also been identified by the enhanced population of thermal electrons [Gurnett et al., 1980]. The properties of this boundary layer resemble in many respects the outer magnetosphere in the subsolar direction. The region just inside the magnetopause as observed with Voyager 2 has also been interpreted in terms of a planetary wind [Krimigis et al., 1979b] because the first order anisotropy indicates a convection velocity away from Jupiter (Chap. 4) and the magnetopause does not constitute a sharp barrier to the particles.

Rotational modulation

A rotational modulation of the energetic particle flux can be accounted for in principle by three models. As so often occurs in physics, reality requires different models in different regions of the magnetosphere.

The clock model [McKibben and Simpson, 1974] asserts that the modulation is due to a global 10-hr modulation of the magnetosphere, either its size, its trapped particle content or its energy. The observed time variation of particle flux j is taken to be a real-time derivative

$$\text{clock modulation:} \quad \frac{dj}{dt} = \frac{\partial j}{\partial t} \qquad (5.14a)$$

The magnetic anomaly model [Dessler and Hill, 1975] asserts that the 10-hr modulation is due to a preferred longitude (related to a surface magnetic anomaly) sweeping past the spacecraft. The observed time variation is due to the rotation of a longitudinal asymmetry

$$\text{magnetic anomaly:} \quad \frac{dj}{dt} = \Omega \frac{\partial j}{\partial \lambda} \qquad (5.14b)$$

where Ω is Jupiter's angular velocity and λ the longitude of the spacecraft.

The disc model [Van Allen et al., 1974a] asserts that the modulation is due to a rotational modulation of the spacecraft's magnetic latitude ψ_m. Such a modulation is due to the fact that the magnetic dipole is tilted with respect to the rotational axis. In this case

$$\text{disc modulation:} \quad \frac{dj}{dt} = \Omega \frac{\partial j}{\partial \psi_m} \frac{\partial \psi_m}{\partial \lambda} \qquad (5.14c)$$

Note that this model can be formulated in different versions depending on whether ψ_m is measured from the surface of minimum magnetic field or the magnetic dipole equator.

In its pure form the clock model postulates that the modulation phase and period are independent of the observer's zenographic latitude or local time. The magnetic anomaly model postulates that the period is independent of latitude and local time but that the phase is dependent on local time. In contrast to these, the disc model predicts a dependence of both phase and period on the observer's location. A spacecraft in the zenographic rotational equator should see a 5-hr modulation if the plausible assumption is made that j is symmetric about $\psi_m = 0$. In analogy to the Earth's magnetotail [Ness, 1965], an additional mechanism may have to be considered in the predawn magnetotail. This is a rocking motion of the neutral sheet about the Sun–Jupiter line in response to solar-wind interaction with the dipole [Behannon, Burlaga, and Ness, 1981; see also Chap. 1]. Because of the large distances at Jupiter, however, propagation delays become quite significant, and it is not yet clear what neutral sheet configuration this process would produce.

Figure 5.17 shows the phase of the Pioneer 10 electron (6- to 30-MeV) modulation relative to the expected position in the basic disc model, where the disc is represented by the dipole equator. As can be seen from this figure, this model holds in the middle magnetosphere out to about 35 R_J. It should be noted that a 180° phase shift relative to the disc model would have been introduced at closest approach if the modulation followed either the clock or anomaly model. The reason is that Pioneer 10 crossed the equator at periapsis and, therefore, the inbound and outbound passes were on opposite sides of the dipole equator.

In the outer magnetosphere (inbound pass ~ 60 to 90 R_J) the phase is advanced by ~ 110°; outbound it lags by 220 to 260° beyond 90 R_J (Fig. 5.17). Thus, the phase shift is one period. The 10-hr modulation of Jovian electrons in the plasma sheath surrounding the magnetosphere and in interplanetary space is also in phase [Chenette, Conlon, and Simpson, 1974]; thus, it appears that the electron flux varies simultaneously in these regions. The Voyager 2 data taken above, rather than below, the magnetic equator (Fig. 5.8) are in phase (synodic period) with Pioneer observations below the equator [Schardt, McDonald, and Trainor, 1981]. The clock model applies in these regions because the disc model requires a 180° phase shift between Pioneer 10-in and Voyager 2-in data, and the magnetic anomaly model would not predict the observed 360° phase shift between the Pioneer 10 inbound and outbound passes. The reader should note that these are the regions where no periodic modulation was observed in the ion and low-energy electron fluxes.

From 40 to 90 R_J outbound, Figure 5.17 shows a linearly increasing lag in the phase of the particle modulation. Such a lag could be the result of finite propagation velocity of the disturbance produced by the rotating tilted dipole [Northrop, Goertz, and Thomsen, 1974]. In deriving a mathematical expression for the disc model that includes the delay, Kivelson et al. [1978] also included the possibility that the disc may bend toward the equator at larger distances. In terms of r, δ, and λ coordinates (radial distance, latitude and System III (1965) longitude), the plasma sheet position is given by:

$$\lambda = 20.8 \mp \cos^{-1} \frac{\tan \delta}{\tan \delta_0} + 36.27 \frac{r - r_0}{v} \tag{5.15}$$

In this expression δ_0 is the inclination of the plasma sheet at distance r, the delay "velocity" is represented by v, and, r_0 is the distance at which the rigid disc model breaks down.

Figure 5.18 illustrates the evolution of a rigid disc into the bent twisted disc. This is accomplished by twisting the disk through an angle $\theta = 36.27 (r - r_0)/v$ around the spin axis and then bending it toward the equator. In Figure 5.18b, the parameters $\delta_0, r_0,$

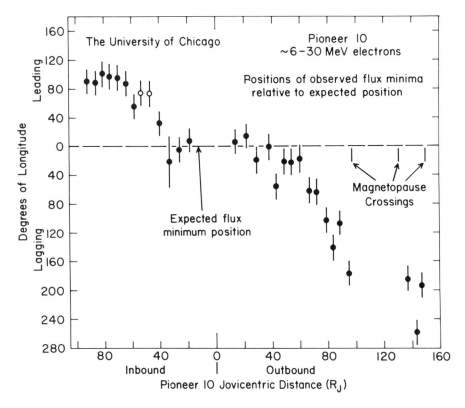

Fig. 5.17. Phase of flux minima of ~ 6- to 30-MeV electrons relative to the position expected from rigid corotation with the magnetic equator at the dipole equator (rigid disc model). Data were taken by the University of Chicago experiment on Pioneer 10. Solid circles correspond to periods when the spacecraft was in the magnetosphere; open circles indicate a period of possible reentry into the magnetosheath [from McKibben and Simpson, 1974].

and v are independent of solar aspect and of longitude; however, this is not a basic feature of the model. As a matter of fact, we know that solar aspect affects the radial extent of the plasma sheet, and one might expect that δ_o would be smallest in the antisolar direction because the plasma sheet will eventually merge into the neutral sheet of the magnetotail; therefore, Figure 5.18b does not illustrate a snapshot of the plasma sheet position but the configuration that appears to rotate past an observer at a specific local time.

The application of this model to Pioneer 10-out data is simple because the latitude of the trajectory was approximately the same as the dipole tilt (9° vs. 10°); thus, only one flux maximum was observed and the $\pm \cos^{-1}(\tan\delta/\tan\delta_o)$ term in Equation 5.15 could be ignored. Little bending towards the equator could have occurred; otherwise, flux maxima (Fig. 5.14) would not have been observed out to the magnetopause. Bending due to the centrifugal stress had been expected [Hill, Dessler, and Michel, 1974; Smith et al., 1974]; however, this will not occur if the plasma temperature is equal to or larger than the corotation energy of a few keV [Goertz, 1976b], and Voyager measurements have confirmed that this is, indeed, the case (Chap. 4). The phase lag (Fig. 5.17) can then be analyzed as a "velocity"; and values from 29 to 43 R_J/hr were obtained, depending on fitting criteria used [Kivelson et al., 1978].

Fig. 5.18. Isomeric projection of the neutral sheet configuration as described by the rigid disc (Fig. 5.18a) and by the bent twisted disk (Fig. 5.18b). The disc is illustrated by equally spaced concentric circles, and the z dimension has been multiplied by three to enhance the dipole tilt. $\theta(R)$, in Figure 5.18a shown in the equatorial plane, is the delay angle used in computing Figure 5.18b. In both figures, the inclination of the inner ring is 10.4°, but decreases to 6.4° for the outer ring in Figure 5.18b [from Schardt, McDonald, and Trainor, 1981].

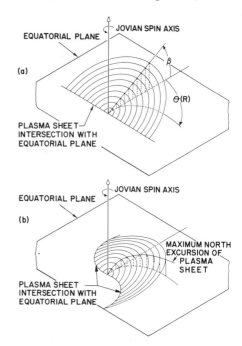

For the lower latitude trajectories of Voyagers 1 and 2 outbound, the disc model predicts two crossings per 10 hr as long as $\delta < \delta_o$; and, indeed, two were found consistently out to 70 R_J and frequently beyond that distance (Fig. 5.15). The analysis of these data is more complex because the Voyager trajectory covered the latitude range from $\delta = -1°$ to 5°. The contributions of the $\cos^{-1}$ term in Equation 5.15 becomes significant but has not been taken fully into account in some published work. As can be seen from Figure 5.15, the two peaks move together as the radial distance increases from 20 to 70 R_J. The decrease in longitudinal separation is substantially greater than can be accounted for by the increase in spacecraft latitude if the magnetic equator stays at the dipole equator, $\delta_o \simeq 10°$. This effect can be explained in terms of a gradual bending of the plasma sheet toward the equator, or as a difference in propagation speed at the longitudes of the two crossings. The latter explanation is employed in a model combining features of the disc and magnetic anomaly models and will be discussed later.

If the propagation velocity is independent of longitude, then Equation 5.15 can be solved for δ_o as a function of the difference in longitude, $\Delta\lambda$, of the two crossings:

$$\delta_o = \tan^{-1}(\tan \delta/\cos 1/2 \ \Delta\lambda) \tag{5.16}$$

Voyager 1 data show that the bending away from the dipole equator ($\delta_o = 10°$) started between 20 and 30 R_J, and that the local inclination, δ_o, near 140 R_J was about 5° [Bridge et al., 1979b]. The bending during the Voyager 2 mission was somewhat less because double peaks occurred out to 200 R_J (Fig. 5.15). This version of the bent twisted disc organizes the data quite well, as can be seen in Figure 5.19, which shows the phase delay vs. distance after the raw data have been corrected for the $\cos^{-1}(\tan \delta/\tan \delta_o)$ term. The resulting "velocities" between 20 and 55 R_J are about 20 R_J/hr, which is somewhat smaller than observed with Pioneer 10. Beyond 70 R_J, the "velocity" increased into the 50 to 200 R_J/hr range.

One explanation for the difference in δ_o between the Pioneer 10 and Voyager missions invokes a local time effect [Thomsen and Goertz, 1981a; Behannon, Burlaga, and Ness, 1981], which is analogous to the behavior of the neutral sheet in the terres-

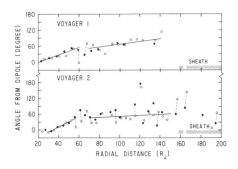

Fig. 5.19. Changes in plasma sheet crossing corrected for lead and lag relative to maximum excursion of bent twisted disc. The angle plotted is $\lambda_{III} - 20.8 \mp \cos^{-1} (\tan \delta / \tan \delta_o)$, where δ is the local spacecraft latitude and δ_o is fit to $\Delta\lambda$ as derived from Equation (5.16). Solid circles refer to crossings or approaches from north to south which occur in the active hemisphere and open circles refer to south to north crossings or approaches [from Schardt, McDonald, and Trainor, 1981].

trial magnetosphere. At the Earth, the full wobble of the magnetic dipole can be seen in the dusk–dawn direction as reflected in the solar-magnetospheric coordinate system [Ness, 1965]. At local times from 1800 to 0600, field lines are pulled back parallel to the plane of the ecliptic and the plasma sheet performs, in essence, a rocking motion about the Sun–Earth line (see Chap. 1 for further discussion). This picture is consistent with $\delta_o \sim 6°$ at 150 R_J for the Voyager 2 trajectory. The Voyager 1 trajectory, however, was closer to dawn (Fig. 5.1a) and should have observed a larger, rather than smaller, value of δ_o. The modified disc model would, therefore, require that the full rocking motion of the dusk–dawn plasma sheet does not necessarily propagate throughout the tail. A model for the modulation based on the rocking motion of the Jovian solar-magnetosphere coordinate system [Behannon, Burlaga, and Ness, 1981] is described in Chapter 1.

In the magnetic anomaly model, the plasma density is higher in the active sector and the Alfvén velocity is presumably lower. Thus, the disturbance would propagate more slowly for the north-to-south crossings of Voyagers 1 and 2 than for the south to north crossings. The phase of the modulation can be accounted for on this basis [Schardt, McDonald, and Trainor, 1981] without requiring a bending of the plasma sheet towards the ecliptic (the reader is referred to Vasyliunas and Dessler, 1981, for a further discussion of this model). The major problem with the disc-anomaly model is that the plasma sheet did not cross Voyager 1 beyond 80 R_J [Ness et al., 1979a] as would be required by the 10° tilt. Similarly, Voyager 2 did not cross the sheet consistently beyond 100 R_J. Thus, the disc-anomaly model requires drastic changes of the current sheet topology, and, hence, in current sheet properties beyond about 80 R_J. Such changes are not apparent in the plasma data (Chaps. 3 and 4).

In summary, we can conclude that the strong periodic modulation is caused by the motion of the plasma sheet in response to the disturbance set up by the tilted spinning dipole. The overall modulation of the electrons, protons, and heavier ions is in phase and extends to $\sim 45\ R_J$ in the subsolar hemisphere and within $\sim 20\ R_J$ of the magnetopause in the predawn magnetosphere. Available data are inadequate to specify uniquely the plasma-sheet configuration. The bent, twisted disc model permits an excellent fit to the data [see also Carbary, 1980]; however, the disc-anomaly model (Chap. 10), and in the tail, the rocking plane model (Chap. 1), have not been ruled out. Knowledge of the current sheet motion is required for calculating scale heights in the plasma sheet from the time variation of particle fluxes; and beyond 30 R_J, a thicker plasma sheet is obtained from the anomaly model than from the bent twisted disc model. In the outer magnetosphere and boundary layer of the predawn magnetosphere, electrons show a small 10-hr modulation that apparently does not affect protons or ions. This modulation is also seen in interplanetary electrons of Jovian origin and is discussed in the next section.

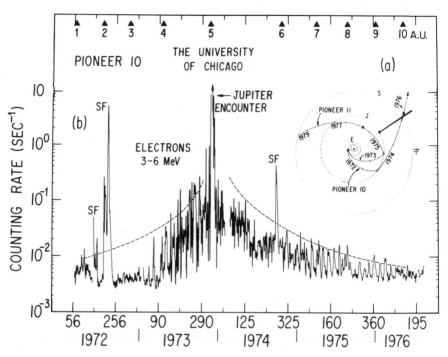

Fig. 5.20. Twenty-four-hour averages of the 3- to 6-MeV electron counting rate observed with the University of Chicago instrument on Pioneer 10. Features identified as SF are of solar flare origin. An intensity dependence of $1/r$ (distance from Jupiter's orbit) is indicated by the dashed line. The solid triangles at the top of the figure refer to Pioneer 10–Sun distances in a.u. The insert shows the Pioneer 10 and 11 trajectories [from Pyle and Simpson, 1977].

5.5. Jovian cosmic rays

Jupiter is an important source of interplanetary electrons in the range 0.2 to 40 MeV. Jupiter also may be a source of interplanetary protons in the MeV range [Simpson et al., 1975]; however, these protons make at most a minor contribution to the ambient flux, and for this reason it has been impossible to establish whether or not protons are emitted. Low-energy ions (<0.5 MeV) of Jovian origin have also been found and resemble the particles observed upstream from the Earth's bow shock (Chap. 4).

The 3- to 6-MeV electron flux in interplanetary space is shown in Figure 5.20. Solar electron events can be identified by their association with solar activity and their softer spectra. The intensity of Jovian electrons falls off as $\sim 1/r$ and is strongly modulated by interplanetary conditions. The regular modulation, especially obvious between 7 and 9 a.u. in Figure 5.20, is due to corotating interaction regions (CIR) that are formed by fast solar wind streams. These regions form an effective barrier and may even influence the release of electrons from Jupiter's magnetosphere. Because electrons move more easily along magnetic field lines than diffuse across them, higher electron fluxes are observed on field lines that connect with the Jovian magnetosphere. For a discussion of interplanetary propagation, the reader is referred to Conlon [1978] and references therein.

The electron spectrum between 0.2 and ~ 10 MeV can be represented as a power law, $E^{-\gamma}$, with $\gamma = 1.5$ [Teegarden et al., 1974]. The spectrum starts to become steeper above 6 MeV and falls into the range $\gamma = 2.5$ to 4. As a matter of fact, the spectral index is modulated with Jupiter's synodic period, and reaches its maximum ($\gamma = 3$ to 4) when the System III (1965) longitude of 240° is at the subsolar point. This modulation has been detected out to 10^8 km from Jupiter and is associated with an intensity modulation of >6 MeV electrons [Chenett, Conlon, and Simpson, 1974]. The electron modulation was also observed by the Voyager missions and softest spectra still occurred at the same subsolar longitude. Preliminary analysis of the Voyager data demonstrated a clear modulation at distances between 5 and 7×10^7 km from Jupiter and showed that the depth of the modulation is changable [Schardt, McDonald, and Trainor, 1981].

The exact nature of this "clock" modulation of the energetic electron flux in the outer magnetosphere and interplanetary space has not been established. McKibben and Simpson [1974] suggested that the energetic electron flux varies simultaneously throughout the outer magnetosphere, especially near the magnetopause. Clearly, the modulation depends on the aspect of the Jovian magnetic field to the solar wind direction. The magnetic anomaly model [Dessler and Hill, 1975 and 1979] explains the modulation by an increased ionospheric conductivity that is produced by precipitating electrons at a magnetic anomaly between $\lambda_{III} = 200$ and 270°. The increased plasma pressure at these longitudes opens additional field lines when the plasma expands into the tail and thus releases a larger electron flux.

An examination of the time delay associated with the expansion of magnetospheric plasma and the phase of the modulation places the most likely release point along the dawn magnetopause at local times from 0200 to 0800. The most characteristic feature of the "clock" modulation is a sharp decrease in the electron flux ($E > 6$ MeV) that lasts only about two hours during each cycle. Because electron diffusion across the magnetopause is enhanced by plasma loading, this flux minimum should occur at a time when a minimum density of the expanding plasma corotates into the dawn magnetopause. Such a minimum would occur [Schardt, McDonald, and Trainor, 1981] at the boundary between the active and inactive hemispheres, if the combination of expansion and partial corotation leads to a rarefaction region between the two plasma streams. Beyond 20 R_J, the active hemisphere, because of its greater plasma loading, is expected to move at a smaller fraction of the corotation velocity than the inactive hemisphere, thus producing a separation as the two regions expand into the tail and rotate across it (see also Chap. 10 and the discussion of rotational modulation in this chapter).

The source strength of Jupiter as a cosmic-ray electron source can be estimated in several ways. The simplest method uses the total electron flux and its first-order anisotropy near the magnetopause. This gives the local loss rate and one has to assume that the loss rate is the same over an extended surface. A loss rate of $\sim 10^{24}$ electrons s^{-1} above 6 MeV was derived based on data from the Pioneer 10-out pass and a surface area equal to that of a 100 R_J sphere [Fillius, Ip, and Knickerbocker, 1977]. Another method is to estimate the total number of Jovian electrons in the heliosphere and assume that their residence lifetime is the same as the known lifetime of solar electrons. This gives a source strength of 10^{25} to 10^{26} electrons s^{-1} [Fillius, Ip, and Knickerbocker et al., 1977]. Conlon [1978] fitted a diffusion-convection model to the observed distribution of Jovian electrons in the heliosphere and used best estimates of the diffusion coefficients to obtain a differential source strength in the range 0.2 to 10 MeV:

$$j(E) = 1.6 \times 10^{26} \, E^{-1.5} \, \text{MeV}^{-1} \, \text{s}^{-1} \tag{5.17}$$

The total power required to produce these electrons can be calculated from the known spectra and estimates of the source strength. For the total flux above 0.2 MeV, this amounts to 10^{13} to 10^{15} W, based on the estimates by Fillius, Ip, and Knickerbocker [1977] and 2×10^{14} W, based on Conlon's [1978] estimate. It is not obvious where this energy comes from. It is often said that Jupiter's rotational energy is sufficient; however, if this rotational energy is used, one must also find a mechanism for removing angular momentum. Whether the torque that can be exerted on the planet by a magnetosphere–ionosphere–atmosphere coupling is large enough is discussed in Chapter 11.

5.6. Summary and discussion

Processes discovered in the Earth's magnetosphere occur, in general, also at Jupiter. The major exception is the decay of albedo neutrons as a significant source of energetic protons, which is important in the terrestrial magnetosphere. Because of the strong Jovian magnetic field, and hydrogen atmosphere, fewer albedo neutrons are produced. Furthermore, other processes accelerate protons to high enough energies (10–200 MeV) to mask the small contribution from albedo neutrons. In contrast, the adiabatic theory of particle motion [Northrop, 1963] and theories of particle diffusion through the magnetosphere [Schultz and Lanzerotti, 1974] are directly applicable if account is taken of the electric fields due to Jupiter's rapid rotation. Thus, fluxes, energies and first-order anisotropies of energetic ions, protons, and electrons can be interpreted in terms of inward diffusion with the conservation of the first and second adiabatic invariants. In this process, the energy gain is approximately proportional to the increase in field strength; thus, the higher intensity of energetic particles in the inner Jovian magnetosphere is at least in part due to the tenfold higher magnetic field strength. Another result of the high field strength is that Jupiter's inner magnetosphere is well insulated from solar wind fluctuations. It is, therefore, believed that the fluctuations responsible for the particle diffusion are not due to the solar wind but to winds in Jupiter's ionosphere. Another important source of fluctuating electric fields is the centrifugal gradient drift instability at the outer edge of the Io torus (Chap. 3). This instability could cause enhanced diffusive transport in the outer Io torus. As in the radiation belts of the earth, particles are lost into the planet's atmosphere by pitch-angle scattering due to resonant interaction with electromagnetic whistler waves.

A new, not yet completely understood, phenomenon is the absorption of energetic particles by the ring and satellites of Jupiter. The original theories of absorption by moons, based on geometric absorption, are inadequate to explain observations [Thomsen, 1979]. Because of the moon's finite conductivity, a current system is set up in the interaction with the corotating magnetosphere; and it is probably the interaction with this disturbance that is the primary mechanism for particle loss. Near Io, interactions with the plasma torus further complicate the situation.

Given the particle population at the boundary between the inner and middle magnetospheres, the "lossy radial-diffusion model" can explain the intensities and energies of particles in the inner magnetosphere; however, solar-wind particles diffusing into the magnetosphere cannot gain enough energy by this process to supply the observed fluxes at this boundary. Thus, an additional acceleration mechanism must be active in the middle and outer magnetosphere. The first model to specifically address this problem was the recirculation model of Nishida [1976] and Sentman, Van Allen, and Goertz [1975, 1978], in which particles diffuse to large L values at a high latitude and then gain further energy as they diffuse back in the equatorial region. In the middle magnetosphere, the electron population follows, in general, stable trapping theory, and the angular distribution (dumbbell) for $R \leq 25 \; R_J$ conforms to the recirculation model. In contrast, the energetic proton population is affected more strongly by dynamic processes and has a pancake angular distribution in the middle magnetosphere. Therefore, the recirculation process probably plays, at most, a minor role in the energization of ions.

A characteristic feature of the middle magnetosphere is the current sheet or plasma disc encircling the planet. It is presumably a direct consequence of the centrifugal stress on the thermal plasma population (Chap. 11). The motion of the current sheet leads to a modulation of particle intensities with two peaks per rotation observed on the low-latitude Voyager spacecraft and only one peak per rotation observed by the Pioneer spacecraft. The disk merges with the tail in the night-side magnetosphere.

Unlike the terrestrial magnetosphere, the β of the magnetospheric plasma is large in the middle and outer magnetospheres. Consequently, these regions are compressible and their radial extent depends sensitively on solar wind pressure (Chap. 3) and on aspect relative to the solar wind direction. The resulting large asymmetry between the subsolar and antisolar hemispheres forms the basis for magnetic pumping [Goertz, 1978; Chap. 10], which could easily supply the required energization. Magnetic pumping is active also in the middle magnetosphere where a day–night asymmetry also exists. However, neither the recirculation nor the magnetic pumping processes can account for the short time variability of fluxes in the middle and outer magnetosphere (~ 1 min. timescale). For an explanation of these, we have to look for instabilities in the magnetically confined high β plasma that fills these regions. As has been found, both in the laboratory and in the geomagnetic tail, some instabilities can rapidly release energy stored in the magnetic field and accelerate some particles well above thermal energies. The plasma waves observed throughout the magnetosphere (Chap. 8) are a great aid for identifying the specific processes involved, although one would expect the most important waves to have frequencies below the threshold of the wave experiments flown as yet.

Unique among the planets investigated so far, Jupiter is a strong source of interplanetary electrons. The energy required to accelerate these electrons can be estimated from the source strength and electron spectra. Based upon the limits on the source strength derived by Fillius, Ip, and Knickerbocker [1977], the required power falls into the range of 10^{13} to 10^{15} W. This covers the 2×10^{14} W based on another estimate [Conlon, 1978]. This power can be compared to the total power in the solar wind of $\sim 2 \times 10^{15}$ W striking Jupiter's magnetosphere (80 R_J radius for magnetosphere); one would expect that only a small fraction of this power is available for accelerating electrons. The other potential source of energy, the Jovian angular momentum, is limited by the maximum torque that can be coupled through the ionosphere. For a discussion of this, we refer the reader to Chapters 10 and 11, where a number of models are described that can yield an adequate amount of power derived from the planetary rotation. However, the unique observable consequences of each model are not well defined.

The 9 hr 55 min. modulation of the Jovian interplanetary electron flux requires an asymmetry in the magnetosphere that rotates with Jupiter. The only cause for such an asymmetry found so far is based on the magnetic field configuration and has given rise to the magnetic anomaly model (Chap. 10). Still, the mechanism by which these electrons are accelerated and released into the outer magnetosphere and interplanetary space is poorly understood.

In summary, to account for the properties of the energetic particle population, magnetospheric models (Chap. 10) have to explain the following points:

1. interaction mechanisms with satellites and Io's torus; these mechanisms must account for the characteristics of the Io absorption features as a function of particle species and energy;
2. acceleration of electrons and protons in the middle and outer magnetospheres, both time averaged and on a timescale of a few minutes;
3. the structure and motion of the plasma disc in the middle magnetosphere;
4. access of ions from the Io torus and interplanetary space into the outer magnetosphere and their subsequent acceleration;
5. the clock modulation of energetic electron fluxes in the outer magnetosphere and in the interplanetary medium; and
6. the mechanism for releasing energetic electrons into interplanetary space and modulating this release with the Jovian rotation period.

Answers to these questions will require a better definition of the properties of the energetic-particle population. Some of this information will emerge from further evaluation of Voyager and Pioneer data; however, new observations with more sophisticated instruments are needed. Crucially important are a greater local-time coverage, a larger time base to distinguish spatial from temporal variations, an extended coverage of lower-energy particles, and an extension of the frequency range of wave measurements.

6

SPECTROPHOTOMETRIC STUDIES OF THE IO TORUS

Robert A. Brown, Carl B. Pilcher, and Darrell F. Strobel

6.1. Introduction

A toroidal volume near Io's orbit is made luminous by multiple optical and ultraviolet line emissions excited by resonant scattering of sunlight and by electron collisions. These emitting atoms and ions have been lost from Io. Table 6.1 summarizes the species and detected transitions as of early 1982. In this chapter we focus on spectrophotometric measurements of these emissions and their physical interpretation. The reader is referred to Pilcher and Strobel [1982] for a more general view of torus emission phenomenology.

The concept of circumplanetary atoms of satellite origin was first proposed by McDonough and Brice [1973] for Titan, a satellite with a dense atmosphere from which Jeans escape is probably important. The Io phenomenon was not anticipated owing to Io's low atmospheric pressure [Smith and Smith, 1972; Pearl et al., 1979]; nevertheless, the discovery by R. A. Brown [1974] of sodium optical emission from Io's vicinity established the first example of a satellite that is a rich, continuing source of material for a planetary environment. We know now that the flow of material from Io dominates the particle and energy budgets of the Jovian magnetosphere.

The primary spatial reference here is to a toroidal volume of $\sim 4 \times 10^{31}$ cm^3 between about 5 and 7 R_J from Jupiter, which includes Io's orbit but lies near the magnetic equatorial plane or centrifugal symmetry surface. This torus contains the bulk of Io's neutral atom clouds and coincides roughly with the UV source region seen by Voyager.

We refer here approximately to the Voyager epoch, the period of the most comprehensive observations. There is intriguing but sometimes conflicting evidence about the plasmasphere's long-term stability, which casts a shadow across the deductions that follow. Spatial inhomogeneity and short-term variability undoubtedly exist. Our simplified picture, based on average properties, is inadequate for such detail and presents some danger of missing a fundamental point associated with, say, nonlinearity or disequilibrium. Nevertheless, we are led by many lines of inference to well-delineated spatial, energetic, and compositional regimes for Jupiter's thermal plasma.

Eight years of study of the manifold emission phenomena associated with the Io torus have produced two kinds of results: (1) demonstrations that certain physical processes obtain in the torus and (2) inferences about the composition and physical state of the torus based on theoretical understanding of the various mechanisms of light production. It is our purpose in this chapter to review these observations, their rationale and implications, plus their relationship to in situ studies of the Jovian environment by the Pioneer and Voyager spacecraft.

6.2. Observational basis: apparent emission rates

This section states the geometrical relationship between the emission and the remote observation of light, incidentally defining the Rayleigh, the photometric unit for

197

Table 6.1. *Observed emitting species in the Io torus*

Species[a]	Source of detection[b]	Excitation mode[c]	Transition type
Na I	GB	RS	Allowed
K I	GB	RS	Allowed
S II	IUE	Coll	Allowed
	Voyager	Coll	Allowed
	GB	Coll	Forbidden
S III	Voyager	Coll	Allowed
	IUE	Coll	Allowed
	GB	Coll	Forbidden
S IV	Voyager	Coll	Allowed
O I	GB	Coll	Forbidden
O II	Voyager	Coll	Allowed
	GB	Coll	Forbidden
O III	Voyager	Coll	Allowed
	IUE[d]	Coll	Allowed
	(GB)[e]	Coll	Forbidden

[a] I indicates the neutral, II indicates single ionization, III indicates double ionization, and so forth.
[b] GB = Ground-based; IUE = international ultraviolet explorer.
[c] RS = resonant scattering; Coll = collisional excitation.
[d] Marginal detection by Moos and Clarke (1981).
[e] Stringent upper limit to [O III] 5007A emission by Brown, Shemansky, and Johnson (1982).

expressing the brightness of line emission [Chamberlain, 1961]. This formalism is the context for decoding the light received from the Io torus in terms of underlying physical and compositional parameters.

Figure 6.1 depicts a telescope having collecting area $A(\text{cm}^2)$ and field of view Ω (sr) observing an emitting layer that is differentially thin and oriented perpendicular to the line of sight. The surface brightness is dJ_λ (photons/cm^2/sr/s) owing to light emission at nominal wavelength λ by particles in the layer. The photon count rate (for unit instrumental efficiency) is

$$\text{count rate (s}^{-1}) = dJ_\lambda A \Omega = dJ_\lambda A \, \frac{a}{r^2} \tag{6.1}$$

In the differential volume, adr, the number density of emitter is n (cm^{-3}) and the average number of photons emitted in the line by each atom or ion per second is ϵ_λ, the unit emission rate. If the emission is isotropic, the photon count rate at the telescope is

$$\text{count rate (s}^{-1}) = nadr\epsilon_\lambda \, \frac{A}{4\pi r^2}$$

which in combination with (6.1) yields:

$$4\pi dJ_\lambda = n\epsilon_\lambda dr$$

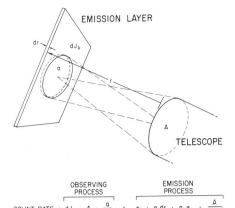

Fig. 6.1. The geometry for remote observations of emitting regions. The count rate for photons at the telescope is computed from two perspectives to derive the result (6.2) relating the observed brightness to the physical process of emission along the line of sight.

By integrating along the line of sight through the source region, we obtain

$$4\pi J_\lambda = \int n(r)\epsilon_\lambda(r)\,dr \qquad\qquad (6.2)$$

The left side of (6.2) is the observed quantity, known as "the apparent emission rate"; if J_λ is expressed in units of 10^6 photon/cm^2/sr/s, then $4\pi J_\lambda$ is in Rayleighs.

The right side of (6.2) is the sum along the line of sight of the number density of emitters weighted by their propensity to emit. The density factor reflects composition; the unit emission rate factor responds to the radiation or charged particle environment of the atoms or ions. When observations are interpreted, this reciprocity in the density-emission rate product causes an ambiguity between composition and physical conditions that can be resolved only by means of external information or assumptions.

In the case of the alkali metal atoms in the Io torus, sunlight is resonantly scattered (absorbed and reemitted) in permitted atomic transitions such as the sodium D-lines; ϵ_λ is proportional to the solar flux at λ and the atomic oscillator strength. For sulfur and oxygen atoms and ions, electrons populate upper atomic levels by means of inelastic collisions, and the atom or ion decays to the ground state by the emission of one or more photons. In this case, ϵ_λ is the product of the electron number density n_e and a rate coefficient $\alpha_\lambda(T_e,n_e)$ which is a function not only of electron temperature but also of electron number density, owing to collisional deexcitation.

When little supporting information is available about the variations of $n(r)$ and $\epsilon_\lambda(r)$ along the line of sight, it is customary to simplify (6.2) to derive the average or characteristic parameters implied by an observations. If the excitation mechanism is known and can be assumed invariant along the line of sight, then the column abundance of the emitting species is

$$N \equiv \int n(r)\,dr = \frac{4\pi J_\lambda}{\epsilon_\lambda}\;(\text{cm}^{-2}) \qquad\qquad (6.3)$$

This expression is the basis, for example, of neutral sodium atom inventories around Io. For an assumed lifetime against loss, that inventory yields an estimate of the sodium production rate. Calculations of this nature are discussed in Section 6.3.

If the depth (D) of the emitting region can be estimated and uniform number density and excitation conditions are assumed, a useful form of (6.2) is

$$\overline{n\epsilon_\lambda} = \frac{4\pi J_\lambda}{D}\;(\text{photons/cm}^3/\text{s}) \qquad\qquad (6.4)$$

This expression is the basis for estimates of the plasma energy loss rate per unit volume through collisionally excited emissions. If the electron number density and temperature can be estimated, then (6.4) yields the average or characteristic atomic or ionic number density.

$$\bar{n} = \frac{4\pi J_\lambda}{n_e \alpha_\lambda (T_e, n_e) D} \qquad (6.5)$$

For two collisionally excited lines originating with the same emitting species, the apparent emission rate ratio,

$$\frac{4\pi J_{\lambda 1}}{4\pi J_{\lambda 2}} = \frac{\overline{\alpha_{\lambda 1} (T_e, n_e)}}{\overline{\alpha_{\lambda 2} (T_e, n_e)}} \qquad (6.6)$$

is a diagnostic indicator of electron temperature or number density or a combination of the two. Two lines in the same multiplet may have closely spaced upper levels that are excited in a constant ratio independent of electron temperature. Collisional deexcitation from these levels competes increasingly with radiative decay as the electron number density increases. Because the excited levels have different lifetimes, this effect becomes important for one level before the other, and the ratio of the two emissions is a function of n_e in that transition range of electron density. When two lines originate in different multiplets, that is, from upper levels belonging to different terms, the emission ratio is a function of the electron temperature. The upper levels are excited in proportion to the fraction of electrons with greater than threshold energy, which is different for the two terms.

The red doublet of singlet ionized sulfur (6716 A, 6731 A) is an example of a case with $\alpha_{\lambda i}$ effectively temperature independent [Brown, 1976], permitting the electron density to be determined from the line ratio. The rate coefficients for 1304 and 6300 A emission from neutral oxygen and 685 and 6312 A emission from doubly ionized sulfur are effectively independent of electron number density under Io torus conditions [Brown, 1981a]; both line ratios yield direct estimates of electron temperature.

Equation (6.2) and its daughter expressions are powerful and convenient tools for inverting photometric observations of line emission in terms of the underlying composition and physical state. However, the observed volume is vast, complex, and detailed; assumptions of uniformity are generally false. Derived results are not literal, they are indicative. They join with other observations and inferences to improve our understanding of the Io torus.

6.3. The atomic clouds

Na D-line optical emission (5890 A, 5896 A) from the near-Io cloud is a bright, accessible phenomenon, and faint D-line emission is observed throughout the Jupiter magnetosphere. The near-Io cloud and one form of remote sodium appear to be kinematically related, while a second remote form requires an episodic acceleration mechanism [Brown and Schneider, 1981]. A sodium source at Io of $\sim 10^{27}$ atoms/s is required to sustain this tableau. Sodium is the best studied atomic species and serves as an archetype for our discussion of the neutral clouds.

Neutral cloud investigations have been of two types: (1) column abundance determinations in which (6.3) is used to invert brightness measurements, and (2) kinematic studies in which measurements of Doppler shifts are used to determine line-of-sight velocities. We present the historical interpretation of these observations, but we caution the reader of a general failure to account adequately for modification of the observed morphological and kinematic structures by ionization and collisions. Although those processes have demonstrated importance for sodium cloud analysis, they have not been incorporated into superseding models. In following a chronological line with simplified but outdated conclusions, we mean to preserve the overall intellectual fabric while exposing the points ready for patches or replacement.

In simple form, the story told by the Na *D*-line observations is that of vast numbers of atoms traversing ballistic trajectories threading the entire Jovian magnetosphere. If the atomic lifetime is spatially invariant, the spatial distribution of these atoms is isomorphic to the source of same-element ions, because each atom may be ionized at a random time and place in its orbit. Before the discovery of an extended neutral oxygen cloud [Brown, 1981a], it seemed that the distribution of alkali metal atoms might be fundamentally different from that of the dominant plasma consituents, sulfur and oxygen probably originating in Io's sulfur dioxide atmosphere. Brown and Ip [1981] have argued that all heavy elements in the Jovian magnetosphere may arise and disperse from Io in a similar fashion. This neutral-phase generation has important implications for the energetics and spatial distribution of the thermal plasma.

Morphology and kinematics

Following the initial demonstration that *D*-line radiation emanates from a substantial volume of space surrounding Io [Trafton et al., 1974; Mekler and Eviatar, 1974], it was shown that the brightness of the sodium near Io varies sinusoidally with Io's orbital longitude, exhibiting a maximum when Io is at elongation [Bergstralh et al., 1975, 1977; see also Macy and Trafton, 1975a,b]. This is the signature of excitation by resonant scattering of sunlight. The modulation is caused by the oscillating heliocentric Doppler shift, which causes the Io sodium atoms to sample the deep Fraunhofer *D* lines at varying height.

Images of the near-Io sodium cloud [Matson et al., 1978; Murcray and Goody, 1978] led to the conclusion that sodium atoms escape Io with a residual velocity of a few km/s [Matson et al., 1978; Smyth and McElroy, 1978]. It was inferred from the asymmetric spatial distribution of this cloud that atoms originate from an extensive portion of the satellite near the sub-Jupiter point. In their imaging study, Murcray and Goody [1978] concluded that the near-Io sodium cloud is mostly inside Io's orbit, extending forward of the satellite about 70° in orbital longitude. Smyth and McElroy [1977] showed that for ejection velocities of a few km/s, atoms from Io spread about 3.5° of longitude per hour into a torus centered on the satellite's orbit. The cloud's angular extent thus implies a sodium lifetime of about 20 hr at the periphery. However, a much shorter lifetime may apply to most Io-generated sodium, as discussed in Section 6.3. The distribution of the near-Io sodium cloud is shown schematically in Figure 6.2.

Systematic variations other than the Doppler brightness modulation have been observed in the near-Io sodium cloud. Trafton and Macy [1975; Trafton, 1977] reported that the sodium emission to the north or south of Io is strongest on the side away from the Jovian magnetic equator. This effect was also observed by Münch and Bergstralh [1977] and Pilcher and Schempp [1979]. It appears to be caused by an interaction between the atoms and the plasma torus, and electron-impact ionization and ion–atom collisions are candidate processes. Morphological variations [Goldberg et al., 1978, 1981] and second-order brightness variations with Io's orbital longitude

Fig. 6.2. A model of the Io Region B
sodium cloud as seen from above
Jupiter's north pole [from Smyth and
McElroy 1978].

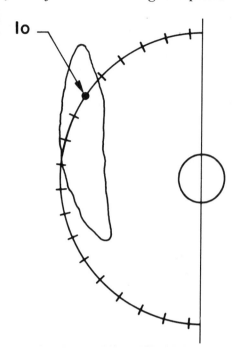

[Bergstralh et al., 1975, 1977] have been attributed by Smyth [1979] to the effects of solar radiation pressure.

There have been many reports of faint sodium emission at great distances from Io [Mekler and Eviatar, 1974; Wehinger and Wyckoff, 1974; Mekler et al., 1976; Wehinger et al., 1976; Trafton and Macy, 1978; Pilcher and Schempp, 1979]. Because of the difficulty in distinguishing this phenomenon from telluric airglow, the nature and even existence of remote sodium had been controversial [Goody and Apt, 1977]. Brown and Schneider [1981] found two distinct kinematical components. The "normal" signature of atoms on bound orbits with large apojoves seems always to be present; these atoms are probably an extension of the bright, near-Io sodium cloud. The signature of "fast" atoms with speeds up to at least 100 km sec^{-1} is seen only occasionally. Brown and Schneider [1981] suggest it is due to sodium produced in the near-Io cloud by collisions between atoms and heavy ions in the corotating plasma. Ion–atom collisions are discussed in Section 6.3.

Other energetic phenomena have been observed in the sodium cloud near Io that may also be caused by ion–atom collisions. Trafton [1975a] and Trafton and Macy [1977, 1978] found asymmetric D-line spectral profiles for the near-Io sodium cloud showing velocities as high as 18 km/s with respect to Io. They concluded that sodium is "streaming" away from Io in the direction of its orbital motion. Similar profiles and conclusions were reported by Carlson et al. [1978]. Transient and anomalous directional features in the Io sodium cloud that have been reported in images [Pilcher, 1980b,c; Pilcher and Strobel, 1982], may be collisionally generated streams of sodium.

The brightness of Jovian sodium emission at large distances from Io has been used to infer the plasma electron density under the assumption of depletion by electron-impact ionization [Mekler and Eviatar, 1978, 1980; Goldberg et al., 1981; Eviatar et al., 1981a]. These authors propose that a decrease in distant sodium emission indicates an increase in plasma density. However, this is not necessarily true. The ion and electron densities are locked together by the requirement of charge neutrality; a transient increase in plasma density would produce both more ionization and more "fast" remote sodium if the Brown and Schneider [1981] mechanism applies. For a fixed Io source rate, the equilibrium production of "fast" remote sodium would be fixed only by the branching ratio for electron-impact ionization and ion–atom elastic collision. It would be independent of the plasma density that would, however, determine the residence time for atoms in the near-Io cloud. The density of distant "normal" sodium, thought to be an extension of this cloud, could therefore be related to the plasma concentration near Io. These considerations suggest that although there is probably a relationship between the plasma density near Io and the brightness of distant sodium, a thorough analysis is necessary before this relationship can be understood. Conclusions [Mekler and Eviatar, 1978, 1980] based on models in which the complexity of the relationship is not taken into account should be viewed with some skepticism.

Resonance emission lines of potassium (7665 A, 7699 A) have also been observed around Io [Trafton 1975b, 1977, 1981]. The distribution of this element seems to be similar to that of sodium. An early suggestion that the potassium lines are more symmetric than those of sodium [Münch et al., 1976] has been superseded by more recent evidence of the kinematic similarities of the two alkali metal clouds (Trauger, private communication, 1981).

Brown [1981a] has reported finding atomic oxygen emission near Io's orbit. His visible detection, $4\pi J$ (6300 A) $= 8 \pm 4\ R$ is compatible with the upper limit $4\pi J$ (1304 A) $< 6.4\ R$ from IUE [Moos and Clarke, 1981]. The spatial distribution of atomic oxygen in the Io torus has not been measured.

The mechanisms by which specific elements are selected and injected from Io into circum-Jovian space have not yet been determined. The ejection speeds implied by the shapes of Io's volcanic plumes are less than the satellite's gravitational escape speed [Cook et al., 1979] and hence inadequate to populate the sodium cloud. Kumar and Hunten [1981] have reviewed sputtering by ion impact, which is possible both at the surface and at the top of the atmosphere [Matson et al., 1974; Haff and Watson, 1979; Haff, Watson, and Yung, 1981]. This mechanism can supply neutral atoms with sufficient speed and also break down low-volatility alkali metals to atomic scale. However, an Io atmosphere poses difficulties for surface sputtering because it may block incoming ions and inhibit the direct escape to space of sputtered atoms [Brown and Yung, 1976]. Furthermore, if atmospheric shielding and surface sputtering are both important on Io, a modulation of the neutral atomic clouds should result from the changing orientation of the corotating ion flux with respect to the highly asymmetrical, solar-fixed distribution of a condensible SO_2 atmosphere [Kumar, 1980]; this effect is not observed [Murcray and Goody, 1978].

Atomic cloud supply rates

The neutral atoms in each region of the Jovian magnetosphere are subject to ionization and so must be replenished constantly. The local sink rates can be estimated from (1) an inventory of atoms based on Equation (6.3), and (2) knowledge of the local lifetime against ionization. Under the reasonable assumption that these atoms originate at Io, the total Io supply rate is the sum of all local sinks. This calculation can be attempted only for the best-studied species, sodium, but a total Io atomic source rate follows from an assumed sodium mixing ratio.

Brown and Schneider [1981] find the "normal" sodium population remote from Io represents a sink $\sim 10^{26}$ atoms/s. The episodic "fast" remote sodium has not been inventoried. These components occupy a region of low plasma density, and the governing loss process is photoionization with lifetime ~ 400 hr [Carlson et al., 1975].

Table 6.2. *Io torus electron density measurements*

Source	$n_e(cm^{-3})$	Reference
Ground-based [S II] observations	2000[a]	Brown (1976)
Ground-based [S II] observations	3000[a]	Brown (1978)
Voyager 1 ultraviolet spectrometer	≥ 2100	Broadfoot et al. (1979)
Voyager I planetary radio astronomy experiment	2000[b]	Warwick et al. (1979a)
Voyager 1 plasma science experiment	2000[b]	Bagenal et al. (1980)
Ground-based [S II] observations	4000[a,c]	Trafton (1980)
Ground-based [S II] observations	1000[a]	Trauger et al. (1980)
Ground-based [S II] observations	5000[d]	Morgan and Pilcher (1982)
Ground-based [O II] observations	2000	Morgan and Pilcher (1982)

[a] These values have been adjusted by Pilcher and Strobel (1982) from those reported by the individual investigators to remove the effects of differences in assumed collision strengths and electron temperatures.
[b] Variations of a factor of 2 about this value were observed between 5 and 7 R_J.
[c] This value corresponds to the characteristic reported line ratio. Trafton reported large variations about this value.
[d] This value corresponds to the reported mean line ratio; the mean electron density reported by Morgan and Pilcher from the [S II] line observations was $\bar{n}_e = 4000$ cm^{-3}.

Images of the near-Io sodium cloud, but excluding the satellite's immediate vicinity, have yielded both an atom count and a lifetime, as discussed above. Smyth and McElroy [1978] find that a source $\sim 2 \times 10^{25}$ atom/s is required for this region in the absence of a shortened sodium lifetime near Io.

Sodium in the immediate vicinity of Io is difficult to observe because the very high surface brightness of Io's telescopic image reduces contrast [Brown et al., 1975]. Furthermore, the sodium lifetime against electron-impact ionization must be short, so a large atomic sink is probably hidden in this region [Brown, 1981b]. Independent measurements of the Io torus electron temperature and number density are available both from Voyager 1 instruments and from ground-based photometry of forbidden line emission. These results are discussed in Section 6.4 and summarized in Tables 6.2 and 6.3. The major electron component is well characterized by a 4–6 eV temperature and number density $n_e \simeq 2000$ cm^{-3}, with a possible small admixture of hotter (100–1000 eV) electrons. Under these conditions, the sodium lifetime is 2–3 hr almost independent of the hot electron component (see Fig. 6.3). We suggest that the remote sodium population and much of the imaged near-Io cloud are atoms that have survived this region of short lifetime. This requires traveling ~ 0.5 R_J with a radial velocity 1–2 km/s, which requires 5–10 hr and implies an approximate order-of-magnitude loss in the hot, poorly observed region. The total sodium sink is then the same factor times the observed, exterior sinks or $\sim 10^{27}$ sodium atoms/s. Brown and Schneider [1981] have argued that the absence of the kinematical signature of remote sodium produced by electron-capture recombination is consistent with a much larger sodium atom population near Io than heretofore has been observed.

The total Io atomic source can be computed from an estimate of the fractional abundance of sodium. Voyager 1 found $\sim 10\%$ sodium ions in the middle magneto-

Table 6.3. *Io torus electron temperature measurements*

	T_e^c		T_e^h			
	10^4K	eV	10^4K	eV	n_e^h/n_e^c	Reference
Voyager 2 ultraviolet spectrometer	6	5	–	–	–	Sandel et. al. (1979)
Voyager 1 ultraviolet spectrometer	10	9	–	–	–	Shemansky (1980a)[a]
Voyager 1 ultraviolet spectrometer	5	4	~120	~100	$2-5 \times 10^{-2}$	Strobel and Davis (1980)[a]
Ground-based measurements of [O II] and [S III]	5	4	–	–	–	Brown (1981a)
Voyager 1 plasma science experiment[b]	5.8	5.0	726	626	2×10^{-4}	Scudder et al. (1981)
Voyager 1 ultraviolet spectrometer	8	7	726	626	$\leqslant 1 \times 10^{-2}$	Shemansky and Smith (1981)

[a] These analyses used less accurate collision strengths.
[b] Values reported for 5.5 R_J.
[c] Superscripts c and h refer to the cold and hot electron components, respectively.

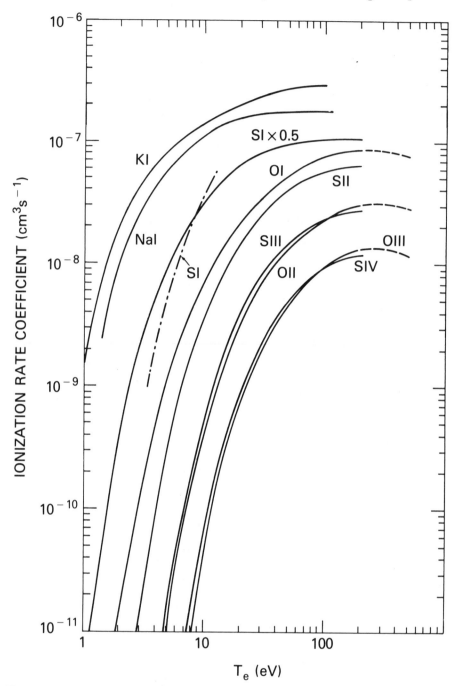

Fig. 6.3. Electron-impact ionization rate coefficients for species present in the Io torus. The coefficients were determined from laboratory cross-section meaurements for O I by Brook et al. [1978], for O II and O III by Aitken and Harrison [1971], and for Na I and K I by Lotz [1967]. The S I coefficients shown by the dash–dot line are based on the theoretical cross sections of Peach [1968, 1971]. All other sulfur coefficients are from the semiempirical expressions of Jacobs et al. [1979]. These tend to be reasonably accurate for ions, but are less accurate for atoms at low temperatures owing to extreme sensitivity to threshold cross-section behavior.

sphere and ~5% in the inner, cool torus (if the gap between the charge-to-mass peaks 16 and 32 is assumed to be filled in by Na^+ [Bagenal and Sullivan, 1981]). If a 5–10% mixing ratio applies to the Io source, the required total atom ejection rate is ~ $(1-2) \times 10^{28}$ atoms/s. From his determinations of the electron temperature and the neutral oxygen density, Brown [1981a], assuming that electron-impact ionization dominated, concluded that the neutral oxygen ionization rate $d(n_{O_1})/dt \simeq 2 \times 10^{-4}$ cm^{-3} s^{-1}. If Brown's assumption is correct, neutral oxygen would be expected to form a more nearly complete torus than sodium, centered near Io's orbit [Smyth and McElroy, 1977]. A torus cross-sectional radius of 1 R_J and Brown's value for $d(n_{O_1})/dt$ leads to a total O I ionization rate of 5×10^{27} s^{-1}. This value is relatively insensitive to the downward revision of the neutral oxygen lifetime when charge exchange is introduced (Section 6.3). The supply rate is proportional to the ratio of the occupied volume and the lifetime, but the former grows proportional to the latter [Smyth and McElroy, 1977]. Ionization rates ~ 3×10^{28} s^{-1} have been estimated from measurements of the total power radiated from the torus, as discussed in Section 6.6.

Ion–atom collisions

Heretofore little consideration has been given to interactions between the corotating heavy ions in the plasma torus and the atoms escaped from Io and forming the neutral atom clouds. In this section we examine elastic and charge exchange collisions, which appear relevant to certain observed torus properties or emission phenomena.

The distribution of encounter speeds for collisions between cloud atoms and plasma ions results from combining the bulk corotational motions with the stochastic distribution of ion velocities in the corotating frame. The former contribution dominates the latter, so we postpone consideration of the ion thermal motions to Section 6.4. At axial distance r from Jupiter, the speed of the corotating ions's cyclotron-guiding center minus the ballistic atom's tangential speed [Brown, 1981a] is

$$v_0(r) = 12.5 \, r - (100/r) \text{ (km/s)} \tag{6.7}$$

which has an average value $\bar{v} \simeq 60$ km s^{-1} in the primary region of plasma/atomic cloud overlap near Io's orbit.

An ion atom collision is described by a cross-section $\sigma(v)$. Practical application requires the rate coefficient,

$$k \simeq \sigma(\bar{v}) \, \bar{v}$$

which can be used to compute species' lifetimes and population balances.

Elastic collisions. "Hard" collisions engaging the repulsive core of the interatomic potential will dominate ion–atom elastic reactions when the atom has negligible polarizability. The momentum transfer (diffusion) cross section can be computed from the Thomas–Fermi interaction [see Torrens, 1972]; it is probably a few times 10^{-16} cm^2 for oxygen and sulfur and is a slow function of velocity in the energy range of interest here (Lindhard and Scharff, 1961). The lifetime against such collisions is then:

$$\tau_{TF} = (n\bar{v}\sigma)^{-1} \simeq 100 \text{ hr (sulfur, oxygen)} \tag{6.8}$$

where we have taken $n = 1500$ cm^{-3} as the total ion number density, $\bar{v} = 60$ km/s and $\sigma = 3 \times 10^{-16}$ cm^2.

For an atom of polarizability α, an attractive force results from the dipole moment induced by a passing ionic charge. The momentum transfer cross section for this "soft" interaction [Dalgarno et al., 1958; Dalgarno, 1962] is

$$\sigma_{pol} = 1.0 \times 10^{-14} \frac{Z\alpha^{1/2}}{\mu^{1/2} v} \ (cm^2)$$

where μ is the reduced mass in amu, Z is the ionic charge, α has units a_0^3, (a_0 = Bohr radius), and v is the encounter speed in km/s. The polarizabilities of Na and K are, respectively, $182 \ a_0^3$ and $257 \ a_0^3$ [Allen, 1973], while those for O and S are about two orders of magnitude smaller. Computation of the lifetime against induced polarization collisions is given by a summation over the various ions with associated charges, reduced mass and number densities $= Z_i, \mu_i$, and n_i:

$$\tau_{pol} = \left[1.0 \times 10^{-9} \alpha^{1/2} \sum \frac{Z_i n_i}{\mu_i^{1/2}} \right]^{-1} s$$

independent of ion temperature. For the ion densities from the constant thermal speed model of Bagenal and Sullivan [1981], for example, the induced-polarization collision lifetime would be about 20 hr for sodium and potassium.

The upper limit to the kinetic energy transfer to the atom in an elastic collision is

$$KE_{max} = \frac{4MM_a}{(M + M_a)^2} KE_0$$

where M and M_a are the ion and atom masses, respectively, and KE_0 is the incident kinetic energy. For classical hard spheres, all transfers between zero and KE_{max} are equally probable. In the Thomas–Fermi case, smaller energy transfers are weakly favored because the differential cross section varies approximately as the transferred energy raised to a power between -1 and -1.5 [Lindhard and Scharff, 1961]. Because $KE_0 \simeq (17M)$ eV at Io's orbit (M in amu) and only about 1.5 eV/amu are required for a cloud atom to gravitationally escape the Jupiter system, virtually all hard elastic collisions result in loss from the Io-connected atomic clouds. The soft collisions at larger impact distances will result in much smaller momentum transfers.

The episodic "fast" remote sodium may be produced by hard ion-sodium atom collisions or by charge-exchange (discussed below) [Brown and Schneider, 1981]. The directional sodium cloud features [Pilcher, 1980b,c; Pilcher and Strobel, 1982] may also result from a combination of these processes. The 20 hr lifetime against soft collisions is comparable to the sodium ionization lifetime at the near-Io cloud periphery, which implies the atoms observed in the near-Io cloud have received significant momentum from the plasma ions. This perturbation may be larger than radiation pressure which has been suggested as an explanation of the observed east–west asymmetry [Smyth, 1979]. We believe a reassessment is in order for those conclusions about the Io source magnitude and anisotropy and the launch mechanism which have been based on purely ballistic interpretations of the kinematical and morphological structures of the near-Io sodium cloud.

Charge-exchange collisions. This process has been discussed recently by Cheng [1980], Kunc and Judge [1981], and Brown and Schneider [1981]. The reactions of particular interest are shown in Table 6.4. The cross sections are evaluated for 60 km s^{-1} encounter speed; lifetimes are computed from (6.2), and an assumed 300 cm^{-3} number density for S II and O II and 100 cm^{-3} for the Na II. The combined charge exchange lifetimes of neutral oxygen and sulfur are 80 and 33 hr, respectively, which are competitive with

Table 6.4. *Charge-exchange reactions at 60 km/s encounter speed*

Reaction[a]	$\sigma(E)$ (cm^2)	τ(hr)[c]
$O^+ + O \rightarrow O^* + O^+$	2×10^{-15} [d]	80
$O^+ + S \rightarrow O^* + S^+(^2P^0)$	1.5×10^{-15} [b]	100
$S^+ + S \rightarrow S^* + S^+$	3×10^{-15} [b]	50
$S^+ + O \rightarrow S^* + O^+$	$\leqslant 1 \times 10^{-17}$ [b]	2×10^4
$Na^+ + Na \rightarrow Na^* + Na^+$	1×10^{-14} [e]	50

[a] The asterisk indicates a fast neutral atom. Reactants and products are in the ground term unless otherwise indicated.
[b] Cross section from R. E. Johnson (private communication), evaluated at the average ion–atom encounter energy, about 310 eV for OII and 580 eV for S II.
[c] Atomic lifetime computed from (III-4) using ion number density 3×10^2 cm^{-3} for O II and S II and 1×10^2 for Na II.
[d] Experimental value of Stebbings et al. (1964).
[e] Empirical value using Figure 9 in Hasted (1962).

electron impact ionization of these species. Substantially larger O II and S II number densities are possible (see Sec. 6.4 and Table 6.6), which would significantly reduce the stated charge-exchange lifetimes. The atomic sodium charge-exchange lifetime is longer than the electron-impact lifetime. The speed of the neutral atom resulting from charge exchange always exceeds that required for direct escape from Jupiter's gravity; thus a sizable, perhaps the major, fraction of the torus particles are lost as fast neutrals rather than as radially diffusing ions captured by the solar wind.

The ionization rate of O II is much slower than that of S II (see Fig. 6.3). Thus with the reactions of Table 6.4, preferential conversion of O II to S II occurs with fast O* escape; subsequent ionization of S II would be the net sink since the reaction of S II with O is extremely slow. As a consequence, the sulfur to oxygen ion ratio is enriched relative to what would be expected from SO$_2$, in agreement with Voyager measurements [Bagenal and Sullivan, 1981; Shemansky and Smith, 1981]. Similarly the atomic S/O ratio should be increased as a result of O* escape. In addition, with O II preferentially converted to S II, the O III/O II density ratio would be significantly less than 1, in agreement with observations discussed in Section 6.4.

Each charge exchange event produces an ion that will subsequently be reaccelerated to corotation. If these events occur frequently enough in comparison to the cooling rate of the ions, only hot ions would occupy the torus. Based on current measurements of the neutral densities, a typical time constant for ion charge transfer is one month, approximately the Coulomb collisional cooling time constant for ions.

6.4. The plasma torus

Two types of ionic emission have been observed from the plasma torus, both of which are excited by collisions with electrons. The forbidden lines at wavelengths between 3000 A and 10,000 A arise from transitions between terms in the p^2 and p^3 ground electron configurations of sulfur and oxygen [Figs. 6.4–6.8). The more energetic lines are observed only from spacecraft (Pioneer, Voyager, IUE) and are due to allowed transitions that occur at wavelengths ($\lambda < 1800$ A) too short for resonant scattering of sunlight to be an important effect.

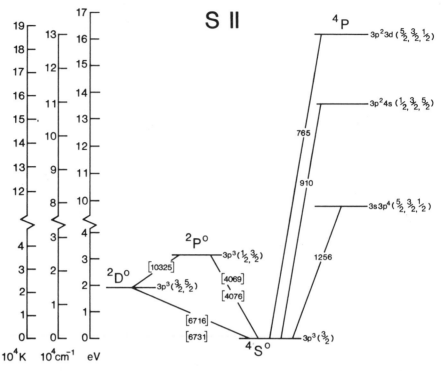

Fig. 6.4. Grotrian diagram for S II. Levels with the same term are shown vertically. The electron configuration and *J* values (in order of increasing energy) are shown for each level.

The forbidden lines

Optical emission from the torus in the red [S II] lines was discovered by Kupo, Mekler, and Eviatar [1976], and its plasma-diagnostic power was pointed out and first applied to the Jupiter torus by Brown [1976]. The first excited levels of that ion $^2D°_{3/2}$ (1.842 eV), $^2D°_{5/2}$ (1.846 eV), $^2P°_{1/2}$ (3.042 eV), and $^2P°_{3/2}$ (3.047 eV) are connected to the ground state $^4S°_{3/2}$, only by electric quadrupole and magnetic dipole matrix elements (Fig. 6.4). These upper levels are therefore populated only by electron collisions, not by photon absorption. Slow decay by photon emission occurs from the $^2D°_{3/2}$ and $^2D°_{5/2}$ levels with lifetimes 6×10^2 s (6731 A) and 2×10^3 s (6716 A), respectively. The $^2P°_{3/2}$ level radiative lifetime is 1.4 s leading to lines at 4068.6, 10320.5, and 10286.7 A in the ratio 1:0.62:0.51. The $^2P°_{1/2}$ level has a 2.4 s lifetime and yields 4076.4, 10370.5, and 10336.4 A emission in the ratio 1:0.68:1.5 [Osterbrock, 1974]. The differing lifetimes for levels in the same term causes the relative emission rate for lines in a multiplet to depend on the electron density, Eq. (6.6). Collisions will populate the $^2D°$ levels, say, approximately in proportion to their multiplicities, 6 and 4 (neglecting the difference in Boltzmann factor between the $^2D°$ levels and radiative cascades from the $^2P°$ levels). At sufficiently low electron number density, all $^2D°$ ions radiatively decay in the excitation ratio: $[4\pi J (6716 \text{ A})/4\pi J (6731 \text{ A})]_{n_{e\cdot \text{low}}} \simeq 1.5$. At sufficiently high electron number density, collisional deexcitation becomes important. The shorter radiative lifetime of the $^2D°_{3/2}$ level enhances 6731 A emission relative to 6716 A, and the line intensity ratio asymptotically approaches the ratio of the equilibrium populations divided by the corresponding radiative lifetimes: $[4\pi J (6716 \text{ A})/4\pi J (6731 \text{ A})]_{n_{e\cdot \text{high}}} \simeq 0.4$. The transi-

S III

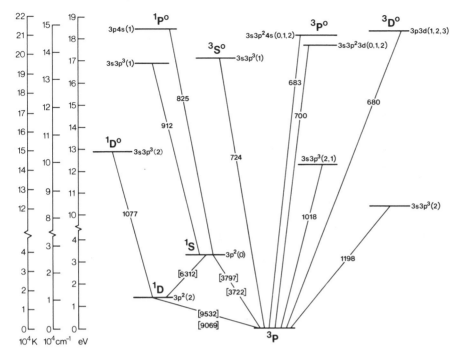

Fig. 6.5. Grotrian diagram for S III.

tion from one limiting case to the other occurs over that range of electron densities for which the collision time is comparable to the radiative lifetime. At the electron temperatures in the Io torus (Table 6.3), the range is $10^2 < n_e < 10^4 \, \text{cm}^{-3}$. The torus electron density is indeed in this range, so $4\pi J \, (6716 \, \text{A})/4\pi J \, (6731 \, \text{A})$ is a useful indicator of n_e. O II has the same term structure as S II; the corresponding [O II] line ratio, $4\pi J \, (3729 \, \text{A})/4\pi J \, (3726 \, \text{A})$, is also sensitive to n_e over the same range. Typical electron densities determined from these line ratios are shown in Table 6.2.

Because of the much shorter lifetimes, the [S II] ratio $4\pi J \, (4069 \, \text{A})/4\pi J \, (4076 \, \text{A})$ is sensitive to n_e only for $n_e > 10^5 \, \text{cm}^{-3}$, higher density than has been observed in the Io torus. Since the small energy difference between the $^2P^o$ levels leads to virtually identical dependences of the rate coefficients of these lines on T_e, this line ratio as observed around Jupiter is essentially independent of the distribution of electron speeds.

The intensity ratio of either blue [S II] line to either red line (for example, $4\pi J \, (4069 \, \text{A})/4\pi J \, (6731 \, \text{A})$) is sensitive primarily to T_e for $n_e < 10^4 \, \text{cm}^{-3}$, but a dependence on electron number density appears for $n_e > 10^4 \, \text{cm}^{-3}$. Morgan and Pilcher [1981] found occasional values of this line ratio not consistent with the electron densities and temperatures of Tables 6.2 and 6.3. These anomalous high blue-to-red emission ratios appear to imply transient high electron density ($n_e > 3 \times 10^4 \, \text{cm}^{-3}$) and low electron temperature ($T_e < 1 \times 10^4 \, \text{K}$).

The extreme ultraviolet lines

In addition to the visible and near UV radiation detected by ground-based instruments, the Io plasma torus emits substantial radiation in the extreme ultraviolet (EUV). Weak emission from an incomplete torus was first detected by the Pioneer 10 UV photo-

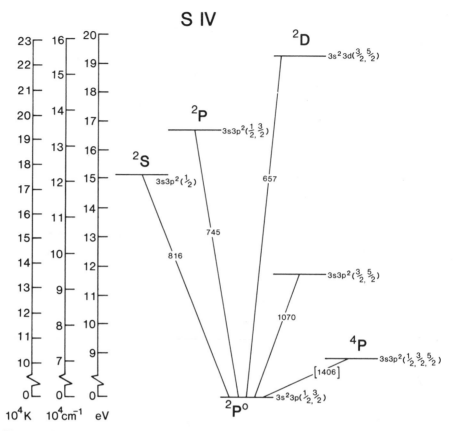

Fig. 6.6. Grotrian diagram for S IV.

meter in both its short ($\sim 40\,R$ for $\lambda < 800$ A) and long (~ 300 R for $\lambda < 1600$ A) channels [Judge et al., 1976]. The ultraviolet spectrometer (UVS) on Voyager 1 measured strong EUV emission (2–3×10^{12} W) from a complete torus [Broadfoot et al., 1979]. The EUV power is about two orders of magnitude greater than the power emitted in the forbidden lines observed from the ground. Observations from rockets and Earth-orbiting telescopes provide continuing ultraviolet measurements [e.g., Moos and Clarke, 1981].

The general characteristics of the Voyager UVS spectra (Fig. 6.9) have been described by Pilcher and Strobel [1982] and can be summarized as follows. The spectra are dominated by features centered at 685 and 833 A. The former is primarily due to overlapping multiplets of S III plus a contribution from O III. The 833 A feature is due to nearly-coincident multiplets of O II and O III plus smaller contributions from S III and S IV. The 685 A feature is surrounded by multiplets of S III (724 A, 729 A) and S IV (657 A, 745 A) that contribute to the wing intensities. There is a persistent feature of S II at 765 A and a variable emission near 910 A that may be due to a blend of an S II multiplet at 910 A with an unknown feature at ~ 900 A. At longer wavelengths there are multiplets of S III (1018, 1077, 1198 A) and S IV (1070 A). Unlike the forbidden lines observed from the ground, none of these multiplets are resolved in the UVS spectra since the instrument has only 30 A resolution. Thus, the fractional contribution of each multiplet is difficult to determine separately.

Emission from S II has been detected at 1256 A with higher spectral resolution by means of IUE [Moos and Clarke, 1981]. IUE spectra also show emission from S III

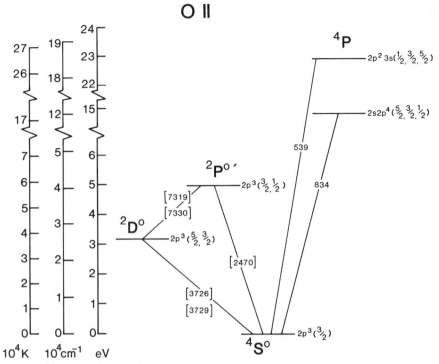

Fig. 6.7. Grotrian diagram for O II.

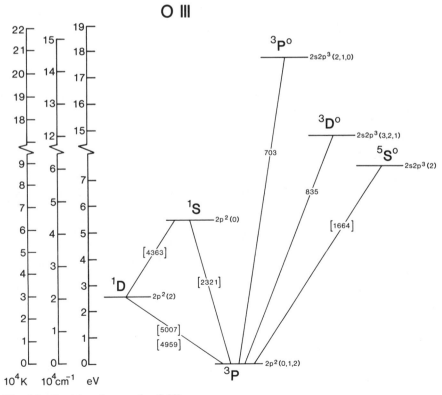

Fig. 6.8. Grotrian diagram for O III.

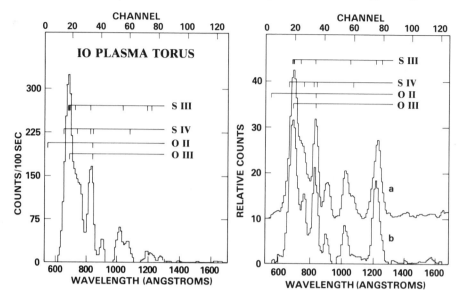

Fig. 6.9. (Left) Voyager 1 EUV spectrum of the plasma torus at elongation; sky background and instrumental scattering have been removed. Multiplets of identified species are indicated. The feature at 685 A is dominated by S III and has a brightness of 200 R [from Broadfoot et al., 1979]. (Right) Voyager 2 EUV spectra of the Io plasma torus showing differences in spectral content indicating composition and/or electron temperature changes. The spectra have been corrected for instrumental scattering. A large fraction of the feature near 1200 A arises from interstellar H Ly-α emission at 1216 A. The spectrum labeled "a" has been displaced upward by ten units [from Sandel et al., 1979].

(1198 A) and possible weak emission from O III (1664 A) and S IV (1406 A). Moos and Clarke [1981] have proposed that a feature at 1729 A is an intercombination line of S III [Moos and Clarke, 1981].

The allowed transitions observed in the EUV from Voyager and IUE are the result of excitation by electron–ion collisions and prompt radiative decay. The emission rates are sensitive to the electron density and temperature, but collisional deexcitation is unimportant and the rate coefficient α_λ (Sec. 6.2) is a function only of T_e. The unit emission rate in this case is given by [Osterbrock, 1974]

$$\epsilon_\lambda = n_e \alpha_\lambda(T_e) = n_e \frac{8 \times 10^{-14}}{\omega} \frac{\overline{\Omega}}{T_e^{1/2}} \exp\left(-\frac{\Delta E}{T_e}\right) (R \text{ cm}^2) \tag{6.9}$$

where ω is the statistical weight of the ground term, $\overline{\Omega}$ is the thermally averaged collision strength, T_e is in eV, and ΔE is energy of the emitted photon in eV. A group of important collision strengths is given in Table 6.5.

In existing analyses of Voyager UVS spectra it has been assumed that conditions are uniform along the line of sight, and the simplified forms (6.4) and (6.5) have been used in deriving plasma parameters. Synthetic spectra are calculated with various values of the electron temperature and density and the ionic densities n_i until a best fit or a group of best fit spectra is obtained. The uncertainties introduced into these calculations by uncertain atomic parameters have been discussed by Pilcher and Strobel [1981]. As of late 1981, the only spectrum that had been subjected to detailed analysis was the

Table 6.5. *Thermally averaged collision strengths*

Ion	Multiplet	λ(A)	Thermally averaged collision strengths[a]	
			$T_e = 5\text{--}10\,\text{eV}$	$100\,\text{eV}$
S II	$g3p^3\,{}^4S^\circ - 3p^4\,{}^4P$	1256	2.2	2.7
	$- 4s\,{}^4P$	910	1.6	6.4
	$- 3d\,{}^4P$	765	7.25	38
S III	$g3p^2\,{}^3P - 3d^3\,{}^3D^\circ$	680	29	75
	$- 4s\,{}^3P^\circ$	683	6.7^c	20^c
	$- 3d\,{}^3P^\circ$	700	10^c	22^c
	$- 3p^3\,{}^3S^\circ$	724	12	21
	$3p^2\,{}^1D - 4s\,{}^1P^\circ$	729	1.5^b	2.5^b
	$3p^2\,{}^1S - 4s\,{}^1P^\circ$	825	0.5^b	0.8^b
	$g3p^2\,{}^3P - 3p^3\,{}^3P^\circ$	1018	3	4.7
	$- 3p^3\,{}^3D^\circ$	1198	4	4.5
S IV	$g3p\,{}^2P^\circ - 3d\,{}^2D$	657	19	27
	$- 3p^2\,{}^2P$	745	16	21
	$- 3p^2\,{}^2S$	816	2.2	4.2
	$- 3p^2\,{}^2D$	1070	4.5	4.5
	$- 3p^2\,{}^4P$	1406	4.5	0.4
O II	$g2p^3\,{}^4S^\circ - 3s\,{}^4P$	539	0.5	0.9
	$-2p\,{}^4P$	834	4.3	8.8
O III	$g2p^2\,{}^3P - 2p^3\,{}^3P^\circ$	703	6.0	9.9
	$- 2p^3\,{}^3D^\circ$	834	7.4	9.9
	$- 2p^3\,{}^5S^\circ$	1664	1.1	0.45

[a] Values for S IV from Bhadra and Henry (1980) and Dufton and Kingston (1980). Values for O III (1664 A) from Baluga et al. (1980). Values for S II, S III, O II and O III (except 1664A) from Ho and Henry (1982). S III (729 A and 825 A) values from calculations of Strobel and Davis (1980) with empirical adjustments where experimental oscillator strengths are available. Possibly important resonance effects near threshold included only in calculations for O III (1664 A) and S IV (1406 A).
[b] Accuracy may be less than ±50%.
[c] The sum of these collision strengths is more accurate than their relative values.

Voyager 1 spectrum presented by Broadfoot et al. [1979] and shown in Figure 6.9. The results of the most recent analysis of this spectrum [Shemansky and Smith, 1981] are given in Table 6.6. The electron temperatures derived from the Voyager UVS data are shown in Table 6.3.

The determination of the densities of O II and O III from the UVS data is complicated by the superposition of their primary emissions in the 833 A feature. Shemansky and Smith [1981] dealt with this problem by determining the O III contribution from the intensity of its strong but blended multiplet at 703 A; they then assumed that the remainder of the 833 A feature was due to O II. An alternate approach uses the IUE upper limit to O III (1664 A) and the absence of O II (539 A) in the Voyager spectra.

Table 6.6. *Number density (cm^{-3})*

Ion	UVS[a]	UVS (this paper)	IUE[b]	Plasma science (6 R_J)[c]
O II	50	35–4000[e]		130/1100
O III	340	< 110	110	160
O IV	< 17			
S II	44[d]	120[g]	140	430/470
S III	160		240[f]	430/560
S IV	220		90	27/170
S V	≤ 11			
n_e	>1850			

[a] From Shemansky and Smith (1981). Uncertainty 20%. $T_e = 8 \times 10^4$K (7eV) and a 2 R_J thick homogeneous torus are assumed.
[b] From Moos and Clarke (1981).
[c] From Bagenal and Sullivan (1981). Single values or those preceding a slash from constant temperature models; values following a slash from constant thermal speed models.
[d] The uncertainty in the SII density is ± 50%.
[e] See text.
[f] Brightness variations observed corresponding to a range of 160–320 cm^{-3}.
[g] Brown and Shemansky (1982).

From the O III collision strengths of Baluja et al. [1980] and others listed in Table 6.5, we derive (by means of Equations (6.6) and (6.9)) the following multiplet intensity ratios as a function of T_e:

T_e (eV)	3	6	9	30	100
$\dfrac{4\pi J(\text{OII, 539 A})}{4\pi J(\text{OII, 834 A})}$	0.0061	0.028	0.058	0.07	0.10
$\dfrac{4\pi J(\text{OIII, 1664 A})}{4\pi J(\text{OIII, 835 A})}$	2.1	0.52	0.32	0.11	0.05

Around the time of the Voyager 1 encounter, $4\pi J(\text{OIII, 1664 A})$ was less than 15 R (Moos, private communication). The Voyager 1 UVS data indicate $4\pi J(\text{OII, 539 A}) <$ 3 R (Shemansky and Smith, 1981). These values lead to the following intensity upper limits as a function of T_e:

T_e (eV)		3	6	9	30	100
$4\pi J(\text{OII, 834 A})$	<	164 R	36 R	17 R	14 R	10 R
$4\pi J(\text{OIII, 835 A})$	<	7 R	29 R	47 R	135 R	285 R

Thus, the Voyager 1 observed intensity at 833 A of 190 R [Shemansky, 1980a] implies either that the source region is cold ($T_e \lesssim 3$ eV) with O II the principle emitter or hot ($T_e \gtrsim 30$ eV) with O III dominant. For regions where T_e is 3 to 10 eV, the $4\pi J$ (O III, 1664 A) upper limit constrains the O III concentration to <110 cm^{-3}. Similarly, the O II concentration must be < 35, 140, and 4400 cm^{-3} for T_e equal to 9, 6, and 3 eV, respectively.

To account for 200 R of 833 A emission when $T_e < 3$ eV, a highly improbable O II density is required which exceeds independent measures of electron density (Table 6.2). If the 833 A feature originates with O III in a region having effective electron temperature ~ 30 eV, the plasma measurements may suggest the emitting region is at the outer edge of the torus ($\sim 7 R_J$). Recent observations have placed an upper limit to $4\pi J$ (O III, 5007 A) $\lesssim 3 R$ [Brown, Shemansky, and Johnson, 1982]. This corresponds to an O III density upper limit significantly below that derived above, and this implies that the suite of emissions now assigned to O II and/or O III cannot currently be reconciled. The most severe constraints on the O II and O III densities are upper limits to O II (539 A) and O III (5007 A).

There are also problems associated with S IV ultraviolet emission. The Voyager UVS intensities for S IV (745 A) and S IV (1070 A) are 150 R and 100 R, respectively [Shemansky and Smith, 1981]. From IUE the S IV (1406 A) intensity is at most 17 R [Moos and Clarke, 1981], which is consistent with the Voyager upper limit of 50 R. Given the collision strengths in Table 6.5, no single consistent effective electron temperature can be inferred from this data set. However, if it is admitted that the 1070 A feature in the Voyager spectra (see Fig. 6.9) is really a blend of the S III (1077 A) and S IV (1070 A) multiplets as would be expected from the relative intensities of S III (1018 A) and S III (1077 A) multiplets in the solar spectrum, then an effective electron temperature $T_e \sim 20$–30 eV is consistent with the data and suggests that S IV emission occurs primarily at the outer fringes of the torus ($\sim 7 R_J$) where the electron temperature is high [Scudder et al., 1981].

Brown and Shemansky [1982] have recently reviewed the observed visible and UV spectrum of S II. The S II $3p^2 3d^4 P$ term, misplaced by Moore [1971], has been found by Petterson and Martinson [1982] at 16.2 eV and the corresponding ground state transition multiplet occurs at 765 A. The Voyager UVS spectra display a $\sim 60 R$ emission feature at this position that was previously assigned to K III [Shemansky and Smith, 1981]. Brown and Shemansky [1982] find the S II emissions from the vicinity of Io's orbit at the epoch of Voyager 1 are consistent with $T_e \sim 7$ eV, $n_e \sim 1.6 \times 10^3$ cm^3 and a S II density ~ 120 cm^{-3}. The corresponding values at the epoch of Voyager 2 are 5 eV, 2.3×10^3 and 170 cm^{-3}. The implications of differential varibility in S II emission brightness at UV versus visible wavelengths have yet to be adequately addressed.

The characteristic electron temperature for S III emission is presumably intermediate between S II and S IV ($5 < T_e < 30$ eV). Unfortunately, attempts to accurately calculate the relative collision strengths of the two S III multiplets at 683 and 700 A (the strongest EUV feature) with state-of-the-art computer codes by Ho and Henry (private communication, 1981) and Davis and Strobel (private communication, 1981) have failed because of the extremely strong configuration mixing. In addition it appears that the S III (1018 A) feature may be contaminated by another emitter, which precludes inference of an accurate electron temperature from the intensity ratio of the 685 and 1018 A features. (See Fig. 6.9.)

We are led by the foregoing discussions of intensity ratios to the view that disjoint regions are addressed by remote spectroscopic studies of the Io plasma torus. The singly ionized species are predominant in low electron temperature regions (< 3 eV), whereas O III and S IV emit primarily in regions where $T_e \sim 30$ eV. This picture is consistent with the ionization energies, ion residence times and the radial variation of electron temperature observed from Voyager. It appears that the Io torus is inhomogeneous along the line of sight and that equivalent homogeneous models have fundamental difficulties in accurately inverting remote measurements in terms of local plasma properties.

Fig. 6.10. The data points show the
measured intensity of the 685 A feature
as a function of distance from Jupiter
measured in the orbital plane of the
satellites. A model torus used to fit the
data is shown to scale above the data; the
intensity predicted by this model is
shown by the solid line. Other
observations (not shown) indicate that
the intensities at the eastern and western
elongation points may differ by as much
as a factor of 2 [from Broadfoot et al.,
1979].

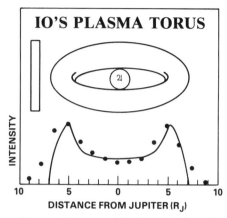

The EUV emissions observed in 1974 from Pioneer 10 were originally thought to be
due in part to atomic hydrogen [Carlson and Judge, 1974; Judge et al., 1976]. If they
were in fact primarily due to oxygen and sulfur, then O I (1304 A, 1356 A), O II
(834 A), and S II (1256 A) may have been primarily responsible [Mekler and Eviatar,
1980]. The torus seen by Voyager may have been 10 times brighter in the long
wavelength channel and 4.5 times brighter in the short wavelength channel of the
Pioneer UV photometer than was recorded during the Pioneer encounter, according
to a preliminary analysis (Judge, private communication, 1980). At the time of the
Pioneer encounter the torus was reported to have an angular extent of ∼ 1/3 of Io's
orbit and to be revolving with Io's orbital velocity, implying a neutral composition.
This is in sharp contrast to the complete, ionized torus observed by Voyager (see Fig.
6.10). Observations of Jupiter and the interplanetary medium by these UV instruments
have been shown to be consistent with measured intensities by Earth orbiting satellites.
Thus, we are forced to conclude that certain plasma characteristics of the torus have
undergone significant changes in the past few years. The interpretation of Pioneer 10
plasma analyzer results by Intrilligator and Miller [1981] suggests that S III, O II, and
O III were present in the torus during the Pioneer encounter at as yet undetermined
concentrations. However, Mekler and Eviatar [1980] have interpreted the Pioneer 10
UV observations as originating from a predominantly O II and S II plasma. Further
analysis of Pioneer data is clearly required before definitive conclusions on compositional differences between the Pioneer and Voyager encounters can be established.
This apparent long-term variability of the plasma torus – reflected to a lesser degree in
differences between Voyager 1 and Voyager 2 results [Sandel et al., 1979; cf. Fig. 6.10)
– is difficult to reconcile with the observed long-term stability of the neutral sodium
cloud, which should respond to changing plasma parameters (see Sec. 6.3).

Ion temperature and spatial distribution

Inside 5.7 R_J, the Voyager 1 plasma experiment [Bridge et al., 1979; Bagenal and
Sullivan, 1981] found resolved but overlapping peaks in the positive ion energy-per-
charge spectra, which were identified as S II, S III, O II, O III, and SO_2^+. These "cool,"
inner-torus spectra were consistent with locally-equilibrated plasma ions at temperatures a few eV or less, with the temperature falling sharply with decreasing radial distance from Jupiter, down to ∼ 0.5 eV at 4.9R_J. These findings are consistent with the
[S II] line-width analyses of Trauger et al. [1979, 1980], which refer to the same region.

Outside 5.7 R_J, the Voyager 1 plasma experiment recorded broad, unresolved spectra. Temperature analysis required external assumptions about the composition and

thermal state of the plasma, but values in the range 40 to 60 eV were found for extreme models [Bagenal and Sullivan, 1981]. Using optical spectroscopy on this "hot" outer torus, Trauger et al. [1979] found ~ 30 eV from the [S III] λ9531 A emission width, and Brown [1981a] found ~ 90 eV from the [S III] λ6312 A emission width at a different epoch. Brown and Ip [1981] have shown that the [S II] 6716 and 6731 A line profiles obtained near Io's orbit have extensive, energetic wings. Brown [1982] found from a detailed line shape analysis that the average S II energy is about 60 eV, whereas a simple interpretation of the line width alone implies 5–10 eV. By comparing the optical line widths, we conclude that outside 5.7 R_J S III is probably hotter than S II, and that more energy may be hidden in possible high-speed tails on the distribution functions of S III and other ions. If the ions occupy the same volume, kinetic thermodynamic equilibrium exists neither within nor between ion species.

The temperatures of the ions are an important factor determining their spatial distributions along the magnetic field lines. At the temperatures of the torus ions, the field-aligned component of the centrifugal force on the ions due to Jupiter's rotation dominates the magnetic mirror force. As a result, the ions have a symmetry plane not at the magnetic equator, but near the centrifugal equator where the centrifugal potential energy of an ion along a field line is a minimum [Hill and Michel, 1976; Cummings et al., 1980]. For a dipolar field this surface is a plane whose tilt relative to the rotational equator is about two-thirds of the tilt of the magnetic equator or, for Jupiter, 7°. The height of an ion above or below the centrifugal equator is a measure of its thermal energy along a field line.

Ground-based observations imply that the spatial distribution of S II inside Io's orbit is generally wedge-shaped, the apex of the wedge lying on or near the centrifugal equator [Hill, Dessler, and Michel, 1974] and about 5 R_J from the center of Jupiter [Nash, 1979; Pilcher, 1980a; Pilcher et al., 1981]. This wedge shape agrees well with the ion distribution expected from the strong radial gradient in the ion temperature determined from Voyager data [Bagenal and Sullivan, 1981]. S II formed near Io's orbit may cool during the slow inward diffusion (see Sec. 6.5), collapsing to the centrifugal equator and forming the wedge. The apex of the wedge as oberved from the ground corresponds to the maximum S II density measured from Voyager [Pilcher, 1980a; Bagenal and Sullivan, 1981].

The strong radial gradient in the ion temperature and temporal variations in the torus structure have led to some confusion as to the isotropy of the ion temperature. Eviatar et al. [1979] presented slit spectra of the red S II lines acquired with the slit perpendicular to the Jovian equatorial plane and 5 R_J from the center of the planet's disk. One of the spectra was acquired when the tilted plasma torus was seen approximately edge-on, thus providing a measure of the thickness of the torus. The parallel temperature of 38 eV derived from this measurement was compared with the perpendicular ion temperature of ~ 2 eV derived from line width measurements by Trauger et al. [1980]. Because both sets of data were acquired at approximately the same apparent radial separation on the sky from Jupiter, they concluded that the ion temperature exhibits a substantial anisotropy. The line of sight of Eviatar et al. passed through the entire torus, but the results are weighted geometrically toward the emitting region near 5 R_J. They noted that a second spectrum obtained the following night "indicates that the nebula has a fairly large radial extent in addition to its thickness." It seems likely, therefore, that they were looking through the wedge, accounting at least in part for the substantial thickness they measured. The measurements of Trauger et al. were apparently acquired on a night when the wedge was absent, and therefore refer entirely to the cool plasma inside Io's orbit. Subsequent line shape measurements [Brown, 1982] have shown hotter S II at larger distances from Jupiter, consistent with the presence of a radial ion temperature gradient. We therefore conclude that there is no convincing evidence for ion temperature anisotropy in the plasma torus.

Pilcher et al. [1981] have shown that the structure of the S II component of the plasma torus varies strongly with both magnetic longitude and time. They found the [S II] 6731 A wedge on a particular night (13 UT March 1981) only between magnetic longitudes System III (1965) = 140°–230°. At higher longitudes (up to at least 325°) the S II emission was concentrated near 5.7 R_J, the distance from Jupiter at which both the Voyager plasma science and planetary radio astronomy experiments detected a charge density maximum [Bagenal and Sullivan, 1981]. At lower longitudes (to at least 60°) there was evidence only of the concentration of S II near 5 R_J. This structural variation with magnetic longitude was far less pronounced a month later in April. In the March data, S II emission extended well beyond Io's orbit (out to 7–7.5 R_J) for longitudes between 150° and at least 325°. Other observations showing S II emission beyond Io's orbit and/or intensity variations with magnetic longitude have been presented by Trafton [1980], Trauger et al. [1980], Pilcher and Morgan [1980], Morgan and Pilcher [1982], and Brown and Shemansky [1982].

These variations of S II forbidden line brightness with magnetic longitude must be due to a corresponding variations of electron or S II density or both. In the EUV-emitting torus, where S II is a minor constituent, the electron density appears to be independent of magnetic longitude (Shemansky and Sandel, 1981; Brown and Shemansky, 1982). Longitude-systematic variations in the S II forbidden line brightness near Io's orbit must therefore be due to variations in the S II number density. S II is a dominant component of the plasma well inside Io's orbit, where it comprises as much as half or more of the total charge density (Broadfoot et al., 1979). In this region variations of the forbidden line brightness probably reflect a variation in plasma mass and charge density with magnetic longitude.

Although S II emission is often observed outside of Io's orbit, most of the sulfur at these distances is more highly ionized. The Voyager 1 plasma science experiment found S III was largely absent in the relatively cool plasma inside of 5.5 R_J where S II was abundant. The S III was found to extend to ~ 7.5 R_J with maximum density at Io's orbit [Bagenal and Sullivan, 1981].

The spatial distribution of S III EUV emission measured from Voyager 1 before encounter is shown in Figure 6.10 [Broadfoot et al., 1979]. The data were fit with the model torus shown in the figure: a radius of symmetry of 5.9 ± 0.3 R_J and a cross-sectional radius of 1.0 ± 0.3 R_J centered on the magnetic equator (although it was not possible to distinguish on the basis of the data between this plane and the centrifugal equator. The intensity distribution corresponding to this model is shown as a solid line superimposed on the data in Figure 6.10. The EUV emission extends beyond the outer boundary of this homogeneous torus, but is consistent with a rather well defined inner edge near 5 R_J.

These data combined show that although there is some spatial separation of S II and S III in the torus, there is a substantial region of overlap. This leads to an inconsistency, as pointed out by Shemansky and Smith [1981]. The S II emission observed from the ground shows a marked variation with longitude and time. The EUV emissions, on the other hand, show no long-term magnetic longitude dependence, only a Jupiter local-time asymmetry, and they are much less temporally variable [Sandel et al., 1979; Shemansky and Sandel, 1981; Sandel and Broadfoot, 1982]. Because S II is presumably the source of S III, the brightness variations of the former are somehow suppressed from manifesting themselves in the emission characteristics of the latter. Shemansky and Smith [1981] have suggested that this suppression might result from a long diffusive loss time. Alternatively, the S II emission may reflect the production rate rather than the spatial density of the ion. A third possibility is that the longitude-variable S II is decoupled from the EUV-emitting and longitude-invariant S III (Brown and Shemansky, 1982). The UV local time asymmetry is interpreted by Shemansky and Sandel [1981] as a 10 hr periodicity in electron temperature.

Pilcher et al. [1981] have suggested that the S II emission from the hot torus (i.e., outside of Io's orbit) may be the product of plasma that was created at or just inside of Io's orbit and is observed as it streams rapidly outward. Radial velocities of order 1 km sec^{-1} are suggested by the short lifetime of S II in the hot plasma. This apparently rapid plasma transport over a limited range of magnetic longitudes [Pilcher et al., 1981] is qualitatively consistent with the description of corotating magnetospheric convection presented by Hill, Dessler, and Maher [1981]. However, there is no reported evidence of this rapid outward transport in the Voyager EUV data.

A difference between the spatial distributions of ionized sulfur and oxygen has been detected both from Voyager [Bagenal and Sullivan, 1981] and from the ground [Pilcher and Morgan, 1979, 1981]. Oxygen is present at greater distances from the symmetry plane of the torus, an effect that appears to be a result of its smaller mass [Bagenal and Sullivan, 1981].

6.5. Radial transport

Two types of radial transport are present in the Io torus; the ballistic motion of neutrals escaping from Io and the cross-L transport of ions that, at least in part, were formed from these neutrals. Both of these transport mechanisms manifest themselves in spatial distributions of the transported material that are larger than their immediate sources: for example, the radial extent of the neutral sodium cloud is large compared to the diameter of Io; the radial extent of the sulfur and oxygen plasma is large compared to the expected dimensions of the neutral clouds of these elements. The spatial distribution of neutrals has been extensively studied [Murcray and Goody, 1978; Smyth and McElroy, 1978; Goldberg et al., 1978; Pilcher and Schempp, 1979] as a means of elucidating the characteristics of the neutral transport mechanism, that is, the directions and velocities of ballistic ejection from Io (see Sec. 6.3). Similarly, the radial distribution of ions determined from the Voyager plasma science experiment has been analyzed in terms of possible characteristics of the source and the radial transport mechanism. Brown and Ip [1981] have argued that extensive neutral clouds are the major ion source, which would have important consequences for the dynamics and structure of the Io torus.

Flux tube interchange diffusion is the cross-L transport mechanism that has been most extensively considered for the plasma torus. Two types of interchange that have been considered for Jupiter are those driven by fluctuating dynamo electric fields in Jupiter's upper atmosphere that map into the magnetosphere [Brice and McDonough, 1973; Coroniti, 1974] and by centrifugal instability resulting in a rapid decrease in plasma density with increasing L-shell [Ioannidis and Brice, 1971; Richardson and Siscoe, 1981; Siscoe and Summer, 1981]. In the presence of plasma distributed nonuniformly in magnetic longitude, centrifugal instability may lead to corotating convection [Hill et al., 1981; see Chap. 10]. Froidevaux [1980] concluded that both types of interchange were possible, whereas Richardson et al. [1980] concluded that only a time-dependent ion ejection rate could account for the Voyager observations, with the rate of centrifugally driven outward diffusion a factor of 50 greater than the inward diffusion rate. (See Chap. 10 for an alternate view.)

In these analyses it was assumed that the ion source term is zero everywhere except at the L-shell of Io. However, this seems highly improbable since the observed extensive atomic clouds of sodium, potassium, and oxygen (Sec. 6.3), and the likely cloud of neutral sulfur, are expected to correspond to ion sources. Further, the absence of enhanced EUV emission when Io is within the field of view of the Voyager spectrometer suggests that copious ionization of oxygen and sulfur atoms is not taking place in the immediate vicinity of the satellite [Shemansky, 1980b], although ionization by Alfvén's [1954] critical-velocity mechanism may not produce such radiation. Attempts to fit the Voyager data with radially extensive ion sources may therefore lead to conclusions significantly different from those summarized above. Nonetheless, it seems clear that ion radial transport rates of the order-of-magnitude of those deduced in the studies cited above are necessary to account for the observed distribution of ionization states.

6.6. Ionization and recombination

Carlson et al. [1975] showed that electron impact ionization is much more rapid than photoionization in the Io torus even for the low ionization potential of neutral sodium (5.14 eV). Rate coefficients for electron-impact ionization for the principal species in the torus are shown in Figure 6.3. In Section 6.3.C we discussed the importance of charge exchange in the ionization of oxygen and sulfur.

The total ionization rate can be calculated under the assumption that the EUV radiation is powered by the magnetic field sweeping of ions created in an extended neutral cloud [Broadfoot et al., 1979; Brown, 1981a]. The acceleration of newly created ions to the corotational velocity results in ion cyclotron-energies of ~ 280 eV and ~ 550 eV for oxygen and sulfur, respectively. This energy may be transferred to the ambient electrons by means of Coulomb interactions or plasma instabilities, and then back to the ions by collisional electronic excitation leading to the emission of radiation [Brown, 1981a,b]. The observed total emission rate of $\sim 2 \times 10^{12}$ W [Broadfoot et al., 1979; Shemansky, 1980a] implies a total ionization rate (for a sulfur-to-oxygen ratio of 0.5) of $\sim 3 \times 10^{28}$ sec^{-1}. In Section 6.3.B we showed that other considerations lead to an estimate of the total ionization rate in the range $(1-2) \times 10^{28}$ sec^{-1}. The comparability of these rates implied by the radiated EUV power supports the assumption made above that the original source of the radiated power may be Jupiter's rotation transferred by the acceleration of newly created ions.

Shemansky and Sandel [1981] report a 40% EUV brightness modulation pattern stationary in Jupiter local time and stable for at least a half-year. Sandel and Broadfoot [1982] have reported a secondary brightening associated with Io's position. Brightness variations in the EUV-emitting torus must be caused by changes in electron temperature, since significant fluctuations in plasma mass are ruled out by the absence of a magnetic longitude dependence in both the EUV luminosity [Shemansky and Sandel, 1981] and the [S II] ratio $4\pi J (6716$ A$)/4\pi J (6731$ A$)$ at 5.9 R_J [Brown and Shemansky, 1982]. Since EUV emission is the main plasma energy sink, these variations seem to reflect modulation of the energy transfer channel(s), the nature of which remains obscure. The local-time and Io-correlated sources apparently supply $\sim 80\%$ and $\sim 20\%$, respectively, of the EUV radiated power.

In the inner torus ($L < 5.8$) radial transport is probably slow (Richardson et al., 1980; Bagenal et al., 1980; cf. Secs. 6.4, 6.5) and ion-electron recombination may affect the ionization distribution. The principal mechanism for recombination in a high-temperature plasma is dielectronic recombination, a process involving the simultaneous change in quantum states of two electrons. It may be viewed as two steps: first, the radiationless capture of an electron with simultaneous excitation of an already bound electron

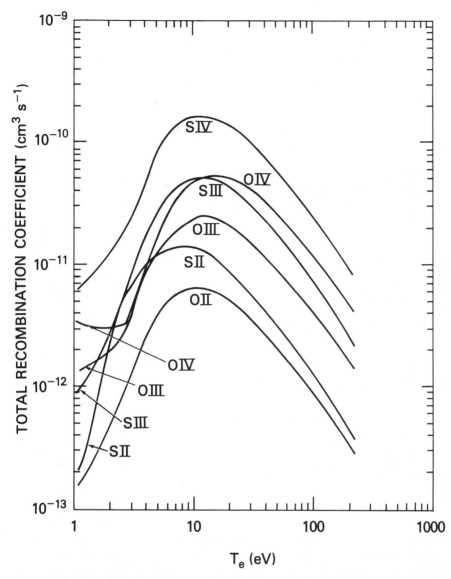

Fig. 6.11. Total (sum of radiative and dielectronic) recombination coefficients [from Jacobs et al. 1978, 1979].

$X^{z+} + e \rightarrow X^{(z-1)+}$ (doubly excited)

followed by a stabilizing radiative transition to a final state below the ionization threshold,

$X^{(z-1)+}$ (doubly excited) $\rightarrow X^{(z-1)+}$ (singly excited) $+ h\nu$

and a subsequent radiative transition. The recombination rate dn_i/dt is given by

$$dn_i/dt = \alpha\, n_i n_e$$

where n_i is the density of the recombining ion and α is the total recombination coefficient, including both radiative and dielectronic components. In Figure 6.11 we show α for the principal Io torus ions as a function of T_e. At low electron temperature ($T_e < 1$ eV) radiative recombination dominates, its rate decreasing with increasing T_e. The dielectronic recombination coefficients peak near $T_e \simeq 10$ eV.

6.7. Concluding remarks

Since the discovery of sodium emission by R. A. Brown [1974], our understanding of Jupiter's heavy-ion environment has improved greatly through the union of the Pioneer and Voyager measurements with continuous spectral observations from the vicinity of Earth. A vast and complex geophysical entity has been identified and is now coming into focus. We know Io is the source of a peculiar mix of both neutral and charged material for Jupiter's magnetosphere. Sputtering may be the escape mechanism from Io, but neither this nor any of the details of timing, isotropy, and species selection are established. The relationships between the torus and the active and bizarre properties of Io's solid body are also not known. We believe energy from Jupiter's rotation drives the torus emissions, but adequate models do not yet exist. We see evidence for thermal and convective ion motions and for collisional acceleration of the ballistic neutral atoms, but a coherent picture has yet to emerge of the alternate avenues along which Io-derived particles are transported from source to sink. There is conflicting evidence concerning the torus' temporal variability from Pioneer/Voyager differences and the remarkable stability of the sodium cloud as seen from Earth.

The Io torus studies are contributing to the refinement of theoretical models and diagnostic tools that have direct application to other astrophysical problems. The detection and interpretation of spectral emissions is the classical basis for our view of the physical state and processes that obtain in active galaxies, stellar coronae, and planetary nebulae. The Io torus is by far the most studied and best characterized astrophysical plasma, and it is the first to be intensively observed from near infrared to vacuum ultraviolet wavelengths. Working out the difficult relationship between torus observations at those greatly different photon energies is an activity relevant to interpretations of ultraviolet emission from extra-solar-system objects obtained from Earth orbit.

In this chapter we have summarized first-order physical characteristics of Jupiter's luminous magnetosphere deduced from optical and ultraviolet spectral observations. In concluding, we wish to point out two lines of development that will improve the specificity of future remote studies of the Jupiter plasmasphere. The first is improvement in values of the atomic parameters, such as transition probabilities, collision strengths, and the cross sections for ion/ion and ion/atom reactions, which are vital for modeling emission line observations in terms of intrinsic plasma properties. The second is a better three-dimensional view of the emitting regions. Central to existing analyses has been the simplifying assumption, when one is required, that the emitting region is homogeneous in depth. Variations in radius and longitude are well documented, so this assumption is patently false. In more sophisticated research geometrical factors will be assessed and corrections incorporated into observing techniques and model interpretations. The data base can respond to this requirement: the torus rotates rapidly and so is unfolded in its third dimension by the changing perspective. Through these and other research developments, we expect telescopes to provide continuing and improving observational contact with Jupiter's remarkable plasma environment.

ACKNOWLEDGMENTS

The contribution of R.A.B. was supported by NASA grant NSG-7634. The contribution of C.B.P. was supported by NASA grants NGL 12-001-057, NSG 7403, and NAGW 153. D.F.S. was supported in part by a grant from NASA Office of Planetary Atmospheres; he would like to thank R. J. W. Henry for collision strengths prior to publication, Davis for unpublished collision strength calculations and V. Jacobs for detailed computer output to construct Figs. 6.3 and 6.11.

7

PHENOMENOLOGY OF MAGNETOSPHERIC RADIO EMISSIONS

T. D. Carr, M. D. Desch, and J. K. Alexander

The radio spectrum of Jupiter spanning the frequency range from below 10 kHz to above 3 GHz is dominated by strong nonthermal radiation generated in the planet's inner magnetosphere and probably upper ionosphere. At frequencies above about 100 MHz, a continuous component of emission is generated by synchrotron radiation from trapped electrons between equatorial distances of about 1.3 and 3 R_J. This component exhibits a broad spectral peak at decimetric (DIM) wavelengths, distinct longitudinal asymmetries arising from asymmetries in Jupiter's magnetic field, and slow intensity variations that are presumably related to temporal changes in the energy, pitch angle, or spatial distributions of the radiating electrons. High resolution mapping of this component will probably continue to provide detailed information on the inner magnetosphere structure that is presently unobtainable by other means. Jupiter's most intense radio emissions occur in the frequency range between a few tenths of a MHz and 39.5 MHz. This decameter-wavelength (DAM) component is characterized by complex, highly organized structure in the frequency-time domain and by a strong dependence on the longitude of the observer and in some cases, of Io. The DAM component is thought to be generated near the electron cyclotron frequency in and above the ionosphere on magnetic field lines that thread the Io plasma torus, but neither the specific location(s) of the radio source(s) nor the specific plasma emission process are firmly established. At frequencies below about 1 MHz there exist two independent components of emission that have spectral peaks at kilometer (KOM) wavelengths. One is bursty, relatively broadbanded (typically covering 10 to 1000 kHz), and strongly modulated by planetary rotation. The properties of this bKOM component are consistent with a source confined to high latitudes on the dayside hemisphere of Jupiter. The other kilometric component (nKOM) is narrow banded, relatively weak, and exhibits a spectral peak near 100 kHz. The nKOM also occurs periodically but at a repetition rate that is a few percent slower than that corresponding to the planetary rotation rate. This component is thought to originate at a frequency near the electron plasma frequency in the outer part of the Io plasma torus (8 to 10 R_J) and to reflect the small departures from perfect corotation experienced by plasma there.

7.1. Introduction

The accidental discovery of the low-frequency Jovian radio emission by Burke and Franklin [1955] holds a special place in the history of solar system exploration. This discovery antedated many of the established historical landmarks, including the discovery of the Earth's radiation belts [Van Allen et al., 1958], the verification of the existence of a solar wind [Neugebauer and Snyder, 1962], and even the dawn of the space age itself in 1957. It foreshadowed a revolution in the thinking about the nature of planetary bodies and especially of their magneto-plasma environments. Considered at one time to be a 100 K gas giant surrounded by only the vacuum of space, Jupiter was soon realized to have a substantial magnetic field containing extensive energetic plasma capable of sustaining radio emission with brightness temperatures in excess of 10^{12} K. A discovery as fundamental as that of Burke and Franklin, but one which this time was not accidental, came in 1964 with the announcement by Bigg [1964] that some of the Jovian emission was directly influenced by the Galilean satellite Io. We now recognize Io as one of the solar system's most unusual members, but this discovery was the first evidence that Jupiter and Io constitute a complicated electro-dynamic system.

Jupiter has now been observed over 24 octaves of the radio spectrum, from about 0.01 MHz (10 kHz) to 300,000 MHz (300 GHz). Its radio emissions dramatically fill the entire spectral region where interplanetary electromagnetic propagation is possible at wavelengths longer than infrared. Three distinct types of radiation are responsible for this impressive radio spectrum. Thermal emission from the atmosphere accounts for virtually all the radiation at the high frequency end. Synchrotron emission from the trapped high-energy particle belt deep within the inner magnetosphere is the dominant spectral component from about 4000 to 40 MHz. The third class of radiation consists of several distinct components of sporadic low frequency emission below 40 MHz, much of which is presumably due to plasma instabilities in the inner magnetosphere that generate radiation slightly above the local electron cyclotron frequency. The nonthermal radiations led to the discovery of the magnetosphere over two decades ago, and they will remain for some time to come major sources of information regarding the magnetosphere and magnetic field of the planet. Although the thermal radio emission has yielded important information on the structure of the atmosphere underlying the magnetosphere [Berge and Gulkis, 1976], the interpretation of the thermal component lies outside the scope of this book.

In this chapter, our interest in the synchrotron and low-frequency components is twofold. First, they provide a means for probing the inner magnetospheric regions (inside $L = 3$) for which the only available in situ measurements are the very few given to us by Pioneer 11, with no prospect for additional ones for many years. Because the theory of the synchrotron emission process is well understood (unlike the low frequency emissions), the synchrotron component is an indispensable diagnostic tool for the study of the radiation belts within a few Jovian radii of the planetary body. Our second interest lies in the elucidation of the largely unexplained intricate phenomenology of the low frequency emission, in the hope of stimulating progress toward its eventual explanation in terms of one or more types of plasma instabilities. When that is accomplished, and the essential related details of emission and propagation are worked out, the low-frequency radiation too will become a useful diagnostic tool for the investigation of the inner Jovian magnetosphere.

Important historical milestones in the radio astronomical investigation of Jupiter are listed in Table 7.1, and a summary plot of the average power spectrum of its nonthermal emissions over the entire radio frequency range, with explanatory labeling, is presented in Figure 7.1. Some salient properties of the various Jovian nonthermal radio components are summarized in Table 7.2. The nomenclature and coverage of the standard radio engineering bands in which the emissions have been found are included in Figure 7.1. Each new component has been named for the band in which it was discovered, but none of the emissions are confined to a single band. There are two distinct subcomponents of the kilometric radiation, as will be explained in Section 7.4. Although the kilometric radiation is most likely distinct from the hectometric, it is not yet clear whether the latter should be considered a separate component from the decametric or merely an extension of it.

7.2. The decimeter wavelength emission

Early evidence for a Jovian magnetosphere: discovery of the synchrotron emission

The high degree of circular polarization of Jupiter's sporadic decametric bursts [Franklin and Burke, 1956] was the first indication of the Jovian magnetic field, although the significance of this early evidence was not widely recognized at the time.

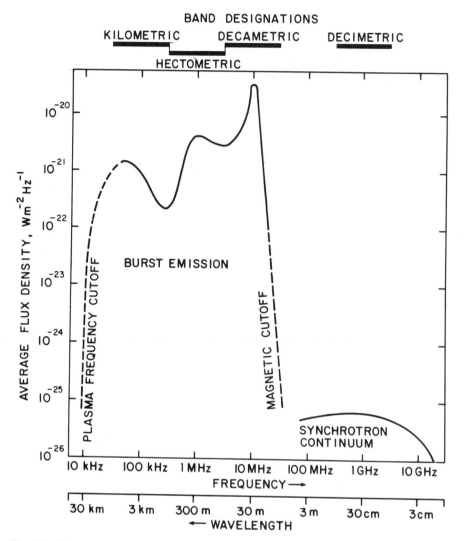

Fig. 7.1. The average power flux density spectrum of Jupiter's non-thermal magnetospheric radio emissions. Burst-component flux densities were averaged over inactive as well as active periods; the instantaneous spectrum may appear considerably different. The highest burst peaks attained values one to two orders of magnitude above the curve. The solid-line part of the burst-component curve is from Schauble and Carr (unpublished). The synchrotron-component curve is the difference between the total and thermal curves in Figure 7.2. Flux densities are normalized to a distance of 4.04 A.U.

For electromagnetic radiation to be polarized, the motions of the emitting particles must be subject to a systematic constraint. It is difficult to conceive of any such constraint acting on the source of naturally occurring radio emissions other than that exerted by an ambient magnetic field. The discovery of the Jovian nonthermal decimetric component [Sloanaker, 1959; McClain and Sloanaker, 1959] and the subsequent revelation by a number of observers of its more conspicuous properties, however, left no room for doubt of the existence of a planetary field. The properties are:

1. Continuous radiation rather than the sporadic bursts characteristic of the decametric component.

Table 7.1. *Historical milestones in Jupiter radio astronomy*

1950 – Decametric radiation (DAM) recorded by Shain at 18.3 MHz but not initially recognized.

1955 – DAM discovered by Burke and Franklin at 22.2 MHz.

1955–58 – first theories for the origin of DAM, including:
- electrostatic discharge noise from lightning in Jovian thunderstorms,
- ionospheric plasma waves generated by atmospheric turbulence, and
- ionospheric oscillations generated by volcanic shock waves.

1958 – nonthermal decimetric radiation (DIM) detected at wavelengths of 10 cm and longer.
    – trapped energetic particles (Van Allen belts) detected in the terrestrial magnetosphere by Explorer-1 & 3 satellites.

1959–61 – DIM attributed to synchrotron radiation from energetic particles trapped in the tilted dipole field of the Jovian magnetosphere.
    – DAM attributed to cyclotron radiation or cerenkov radiation by energetic electrons in the Jovian ionosphere.

1964 – Io influence on DAM discovered by Bigg.

1965 – first high resolution map of DIM obtained by Berge.
    – terrestrial kilometric radiation (TKR) detected by USSR Elektron-2 satellite.

1963–69 – first applications of plasma kinetic theory to explanation of DAM.

1973–74 – first detection of Jupiter below 1 MHz with RAE-1 and IMP-6 satellites.
    – first detailed studies of TKR with IMP-6 satellite.

1978–79 – full spectrum of Jupiter's low frequency emissions measured by Voyager.

2. A nonthermal spectrum that is nearly flat in the decimeter-wavelength band, but extends well into the microwave region where it overlaps the thermal component.
3. A distributed emission region that is centered on the planet and is roughly three Jovian diameters wide east-west and one diameter high north-south.
4. A linearly polarized component, amounting to 20 to 25% of the total intensity, most of the remainder being unpolarized.
5. A weak circularly polarized component that alternates between right- and left-hand polarization as the planet rotates.
6. An almost sinusoidal rocking of the plane of linear polarization of about $\pm 10°$ as the planet rotates, with the average orientation of the polarization plane perpendicular to the rotation axis.
7. A periodic variation in the total flux density as the planet rotates, generally with two maxima and two minima per rotation.

These clues clearly pointed to synchrotron emission by high energy electrons trapped in a Jovian Van Allen Belt [Drake and Hvatum, 1959; Roberts and Stanley, 1959; Field, 1959; Radhakrishnan and Roberts, 1960; Morris and Berge, 1962; Gary,

Table 7.2. *Components of Jupiter's radio spectrum*

Emission component	Decimetric (DIM)	Decametric–hectometric (DAM–HOM)	Broadband kilometric (bKOM)	Narrowband kilometric (nKOM)
Frequency range	<80 MHz–300 GHz	~0.2–39.5 MHz	~10–1000 kHz	40–200 kHz
Polarization	~25% linear from 600 MHz to 5 GHz; ~1% circular at 1.4 GHz	Circular or elliptical; primarily RH above 20 MHz but mixed at lower frequencies	Circular or elliptical; emission centered near 200° CML is LH from above dayside and RH from over nightside	LH for northern magnetic latitudes, RH for southern magnetic latitudes
Peak flux density (4 A.U. equiv.)	~6 × 10^{-26} W/m^2 Hz at 800 MHz	3 × 10^{-19} W/m^2 Hz at 10 MHz	~10^{-20} W/m^2 Hz at 60 kHz	10^{-20} W/m^2 Hz at 100 kHz
Ave. isotropic equiv. total power	2 × 10^9 W	4 × 10^{11} W	~5 × 10^8 W	~10^8 W
Dynamic spectrum	Smooth, continuous	Arc-shaped drifting features with duration of few min. at each frequency; occasional ms bursts at certain γ_I-CML	Short impulsive bursts superposed on slower drifts; longer duration at lower frequencies	Smooth, narrow-band emission repeatable at 3–5% slower than System III period
Origin	Synchrotron radiation from trapped electrons between 1.6 and 3 R_J	Plasma instability in auroral ionosphere and inner magnetosphere. Possibly R-X mode from Io plasma torus L-shells	Plasma instability in auroral ionosphere or inner magnetosphere. Possibly constrained to daytime sector with L-O mode	Plasma instability at low latitudes in inner magnetosphere near outer edge of Io plasma torus

1963; Chang and Davis, 1962; and others. See Berge and Gulkis, 1976]. Before the reasons for this interpretation are presented, however, we shall briefly review some basic facts regarding such radiation.

Radiation by high-energy electrons in a magnetic field

A detailed treatment of synchrotron radiation theory as it is applied to astrophysics, with an extensive annotated reference list, is given by Pacholczyk, [1970]. Developments in the theory making it more readily applicable to the Jupiter problem have been made by Chang and Davis [1962], Korchak [1962], Thorne [1963], Ortwein, Chang, and Davis [1966], Legg and Westfold [1968], Clark [1970], Gleeson, Legg, and Westfold [1970], Degioanni [1974], and others. A qualitative understanding of most of the observed Jovian synchrotron emission phenomena, however, requires only a knowledge of the radiation characteristics of a single high energy electron in a magnetic field, a brief discussion of which is presented in this section.

If an isolated electron (of charge $-e$ and mass m_e) in a uniform magnetic field (**B**) is given an initial velocity (v) perpendicular to the field, where v is sufficiently small that the relativistic increase in mass with velocity is negligible, the electron will revolve in an almost circular path at $Be/(2\pi m_e)$ revolutions per second, in the sense of rotation of a right-hand screw advancing in the direction of **B**. Its orbit will not be precisely circular because of a relatively slow decrease in the radius of curvature due to the loss of energy (and velocity) by radiation. The frequency of revolution, however, will remain very nearly constant as the velocity and radius of curvature both decrease toward zero. Nearly monochromatic radiation will be emitted at the same frequency at which the electron is revolving. This frequency is the electron cyclotron frequency (or gyrofrequency), and is given by

$$f_c = \frac{Be}{2\pi m_e} \approx 2.8 \; B(\text{Gauss}) \; \text{MHz} = 2.8 \times 10^4 \; B \; (\text{Telsa}) \; \text{MHz} \qquad (7.1)$$

The radiation will be emitted in all directions, but its polarization will depend on the direction. Radiation propagating in the same direction as **B** will be right-hand (RH) circularly polarized, and in the opposite direction left-hand (LH) circularly polarized.* For all propagation directions within the plane of the orbit of the electron the polarization will be linear, with the polarization plane perpendicular to **B**. In other directions of propagation, the polarization will be RH or LH elliptical.

If the electron velocity is close enough to c that the mass increase due to relativity is appreciable, a relativistic beaming of the emitted radiation in the instantaneous direction of the velocity vector will also occur [Jackson, 1975]. This beaming is narrower the higher the electron energy, most of the power being radiated within an angle of $(1 - v^2/c^2)^{1/2}$ radians with respect to v. The half-power width of the instantaneous emission beam is given approximately by

$$\theta \approx 56/E \qquad (7.2)$$

---

* We use the right-hand screw rule for defining the RH polarization sense: The electric vector at each point in a fixed plane perpendicular to the propagation direction rotates in the same sense as a RH screw advancing in the direction of propagation. That is, rotation is counterclockwise when propagation is toward and viewed by the observer. For LH polarization, the electric vector of course rotates in the opposite sense. This is the convention that is generally used in radio astronomy and plasma physics, but is opposite to the one that until recently was used in physical optics.

where θ is in degrees and E is the electron energy in MeV. A distant observer in the plane of the electron orbit would see one pulse each revolution, as the beam swept past (if such an observation were possible). The emission beam pattern averaged over a full revolution (or many revolutions) would have a maximum in the orbital plane but would be constant in azimuth. The angular thickness of this radiation sheet would depend on the electron energy. For example, for a 20 MeV electron the half-power beam thickness of the sheet is about 2.8°, according to (7.2).

The instantaneous intensity seen by a distant observer in the orbital plane as a function of time can be represented by a Fourier series in terms of consecutive-integer harmonics of f_c (calculated from the relativistic mass of the electron rather than its rest mass). The power spectrum of the radiation from this single electon consists of a large number of narrow spikes separated by f_c. The envelope of these harmonics has a broad maximum at a frequency that is proportional to BE^2. As in the case of a low energy radiating electron, the polarization is linear when the electron orbit is seen edge-on, that is, when the line of sight is perpendicular to **B**. If the line of sight is along **B** (i.e., a face-on orbit) the radiation intensity is at a minimum because of the beaming effect, but whatever radiation is observed is RH circularly polarized if **B** points toward the observer. It is LH if **B** points away from the observer. For propagation in any other direction, the radiation is elliptically polarized; it contains both linearly and circularly polarized components (but no unpolarized component).

If the velocity of the electron is not perpendicular to **B**, its path is helical. The longitudinal and transverse components of **v** are $v \cos \alpha$ and $v \sin \alpha$, respectively, where α is the pitch angle. In this case, the direction of maximum radiation intensity is along the surface of a cone of opening angle 2α, and the synchrotron emission consists of integral harmonics of $f_c / \sin^2 \alpha$ rather than of f_c. The latter fact was not recognized until 1967 (see e.g., Scheuer [1968]), and as a result, a number of theoretical treatments of synchrotron emission published earlier are in error in some respects [Pacholczyk, 1970]. An observer in the path of the beam from a relativistic electron that is traveling along a helix with pitch angle α sees a broad spectrum (the envelope of the harmonics) with a maximum at approximately

$$f_{max} \approx 4.8\, E^2 B \sin \alpha \qquad\qquad (7.3)$$

where f_{max}, E, and B are in MHz, MeV, and gauss (G), respectively. Thus, a 20 MeV electron with a 90° pitch angle in a magnetic field of 0.5 G emits a spectrum having a maximum close to 1000 MHz. The total power radiated by the electron, in watts, is approximately

$$p \approx 6 \times 10^{-22} E^2 B^2 \sin^2 \alpha \qquad\qquad (7.4)$$

The observed synchrotron radiation comes, of course, from vast aggregates of electrons, rather than from a single one. The synchrotron-emitting electrons in a planetary magnetosphere with a primarily dipole magnetic field configuration are magnetically trapped, and there may be wide distributions of electron energy and equatorial pitch angle and magnetic field intensity and direction. All traces of the harmonic spike structure characteristic of the single-electron spectrum are of course smoothed out.

First-order explanation of the synchrotron emission phenomena

A "first order" explanation of the synchrotron emission phenomena, accounting for the early-observed effects listed previously, is given in this section. The explanation is based on the changing viewing geometry for trapped relativistic electron orbits in the

Jovian magnetic field as the planet rotates, assuming somewhat simpified orbits and field configuration.

The simplest useful approximation of Jupiter's magnteic field is that of a magnetic dipole located at the center of the planet, its axis tipped about 10° with respect to the planetary spin axis. The magnetic equator is thus tilted about 10° with respect to the spin equator. The belt of trapped charged particles is approximated by a shell about 2 R_J above the cloud-top level and centered on the magnetic equator. The synchrotron-emitting electrons follow paths that are relatively flat helices, that is, their equatorial pitch angles are not far from 90°. The mirror points of most of the electrons, at which the instantaneous pitch angles reach exactly 90°, are within about 1 R_J above and below the magnetic equatorial plane. The observed synchrotron emitting region is thus the projection upon the plane of the sky* of a toroidal source surrounding the planet. The width and height of the region are roughly 6 and 2 R_J, respectively. The long axis of the emitting region actually rocks ±10° with respect to the rotational equator as the planet rotates, but this effect was not observed until considerably later when aperture-synthesis arrays of relatively high angular resolution became available [Branson, 1968].

The radiation seen along any line of sight is predominantly randomly polarized, with some residual linear component. When the line of sight to an element of volume in the emitting region is perpendicular to **B** at the volume element, the radiation from it will be linearly polarized in a plane that is perpendicular to **B**. However, **B** is not uniform. For the majority of the volume elements, the line of sight is not perpendicular to **B**, and the polarization will be elliptical. The polarization ellipse can have either sense and any of a wide range of axial ratios and major axis orientations. The superposition of the contributions from all the volume elements thus leads to predominantly unpolarized radiation with a residual linearly polarized component. (There is also a very small residual circular component, provided the resultant of the intensity contributions in one sense of elliptical polarization is greater than that of the other.) The polarization plane of the linear component is parallel to the plane of the magnetic equator. For qualitative visualization, it is convenient to think of the polarized component as having been radiated by electrons in circular orbits lying within the plane of the magnetic equator.

As the planet rotates, the intersection of the plane of the magnetic equator with the plane of the sky rocks back and forth by ±10° relative to the spin equator, which remains fixed. This accounts for the observed ±10° rocking of the plane of the linearly polarized component. The much weaker circularly polarized component was not investigated extensively until after these first-order effects were; it will be discussed later.

The observed variation in total flux density as the planet rotates is a manifestation of the beaming of the emitted radiation. For the simple dipole field model the beaming is maximum within the plane of the magnetic equator. When the *Jovicentric declination of the earth* D_E is zero, the earth lies in the plane of Jupiter's spin equator, and the magnetic equator would be seen on edge every 180° of rotation. (D_E is the angular distance of the earth from Jupiter's spin equator.) This would give rise to two flux density maxima of equal magnitude and spacing as the planet rotates. On the other hand, when D_E is positive (its possible values over Jupiter's 11.9 year orbital period range between +3.3° and −3.3°), the two flux density maxima are separated by rotation intervals alternately less than and greater than 180°. When D_E is negative, the phase of alterna-

---

* For an observed region such as Jupiter's magnetosphere, which subtends a relatively small solid angle from Earth, the *plane of the sky* is perpendicular to the radius vector from an observer to an arbitrary reference point within the region.

tion of the narrower and wider rotation intervals is reversed. To a first approximation, this is what is observed. However, there are important departures from the predictions of this simple model; these will be discussed later.

Finally, the shape of the power spectrum of the synchrotron emission in the decimeter wavelength range is a consequence of the combined effects of several factors. These are:

1. the distribution of radiated power by a single electron as a function of frequency, direction, electron energy, electron pitch angle, and **B**;
2. the distribution of electron energies;
3. their pitch angle distribution;
4. their spatial distribution.

The electron energy distribution is one of the dominant factors. The spectral shape can be approximated from the power spectra of individual electrons in circular orbits seen edge-on as a function of their energies, together with the energy distribution of the electrons.

Present status of Jovian synchrotron phenomenology

It is not surprising that the great increase in resolution, sensitivity, and versatility of radio telescopes within the past decade has resulted in a greatly increased knowledge of the synchrotron emission phenomenology. There has also been considerable progress in emission region modeling. In the light of these relatively recent developments, we present in this section a brief overview of the phenomenology of Jupiter's synchrotron radiation, with emphasis on progress that has been made since the comprehensive review by Berge and Gulkis [1976]. The relatively large number of papers referenced in this review deal mainly with the synchrotron flux density spectrum and its variability, the central meridian longitude and magnetic latitude dependence of flux density, studies of the linearly and circularly polarized components, decimetric rotation period measurements, and relatively high-resolution mapping of the emitting region. Despite the fact that the emission mechanism is well understood and much is known of the general properties of the source, the detailed picture is still far from complete. As in the case of most reviews of active fields of research, recent results do not uniformly fall into neat and consistent patterns. The filtering action of time is usually needed to reconcile and refine new results.

Flux density spectrum. Figure 7.2 shows a plot of the flux density spectrum of the synchrotron emission and its region of overlap with the thermal component. The points were obtained from the compilations by Dickel et al. [1970] and Berge and Gulkis [1976], with additions from Gerard [1976], Shawhan et al. [1977], Gardner and Whiteoak [1977], Neidhofer et al. [1977], Werner et al. [1978] and de Pater [1980a, 1981c]. Much of the scatter is due to real long-term intensity variations.

An accurate separation of the synchrotron and thermal components of the radiation in the spectral region between about 1000 and 15 000 MHz is difficult. An accurate model accounting for one or the other component is required before a separation can be effected [de Pater, 1980b]. In the absence of such a model, several more approximate methods have been used [Berge and Gulkis, 1976]. In one such method, utilizing radio telescopes that are incapable of resolving the source region, the separation has been made on the basis of the polarization properties of the components [Roberts and Komesaroff, 1965; Dickel, 1967; Morris et al., 1968; de Pater, 1981c]. The thermal

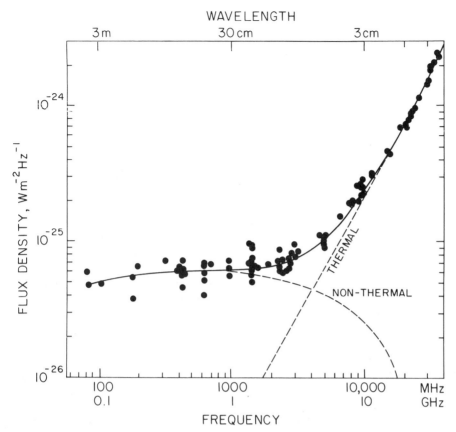

Fig. 7.2. Measurements of Jupiter's average flux density above 80 MHz. The dashed-line portion of the thermal component curve was made to fit points from de Pater [1980a, 1981c] corresponding to 320 K at 1410 MHz and 220 K at 4900 MHz, and to merge with the total flux density curve at higher frequencies. The dashed-line portion of the nonthermal curve was obtained by subtracting the thermal curve from the total curve. See text for references from which indicated points were taken.

component is assumed to be completely unpolarized; the overlapping nonthermal component is assumed to have the same degree of linear polarization that it has at lower frequencies where the thermal component is negligible. Then the flux densities of the two components separately can be calculated from the measured flux density and degree of linear polarization of the two combined. The deficiency in this method lies in the fact that the constancy of the degree of linear polarization of the nonthermal component becomes progressively more uncertain for frequencies above about 2000 MHz.

In other separation methods, employed with radio telescopes that at least partially resolve the sources, an approximate model is developed for (1) the entire synchrotron emitting region, (2) the part of the synchrotron emitting region in front of the visible disc, or (3) the thermal disc source alone. The approximate fractions of the total flux density due to the synchrotron and thermal components can then be ascertained [Berge, 1966; Branson, 1968; Beard and Luthey, 1973; Degioanni and Dickel, 1974; de Pater, 1980b].

The smooth curve in Figure 7.2 is an approximate fit to the points representing total flux density. The dashed curves approximate the separate flux densities of the two

Fig. 7.3. Frequency of maximum
synchrotron emission power for 90°
pitch angle electrons as a function of
electron energy and source magnetic
field.

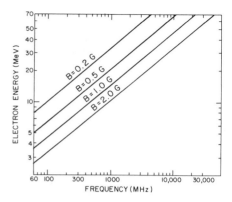

components in the region of overlap. The dashed thermal component curve was made
to pass through points corresponding to disc temperatures of 320 K at 1410 MHz
[de Pater, 1980a] and 200 K at 4900 MHz [de Pater, 1981c], and to merge with the
total flux density curve at still higher frequencies. This dashed curve was then subtracted from the total intensity curve to obtain the dashed portion of the nonthermal
curve in Figure 7.2, and also the corresponding portion of the curve in Figure 7.1. The
unexplained long-term variations that are observed in the total flux density occur in
the nonthermal component; the thermal component is presumably of much more
nearly constant intensity (in time but not in frequency).

Our separation of the two components in Figure 7.2 depends critically on de Pater's
estimates of the disc temperatures at 1410 and 4900 MHz. De Pater obtained the 1410
MHz temperature by developing a synchrotron source model based on her high-resolution maps (to be discussed later), and subtracting its contribution from the total
observed flux density. She obtained the 4900 MHz temperature by assuming that the
degree of linear polarization of the nonthermal component at 4900 MHz is the same as
it is at 1410 MHz, at which frequency the thermal component is negligible, and then
deducing the amounts of thermal and nonthermal radiation at 4900 MHz from the
measured degree of linear polarization and flux density of the two combined.

In Figure 7.3 we have plotted the energy of a synchrotron-emitting electron with
90° pitch angle as a function of the frequency at which its radiated power is maximum,
using (7.3), for each of a series of magnetic field values. Fields within this range are
encountered in the actual emission belt, as mapped between about 1400 and 5000
MHz [e.g., Berge, 1966; de Pater, 1981] and assuming a dipole field with a magnetic
moment of $4 \, G R_J^3 (4 \times 10^{-4} \, \text{T} \, R_J^3)$ [Acuña and Ness, 1976c]. Thus in the 1 G region,
most of the 1000 MHz radiation is emitted by approximately 14 MeV electrons, while
that at 3000 MHz comes from electrons in the vicinity of 24 MeV.

The lowest-frequency measurements of the synchrotron radiation are those by
Gower [1968] at 81.5 MHz and by Slee and Dulk [1972] at 80.0 MHz. The flux density
in this region is only slightly less than at higher frequencies. Here most of the radiation
comes from electrons of lower energies. Because mapping has not yet been possible at
the lower frequencies, the magnetic field range in the 80 MHz emission region is not
known. However, if 1 G can still be considered a typical value, then the majority of the
electrons must have about 4 MeV energy. More likely, the field in most of the 80 MHz
emitting region is somewhat less than 1 G, in which case the predominant electron
energies are somewhat higher than 4 MeV. Although it has not been detected there,
the synchrotron emission spectrum probably extends below 40 MHz, into the realm of
the decametric bursts.

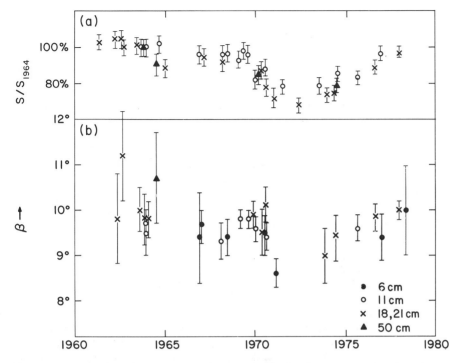

Fig. 7.4. (*a*) Variation of the total peak flux density *S* with time compared to the value measured in 1964.0. (*b*) Variation of β (as determined by de Pater's model-independent method) with time. Both compilations were made by de Pater [1981c].

Intensity fluctuations. During the first few years after the discovery of the Jovian synchrotron emission, relatively large and erratic intensity fluctuations on timescales of days or weeks were believed to be characteristic of the radiation. It was later concluded that they were not real but must have been due to background confusion effects or instrumental instabilities. However, in subsequent years occasional reports of longer period fluctuations continued to appear [e.g., Gerard, 1970; Klein et al., 1972]. Although there was at first some reluctance to return to the idea of variability, there is now no doubt of the existence of relatively large fluctuations in flux density on a timescale of years [Gulkis et al., 1973; Gerard, 1976; Berge, 1974; Klein, 1976; Hide and Stannard, 1976; Shawhan et al., 1977; de Pater, 1981c].

Figure 7.4*a* shows for several frequencies the variability of the ratio of the total flux density to that occurring at 1964.0 over a 17-year interval (after Hide and Stannard [1976] as updated by de Pater [1981c]). The data have been corrected for differences in distance and for systematic intensity variations with respect to central meridian longitude and magnetic latitude. The flux density decreased by 30% in a decade, clearly going through a minimum in 1972–1973. It is the synchrotron rather than the thermal component that has varied, for otherwise the brightness temperature of the thermal component would have had to change by an enormous amount. Klein [1976] states that there was no apparent correlation with solar activity. Moreover, the variations do not appear to be related to changes in D_E. According to Shawhan et al. [1977] the flux density varies in such a way that the shape of the spectrum changes. For example, their measurements reveal a pronounced spectral dip of about 15% near 1400 MHz which was not previously present. The data in Figure 7.4*a* do not bear this out, however,

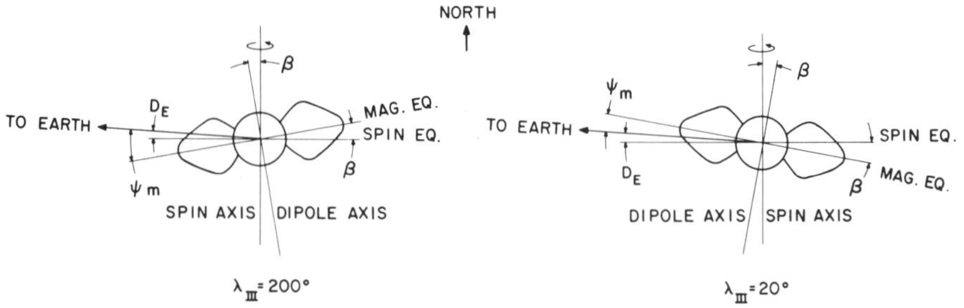

Fig. 7.5. Geometrical relationship between the dipole tilt angle β, the jovicentric declination of Earth D_E, and the magnetic latitude of the observer ψ_m, for the extreme cases when the observer's central meridian longitude λ_{III} is 200° and 20°.

perhaps because the points for the different frequencies in this figure are too sparsely distributed to show the effect.

In addition to the very slow intensity fluctuations that have been discussed, Gerard [1970] reported relatively rapid fluctuations occurring on time scales of days or weeks. During a 3 1/2-month period of intensive monitoring at 1408 MHz in 1974, Gerard [1976] observed two surges in intensity reaching a level 9% higher than the background, each with a half-intensity duration of about a week. Two or three smaller surges were also noted. He points out that the radial diffusion models of the magnetosphere cannot account for such rapid fluctuations, because the time constant for changes in the inner, stably trapped electron zone is predicted from the theory to be a few years [Coroniti, 1975]. Instead, he attributed them to some transient magnetospheric phenomenon such as substorms acting in addition to radial diffusion. Hill, Dessler, and Maher [1981] believe that significant changes in corotating convection patterns can occur over time intervals as short as 30 hr (see Chap. 10). Such changes would probably affect the synchrotron-emitting electrons and might therefore be related to Gerard's rapid fluctuations.

Whether or not Gerard's short-term fluctuations can be established as fact, the oberved long-term ones are undoubtedly manifestations of changes in the energy, pitch angle, or spatial distributions of the radiating electrons. Despite the fact that the deepest minimum in the curve in Figure 7.4a occurs within a year of the minimum in the 11.9 yr D_E cycle, the two curves are dissimilar elsewhere and are very likely unrelated. De Pater [1981c] has found some evidence in support of an earlier suggestion by E. T. Olson [de Pater and Dames, 1979] that the observed flux density decrease was due to a contraction of the radiation belts, or at least that the two effects were related. The discovery of the temporal spectral variations are of especial importance because of their potential utility as a means for monitoring one or more parameters of the radiation belt electrons. The fact that there is no apparent correlation with solar activity (see, however, the discussion in Sec. 7.3 of reported solar influence on low-frequency burst emission) should not be discouraging, because loading of the radiation belts is probably only very weakly dependent on the solar wind. Further observations of variability will be awaited with keen interest, and we anticipate increased theoretical activity stimulated by the need for a detailed explanation of the temporal effects.

The variation of flux density with central meridian longitude and magnetic latitude. Several of the parameters of the synchrotron radiation vary systematically as the planet rotates and are also influenced by the much slower periodic variation of the Jovicentric declination of the earth, D_E. As explained in Section 7.3, these effects are attributable to a predominantly dipolar configuration of the magnetic field in the radia-

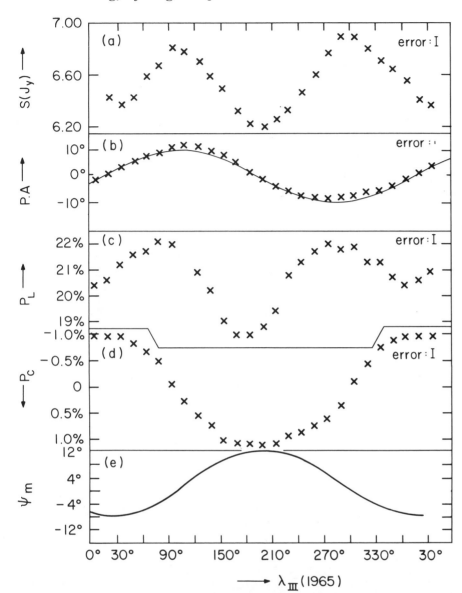

Fig. 7.6. Various parameters of the integrated radiation of Jupiter at 21 cm as functions of System III longitude. The north magnetic pole is assumed to be at 200°. (*a*) Total integrated flux density in Janskys, where 1 Jy = 10^{-26} W m^{-2} Hz^{-1}. (*b*) Position angle of the electric vector measured eastward from north in the sky. The smooth curve is defined by P.A. = $10°\sin(\lambda_{\mathrm{III}} - 20°)$. (*c*) Degree of linear polarization. (*d*) Degree of circular polarization. (*e*) Magnetic latitude of Earth. Adapted from de Pater [1980a].

tion zone, the dipole axis being tipped about 10° with respect to the rotation axis. The clearest evidence for the 10° tipping is derived from the measurements of the linearly polarized component and will be discussed later. These effects are most easily described in terms of the central meridian longitude (CML) of the planetary disc as seen by the observer at a given time, and the magnetic latitude of the observer (ψ_m). The longitude system currently in use for Jovian radio and magnetospheric phenomena is designated System III (1965) and is based on a sidereal rotation rate of 870.536°/day (rotation

Fig. 7.7. The parameters of the integrated radiation of Jupiter as functions of the magnetic latitude of Earth with respect to Jupiter. (*a*) Total integrated flux density in Janskys. (*b*) Polarized flux density, $(Q^2 + U^2)^{1/2}$, in Janskys. (*c*) Position angle of the electric vector. (*d*) Degree of linear polarization. (*e*) Degree of circular polarization. After de Pater [1980a]. Positive P_c signifies the LH sense, and positive ψ_m indicates magnetic latitudes north of the magnetic equator. (In de Pater's original figures, the opposite sign convention for LH and RH is used.)

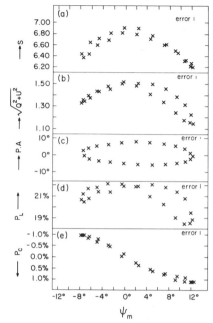

period ~9 hr 55 min 29.71 s). The value of the System III (1965) CML at a given time is designated by the symbol λ_{III}. (In addition to its use in specifying CML, the System III concept is extended to specify meridians that corotate with the planet and are not necessarily on the central meridian at the time under consideration.) The magnetic latitude of the observer is the angle between the magnetic equatorial plane, which is perpendicular to the assumed location of the dipole axis, and the radius vector from the center of the planet to the observer. For observations from Earth, the magnetic latitude of the observer is, to a good approximation,

$$\psi_m = D_E + \beta \cos (\lambda_{III} - \lambda_p) \tag{7.5}$$

where D_E is the Jovicentric declination of Earth (given in the *American Ephemeris and Nautical Almanac*), β is the assumed tilt of the dipole axis with respect to the axis of rotation, and λ_p is the value of λ_{III} toward which the northern-hemisphere pole of the magnetic dipole is tilted. Figure 7.5 illustrates the geometrical relationships between the quantities. Figure 7.6e shows a plot of ψ_m vs. λ_{III} as the planet rotates, for a time at which $D_E = 2.24°$. The values of λ_p and β for this plot are 200° and 10°, respectively. The choice of λ_p and β are model dependent, of course.

It appears that the measurements of integrated polarization parameters as a function of CLM that show the least statistical scatter are those of de Pater [1980a] obtained at 1412 MHz with the Westerbork Synthesis Radio Telescope in the Netherlands. The results of de Pater presented in Figures 7.6 and 7.7 will be used to illustrate the several longitude-correlated effects. Figure 7.6a shows the variation of total flux density with λ_{III}. This effect is most apparent at frequencies between about 1000 and 4000 MHz. It is caused by beaming of the emission [Morris and Berge, 1962; Gary, 1963] and has been extensively investigated by Roberts and Komesaroff [1965], Barber [1966], Roberts and Ekers [1968], Gulkis et al. [1973], Neidhofer et al. [1977] and de Pater [1980a, 1981c]. As mentioned previously, the radiation is assumed to a first approximation to be beamed in the plane of the magnetic equator, that is, at $\psi_m = 0°$. On the basis of the simple model, one would expect two maxima of equal heights in the

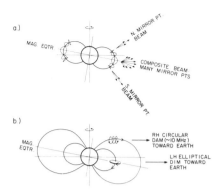

Fig. 7.8. (a) Beaming and (b) polarization geometry of emission by energetic trapped electrons gyrating in Jupiter's magnetic field.

curve of Figure 7.6a, with the peaks centered on the values of λ_{III} for which $\psi_m = 0°$, but with the two minima at different levels (except when $D_E = 0°$). It is apparent from (7.4) or Figure 7.6e that $\psi_m = 0°$ at $\lambda_{III} = 98°$ and $304°$. Figure 7.6a indeed displays two peaks, centered at about $\lambda_{III} = 100°$ and $300°$, in close agreement with the theory, but the peaks are of unequal shapes and heights. (It is only when $D_E = 0°$ that the two peaks are $180°$ apart.) It has been widely reported that the higher-longitude peak is the higher of the two. This is one of many second-order effects not accounted for by the simple tilted-dipole field model.

In Figure 7.7a the total flux density is plotted as a function of ψ_m instead of λ_{III}. For a dipole model in which none of the radiation is blocked by the planet, the points would like on a single arch-shaped curve that is symmetrical except it would extend farther for positive than for negative ψ_m values, because $D_E > 0°$ (see Fig. 7.5). If the electrons in such a model were monoenergetic, all with an energy of, say, 20 MeV, and if they had equatorial pitch angles of exactly $90°$, the half-flux- density width of the flux density vs. ψ_m curve would be about $3°$, as given by (7.3). The fact that the curve in Figure 7.7a is much wider than this is due to several factors, the principal ones of which are that there are distributions rather than single values of equatorial pitch angle (α_e) and electron energy. Those electrons with α_e considerably different from $90°$ make helical excursions to and from mirror points well above and below the magnetic equator. Most parts of the helical paths of these electrons are appreciably inclined with respect to the equatorial plane and therefore emit radiation that is not beamed closely to it. The width of the beam must be at least as great as the angle of intersection of the electron orbital planes at pairs of conjugate mirror points that are farthest above and below the magnetic equator (not considering shadowing by the planetary body). This effect is illustrated in Figure 7.8a. The presence of relatively low energy synchrotron-emitting electrons also tends to widen the beam, as does the presence of multipole field components.

The function indicated by the points in Figure 7.7a is actually double valued, neither branch of which is symmetrical about $\psi_m = 0$. Thus, the shapes of the parts of the beam above and below the magnetic equator are different, and the latitudinal profile of the beam changes somewhat with longitude. For the dipole model on which the plot in Figure 7.6e is based, the System III (1965) longitudes $200°$ and $20°$ define a plane of symmetry. The points in Figure 7.6a are not symmetrical about longitude $20°$ (where ψ_m is minimum), although they are more nearly so about $200°$ (where ψ_m is maximum); this is the same effect as the splitting into two arches seen in Figure 7.7a.

The difference in the beam profile for northern and southern magnetic latitudes has long been observed. This difference depends on D_E. When $D_E > 0°$, the decrease in S with increase in $|\psi_m|$ northward from the magnetic equator is more steep than it is when $|\psi_m|$ increases southward from the equator. The reverse is true when $D_E < 0°$ [Roberts and Komesaroff, 1965; Gulkis et al., 1973]. Degioanni [1974] showed that this effect is probably caused by the asymmetrical blocking by the planet of much of the radiation from the back side of the emission region when $D_E \neq 0°$. Figure 7.5 will aid in visualizing it. The other asymmetries that are observed are attributed to the multipole field components, in combination with planetary shadowing, as discussed later in this section.

Polarization. Over the frequency range from 600 MHz or below to at least 5000 MHz, the nonthermal component is about 75%–80% unpolarized and about 20%–25% elliptically polarized. The shape of the polarization ellipse changes as the planet rotates, and its polarization sense reverses twice each rotation. (The polarization ellipse is the locus of the tip of the electric vector of the polarized component during each cycle as the wave passes.) The ellipse for this radiation is always thin, the ratio of its major to minor axis lengths (axial ratio) never being less than about 40. The minor axis momentarily becomes zero at each reversal of sense. The major axis remains very nearly perpendicular to the magnetic dipole axis during rotation (to within $\pm 1°$ or less).

Before proceeding with the discussion of the linearly polarized component, we shall dwell briefly on general methods for specifying polarization. A set of four parameters is required to specify the intensity and polarization state of partially polarized radio emission. Many equivalent sets are possible. The two sets which are used most often for Jupiter are as follows:

1. Total flux density (S), degree of linear polarization (P_L), degree of circular polarization (P_C) with sign to indicate sense, and position angle of the plane of linear polarization (P.A.).
2. The four Stokes parameters, I, Q, U, and V.

P_L and P_C are the fractions of the total flux density that are linearly and circularly polarized, respectively. They pertain to the linear and circular parts of the polarization ellipse. The Stokes parameter I is simply the total flux density, S. The polarized part of S is $(Q^2 + U^2 + V^2)^{1/2}$; the linearly and circularly polarized parts are $(Q^2 + U^2)^{1/2}$ and V, respectively. The sign of V specifies the polarization sense; $-$ is for right hand (RH) and $+$ is for left hand (LH). The tilt angle of the major axis of the polarization ellipse, with respect to the x axis, is $(1/2) \tan^{-1} (U/Q)$. (In the coordinate system that is assumed, propagation is in the $+z$ direction, and the x axis is in such a direction that $Q > 0$ and $U = 0$ for a wave that is linearly polarized in the xz plane.) The axial ratio of the ellipse is the cotangent of $(1/2) \sin^{-1} [V/(Q^2 + U^2 + V^2)^{1/2}]$. Typical values of P_L and P_C for Jupiter's nonthermal decimetric radiation are 0.25 and 0.01, respectively, so that $(Q^2 + U^2)^{1/2} = 0.25S$ and $V = 0.01S$, where S is the total flux density of only the nonthermal part. If the thermal component were included with the nonthermal, P_L and P_C would of course be smaller. Thus, the axial ratio of the polarization ellipse in this case is approximately $\cot [0.5 \sin^{-1} (0.01/0.25)]$, or 50. The principal advantage in using Stokes parameters instead of a more easily visualized set of quantities results from their additive property: if several independent waves propagating in the same direction are superimposed, the Stokes parameters of the resultant wave are the sums of the corresponding Stokes parameters of the individual waves. This property greatly facilitates the matching of assumed models of a distributed source of partially polarized radiation, such as Jupiter's magnetosphere, to the observed integrated Stoke's parameters. For additional information on Stokes parameters see Kraus [1966] and references cited therein.

The linearly polarized component. Figure 7.7*b* shows the linearly polarized component of the flux density, which in terms of Stokes parameters is $(Q^2 + U^2)^{1/2}$, plotted as a function of the magnetic latitude ψ_m. The variation with ψ_m is significantly different from that for the total flux density in Figure 7.7*a*. A difference is to be expected because most of the linearly polarized radiation comes from the magnetic equatorial region and is relatively sharply beamed within the equatorial plane whereas the total flux density contains the radiation from outside the equatorial plane as well. The latter is the sum of elementary contributions having a wide range of emission beam directions and polarizations, and therefore the resultant beam is broadened and has a lower degree of linear polarization.

Measurements of the variation of the plane of linear polarization with CML have been made by Morris and Berge [1962], Gary [1963], Roberts and Komesaroff [1965], Komesaroff and McCulloch [1967], Morris et al. [1968], Roberts and Ekers [1968], Whiteoak et al. [1969], Berge [1974], McCulloch [1975], Stannard and Conway [1976], Gardner and Whiteoak [1977], Neidhofer et al. [1977, 1980], de Pater [1980a, 1981c], and Komesaroff et al. [1981]. Such meaurements provide the best means for determining the tilt β of the equivalent magnetic dipole, and they have the potential for yielding more detailed information on the structure of the field in the vicinity of the magnetic equator than do the other CML-related quantities. Figure 7.6*b* shows, for the 1412 MHz data of de Pater, the position angle (P.A.) of the plane of linear polarization relative to the plane of the rotational equator, plotted as a function of λ_{III}. It is clear from this figure that the polarization plane rocks back and forth by approximately $\pm 10°$ with respect to the rotational equator. Because the polarization plane is parallel to the magnetic equatorial plane in the simple dipole model, the value of the tilt angle β of the dipole axis with respect to the rotation axis must also be approximately $10°$ (see Fig. 7.5). The smooth curve in Figure 7.6*b* is the calculated position angle of the dipole model magnetic equatorial plane, relative to that of the spin equator, plotted as a function of λ_{III}. The departure of the measured points from the calculated curve is a clear indication of the limitations of the model. Figure 7.7*c* shows the variation of P.A. with respect to ψ_m. The quasielliptical curve defined by the points in this figure is approximately symmetrical about the vertical line $\psi_m = D_E = 2.24°$, but this symmetry and that about the horizontal line P.A. $= 0°$ are both imperfect.

Roberts and Komesaroff [1965] and most of the other authors referenced above have found it useful to evaluate the amplitudes and phases of the least squares Fourier series fit to the measured P.A. vs. λ_{III} curves. The series can be expressed in the form

$$\text{P.A.} = A_0 - A_1 \sin (\lambda_{III} - \ell_1) - A_2 \sin 2 (\lambda_{III} - \ell_2) - A_3 \sin 3 (\lambda_{III} - \ell_3) \qquad (7.6)$$

Here the amplitude A_1 can be identified with the tilt angle (β) of the dipole axis of the dipole-approximation field model, and the phase ℓ_1 with the System III longitude toward which the northern-hemisphere end of the dipole axis is tipped. Significant values of amplitudes and phases of at least the first two sinusoidal terms can generally be determined when the measurements are made under suitable circumstances. For example, for measurements made during October 1975 at 2700 MHz [Komesaroff and McCulloch, 1975] the amplitudes and phases of the first three and the fifth sinusoidal terms were as given in Table 7.3. Values for the other terms were not of sufficient statistical significance to be listed.

Gardner and Whiteoak [1977] and Neidhofer et al. [1980] report a slight change in the indicated value of A_1 as a function of frequency, the value ranging from about $10°$ at 1400 MHz to about $9°$ at 5000 MHz. This effect, Gardner and Whiteoak suggest, may indicate that the overall distribution of synchrotron radiation changes with frequency. This would be consistent, they point out, with theoretical predictions

Table 7.3. *Fourier amplitudes and phases of polarization position angle as a function of CML; measurements at about 2700 MHz during October 1975 [Komesaroff et al., 1980]. All quantities are in degrees.*

Amplitude	Std. error	Phase	Std. error
$A_1 = 9.85$	0.04	$\ell_1 = 195.3$	0.2
$A_2 = 1.16$	0.04	$\ell_2 = 174.9$	1.1
$A_3 = 0.19$	0.04	$\ell_3 = 202.0$	4.3
$A_5 = 0.16$	0.04	$\ell_5 = 177.7$	3.0

[e.g., Beck, 1972] that the most energetic electrons will be found to be more concentrated toward the planetary surface, and with the Pioneer in situ measurements, which seem to support the prediction [e.g., Fillius, 1976]. More recently, Komesaroff et al. [1981] also found an apparent decrease in A_1 with increasing frequency, but it was only half as large as that reported previously and was of marginal statistical significance.

De Pater [1980b] has reached a quite different conclusion regarding the variability of β. She found the emission regions at 1412 MHz and 4885 MHz to be of essentially the same size and shape, and concluded that the reported changes in β with frequency may be temporal changes instead. She recalculated β from all available previously published data on the basis of the maximum peak-to-trough excursion of the P.A. vs. λ_{III} curve, instead of using the customary least squares fit of the fundamental Fourier term to the data. The resulting curve of β vs. time over a 16-year interval, including de Pater's own data, is shown in Figure 7.4b. A comparison of the curves in Figures 7.4a and 7.4b strongly suggests a correlation of β with flux density. De Pater points out that the two brightest regions on synchrotron source maps were apparently closer together over a span of a few years that included the time of the observed flux density reduction than they were either before or after that interval [de Pater, 1980b]. She offers as a possibility the suggestion that the minima in the curves of both Figures 7.4a and 7.4b result from a contraction of the emission region. The fact that during the flux density minimum a larger proportion of the radiation seems to have come from closer to the planetary body, where the multipolar field effects are more pronounced, was perhaps the cause of the observed change in β. This is a most interesting idea. The accumulating evidence for the long-term aperiodic variability of the synchrotron emission parameters will no doubt stimulate new measurement programs.

Progress has also been made in accounting for the Fourier harmonics above the fundamental. The earliest efforts in this regard were those of Conway and Stannard [1972], Komesaroff and McCulloch [1975], Gerard [1976], and Gardner and Whiteoak [1977]. Gerard showed that when the shadowing of radiation from behind the planet is taken into consideration, the quadrupole field component can give rise to the second harmonic term. He suggested that the "charged particle equator" (i.e., the surface of minimum B within which particles having 90° equatorial pitch angle would lie) is warped rather than planar in a particular longitude sector. He succeeded in calculating the λ_{III} values at which the nulls in the second harmonic term occur from the magnetic quadrupole moment as measured in situ by Pioneer 11 [Acuña and Ness, 1976c].

Smith and Gulkis [1979] were able to derive a curve of P.A. vs. λ_{III} from the Pioneer 11 magnetic field measurements which was in good agreement with the measured curve. They assumed that only electrons lying in the non-planar equatorial surface (charged particle equator) at a distance of about $2\ R_J$ from the center of the planet need be considered, and that beaming is unimportant but that planetary shadowing must be accounted for. They concluded that the prospects for using radio data to improve the magnetic field model are promising.

Komesaroff and McCulloch [1981] showed that a good approximation to the amplitudes and phases of the first two sinusoidal terms of the Fourier series fit to the 16-year Komesaroff et al. [1981] data set could be derived from Pioneer 11 magnetic field measurements. The dipole and quadrupole field components were used, and planetary shadowing of emitted radiation was taken into account. There was good agreement in the case of the second harmonic amplitude only when it was assumed that the distance of the emitting shell of electrons was between 1.5 R_J and 2 R_J from the center of the planet. They concluded that the form of the position angle curve is relatively insensitive to any variation in the pitch angle distribution, but it depends sensitively on the magnetic field structure, at least on the dipole and quadrupole components in the region between 1 R_J and 2 R_J.

Birmingham [1981] followed the approach taken by Smith and Gulkis [1979], starting with basically the same assumptions, but his calculations were more detailed. For both the 0_4 magnetic field model and the P10-11 model derived from Pioneer 10 and 11 field measurements combined [Smith and Gulkis, 1979], he calculated the Fourier amplitudes and phases for each of a series of values of the effective distance of the radiating electrons (in the warped magnetic equatorial surface) from the center of the planet. Birmingham stated that although there was good qualitative agreement of his calculated results with the observations, there were several types of quantitative disagreements. It was apparently not possible to determine definitely which was the most realistic of the assumed distances of the equivalent ring of emitting electrons from the center of the planet, nor which of the two magnetic field models gave the better result. He concluded that although the model can be refined in several ways, further effort employing this particular approach is probably not justified at present because of the uncertainties in the radio, magnetic field, and energetic electron data used.

In summary, the dependence of linear polarization position angle on CML as modeled from the Pioneer 11 magnetic field measurements by Gerard, Smith, and Gulkis, Komesaroff and McCulloch, and Birmingham is in each case in good qualitative agreement with observation. However, only Birmingham's work appears sufficiently detailed to reveal the severe limitations on the use of this type of radio data for improving existing models of the magnetic field and the high energy particle distributions. Clearly, the high-resolution intensity and polarization parameter maps of the emission region would provide a sounder basis for model development than do the measurements of integrated polarization parameters utilized in the preceding papers. As we shall see, relatively high-resolution maps have recently been made and have indeed been used with apparent success in model development.

Figure 7.6c shows the variation of the *degree* of linear polarization P_L (ratio of linearly polarized flux density to total flux density) with respect to λ_{III}. The two maxima in P_L occur very nearly at the zero crossings of ψ_m, and hence they coincide with the maxima in the total flux density curve in Figure 7.6a. P_L has minima when Earth is farthest above and below the magnetic equatorial plane. If D_E had been zero the two minima, occurring at $\lambda_{III} = 190°$ and $360°$, would probably have reached nearly equal levels. Figure 7.7d shows the corresponding variation with respect to ψ_m. As Berge and Gulkis [1976] have pointed out, the similarity of the curve of P_L vs. λ_{III} to that of total flux density vs. λ_{III} (Fig. 7.6a) indicates that the linearly polarized component is more sharply beamed than is the total emission. If the beaming of the two were the same, their intensities would vary proportionally, and the ratio P_L would remain constant instead of varying as it actually does. The total radiation thus consists of (1) an unpolarized nonthermal component that is beamed, (2) a polarized nonthermal component that is more sharply beamed, and (3) a thermal component that is unpolarized, not beamed, and is relatively small at frequencies below 2000 MHz. No variation in the degree of linear polarization of the nonthermal component with respect to frequency has yet been found, although such a variation is probable in both the lowest and highest frequency regions of the observable synchrotron spectrum.

The circularly polarized component. High energy electrons in a tilted Jovian dipole field with equatorial pitch angles (α_e) near 90° direct most of their synchrotron radiation close to the plane of the magnetic equator. In these directions the polarization is linear, or at least elliptical with a high axial ratio. For propagation directions farther out of the equatorial plane, where the intensity is less, the axial ratio is also less, making the polarization ellipse more nearly circular. Thus, although the degree of linear polarization (P_L) is greatest for propagation directions parallel to the equatorial plane, the absolute value of the degree of circular polarization ($|P_C|$) is greatest in propagation directions which are most nearly perpendicular to it. Now if in addition there are other electrons having an extended range of α_e values both less and greater than 90°, major parts of their helical paths are more or less steeply inclined with respect to the magnetic equatorial plane, as suggested in Figure 7.8a, causing (1) a broadening of the curve of S vs. ψ_m, (2) a reduction in the amplitude of the curve of P_L vs. ψ_m, with an increase in the relative amount of unpolarized radiation, and (3) an increase in the amplitude of the curve of P_C vs. ψ_m. The reason for the latter is that the farther the mirror points of electrons near the central meridian are from the magnetic equator (and hence from the Earth direction), the more nearly perpendicular is the line of sight from Earth to the orbital planes of the mirroring electrons. Each electron emits most of its radiation from the vicinity of the mirror points, where it spends most of its time. The circularly polarized components of opposite sense from conjugate pairs of mirror points are of equal intensity when $\psi_m = 0$, so that their net contribution to the circular polarization is zero. However, when $|\psi_m|$ is maximum, one of the mirroring orbits is seen more nearly face-on than the other, and $|P_C|$ then reaches a maximum.

The sign of P_C is positive, corresponding to the LH sense, if **B** projected onto the line of sight is away from the observer; it is negative (RH) if the **B** component is toward the observer.

Measurements of the degree of circular polarization of Jupiter's synchrotron radiation have been made by Berge [1965, 1974], Seaquist [1969], Komesaroff et al. [1970], Stannard and Conway [1976], Roberts and Komesaroff [1976], Biraud et al. [1977], Neidhofer et al. [1977, 1980], and de Pater [1980a]. Figure 7.6d, from de Pater's 1412 MHz measurements, shows the degree of circular polarization, P_C, as a function of λ_{III}. The two zeros in the P_C curve occur at λ_{III} values of approximately 100° and 300°, about the same longitudes at which the maxima in the S (Fig. 7.6a) and the P_L curves (Fig. 7.6c) occur. It is apparent from Figure 7.7e that the zeros of P_C lie close to $\psi_m = 0°$, in agreement with the qualitative picture of the emission process presented above. The two absolute-value maxima of the P_C curve occur very nearly at the propagation directions which are farthest above and below the magnetic equatorial plane, nearly in coincidence with the S and P_L minima, again in close agreement with our qualitative picture.

The tendency for the curve of P_C vs. ψ_m (Fig. 7.7e) to become flattened, that is, more nearly horizontal, the farther ψ_m is from zero was first observed by Biraud et al. [1977] and Roberts and Komesaroff [1976], and has subsequently been observed by others.

Roberts and Komesaroff attribute this characteristic shape to the pitch angle distribution of the radiating electrons. Using a thin L-shell emission model, they were able to adjust the *shape* of the theoretically derived curve of P_c vs. ψ_m by altering the pitch angle distribution function, and to adjust the *amplitude* of the curve by manipulating the assumed value of the equatorial magnetic field **B** within the L-shell. The value they obtained is 0.3 G. This can be considered a first approximation to the mean value of **B** for the actual distributed emitting region weighted by the brightness. The high resolution mapping of de Pater and Dames [1979] indicated, however, that the synchrotron emission peak in the equatorial plane was at about 1.3 R_J, decreasing to half brightness at about 3.4 R_J. Thus, the actual emission region is far more extended and complex than the thin shell model upon which the radio determination of **B** was based. In order to obtain radio measurements of **B** at a particular point to within accuracies better than, say $\pm 50\%$, a more realistic model of the emission region will have to be used.

In a discussion of the same radio observations, Roberts [1976] concluded that theoretically derived curves of S vs. ψ_m and P_L vs. ψ_m are much less sensitive to the nature of the pitch angle distribution than is the P_c vs. ψ_m curve. He also pointed out that only the latter curve provides a means for measuring **B**.

Still another important piece of information that can be deduced from measurements of P_c as a function of ψ_m is the sense of the magnetic dipole, that is, whether the majority of **B** field lines emerge from the northern or the southern Jovigraphic hemisphere. Berge [1965], who first measured the circularly polarized component, found that it emerges from the northern hemisphere, opposite to the direction of the terrestrial field relative to the rotational angular velocity vector of the planet. This same conclusion had previously been reached by Warwick [1963a] and Dowden [1963] on the basis of measurements of the decametric radiation, after having made certain unverified assumptions regarding the unknown decametric emission mechanism. In arriving at his conclusion, Berge found that the circularly polarized component is LH when Earth is north of the magnetic equator and RH when south of it. This is apparent in Figure 7.7e. When Earth is at its maximum northern magnetic latitude, for which λ_{III} is 201°, the northern-hemisphere magnetic pole is tipped toward the observer. As we have previously stated, most of the circularly polarized radiation comes from the vicinity of electron mirror points in the region near the central meridian. The polarization sense of the integrated wave contributions of all these electrons, which have a symmetrical distribution of α_e values centered on 90°, is the same as it would have been if they all remained in the magnetic equator, with α_e values of exactly 90°. Thus, we can use the circular orbit shown in the magnetic equator in Figure 7.8b to deduce the direction of **B**. The pertinent facts are (1) in an orbit seen obliquely from its northern face the electron emits LH elliptical radiation and (2) for radiation having the LH sense, the direction of **B** projected onto the line of sight must be away from the observer. It follows that field lines must penetrate the electron orbital plane from its northern side, as shown in Figure 7.8b, and therefore that the majority of **B** lines emerge from the northern Jovigraphic hemisphere of the planet and enter the southern hemisphere. This was essentially Berge's line of reasoning. As we discuss further in Section 7.3, an earlier analysis of the low-frequency burst emissions led to the same conclusion.

Decimetric rotation period measurements. The rotation period of Jupiter's inner magnetosphere can be determined from both the decametric burst emission (below 40 MHz) and the decimetric synchrotron emission. Such measurements by both methods have been made for a number of years. As techniques have improved, the decimetric and decametric measurements have tended to converge, providing the basis for establishing the currently accepted value, designated System III (1965). This value is 9 hr 55 min 29.71 s [Riddle and Warwick, 1976]. Although the mean of the better decimetric measurements [Berge and Gulkis, 1976] appeared to be slightly different from the best at decametric wavelengths [Carr and Desch, 1976 and references therein], more recent results have essentially eliminated the discrepancy.

The method for arriving at the rotation period from synchrotron emission data involves the determination of fundamental periodicities in measurements of the linear-polarization-plane position angle as a function of time or in measurements of the degree of circular polarization as a function of time. The position angle method has been used much more extensively than that based on circular polarization. One way of calculating the period from decimetric position angle measurements is to vary (in small increments) the assumed period upon which λ_{III} in the Fourier series fit (7.6) is based, until the phase ℓ_1 of the fundamental term attains a maximum or minimum value. The best decimetric rotation period measurement is that of Komesaroff et al. [1981], based on over 16 years of linear polarization position angle measurements at three observatories. This value agrees with the System III (1965) value to within 0.02 s. Very nearly the same result has been obtained from measurement of the circularly polarized component. Biraud et al. [1977], using their own measurements of this component at 1400 MHz together with the earlier data of Komesaroff et al. [1970], Berge [1974], and Seaquist [1969], obtained the value 9 hr 55 min 29.69 s ± 0.12 s. This is not significantly different from the System III (1965) value nor from the above determined from position angle measurements.

The rotation period measured by Komesaroff et al. [1981] at decimetric wavelengths and that by May, Carr, and Desch [1979] at decameter wavelengths (see Sec. 7.3) both differ from the System III period by only 0.02 s. An error of 0.02 s in the value assumed would cause the apparent longitudes of features corotating with the inner magnetosphere to drift only 10° in 50 yr. Differential magnetic rotation between the synchrotron and decametric regions ($r < 4 R_J$) must be very slight if it exists at all.

Source mapping. The first measurements yielding information on the spatial distribution of the Jovian decimetric source were the interferometric observations of Radhakrishnan and Roberts [1960] and Morris and Berge [1962]. Berge [1966], by fitting a model to data obtained with the Owens Valley Radio Observatory interferometer, produced a two-dimensional brightness distribution map of the synchrotron-emitting region which in its gross structure is remarkably similar to the best available today. Berge had assumed a highly symmetrical model, and as a result his map does not display irregularities attributable to nondipolar field components; a 10° tilt of the dipole was built into his model. Barber [1966] and McAdam [1966] also made interferometric measurements of the emission region. Branson [1968] used the method of aperture synthesis to produce three two-dimensional maps at equally spaced intervals of CML. His maps clearly show the wobble of the magnetic equator with rotation and also effects due to nondipolar field components. Branson, and also Conway and Stannard [1972], found evidence for a "hot spot" at a System III (1965) longitude of about 200°. De Pater and Dames [1979] confirmed the existence of the hot spot in 1973, but found it to be located at a longitude about 60° higher than reported by Branson. Ground observations have revealed a striking enhancement of UV radiation from sulfur ions in the inner part of the Io torus (see Chap. 6) at about the same longitude as the radio hot spot. Dessler [1980a] has suggested that the two effects are related.

De Pater, using the Westerbork aperture synthesis array, has extensively mapped the synchrotron source region at 1412 MHz [de Pater, 1980a,b] and 4885 MHz

LINEARLY POLARIZED FLUX DENSITY	MAGNETIC FIELD ORIENTATION	EQUATORIAL SCAN, POLARIZED FLUX

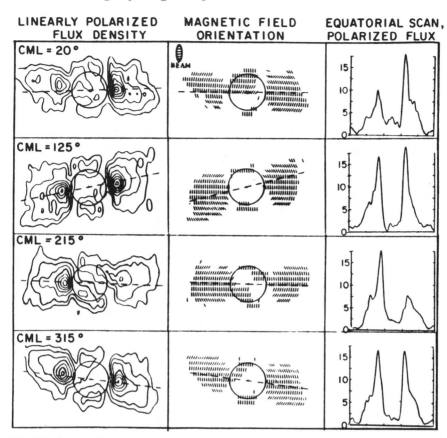

Fig. 7.9. Maps of the linearly polarized flux density at 6 cm wavelength (left-hand panels), magnetic field orientation (center panels) and polarized flux density scans (in Jy) across the magnetic equator. The value of the CML is given in the upper left-hand corner of the first panel for each of the four rows of maps. The projected average magnetic field directions are drawn perpendicular to the measured orientation of the linearly polarized component of the radio emission. Adapted from de Pater [1981].

[de Pater, 1981c]. At 4885 MHz, she constructed for each 15° strip of CML a brightness distribution map of the linearly polarized component, a map of the projected average magnetic field directions (as deduced from the linear polarization direction) and a cross-scan plot of the linearly polarized brightness along the magnetic equator. Examples selected from this series of 24 sets of maps are shown in Figure 7.9.

Assuming the simple dipole model, one would expect the maxima of the total intensity and those of the linearly polarized component to occur at the λ_{III} values of 110° and 290°, and the circularly polarized maxima (of opposite senses) to be at 20° and 200°. As we have seen from Figure 7.6, this is approximately the case. However, it is apparent from de Pater's complete series of maps, that this is far from true in the vicinity of the magnetic equator. Conspicuous anomalies are present in the distribution of both the linear and circularly polarized components. For example, in Figure 7.9, at $\lambda_{III} = 20°$ the brightness center to the right of the planetary disc is much stronger than the one to the left for the linearly polarized intensity. The reverse is true at $\lambda_{III} = 215°$, about 180° of rotation later. Such effects are manifestations of the multipolar magnetic field structure and are related to the hot spot of Branson [1968] and the warped equator of Gerard [1976]. Another curiosity is the previously undetected emission regions can be seen near the poles at the top and bottom of the planetary disk in Figure 7.9. They are perhaps due to mirroring electrons relatively close to the poles whose orbits are viewed nearly edge-on (i.e., with the line of sight nearly perpendicular to **B**), according to de Pater.

Insofar as Jupiter's magnetosphere is concerned (and hence this book), the ultimate goal of studies of the synchrotron radiation is the creation of the best possible source model incorporating detailed information on the spatial, energy, and pitch-angle distributions of the radiating energetic electrons and how and why these distributions vary with time. The high-resolution measurements of de Pater have made possible the construction of such a model in considerable detail. This model is described in two papers by de Pater [1981a,b] and in her thesis [de Pater, 1980b]. It is undoubtedly the most advanced treatment of the inner magnetosphere that has yet been published and may not be surpassed until measurements made at still higher resolution with the Very Large Array (VLA) in New Mexico are incorporated into a new model.

De Pater's model is discussed in Chapter 5.

7.3. The decameter and hectometer wavelength emission

In the remainder of this chapter we shall be concerned with the complex phenomenology of Jupiter's radio bursts, which are confined to the frequency range below 40 MHz. Despite the fact that the decametric bursts were discovered before the decimetric radiation, the emission mechanism is still only partially understood. On the basis of the large power flux density observed (Fig. 7.1), it is clear that whatever the mechanism (mechanisms) may be, collective interactions between the emitting electrons play a most important role. This is not the case with the synchrotron-emitting electrons. Decametric emission is believed to be near the fundamental of the cyclotron frequency, rather than at its high harmonics as in the case of the synchrotron radiation. The 40 MHz high-frequency cut-off is thus believed to be approximately the electron cyclotron frequency in the region of strongest magnetic field encountered by the emitting electrons. Since there is no evidence of second harmonic emission, the energies of the electrons must be very low in comparison with the synchrotron emitters. These energies are generally believed to be in the tens of keV range.

Another important difference between the two types of emission is in the effect of the surrounding medium. Although the propagation and polarization of the synchrotron radiation are only slightly affected by the plasma adjacent to the emission site, this is not true of the burst radiation. If the emission is in the extraordinary mode, as is generally believed, then it is necessary to assume that the emitted radiation is slightly above the cut-off frequency in order for it to escape the stop zone (see Chap. 9). During escape, its propagation path and polarization may be strongly affected by the plasma and fields it encounters.

The lack of a comprehensive theory to date accounting for the observed radiation phenomenology below 40 MHz can be attributed to several factors. From an observational perspective these include the complex observed phenomenology, and the difficulty of untangling first order from second order effects and of separating out propagation effects and artifacts of observation. All of these have contributed in one form or another to prevent us from answering the fundamental question of locating the radio source. From the point of view of the theorist, the principal obstacles have been the relatively large number of possible plasma instabilities involving various types of interactions in which emitting electrons might participate, the inherent difficulty in formulating mathematical descriptions of these processes and in choosing the important ones, and inadequate knowledge of the characteristics of the magnetospheric medium. Chapters 9 and 12 are concerned with some of these problems. In the remainder of this chapter we present the extremely diverse phenomenology of the burst radiation, one of our foremost aims being to make it more accessible to those searching for observational constraints with which to limit the theoretical possibilities.

Historical background: from initial discovery to Voyager encounter

We begin this section with a general pre-Voyager overview of some of the major properties of the Jovian emission, specifically of those radio components now commonly referred to as DAM and HOM (for decameter- and hectometer-wavelength emissions, respectively), or collectively as just DAM. (Detailed reviews of the pre-Voyager descriptions may be found in Warwick [1967], Carr and Gulkis [1969], Warwick [1970], Carr and Desch [1976].) Initial observations showed that in addition to being very intense the emission is also elliptically polarized, sometimes in the left-hand sense but more commonly right-hand. The radiation was also observed to be quite sporadic, exhibiting large intensity fluctuations on seconds to tens-of-seconds timescales. This is in sharp contrast to the smoothly and only slightly varying decimeter-wavelength emission (DIM) discussed previously. In addition, observers noted that DAM was invariably confined to frequencies below 40 MHz. The low-frequency emission cut-off, if indeed there was one to be found, was not directly observable owing to the opacity of the Earth's ionosphere at long wavelengths. Ground-based observers were able to establish the existence of DAM at frequencies as low at 4 MHz [Carr et al., 1964; Ellis, 1965; Zabriskie, 1970]. Subsequent earth- and lunar-orbiting satellite observations in the late 1960s and early 1970s [Desch and Carr, 1974; Brown, L. W., 1974; Kaiser, 1977; Desch and Carr, 1978]

1. confirmed the existence of sporadic Jovian emission at frequencies as low as 450 kHz,
2. revealed the broad spectral peak centered at about 8 MHz, and
3. showed that some Io influence over the emission extended down as far as 2 MHz.

Using ground-based telescopes of low spatial, time, and frequency resolution, astronomers recognized that Jupiter possesses a "permanent" DAM dynamic spectrum, that is, landmark emission patterns in the frequency-time domain that reappear when specific central meridian longitudes and Io orbital phases recur. Later, new hierarchies of dynamic spectral pattern organization were found to exist when higher time and frequency resolutions became available. For example, the Io-controlled emission exhibited highly structured and rapidly drifting bursts, called *S* bursts or millisecond bursts, in which there is structure on timescales down to about a millisecond. Continuous monitoring of DAM by fixed-frequency and swept-frequency receivers at a number of observatories

1. revealed the characteristic Jovian-longitude emission zones known as sources A, B, C, and D, and their respective dependence on Io phase,
2. yielded precise measurements of Jupiter's magnetic rotation period, and
3. demonstrated the existence of 11.9-yr periodicities in emission phenomena attributable to changes in the Jovicentric declination of Earth (D_E).

The stability of the source occurrence probability distributions and of dynamic spectral landmarks within sources, the approximate symmetry of sources A and B relative to the tilted magnetic dipole, and the correlations with respect to D_E were generally accepted as due to the geometrical effects of relatively narrow emission beams. These same synoptic monitoring programs also provided the long data spans necessary to search, with marginal success, for correlations with solar activity, and with no success for effects due to satellites other than Io (e.g., Kaiser and Alexander [1973]).

In short, by the late 1960s the following picture (oversimplified here) was developed from these and other observations and certain assumptions. It was concluded that Jupiter must have several strongly beamed radio sources in both northern and southern hemispheres emitting mostly in the extraordinary mode at or near the local electron cyclotron frequency. Jupiter's global magnetic field must strongly influence the emission properties. This field must have a peak intensity of about 14 G in the northern hemisphere. The field lines must be directed from the vicinity of Jupiter's northern geographic pole toward the southern geographic pole; that is, Jupiter's magnetic field must be flipped relative to Earth's. In fact, these points concerning the field intensity and topology were later confirmed by Pioneer 10 and 11 observations [Smith et al., 1975; Acuña and Ness, 1976c]. Further, in the case of that part of the emission directly stimulated by Io, it was necessary to assume that communication by means of particles or waves occurs between Io and the distant place at which emission occurs. Other radio components independent of Io orbital phase suggested one or more additional controlling influences unrelated to Io.

The Voyager spacecraft observations have been invaluable in complementing and extending the earlier ground-based and satellite efforts in a number of ways. (An overview of the Voyager Planetary Radio Astronomy results can be found in Boischot et al. [1981].) Briefly, because the Voyager observations were free of ionospheric scintillation effects and were of sufficient sensitivity and frequency resolution, another level of structure became apparent in the frequency-time dynamic spectra. For example, the frequency-time landmark features were resolved into many nested spectral arcs, which appeared like rows of opening and closing parentheses of smoothly varying size and curvature. Two distinct classes of arcs became apparent: *greater arcs* and *lesser arcs*, generally distinguishable by their frequency extent and curvatures. Greater arcs can extend from ~ 1 MHz up to 39.5 MHz; the lesser arcs are more limited in frequency extent, rarely exceeding 15 MHz peak frequency. In addition, Voyager was able to provide the first information on the behavior of the emissions as observed from above Jupiter's nightside hemisphere.

General properties

Spectral characteristics. As has been mentioned, early ground-based and satellite studies had established a high frequency cut-off at about 40 MHz and a broad peak centered at about 8-MHz in the time-averaged flux spectrum. The Voyager observations confirmed these features, and by virtue of greatly increased sensitivity and frequency resolution, made possible an accurate determination of the entire spectrum of the burst emission from 40 MHz down to the lowest frequencies at which radio propagation is possible. This spectrum together with that of the synchrotron component constitute Jupiter's nonthermal emission; the spectra are shown in Figure 7.1. In this figure the peak in the DAM range is broad, extending over a band about 4 MHz wide centered on about 10 MHz. The high-frequency shoulder is much steeper than that at low frequencies and corresponds to a spectral index (i.e., the exponent n in $S \propto f^n$) of about -6. The Voyager results confirmed that there is indeed a high-frequency cut-off at 39.5 MHz, and that it is of natural origin, not simply an artifact due to inadequate sampling of weak events. This spectral boundary, labeled "magnetic cut-off" in Figure 7.1, has long been suspected to represent the maximum electron cyclotron frequency occurring in the region of emission, an idea that is strongly supported by the Pioneer and Voyager magnetometer results. The drop-off below about 2 MHz, on the other hand, may be due in part to propagation effects induced by the dense Io plasma torus.

The Voyager observations have extended the range to lower frequencies so that a minor peak is apparent at about 100 kHz. This peak has an intensity level 1.5 orders of magnitude below the principal peak, and the shoulder on the low-frequency side extends to the boundary for freely propagating electromagnetic radiation, labeled "plasma cut-off" in Figure 7.1. This cut-off varies with both the propagation path and with time; it is essentially the maximum plasma frequency between source and observer and is believed to occur in the 1 to 10 kHz range. Below the plasma cutoff lies the realm of the plasma waves – waves that are bound to the medium. They are the subject of Chapter 8.

Examination of frequency-time dynamic spectra, discussed in detail below, shows that several distinct types of emission make up the spectrum shown in Figure 7.1. The "greater arcs" can extend from at least 1 MHz up to the 39.5-MHz cut-off, whereas the "lesser arcs" are more limited in frequency extent. It is these two components that collectively make up the single flux peak near 10 MHz. The low-frequency secondary hump is also made up of two separate sources, the narrow band (n) and broadband (b) kilometric (KOM) radiation, discussed later.

By integrating the power flux density curve over the entire frequency range and assuming that the radiation is emitted uniformly and isotropically over $4\pi sr$, we can arrive at a crude estimate of the total (average) radiated power. The value obtained is 3 $\times$ 10^{11} W, a number that is very likely an overestimate since the DAM component, at least, is not emitted isotropically. A lower-limit estimate of the average value of the total radiated power is probably 6 $\times$ 10^{9} W, arrived at by assuming an emission cone conical-sheet thickness of 5°.

Temporal structure and dynamic spectrum. Frequency-time dynamic spectra of Jovian emission are extremely complex, although well ordered and even predictable. Figure 7.10 shows a 10-h spectrum from Voyager observations in which increasing darkness is proportional to increasing radio intensity. The spectrum covers the frequency range from 20 kHz to 40 MHz. The emission is organized into several hierarchies according to timescale. Most apparent is the organization into individual *storms*; for example, the event from 03 hr to 05 hr SCET (spacecraft event time) centered at 20 MHz is a single storm. Later we shall see that because this storm is associated with a particular longitude region and Io's orbital phase was not within two special zones, it is referred to as a non-Io-A storm (see Table 7.4). These active periods are quite episodic. Storms in some frequency ranges may last up to several hours with long periods of complete radio quiet separating them. It is on this timescale that the permanent Jovian dynamic spectrum appears.

Detailed structure is also apparent within each storm in these relatively low-resolution spectra and here a second hierarchy appears. On a timescale of minutes, we observe the ubiquitous *spectral arcs*, one of the principal Voyager radio astronomy discoveries. Note the well-defined closing parenthesis-shaped arcs which make up the non-Io-A storm just described. In contrast, at 22 hr to 00 hr a non-Io-B storm occurred in which the arcs are oriented in opening-parenthesis fashion. This is another example of the remarkable repeatability of the Jovian radio morphology: the arcs are always oriented in the same way within a given storm class so that they constitute another level of permanent dynamic spectra. The detailed morphology of these landmark features is described later.

Modulations appear on a timescale of seconds, some of which are intrinsic to the source itself, and some of which are not. An example is shown in Figure 7.11 which has

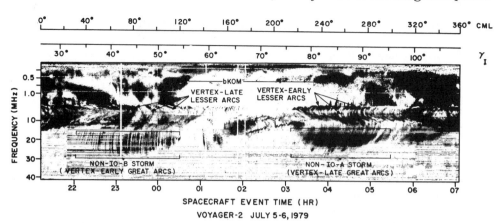

Fig. 7.10. A representative dynamic spectrum of Jupiter's low frequency radio emissions
during one complete rotation of the planet. The darkness of the grey shading is
proportional to signal intensity, plotted here as a function of frequency and time. The
frequency axis is actually divided into two distinct bands that adjoin at about 1.3 MHz;
the low frequency band consists of 70 frequency channels spaced at 19 kHz intervals
between 1 and 1326 kHz, and the high frequency band is composed of 128 channels
spaced 307 kHz apart between 1.2 and 40.5 MHz. At frequencies above 20 MHz, two
major DAM "great arc" storms can be seen centered on about 23 hr and 04 hr spacecraft
event time. At frequencies between a few MHz and about 15 MHz the "lesser arc" DAM
can be seen with "vertex-late" curvature after 03 hr. Some of the DAM arcs appear to
extend to below 0.5 MHz in the low frequency band where the frequency scale is greatly
expanded.

been adapted from work by Riihimaa [1971]. This figure is again in the format of a
dynamic spectrum, but at much higher frequency and time resolution than those
shown previously. The negatively drifting (i.e., $df/dt < 0$) emission features are called
modulation lanes. Generally, modulation lanes can have drift rates of from -150 to
$+150$ kHz/s; those shown in the figure drift at about -150 kHz/s. By way of contrast,
the greater arcs in this frequency range drift about an order of magnitude slower on the
average. Vertical gaps, centered at about 16 and 20 s, break up the envelopes of the
modulation lanes. Intensity structure on this 1 to 5 s timescale, as illustrated by these
envelopes and gaps in Figure 7.11, are commonplace in groundbased data and are
referred to as L (long) *bursts*. Douglas and Smith [1967] first demonstrated that these
temporal signatures are due primarily to the infuence of interplanetary scintillation.
This modulation is most apparent in data taken at significant distances from Jupiter.
L-burst groups also experience a 30–60 s modulation caused by the Earth's ionosphere.
Thus we might expect, and indeed there is some evidence, that close to Jupiter most
but not all of the radiation is in the form of relatively long bursts of emission. These
long bursts are subsequently broken up into L-burst trains by scintillation during
propagation to Earth. The implication here is that the fundamental timescale of the
original burst from which the L bursts is formed is actually tens of seconds or minutes
and may simply correspond to the 3- and 6-min timescale of the arc structure [Genova
and Leblanc, 1981]. The observed modulation-lane structure must be primarily intrinsic to the source, or at least to propagation effects in the immediate neighborhood of
the source, because the drift rates are principally dependent on Jovian longitude
[Riihimaa, 1974]. However, a secondary dependence on the solar elongation at the
time of the observation has been reported [Riihimaa, 1979]. These conclusions have of
necessity been drawn from Earth-based observations because the Voyager radio
astronomy instrument is incapable of resolving modulation-lane structure.

Table 7.4. *Components of Jupiter's decametric emissions*

Source designation	CML range[a]	γ_{Io} range[a]	Maximum frequency (MHz)	Dominant polarization	Arc curvature (vertex)	Notes
Io-D	0°–200°	95°–130°	18	LH	Early	Also called "fourth source"
Io-B	15°–240° (105°–185°)	40°–110° (80°–110°)	39.5	RH	Early	Also called "early source"
non-Io-B	80°–200°	0°–360°	38	RH	Early	Weak from above day hemisphere but strong when viewed from above night hemisphere
Io-A	180°–300° (200°–270°)	180°–260° (205°–260°)	38	RH	Late	Also called "main source"
non-Io-A	200°–300° (230°–280°)	0°–360°	38	RH	Late	Strong from above day hemisphere but weak when viewed from above night hemisphere
Io-C	280°–60° (300°–20°)	200°–260° (225°–260°)	36	RH and LH	Late	Also called "third source"
non-Io-C	300°–360°	0°–360°	32	RH and LH	Late	Moderately strong from above day hemisphere and very weak when viewed from above night side

[a] Numbers in parentheses give widths at half-maximum for the major sources as observed from Earth at frequencies near 20 MHz.

Fig. 7.11. An idealized sketch
illustrating *L*-burst envelopes caused by
interplanetary scintillations and drifting
modulation lanes as they would appear
in a dynamic spectrum. Adapted from
Riihimaa [1971].

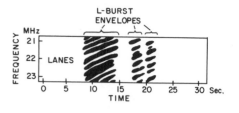

Fig. 7.12. A dynamic spectrum showing
a train of simple *S* bursts recorded near
32 MHz by Desch, Flagg, and May
[1978]. Notice that the full display covers
a time interval of less than 200 ms, with
an effective frequency and time
resolution of 3 kHz and 0.3 ms,
respectively.

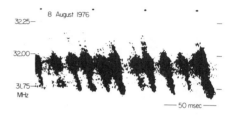

At time resolutions shorter than about 0.01 s, *S* (short) bursts become apparent.
These short-lived and rapidly drifting emission features are shown in Figure 7.12. The
S-burst modulation is also a source-related phenomenon, appearing only in association
with Io-stimulated emission. They will be discussed in detail.

Any or all of the above phenomenology is observable during an active period. Thus,
a given storm always consists of spectral arcs of one variety or another but the nature
and even occurrence of *S* bursts and modulation lanes depend on the CML, Io phase
and observing frequency.

On much longer timescales, years for example, variations in activity become
apparent following statistical analysis of the data. The best known of these is the
11.9-yr modulation, corresponding approximately to the length of a Jovian year. This
modulation [Carr et al., 1970] has been shown to correlate best with the Jovicentric
declination of the observer (D_E for an Earth-based observer), indicating that the radiation is strongly beamed latitudinally. Striking evidence for this (see, e.g., Fig. 7.13) is
the almost complete disappearance of DAM at some frequencies during Jovian
apparitions for which D_E is near its most negative value ($-3.3°$).

Finally, there are more subtle variations in the activity level, which appear on a scale
of days to weeks. These are partially geometrical; that is, they are due to changes in the
CML and Io phase required to observe storms associated with particular source
regions. However, some of these variations are possibly due to some form of solar influence, such as to fluctuations in the solar-wind pressure and/or interplanetary magnetic
field orientation across the Jovian magnetosphere. The latest and most sophisticated
work on this problem has been done by Terasawa, Maezawa, and Machida [1978],
Barrow [1979], and Levitskii and Vladimirski [1979]. All of these authors have found
solar influences effective in modulating DAM; the solar variable examined ranged
from sector boundary and stream–stream interactions in the solar wind to the geomagnetic A_p index. These studies indicate that decametric activity can be enhanced in
direct response to a disturbance propagating with the solar wind. It should be understood, however, that these studies pose a complicated analysis problem involving the
complete removal of the Io-controlled (geometry-related) events and the necessity of
working with incomplete data sets. The arguments are necessarily statistical and there
is not yet general agreement as to the nature and, indeed, even the existence of a solar
influence. Not yet examined is the possibility that Io vulcanism has some influence on
these variations.

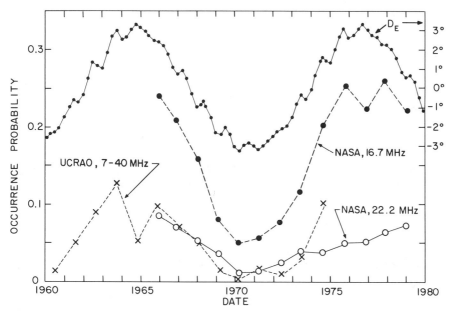

Fig. 7.13. A plot of the apparition-averaged probability of occurrence of DAM (three lower curves) and the jovicentric declination of the Earth, D_E, from 1960 to 1980. The x's correspond to swept-frequency measurements of DAM between 7 and 40 MHz obtained at the University of Colorado Radio Astronomy Observatory [Warwick et al., 1975], and the large circles and dots pertain to fixed-frequency measurements obtained with the NASA multistation Jupiter Monitor Network. Notice that the long-term variations in DAM activity correlate well with D_E in spite of the fact that D_E varies only by $\pm 3.3°$.

CML and Io phase control. The distribution of Jovian storms in time represents a well-organized, nonstochastic process. This is manifestly evident when active periods are superposed in histogram form using starting epochs separated by about 10 h. An example is shown in Figure 7.14, which shows results at 18–22 MHz [Thieman, 1977]. A three-peaked distribution is generally seen above about 10 MHz. Each individual peak in the DAM range has come to be called a radio "source" with varying nomenclatures (see Table 7.4). The sources can be maintained at the same longitude over long periods of time by choosing the proper interval between epochs [Carr, 1972b]. Such efforts have led to very accurate determinations of the rotation period of Jupiter's magnetic field and to the establishment of a strict upper limit of < 0.03 s/yr on possible linear secular changes in that period (see, e.g., May, Carr, and Desch [1979]). The latter determination indicates that Jupiter's rotation is very stable over a timescale of at least a decade. Other methods of determining Jupiter's rotation period have also been employed successfully. Duncan [1971] used a two-dimensional periodogram analysis, solving for Jupiter's period by sorting in both CML and Io phase space. Kaiser and Alexander [1972] used the method of power spectral analysis. These measurements of DAM and other period determinations at DIM wavelengths (e.g., Berge [1974]) led to the official adoption of a radio longitude system, called System III (epoch 1965). This system is based on a rotation rate of 870.536 deg/day (9 hr 55 min 29.71 s) [Riddle and Warwick, 1976; Seidelman and Divine, 1977]. It is used universally for specifying both the central meridian longitude, or CML (1965), and the longitude of meridians which corotate with the planet (see discussion of decimetric rotation period measurements in Sec 7.2).

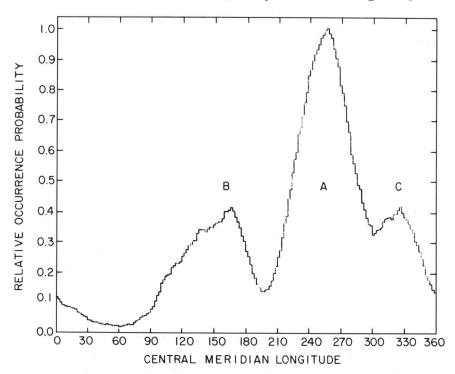

Fig. 7.14. Variation of the relative probability of occurrence of DAM as a function of central meridian longitude. The data were compiled by Thieman [1977] from observations at 18, 20, and 22 MHz carried out between 1957 and 1975 at the University of Florida and University of Texas radio observatories.

If we take the same data used to make Figure 7.14 but replot the events as a function of two variables, CML and Io phase, we obtain a plot like that shown in Figure 7.15. Here Io phase is plotted, as usual, in terms of angular departure from superior conjunction relative to the observer. This figure illustrates several important morphological features of DAM, most notable among them the control of the radiation by Io. Io modulation of DAM is not a subtle effect. Note that the B source, $80° <$ CML $<$ $200°$, is observed only infrequently, except when $80° < \gamma_1 < 110°$. When Io is in the preferred location, centered at about $90°$ γ_1, the emission "turns on," increasing flux levels a hundredfold over average source B intensity levels [Desch, Carr, and Levy, 1975; Desch, 1980]. We refer to the intense component as the Io-related source B, or just Io-B, and to the remainder of the emission as the non-Io-B source. Similar remarks apply to sources A and C, except that the preferred Io phase is centered on $240°$ γ_1. Source D seems to have only an Io-dependent component, which is shifted in Io phase relative to Io-B. The preferred Io phase for source D is about $105°$ γ_1.

Each of these principal sources has a characteristic frequency-time dynamic spectrum that is illustrated schematically in the smaller panels in this figure. The detailed behavior of the frequency ranges, arc shapes, and nose frequencies will be described.

At frequencies below about 10–15 MHz the CML-γ_1 morphology is different from that just described. In the 5-MHz-wide band centered on the 10 MHz spectral peak, the emission is more continuous as a function of CML than it is at higher frequencies. This emission is made up principally of lesser arcs as mentioned previously. Below 2 MHz, the HOM once again exhibits strong CML modulation as suggested by the polarization

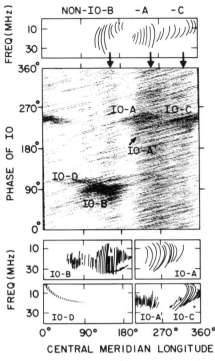

Fig. 7.15. Two-dimensional dependence of DAM activity on longitude (CML) and Io phase (γ_I) at about 20 MHz (J. R. Thieman, personal communication). The Io-related sources are each labeled in the CML-Io phase plot and schematic illustrations of the Io-dependent dynamic spectra are shown in the four lower panels. The top panel illustrates the dynamic spectral behavior of the Io-independent emissions. See text for details.

plot in Figure 7.17. In this frequency band HOM is rarely seen in the CML range 150° to 250°, that is, when the longitude meridian containing Jupiter's north magnetic dipole tip is facing, or nearly facing, the spacecraft. At still lower frequencies the kilometric-wavelength emission is evident and will be discussed separately later.

Polarization. It has long been known that the polarization sense of radiation from sources A and B (the principal ones at the higher frequencies) is predominantly right hand (RH), but as the frequency is decreased below about 20 MHz, progressively more left hand (LH) polarization becomes apparent outside the A–B longitude region (see Carr and Desch [1976], and references therein; see footnote in Section 7.2 of this chapter for definition of LH and RH polarization senses). It was recognized that these observations are consistent with the tilted-dipole field model if it is assumed that emission is in the extraordinary mode, and that the sources are relatively close to the planet. The evidence for this is that most RH polarization occurs within approximately ±90° of the CML toward which the northern-hemisphere end of the magnetic dipole is tipped (200° λ_{III}) and that most LH occurrences are within the other 180° of CML. This implies that the magnetic dipole moment is approximately parallel rather than antiparallel to the rotation vector; that is, that the field lines emerge from the northern Jovigraphic hemisphere, as was later verified by DIM observations and Pioneer magnetometer measurements. The polarization senses of DAM and of the circular component of DIM are usually opposite; the reason for this, on the basis of the above model, is clear from Figure 7.8*b*.

These general conclusions regarding the polarization sense of DAM were based on fixed-frequency observations made at a relatively few select frequencies. The Voyager radio astronomy measurements of polarization sense extend over the entire frequency band of the Jovian emission, however, and we illustrate some of these observations in Figure 7.16. The polarization sense during four consecutive rotations is shown in the

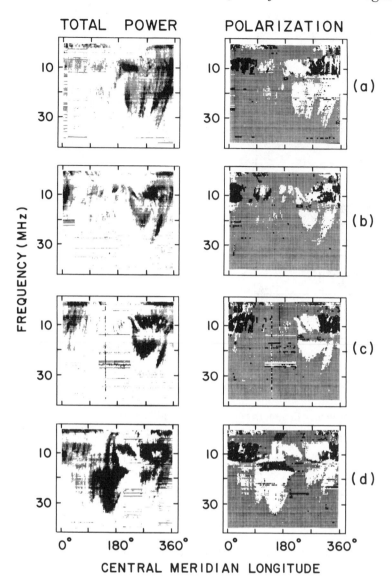

Fig. 7.16. Dynamic-spectral displays of the total received power (left-hand panels) and sense of polarization (right-hand panels) of DAM observed by Voyager 2 during four successive rotations of Jupiter. The top panels begin at 2330 spacecraft event time on July 2, 1979 and the bottom panels end 40 hours later at 1518 on July 4, 1979. The polarization display is coded so that right-hand polarized emission is white, left-hand emission is black, and intervals with no emission or unpolarized emission are grey. The Voyager antenna and impedance patterns produce bands of low sensitivity centered near 5 and 15 MHz, and occasionally the indicated polarization is artifically reversed. This effect can be seen in panel (*d*) between longitudes of about 100° and 210° where the polarization appears to be left-hand in spite of the overwhelming presence of right-hand emission at surrounding frequencies. The high frequency event at high longitudes in panel (*a*) is an Io-C storm, and the event centered near 150° in panel (*d*) is an Io-B storm. Notice the pattern of left-hand emission below 90° and above 270° at frequencies near 10 MHz. Note also the predominance of right-hand polarization at all longitudes for frequencies above 15 MHz.

right-side panels with the convention that white is RH polarized, black is LH polarized, and gray is unpolarized or no emission. Note the predominance of RH (white) polarization for each storm at all frequencies above about 12 to 14 MHz. Exceptions to this rule (not shown here) are the Io-D source, which is often LH and Io-C source, which can be LH polarized up to 20 MHz but rarely so at higher frequencies.

Three parameters are needed to specify the polarization state of a wave if total intensity is selected as the fourth (see Sec. 7.2 for a discussion of polarization parameters). The three most often used in DAM polarization measurements are (1) the axial ratio, with sign to indicate left-hand or right-hand sense, (2) the tilt angle of the major axis of the polarization ellipse, and (3) the degree of polarization, that is, the ratio of the polarized to the sum of the polarized and unpolarized power. Because of the effect of terrestrial ionospheric instability, in only one paper [Parker, Dulk, and Warwick, 1969] have ground-based measurements of the tilt angle been considered reliable. The degree of polarization was found by Barrow and Morrow [1968] to be > 0.70 for about 80% of the time at 18 MHz. Similar results were obtained at other frequencies (e.g., Sherril [1965]), and the generalization is usually made that the Jovian emission is nearly completely polarized across its entire frequency range. Measurements of the axial ratio r have indicated that DAM is elliptically polarized with $1 < r < \infty$ at the higher frequencies, becoming more nearly circular ($r = 1$) toward 10 MHz [Kennedy, 1969].

Because the only unambiguous polarization information provided by the Voyager radio astronomy receiver is the polarization sense, that is, whether the RH or LH circular component is the larger, we must continue to rely on the earlier fixed-frequency ground-based measurements. Between 5 and 15 MHz, where the lesser arcs are often almost continuous, agreement between Voyager and earlier ground-based [Dowden, 1963; Kennedy, 1969] polarization sense measurements appears to be good. LH polarization is observed from 0° to 135° CML, RH polarization from 135° to 300° CML. Between 300° and 360°, both senses are observed.

Polarization measurements below 2 MHz have been made only by the Voyager radio astronomy experiment. In Figure 7.17 we show 10 rotations of the planet with the time for 0° CML indicated in each rotation. The hectometric (HOM) emission is the black and white striped activity centered on 0° CML. The emission at the lowest frequencies, centered at about 200° CML, is kilometric emission (KOM), to be discussed later. The HOM displays fairly clear polarization reversals as a function of CML. For example in Figure 7.17, RH (white) polarization is generally evident between 0° and 90° CML where it is flanked by LH (black) emission centered at 330° and 100° CML. The LH polarization is apparently of greater absolute intensity because relatively far from the planet the RH component is below the receiver detection threshold and only the LH polarized emission was detected [Kaiser et al., 1979].

Although the DAM polarization pattern remains unchanged as viewed on the Voyager outbound trajectory (relative to inbound), the polarization pattern below 1 MHz is not as clearly defined outbound as it is inbound. Some limited CML ranges, for example the early longitude RH band, may even have reversed polarization (see Alexander et al. [1981] for details). Additional work is needed to clarify this phenomenon.

Detailed properties

Dynamic spectral landmarks. One of the most distinctive characteristics of the DAM concerns the repeatability of both coarse and fine emission features within the dynamic spectra. As first discussed by Warwick [1964] and Dulk [1965], both the Io-dependent and Io-independent emission sources have unique morphologies in

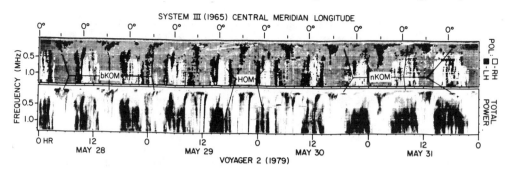

Fig. 7.17. Dynamic-spectral plots showing the polarization (upper panel) and total intensity of HOM and KOM (kilometer-wavelength emission) for ten consecutive rotations of Jupiter. Polarization sense is coded such that RH is white and LH is black. Episodes of HOM recur at 10-hr intervals centered near 0° central meridian longitude and tend to show both right- and left-hand polarization in a left-right-left (black-white-black) sequence. Left-hand polarized bKOM (broadband KOM) can be seen centered at about 200° CML in between each HOM episode.

frequency-time spectra that repeat on a rotation-by-rotation basis or, in the case of the Io-dependent sources, when the same CML and Io phase coordinates are presented to the observer. It is the latter dependence that is particularly striking. Furthermore, the features are apparently so stable over very long periods of time that comparison of landmark events separated by a Jovian year (~ 11.9 yr) may be used to measure the Jovian rotation period with considerable precision [Alexander, 1975].

This "permanent" spectrum, as it is called, is apparent on two timescales. On a scale of tens of minutes to hours, each source has a characteristic emission envelope, defined by its upper and lower frequency limits vs. time. And on a timescale of minutes, the spectral arcs, which form a kind of skeleton within each source, have a characteristic curvature, repetition rate, and vertex frequency (i.e., the frequency at which a vertical line is tangent to a given arc). The envelope landmarks were first described by Warwick [1964] and Dulk [1965] from ground-based studies. Using the notation of Table 7.4, Io-B is the most broadbanded of the sources and, as viewed from Earth, extends from frequencies below the ionospheric cut-off up to the magnetic cut-off at 39.5 MHz. In the best-developed cases the early phase of Io-B storms exhibits a drift of the high-frequency envelope from lower to higher frequencies, followed by a long, narrow-band tail that drifts down in frequency. The fabled high-frequency cut-off can be reached either during the development of the upward drifting envelope or by the narrow-band tail which follows it. The Voyager-PRA records (see Fig. 7.15) show that Io-B is made up of vertex-early arcs, closely spaced, with large radii of curvature – that is, they almost "stand up." When the CML-Io phase geometry is appropriate for Io-B (150° CML, 90° γ_1), the arcs can extend from about 1 MHz up to 39.5 MHz. For a given range of CML, the emission features evolve in a systematic fashion as the Io phase progresses from 60° to 120° [Dulk, 1965; Boischot et al., 1981].

The Io-A source is nearly the mirror image of Io-B. The sample spectrum in Figure 7.18 shows how the upper-frequency envelope drifts up and then down in frequency with time, like Io-B, but the spectral arcs are oriented vertex late. Generally, the Io-A arcs have smaller radii of curvature with lower vertex frequencies than those of Io-B, and they generally do not extend quite as high in frequency. Vertex frequencies are in the range of 10 to 12 MHz.

In addition to the traditional Io-A source, observed for 230° < γ_1 < 250°, Leblanc [1981] has noted that an additional Io-related-A (Io-A') emission extends to slightly

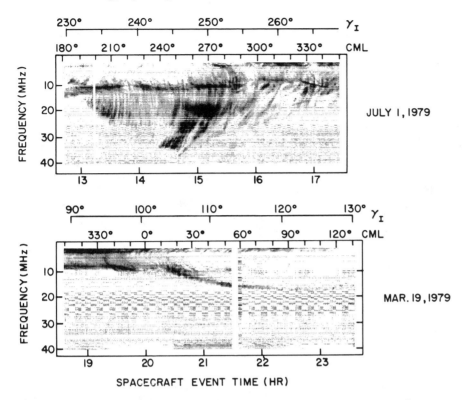

Fig. 7.18. Typical dynamic spectra of Io-A (upper panel) and Io-D (lower panel) events from Voyager Planetary Radio Astronomy experiment.

lower Io phase. The identification is based on characteristic spectral forms which, like Io-A, are vertex late arcs but with very little curvature and very compressed frequency range. This emission rarely extends below 10 MHz or above 25 MHz.

The Io-C source (at CML > 270°) in Figure 7.16a displays a characteristic negatively drifting envelope made up of vertex late arcs with small curvature. Vertex frequencies are commonly near 10–15 MHz or below. The upper-frequency limit of the Io-C source is only rarely greater than 30–32 MHz.

The Io-D source generally corresponds to only one slowly drifting narrow band of emission. The event shown in Figure 7.18, for example, drifts slowly up in frequency from about 5 MHz to 18 MHz over a period of slightly over 2 h. This source often has the appearance of a single, long, vertex-early arc, but it may actually consist of finely structured sections of arcs. The vertex frequency, when visible, is always below 10 MHz.

The Io-independent A- and B-source spectra often appear as less well-developed versions of the Io-dependent components. The intensities are much smaller, the frequency range of the arcs appears somewhat compressed, and the curvature is greater, that is, the Io-independent arcs are not as nearly vertical as the Io-related arcs. The restricted frequency range is probably an intensity threshold effect, however, as originally suggested by ground-based observations [Desch et al., 1975] and confirmed by recent Voyager observations [Barrow and Desch, 1980; Barrow and Alexander, 1980]. The high frequency limit of non-Io-B, for example, is at least 36 to 38 MHz; of non-Io-A, 32 to 34 MHz; and of non-Io-C, 30 to 32 MHz. The values are not much less than the Io-related source high-frequency cut-offs. The strong implication here is that, if emission is at or near the local electron cyclotron frequency, then stably trapped particles alone cannot account for either the Io or non-Io radiation because the maximum magnetic field strengths in the southern hemisphere are not high enough to support these peak frequencies. That is, the conjugate mirror points at these relatively high frequencies in the southern hemisphere would be below the ionosphere.

All of the traditional sources described up to now have consisted of greater-arc emission. However, often superposed on the greater arcs in the 2 to 20 MHz band are intense lesser-arc emissions, so called for their restricted frequency range. Some of this emission is evident in Figure 7.10. Lesser-arc emission appears more nearly continuous compared to the episodic occurrence of the greater-arc events. This radiation is independent of Io control.

Below about 2 MHz, the spectra seem more amorphous than at higher frequencies. There is some evidence that most, if not all, of this radition is simply the lower-frequency extension of the DAM greater and/or lesser arcs.

All of the DAM arcs seem to share certain statistical properties. For example, the average single-frequency duration and time interval between arcs is 3 to 6 min; however, the arcs are not periodic in time. The depth of modulation between arcs can be 20 dB (factor of 100 in power) or greater. The arc vertex frequencies seem to vary in a systematic way as a function of CML. Staelin [1981] has related this variation to the O_4 magnetic field model [Acuña and Ness, 1976c] in the context of a conical beam pattern.

Local time effects. We noted earlier with regard to Io control of DAM that Io phase is almost always plotted relative to the observer rather than, say, the Sun. Early ground-based studies showed that the Io effect is most pronounced in geocentric as opposed to heliocentric coordinates, implying that the observed morphology is due to the passage of a narrowly filled radiation cone past the observer [Bigg, 1964; Dulk, 1967]. According to this view, the preferred coordinates should remain at 90° and 240° γ_i, even in the event of large changes in the observer's local time relative to Jupiter. (Local time is defined here as the Sun–Jupiter–observer angle, measured in degrees or hours, where 12 h corresponds to 0° local time.) As viewed from Earth, the Sun–Jupiter–observer angle varies only between ±12°, so a definitive test of this model could not be made until Voyager was on its outbound flight path after Jupiter encounter. The results are shown in Figure 7.19. Here the preencounter and post-encounter morphologies as observed from Voyager 1 and Voyager 2 are compared at 20 MHz, and it is evident that the concept of radiation beams rotating with the planet is substantiated [Alexander et al., 1981]. The local times are about 10 and 4 hr, respectively, corresponding to an angular change of about 90° between the two observing geometries. The Io-controlled sources B, A, and C are clear in both panels and their preferred Io phase and CML locations are unchanged to within the effective grid size of the plots. Alexander et al. also stressed that the occurrence probability levels did not appear to change after encounter, both were consistent with 100% probability of detection. Also as expected from the corotating beam model, the polarization sense of the sources did not change after encounter.

On the other hand, Alexander et al. show that the non-Io emission undergoes a dramatic change in occurrence probability level between pre- and post-encounter observing intervals, in contrast to the Io-related emission (see Fig. 7.19). Note that the non- Io-B events (CML < 200°) become much more common after both the Voyager 1 and Voyager 2 encounters, whereas the non-Io-A events (CML > 200°) are greatly enhanced before encounter. Examination of the source flux density levels indicates

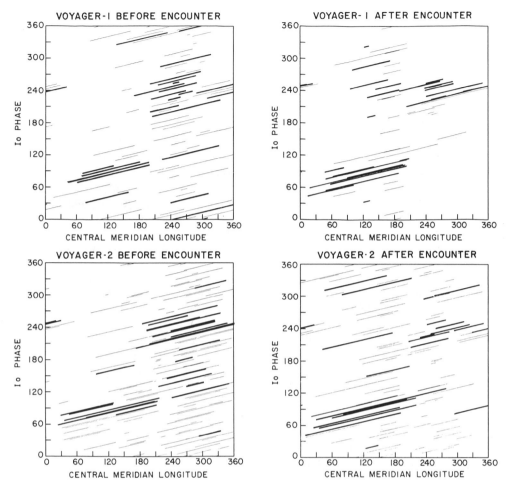

Fig. 7.19. Plots of the occurrence of DAM at frequencies above 20 MHz observed by Voyager 1 (above) and Voyager 2 (below) for one-month intervals before (left) and after (right) encounter as a function of CML and Io phase (γ_1). The heavy lines denote those events that extended to 30 MHz or higher, and so they tend to concentrate in regions of the diagram associated with Io-related activity. After encounter there is a shift in the preferred longitude for Io-independent DAM but no change in Io-related DAM. From Alexander et al. [1981].

that these changes may primarily be a consequence of changes in emission intensity, rather than of the total absence of one component. In basing non-Io source A and B identifications on dynamic spectral properties, that is, vertex-late vs. vertex-early are curvature, respectively, Leblanc [1981] independently came to the same conclusions as Alexander et al. regarding the intensity changes of the two sources after encounter.

Below 15 MHz and above 1 MHz the lesser arc emission is dominant, and here there is also evidence of a local time effect in the data [Alexander et al., 1981; Leblanc, 1981]. The occurrence probability of emission at longitudes below about 140° decreases after encounter, and the lesser arcs with vertex-late curvature become weaker and rare.

The important question arises as to how the local time effect is caused by the Sun. Two types of solar influence that could conceivably produce such an effect are solar photons and the solar wind. If it is the latter, the local time effect might be expected to depend on the solar wind activity level in some observable way. But if solar photons are responsible, the effect may be due somehow to the ionizing action of ultraviolet radiation in the ionosphere. If so, new clues to the emission or initial propagation of DAM will be at hand. In any event, the recent discoveries of the Jovian local time effect and the dynamic spectral arcs are providing new impetus to the search for the origins of DAM.

S bursts. The Jovian *S* bursts, or millisecond bursts as they are sometimes called, were first recognized by Gallet [1961] who noted a clear distinction between *L*- (long) bursts and *S*-(short) bursts, a distinction that is still made today. The fundamental differences between the two types of emission were pointed out earlier: the *L*-burst modulation envelope originates largely in the solar wind and has a timescale of a few seconds; the *S*-burst modulation is intrinsic to the source or to its immediate surroundings and has a timescale of milliseconds. Although the *S* bursts account for a relatively small fraction of the DAM emission – probably less than 10% – their source-intrinsic nature makes them of such interest that they have undergone intense scrutiny since their discovery.

A sample *S*-burst dynamic spectrum is shown in Figure 7.12. (Readers wishing to examine many more *S* bursts are referred to an atlas of spectra by Ellis [1979]). These are so-called simple *S* bursts whose phenomenology is easily described: they usually have narrow (< 200 kHz) instantaneous bandwidths, single-frequency durations of 10 to 100 ms, and rapid frequency drift rates (df/dt) ranging from -5 to -45 MHz/s. It is important to note that the drift rates are always negative; that is, simple *S* bursts invariably drift from higher to lower frequencies with time. The total band through which the bursts drift is generally less than a few MHz, which explains why Gallet [1961] never saw bursts simultaneously on his two channels, which were separated by 2 MHz. A further important characteristic of the *S* bursts is their strict association with Io-dependent emission sources and, in fact, with only certain CML and Io-phase values within a given Io-related source [Riihimaa, Dulk, and Warwick, 1970; Leblanc and Genova, 1981]. More complex *S* bursts, which nearly defy classification, have also been described in the literature (see, e.g., Ellis [1975], Krausche et al. [1976], Flagg, Krausche and Lebo [1976], Riihimaa [1977]). However, in the model discussion that follows, only the simple-burst morphology described above and illustrated in Figure 7.12 is considered.

A major clue in the attempt to understand *S* bursts came with the discovery that the drift rates were frequency dependent, increasing in absolute value nearly monotonically with frequency between about 5 and 20 MHz. Ellis [1974] showed that electron cyclotron radiation from electrons that conserved the first adiabatic invariant of motion about field lines would drift in frequency with monotonically increasing rates, and that the observations could be matched provided nearly monoenergetic (3-keV) electrons with narrow, that is, conical (3.5° equatorial) pitch angle distributions are assumed. In view of the manifestly Io-related nature of *S* bursts, the initial electron acceleration is assumed to take place at Io in this model. Further, because the drift rates are always negative, it is presumed that the radiation is only observed following reflection of the electrons from their mirror points that are near the top of Jupiter's ionosphere. That is, outward streaming electrons generate negatively drifting emission because of the constantly decreasing field magnitude and hence cyclotron frequency. These, then, are trapped or quasitrapped electrons.

A key feature of this model was the prediction of (1) a turnover in the drift rate vs. frequency curve (near 27 MHz, see Ellis [1974]), and (2) virtually zero drift rate at the mirror point (i.e., the high frequency cut-off) where the electron velocity parallel to **B** is zero. Thus, the critical test of the model was provided by observations at 32 MHz [Desch, Flagg, and May, 1978]; drift rates, instead of being zero, exceeding any previously recorded at lower frequencies. Subsequent observations at 34 MHz, within a few MHz of the mirror-point frequency (based on the magnetic field model), confirmed this result [Flagg and Desch, 1979]. Because no turnover in the spectrum was seen, this was clearly in conflict with model predictions.

Desch et al. argued that the model was basically sound but that, in order to explain the large drift rates they measured at high frequency, the electron acceleration responsible for the *S* bursts must take place in Jupiter's ionosphere, not at Io. Implicit in this observation was the additional requirement that the initial pitch angles be relatively far removed from 90°. These are decidedly nonmirroring, untrapped electrons. Subsequent observations of a similar nature by Leblanc et al. [1980] and by Ellis [1980] yielded the same conclusions, although Ellis in addition showed that, over the frequency spectrum below 30 MHz the best agreement with observations was obtained with 3 keV electrons possessing a broad range of initial pitch angles at the acceleration point in Jupiter's ionosphere.

At the same time that the high frequency observations were being made, Riihimaa [1979] challenged the evidence that the drift rates were frequency dependent at all. In observations conducted at 22 MHz over many years, he showed that average drift rates could vary widely from storm to storm. The controlling factor in determining drift rates, he contended, was not observing frequency but observer viewing geometry, namely D_E. This particular issue was resolved by Flagg and Desch [1979] who showed that when simultaneous, multifrequency observations are made the frequency dependence is still evident. Thus, there seemed to be a strong short-term dependence on frequency, and for reasons as yet unexplained, a weak long-term (11.9-yr) dependence on D_E.

Narrow-band events (*N* events). A third class of emission events, which are apparently distinct from the *L* and *S* bursts, has been studied in detail by Riihimaa [1968, 1977]. These are the narrow-band events, or *N* events (this designation has not previously been used in the literature), frequently seen in the source B region between 21 and 30 MHz. They have also been observed by Krausche et al. [1976], Flagg, Krausche, and Lebo [1976], Ellis [1979], and Leblanc, Genova and de la Noe [1980]. As mentioned earlier, narrow-band emissions were also observed with relatively low time resolution by Warwick [1963a] and by Dulk [1965] as the extended tails of the spectral landmarks. However, the events observed by Riihimaa do not seem to be part of known spectral landmarks. They have a bandwidth of about 200 kHz and durations anywhere from a few seconds, characteristic of *L* bursts, up to tens of minutes. Bandwidths of *L* bursts are usually about 2 to 5 MHz. An example of an *N* event of at least 9-min duration is shown in Figure 7.20. A question which immediately comes to mind is that if *L* bursts represent what is left over when longer duration bursts are broken up by interplanetary scintillation, how do the *N* events survive being broken up too? Although this question has not yet been answered, Riihimaa's statement that they are among the most intense of Jupiter's bursts may be a clue. Perhaps they are so intense that the scintillation minima are still strong while the maxima saturate the receiving-recording system.

Although the center frequency of an *N* event remains nearly constant, or at most displays a slow but random drift, sometimes it flutters rapidly back and forth between two frequencies. Often a single *N* event flares out into a train of *S* bursts with individual drift ranges much wider than that of the original *N* event.

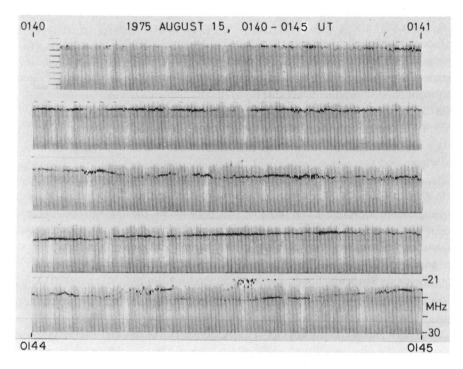

Fig. 7.20. An *N* event is shown starting in the upper left-hand corner and progressing in 1-min sections from left to right and from top to bottom. Five minutes of the original 9-min event are shown. From Riihimaa [1977].

Sometimes interactions occur between an *N* event and *S* bursts that originate well outside the *N* event but drift into it [Riihimaa and Carr, 1981]. In a typical interaction, the drifting *S* burst crosses the *N* event, leaving a "tilted-V-shaped" gap in the latter. The gap represents a quenching of the *N* event for a substantial fraction of a second – far longer than the duration of the *S* burst that triggered it. Much more intricate inter-action patterns are produced when the *N* events are fluttering in frequency simultaneously with being cut across by *S* bursts from outside.

Riihimaa [1968] and Boischot et al. [1980] have observed split *N* events, consisting of long enduring parallel pairs of narrow emission bands. They are separated by a sharply delineated narrow band containing no activity. Excellent examples of such events are shown in Figure 7.3 of Riihimaa [1968].

Riihimaa and Carr [1981] consider the possibility of explaining (1) the *S* bursts as arising from bunches of electrons emitting Doppler-shifted cyclotron emission as they ascend a flux tube, (2) the *N* events as gyroemission from a fixed field location by a different population of electrons streaming through it, and (3) the observed interaction as the result of the invasion of the stationary *N*-event emission region by the ascending *S*-burst group. They concluded that it is difficult to explain the more complex interac-tions by means of such a simple model.

Beaming effects. Evidence for sharp beaming of DAM was provided initially by ground-based observations of the variations in several emission parameters with observer viewing geometry [Douglas, 1964]. Gulkis and Carr [1966] showed, for example, that the centroid of source A varied almost sinusoidally in System III longitude with a period of 11.9 yr, a time interval corresponding to the planet's orbital period. The important parameter that varies over this period is the observer's jovigraphic latitude, or D_E. (The jovigraphic latitude of the Sun, D_s, also varies almost in phase with D_E during this period but the possibility of a D_s dependence has been largely ruled out by the observations that show a slightly better correlation with D_E.) D_E changes by only $\pm 3.3°$ over 11.9 yr; however, the radiation is apparently so strongly beamed in latitude that this excursion is enough to produce the observed results. Changes in occurrence probability with D_E are shown in Figure 7.13. Carr [1972b] showed that both the longitude oscillations and the variations in overall occurrence probability of source A were consistent with a rotating radiation beam having a leading edge parallel to the spin axis and a wedge-shaped trailing edge. The radiation escapes in a direction that favors the northern hemisphere.

Since this early work, additional D_E-dependent phenomena have been found, notably the long-term variations in the Io- controlled source centroids in Io phase [Lecacheux, 1974; Thieman, Smith, and May, 1975] and also in CML [Boyzan and Douglas, 1976]. The effect is strongest for the Io-B and Io-C sources whose oscillations in γ_1 are in phase and of peak-to-peak amplitude 5° to 10°. There is not yet general agreement as to the behavior of Io-A, probably owing to confusion between Io- related and non-Io-related components. Lecacheux [1974] argued that the strong latitudinal beaming implicit in the observed D_E control of the preferred Io phases is inadequately accounted for by the conical sheet model of Dulk [1965] (but see Goldstein and Eviatar [1979]). However, Desch [1978] obtained quantitative agreement with the observations by simply requiring that the emission be radiated from the Io flux tube at a preferred angle to the ambient field regardless of observer vantage point. The results agreed with and lent credence to the source location assignments inferred from polarimetry, namely Io-B in the northern and Io-C in the southern hemisphere.

Estimates of the actual emission beamwidth have been made through examination of dynamic spectra [Warwick, 1963a; Alexander, 1975] and through direct stereoscopic measurement from space [Poquerusse and Lecacheux, 1978]. (However, observations made at the same time by Reyes and May [1981] appeared not to be entirely consistent with the latter result.) Both stereoscopic measurement and examination of dynamic spectra indicate radiation patterns no greater than 10° wide. An independent estimate comes from observations of the 3 to 6 min timescale inherent in the spectral arc emission. If the beam responsible for the arc is carried past the observer at Jupiter's diurnal rate then the indicated beamwidth can be no greater than 2° to 4°. If the beam is carried with Io's motion, the beamwidth is a factor of four smaller.

Emission in the HOM band also exhibits dramatic changes in morphology as a function of observer latitude. These have been modeled by Alexander et al. [1979] in terms of a 10° wide beam centered at a fixed magnetic latitude of +3°. These same studies showed that simultaneous observations of radio events by both Voyager spacecraft showed significant decorrelations when observer angular separations exceeded 3°. This latter result is consistent with the 2° to 4° beamwidth inferred from the spectral arc timescales.

Source size, motion, and structure. Both direct and indirect methods have been used to estimate the size of the radiating source of DAM, and both lead to approximately the same upper limit. Very long baseline interferometry (VLBI) yields a direct measure of source size, or an upper limit if the source is unresolved at the longest baselines. Using baselines of over 400 000 wavelengths (~7000 km) neither Dulk [1970] nor Lynch, Carr, and May [1976] succeeded in resolving sources responsible for either *S* bursts or

L bursts, indicating an angular diameter for these sources of < 0.1 arc s, or a source linear dimension of < 400 km. One usually assumes that the source is spatially incoherent in deriving this upper bound to the size; however, Ratner [1976] has argued that the limits apply even if the source is spatially coherent.

The other model-dependent estimate of *S*-burst source size is derived from measurements of the instantaneous bandwidth of the emission. Adopting the model described previously, the instantaneous bandwidth is a consequence of the spatial extent along the Io flux tube (IFT) of an emitting electron bunch. This corresponds to the instantaneous linear physical dimension of the source. Near the surface of Jupiter, the magnetic field gradient in terms of electron cyclotron frequency is about 1 kHz/km, indicating a source size < 200 km for bandwidths < 200 kHz. This is consistent with the direct VLBI measurements.

VLBI also provides a possible means of detecting source motion in certain directions. Analysis by Lynch, Carr, and May [1976] of the interferometer fringe stability over short time scales (< 0.5 s) has shown that the fluctuations in source position parallel to the interferometer baseline is less than 0.05 arc s, or < 150 km. They concluded that all the members of a sequence of *S* bursts, at a given frequency, must come from the same source. A similar investigation of fringe stability by Dulk [1970] led to the conclusion that the *L*-burst source cannot fluctuate spatially by more than 0.20 arc s.

VLBI has also been used to search for direct evidence of sweeping effects in the radiation beams responsible for *S* bursts. Based on the simultaneous arrival, within experimental error, of the burst envelopes at two widely-separated stations (after accounting for light travel time), Lynch, Carr, and May were able to rule out the possibility of sweeping beams carried along either with Jupiter's diurnal motion or with Io's orbital motion. They also found no evidence of the beam sweep that would result from a group of radiating electrons moving along a curved magnetic field line. This would seem to contradict the *S*-burst model described earlier. However, if the electron energy were greater than about 3 keV with near-zero pitch angle (or higher energies with commensurately larger pitch angles), the beam-sweep effect would have been undetectable. Using a refined statistical approach applied to 26 MHz VLBI data, Ratner [1976] came to much the same conclusions regarding the lack of evidence for sweeping-beam effects. All investigators agree that the millisecond-burst structure must result from the emission mechanism itself rather than from the passage of a beam pattern across the observer.

Concluding remarks concerning DAM

In this section we review what is known and what is not well understood about the phenomenology of DAM. The more important of the first-order observational features of DAM that are generally recognized are:

1. The flux density spectrum has a broad peak near 10 MHz and an intrinsic high-frequency cut-off near 39.5 MHz.
2. There is a strong, periodic organization of the activity into storms, with durations of hours, that are modulated on timescales from min to ms. The detailed emission patterns that make up the storms are predictable functions of observing frequency, CML and γ_1.
3. Activity above 20 MHz is predominantly RH elliptically polarized; at lower frequencies either polarization sense is observed depending on CML.
4. Although Io acts to enhance the emission intensity and alter the radius of curvature of the arcs, the basic character of the arcs is principally a function of the CML.

5. The Io-related emission is independent of the solar hour angle of the source, but the Io-independent emission shows diurnal as well as CML control.
6. The strong dependence of the DAM emission on observer latitude is evidence of sharp latitudinal beaming of the radiation.
7. At any given frequency the source size is limited to no more than several hundred km.

Apart from the questions concerning the nature of the processes that give rise to DAM, more properly the subject of Chapter 9, there are a number of observational issues and matters of interpretation of the DAM phenomena that remain unsettled. Some of these questions are long standing, such as the location of the radio sources; others have only recently been raised as a direct consequence of the Voyager observations. We will devote some discussion to the important question of the radio source location and, finally, itemize those questions that should be addressed in the near future regarding the observations.

As stated previously, the observations have long been interpreted to indicate the emission takes place at or near the local electron cyclotron frequency [e.g., Warwick, 1970] and, for the Io-related sources at least, on the IFT. If one accepts these conclusions, then only the source magnetic hemisphere is left unspecified, and here a variety of observations have yielded self-consistent answers in all cases except one. For example, few would argue with the conclusion that the Io-B and Io-A sources are in the northern hemisphere. Their peak frequencies, 39.5 and 38 MHz, respectively, are attainable only in the northern hemisphere. The predominant polarization is RH, consistent with extraordinary-mode emission from Jupiter's northern hemisphere. And the measured long-term oscillation of Io-B in γ_i is in quantitative agreement with a northern-IFT source. The dynamic spectra of the non-Io counterparts of these two sources have great arcs which are of essentially the same shape, occur at the same longitude and display the same polarization. In addition, both have peak frequencies as great or nearly as great as their Io-dependent complements. Non-Io-A, in addition, has been modeled with some success in terms of a northward-tipped beam. All of these similarities imply the same hemisphere is active for the non-Io-A and B sources as for the Io-related sources.

There is also probably widespread agreement concerning the Io-D source. Io-D emission has characteristics which are in agreement with a southern-hemisphere origin: LH polarization and a modest peak frequency of 18 to 20 MHz.

The Io-C source presents some difficulties, however. Both LH and RH polarizations are evident, with the former often dominant below 20 or 22 MHz. Barring unknown propagation effects, extraordinary-mode emission requires that the LH polarized emission originate in the southern hemisphere, where the wave vector is initially antiparallel to $\boldsymbol{B}$. This source exhibits a peak frequency generally not in excess of 30 to 32 MHz, nearly consistent with either north or south, but on some occasions events reach as high as 36 MHz, a fact probably only consistent with a northern hemisphere origin. The 11.9-yr oscillation of the source in γ_i (measured at 22 MHz) is consistent with a southern hemisphere source only, however. Similar comments apply to the non-Io component of the C source, except that the peak frequency has never been observed to exceed 32 MHz, which is probably in the range of frequencies attainable in the southern hemisphere. It is likely that the final answer to the question of source location will have to await direct measurement by ground-based interferometry, if this proves feasible.

Some other points to be settled from either ground-based or Voyager observations are:

1. Is there a distinct component of HOM in addition to the low-frequency extension of DAM? A fraction of the emission below 2 MHz is undoubtedly the extension of greater and lesser arcs into the HOM band. Some of it, however, may represent an independent component.

2. How are the arcs produced? Several theories (see e.g., Chapt. 9) have been advanced to explain arcs; the data must be examined in ways that provide definitive constraints for the various models.

3. What is the relationship between greater and lesser arcs? Unlike much of the greater-arc emission, the lesser-arc emission appears to be more nearly continuous and to be entirely independent of Io control. This suggests an independent origin.

4. How and where do the modulation lanes originate?

5. What aspects of the emission mechanism account for the pronounced differences between L bursts, S bursts, and N events. The observed interactions of S bursts with N events (and with L bursts) are undoubtedly an important new source of information relating to this question, and should provide a stimulus to theoretical activity.

6. Is the interpretation of the S-burst observations correct in indicating a source of keV electrons accelerated outward from Jupiter's ionosphere? Most theories provide solely for Jupiterward, that is, Io-accelerated electrons (see, however, Sharp et al. [1978] and Smith and Goertz [1978]).

7. Which interplanetary solar parameter best correlates with the occurrence of non-Io DAM? The weight of evidence now seems to favor the existence of some solar control over the non-Io component of the emission, but the nature of the interaction is unclear. Comparison of in situ measurements by Voyager of solar-wind and interplanetary magnetic field quantities with the occurrence of DAM will be helpful.

7.4. Emissions at kilometric wavelengths

The first evidence for a kilometer-wavelength component of Jupiter's radio spectrum appeared in the Voyager Planetary Radio Astronomy data in April, 1978, when the spacecraft were still more than 2 AU away from their epochal encounters with the planet. By the end of 1978, distinct episodes of LH polarized emissions were routinely being detected at frequencies between 60 kHz and a few hundred kHz. Activity at even lower frequencies in the range covered by the Plasma Wave Science instrument also became commonplace. Such events would be nearly impossible to detect from a spacecraft near Earth because of the overwhelming presence of intense terrestrial kilometric radiation. However, for an observer situated well away from Earth, the Jovian kilometric emissions are often readily recognized by virtue of their distinctive dynamic spectral characteristics, their high degree of polarization, and their tendency to occur at CML values between 120° and 270° where HOM activity is rare.

The first brief reports of the new low frequency emissions appeared in the initial accounts of the Voyager 1 and 2 Jupiter encounters [Scarf, Gurnett, and Kurth, 1979; Warwick et al., 1979a,b; Gurnett, Kurth, and Scarf, 1979a]. More detailed discussions have been given by Kurth et al. [1979a, 1980] who described spectral, spatial, and temporal characteristics observed between about 10 and 56 kHz and by Desch and Kaiser [1980] who studied the occurrence patterns, polarization, and intensity spectra for the frequency range between 20 kHz and about 1 MHz. Green and Gurnett [1980] used a ray-tracing calculation to model the refraction of the kilometric waves in the Io plasma torus to demonstrate that the source region is probably located on auroral field

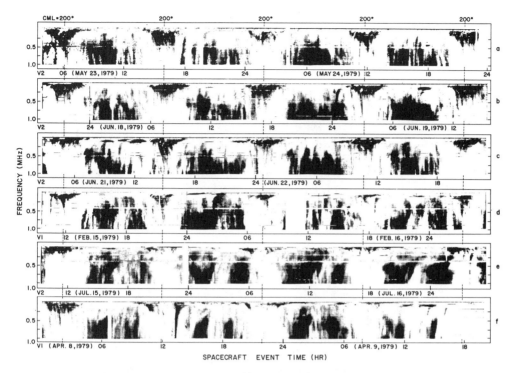

Fig. 7.21. Examples of dynamic spectra of HOM and KOM emissions. Each panel is a frequency-time display of signal intensity recorded between 20 kHz and 1.0 MHz over an interval of nearly two days, and all are aligned in System III Central Meridian Longitude (CML). The bKOM is usually seen at frequencies below about 0.5 MHz when the CML is near 200°. The higher frequency emission that occurs outside the longitude band 200° ± 90° is the HOM. The top four panels are examples of preencounter, dayside obervations of Voyager 2 (*a*)–(*c*) and Voyager 1 (*d*). The two bottom panels are examples of post-encounter, nightside observations from Voyager 2 (*e*) and Voyager 1 (*f*). Examples of nKOM can be seen in between the bKOM events in panels (*c*) and (*d*).

lines between the Io torus and the Jovian ionosphere. Desch and Kaiser [1980] further suggested that the source appears to be fixed in local time rather than to rotate with the planet. Thus the 10-h periodicity was attributed to variations in the observer's magnetic latitude.

Some investigators have referred to the kilometer-wave emissions as "Jovian kilometric radiation" in analogy to the "terrestrial kilometric radiation" or "TKR" first discussed by Gurnett [1974]. We adopt the convention initiated by the Voyager planetary radio astronomy investigators in which the Jovian kilometric radiation is denoted KOM in order to distinguish it from its higher frequency spectral cousins DAM, DIM, etc. As we have noted early in this chapter, there are two apparently distinct kilometric emissions – the broad-band kilometric radiation (bKOM) and the narrow-band kilometric radiation (nKOM) – and we will discuss them below in that order.

Broadband kilometric radiation

Several representative examples of dynamic spectra of individual bKOM events are shown in Figure 7.21. In each case we see a period of emission extending over an hour or more and covering a frequency range of several hundred kHz. The events tend to last longer at low frequencies than at high frequencies. The dynamic spectra also show evidence of slow drifts in frequency on a scale of ∼ 10 min, and we shall return to these later. Notice also that both the high frequency and low frequency limits to the emission vary from event to event. For example, the first bKOM event in Figure 7.21c (∼ 05 hr on June 21, 1979) extends in frequency only up to about 300 kHz, but in the next event 10 hr later we see emission reaching up to at least 800 kHz. Similarly, in the events displayed in the top panel (Fig. 7.21a) we see low frequency cut-offs that range from ≤ 20 kHz to about 100 kHz. The examples in Fig. 7.21 are ordered according to the jovigraphic latitude and local time of the observing spacecraft. The events in Figure 7.21a–c were obtained by Voyager 2 prior to closest approach from a jovigraphic latitude of + 7° and a local time meridian of 9.5 hr. The events in Figure 7.21d were obtained with Voyager 1 from a slightly lower latitude (+ 3°) and from nearly the same dayside local time sector (10.5 hr). The events illustrated in Figure 21e and f were obtained from above the post-midnight local time sector and at a latitude (+ 5°) that was intermediate to that for the other panels.

The events recorded from above the daytime hemisphere (Fig. 7.21a–d) are all similar in the sense that they show a characteristic tapered shape in which the higher frequency portions occur only during a limited longitude interval centered on 200° CML. The events recorded by Voyager 2 at the higher latitude tend to extend over a wider longitude range than the Voyager 1 dayside events, and this is typical of most (but not all) bKOM activity observed by the two spacecraft before closest approach. Most of the post-midnight sector examples (Fig. 7.21e, f) are distinctly different. The high frequency activity peak near 200° longitude is not observed, and instead we see two periods of emission over a limited frequency range and centered at about 150° and 240° CML. Although there is considerable variability in all aspects of the individual dynamic spectra observed from above the sunlit and nightside hemispheres of Jupiter, the trend illustrated in Figure 7.21 is typical of most of the Voyager data. Namely, bKOM events tend to build toward a maximum at 200° CML when observed from the morning sector and to display a "bite" in the dynamic spectrum at 200° CML when observed from above the post-midnight sector.

The drifting structure seen in the individual dynamic spectra is a common feature of bKOM. Both positively and negatively drifting features can occur – sometimes simultaneously – although both Kurth et al. [1979a] and Warwick et al. [1979b] reported that negatively drifting structures appeared to be prevalent in events observed before encounter. Subsequent studies have shown the reverse to be true after encounter. The drift rates are generally greater at high frequencies than at low frequencies and are typically ∼ 1–10 kHz/min at 100 kHz and below.

If the frequency drifts are related to the motion of an exciter along a gradient in the density or magnetic field in the source region then the overlapping occurrence of features of opposite senses of drift must imply the simultaneous occurrence of "ascending" and "descending" sources. The measured drift rates correspond to significantly lower velocities than would be expected for either energetic particles or Alfvén waves in any region between Io's orbit and the Jovian ionosphere. Thus, the drifting bKOM features are not easily explained by simple analogy to drifting solar radio bursts. Both the typical durations of the drifting bKOM features at a single frequency and their frequency-time slopes are roughly comparable to the durations and slopes of the DAM arcs at low frequencies. However, no evidence has been presented to date to demonstrate any physical connection between the two phenomena.

On the shortest timescale at which the bKOM activity has been studied (∼ 60 ms) the emission is found to be very impulsive. An example of this property of the bKOM events is shown in Figure 7.22 where we can see large amplitude fluctuations in mea-

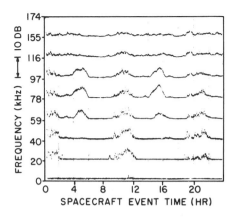

Fig. 7.22. Plots of relative signal intensity at eight frequencies between 20 and 174 kHz as measured by Voyager 2 on May 27, 1979. bKOM events occur at approximately 01, 11, and 21 h spacecraft event time and nKOM events can be seen between 78 and 116 kHz at about 05 and 15 h. Notice that when examined with 6-s time resolution the bKOM is bursty and nKOM is relatively smooth. Adapted from Boischot et al. [1981].

surements taken every 6 s. Kurth et al. [1979a] found that in the frequency range between about 5 and 12 kHz the short, impulsive bursts appeared to be superposed on a relatively smooth background. Moreover, the short bursts drifted systematically at rates of the same order of magnitude as the drift rates for the longer duration features evident in Figure 7.21.

Scarf et al. [1979] noted that bKOM could often be detected down to a minimum frequency between 5 and 10 kHz. Because such a low frequency cut-off is still significantly above the cut-off frequency for electromagnetic waves propagating in the solar wind in the outer solar system, the bKOM low frequency cut-off must be imposed near Jupiter. Scarf et al. suggested that the electron plasma frequency of the Jovian magnetosheath (typically ~ 5 kHz) could determine the lowest frequency at which an observer outside the Jovian magnetosphere can detect bKOM. On many occasions the low frequency cut-off is greater than 20 kHz (see discussion of Fig. 7.23), and Warwick et al. [1979b] reported that the cut-off frequency extent of the bKOM was probably set by propagation through the Io plasma torus.

Representative power flux densities of bKOM emissions are shown in Figure 7.23. The data are taken from the statistical analysis of 12-day spans of Voyager observations by Desch and Kaiser [1980]. The two solid curves display the most commonly occurring flux density as a function of frequency (mode spectrum) and the flux density that was exceeded only 10% of the time at each frequency (peak spectrum) as observed by Voyager 2 before Jupiter encounter. Only periods of time when the bKOM source was active were included. Spectra of individual events can vary considerably, both within the event and from event to event, but the curves in Figure 7.23 provide a good statistical summary of the average bKOM spectrum. The greatest flux density occurs most often between 60 and 80 kHz, and the half power bandwidth is approximately equal to the frequency at the peak. Only about 10% of the events are observed to extend below 20 kHz or above 1 MHz.

The dashed curve in Figure 7.23 shows the mode spectrum from the post-encounter Voyager 1 data. We see that after encounter the measured flux density levels are generally lower than observed before encounter, the peak frequency is lower, and the emission does not often extend to frequencies as high as are observed in the preencounter data set. Thus the average spectral properties of bKOM resemble the pattern observed for the terrestrial kilometric radiation [Kaiser and Alexander, 1977] where high flux densities and higher frequencies are observed from above one hemisphere and lower flux levels and slightly lower peak frequencies are detected from above the opposite hemisphere. In the case of TKR, the data clearly indicate a source region in the midnight sector that radiates preferentially in directions over the nighttime

Fig. 7.23. Power flux density spectra of
bKOM showing the spectrum most
commonly observed before and after
Voyager 2 encounter (mode spectrum)
and the spectrum representative of the
strongest bKOM events (peak spectrum).
From Desch and Kaiser [1980].

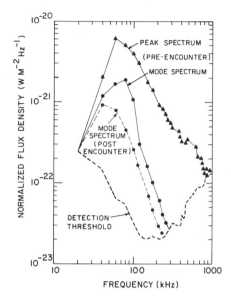

hemisphere. On the other hand, Desch and Kaiser [1980] argue that the
bKOM measurements are most simply explained by a source that favors the daytime
hemisphere.

If the radiation is assumed to be emitted uniformly over 4π sr, the isotropic
equivalent total radiated power derived by integration of the preencounter mode spectrum is 8×10^8 W. Using a slightly different method, Alexander et al. [1981] find an
average power of 3×10^8 W in the kilometric band. The peak power radiated by
bKOM as derived from the peak spectrum in Figure 7.23 is 2×10^9 W, and Desch and
Kaiser [1980] report that a level of 5×10^9 W is exceeded only 1% of the time. After
encounter, the total integrated flux is lower by about a factor of 2.5.

The results of a statistical analysis by Desch and Kaiser [1980] of the variation of
occurrence probability of bKOM as a function of CML and Io phase are illustrated in
Figure 7.24. The emissions are tightly confined in CML, but the two-dimensional time-
line plots give no indication for any dependence on Io phase. The histograms show the
variation of 97 kHz activity with CML for three different time intervals. The two
preencounter histograms have a single longitude maximum centered between about
190° and 240°, but after encounter when the Voyager 1 spacecraft was above the post-
midnight local time sector the CML histogram shows a bimodal distribution with
peaks near 150° and 240°. The Voyager 2 data, not displayed in Figure 7.23, also
showed a similar split in the post-encounter CML profile. Thus, on a statistical basis,
one finds a pattern in the occurrence of bKOM that is typified by the individual events
illustrated in Figure 7.21. In particular, the shape of the longitude distribution of
bKOM depends on the observer–Jupiter–Sun angle.

In addition to changes in the shape of the CML profile of occurrence probability, the
absolute levels of activity appear to depend on observer–Jupiter geometry. Desch and
Kaiser [1980] noted that at 97 kHz both the peak occurrence probability and the average occurrence probability over the 120°–270° CML sector of bKOM activity were
significantly higher for the Voyager 2 preencounter data than for the other data sets
illustrated in Figure 7.24. This seems to be primarily a consequence of the higher jovigraphic latitude at which the dayside Voyager 2 data were collected. Gurnett et al.
[1979] also noted that at frequencies between 10 and 56 kHz the bKOM was clearly

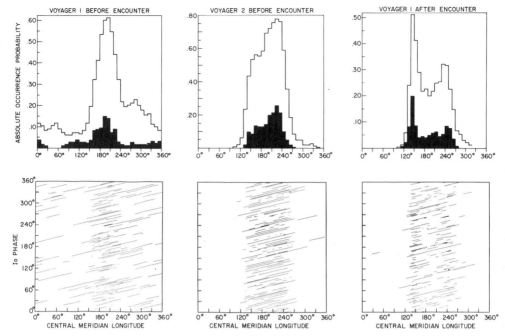

Fig. 7.24. Plots of the occurrence of bKOM at 97 kHz as a function of Central Meridian Longitude (upper panels) and as a joint function of longitude and Io phase (lower panels). The unshaded (shaded) histograms include only those events exceeding a normalized flux density of 2.5×10^{-22} (2.5×10^{-21}) W/m^2/Hz. The activity does not appear to be as tightly organized in longitude in the Voyager 1 preencounter plot because the distinct nKOM events were not purged from the data set. Notice (1) the bifurcation of the longitude peak after encounter and (2) the lack of evidence for any influence by Io. Adapted from Desch and Kaiser [1980].

more intense and more regular in the higher latitude Voyager 2 preencounter data than in the Voyager 1 preencounter data set. Kurth, Gurnett, and Scarf [1980] suggested that at 56 kHz there is a shadow zone of half-width $\sim 10°$ centered on the magnetic equator due to refraction by the Io plasma torus. Thus the Voyager 2 preencounter observations from a latitude of $+7°$ (versus $+3°$ for Voyager 1 before encounter) would place the observer farther outside of the shadow zone for a greater period of time during each rotation of the planet. Arguments favoring a low latitude shadow zone were also made by Warwick et al. [1979b] on the basis of Voyager 2 near encounter data and by Green and Gurnett [1980] on the basis of a ray tracing model.

On the other hand, variations in shadowing due to the periodic rocking of the Io plasma torus can not readily explain the bifurcation of the CML histograms in the post-encounter Voyager data. In spite of the fact that the Voyager post-encounter latitude is intermediate to the two preencounter cases in Figure 7.24, Desch and Kaiser [1980] found a slightly lower absolute occurrence probability in their post-encounter survey than in the Voyager 1 preencounter data set. Thus, the level of activity appears to be diminished when viewed from the nightside of Jupiter in addition to being positively correlated with observer's latitude. Such a local time dependence in the occurrence probabilities is consistent with the preencounter/post-encounter differences in flux densities illustrated in Figure 7.23.

In addition to the major concentration of bKOM activity that occurs near CML = 200° when the northern tip of Jupiter's magnetic dipole is tilted toward the observer,

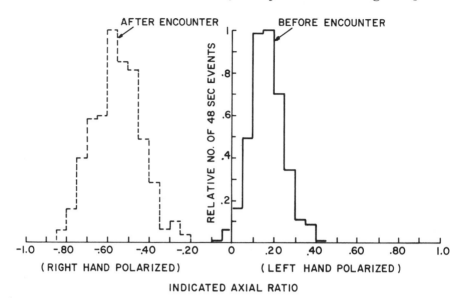

Fig. 7.25. Histograms of the polarization axial ratio of bKOM at 97 kHz as measured simultaneously from above Jupiter's day hemisphere by Voyager 2 (solid lines) and from the night hemisphere by Voyager 1 (dashed lines). Notice that the polarization is almost exclusively left-hand in the first case and right-hand in the latter. Adapted from Desch and Kaiser [1980].

both Scarf et al. [1979] and Warwick et al. [1979a] noted that there exists a subsidiary concentration of emission at longitudes between about 20° and 40° or in roughly the longitude range of the Southern dipole tip. This component is considerably weaker and more intermittent than the emission near 200°. In contrast to the positive latitude correlation in the 200° bKOM activity, the bKOM events at early longitudes were more common in the Voyager 1 inbound data than in the corresponding Voyager 2. This inverse correlation with observer's latitude lends support to the physical association of the 200° emission with a northern hemisphere source and the 20°–40° emission with a southern source.

The polarization properties of bKOM are particularly interesting. By examining the long-term statistical behavior of the bKOM polarization and by interpreting the results as time-averaged properties, Desch and Kaiser [1980] were able to specify more than just the polarization sense of the waves. They found that the emission appears to be strongly polarized over essentially its full frequency range (~ 20 kHz to 1 MHz). The degree of polarization must be at least 75% and could be as high as 100% polarized. Statistical studies of the measured wave axial ratio also suggest that the waves must be more nearly circular rather than linearly polarized.

Desch and Kaiser [1980] also reported that the sense of polarization of bKOM reverses depending on whether the observer is situated over the dayside or the nightside of Jupiter. When viewed from above the day hemisphere, bKOM is LH polarized; when viewed from above the night hemisphere, the polarization is RH. This remarkable result is illustrated in Figure 7.25, which shows histograms of indicated axial ratio at 97 kHz observed by Voyager 1 and Voyager 2 at the same time but from two different local time locations. Voyager 1 was in the post-midnight sector after encounter and measured the bKOM to be exclusively RH. During the same period of time Voyager 2 was approaching Jupiter from above the prenoon sector and detected bKOM that was

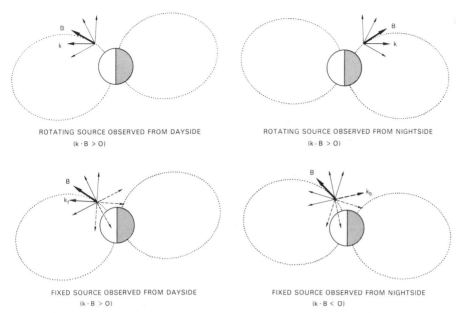

ROTATING SOURCE OBSERVED FROM DAYSIDE
(k · B > 0)

ROTATING SOURCE OBSERVED FROM NIGHTSIDE
(k · B > 0)

FIXED SOURCE OBSERVED FROM DAYSIDE
(k · B > 0)

FIXED SOURCE OBSERVED FROM NIGHTSIDE
(k · B < 0)

Fig. 7.26. A schematic illustration of some differences between a rotating, light-house-like radio source (upper panels) and a source that remains fixed in local time and emits into a forward (f) and back (b) lobe (lower panels). The source rotating with the planet would always appear to have the same polarization sense, because the relative orientation of the magnetic field B and the wave vector **k** would always be the same with respect to the observer. For a fixed source, the forward lobe wave vector $\mathbf{k}_f$ and the backward lobe wave vector $\mathbf{k}_b$ are reversed with respect to B. Hence a viewer on the nightside would measure the opposite polarization compared to a dayside observer if the source remains near a particular local time meridian.

predominantly LH polarized. Both spacecraft observed the polarization reversal with respect to the time of passage from the sunward hemisphere to the nightside hemisphere at encounter. Although very little information exists on the polarization of the secondary component of bKOM observed at longitudes around 20°–40°, Desch and Kaiser [1980] report that just before Voyager 2 encounter, when the 200° emission was LH polarized, activity at the earlier longitudes was clearly RH.

Desch and Kaiser [1980] argued that the polarization data are strongly suggestive of a radio source that remains fixed in some local time zone rather than one that rotates around the planet. The base mode for a particular electromagnetic wave, ordinary or extraordinary (O or X), will have a particular sense of polarization that will depend on whether the wave normal vector, **k**, has a component that is parallel or antiparallel to the magnetic field vector, **B**, at the source. A source that is constrained to a fixed range of longitudes and that rotates around the planet in synchronism with the rotation of the magnetic field will always have **k** oriented in the same sense with respect to **B** when observed at the same CML regardless of the observer's local time. However, for a source that does not rotate with the planet an observer above the sunlit hemisphere would detect emission of one polarization, and an observer above the opposite hemisphere would detect the opposite polarization because the waves escaping in the two opposite directions would have **k** in the opposite sense with respect to **B**.

This situation is illustrated in Figure 7.26. For a northern hemisphere source located near the planet (as opposed to near the Io plasma torus), a "forward lobe" will have a wave vector component $\mathbf{k}_f$ parallel to $\mathbf{B}$, and a "back lobe" of the same source will have a wave vector component $\mathbf{k}_b$ antiparallel to $\mathbf{B}$. Thus the two lobes will be seen to have opposite polarizations. If the observed polarization is unchanged when viewed from both day and night hemispheres, then we always observe forward lobe radiation. For the polarization to reverse, we must be seeing foward lobe radiation when above one hemisphere but back lobe radiation when above the opposite hemisphere. Absorption or refraction effects associated with the propagation of back lobe waves over the polar cap toward the night hemisphere are likely to contribute to the lower intensity and occurrence probability observed for the nightside data.

The data do suggest a northern hemisphere source for CML $\sim 200°$ because the profiles of occurrence probability show a direct correlation with latitude. The maximum in the histograms produced by both Desch and Kaiser [1980] and Kurth et al. [1979a, 1980] occur when the observer's magnetic latitude reaches its greatest northern value. This corresponds to the greatest amount of tipping of an obscuring Io plasma torus out of the way of the line of sight to a source in the northern hemisphere. Furthermore, the higher flux densities and occurrence probabilities observed in the preencounter dayside data sets suggest a source located in the day hemisphere.

If we accept the evidence for a northern hemisphere, dayside source as sketched in Figure 7.26, then we can also infer the electromagnetic wave mode. Because we observe LH polarization for the forward lobe when $\mathbf{k}$ is parallel to $\mathbf{B}$ in the north, the radiation must be in the ordinary mode. Green and Gurnett [1980] also noted that if the bKOM is associated with the Io plasma torus or with magnetic field lines that thread the torus, then a comparison of the low frequency limit of bKOM with the range of propagation cutoffs inside $\sim 10 R_J$ implies left-hand ordinary (L-O) rather than right-hand extraordinary (R-X) mode emission. They pointed out that since bKOM is observed to frequencies as low as 10 kHz, the propagation cut-off frequency at the source must be below 10 kHz. Because the R-X cut-off is determined by the upper hybrid frequency, which is greater than about 50 kHz everywhere inside the likely source region, the R-X mode can be ruled out.

The bKOM is sporadic on a timescale of hours and longer. Successive individual events do not necessarily occur at the same flux density levels or with identical dynamic spectral shapes. Both the high frequency and the low frequency cut-offs can change by a factor of ~ 3 in just a few rotations of Jupiter, and the emission can occasionally disappear entirely for several rotations. Kurth et al. [1980] developed a simple bKOM activity index based on the total intensity observed at 56 kHz over one rotation of Jupiter. When plotted as a function of time, this activity index exhibited marked fluctuations on a timescale of ~ 1–15 days. Kurth et al. [1980] pointed out that these time variations could be indicative of changes in the properties of the Io plasma torus, the injection of fresh torus ions from Io, the interaction of the solar wind with Jupiter's magnetosphere, or some other dynamic process analogous to terrestrial magnetospheric substorms. The analysis of these long term fluctuations in bKOM activity is just beginning to be used as a remote diagnostic of temporal variations in Jupiter's magnetosphere.

Narrow band kilometric radiation

A second, quite distinctive component of Jupiter's kilometer wavelength emissions is the narrow band KOM, or nKOM. This class of emission was first noted in the initial report of Voyager 2 Jupiter encounter results [Warwick et al., 1979b]. Most of the information in the discussion below is taken from the subsequent definitive study of nKOM by Kaiser and Desch [1980].

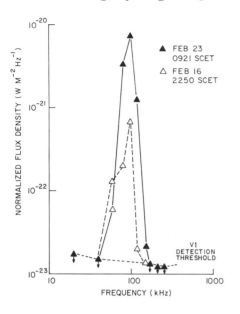

Fig. 7.27. Typical power flux density spectra of nKOM. From Kaiser and Desch [1980].

An example of the dynamic spectral appearance of nKOM can be found in Figure 7.21*c* (0900-1300, 2100-2300 on June 21) and Fig. 7.21*d* (0000-0200 on February 16). The emission occurs at frequencies between about 50 and 180 kHz in events that typically last for a few hours or less. Individual events tend to be composed of a smooth rise and fall in signal level with little or no burstiness noticeable of the sort that is characteristic of the bKOM.

The remarkably narrow instantaneous bandwidth observed for nKOM is illustrated in Figure 7.27. The figure, taken from the report by Kaiser and Desch [1980], shows two examples of power flux density spectra. In each case, we see a peak at about 100 kHz and a very sharp cut-off at both high and low frequencies. Bandwidths of only 40 to 80 kHz are commonly observed. According to Kaiser and Desch, the isotropic equivalent total radiated power in nKOM can occasionally reach $\sim 10^9$ W but is more typically $\sim 10^8$ W.

Polarization measurements of nKOM show that the emission is usually observed to be LH when the observer is situated north of the magnetic equator and is RH when observed from above the southern magnetic hemisphere. Kaiser and Desch [1980] noted that polarization reversals sometimes occur during the course of an individual nKOM event at the time of Voyager spacecraft crossings of the magnetospheric current sheet, but that this was not an invariant property of nKOM.

Initial studies of the Central Meridian Longitude distribution of nKOM were inconclusive because CML histograms of occurrence probability calculated from nKOM event catalogs showed no organized pattern. (In fact, most of the CML-independent activity that appears in the left-hand panel of Figure 7.24 is due to nKOM.) This was considered highly unusual because virtually every other component of the Jovian emission exhibits at least some organization in CML. This puzzle was solved when Kaiser and Desch [1980] plotted the longitudes where nKOM was observed on a rotation-by-rotation basis, and their results are shown in Figure 7.28. We see that the emission tends to occur at slightly later longitudes on each successive rotation of the planet, and the amount of slippage is about the same from one rotation to the next. In other words, the nKOM events do tend to recur periodically but at a rate that is slightly slower than the rotation rate of the planetary magnetic field.

Fig. 7.28. Plots of the occurrence of
bKOM (shaded patches) and nKOM (line
segments) as a function of date and
central meridian longitude. The data are
displayed on a Jupiter rotation-by-
rotation basis and time, as measured by
the number of 10-hr planetary rotations,
increases downward. Voyager 1
observations are shown in the top panel
and Voyager 2 observations are displayed
below. Notice that the bKOM always
appears in the longitude band between
about 120° and 280° but that nKOM
drifts in System III longitude due to its
slower recurrence period. Adapted from
Boischot et al. [1981].

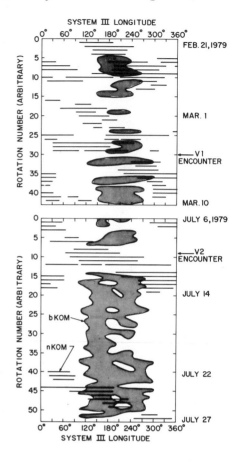

Kaiser and Desch [1980] found that the best-fit recurrence period for the nKOM
activity corresponds to a rate that is 3% to 5% slower than what would be expected for
a rigidly corotating source. Based on Hill's [1980] model of the radial breakdown in
magnetospheric corotation (see also Chap. 10) this amount of corotation lag would correspond to a source at a radial distance of 8 to 9 R_J, and Bagenal and Sullivan [1981]
did observe evidence for 5% to 10% deviations from corotation in plasma ion data
taken near the outer edge of the torus. This distance range corresponds to the outer
periphery of the Io plasma torus where measured upper hybrid resonance frequency
and plasma frequency of the torus electron plasma are essentially identical to the
observed nKOM emission frequency, ~ 100 kHz [Birmingham et al., 1981]. Thus, the
outer portions of the torus where departures from corotation are beginning to become
measurable are plausible nKOM source locations.

The Voyager data near the times of closest approach provide additional evidence for
a source location near or just inside of 10 R_J. Kaiser and Desch [1980] noted that
nKOM events were detected before and after Voyager 1 closest approach when the
spacecraft was at a radial distance of ~ 12 R_J and also near the time of Voyager 2 closest approach at 10 R_J. They pointed out that when an observer is situated that close to
the Io torus, whose vertical dimensions are ~ ± 1 R_J, refraction makes impossible the
detection of a 100 kHz source planetward of the torus. They noted further that there
was no evidence in the Voyager data to suggest that the source was situated farther
from Jupiter than 10 to 12 R_J, and that the range of electron cyclotron frequencies and

plasma frequencies extant in the magnetosphere beyond this distance are too low to support emission above ~ 50 kHz. Thus there are at least four lines of evidence for an nKOM source near the outer portion of the Io torus:

1. the detection of a 3% to 5% departure from corotation,
2. the limited range of directions visible during events detected near Voyager closest approach,
3. the unique coincidence of torus electron plasma frequencies and nKOM emission frequencies, and
4. the existence of strong upper hybrid resonance emissions in that region of space that might be the source of the nKOM.

The records of the occurrence of nKOM plotted in Figure 7.28 show that emission is not always observed on every rotation, but instead nKOM is intermittent on a scale of several days. Moreover, a source can disappear altogether and then reappear many rotations later still in phase with its predecessor. For example, in the Voyager 2 plot, emission was observed from a source that drifted across the full 360° range of System III longitudes between about July 6 and July 14, 1979, and then no nKOM was detected during the next 20 rotations of Jupiter. When nKOM emissions appeared again on July 22, they occurred at the same longitude as would be expected if the activity observed prior to July 14 had continued at the same repetition rate.

Two source regions appear to be active in the Voyager 1 plot in Figure 7.28. One region is observed at longitudes from about 100°–190° on February 21, 1979, which then drifts to about 300° CML after a dozen rotations. This region then ceases to be active except for a few rotations just before the time of Voyager 1 encounter when it appears at longitudes between ~ 30° and 180° still in phase with the earlier activity. A second source begins at about 360° CML on February 23 and drifts through a full 360° range of longitudes over the remaining 40 rotations displayed in the plot. Although there are occasional interruptions in activity for a day or so, there is no apparent change in its location or longitude extent even though Voyager 1's closest approach occurred in the middle of the observing span.

Thus, we are presented with a picture of an nKOM source whose physical extent or radiation pattern near the ecliptic plane are confined to a limited longitude range. It rotates at a rate slightly slower than the planetary magnetic field and maintains its identity over long periods of time even though the radio-wave activity visible from near the ecliptic varies episodically, either waxing or waning for days at a time. The emission may be similar to that responsible for terrestrial nonthermal continuum radiation in which narrow-band electromagnetic waves are generated via coupling with electrostatic waves near the upper hybrid frequency [Jones, 1980; Kurth, Gurnett, and Anderson, 1981].

7.5. Concluding remarks

We conclude this discussion of the kilometer-wave radio emissions from Jupiter with a recapitulation of what we believe we know and what we view as the outstanding questions concerning KOM at present (1981).

For bKOM, we know:

1. dynamic spectra, power flux density spectra and occurrence probability profiles as a function of CML all show evidence of a dependence on observer–Jupiter–Sun angle and on observer's latitude;

2. emissions near 200° CML are from a northern-hemisphere source and are LH polarized when viewed from above the sunlit hemisphere but are RH polarized when viewed from above Jupiter's nightside;
3. there exists a complementary southern hemisphere source near CML ~ 20° for an observer at southern jovigraphic latitudes;
4. a model of the northern-hemisphere source that emits in the left-hand-ordinary mode from a region above Jupiter's dayside ionosphere in the northern hemisphere is consistent with all the observations; and
5. refraction of bKOM waves as they propagate through or over the Io plasma torus is likely to influence the spectral properties observed for bKOM.

For bKOM, we do not yet know:

1. how far above 1 MHz the emission can extend;
2. what is the origin of the drifting patterns seen in dynamic spectra;
3. what is the origin of the observed temporal variations in bandwidth and intensity; and most importantly,
4. what is the emission mechanism.

For nKOM, we know:

1. the emission occurs only over a narrow band of frequencies comparable to the electron plasma frequency near the edge of the Io plasma torus;
2. the repetition rate of the emission is slower than the System III rate by an amount comparable to the corotation lag in the outer edge of the Io plasma torus; and
3. the emission is intermittent, being active for days at a time then disappearing for several days and then reappearing without apparently changing its location in the "not-quite-corotating" frame.

For nKOM, we do not yet know:

1. what causes the polarization reversals often seen during an event;
2. what causes the episodic appearance and disappearance of the emission; and fundamentally,
3. what is the emission mechanism.

ACKNOWLEDGMENTS

The authors are grateful to their colleagues, T. J. Birmingham, M. L. Goldstein, S. Gulkis, M. L. Kaiser, and W. S. Kurth, for their very thorough review and constructive comments on earlier versions of this chapter. We thank J. J. Schauble and J. R. Thieman for providing unpublished data for Figures 7.1 and 7.15, respectively. We also thank B. C. Holland, F. H. Hunsaker, and L. A. White for their untiring efforts in producing the original typescript and most of the illustrations. We acknowledge the partial support of NASA through grant NAGW-196.

8

PLASMA WAVES IN THE JOVIAN MAGNETOSPHERE

D. A. Gurnett and F. L. Scarf

The recent Voyager encounters with Jupiter have now provided us with the first comprehensive investigation of plasma waves in the magnetosphere of Jupiter. The most striking feature of the Jovian plasma wave observations is the close similarity to the plasma-wave phenomena observed in the Earth's magnetosphere. Essentially, all major types of plasma waves detected in the Jovian magnetosphere have analogs in the Earth's magnetosphere. These include, for example, electrostatic waves near and upstream of the bow shock, electromagnetic continuum radiation, lightning- generated whistlers, whistler-mode chorus and hiss, electrostatic electron cyclotron and upper hybrid emissions, and broadband electrostatic noise.

8.1. Introduction

Any waves that are influenced by the presence of a plasma are called plasma waves. Because wave-particle interactions in a collisionless plasma produce scattering and thermalization effects somewhat similar to collisions in an ordinary gas, plasma waves are now recognized as being of fundamental importance for understanding the equilibrium state of planetary magnetospheres. In the Earth's magnetosphere, wave-particle interactions are known to be responsible for heating the solar wind at the bow shock, for the diffusion that allows plasma to enter the magnetosphere, and for the pitch-angle scattering that causes the loss of energetic particles trapped in the magnetic field. A large number of different types of waves can occur in planetary magnetospheres. The properties of some of the more important plasma wave modes are summarized in Table 8.1. In general, plasma waves can be classified as either electromagnetic, which have both electric and magnetic fields, or electrostatic, which have no magnetic field. The electromagnetic modes include two free space modes that can escape from the plasma and be detected remotely and several internal modes, such as the whistler mode, that cannot escape from the plasma. Typically, the electromagnetic modes tend to have propagation velocities near the speed of light. The electrostatic modes, on the other hand, tend to have much lower propagation velocities with properties somewhat similar to sound waves in an ordinary gas. At low frequencies, below the ion cyclotron frequency, plasma waves display a fluidlike behavior involving the bulk motion of the entire plasma. Two such modes, called the shear and compressional Alfvén waves, are the principal mechanisms by which all fluid disturbances are propagated through a plasma. For a further review of the types of wave modes that can exist in a plasma, the reader is referred to one of the standard texts on the subject, for example, Stix [1962] or Krall and Trivelpiece [1973].

The recent Voyager encounters with Jupiter now provide us with the first comprehensive investigation of plasma waves in the magnetosphere of Jupiter. The most striking feature of the Jovian plasma-wave observations is the very close similarity to the plasma-wave phenomena observed in the Earth's magnetosphere. Essentially, all of the major types of plasma waves detected in the Jovian magnetosphere have analogs in the Earth's magnetosphere. These include, for example, electrostatic waves near and upstream of the bow shock, electromagnetic continuum radiation, lightning-generated

Table 8.1. *Plasma wave modes*

Plasma wave mode	Frequency range	Electromagnetic/electrostatic	Free energy source
Langmuir mode (electron plasma oscillation)	$\omega \simeq \omega_{pe}$	Electrostatic	Beam $(\partial f / \partial v_{\parallel} > 0)$
Ion-acoustic mode	$\omega \lesssim \omega_{pi}$	Electrostatic	Drift between electrons and ions
Electron Bernstein modes (electron cyclotron waves)	Bands near $\omega \simeq (n + 1/2)\omega_{ce}$	Electrostatic	Ring distribution (electrons) $(\partial f_e / \partial v_{\perp} > 0)$
Ion Bernstein modes (ion cyclotron waves)	Bands near $\omega \simeq (n + 1/2)\omega_{ci}$	Electrostatic	Ring distribution (ions), field-aligned currents
Free space (R,X) mode	$\omega > \omega_{R=0}$	Electromagnetic	Beam, loss cone
Free space (L,O) mode	$\omega > \omega_{pe}$	Electromagnetic	Beam, loss cone, coupling to electrostatic waves
Z-mode	$\omega_{UHR} > \omega > \omega_{L=0}$	Electromagnetic, electrostatic near ω_{UHR}	Beam $(\partial f / \partial v_{\parallel} > 0)$
Whistler mode	$\omega < \mathrm{Min}\ \{\omega_{ce}\ ,\ \omega_{pe}\}$	Electromagnetic, electrostatic near ω_{LHR}	Loss cone, beam above ω_{LHR}.
Alfvén modes (shear and compressional)	$\omega \ll \omega_{ci}$	Electromagnetic	Pressure anisotropies

Note: $\omega_{R=0} = \omega_{ce}/2 + \sqrt{(\omega_{ce}/2)^2 + \omega_{pe}^2}$ $\qquad \omega_{UHR} = \sqrt{\omega_{ce}^2 + \omega_{pe}^2}$

$\omega_{L=0} = -\omega_{ce}/2 + \sqrt{(\omega_{ce}/2)^2 + \omega_{pe}^2}$ $\qquad \omega_{LHR} \cong \sqrt{\omega_{ce}\,\omega_{ci}}$, if $\omega_{pe} \gg \omega_{ce}$

whistlers, whistler-mode chorus and hiss, electrostatic electron cyclotron and upper hybrid emissions, and broadband electrostatic noise. Because all of these plasma-wave phenomena have received extensive study in the Earth's magnetosphere over the past few years, it has been possible to make rapid progress in the initial identification and first-order understanding of nearly all of the plasma waves observed in the Jovian magnetosphere. In this section, we review the Jovian plasma-wave observations and discuss the relationships and similarities to plasma-wave observations in the terrestrial magnetosphere.

The first indication of important nonthermal wave emission processes at Jupiter was obtained by Burke and Franklin [1955], who discovered that Jupiter was a strong radio emitter in the decameter wavelength range. A few years later, the detection of synchrotron radiation at much shorter wavelengths, in the microwave region of the spectrum, demonstrated that Jupiter had a radiation belt with a substantial population of trapped electrons. Even though no in situ plasma-wave measurements were available prior to the Voyager observations, strong indirect evidence was available that other local plasma-wave emissions existed that could not be detected remotely from the planet. In developing models of the Jovian radiation belts before the Pioneer encounters with Jupiter, several investigators [Kennel, 1972; Thorne and Coroniti, 1972; Coroniti, 1974] considered the effect of unstable whistler-mode and ion-cyclotron waves on the equilibrium energetic particle distributions in the Jovian magnetosphere. The wave generation processes were expected to be similar to the terrestrial radiation belts, where the loss cone imposed by the atmosphere assures that the whistler mode and ion-cyclotron mode will be unstable if the trapped particle fluxes are sufficiently high. According to the limiting flux theory of Kennel and Petschek [1966] pitch-angle scattering by whistler and ion-cyclotron waves produces an upper limit to the trapped particle intensities that can occur in a radiation belt.

The case for the existence of whistler-mode emissions in the inner magnetosphere of Jupiter was strengthened considerably by the Pioneer 10 and 11 energetic electron measurements, which showed the occurrence of hat-shaped pitch-angle distributions with maximum intensities perpendicular to the magnetic field [Van Allen et al., 1975; Fillius et al., 1976]. Pitch-angle distributions of this type are observed in the terrestrial magnetosphere and are generally believed to arise from pitch-angle scattering by whistler-mode waves [Lyons, Thorne, and Kennel, 1971, 1972]. Further detailed analyses of the Pioneer data by several investigators, including Scarf and Sanders [1976], Coroniti [1975], Baker and Goertz [1976], Barbosa and Coroniti [1976], and Sentman and Goertz [1978], provided specific estimates of the frequency range and intensity of the whistler-mode emissions. These studies suggested, for example, that whistler-mode magnetic field intensities of a few milligammas could be expected in the frequency range of a few kilohertz. Predictions of other types of plasma waves were less specific and were based more on analogies with the Earth's magnetosphere. Scarf [1976] suggested, for example, that electrostatic waves driven by electrons and ions streaming into the solar wind should be observed upstream of the bow shock and that current-driven instabilities should be observed in the field-aligned current system produced by the Io interaction. In our original proposal for the Voyager plasma-wave investigation we also suggested that the local electron concentration could be determined from the cutoff of trapped continuum radiation at the local electron plasma frequency, similar to the continuum radiation cutoff observed in the Earth's magnetosphere [Gurnett and Shaw, 1973]. As it turns out, continuum radiation was detected at Jupiter and the plasma-frequency cutoff has proven to be a valuable method of providing absolute, sheath-independent measurements of the electron concentration in the Jovian magnetosphere.

A complete discussion of the Voyager plasma wave observations involves results that depend somewhat on the instrument characteristics; therefore, a brief review of the instrumentation is useful. For a complete description of the Voyager plasma wave instrument, see Scarf and Gurnett [1977]. Because the plasma-wave investigation was added to the Voyager mission relatively late in the development phase of the project, the available mass, power, and telemetry allocations were minimal. Consequently, even though it was desirable to measure both the electric and magnetic fields of plasma waves, only electric field measurements were possible. Electric-field measurements are obtained using two orthogonal 10 m antennas that are shared with the planetary radio-astronomy experiment [see Warwick et al., 1977]. Voltages from the 10 m elements are combined differentially to form an electric dipole antenna with an effective length of about 7 m. Two methods are used to process the signals from the electric antenna. First, for survey analyses, a 16-channel spectrum analyzer is used to make absolute electric-field spectral-density measurements at 16 logarithmically spaced frequencies from 10 Hz to 56 kHz. Typically, this spectrum analyzer provides one complete frequency scan every 4 s with a dynamic range of 100 db. Second, for detailed high-resolution spectrum analyses, the electric-field waveform in the frequency range from about 50 Hz to 14 kHz is transmitted to the ground using the high rate (115 kbs) imaging telemetry system. Because the wideband measurement must be time shared with the imaging system, wideband high-resolution spectrum measurements are only available for short intervals, usually consisting of isolated 48-s bursts. Because the wideband data transmission employs an automatic gain control receiver to maintain the signal within the dynamic range of the telemetry system, no absolute field-strength measurements can be obtained from the wideband data alone. The wideband wave-form measurements do, however, provide a valuable capability for analyzing complex plasma-wave phenomena, because the broadband measurements give essentially continuous coverage in frequency and time, subject only to the constraints imposed by $\Delta f \Delta t \sim 1$.

In describing the plasma-wave results obtained by Voyager we will discuss the observed phenomena in several broad categories, ordered more or less according to the radial distance at which each phenomenon was first observed. In general, the results presented are limited to discussions of plasma waves confined to the near vicinity of the planet. For a discussion of freely escaping electromagnetic emissions, the reader is referred to Chapter 7, which describes the phenomenology of Jovian radio emissions detected by Voyager.

8.2. Upstream waves and bow shock

During the approach to Jupiter two primary types of plasma waves, ion-acoustic-like waves and electron plasma oscillations were detected by the plasma-wave instrument in the solar wind upstream of the bow shock. A wideband frequency-time spectrogram illustrating the detailed structure of a series of ion-acoustic wave bursts observed upstream of Jupiter at a radial distance of 198.5 R_J is shown in Figure 8.1. Typically, these waves consist of an intense narrowband emission in the frequency range from about 500 Hz to 2 kHz. The center frequency of the emission varies erratically, sometimes remaining nearly constant for several seconds, as near the beginning of the spectrogram, and at other times displaying an inverted U-shaped feature lasting only 1 or 2 s, as near the end of the spectrogram. Typical broadband electric-field amplitudes for these bursts range from about 10 to 30 μV/m.

Comparisons of the Voyager spectrograms with comparable spectrograms obtained by Earth-orbiting spacecraft [Kurth, Gurnett, and Scarf, 1979] provide a strong indica-

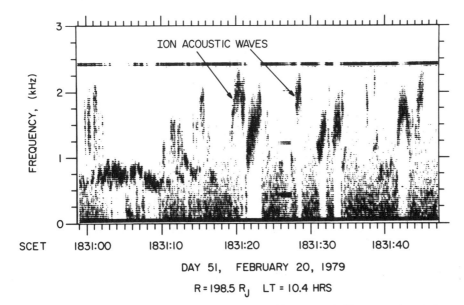

Fig. 8.1. A wideband frequency-time spectrogram of ion-acoustic-like waves detected upstream of the bow shock at Jupiter by the Voyager 1 Plasma Wave Experiment. Similar waves are observed upstream of the Earth's magnetosphere and are produced by ion beams streaming into the solar wind from the bow shock.

tion that these waves are essentially identical to the waves identified by Gurnett and Frank [1978a] as ion-acoustic waves upstream of the Earth's bow shock. These waves were first observed by Scarf et al. [1970] and are known to be closely correlated with suprathermal protons streaming into the solar wind from the Earth's bow shock. The identification of the mode of propagation as the ion-acoustic mode by Gurnett and Frank [1978a] is based on several factors including (1) the wavelength, which extends down to only a few times the Debye length λ_D, (2) the polarization, which is linear and approximately parallel to the static magnetic field, and (3) the absence of a wave magnetic field, which indicates that the wave is electrostatic. These conclusions have recently been further confirmed and amplified by Anderson et al. [1981]. The short wavelengths imply that the frequency observed in the spacecraft frame of reference is mainly determined by Doppler shifts, so that

$$f = (V_s/\lambda) \cos \theta_{kv} \tag{8.1}$$

where V_s is the solar wind velocity, λ is the wavelength, and θ_{kv} is the angle between the propagation vector and the solar wind velocity. The large solar wind Doppler shift explains why the observed frequency extends well above the ion plasma frequency ($f_{pi} \simeq 100$ Hz at 5 AU), which is the upper frequency limit of the ion-acoustic mode in the rest frame of the plasma. Although all the present indications support the identification of these waves as ion-acoustic waves, a complete theory for the generation of these waves by the upstream ion beams has not yet been established. For this reason, Gary [1978] and others have referred to these waves as ion-acoustic-like waves, because the mode involved may be a beam-related mode that is formally distinct from the ion-acoustic mode, although it has many similar characteristics. It is also interesting to note that these ion-acousticlike waves are also observed in the solar wind at locations remote from any possible bow-shock effects [Gurnett and Anderson, 1977], so that the conditions required for their generation are not uniquely related to magnetospheric bow shock interactions. In addition, Kurth, Gurnett, and Scarf [1979] have shown that the same type of ion-acousticlike waves are observed ahead of interplanetary shocks.

The second principal type of plasma wave observed upstream of the bow shock consists of electron plasma oscillations, also sometimes called Langmuir waves. A wideband frequency-time spectrogram of electron-plasma oscillations detected at a radial distance of 82.3 R_j is shown in Figure 8.2. The primary Langmuir wave emission is evident as the narrow dark line at about 5 kHz. This frequency corresponds very closely to the local electron plasma frequency, f_{pe}, in the solar wind. Because the wavelength of the main emission is believed to be very long, $\lambda \simeq 1$ to 10 km, the Doppler shift of the primary emission line is quite small, $\Delta f \simeq 100$ Hz [Gurnett et al., 1981a]. In addition to the primary Langmuir wave emission, many impulsive secondary emissions can also be seen extending upward and downward in frequency from the primary emission line. These upper and lower sideband emissions are believed to be short wavelength Langmuir waves produced from the primary "pump" wave by non-linear parametric interactions. The frequency shift is due to a combination of the Doppler shift, which can be of either sign depending on the direction of propagation, and the intrinsic frequency shift caused by the wave number dependence in the Langmuir wave dispersion relation,

$$f^2 = f_{pe}^2 (1 + 3k^2 \lambda_D^2) \tag{8.2}$$

which always produces an upward frequency shift. Detailed analyses by Gurnett et al. [1981a] indicate that the sideband emissions have wave numbers of approximately $k\lambda_D \simeq 0.15$, whereas the primary pump wave has a wave number about a factor of 10 smaller, with $k_0\lambda_D \simeq 0.015$. The electric field strength of the primary emission line is typically in the range from about 10 to 100 μV/m and tends to increase in intensity as the spacecraft approaches the shock. Because of the bursty character of the sideband emissions, it has not been possible to accurately determine the intensity of these emissions. However, it is believed that the sideband bursts are considerably more intense than the main emission line because the bursts frequently saturate the wideband receiver. The short duration of the sideband bursts, sometimes lasting only a few milliseconds, has led to the suggestion that these bursts may be so intense that nonlinear effects cause the wave to intensify and collapse into discrete spatial structures with scale sizes of only a few Debye lengths [Gurnett et al., 1981a]. Such collapsed spatial structures are called solitons. For a review of the theory of Langmuir wave solitons, see Nicholson et al. [1978].

Comparisons with plasma wave observations near the Earth show that the electron plasma oscillations detected by Voyager have characteristics very similar to electron-plasma oscillations observed upstream of the Earth's bow shock, thereby suggesting that they are generated by essentially the same mechanism. Scarf et al. [1971] showed that the electron plasma oscillations observed upstream of the Earth's bow shock are correlated with the occurrence of beams of nonthermal electrons arriving from the bow shock. Further studies by Gurnett and Frank [1975] and Filbert and Kellogg [1979] showed that the plasma oscillations occur when the beam of nonthermal electrons produces a double hump in the electron distribution function. The double hump arises because of time-of-flight considerations involving the propagation of the electron beam from the bow shock to the spacecraft. These considerations all strongly suggest that the electron plasma oscillations upstream of the terrestrial and Jovian bow shock are produced by the classic electrostatic two-stream instability that occurs when a superthermal electron beam streams through a stationary background

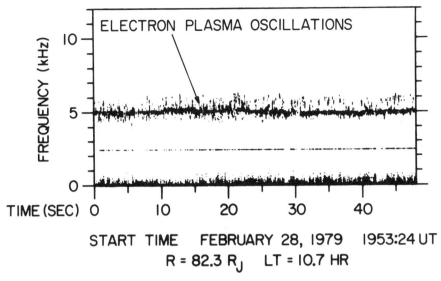

Fig. 8.2. A wideband frequency–time spectrogram of Langmuir waves detected by Voyager 1 in the solar wind upstream of the Jovian bow shock. These waves are generated by low energy, 1–10 keV, electrons streaming into the solar wind from the bow shock. Note the fine structure, which is believed to arise from nonlinear parametric decay processes.

plasma [Stix, 1962]. The principal scientific questions that remain have to do mainly with the nonlinear beam stabilization processes that prevent the instability from completely disrupting the beam [Papadapoulos, Goldstein, and Smith, 1974]. Both the terrestrial and Jovian observations show that the electron beam that generates the plasma oscillations can propagate large distances ($> 10^5$ km) through the solar wind plasma. The nonlinear parametric interactions and soliton collapse effects observed upstream of the Jovian bow shock provide the first space-plasma measurements of the nonlinear effects involved in the beam stabilization process.

At the Jovian bow shock, an abrupt burst of low frequency electric field noise occurs with characteristics similar to the terrestrial bow shock. Because of the fluctuations in the position of the bow shock, a total of at least 23 shock crossings were observed during the inbound and outbound passes of Voyagers 1 and 2. These shocks display a variety of characteristics, ranging from narrow laminar structures characteristic of quasiperpendicular shocks to extended structures characteristic of quasiparallel shocks. The frequency-time spectrogram of one shock for which we obtained exceptionally good wideband electric field spectrum measurements is shown in Figure 8.3. The abrupt increase in both the intensity and frequency of the low-frequency electric field noise at the shock is clearly evident about 39 s after the start of the spectrogram. At peak intensity, the broadband electric-field strength is about 1.3 mV/m. This electric field noise is thought to be responsible for the ion and electron heating that takes place at the shock. Inspection of Figure 8.3 also shows that the main shock disturbance is preceded by a low-frequency precursor that starts about 10 s before the shock and gradually rises in frequency and intensity as the shock approaches. The onset of the low frequency precursor occurs at about the same time that the intensity of the upstream electron plasma oscillations starts to decrease. The rapid decrease in the electron plasma oscillation intensities just ahead of the shock is believed to be due to an increase in Landau damping as the electron temperature increases just ahead of the

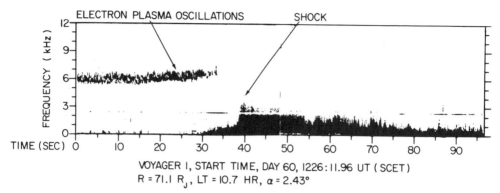

VOYAGER 1, START TIME, DAY 60, 1226:11.96 UT (SCET)
R = 71.1 R$_J$, LT = 10.7 HR, α = 2.43°

Fig. 8.3. A wideband spectrogram of the electric field turbulence at one of several bow-shock crossings observed upstream of the Jovian magnetosphere. Note the abrupt termination of the electron plasma oscillations just ahead of the shock and the ramp of low frequency noise starting about 10 s ahead of the shock. The abrupt burst of electric field noise at the shock is probably ion-acoustic or Buneman-mode turbulence driven by currents at the shock front. The electric field turbulence extends for several minutes into the region downstream of the shock [from Scarf, Gurnett, and Kurth, 1979].

shock. Downstream of the shock, the electric field turbulence decays rapidly in intensity and is no longer detectable a few minutes after the shock crossing. For further details of this shock, see Scarf, Gurnett, and Kurth [1979].

To compare the electric-field turbulence at the terrestrial and Jovian bow shock, the left panel of Figure 8.4 shows a composite summary of the electric field spectra (one 4 s scan at the time of maximum intensity) for nine bow shock crossings detected during the dayside inbound passes of Voyagers 1 and 2. The corresponding panel on the right shows a composite summary of the electric field spectra from 36 terrestrial bow shock crossings selected at random from the IMP 6 data [Rodriguez and Gurnett, 1975]. As can be seen, the general range of electric field spectral densities in the shock are similar at Jupiter and Earth. In both cases the spectrum shows a broad peak at low frequencies, with a rapid decrease in intensity as the frequency approaches the electron plasma frequency. For comparison, the nominal electron plasma frequencies in the solar wind are indicated in Figure 8.4 at Jupiter and Earth. It is evident that the electric-field spectrum at Jupiter is shifted downward in frequency compared to the Earth, approximately in proportion to the electron plasma frequency in the solar wind. Because of the possibility of large Doppler shifts, it is not known whether the upper cutoff near the plasma frequency is directly related to f_{pe} or is an upper limit on the Doppler shift, which also varies in direct proportion to f_{pe} because of the density dependence of the minimum wavelength, $\lambda_{min} = 2\pi\lambda_D$, [Gurnett and Anderson, 1977]. According to long standing ideas [Fredricks et al., 1968; Tidman and Krall, 1971; Greenstadt and Fredricks, 1979] the electric-field turbulence is thought to consist of ion-acoustic or Buneman mode waves driven by currents in the shock front. Although the main instability takes place near the ion plasma frequency, Doppler shifts caused by the solar wind motion tend to shift the frequency upward as detected in the spacecraft frame of reference. The broad frequency spectrum presumably arises because of the wide range of wave numbers and wave normal directions that are present in the electrostatic turbulence.

Although the general features of the bow shock turbulence appear to be quite similar at Earth and Jupiter, several important differences are evident. As can be seen from Figure 8.4, the rapid monotonic rise in the electric field intensities at low frequencies, less than 100 Hz, is not as evident at Jupiter. In the Earth's bow shock this low-

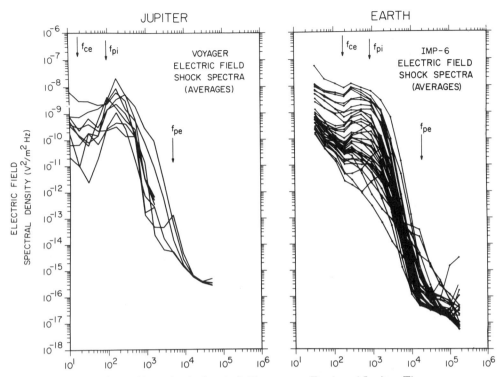

Fig. 8.4. A comparison of bow-shock electric field spectra at Earth and Jupiter. The peak electric field intensities are quite similar; however, the frequencies at the Jovian shocks appear to be shifted downward somewhat compared to the terrestrial shocks, probably owing to the lower solar wind plasma frequency. A notable difference at Jupiter is the near absence of the whistler-mode component that is apparent at low frequencies in the terrestrial shock spectra [from Rodriguez and Gurnett, 1975].

frequency electric field noise is believed to be due to whistler-mode waves generated in the shock [Rodriguez and Gurnett, 1975; Greenstadt et al., 1981]. It appears, therefore, that whistler-mode noise may not be as important in the Jovian bow shock as in the Earth's bow shock. This difference may be related to the average magnetic field direction, which tends to be more nearly perpendicular to the solar wind flow at Jupiter. It is interesting to note that at Venus, where the nominal magnetic field direction is even more radial than at Earth, the low-frequency noise is elevated with respect to the terrestrial case [Scarf et al., 1980]. However, the magnetic field direction is not the only important solar-wind characteristic that changes significantly with heliocentric radial distance, and it is possible that variations in the plasma β-value ($\beta = 8\pi NkT/B^2$), the thermal anisotropy ($T_{\parallel}/T_{\perp}$) or other parameters could produce the observed changes in the low-frequency turbulence spectrum.

Another feature of the Jovian bow shock interaction that seems to be different than the Earth is the turbulence downstream of the shock. At Earth, the entire magnetosheath region downstream of the shock is very turbulent, with intense electrostatic turbulence extending several hundred thousand kilometers into the downstream region [Rodriguez, 1979]. At Jupiter the electrostatic turbulence apparently decays in intensity very rapidly in the region downstream of the shock. Consequently, the electric-field noise levels are very low throughout most of the magnetosheath. The reason for this large difference in the magnetosheath electrostatic turbulence levels is not presently understood.

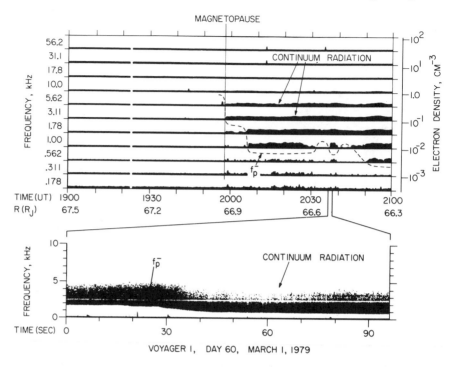

Fig. 8.5. A Voyager 1 magnetopause crossing that shows the electromagnetic continuum radiation trapped in the low density cavity inside the magnetosphere. The sharp low frequency cutoff of the continuum radiation is at the local electron plasma frequency, which gives a direct measurement of the electron concentration.

8.3. Trapped continuum radiation

Proceeding inward closer to Jupiter the next principal type of plasma wave detected was continuum radiation trapped in the low density cavity of the magnetosphere. Figure 8.5 shows the initial magnetopause crossing at which this radiation was first discovered. Immediately after the magnetopause crossing at 1957 UT, a relatively steady level of electric field noise is evident at frequencies below about 10 kHz. This noise has a relatively smooth spectrum with a sharp low-frequency cutoff. A wideband spectrogram showing the extremely sharp low-frequency cutoff is shown in the bottom panel of Figure 8.5. On the basis of similar observations of the same radiation in the Earth's magnetosphere [Gurnett and Shaw, 1973] the low frequency cutoff of the continuum radiation is identified as the local electron plasma frequency. Because the electron plasma frequency is determined by the electron concentration, $f_{pe} \simeq 9 (n_e)^{1/2}$, where n_e is in units of cm^{-3} and f_{pe} is in kHz, the electron concentration can be inferred from this cutoff. The inferred electron concentration variations are indicated by the dashed lines in the top panel, with the electron concentration scale shown to the right. Because of the relatively coarse frequency resolution of the 16-channel spectrum analyzer, the electron concentration is determined only to within a factor of three using these data. More accurate electron concentration measurements, to an accuracy of a few percent, can be obtained from the wideband spectrograms, as in the bottom panel. However, these high-resolution wideband measurements are only available at isolated points. A complete survey of all the electron concentration measurements obtained using this technique is given by Gurnett et al. [1981b].

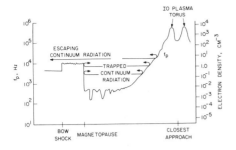

Fig. 8.6. A schematic illustration showing the reflection and trapping of continuum radiation in the low-density cavity inside the magnetosphere.

The mechanism by which the continuum radiation is trapped in the magnetosphere is illustrated in Figure 8.6. Because the electron concentration in the magnetosphere is much lower than in the solar wind, a large region exists where electromagnetic radiation can be trapped within the magnetosphere at frequencies below the solar-wind plasma frequency. Because the plasma is essentially collisionless, the reflection coefficient at the boundaries of the cavity is very close to unity, so that a large number of reflections can occur. The low density cavity not only includes the region around the nose of the magnetosphere but also extends into the tail of the magnetosphere, extending several AU or more downstream of Jupiter [Scarf, et al., 1981]. Because many reflections can occur, the trapped continuum radiation fills essentially the entire magnetospheric cavity. Representative ray paths illustrating the multiple reflections that can occur are shown in Figure 8.7. Because the reflection point at a specific frequency depends on the electron concentration distribution in the magnetosphere, the region accessible to the radiation varies with the wave frequency. At frequencies below the maximum plasma frequency in the distant plasma sheet, which is about 500 Hz, the radiation is trapped between the magnetopause and plasma sheet as shown in Figure 8.7. At higher frequencies, but still below the solar-wind plasma frequency, the radiation penetrates through the plasma sheet and is reflected at the magnetopause. At even higher frequencies, above the magnetosheath and solar-wind plasma frequency, the continuum radiation can escape freely from the magnetosphere as illustrated in Figure 8.6. Escaping continuum radiation, at frequencies above the solar wind plasma frequency, was detected in the solar wind during both the inbound and outbound passes of Voyagers 1 and 2. In this frequency range the continuum radiation changes into a discrete spectrum consisting of many closely spaced narrowband emissions. The pervasive character of the trapped continuum radiation is further demonstrated by the close similarity of the radiation spectrum at widely separated points in the magnetosphere. The left-hand panel of Figure 8.8, for example, shows three representative power spectra from the dayside plasma sheet, the nightside plasma sheet, and the tail lobe regions of the magnetosphere. With the exception of the low-frequency cutoff, which depends on the local plasma frequency, the high frequency parts of spectra are essentially identical in all three regions. The spectrum of the radiation is very steep, $\sim f^{-4}$, and extends down to extremely low frequencies, ≤ 20 Hz, in the tail lobe, implying electron concentrations less than 3×10^{-6} cm^{-3} in this region.

Comparisons of the Jovian and terrestrial continuum radiation provide strong evidence that essentially the same type of radiation is present at both planets. In both cases the spectrum of the trapped component is relatively smooth and steady, and a sharp low-frequency cutoff is present at the local electron plasma frequency. For comparison, the right-hand panel of Figure 8.8 shows some representative continuum radiation spectra in the Earth's magnetosphere. As can be seen, the general shape of the terrestrial continuum radiation spectrum is very similar to the Jovian continuum radia-

Fig. 8.7. Typical ray paths for the
trapped continuum radiation showing
that the radiation can propagate long
distances in the magnetospheric cavity.
At sufficiently low frequencies, the
radiation is trapped as shown between
the magnetopause and the plasma sheet,
whereas at somewhat higher frequencies
the radiation can penetrate through
the plasma sheet thereby allowing
propagation between the northern and
southern tail lobes.

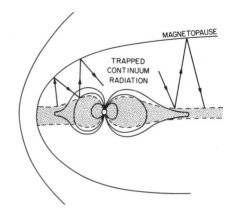

tion spectrum. However, at Jupiter, the intensities are generally higher and the radiation extends down to lower frequencies, mainly because of the much lower plasma concentrations in the Jovian magnetotail. The main argument that the continuum radiation at Jupiter consists of electromagnetic radiation comes from comparison with terrestrial observation. Because the Voyager instrumentation does not include a magnetic antenna, the electromagnetic character of the noise cannot be directly confirmed at Jupiter. However, magnetic field measurements with the IMP 6 spacecraft in the Earth's magnetosphere [Gurnett and Shaw, 1973] show that the continuum radiation consists of electromagnetic radiation with $E \simeq cB$. Terrestrial observations also demonstrate [Gurnett, 1975; Kurth, Gurnett, and Anderson, 1981] that the continuum radiation consists of both a trapped and escaping component, with the spectrum changing to a series of narrowband emissions above the solar wind plasma frequency very similar to the Jovian observations.

Two basic mechanisms have been proposed for generating the continuum radiation observed in planetary magnetospheres. First, Frankel [1973] suggested that continuum radiation escaping from the terrestrial magnetosphere may be generated by synchrotron radiation from energetic electrons; and second, Gurnett [1975] and Gurnett and Frank [1976] suggested on the basis of observations obtained in the Earth's magnetosphere that the continuum radiation is generated by mode-coupling from electrostatic waves near the local electron plasma frequency. The electrostatic mode-coupling mechanism has been further refined and discussed by a number of investigators including Jones [1976], Kurth et al. [1979b], Kurth, Gurnett, and Anderson [1981], and Melrose [1981], and it now appears to be the favored mechanism for generating the continuum radiation. The specific electrostatic modes involved in the generation of the continuum radiation are discussed in the next section.

Recently, Barbosa [1981] has addressed the question of how the narrowband emission generated by a mode-coupling process could be converted into a nearly continuous spectrum. Although the trapped continuum radiation spectrum is usually relatively smooth and continuous, as in Figure 8.5, occasionally narrowband features are evident [see Figure 8 from Gurnett and Shaw, 1973; or Figure 3 from Gurnett, Kurth, and Scarf, 1979a]. This narrowband structure strongly suggests that the radiation originates as a superposition of many distinct narrowband emissions. Barbosa's mechanism for spreading the spectrum of the trapped continuum radiation involves the Doppler shift that occurs each time the radiation reflects from the magnetopause boundary. Because of the random fluctuations in the velocity of the magnetopause in response to variations in the solar wind pressure, this mechanism converts the nar-

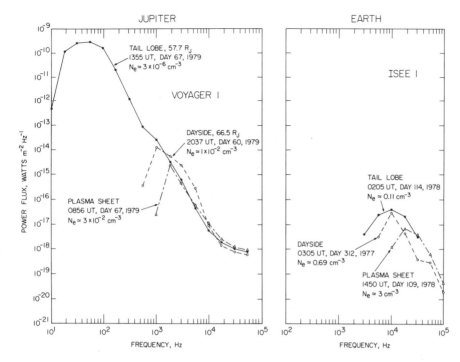

Fig. 8.8. A comparison of continuum radiation spectra in the Jovian and terrestrial magnetospheres. Although the overall features are quite similar, the continuum radiation extends to much lower frequencies at Jupiter and is much more intense, apparently owing to the steep spectrum and extremely low electron concentrations in the Jovian tail lobes. Note that the spectral intensities are similar in the upper part of the frequency range.

rowband emissions generated by the mode-coupling process to an essentially continuous spectrum at frequencies below the solar wind plasma frequency. At frequencies above the solar wind plasma frequency reflections do not occur at the magnetopause, which explains why the spectrum switches from a continuum to series of narrowband emissions at this frequency.

8.4. Upper hybrid and electron cyclotron waves

Within the Jovian magnetosphere, numerous narrowband electrostatic emissions were detected by both Voyagers 1 and 2 over a wide range of radial distances. On the basis of similar observations in the Earth's magnetosphere these emissions are identified as being of two related types: upper-hybrid resonance waves at $f_{UHR} = (f_{ce}^2 + f_{pe}^2)$, and electron cyclotron waves near half-integral harmonics of the electron cyclotron frequency, $(n + 1/2) f_{ce}$. These two types of waves are shown circled in Figure 8.9. The electron cyclotron waves occur just above the electron cyclotron frequency, at about $3 f_{ce}/2$ and $5 f_{ce}/2$, and the upper-hybrid resonance waves occur at slightly higher frequencies, a specific example being the isolated intense emission in the 31.1-kHz channel at about 0000 UT on March 5. As can be seen from the magnetic latitude given in the lower panel of Figure 8.9, the upper-hybrid and electron cyclotron emissions occur simultaneously, however, occasionally only one of the two types may be present at a time. The intensities of the upper-hybrid and electron cyclotron waves vary over a

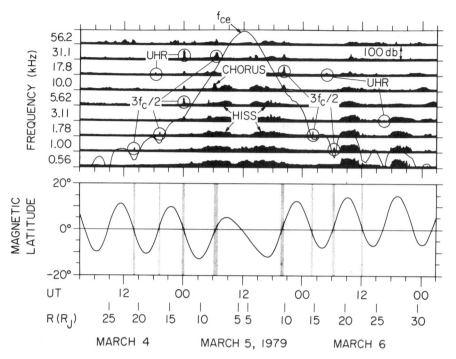

Fig. 8.9. The 16 channel spectrum analyzer data for the Voyager 1 pass by Jupiter showing the occurrence of low order, $3 f_{ce}/2$ and $5 f_{ce}/2$, electron cyclotron emissions and *UHR* emissions near the equatorial plane. Whistler-mode chorus and hiss emissions are also evident in the inner region of the magnetosphere [from Kurth et al., 1980b].

wide range, from only a few microvolts per meter to a few millivolts per meter. Generally, the most intense emissions occurred in the inner magnetosphere, near closest approach.

A high-resolution wideband spectrogram illustrating the detailed frequency-time structure of the upper-hybrid waves is shown in Figure 8.10. As can be seen, the upper-hybrid waves consist of intense narrowband emissions near the low-frequency cutoff of the trapped continuum radiation, at approximately f_{pe}. The frequency of these emissions, near the low-frequency cutoff of the continuum radiation, is consistent with the fact that the electron cyclotron frequency is much less than the electron plasma frequency, which places the upper-hybrid resonance frequency very close to the plasma frequency, $f_{UHR} \simeq f_{pe}$. A striking characteristic of the upper-hybrid resonance emissions shown in Figure 8.10 is the fact that the emission frequency is not continuous, but rather consists of a series of discrete lines that have a frequency spacing corresponding to the local electron cyclotron frequency. Careful comparisons indicate that the individual emission lines occur when $(n + 1/2) f_{ce} \simeq f_{UHR}$. As can be seen from the continuum radiation cutoff, the variations in plasma concentration cause the upper-hybrid resonance frequency to sweep across successive half-integral harmonics of the electron cyclotron frequency. Intense upper-hybrid emissions occur whenever the plasma concentration is such that the condition $(n + 1/2) f_{ce} \simeq f_{UHR}$ is satisfied.

A somewhat similar relationship between the upper-hybrid emissions and the half-integral electron cyclotron harmonics is evident in the measurements from the planetary radio astronomy experiment near closest approach [Warwick et al., 1979a]. Figure 8.11 summarizes the frequency structure of the upper-hybrid and electron cyclotron

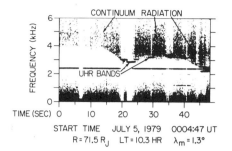

Fig. 8.10. A wideband spectrogram showing the details of the upper-hybrid resonance emissions. Note that this emission is not a continuous band, but consists of a series of discrete emissions that occur near half-integral harmonics of the electron cyclotron frequency, when $(n + 1/2) f_{ce} \simeq f_{UHR}$.

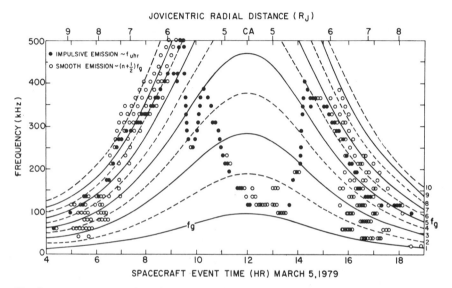

Fig. 8.11. A summary of the frequencies of upper-hybrid and electron cyclotron emissions observed by the planetary radio astronomy experiment during the pass through the Io plasma torus on March 5, 1979 [Birmingham et al., 1981]. The black circles signify impulsive emissions near f_{UHR} while the open circles denote smooth emissions near $(n + 1/2) f_{ce}$.

waves observed during the pass through the Io torus [Birmingham et al., 1981]. Two types of emissions can be identified in this region: relatively intense impulsive emissions at f_{UHR}, and somewhat weaker smooth emissions at $(n + 1/2) f_{ce}$. The impulsive emissions at f_{UHR} are probably analogous to the upper-hybrid waves shown in Figure 8.10, although the splitting into discrete lines is not as evident, possibly because of the more limited frequency resolution. This upper-hybrid resonance emission was used by Warwick et al. [1979a] to obtain the electron concentration profile shown in Figure 8.12. The two broad peaks in the concentration profile at about 0930 and 1430 UT correspond to the inbound and outbound passes through the Io torus. The smooth emissions which tend to occur at $(n + 1/2) f_{ce}$ in Figure 8.10 appear to be yet another type of electron cyclotron emission. These emissions are characterized by relatively smooth diffuse frequency structure that contrasts sharply with the more impulsive low-order electron cyclotron emission observed farther out in the magnetosphere. These smooth diffuse bands also tend to reach maximum intensity near f_{UHR}, rather than at the low order half-integral harmonics.

   Upper-hybrid resonance emissions displaying somewhat similar quantized frequency structure at half-integral harmonics of the electron cyclotron frequency have

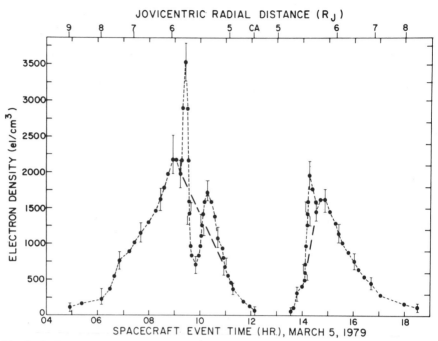

Fig. 8.12. The electron concentration profile obtained from the upper-hybrid emissions in Figure 11. The two broad peaks correspond to the inbound and outbound passes through the Io torus. The dashed line indicates a possible smoothing according to Cummings et al. [1980]. The peaks farthest left and farthest right correspond to the Io *L*-shell.

also been observed in the Earth's magnetosphere [Shaw and Gurnett, 1975] and have been discussed in detail by Kurth et al. [1979b]. The occurrence of half-integral cyclotron harmonic effects at the upper-hybrid resonance demonstrates that these emissions are related to a general class of electron cyclotron emissions at $(n + 1/2)f_{ce}$ which were first observed near $3f_{ce}/2$ and $5f_{ce}/2$ in the Earth's magnetosphere by Kennel et al. [1970]. When the electron plasma frequency is well above the cyclotron frequency, $f_{pe}/f_{ce} \gg 1$, these emissions split into two well-defined groups consisting of one prominent emission line near f_{UHR}, and a set of low order lines: at $(n + 1/2)f_{ce}$, where n usually does not extend above 2 or 3. These emissions have been extensively studied in the Earth's magnetosphere by many investigators, including Fredricks and Scarf [1973], Scarf et al. [1973], Shaw and Gurnett [1975], Christiansen et al. [1978], and Kurth et al. [1979b]. Relatively smooth diffuse electron cyclotron emissions possibly related to the smooth $(n + 1/2)f_{ce}$ emissions in the Io torus have also been observed in the Earth's magnetosphere [Shaw and Gurnett, 1975; Hubbard and Birmingham, 1978]. Thus, a wide body of research exists in the Earth's magnetosphere that is directly applicable to the interpretation of the Jovian upper hybrid and electron cyclotron emissions.

Theoretical studies of the upper hybrid and electron cyclotron waves by numerous investigators including, for example, Fredricks [1971], Young, Callen, and McCune [1973], Hubbard and Birmingham [1978], Ashour-Abdalla and Kennel [1978a,b], and others have accounted for the main features of these emissions. Studies of the Harris dispersion relation [Harris, 1959] for electrostatic waves propagating in a hot plasma with a magnetic field indicate that the underlying requirement for instability is that the electron velocity distribution function must have a region of positive slope, $\partial f/\partial v_\perp > 0$, with respect to the component of velocity perpendicular to the magnetic field, $v_\perp$. Cyclotron damping strongly attenuates waves near harmonics of the cyclotron frequency; therefore, the instability always occurs between an adjacent pair of cyclotron harmonics, but not necessarily at an exact half-integral harmonic. The identification of a given emission band with a specific half-integral harmonic is, therefore, only an identifying name, and does not necessarily imply that the emission frequency is exactly at the half-integral harmonic. The detailed growth rate of the electrostatic instability, and the number of unstable bands depends on a large number of parameters, the most important of which are the cold-to-hot electron concentration ratio and the cold electron temperature. Comparisons of wave intensities with electron distribution function measurements in the Earth's magnetosphere by Rönnmark et al. [1978] and Kurth et al. [1979b, 1980a] show good general agreement between theory and observations, although in some cases the source of the free energy is not completely obvious, possibly owing to modifications of the distribution function by the waves.

The existence of upper hybrid and electron cyclotron waves near the magnetic equator implies that electron distribution functions with a significant positive $\partial f / \partial v_\perp$ occur in this region of the Jovian magnetosphere. However, this relationship has not been confirmed with particle measurements, mainly because no electron measurements are available from Voyager with sufficient angular resolution in the proper energy range, 100 eV to 10 keV, to provide an adequate test of the theories involved.

The electrostatic upper-hybrid and electron cyclotron waves are possibly important for three processes in the Jovian magnetosphere. First, if the waves are sufficiently intense, they can cause pitch-angle diffusion and loss of energetic electrons trapped in the magnetic field; second, they can transfer energy from the hot to the cold electron population; and third, the upper-hybrid emission can couple to a free space electromagnetic mode and produce radio emissions. The possibility that low-order electron cyclotron waves could cause significant pitch-angle diffusion was first proposed by Fredricks and Scarf [1973] to explain the diffusion and energization of auroral electrons in the Earth's magnetosphere. Subsequent studies by Lyons [1974] demonstrate that electron cyclotron waves with intensities of 10 to 100 mV/m can account for strong pitch-angle diffusion of 1 to 20 keV electrons. Electron cyclotron waves with intensities approaching these levels have been reported by Scarf et al. [1973] in association with substorms in the Earth's magnetosphere. At Jupiter it is not known whether the amplitudes of the low order electron cyclotron waves are sufficiently large to produce significant pitch-angle diffusion effects. In the middle and outer regions of the magnetosphere the field strengths of the $3 f_{ce}/2$ and $5 f_{ce}/2$ emissions are very weak, only a few tens of μV/m; thus it is very unlikely that these waves produce important effects. Only in the inner magnetosphere, at radial distances less than about 15 R_J, do the electron cyclotron emissions reach intensities of a few millivolts per meter [Kurth et al., 1980b], which are possibly large enough to cause significant pitch-angle diffusion. In addition to pitch-angle scattering, studies of the nonlinear stabilization of the electron cyclotron harmonic emissions by Ashour-Abdalla and Kennel [1978b] suggest that these waves may cause a significant energy transfer from the hot to the cold electron populations. Such a heating process could, for example, be important in the Io plasma torus, which is known to consist of a relatively cool (10 eV) core component and a much more energetic (1 keV to 1 MeV) magnetospheric component.

The possibility that the upper-hybrid resonance emissions may play an important role in the generation of low frequency radio emissions has been suggested by a number of authors. In an early study of the terrestrial continuum radiation, Gurnett [1975] presented radial intensity profiles that strongly suggested that the continuum radiation originates from intense electrostatic noise bands near the local electron plasma frequency. Jones [1976] later suggested a specific mechanism for converting the electrostatic noise to electromagnetic radiation. Recently, after reviewing a number of mechanisms for generating the continuum radiation at both Jupiter and Earth, Melrose [1981] concluded that the most likely process for generating the continuum radiation involves a nonlinear interaction between the upper-hybrid emissions and a yet unidentified low-frequency mode. Strong evidence for the importance of the upper-hybrid emission as a source of both the trapped and escaping terrestrial continuum radiation has recently been presented by Kurth, Gurnett, and Anderson [1981]. Because the upper-hybrid emissions are observed over a wide range of radial distances inside the Jovian magnetosphere, including the Io torus [Warwick et al., 1979a; Birmingham et al., 1981], it seems quite likely that these emissions are involved in not only the generation of the trapped and escaping continuum radiation at Jupiter, but also other types of Jovian radio emissions at higher frequencies. A prime candidate is the narrowband Jovian kilometric radiation reported by Kaiser and Desch [1980], as this radiation appears to be generated near the equatorial plane in the outer regions of the Io plasma torus, which is a region known to include intense upper-hybrid resonance emissions.

8.5. Whistler-mode waves

Because several types of whistler-mode waves are observed in the Earth's magnetosphere, it was anticipated that some type of whistler-mode noise would be detected during the Voyager passes by Jupiter. In fact, four different types of whistler-mode signals were detected, each related to a type of whistler-mode noise present in the Earth's magnetosphere. The four types are (1) lightning-generated whistlers, (2) chorus, (3) hiss, and (4) auroral hiss.

Whistlers

One of the more striking results from the Voyager 1 plasma wave observations was the discovery of whistlers generated by lightning at Jupiter [Gurnett et al., 1979b]. The occurrence of lightning at Jupiter was not totally unanticipated; several investigators had already suggested that lightning was likely to be present in the Jovian atmosphere [Sagan et al., 1967; Bar-Nun, 1975]. Also, the possibility of detecting whistlers generated by lightning was considered in the overall planning and design of the plasma wave instrument [Scarf and Gurnett, 1977]. Nevertheless, when signals were detected from the Voyager 1 plasma wave instrument with the unmistakable whistling sounds characteristic of lightning-generated whistlers, it was regarded as a remarkable discovery because it was completely unknown whether the signals would be detectable given the long propagation paths involved (about 10 times longer than in the Earth's magnetopshere) and the largely unknown attenuation effects along the propagation path. During the pass by Jupiter, a total of 167 whistlers were identified in the Voyager wideband data. All of the whistlers detected were observed in a narrow range of L-shells near the inner edge of the Io plasma torus, from about 5.2 to 5.9 R_J and within $\pm 10°$ of the magnetic equator. Two of the best examples are shown in Figure 8.13. The characteristic decrease in the frequency with increasing time, which uniquely identifies these signals as whistlers, is clearly evident. Gurnett et al. [1979b] show that these signals fit the well-known equation

$$t - t_0 = D / \sqrt{f} \qquad\qquad (8.3)$$

for the dispersion of whistlers first given by Eckersley [1935], where D is a constant called the dispersion. For the frequencies of interest the whistlers propagate approxi-

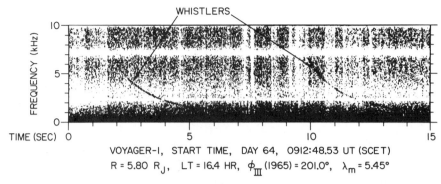

Fig. 8.13. A spectrogram of two long dispersion whistlers detected by Voyager 1 in the Io plasma torus. The whistler dispersion gives a weighted integral of the electron concentration along the propagation path.

mately along the magnetic field line from the base of the ionosphere to the spacecraft. The dispersion is proportional to the integral of $(n_e)^{1/2}/B$ along the ray path. The whistler dispersion varied over a large range, from $D \simeq 35$ to 570 s $\mathrm{Hz}^{1/2}$. The small dispersion whistlers were observed near the edge of the torus and are believed to have reached the spacecraft without passing through the torus, whereas the large dispersion whistlers are believed to have passed through the torus. The very small dispersion of some of these whistlers, $D \simeq 50$ s $\mathrm{Hz}^{1/2}$, demonstrates that the average electron concentration outside of the torus is small, only about 10 to 40 cm^{-3} along the $L = 6$ field line. For the whistlers that have passed through the Io plasma torus the dispersion provides a quantitative measure of its north–south thickness. Typical values for the torus thickness at $L \simeq 6$ from whistler dispersion measurements range from about 1.8 to 5.0 R_J [Gurnett et al., 1981b]. These thicknesses tend to be about a factor of 2 larger than the scale heights estimated from in situ plasma measurements [Bagenal, Sullivan, and Siscoe, 1980; Warwick et al., 1979a; Birmingham et al., 1981], presumably because the in situ plasma measurements do not include the contribution of light ions, such as protons, that would tend to increase the electron concentration away from the equator. Comparisons of the whistler dispersion with in situ measurements of heavy ion concentrations near the equatorial plane [Tokar et al., 1982] show that light ions tend to be the dominant constituent beyond about 2 R_J from the magnetic equator at $L \simeq 6$.

In addition to the scale height and electron concentration determinations, the Voyager whistler observations also place limits on the electron temperature in the torus because of the absence of appreciable Landau damping effects. As shown by Menietti and Gurnett [1980] the attenuation of whistlers due to Landau damping by hot electrons becomes unacceptably large if the electron temperature in the torus exceeds about (2 to 3) $\times 10^5$ K. This upper limit is in good agreement with the results from the Voyager 1 ultraviolet spectrometer [Broadfoot et al., 1979] and plasma instrument [Scudder et al., 1981a], which indicate an average electron temperature of about 10^5 K.

Chorus and hiss

Two other types of whistler-mode waves called chorus and hiss were detected by both Voyagers 1 and 2 as the spacecraft passed through the Io torus. Both of these whistler-mode waves are spontaneously generated within the magnetosphere by interactions with energetic trapped electrons and are named after the corresponding phenomena in

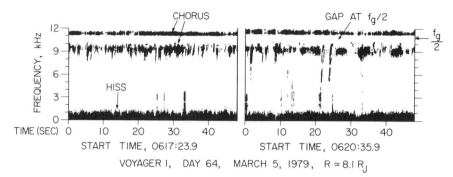

Fig. 8.14. Wideband frequency-time spectrograms of whistler-mode chorus and hiss observed in the Io plasma torus. The hiss consists of a relatively structureless emission below about 1 kHz, and the chorus consists of many discrete narrowband tones generally rising in frequency from about 8 to 12 kHz. Note the gap in the chorus spectrum at approximately $f_{ce}/2$.

the Earth's magnetosphere. The spectral characteristics of the chorus and hiss emissions are illustrated in Figure 8.14, which shows a wideband spectrum obtained from Voyager 1 at about 8.1 R_J. The chorus emissions occur from about 8 to 12 kHz and have a very complex spectral structure, typically consisting of short bursts that increase in frequency with increasing time. The hiss emissions, on the other hand, occur at much lower frequencies (below about 2 kHz) and are nearly structureless. A remarkable feature of the chorus emissions is the distinct gap in the spectrum at one-half the electron cyclotron frequency, $f_{ce}/2$. As can be seen from Figure 8.14, the low frequency cutoff of the upper portion of the chorus band is very close to $f_{ce}/2$. A well-defined gap occurs below this cutoff with a bandwidth of about 1 kHz. The chorus intensities are very similar above and below the gap. Within the gap, the intensities are at least 25 db below the peak intensities on either side. A qualitative difference is also evident in the detailed frequency-time structure above and below the gap; more rising tones and discrete features tend to occur in the lower band.

The spatial distribution of the chorus and hiss emissions detected by Voyager 1 can be seen in Figure 8.9. In the 16 channel spectrum analyzer data, the hiss corresponds to the very broadband emission extending from about 50 Hz to 10 kHz, with maximum intensity at a few hundred Hz [Scarf, Gurnett, and Kurth, 1979]. The hiss occurs from about 5.5 to 10 R_J on both the inbound and outbound legs, in the regions that correspond to the inbound and outbound passes through the Io plasma torus. The chorus emissions, on the other hand, occur only in a single frequency channel (10 kHz) for a brief period around 0615 UT on the inbound leg. This time corresponds to a crossing of the magnetic equator, thereby suggesting that the chorus may be confined to the vicinity of the magnetic equator. At maximum intensity the integrated broadband electric field strength of the hiss is a few millivolts per meter [Scarf, Gurnett, and Kurth, 1979]. The chorus is somewhat less intense than the hiss, and has a maximum broadband electric-field strength of about 0.26 mV/m [Coroniti et al., 1980a].

The whistler-mode hiss and chorus emissions observed at Jupiter have many remarkable similarities to hiss and chorus emissions observed in the Earth's magnetosphere. In fact, our identification of these waves as whistler-mode emissions rests mainly on these close similarities, because no magnetic field measurements are available from Voyager to prove conclusively that these waves are propagating in the whistler mode. In the Earth's magnetosphere whistler-mode emissions called chorus, with spectral characteristics similar to Jovian chorus have been observed for many years [Helliwell, 1965].

The earliest reports of ground-based very low frequency (VLF) radio observations with characteristics resembling what is now called chorus were by Preece [1894] and Burton and Broadman [1933]. The first systematic studies of the phenomena called chorus were conducted by Allcock and Martin [1956] and Allcock [1957] who coined the name "dawn chorus" because the received audio signals sounded like the chorus from a distant rookery at sunrise. After spacecraft measurements were available, it became apparent that the chorus emissions were due to a cyclotron resonance involving electrons trapped in the magnetic field. Tsurutani and Smith [1974] and Burtis and Helliwell [1976], for example, showed that the frequency of chorus emissions were related to the equatorial electron cyclotron frequency along the observer's L-shell. They also found that the chorus emissions frequently have a gap at one-half the electron cyclotron frequency. At about the same time, Russell, Holzer, and Smith [1969] identified another type of whistler-mode emission, later called plasmaspheric hiss by Thorne et al. [1973], which is oberved mainly inside the plasmasphere. In contrast to the chorus emissions, the plasmaspheric hiss has a relatively broad bandwidth with nearly constant intensity and occurs at frequencies well below the equatorial electron cyclotron frequency.

The first comprehensive theory of whistler-mode emissions, with applications to both chorus and hiss, was given by Kennel and Petschek [1966], who identified the electron pitch-angle anisotropy produced by the atmospheric loss cone as the free energy source for driving the whistler-mode instability. By utilizing the principal of conservation of angular momentum Brice [1964] showed that the generation of right-hand polarized whistler-mode waves leads to a reduction in the pitch angle of the resonant electrons. Therefore, the spontaneous emission of whistler-mode waves by a plasma tends to drive the resonant particles toward the loss cone, eventually leading to the loss of the particles. In the Earth's magnetosphere whistler-mode chorus and hiss emissions are now regarded as the primary mechanisms for the scattering and loss of energetic electrons trapped in the magnetosphere [Lyons, Thorne, and Kennel, 1972]. Similarly at Jupiter, chorus and hiss emissions are now thought to be mainly responsible for the scattering and loss of trapped electrons, with the hiss mainly interacting with high energies [Scarf et al., 1979a; Thorne and Tsurutani, 1979] and the chorus interacting with low energies [Coroniti et al., 1980a]. Figure 8.15 [from Coroniti et al., 1980a] for example, shows the resonant energies and pitch-angle scattering times for a typical chorus and hiss spectrum. As can be seen, the hiss tends to resonate with electron energies of a few hundred keV, and the chorus resonates with electrons of a few keV. The occurrence of hiss mainly in the high density region of the Io plasma torus is a consequence of the dependence of the whistler-mode resonance condition on the electron concentration [Kennel and Petschek, 1966]

$$E_{Res} = \frac{B^2}{2\mu_o n_e} \left(1 - \frac{f}{f_c}\right)^3 \frac{f_{ce}}{f} \tag{8.4}$$

where $B^2/2\mu_o n_e$ is an energy that characterizes the whistler-mode interaction. Higher electron number densities lower the resonant energy, thereby increasing the number of resonant electrons and the growth rate of the instability. Representative ray paths for hiss generated in the Io torus are illustrated in Figure 8.16, based on a similar sketch of hiss trapped in the Earth's plasmasphere by Thorne et al. [1973]. Because of the decreasing index of refraction away from the magnetic equator, whistler-mode waves tend to be reflected as soon as the wave frequency drops below the local lower-hybrid resonance frequency f_{LHR}. This lossless reflection process leads to repeated passes

Fig. 8.15. A typical chorus and hiss spectrum and the corresponding electron resonance energies and scattering times for resonant whistler-mode interactions. The hiss scatters relatively energetic, few hundred keV, electrons whereas the chorus interacts with much lower-energy electrons, that have energies of a few keV [from Coroniti et al., 1980a].

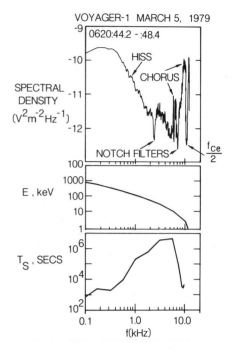

of the whistler-mode wave through the torus and increases the time available for wave-particle interactions.

Although the Kennel–Petschek theory provides a general framework for understanding magnetospheric whistler-mode instabilities, many of the detailed characteristics of the whistler-mode emissions remain difficult to explain. This is particularly true for the chorus bursts, which display very complex frequency-time characteristics. Specific models for the frequency variation of the chorus emissions have, for example, been developed by Helliwell [1967] and Helliwell and Crystal [1973] to account for the nonlinear feedback between the whistler-mode waves and the resonant electrons. However, it is uncertain at present whether these models can account for the detailed temporal behavior of individual chorus bursts. Similarly, even though numerous theories have been proposed to explain the gap in the chorus spectrum at $f_{ce}/2$ [Tsurutani and Smith, 1974; Maeda, Smith, and Anderson, 1976; Burtis and Helliwell, 1976; Curtis, 1978] the exact explanation of this effect remains unresolved.

Auroral hiss

A fourth type of whistler-mode noise called auroral hiss was also observed during the Voyager 1 pass through the Io torus [Gurnett, Kurth, and Scarf, 1979b]. The name of this noise is derived from a type of whistler-mode noise commonly observed in the Earth's auroral zones in association with intense low-energy, 10 eV to 1 keV, auroral electron precipitation. The distinguishing characteristics of the terrestrial auroral hiss are that it has (1) a very broad bandwidth, and (2) a V-shaped low-frequency cutoff [Gurnett, 1966; McEwen and Barrington, 1967]. The upper frequency limit of the emission typically extends up to near the electron cyclotron frequency and the lowest frequency, near the point of the V, is near the local lower-hybrid resonance frequency $f_{LHR} \simeq (f_{ce} f_{ci})^{1/2}$.

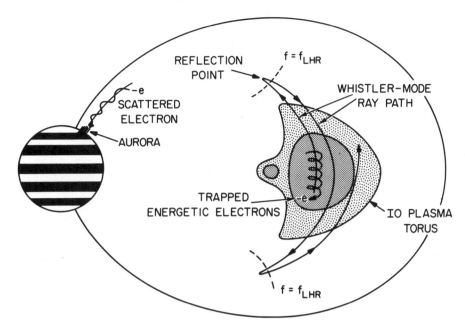

Fig. 8.16. A schematic illustration of typical whistler-mode ray paths for hiss and chorus interacting with trapped electrons in the Io plasma torus. The high plasma density in the torus lowers the resonance velocity and increases the whistler-mode growth rate. As the wave propagates away from the magnetic equator, the emissions are reflected as soon as the wave frequency drops below the local lower-hybrid resonance frequency, which returns the waves to the equatorial plane and further increases the time available for resonant interactions. This reflection process is sometimes compared with the mirrors in a laser, which cause repeated passes of the wave through the amplifying region.

A series of wideband frequency-time spectrograms illustrating the main features of the auroral hiss observed by Voyager 1 is shown in Figure 8.17. Because of the time sharing with the imaging system these spectrograms are not contiguous in time. Nevertheless, a broad band of noise can be identified during this period with a very well-defined low-frequency cutoff that increases linearly in frequency with increasing time. The broadband character of this emission and the smoothly varying low-frequency cutoff provide an almost unmistakable identification of this emission as a V-shaped auroral hiss emission. This event occurs during the inbound pass through the inner edge of the torus, and another comparable emission is evident in the survey data during the outbound pass through the inner edge of the torus. Unfortunately, on both the inbound and outbound passes only one leg of the V can be distinguished, since the leg inside the torus tends to be obscured by the whistler-mode hiss emissions.

From measurements of the cutoff frequency as a function of time and a simple model for the whistler-mode ray paths, a rather accurate determination of the source position can be obtained. For the auroral hiss emissions, which are known to be propagating at angles close to the resonance cone, the limiting ray path angle with respect to the magnetic field is given by

$$\tan^2 \psi = \frac{f^2 - f_{LHR}^2}{f_{ce}^2} \tag{8.5}$$

Because higher frequency waves propagate at larger angles to the magnetic field, it is evident from the above equation that as the spacecraft approaches the field line

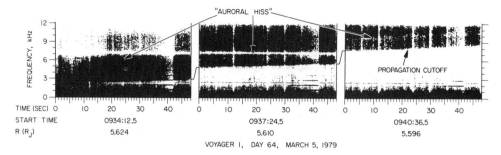

Fig. 8.17. A wideband spectrogram of auroral hiss observed at the inner edge of the
torus. The auroral hiss is characterized by a broadband emission spectrum with a sharply
defined low-frequency cutoff that increases linearly with time. The low frequency cutoff
is produced by a frequency-dependent limiting angle for the whistler-mode ray path
relative to the magnetic field [from Gurnett, Kurth, and Scarf, 1979b].

through the source the highest frequencies are detected first, with progressively lower
and lower frequencies accessible as the spacecraft comes closer to the source field line.
The sharp low-frequency cutoff of the auroral hiss indicates that the source terminates
at a well-defined distance h upward along the magnetic field line. For the observed variation of the cutoff frequency in Figure 8.17, a best fit analysis of Equation 8.1 indicates that the spacecraft crossed the field line through the source at 0930 UT, and that
the distance h to the source was 0.65 R_J [Gurnett, Kurth, and Scarf, 1979b]. Because
the spacecraft was located close to the centrifugal equator at this time [Cummings,
Dessler, and Hill, 1980], the source could be located either to the north or south of the
equator. The approximate location of the auroral hiss source is shown in Figure 8.18 in
relation to the electron number-density contours of the Io torus given by Bagenal,
Sullivan, and Siscoe [1980]. It is evident from this comparison that the auroral hiss
source is located almost exactly coincident with the sharp inner edge of the torus at
$L = 5.6$. Because the auroral hiss was observed on both the inbound and outbound
passes through the inner edge of the torus, it is concluded that the auroral hiss source
extends longitudinally around the torus on the $L = 5.6$ L-shell at a distance of about
0.65 R_J from the equator.

   The electric field intensity of the auroral hiss is small, only about 30 μV/m broadband intensity. Therefore, it is unlikely that the auroral hiss plays a significant role in
the dynamics of the torus plasma. The principal importance of the auroral hiss is as an
indicator of auroral processes occurring at the inner boundary of the torus. In the terrestrial auroral zone, the occurrence of auroral hiss is closely correlated with intense
fluxes, 10^8 to 10^{10} electrons cm^{-2}, of low energy, 10 eV to 1 keV, auroral electrons
[Gurnett and Frank, 1972]. These electrons are believed to be part of the primary field-
aligned current circuit that couples the outer magnetosphere to the auroral ionosphere.
As presently understood, the auroral hiss is generated by Cerenkov radiation from the
suprathermal auroral electron beam, possibly amplified by coherent processes [Maggs,
1976]. Because auroral hiss is generally regarded as a reliable indicator of suprathermal
electron beams in the Earth's auroral zones, the observation of auroral hiss at the edge
of the Io plasma torus is strongly suggestive of intense electron beams in this region of
the Jovian magnetosphere. Most likely these electron beams would be associated with
a Birkeland (field-aligned) current linking the Io torus with the Jovian ionosphere as
illustrated in Figure 8.18. Because the auroral hiss appears to be propagating toward
the equatorial plane, the electron motion would have to be in the same direction, corresponding to a field-aligned current directed away from the equatorial plane. At the

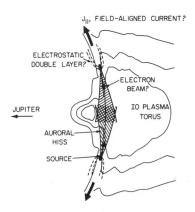

Fig. 8.18. A schematic illustration of the auroral hiss source region inferred from the low-frequency cutoff of the auroral hiss. Based on analogy with auroral hiss observed in the Earth's magnetosphere, this noise indicates the presence of an intense beam of low-energy (10 eV to 1 keV) electrons at the inner edge of the hot torus, possibly accelerated in a region of parallel electric fields (electrostatic double layer) near the edge of the torus.

present time, no evidence is available supporting the existence of such a current. However, the magnetic field is quite strong in this region, and the magnetic field perturbation caused by an electron beam of 10^8 to 10^{10} electrons cm^{-2} would be difficult to detect with the Voyager magnetometer (M. Acuña, personal communication).

If field-aligned currents are present at the inner edge of the Io torus, these currents would probably be caused by a divergence in the perpendicular current system at the torus boundary, with a resultant charge build up and acceleration of particles along the magnetic field lines. The acceleration process would be very similar to the terrestrial auroral electron acceleration, with the torus plasma playing a role similar to the terrestrial ionosphere. The abrupt termination of the auroral hiss source at a distance of about $0.65\ R_J$ from the equator may in fact be directly indicative of the electron acceleration region. For terrestrial auroral hiss the sharp V-shaped low frequency cutoff has been widely interpreted as giving the location of the auroral electron acceleration region [Shawhan, 1979], possibly associated with the formation of an electrostatic double layer as illustrated in Figure 8.18.

8.6. Broadband electrostatic noise

During both the inbound and outbound passes of Voyager 1 and 2 several regions of intense low-frequency electric field noise were encountered near the outer boundary of the plasma sheet [Barbosa et al., 1981] with characteristics very similar to a type of noise observed in the Earth's magnetosphere called broadband electrostatic noise [Gurnett, Frank, and Lepping, 1976]. A representative example of this noise is shown in Figure 8.19. Typically, this noise has a broad frequency spectrum extending more or less monotonically from less than 10 Hz to greater than 1 kHz. A distinguishing characteristic of this noise is the spiky appearance in the wideband frequency-time spectrograms, with many impulsive vertical features, as in the bottom panel of Figure 8.19. It is this spiky appearance that leads us to believe that the noise is electrostatic; the comparable type of noise in the Earth's magnetosphere has a negligible magnetic component and is also spiky. Although the noise occurs in a frequency range where the whistler-mode can propagate, it is unlikely that the noise is propagating entirely in the whistler mode because the magnetic field is very small and the noise sometimes extends to frequencies well above the electron cyclotron frequency. Unfortunately, even in the Earth's magnetosphere we have little precise knowledge of the mode of propagation. The very broad frequency spectrum and spiky appearance suggest that the noise may consist of short wavelength waves that are strongly Doppler shifted.

In the Earth's magnetosphere the broadband electrostatic noise is known to be associated with boundaries that carry large field-aligned currents, such as the bound-

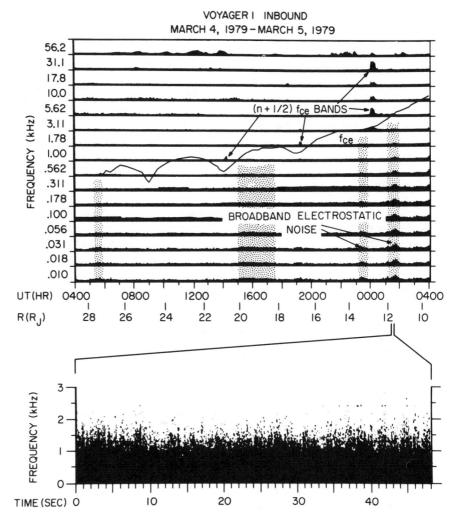

Fig. 8.19. An example of the broadband-electrostatic noise detected near the outer boundary of the plasma sheet. This noise has a broad frequency spectrum and is very spiky, indicating that the noise consists of short-wavelength electrostatic waves.

ary of the plasma sheet [Gurnett, Frank, and Lepping, 1976], the auroral field lines [Gurnett and Frank, 1977], the polar cusp [Gurnett and Frank, 1978b] and the magnetopause [Gurnett et al., 1979a]. At Jupiter, the broadband electric-field noise also appears to be associated with boundaries. Barbosa et al. [1981] demonstrate that broadband electrostatic noise occurs near the outer boundary of the Jovian plasma sheet. For example, the two regions of broadband electric field noise at 2320 and 0140 UT in Figure 8.19 are symmetrically located north and south of the magnetic equator (which can be identified from the $3 f_{ce}/2$ and UHR emissions). The broadband electrostatic noise events occur at peaks in the plasma density that have been interpreted as a bifurcation in the north–south density profile of the plasma sheet [McNutt, Belcher, and Bridge, 1981]. During the dayside crossing of the magnetopause by Voyager 2 a similar type of broadband noise was also identified near the magnetopause [Gurnett, Kurth, and Scarf, 1979a]. The overall morphology of the broadband electrostatic noise is,

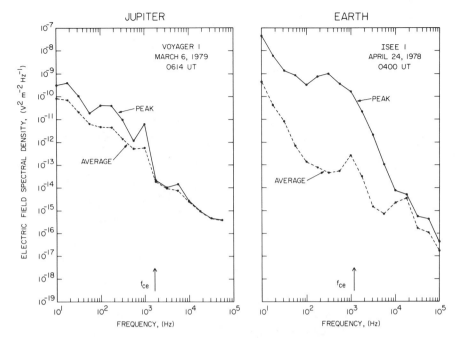

Fig. 8.20. A comparison of the spectra of broadband-electric-field noise observed near the boundary of the Jovian and terrestrial plasma sheets. The Jovian broadband-electrostatic noise generally occurs at lower frequencies and has lower intensities than the comparable noise in the terrestrial magnetosphere [from Barbosa et al., 1981].

therefore, quite similar at Earth and Jupiter, with a strong tendency for the noise to occur near boundaries that carry field-aligned currents.

A comparison of the frequency spectrum of the broadband electrostatic noise at Earth and Jupiter is shown in Figure 8.20. In both cases the spectra were obtained from the outer boundary of the plasma sheet on the evening side of the magnetosphere. As can be seen, the form of the frequency spectrum is very similar at both planets, with the intensity generally decreasing toward increasing frequency approximately as $f^{-2.0}$. In some cases, a distinct break occurs in the slope of the spectrum near the electron gyrofrequency. The electric-field intensities at Jupiter are usually significantly smaller than at Earth, as if the basic spectrum were shifted downward in frequency by about a factor of 10. A downward shift in the frequencies at Jupiter is probably not surprising as all of the characteristic frequencies of the plasma in the Jovian plasma sheet are substantially lower than in the terrestrial plasma sheet.

Among the various magnetospheric plasma wave processes that have been discussed, the broadband electrostatic noise is probably the least well understood. At present, only two mechanisms have been proposed to explain this noise. Ashour-Abdalla and Thorne [1977] have suggested that the noise may be caused by electrostatic ion-cyclotron waves driven by a loss-cone instability, and Huba, Gladd, and Papadopoulos [1978] have suggested that the noise may be caused by a lower-hybrid drift instability driven by the cross-field current. Both of these mechanisms would require large Doppler shifts to explain the observed frequency spectra. Because the noise seems to be associated with field-aligned currents, it also appears that current-driven instabilities, such as the current-driven ion-cyclotron instability [Kindel and Kennel, 1971] should be considered. Electrostatic ion-cyclotron waves have been identified in association with auroral field-aligned currents at relatively low altitudes in

Fig. 8.21. Evidence of intense
electrostatic waves in the Io plasma
torus. The strong impulsive bursts
extending from ~ 3 to 10 kHz in the top
panel are believed to be short wavelength
electrostatic waves. These waves occur
in a region where the energetic ion
phase-space density decreases rapidly,
suggesting that these waves may cause
precipitation of energetic ions.

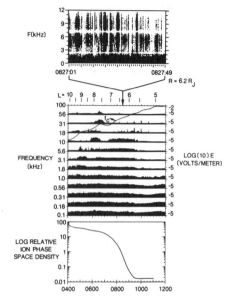

the Earth's magnetosphere [Kintner, Kelley, and Mozer, 1978]. Whether waves of this type can account for the broadband electrostatic noise observed at much higher altitudes in the magnetosphere remains to be determined.

Probably the most important implication of the Jovian broadband electrostatic noise is that this noise provides indirect evidence of substantial field-aligned currents flowing along the boundary of the plasma sheet. In the Earth's magnetosphere a substantial body of evidence suggests that the broadband electrostatic noise is associated with field-aligned currents [Gurnett and Frank, 1977]. Although a full understanding of the plasma instability involved in the generation of this noise is not yet available, it can reasonably be presumed to be a current-driven instability. The observation of broadband electrostatic noise along the outer boundary of the Jovian plasma sheet is in close agreement with terrestrial observations, suggesting that a field-aligned current system links the Jovian plasma sheet to the auroral ionosphere in a manner similar to the auroral current system in the terrestrial magnetosphere.

8.7. Discussion

Although the general characteristics of the Jovian plasma wave phenomena can now be regarded as reasonably well understood, numerous detailed observational and theoretical questions remain. For example, although detailed particle loss rates have been computed, it is still not certain that the chorus and hiss emissions are entirely responsible for the particle precipitation and aurora observed at the foot of the torus field lines. Goertz [1980a] has pointed out that the energetic electron diffusion and energy transport rate into the torus is too small by about a factor of 10 to account for the energy precipitated in the aurora (see Chap. 12). Thus, it is questionable whether electron scattering by whistler-mode waves in the torus can account for the torus aurora. It appears more likely that proton precipitation by electromagnetic ion-cyclotron waves or some other type of wave-particle interaction could account for the observed auroral light intensities. Unfortunately, the Voyager plasma-wave instrument does not extend to sufficiently low frequencies to determine whether ion-cyclotron waves are present. In a possibly related development, Scarf, Gurnett, and Kurth [1981] have recently pre-

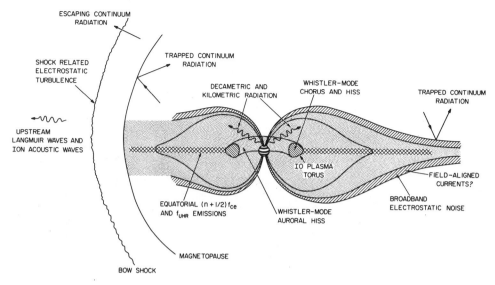

Fig. 8.22. A summary of the principal types of plasma waves detected by Voyager and their regions of occurrence.

sented observations suggesting that significant levels of electrostatic noise may also exist in the torus, superimposed on the whistler-mode hiss. The top panel of Figure 8.21 shows one example of the impulsive spectral characteristics that are believed to be indicative of short wavelength electrostatic waves in the Io plasma torus. The broad bandwidth of the individual bursts could be caused by large Doppler shifts as the corotating plasma sweeps past the spacecraft. The central panel in Figure 8.21 shows part of the 16 channel survey data during the inbound passage through the Io torus. The uniformly enhanced wave levels detected in the 3.1 to 10 kHz channels before about 0927 UT are thought to be caused by these broadband electrostatic noise bursts.

The possible role that this electrostatic noise could play in the torus dynamics is not understood. However, the phase space density plot in the bottom panel in Figure 8.21 [adapted from Armstrong et al., 1981] shows that significant ion precipitation developed in the region where the electrostatic noise is present. Because the ion losses stopped just where the wave turbulence ended, it is quite possible that this electrostatic noise plays a role in precipitating energetic ions and thereby producing the aurora at the foot of the Io field lines.

Another unresolved issue involves the observation of auroral hiss at the inner edge of the Io torus. As discussed earlier, the presence of auroral hiss is thought to indicate the presence of field-aligned particle beams and currents in this region, possibly including the formation of electrostatic double layers along the field lines. Is it possible that these particle beams and currents contribute to the torus aurora? At the present time, essentially nothing is known about the suprathermal particles responsible for the auroral hiss, other than what can be concluded from the general similarity to terrestrial auroral hiss.

Substantial questions still remain concerning the continuum radiation observed in the Jovian magnetosphere. How is this radiation generated? Is the mode conversion discussed by Melrose [1981] adequate for explaining the continuum radiation, or is some other process necessary? What is the Q factor of the magnetospheric cavity for the trapped continuum radiation, and how much radiation escapes down the tail?

Table 8.2. *Jovian plasma waves*

Name	Region of occurrence	Frequency range	Representative broadband electric field strength	Plasma wave mode	Free energy source
Upstream ion-acoustic waves	Upstream of bow shock	$f \simeq (V_s/\lambda)\cos\theta$, 100 Hz to 3 kHz	10 to 30 μV m^{-1}	Ion-acoustic-like mode	Ion beams from bow shock
Upstream electron plasma oscillations	Upstream of bow shock	$f \simeq f_{pe}$, $\sim$3 to 5 kHz	10 to 100 μV m^{-1} (very impulsive)	Langmuir mode	Electron beam from bow shock
Bow shock noise	Bow shock	Broadband 10 Hz to $\sim$10 kHz	1 to 3 mV m^{-1}	Ion-acoustic or Buneman mode, whistler mode and Langmuir mode	Current in bow shock, beams and anisotropies generated in shock
Continuum radiation	Magnetospheric cavity	$f > f_{pe}$, 20 Hz to $\sim$10 kHz	See Fig. 8.8	Free space (R,X) and (L,O) modes	Coupling from electro-static UHR emissions
UHR emissions	Magnetic equator and Io torus	$(n + 1/2)f_{ce} \simeq f_{UHR}$, 1 to 400 kHz	3 to 10 mV m^{-1}	Z-mode, or electron Bernstein mode at UHR	$\partial f_e/\partial v_\perp > 0$, instability depends on n_c/n_H
Electron cyclotron waves (a) Discrete	Magnetic equator	$f = 3/2 f_{ce}$, $5/2 f_{ce}$; 1 to 30 kHz	1 to 3 mV m^{-1}	Electron Bernstein mode	$\partial f_e/\partial v_\perp > 0$, instability depends on n_c/n_H

(b) Diffuse	Io plasma torus	$f = (n + 1/2) f_{ce}$ near f_{LHR}, 50 to 500 kHz		Electron Bernstein mode	Thermal excitation, or $\partial f_e/\partial v_\perp > 0$
Whistlers	Io plasma torus	$f \approx f_{ce}$, 1 to 7 kHz	30 μV m^{-1}	Whistler mode	Lightning
Chorus	Io plasma torus	Near $f_{ce}/2$, ~ 8 to 12 kHz	0.3 mV m^{-1}	Whistler mode	Electron loss cone anisotropy, resonant energy few keV
Hiss	Io plasma torus	$f \lesssim f_{ce}$, 10 Hz to 10 kHz	1 to 2 mV m^{-1}	Whistler mode	Electron loss cone anisotropy, resonant energy few hundred keV
Auroral hiss	Io plasma torus	$f \lesssim f_{ce}$, 1 to 10 kHz	10 to 30 μV m^{-1}	Whistler mode	Electron beam, 10 eV to 10 keV
Broadband electrostatic noise	Outer boundary of plasma sheet	10 Hz to 1 kHz	See Figure 8.20	Unknown, possibly ion cyclotron or LHR drift mode	Unknown, probably field-aligned currents

Could the tail be sufficiently long to allow the "trapped" continuum radiation to escape freely out the tail in regions of low solar wind plasma density downstream of Jupiter?

Because of limitations of the Voyager instrumentation, the electron distribution functions responsible for the $(n + 1/2)f_{ce}$ and f_{UHR} waves have not yet been determined. A basic question, therefore, remains as to why these waves are so closely confined to the magnetic equator. Is the equatorial confinement indicative of a highly anisotropic angular distribution that occurs only at the magnetic equator, or is the equatorial confinement a propagation effect? Can equatorial upper hybrid emissions account for the narrowband kilometric radiation by processes similar to the generation of the continuum radiation? Similarly, in the plasma sheet basic questions remain concerning the broadband electrostatic noise. How is this noise generated? Is the noise indicative of field-aligned currents linking the plasma sheet to the auroral ionosphere? If so, what role does this turbulence play in high latitude auroral processes? Does the noise produce sufficient anamolous resistivity to cause field-aligned potential drops, ion-acceleration, and plasma heating effects similar to the processes believed to be occurring in the terrestrial magnetosphere?

In conclusion, it is clear that the Voyager observations have added greatly to our knowledge of plasma wave and radio emission processes in the Jovian magnetosphere. Table 8.2 and Figure 8.22 summarize the many types of waves observed and their regions of occurrence. Although in most cases a first-order understanding is available to account for the generation of these waves, questions still remain concerning the possible role that these waves play in the physics of the Jovian magnetosphere.

ACKNOWLEDGMENTS

We would like to extend our thanks to W. Kurth, R. R. Anderson, and R. West for their assistance in preparing illustrations of the Voyager and ISEE data and for their helpful discussions concerning the interpretation of the data.

The research at The University of Iowa was supported by NASA through Contract 954013 with the Jet Propulsion Laboratory and through Grants NGL-16-001-002 and NGL-16-001-043 with NASA Headquarters and by the Office of Naval Research. The research at TRW was supported by NASA through Contract 954012 with the Jet Propulsion Laboratory.

9

THEORIES OF RADIO EMISSIONS AND PLASMA WAVES

Melvyn L. Goldstein and C. K. Goertz

A generally accepted theory of the enigmatic phenomenon of planetary radio emission is not yet available. In this chapter, we direct our attention primarily to the question of how the Jovian decameter radiation might be generated via both direct and indirect mechanisms. Direct mechanisms transform the free energy contained in an electron distribution (typically a loss-cone) directly into electromagnetic waves. Indirect mechanisms transform the free energy contained in an electron beam distribution first into electrostatic waves that can then couple, in some manner, to produce electromagnetic waves. The growth rates for the unstable electromagnetic and electrostatic waves are derived. Nonlinear theories are briefly discussed as they apply to the case of Jupiter's decametric radiation. Because most of the Jovian radio emission seems to be controlled by Io, we describe how Io, through the emission of kinetic Alfvén waves, can produce a "beamlike" electron distribution. It is more difficult to understand how Io can enhance or produce a "loss-cone" distribution. Thus we conclude that, at least for Jovian radio phenomena, indirect mechanisms are preferred. We also describe theories and models for the generation of the dynamic spectral arcs that characterize the radio spectrum from hectometric to decametric wavelengths.

9.1. Introduction

Jupiter is the most powerful planetary source of nonthermal electromagnetic radiation in the solar system, with a radio spectrum extending from a few kHz to over 100 MHz. The phenomenology of the decimeter component in the GHz range has been discussed in Chapter 7. Here we concentrate on the complex region of Jupiter's spectrum at decameter wavelengths, the so-called DAM. The relevant observations are described in Chapters 7 and 8 and we reference them freely.

Theoretical interpretations of Jupiter's low frequency radio emissions have been plentiful and imaginative, ranging from the suggestion by Vasil'ev, Volovik, and Zalyubovskii [1972] that extensive air showers of ultrarelativistic cosmic rays impinging on the Jovian atmosphere coherently excite the observed radio frequencies, to more prosaic proposals relating the observations to plasma instabilities well known from laboratory and space physics. However, owing perhaps to the rich and detailed phenomenology and fascinating morphology, or perhaps to our limited but rapidly increasing knowledge of the physical environment of Jupiter's magnetosphere and upper ionosphere, there are as yet no theories of any of the low frequency radio components that are generally accepted as being completely correct. This is in contrast to the situation at decimeter wavelengths, where, as pointed out in Chapter 7, a rather coherent picture of the emission mechanism has been formulated. As we shall see, however, some aspects of the observations at decameter wavelengths do appear consistent with at least some of the theories.

There is something of a renaissance of interest in this area of research as the results of the Voyager missions are digested and assimilated. In the following discussion, we emphasize the basic theoretical ideas that underly both the older and newer theories and models, and thus provide a suitable foundation for understanding the latest literature. In this chapter, we follow Smith [1976a] in using the term "phenomenology" to

Fig. 9.1. A schematic view of the flow of information from the Jovian ionosphere, magnetosphere, and Io plasma torus to a remote observer. In many situations, the physical mechanisms for energy release in the plasma environment must be inferred from electromagnetic radiation that is detected far from the source region. Theoretical attempts to deduce the primary physical processes controlling the release of energy often proceed by trying to simulate nature by hypothesizing the existence of a source of free energy, then calculating the response of the system, and comparing the result to the observations.

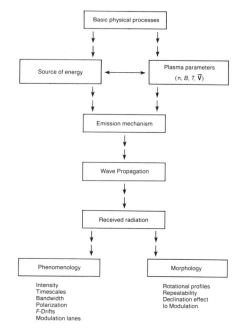

refer to observables (intensity, polarization, frequency bandwidths, etc.), and in reserving "morphology" for interpretive organization of data such as, for example, occurrence probabilities. Generally, we will concentrate on theoretical work completed since Smith's review, referring to earlier work as necessary to make our discussion fairly self-contained.

In principle, the observed radiation may be used to study the relevant physical processes and diagnose the physical conditions (i.e., plasma parameters) at Jupiter. Figure 9.1 shows in a simplified manner the flow of information from Jupiter (its ionosphere, magnetosphere, or Io plasma torus) to a remote observer. By some physical process, a source of energy is released in a plasma. This energy causes radiation by the action of one or more emission mechanisms. If the radiation can propagate to the observer, information about the physical processes, energy exchange, and plasma environment can be obtained in a coded way. The decoding of this information is obtained by running back through the scheme of Figure 9.1. Evidently this inversion process can easily lead to serious errors because the number of unknown variables exceeds the number of observed independent parameters. Thus, theorists often go the other way, trying to simulate what happens in nature. Starting with a physical process, for example, the interaction of Io with the magnetosphere, they calculate the resulting free energy (residing either in spatial or velocity-space inhomogeneities of the plasma distribution). Once the sources of free energy have been identified, one tries to estimate the efficiency by which this energy is transformed into observable radiation. Sometimes, but not always, propagation and refraction of the radiation through the often inhomogeneous medium between the source and observer is calculated.

A few general comments about the decameter observations can be made before getting into detailed theoretical formulations. Some of these comments have been made before [Smith, 1976a], but are worth repeating. The high intensities, limited bandwidth, short timescales, and sporadic nature of the radio emissions indicate that they are due to stimulated emission from plasma microinstabilities. In addition, the remarkable repeatability of many of the decameter and kilometer wavelength spectra imply a long term stability of at least some of the plasma parameters. However, much variability is still present, including source drifts in longitudes by up to 10°. Although the basic plasma parameters may be stable over long periods of time, it must be stressed that the decameter radiation is not always observed even when the viewing geometry is propitious. This is apparently because of variations in the intensities and frequency bandwidths of the various sources; the Planetary Radio Astronomy (PRA) experiment on Voyagers 1 and 2 demonstrated that at sufficiently low signal strengths the occurrence probability in CML is nearly unity at almost all longitudes at some frequency (cf., Fig. 7.11). This is true of both the Io phase-independent and phase-dependent components. The variations in intensity and bandwidth are, however, significant, indicating that there must be some variable and random influence on the propagation characteristics of the medium, emission mechanism, or physical processes that generate the free energy. The distinguishing characteristics of the Io phase dependent and independent sources are described in detail in Chapter 7 (in particular see the discussion of Fig. 7.8).

Nearly all of the decameter radiation appears to be controlled by Io in some manner. Apparently either Io itself, or its plasma torus, modifies the plasma distribution (in configuration or velocity space) in such a way that free energy is converted into electromagnetic waves. In addition, Io probably enhances or creates sources of free energy, thus accounting for the greater intensities of some of the Io-controlled sources in comparison with the Io-phase-independent sources. It has become increasingly clear from both ground-based and space probe studies of the plasma torus that Io is ultimately responsible for many, if not all, of the dynamical phenomena in the inner Jovian magnetosphere. Hence, the traditional division in decameter morphology between Io-independent and Io-controlled sources is something of a misnomer. Evidently the essential distinction is whether or not the particular decameter phenomenon is correlated with the phase of Io. Thus, we will occasionally refer to the DAM sources as being either Io-phase independent or Io-phase dependent.

The repeatability of DAM and HOM spectra suggests that the frequencies are related to a characteristic frequency of the plasma. The possibilities are the electron and ion cyclotron frequencies, plasma frequencies, and upper- and lower-hybrid frequencies. The characteristic ion frequencies and the lower-hybrid frequency, are too low to control the emission of megahertz electromagnetic waves, although in Section 9.3 we will see that these low frequency waves may play a crucial role in the chain of events leading to the observed electromagnetic radiation. The highest electron number densities inferred from radio occultation measurements [Fjeldbo et al., 1975, 1976; Eshleman et al., 1979a,b; Chap. 2] correspond to $f_{pe} \simeq 3$ MHz, which not only excludes Langmuir waves as the source of the decameter emissions, but also implies that the electron cyclotron frequency and upper-hybrid frequencies are very nearly equal. Thus, only the electron cyclotron frequency and upper-hybrid frequency appear to be directly related to the observed spectra.

The argument that the observed radiation is the upper-hybrid or electron cyclotron frequency at the source suggests that the maximum frequency of the Io controlled DAM (39.5 MHz) then corresponds to the maximum magnetic field at the foot of Io's flux tube (IFT), which is inferred from Pioneer 11 observations [Smith, Davis, and Jones, 1976; Acuña and Ness, 1976a; and Chap. 1] to be approximately 14 G in the northern hemisphere. Consequently, any electrons reaching these high fields must have pitch angles inside the steady state loss-cone. We are thus led to the immediate conclusion that stably trapped electrons cannot produce cyclotron frequency radiation at the highest decameter frequencies unless the exciters have been scattered into the loss-cone and/or have been accelerated somewhere along the northern half of the Io flux tube. For this reason, generation mechanisms involving trapped electrons have been proposed only in connection with the Io-phase independent sources [Goldstein and Eviatar, 1979]. Analysis of Voyager PRA data indicates that the historical distinction between the frequency range of the Io-phase independent and Io-phase dependent sources that was based primarily on ground-based observations may actually be a manifestation of differences in the intensities of the two sources as first suggested by Desch [1976]. Barrow and Alexander [1980] and Barrow and Desch [1980] report that the upper-frequency limits of the Io-independent source B can reach 36–38 MHz, very close to the highest observed frequencies of the Io-dependent source (39.5 MHz). For a more complete description of the morphology and phenomenology of these and other aspects of DAM, we refer the reader to Chapter 7.

When the waves are observed, they appear at frequencies much higher than the electron cyclotron frequency at the observer, thus the waves are high frequency waves. In a magnetized plasma there are two high frequency normal modes [Stix, 1962]: the ordinary (O) and the extraordinary (X) mode. The X mode consists of two branches, a slow mode for which the phase velocity is less than the velocity of light (also called the z mode or band II); and a fast mode for which the phase velocity is larger than c (also referred to as band III or simply the X mode). Between the two branches there is a "stop band" where the electromagnetic X-mode wave becomes evanescent and cannot propagate. Consequently, the slow X mode cannot easily escape from Jupiter. For the ordinary mode there is no stop band above the local electron plasma frequency. Therefore, far from Jupiter the observed radiation must be in either the fast X mode or O mode. Observations indicate that the observed radiation is X mode. However, this does not prove that it is generated in that mode because strong mode coupling between the point of observation and the source is possible.

We can summarize the theoretical problem of decameter radio waves as follows: by what sporadic mechanisms is the plasma distribution modified so that its free energy is converted with adequate efficiency into electromagnetic waves near the local electron cyclotron frequency and how does this radiation propagate through the intervening plasma (including the Io plasma torus) to a distant observer? One of the keys to answering this question is the in situ form of the electron distribution function. This is not easily determined, primarily because there are no observations of the electron distribution in the suspected source region, the Jovian ionosphere. Much of our discussion below will concentrate on the question of what clues are contained in the extant observations that permit models of the distribution function to be constructed. Because the Io flux tube appears to be the likely location of the energetic electrons that generate the decameter radiation, we discuss several properties of that region that can give rise to unstable distributions. Several theories for the emission process yield similar observable characteristics, even though they require quite different distribution functions. We try to distinguish those aspects of the various theories that seem more plausible than others.

The various theories can be distinguished by whether they depend on direct or indirect emission processes. The term "direct emission" refers to particle-generated electromagnetic waves; whereas the term "indirect emission" refers to wave-wave interactions and scattering (mode conversion). Until recently most of the literature dealing with Jovian emissions has concentrated on direct linear emission mechanisms. In contrast, the terrestrial kilometric radiation (TKR), which is similar in many respects to Jovian radiation, has spawned several nonlinear theories of its generation. Similar theoretical tools are now being applied to Jupiter's radio sources.

In the following section, we review the linear plasma theory that has been utilized in constructing "direct emission" theories, and follow that with a review of several applications to the DAM problem. In Section 9.3 we review some of the techniques used in nonlinear (indirect) emission theories and emphasize applications to Jupiter. Much of the theoretical formalism we develop is also useful for discussions of a number of other Jovian radio and plasma phenomena besides DAM. In Chapter 12, the linear instability formalism is employed to describe several microscopic plasma processes that give rise to particle diffusion and whistler wave generation.

In several instances, newly discovered components of Jupiter's radio spectrum, in particular, the nKOM and bKOM components described in Chapter 7, are only just beginning to inspire theoretical attention, and thus a "review" would be premature. We have therefore chosen to concentrate on the microscopic conditions that can produce DAM. Furthermore, we emphasize work that has been completed since Smith's excellent review [Smith, 1976a], at the risk of being perhaps too uncritical of work that has been available to us only in preprint form. Where possible, we try to relate the formalism to such macroscopic phenomena as occurrence probability and source location, but we make no claims of completeness.

9.2. Linear theories

Direct emission mechanisms

Plasmas are characterized by a multitude of instabilities that occur if free energy, stored in thermal energy, kinetic energy, ordered motion, potential mechanical energy, electrostatic energy, etc, exceeds a critical level and becomes transformed into other forms including radiation. In a typical situation, a small perturbation in some plasma parameter, or a spontaneously emitted plasma wave, grows exponentially with time ($\sim \exp \gamma t$), that is, the temporal change of the perturbation is proportional to the perturbation itself. When the growth rate for waves is positive ($\gamma > 0$), we speak of coherent or stimulated emission. In this section and in the appendix, we derive general expressions for the linear growth rate of various wave phenomena, both electromagnetic and electrostatic that are then used here, in Section 9.3, and in Chapters 8 and 12. For an alternative point of view to the one we present, the interested reader should refer to Kennel and Wong [1967] and Baldwin, Bernstein, and Weenik [1969]. Both those references proceed from the linearized Vlasov equation and include all wave modes and directions of propagation. The derivation by Melrose [1968] and Harris [1968] is also general in that all wave modes and directions of propagation are included, but is couched in terms of the semiclassical theory of emission and absorption of waves. Thus, both stimulated and spontaneous emission can be included in a straightforward way. In the Vlasov treatment, the inclusion of spontaneous emission is not at all a trivial exercize, and the reader is referred to Wu [1968] for details. In the following discussion, we use the approach and notation of Melrose [1968].

We treat a homogeneous magnetized plasma which is both gyrotropic (i.e., the distribution functions are assumed to be independent of azimuthal direction about the magnetic field [see, for example, Montgomery and Tidman, 1964]) and collisionless. Wave amplitudes are assumed to be small enough to leave the particle orbits essentially unperturbed. The waves are harmonic and are not significantly damped by the background medium. We deal only with the normal modes that exist in a (cold) magnetized plasma containing no free energy. The normal modes of interest to us here and in Section 9.3 include the upper- and lower-hybrid electostatic modes, and the ordinary and extraordinary electromagnetic modes. The addition of a population of particles not in thermal equilibrium introduces a non-Hermitian part to the dielectric of the medium. It is this part that implies growth (or damping). In principle, observations (Chap. 7) could determine which mode the radiation is in, and theory could determine the form of the distribution function needed to amplify that mode. As we shall see, however, the situation is rarely that simple. Various theoretical mechanisms are capable of generating the same wave mode, and often, unless the source location is known with some precision, observations are incapable of determining the original wave mode at the source.

The linear growth rate γ for arbitrary direction of propagation and for all polarizations is derived in the appendix and is given by

$$\frac{\gamma^{\pm}(k)}{\omega} = \sum_{\nu} \frac{2\pi^2 q^2 c^2}{\omega^2 |k_{\parallel}|} \, k_{\parallel} \, n^{\pm} \int dp_{\parallel} \tag{9.1}$$

$$\cdot \int dp_{\perp} p_{\perp}^2 \, \delta(p_{\parallel} - p_{\parallel 0}) \, \frac{\theta_{\nu}^{\pm} \, G_{\nu}^{\pm} \, f^{\pm}(p_{\perp}, p_{\parallel})}{(m^2 c^4 + p^2 c^2)^{1/2}}$$

where

$$\theta_{\nu}^{\pm} = |J_{\nu \pm 1}(z) E_{\mathrm{R}} + J_{\nu \mp 1}(z) E_{\mathrm{L}} \tag{9.2}$$

$$+ (2)^{1/2} \frac{v_{\parallel} k_{\perp}}{v_{\perp} |k_{\perp}|} \, J_{\nu}(z) E_{\parallel}|^2 / W(k_{\perp}, k_{\parallel})$$

indicates the polarization of the particular mode under consideration, and

$$W = (|E(\mathbf{k})|^2 \omega \, \partial \Lambda / \partial \omega) / (4\pi \lambda_{ss})$$

is the wave energy as a function of parallel ($k_{\parallel}$) and perpendicular wave number ($k_{\perp}$). E is the wave electric field strength. The subscripts R and L refer to right-handed and left-handed electromagnetic modes, and the subscript $\parallel$ refers to the longitudinal polarization of the electrostatic modes. Note that the magnitude of $E(\mathbf{k})$ cancels in (9.2), as it should. Λ, λ_{ss} as well as the other notation follows Melrose [1968] and is defined in the appendix.

The sign of the growth rate is determined by the operator G which acts on the particle distribution function. G is given by

$$G_{\nu}^{\pm} = \left[k_{\parallel} p_{\perp} \frac{\partial}{\partial p_{\parallel}} + (\omega \gamma_r m^{\pm} - k_{\parallel} p_{\parallel}) \frac{\partial}{\partial p_{\perp}} \right] \tag{9.3}$$

Amplification ($\gamma > 0$) is only possible if $G_{\nu}^{\pm} f(\mathbf{p}) > 0$ for some values of $\mathbf{p}$. The interpretation of this criterion is given below.

The resonance condition for the parallel momentum in (9.1) can be written as

$$p_{\parallel 0} = \gamma_r^{\pm} m^{\pm} \beta_{\parallel 0} \, c = [1 - \nu \omega_c^{\pm} / \gamma_r^{\pm} \omega] \, (\omega / k_{\parallel}) m^{\pm} \gamma_r^{\pm} \tag{9.4}$$

or

$$v_{\parallel 0} = (\omega / k_{\parallel}) (1 - \nu \omega_c^{\pm} / \gamma_r^{\pm} \omega)$$

where $\gamma_r = (1 - v^2/c^2)^{-1/2}$ is the Lorentz factor and the phase velocity of the wave along the magnetic field is $\omega / k_{\parallel}$. The $\pm$ superscript refers to contributions from ions (+) or electrons (−).

We can now specify general rules concerning properties of the distribution function necessary for amplification. The delta function in (9.1) indicates a resonance condition that requires that the Doppler-shifted wave frequency must approximately equal ν times the relativistic cyclotron frequency of the resonant particle. The integer ν indicates the change of perpendicular energy of a particle upon emission of a photon. If $\nu = 0$, the change is zero, which is the well-known Cerenkov emission, or Landau resonance, in which only the parallel energy of the particle is changed. The resonances for $\nu \neq 0$ are known as cyclotron (or gyro-) resonances. Positive ν, corresponding to a decrease of $p_\perp$, is called the normal Doppler effect; whereas negative ν corresponding to an increase of $p_\perp$, is called the the anomalous Doppler effect. From the form of G we find that Cerenkov emission is amplified only if the distribution function has a positive slope in $p_\parallel$, that is, satisfies a Penrose criterion [Penrose, 1960]. We call these distributions beamlike because such distributions contain free energy in the parallel velocities and more particles can feed energy into waves than absorb energy from them. This situation is equivalent to population inversion in maser theory except that the population levels in the plasma are continuous.

For positive ν we can get amplification from either a beamlike distribution with excess free energy parallel to the magnetic field **B**, or a loss-cone type distribution with more free energy perpendicular to **B**. The loss-cone distributions contain more particles feeding energy into the wave (when $p_\perp$ decreases) than absorbing energy from the wave (when $p_\perp$ increases). Such distributions are equivalent to the population inversion of a maser. In both cases the population levels are quantized, although for all practical purposes in the loss-cone situation the quantization step $\Delta p_\perp^2 = qB/c\hbar$ (see Eq. A.13) is negligible. The distribution can therefore be considered continuous in $p_\perp$. Similarly, for the anomalous Doppler-shifted emission, instability can result from either a beamlike or an antiloss-cone distribution.

If we use the resonance condition (9.4), G can be rewritten as

$$G_\nu^\pm f^\pm (p_\perp, p_\parallel) = \left[k_\parallel p_\perp \frac{\partial}{\partial p_\parallel} + \nu \omega_c^\pm m^\pm \frac{\partial}{\partial p_\perp} \right] f^\pm (p_\perp, p_\parallel) \Bigg|_{p_\parallel = p_{\parallel 0}} \quad (9.5)$$

The instability conditions for electrons can then be explicitly summarized as follows:

For the Landau resonance, $\nu = 0$:

$$\frac{\partial f}{\partial p_\parallel} \Bigg|_{p_\parallel = p_{\parallel 0}} > 0 \qquad \text{"beam"}$$

For the "normal" Doppler resonance, $\nu > 0$:

$$\left[p_\perp \frac{\partial}{\partial p_\parallel} + \frac{\nu \omega_{ce}}{\omega} m v_{ph} \frac{\partial}{\partial p_\perp} \right] f \Bigg|_{p_\parallel = p_{\parallel 0}} > 0 \qquad \begin{array}{l} \text{"loss-cone"} \\ \text{and/or} \\ \text{"beam"} \end{array}$$

For the "anomalous" Doppler resonance, $\nu < 0$:

$$\left[p_\perp \frac{\partial}{\partial p_\parallel} - \frac{|\nu| \omega_{ce}}{\omega} m v_{ph} \frac{\partial}{\partial p_\perp} \right] f \Bigg|_{p_\parallel = p_{\parallel 0}} > 0 \qquad \begin{array}{l} \text{"anti-loss-cone"} \\ \text{and/or} \\ \text{"beam"} \end{array}$$

Whether a distribution is a loss-cone or antiloss-cone is determined by the sign of $\partial f/\partial p_\perp$ at resonance, that is,

$$\frac{\partial f}{\partial p_\perp}\bigg|_{p_\| = p_{\|0}} \begin{array}{l} > 0 \\ < 0 \end{array} \quad \begin{array}{l} \text{loss-cone distribution} \\ \text{antiloss-cone distribution} \end{array} \tag{9.6}$$

Obviously, an extensive study of (9.1) is beyond the scope of any single review. Instead of attempting to describe all possibilities, we restrict ourselves to two extreme cases: exactly parallel propagation ($k_\perp = 0$) and nearly perpendicular propagation ($k_\| \simeq 0$). The discussion of these two cases will serve to illustrate the basic principles of the linear analysis and introduce the sometimes confusing terminology used in the literature. For detailed investigations including obliquely propagating waves the reader is referred to appropriate references. An example of beamlike distributions unstable at both the Landau and normal Doppler resonances to the excitation of electrostatic waves ($E_\|$) is described in Section 9.3. Other examples dealing with electromagnetic waves are referred to below and are also treated in Chapter 12.

Parallel propagation

Let us look at the case of the parallel or quasiparallel propagating waves. In this case the argument of the Bessel functions in $\theta_r^\pm$ equals zero. Therefore, electrons can resonate with longitudinally polarized waves only when $\nu = 0$ (because only J_0 is nonzero if $z = 0$). Similarly, electrons resonate with right hand (RH) polarized waves if $\nu = 1$ and with left hand (LH) polarized waves if $\nu = -1$. In Section 9.3, we discuss an example where the $\nu = 0$ resonance excites obliquely propagating longitudinal waves. However, for exact parallel propagation, γ is zero because the term $\partial \Lambda / \partial \omega$ in W approaches infinity when evaluated at the solutions of the normal mode cold plasma dispersion relation.

From (9.1), (9.2), and (9.4) it would appear that the anomalous Doppler resonance ($\nu = -1$) can give rise to linearly unstable LH polarized ordinary mode waves. However, because $\nu < 0$, the resonance condition shows that the parallel resonance velocity is larger than the phase velocity. The phase velocity of the ordinary wave is itself almost always larger than c for frequencies of interest (i.e., $\omega \geq \omega_{ce}$) so that no resonance is possible. This is most easily seen from the cold plasma dispersion relation for the parallel propagating O mode, which is given by [Stix, 1962]

$$n^2 = k_\|^2 c^2 / \omega^2 = \frac{(\omega - \omega_R)(\omega + \omega_L)}{(\omega - \omega_{ci})(\omega + \omega_{ce})} \tag{9.7}$$

where the left hand cut-off frequency is defined by

$$\omega_L = -(1/2)\omega_{ce}\{1 - [1 + 4(\omega_{ce}\omega_{ci} + \omega_{pe}^2)/\omega_{ce}^2]^{1/2}\}$$

and the right hand cut-off frequency is defined by

$$\omega_R = -(1/2)\omega_{ce}\{1 + [1 + 4(\omega_{ce}\omega_{ci} + \omega_{pe}^2)/\omega_{ce}^2]^{1/2}\}$$

The dispersion relation is illustrated in Figure 9.2.

In contrast, amplification of both the slow and fast branches of the X mode is possible via the normal Doppler resonance, $\nu = 1$. On the slow branch ($\omega < \omega_{ce}$), $p_{\|0}$ and v_{ph} have opposite signs (see 9.4) and the amplified waves propagate backwards relative to the beam. As is true of all waves produced on the slow X-mode branch, this radiation cannot freely escape from the source because of the existence of an evanescent stop zone where the wave frequency becomes imaginary. The existence of this stop zone is apparent from this dispersion relation for the parallel propagating X mode [Stix, 1962]:

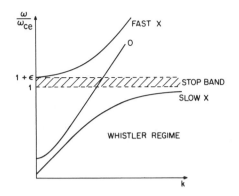

Fig. 9.2a. Dispersion relation for parallel propagating electromagnetic waves. The parameter $\epsilon = \omega_{pe}^2/\omega_{ce}^2$ is assumed to be small.

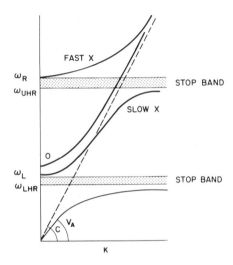

Fig. 9.2b. Dispersion relation for perpendicular propagating modes.

$$n^2 = \frac{k_\parallel^2 c^2}{\omega^2} = \frac{(\omega - \omega_R)^2}{(\omega + \omega_{ci})(\omega - \omega_{ce})} \tag{9.8}$$

This dispersion relation is also illustrated in Figure 9.2. The "whistler" mode lies on the slow branch ($\omega < \omega_{ce}$). Between ω_{ce} and ω_R, the evanescent band ($n^2 < 0$) inhibits the escape of the slow mode. Smith [1976b] has investigated the escape of the slow mode by tunneling through the stop zone and has shown that while this may be possible for radiation below 20 MHz, it is unlikely to occur at higher frequencies.

Another difficulty with backward propagating waves is that because the frequency is close to ω_{ce} they can resonate with thermal electrons and be damped. At the top of the Jovian ionosphere the ambient electron temperature has been estimated by Atreya, Donahue, and Waite [1979] to be 0.1 eV, and the effects of damping in the high frequency source region can be important in some situations (cf. Section 9.3). This mechanism can thus produce whistler mode waves at $\omega < \omega_{ce}$ [see, for example, Chang, 1963]. Observations of whistlers have been reported at Jupiter near $L = 8$ in the equatorial plane at the outer edge of the Io torus [Coroniti et al., 1980a] and are discussed in some detail in Chapters 8 and 12. Growth rates have been calculated using Equation (9.1) [Sentman and Goertz, 1978]. Because the direction of propagation is nearly parallel to **B**, one has no way of knowing if similar radiation is produced near the foot of the Io flux tube without actually placing a receiver in the high-latitude Jovian ionosphere.

From Equation (9.4), we see that on the fast branch ($\omega > \omega_R$) amplified waves propagate forward relative to the particles. Then, because both the density and magnetic field increase with decreasing distance from Jupiter, a wave initially produced above the local cut-off frequency by downward moving electrons will soon propagate into a region where $\omega \simeq \omega_R$, be reflected, and escape the Jovian magnetosphere. Of course, if a particle distribution is directed away from the planet, the wave can escape directly without reflection. However, because **k** is parallel to **B**, such escaping radiation would not in general reach an observer located near the ecliptic plane, and therefore, nearly parallel propagating waves are unlikely to be the source of the decameter radiation.

Perpendicular propagation

For perpendicular or quasiperpendicular propagation, the restrictions on ν are not as severe. This can be seen by setting $k_\parallel = 0$ in (9.1). For the normal Doppler effect ($\nu > 0$), a loss-cone distribution ($\partial f/\partial p_\perp > 0$) is required for amplification, independent of the mode considered. For the anomalous Doppler effect ($\nu < 0$), amplification requires an antiloss-cone distribution ($\partial f/\partial p_\perp < 0$), again independent of mode. However, anomalous Doppler effect interactions at exactly perpendicular propagation are unlikely because the resonance energy required is infinitely large. For the Landau resonance ($\nu = 0$), instability requires a beamlike distribution $\partial f/\partial p_\parallel > 0$).

For a variety of reasons, one of the most frequently hypothesized mechanisms for producing DAM is that of directly amplifying fast X-mode waves propagating nearly perpendicular to **B**. The advantages of such a scheme are obvious: the observed radiation is inferred to be X-mode, compilations of occurrence probability and identification of the two distinct sources (A and B; or Main and Early, respectively – see Chap. 7) certainly suggest (but do not require) that the emission pattern be a thin conical sheet with half angle ψ close to 80° [Dulk, 1965]. This would result naturally from waves excited nearly orthogonal to **B**. Furthermore, direct amplification of the fast X mode eliminates such complications as tunneling and mode conversion. Various versions of such mechanisms have been described by Twiss and Roberts [1958], Twiss [1958], Hirshfield and Bekefi [1963], Fung [1966a,b,c], Goldreich and Lynden-Bell [1969], Melrose [1976], Wu and Freund [1977], and Goldstein and Eviatar [1972, 1979] [Ben-Ari, private communication, has reported an error in the instability analysis in this last reference; consequently we will not discuss that aspect of those two papers]. These references consider instabilities driven by excess perpendicular particle energy arising from either loss-cone distributions, Dirac delta functions in $p_\perp$, or temperature anisotropies.

The cyclotron maser instabilities are examples of the loss-cone driven instabilities discussed above [see, for example, Wu and Lee, 1979; Lee, Kan, and Wu, 1980]. They have been invoked frequently as the excitation mechanism of DAM. However, many of the recent applications of this class of instabilities have been directed toward explaining TKR. In the TKR source it is often assumed that the ambient plasma density n_e is so low that the density of the energetic distribution n_b exceeds it by a large factor [Wu and Lee, 1979; Lee, Kan, and Wu, 1980]. In contrast, while the Jovian source region is also underdense ($\omega_{pe} << \omega_{ce}$), it has generally been assumed that n_e is still large ($\simeq 10^5$ electrons/cm^3), so that $n_e/n_b >> 1$. Lee and Wu [1980] have studied the instability for finite values of n_e/n_b. They computed the growth rate for both the slow and fast extraordinary modes as well as for the ordinary mode from Equation (A.24) or (9.1) using a loss-cone distribution function of the form $F(v_\perp, v_\parallel) = F(v_\perp)F(v_\parallel)$, where

$$F(v_\perp) = \frac{1}{\pi \beta_\perp^2} \left(\frac{v_\perp}{\beta_\perp}\right)^{2\alpha} \exp\left(-\frac{v_\perp^2}{\beta_\perp^2}\right)$$

$$F(v_\parallel) = \frac{1}{(2\pi)^{1/2} V_\parallel} \exp\left[-\left(\frac{v_\parallel - v_0}{2v_\parallel^2}\right)^2\right]$$

and $\beta_\perp^2 = 2T_\perp/\{m_e(\alpha + 1)\}$. The growth rates for both the fast and slow extraordinary modes peak just above ω_R and just below ω_{ce}, respectively. In general, the growth of the slow mode, the one that cannot escape because of the stop zone, greatly exceeds that of the fast mode because the resonance velocity for the slow mode is less than that of the fast mode, which allows many more particles to participate in amplification of the slow mode. The question of whether substantial growth of fast mode waves can actually occur before the instability saturates because of the much faster growth of slow mode waves has not been addressed. The O mode is also unstable, but is not competitive with the two extraordinary mode waves.

The growth of the slow mode waves is related to the excitation of the Weibel instability [Weibel, 1959], which arises from an axial bunching of the energetic electrons (the cyclotron maser instability arises from an azimuthal bunching). Lee and Wu comment that at lower phase velocities the slow mode instabilitiy becomes a Weibel instability. This poses a potential difficulty with the cyclotron maser mechanism. Chu and Hirshfield [1978], in their comparative study of the two types of instabilities, concluded that they were both simultaneously present and always competed with one another. This means that only one branch can be unstable for a given value of $k_\parallel$. In general, the cyclotron maser instability tends to dominate for small $k_\parallel$, which would tend to produce conical emission patterns. Even ignoring for the moment the severe problem of propagating through the stop band, obliquely propagating waves are difficult to reconcile with the conical sheet hypothesis as well as with other observations discussed in Chapter 7. We return to this point in Section 9.4. Nonetheless, in the absence of a detailed comparison between the Weibel and cyclotron mechanisms, it does not appear possible to reach any firm conclusions about the viability of the cyclotron mechanism as the exciter of planetary radio emission.

The loss-cone instability mechanism, taken in a general context to include any instability driven by free energy in perpendicular motion, has several attractive features. For one, in principle it can produce escaping X-mode radiation if it can be shown that growth of the slow mode does not dominate. Loss-cone instabilities are also consistent with Dulk's conical sheet hypothesis, but the other theories also share this property. When combined with models of the global Jovian magnetic field (Chap. 1), this mechanism (as well as others discussed below) provides a possible explanation for the locations of some of the more prominent sources, in particular, sources A and non-Io B [see, for example, Goldstein, Eviatar, and Thieman, 1979; Goldstein and Eviatar, 1979; Oya, Morioka, and Kondo, 1979; Goldstein and Thieman, 1981].

There is a problem, however, with the location of the Io-B source. The origin of this discrepancy could well lie with either the magnetic field models or the emission mechanism. From the analysis by Goldstein, Eviatar, and Thieman [1979] and Goldstein and Thieman [1981], it is clear that neither the Goddard O4 model nor the JPL magnetic field models provide a natural explanation of the Io-B source location within the context of any of the proposed emission mechanisms. In Section 9.4, we return to this problem.

Finally, whereas loss-cone instabilities have of course been invoked to explain observations of TKR, the situation at Jupiter differs in several respects. For example, a satisfactory model of just how Io might drive or enhance a loss-cone process has not been developed.

Summary of direct linear mechanisms

Of the many direct linear mechanisms it is difficult to select one that is the most likely candidate for DAM radiation quite apart from the neglect of nonlinear effects. The properties of the various direct linear mechanisms are summarized in Table 9.1. It appears that processes that generate only the slow X mode must be discarded because of the problem of propagating through the stop zone. The fast X-mode branch will be amplified in the direction of the electron propagation, so that unless the energetic electrons are coming up from the ionosphere, the waves must be reflected before leaving the magnetosphere. The reflection coefficient is certainly less than one, thereby reducing the overall efficiency of the process. The frequency drifts of S-bursts, however, do suggest the possibility of upward propagating electrons (Chap. 7). Any theory that relies on a loss-cone distribution produced by stably trapped particles does not require a downward flux of electrons and hence is not subject to this criticism. However, electrons that reach a 14 G field at the foot of the Io flux tube must be inside the loss-cone. Thus, mechanisms which rely on stably trapped electrons are not applicable to the production of Io controlled DAM.

The highest decameter frequencies can be excited by loss-cone distributions as has been pointed out by Wu and Lee [1979] in their theory of TKR. In their model, precipitating electrons with pitch-angles outside the loss-cone are mirrored so that the albedo distribution shows a depletion near zero pitch angle, that is, the ascending electrons have a loss-conelike distribution. This model has been amplified by Lee, Kan, and Wu [1980] in their paper on TKR. This picture is also a possible candidate for the generation of DAM [Wu and Freund, 1977]. A loss-cone type distribution in the ascending electrons, after mirroring above the ionosphere, could amplify freely escaping quasi-perpendicular fast X-mode waves. Thus, one avoids the problem of reflection alluded to above, but replaces it with another; namely, reduced efficiency because the albedo flux is less than the precipitating flux. As we shall see in the following section, when nonlinear processes are considered, it becomes possible to construct a theory consistent with the conical sheet model without reflecting the particles although still assuming downward propagating beams. This avoids the two losses of efficiency previously discussed. However, the beam driven instabilities require two steps – one to generate electrostatic waves, and a second to up-convert them into fast extraordinary mode waves. This, of course, reduces the overall efficiency of that process. A detailed comparison of the overall efficiencies of loss-cone and beam driven process is yet to be carried out.

The backward propagating ordinary mode in the frequency range $\omega_{ce}/2 < \omega < \omega_{ce}$ requires no reflection and will not encounter a stop band. However, it is elliptically polarized in a LH sense, which is only rarely observed (Chap. 7). Furthermore, as we have pointed out, the growth rates of waves in the ordinary mode are generally less than those of extraordinary waves [Hirshfield and Bekefi, 1963; Lee and Wu, 1980]. At low frequencies (< 1 MHz) the Jovian kilometric radiation (KOM) may well be in the ordinary mode [Warwick et al., 1979b; Green and Gurnett, 1980; Desch and Kaiser, 1980; and Chaps. 7 and 8].

Several of the assumptions made in the derivation of (9.1) impose physical limitations on its application. Interparticle collisions are neglected that could enhance

Table 9.1. *Summary of direct linear mechanisms*

Mode	Frequency range	Direction of propagation	Form of free energy	Comments
Fast x mode or x mode or band III		$k_\parallel \gg k_\perp$	Excess parallel energy (beams)	Does not produce radiation cones
	$\omega > \omega_{ce}$	$k_\parallel \ll k_\perp$	Excess perpendicular energy (loss-cone)	Provides radiation cones. Most commonly assumed linear mechanisms for DAM, HOM
Slow x mode or z mode or band II		$k_\parallel \gg k_\perp$	Loss-cone (and/or beams)	Whistlers encounters stop zone and cannot escape freely
	$\omega < \omega_{ce}$	$k_\parallel \ll k_\perp$	Excess perpendicular energy	Small growth rates Landau damping by thermal particles
Ordinary mode		$k_\parallel \ll k_\perp$	No resonance possible with electrons	
	$\omega_{ci} < \omega < \omega_{ce}$	$k_\parallel \gg k_\perp$	Excess perpendicular energy	Wrong polarization small growth rates, may be responsible for KOM

absorption of electromagnetic waves and would also tend to smooth out velocity-space anisotropies resulting in reduced growth rates. When amplitudes become large, the particle orbit is subject to large deviations from the assumed unperturbed helical trajectory. The linear analysis then must be modified by nonlinear perturbed orbit effects. (For a general discussion of the techniques involved, see Völk [1975].) We have not considered the effects of Landau damping or cyclotron absorption by thermal electrons. Those resonances could quickly damp the slow X mode near the cyclotron frequency. The Landau damping of this (whistler) mode has also been investigated by Menietti and Gurnett [1980] whose results are discussed in Chapter 9.

Indirect emission mechanism

In the previous section, we have seen that an important constraint on the generation mechanisms of DAM radiation is the escape of the radiation. Many linear theories suggest amplification on the slow X branch that cannot escape directly to free space. Oya [1974], Oya, Morioka, and Kondo [1979], Oya, Kondo, and Morioka [1980] have attempted to account for the escape of waves by considering a process which converts the slow extraordinary mode into the escaping ordinary mode. The coupling is found to be especially strong in the vicinity of the point where the wave frequency equals the local plasma frequency (ω_{pe}). The extraordinary mode is obtained as a result of mode coupling between a beam excited upper-hybrid resonance electrostatic wave and plasma inhomogeneities [Zheleznyakov, 1966]. Apart from the severe difficulties inherent in proposing a mechanism whose overall efficiency is proportional to the product of three processes, all of which with intrinsic efficiencies less than one (beam to electrostatic, electrostatic to X mode, and X mode to O mode), Oya's theory requires plasma densities in excess of 10^7 cm^{-3} to meet the condition $\omega = \omega_{pe}$. In the Jovian ionosphere, the maximum plasma frequency is no greater than about 3 MHz (Chap.2). Futhermore, the resulting O-mode radiation requires that the DAM sources lie primarily in the southern hemisphere. This appears contrary to the Voyager observations described in Chapter 7.

We now turn to an investigation of beam-driven instabilities. We will find that some of the difficulties experienced by loss-cone driven processes are absent, only to be replaced by new problems that we have not as yet discussed.

9.3. Nonlinear theories

Indirect emission mechanisms

We have seen that many attempts at explaining the Jovian decameter radiation involve direct excitation of electromagnetic radiation by energetic electrons. Several examples of such theories have been discussed. Common to those theories is the assumption that the energetic electron distribution function contains more free energy in velocity components perpendicular to the magnetic field than parallel to it as is the case for electromagnetic instabilities driven by loss-cone distributions or temperature anisotropies with $T_\perp > T_\parallel$. It is clear from the discussion in Section 9.2 that none of these theories is without its difficulties in accounting for the rich phenomenology of DAM. Recently, it has been suggested that the decameter radiation can arise from indirect processes where the energetic electrons first excite electrostatic waves $(E = E_\parallel)$, which then combine to produce the observed electromagnetic radiation. To some extent these theories are motivated by similar analyses of TKR.

The indirect emission mechanisms start with the assumption that the distribution function of energetic electrons is essentially that of a beam; the excess free energy is parallel to the magnetic field. This is precisely the complement of the examples discussed in Section 9.2.

Application of indirect nonlinear emission mechanisms to the problem of generating DAM has occurred only fairly recently. Earlier, mode coupling theories have been proposed for generation of type III (fast drift) solar radio bursts [Papadopoulos, Goldstein, and Smith, 1974], and generation of TKR [Barbosa, 1976; Maggs, 1978; Roux and Pellat, 1979]. In their 1979 paper, Roux and Pellat suggested that their up-conversion process for the generation of TKR might be adaptable to the Jovian problem. They noted that beam driven instabilities could excite both upper-hybrid and lower-hybrid waves that in turn could couple to produce electromagnetic radiation. In the terrestrial environment, they had concluded that the most appropriate coupling would result in either an ordinary mode wave near ω_{UHR}, or an extraordinary mode wave at $2\omega_{UHR}$ (for a related analysis see Barbosa [1976]). However, in the Jovian magnetosphere, where $\omega_{pe}/\omega_{ce} \ll 1$, Roux and Pellat recognized that the coupling $\omega \geq \omega_{LHR} + \omega_{UHR}$ might produce radiation above ω_R, the right-hand cut-off frequency defined following (9.7); thus giving a freely escaping electromagnetic wave with frequency close to the local gyrofrequency.

Detailed analyses of several nonlinear indirect mechanisms for the production of DAM, including the possibility suggested by Roux and Pellat [1979], have been developed recently. The first is the work of Ben-Ari [1980] who assumed that a beam of electrons was accelerated by the motion of Io (see discussion in Section 9.4). In the source region the density of the beam n_b was assumed less than n_e, in contrast to some of the theories of TKR described above in Section 9.2 . Because the plasma is characterized by $\omega_{ce} \gg \omega_{pe}$, such a "bump-in-tail" distribution is unstable to the excitation of upper-hybrid electrostatic waves. Ben-Ari considered four different resonant three wave interactions that could couple upper-hybrid waves to electromagnetic radiation above ω_R. In a subsequent paper, Goldstein et al. [1982] included the excitation of lower-hybrid waves by the same electron beam and then analyzed the up-conversion of upper-hybrid and lower-hybrid waves in detail.

We first briefly review some aspects of these up-conversion theories. For the purposes of this discussion, we assume the existence of a beam of energetic electrons in the Jovian magnetosphere. In Section 9.4 we discuss a possible way to produce beamlike distributions, but here we proceed without any consideration of the beam generation mechanism and merely explore the consequences of this assumption.

Let $F_e(v_\parallel, v_\perp)$ be the total electron distribution function made up of a cold thermal component (0.1 eV thermal energy), and an energetic beam. It suffices in describing the linear electrostatic instability to approximate the electron motion perpendicular to the magnetic field as cold. Thus, we take $F_e(v_\parallel, v_\perp) = F_e(v_\parallel) \delta(v_\perp) / 2\pi v_\perp$, with $F_e(v_\parallel) = f_e(v_\parallel + (n_b / n_e) f_b (v_\parallel)$. The thermal component with density n_e and the beam with density n_b have distributions denoted by f_e, and f_b, respectively. We represent each with a Maxwellian, so that

$$f_e(v_\parallel) = \frac{1}{(2\pi)^{1/2} V_e} \exp \left[- \frac{v_\parallel^2}{2 V_e^2} \right] \tag{9.9}$$

$$f_b(v_\parallel) = \frac{1}{(2\pi)^{1/2} V_b} \exp \left[- \frac{(v_\parallel - u)^2}{2 V_b^2} \right]$$

where V_e is the thermal velocity, V_b the thermal spread of the beam, and u the beam speed. The phase velocities of the upper-hybrid and lower-hybrid waves can be calculated from the real part of the cold-plasma dielectric function [Harris, 1968]

$$\epsilon = 1 - \omega_{pe}^2 \left(\frac{\sin^2 \theta}{\omega^2 - \omega_{ce}^2} + \frac{\cos^2 \theta}{\omega^2} \right) - \frac{\omega_{pi}^2 \sin^2 \theta}{\omega_{ci}^2} \tag{9.10}$$

where θ is the angle between $\mathbf{k}$ and $\mathbf{B}$.

The solution for the high frequency branch (between the electron cyclotron frequency and the upper-hybrid frequency, is

$$\omega^2 = (1/2) [\omega_{UHR}^2 + (\omega_{UHR}^4 - 4\omega_{ce}^2 \omega_{pe}^2 \cos^2 \theta)^{1/2}] \tag{9.11}$$

The frequency of this mode ranges between ω_{ce} at $\theta = 0$ and ω_{UHR} at $\omega = \pi/2$. The low frequency mode is given approximately by

$$\omega_1^2 \simeq \omega_{pe}^2 \cos^2 \theta + \omega_{pi}^2 \tag{9.12}$$

The growth rate for these electrostatic waves can be found from (9.1) or (A.24) if the electrostatic term proportional to $E_{\parallel}$ is the only one retained and the nonrelativistic approximation is made. Thus from (9.1),

$$\gamma = - \frac{4\pi^2 e^2 n_e}{m_e k^2} \sum_{\nu} \int_0^\infty 2\pi v_\perp \, dv_\perp \, J_\nu \left(\frac{k_\perp v_\perp}{\omega_{ce}} \right) \tag{9.13}$$

$$\times \left[\left(\frac{\partial}{\partial v_\parallel} + \frac{\nu \omega_{ce}}{k_\parallel v_\perp} \frac{\partial}{\partial v_\perp} \right) F_e(v_\parallel, v_\perp) \Bigg|_{v_\parallel = (\omega - \nu\omega_{ce})/k_\parallel} \right] / (\partial \epsilon / \partial \omega)$$

the only important contributions in (9.13) come from the $\nu = 0$ and $\nu = -1$ terms [Goldstein et al., 1982]. If the thermal plasma is assumed sufficiently cold, the contributions from the ambient electrons can be neglected and γ reduces to

$$\gamma = \tag{9.14}$$

$$\frac{\left(\dfrac{\pi}{2} \right)^{1/2} \left(\dfrac{\omega}{2k^2} \right) \left\{ \dfrac{n_b(u - \omega/k_\parallel)}{n_e V_b^3} \exp - \dfrac{(\omega/k_\parallel - u)^2}{2 V_b^2} + \dfrac{k_\perp^2}{k_\parallel \omega_{ce}} \dfrac{n_b}{n_e} f_b \left(\dfrac{\omega + \omega_{ce}}{k_\parallel} \right) \right\}}{\left[\dfrac{\omega^2 \sin^2 \theta}{(\omega^2 - \omega_{ce}^2)^2} + \dfrac{\cos^2 \theta}{\omega^2} + \dfrac{\omega_{pi}^2 \omega^2 \sin^2 \theta}{\omega_{pe}^2 (\omega^2 - \omega_{ci}^2)^2} \right]}$$

where the first term in the denominator represents the interaction of the plasma and beam and the last term represents the anomalous ($\nu = -1$) Doppler contribution from the beam.

The high frequency wave will be excited through the beam-driven Landau resonance with growth rate

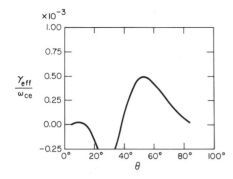

Fig. 9.3a. Growth rate of the upper-hybrid wave as a function of angle of propagation for $\omega_{pe}/\omega_{ce} = 0.1$. Cyclotron damping is included.

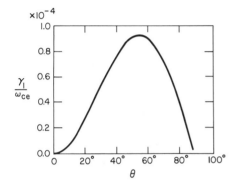

Fig. 9.3b. Similar to Figure 9.3a, but for the lower-hybrid wave.

$$\gamma = \left(\frac{\pi}{2}\right)^{1/2} \frac{n_b}{n_e} \frac{\omega_{ce}}{2} \left(\frac{\omega_{pe}}{\omega_{ce}}\right)^4 \left(\frac{u}{V_b}\right)^2 \cos^2\theta \sin^2\theta \qquad (9.15)$$

$$\times \left(\frac{u}{V_b} - \frac{\omega_{ce}}{k V_b \cos\theta}\right) \exp\left[-(u - \omega_{ce}/k\cos\theta)^2/2 V_b^2\right]$$

Similarly the growth rate of the low-frequency wave, driven unstable by the cyclotron term is

$$\gamma_1 \simeq \frac{\omega_{pe}}{8} \left(\frac{\omega_{pe}}{\omega_{ce}}\right)^2 \left(\frac{u}{V_b}\right) \sin\theta \sin 2\theta \left(\frac{n_b}{n_e}\right) \exp\left[-\frac{(\omega_{ce} - k_1 u \cos\theta)^2}{2 V_b^2 k_1^2 \cos^2\theta}\right] \qquad (9.16)$$

where we have assumed that θ is not too close to $\pi/2$ so that the ion terms can be neglected. Note that for both upper-hybrid and lower-hybrid waves, $k \simeq \omega_{ce}/u \cos\theta$.

In Figure 9.3, taken from Goldstein et al. [1982], we plot the magnitude of these growth rates for $n_b/n_e \simeq 10^{-2}$ and $\omega_{pe}/\omega_{ce} = 0.1$. Both γ and γ_1 peak for propagation near 50°, which is consistent with the claim made in Section 9.2 that beam driven instabilities tend to excite obliquely propagating waves. The choice of 10^{-2} for n_b/n_e is made solely for illustrative purposes as it is by no means certain that any of the energetic electron populations that have been observed in the Jovian magnetosphere are the exciters of DAM, (see Chaps. 5 and 12). The source for the DAM is intimately associated with the current system driven by the motion of Io through the plasma torus, and there have been no in situ observations of the electron distribution function inside the current carrying flux tube. In 9.3*a* cyclotron damping by the thermal plasma has been included. For a detailed discussion of its importance the reader is referred to Goldstein et al. [1982].

Three-wave interactions

The electrostatic waves excited by the beam must be converted into electromagnetic wave modes in order to produce the observed decameter radiation. There are several possibilities. We confine our attention to three-wave processes that can become important if the intensity of the upper-hybrid wave grows to a large amplitude. One possibility is that an intense upper-hybrid wave can decay into two other waves, one of them electromagnetic. That process can be schematically denoted.

$$\sigma \rightarrow \sigma_1 + \sigma_2 \tag{9.17}$$

This notation implies that two resonance conditions must be satisfied, one requiring momentum conservation:

$$\mathbf{k} = \mathbf{k}_1 + \mathbf{k}_2$$

and a second ensuring conservation of energy (frequency matching):

$$\omega = \omega_1 + \omega_2$$

Alternatively, the upper-hybrid wave could combine with a second electrostatic wave to produce an electromagnetic wave. This process is known as up-conversion, which we denote as

$$\sigma + \sigma_1 \rightarrow \sigma_2 \tag{9.18}$$

The resonance conditions become

$$\mathbf{k} + \mathbf{k}_1 = \mathbf{k}_2 \quad \text{and} \quad \omega + \omega_1 = \omega_2$$

Decay instabilities

Ben-Ari [1980] has considered decay interactions in which the σ_1 wave was either a lower-hybrid wave, an ion cyclotron wave, or an ion acoustic wave. She showed that when an upper-hybrid wave decays into a lower-hybrid wave, the resulting electromagnetic wave will be either an ordinary mode wave propagating across the field, or an extraordinary mode wave propagating along the field, antiparallel to the direction of the beam. Alternatively, if the decay produces a low frequency electrostatic ion cyclotron wave, then the resonance conditions restrict the electromagnetic wave to only the ordinary mode. Furthermore, decay into electrostatic ion cyclotron waves is unlikely because a higher threshold is required than for decay into a lower-hybrid electrostatic wave. Another possibility is that the decay product can be an acoustic wave. This can only happen if the thermal electron temperature exceeds the thermal ion temperature by an amount sufficient to allow ion acoustic waves to propagate as a normal mode of the plasma [see Krall and Trivelpiece, 1973]. Ben-Ari found that the largest decay rates produced ordinary mode electromagnetic waves in the direction antiparallel to the beam direction.

From these results it appears unlikely that decay instabilities (9.17) play a dominant role in exciting DAM, primarily because DAM appears to be almost entirely extraordi-

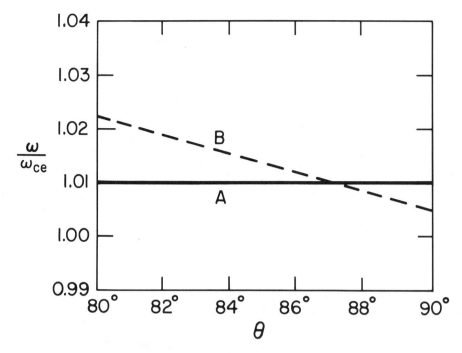

Fig. 9.4. The right-hand cutoff frequency ω_R (curve A) and the upconverted frequency ω (curve B) are shown as functions of θ, the propagation angle. Up-conversion is possible only for angles of propagation for which $\omega > \omega_R$.

nary mode radiation. The one possibility that can generate extraordinary mode radiation requires that the radiation be emitted along the field. In Section 9.2 we discussed the difficulties associated with parallel propagation. This leaves up-conversion as the most promising nonlinear process. We now consider the up-conversion of a lower-hybrid and upper-hybrid wave in greater detail.

Three-wave up-conversion

The frequency matching condition requires that the frequency of the up-converted wave, given by $\omega_2 = \omega + \omega_1$, must be greater than ω_R. The constraints under which this can be satisfied are illustrated in Figure 9.4. Assuming that $\omega_{pe}/\omega_{ce} \simeq 0.1$, ω_2 can exceed ω_R if θ, the angle between **B** and **k** is less than about 88° (implicit in Figure 9.4 is the assumption that θ for both electrostatic waves has the same magnitude). Recall that the growth rates for the lower-hybrid and upper-hybrid waves were greatest near 50°, thus up-conversion is apparently allowed. Ben-Ari [1980] considered this regime and found that the up-converted waves preferentially propagated nearly parallel to **B**. Thus, if one wishes to produce a hollow conical emission pattern, up-conversion cannot take place in the same physical location as the excitation. This has led Goldstein et al. [1982] to reconsider the up-conversion process by allowing for the possibility that once amplified the electrostatic waves propagate and refract so that the conversion to electromagnetic radiation will not occur where the electrostatic waves are first produced [cf., Roux and Pellat, 1979]. In addition, Goldstein et al. assume that the lower-hybrid waves are self-consistently amplified by the same electrons that excite the upper-hybrid waves.

Fig. 9.5. A schematic view of the
excitation, propagation, and refraction
of electrostatic lower-hybrid and
upper-hybrid electrostatic waves.
Up-conversion would produce a hollow
conical emission pattern of fast X mode
radiation.

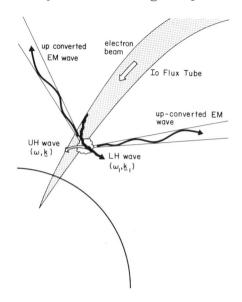

Because the wave number of the electromagnetic wave is small, that is, $|\mathbf{k}_2| \ll |\mathbf{k}|$ or $|\mathbf{k}_1|$, the three-wave resonance for up-conversion requires that $\mathbf{k} \simeq -\mathbf{k}_1$. Consequently, the two electrostatic waves must propagate in opposite directions for the up-conversion to take place. As we have seen, an electron beam is unstable to excitation of both upper-hybrid waves and lower-hybrid waves. If the high frequency wave is excited by the Landau resonance (see Eq. 9.15), and the low frequency wave is excited by the cyclotron resonance (Eq. 9.16), then one is guaranteed that $|\mathbf{k}| \simeq |\mathbf{k}_1| \simeq \omega_{ce}/u \cos\theta$.

Both waves have phase velocities close to the velocity of the beam. However, although the low frequency wave also propagates with a group velocity whose parallel component is in the beam direction, the parallel component of the group velocity of the high frequency wave is antiparallel to the beam. The vector direction of the group velocity of each of the waves is orthogonal to its phase velocity [Stix, 1962].

As the group velocity of the high frequency wave propagates upward, the wave encounters a region where the frequency equals the local upper-hybrid frequency. From (9.10) and (9.11) it is clear that as $\omega \to \omega_{UHR}$, $\theta \to \pi/2$. The group velocity approaches zero at this resonance. On the other hand, the low frequency wave encounters the ion plasma frequency as it propagates downward into higher density regions (cf., Eq. 9.12). The phase velocity of this wave also refracts toward $\pi/2$ and the group velocity again approaches zero. As a result the two waves can have oppositely directed wave vectors as required by the resonance condition. This model requires that although the upper- and lower-hybrid waves are both generated by the electron beam, the waves are not generated in the same place. This is schematically illustrated in Figure 9.5. As long as the wave-particle resonance conditions are satisfied during the refraction, the electrostatic waves will experience a convective amplification.

Recall from Figure 9.4 that frequency matching requires that $\theta < 88°$. A quantitative estimate of how close to $\pi/2$ the electrostatic waves must be to satisfy the wave-number matching condition (9.18) is also necessary. Because the perpendicular components of $\mathbf{k}$ are conserved during propagation of the electrostatic waves, $\mathbf{k}_\perp + \mathbf{k}_{1\perp} \simeq 0 \simeq \mathbf{k}_{2\perp}$ so long as the density gradients in the source region are much greater than the gradients in $\mathbf{B}$. However, both electrostatic waves have parallel components of their wave vectors in the same direction so that $k_{\parallel} + k_{1\parallel} \simeq 2k_{\parallel}$. Up-conversion to an electromagnetic wave propagating nearly orthogonal to **B** then requires that

$$2k_{\parallel} = k_{2\parallel} << k_{21} = k_2 \sin\theta_2 \qquad (9.19)$$

An estimate of $k_{2\parallel}$ can be obtained by noting that close to the reflection point the amplitudes of the electrostatic waves are described by Airy functions [Ginzburg, 1970; Roux and Pellat, 1979]. The wavelength in the vicinity of the reflection point is approximately the distance from the point of reflection to the first maximum of the Airy function, or

$$k_{\parallel} \simeq 2\pi(k_{0\parallel}L_N)^{2/3}/L_N \qquad (9.20)$$

where L_N is the density scale height at the point of reflection and $k_{0\parallel}$ is the initial magnitude of the parallel component of **k** or k_1 at the point of excitation, that is, $k_0 \simeq \omega_{ce}/u \cos(50°)$. The density scale height is estimated to range from 600–900 km in the Jovian ionosphere (see Chap. 2). This implies from (9.19) and (9.20) that $\theta_2 \gtrsim 50°$ for 20 MHz radiation. Taken together with the frequency matching requirement that $\theta \lesssim 88°$, we see that the up-conversion process can generate fast X-mode radiation in a hollow conical emission pattern.

The model described above requires a single electron beam directed downward. In Section 9.4 we describe a model of the coupling between Io and energetic electrons that produces two electron beams; one directed downward and the second directed upward. The two beams are on opposite sides of the Io flux tube. This suggests the possibility that the electrostatic waves could propagate across the flux tube with relatively little refraction and then up-convert. In that case the parallel components of the wave vectors would be oppositely directed and the constraints on the wave-number matching would be even less severe. Whether this geometry would result in a hollow conical emission pattern has not as yet been worked out.

We have so far shown that electron beams can excite electrostatic waves, that these waves can propagate to a geometrical configuration where the three-wave up-conversion matching conditions can be met, and that the resulting electromagnetic radiation will be beamed into a hollow conical emission pattern (see Fig. 9.5). Up-conversion processes can be intrinsically inefficient and it is important to estimate the coupling coefficients and the energy requirements before concluding that the process is viable. Roux and Pellat [1979] have computed the coupling of a coherent three-wave interaction involving two upper-hybrid waves. The coupling efficiency of an upper- and lower-hybrid wave has been computed by Goldstein et al. [1982] who considered only the coupling of incoherent waves. The derivation of the coupling coefficients is a very lengthy exercise that we will not repeat. Rather, we limit ourselves to a discussion of the underlying physics and discuss some of the results. The interested reader is referred to Davidson [1972] and Tsytovich [1970] for more details. The full generalization to electromagnetic waves can be found in Pustovalov and Silin [1975], and Larson and Stenflo [1976]. The application to TKR was discussed by Barbosa [1976] and Roux and Pellat [1979], and application to DAM by Ben-Ari [1980] and Goldstein et al. [1982].

The nonlinear currents that generate the up-converted electromagnetic wave can be computed if one uses fluid equations to describe the motion of the electrons in response to both the electrostatic fields and the magnetic fields. One approach is to linearize the electron fluid equations for conservation of electron mass and momentum. One of the contributions to the nonlinear current can be computed from the terms representing the product of the electron density perturbations driven by the upper-hybrid wave and the electron velocity perturbations driven by the lower-hybrid wave. In a similar manner, the lower-hybrid density fluctuations and the upper-hybrid velocity perturbations produce a second term in the nonlinear current. A third contribution comes from the pondermotive force that arises from the interaction of the perturbation velocities of the upper- and lower-hybrid waves through the $(\mathbf{v} \cdot \nabla)\mathbf{v}$ term in the momentum equation. The resulting force produces a nonlinear velocity perturbation $\mathbf{v}_{NL}$. The current can then be written [Goldstein et al., 1982]

$$J_{NL} = -ne\mathbf{v}_1/2 - n_1 e\mathbf{v}/2 - n_e e\mathbf{v}_{NL} \tag{9.21}$$

This current then enters the right-hand side of Equation (A.1), where now Λ_{ij} and E_j describe the cold plasma dispersion relation and electric field of the up-converted electromagnetic wave. From (9.21) and (A.2), one can derive an equation relating the amplitude of the electromagnetic wave to the product of the amplitudes of the two electrostatic waves. Details can be found in the papers referenced above.

The efficiency of this process can be estimated by asking what fraction of the energy in the beam must be converted into electrostatic energy to produce the observed power in decameter radiation. Such an estimate has been made for the incoherent process described by Goldstein et al. [1982]. They found that the observed radiation could be accounted for if approximately 10^{-3}–10^{-4} of the beam energy was converted into electrostatic waves. These estimates depend on the choice made for the source size, ambient electron density, beam density, etc., and the reader is referred to their paper for more details. Their calculation ignores the geometric amplification of the electrostatic waves known to occur near resonance. This amplification can be estimated from knowledge of the Airy functions mentioned above and has been done by Roux and Pellat [1979] for the upper-hybrid wave in the terrestrial magnetosphere. Computer simulations of the electrostatic simulations described above have shown that as much as 10% of the beam energy can be converted into electrostatic waves [Rowland, Palmadesso, and Papadopoulos, 1981]. Thus, the indirect up-conversion process appears to be a viable candidate for producing DAM.

Direct emission mechanisms

Although we have discussed what must appear to be a plethora of emission mechanisms for exciting the DAM, we have by no means exhausted all of the possibilities considered in the literature. In addition to the indirect wave coupling process described above where the beam first excites two electrostatic waves that in turn up-convert to give the escaping electromagnetic radiation, one can imagine situations where the beam directly amplifies fast X-mode electromagnetic radiation via a direct coupling with a second intense electrostatic wave. Two theories have been proposed for TKR based on this approach. Neither has been applied to the Jovian situation, primarily because it is not clear that the high efficiencies obtained in these direct mode coupling theories are required by the observed emission. In addition, one also would need to assume the existence of some source of intense, coherent, electrostatic waves. Although these have been observed in the Earth's auroral zone, they have not been reported at Jupiter. Of course, such waves would not propagate from the region of excitation, and so would be confined to the presumed DAM source region in the upper ionosphere.

The first of these two direct nonlinear theories [Palmadesso et al., 1976] assumed the existence of electrostatic electron cyclotron waves. For efficient coupling, the escaping electromagnetic wave then had to be in the O mode. Subsequently, Grabbe, Papadopoulos, and Palmadesso [1980] and Grabbe [1981b], motivated by observations of large amplitude coherent electrostatic ion cyclotron (EIC) waves in the auroral zone [see, e.g., Lysak, Hudson, and Temerin, 1981], assumed that the electrostatic waves were EIC. In these papers, the authors showed that interaction between EIC waves and the electron beam could produce escaping fast X-mode radiation just above ω_R in a source region with $\omega_{pe} < \omega_{ce}$. We will not describe these theories here in any greater detail. The interested reader will find more complete descriptions in the references cited above and in a review of TKR theories by Grabbe [1981a].

In the nonlinear theories discussed above, the existence of an electron beam is required. Because it is believed that most DAM is controlled in some way by Io, we are faced with the question of how Io can produce an electron beam. Details of the interaction of Io with the corotating plasma are described in Chapter 10. In the next section we discuss those aspects of this interaction that bear on the generation of electron beams. We also discuss a few aspects of DAM morphology related to this same "coupling mechanism" [Smith, 1976a].

9.4. The Io and plasma torus interaction

Since the discovery by Bigg [1964] of the remarkable control of the orbital position of Io over the DAM radiation, considerable theoretical attention has been given to the explanation of the Io interaction with the Jovian magnetosphere. Several mechanisms including the excitation of large amplitude MHD waves, the generation of electrostatic electric fields parallel to **B** and the effects of the Io wake have been considered. The recent Voyager 1 and 2 encounters have altered our view of the basic Io interaction considerably. The plasma environment of Io is quite different from what had been assumed earlier and the great Io torus affects significantly the interaction of Io with the magnetosphere (Chap. 10). It also changes the propagation characteristics of electromagnetic waves and determines much of the phenomenology and morphology of KOM radiation (see Chap. 7).

In this section we review one mechanism by which Io can accelerate electrons. The end result is an electron distribution that satisfies the requirements discussed in Section 9.3 for beam-driven instabilities. However, because loss-cone distributions may arise from beamlike distributions after magnetic mirroring, we cannot use the Io-magnetosphere coupling mechanism described below to choose indirect over direct emission mechanisms with certainty, although the model is suggestive that beamlike distributions should arise relatively naturally.

Generation of Alfvén waves by Io

The interaction of Io with the corotating torus plasma is described in Chapter 10. It seems that Io's interaction is electromagnetic in nature and related to the induced electric field on Io moving through the magnetized plasma [Neubauer, 1980]. This field drives an electric current through Io (either through its body or through its ionosphere), which produces polarization perturbations of the electric field and these are transferred to the plasma along magnetic field lines. As a result, the plasma velocity in Io's vicinity changes. Far away from Io the plasma corotates at a velocity $v_0 = \omega r$. Near Io the plasma is slowed down to $v_{Io} = v_0 - v_\perp$. The velocity perturbation $v_\perp = v_0 - v_{Io}$ is discussed in Chapter 10. A schematic view of the situation is shown

Fig. 9.6. A sketch of the Alfvén wave
model for the Io interaction as seen in
Io's frame. The top figure shows the
Alfvén wings as dashed lines and the
magnetic field lines as solid lines. The
bottom two figures show the magnetic
field configurations in the $\hat{x}$–$\hat{y}$ plane at
different values of $\hat{y}$. $\hat{y}$ is the direction of
the torus plasma flow relative to Io. The
acceleration of electrons takes place at
the Alfvén wings and the space between
the wings is filled with accelerated
electrons, represented by the dots.
These electrons carry a current parallel
to the Alfvén wings, indicated by the
heavy arrows. The reduction of the
plasma convection velocity behind the
Alfvén wings is also indicated. The
reader is referred to Chapter 10 for more
details.

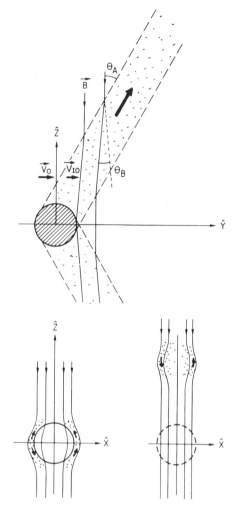

in Figure 9.6. Io interacts with the plasma through what has been called an "inductive interaction" [Gurevich, Krylov, and Fedorov, 1978] by which magnetohydrodynamic waves are produced. The circuit for the current flowing through Io is closed in the plasma by currents flowing in the wave fronts. Note that the waves are not harmonic waves, but comprise a pulselike wave packet. Direct measurements of the perturbation magnetic fields produced by this current system have been reported by Acuña, Neubauer, and Ness [1981]. The only mode that can guide energy along the field lines is the Alfvén wave. Hence, it alone can account for a small source size near the Jovian ionosphere of the Io controlled DAM. The other two MHD modes are not guided and will not be considered here. Under certain conditions, for example, low plasma density and small scale sizes, the field aligned currents in the Alfvén wave are carried by accelerated electrons, i.e., a beam. The energy of these electrons can be estimated from the properties of the electric field of the Alfvén wave.

The Alfvén waves created by Io are shear Alfvén waves. Because the disturbance is spatially localized, one must view it as a wave packet consisting of a Fourier spectrum in the perpendicular wave-number $k_\perp$. The Fourier components of this shear Alfvén wave propagate obliquely with respect to the background magnetic field $\mathbf{B}_0$ although the wave packet itself propagates almost exactly along the field lines in Jupiter's frame of reference. (The group velocity of Alfvén waves is along **B**.) The "quality" of guidance of the wave-packet along the magnetic field depends on two factors: The amount of dispersion and dispersive spreading of the wave packet and, the inhomogeneities of the medium where the wave propagates. The fact that for $k_\perp \neq 0$ Alfvén waves are dispersive is a well documented but little known fact [see, e.g., Fejer and Lee, 1967]. Due to the dispersive nature of obliquely propagating Alfvén waves, a wave packet will spread out as it propagates. The spread is rapid for large values of $k_\perp$ and slow for small values; a critical wave number above which dispersion is strong is ω_{pe}/c for a cold plasma ($\beta < m_e/m_i$) and ω_{ci}/V_i for a warm plasma ($1 > \beta > m_e/m_i$). In the center of the Io torus $T_i \simeq 20$ eV, $n \simeq 2 \times 10^3$ electrons/cm³, $B_0 \simeq 2 \times 0^{-2}$ G, and $\beta \simeq 4 \times 10^{-3}$. But because β rapidly decreases along a field line as the wave propagates away from the torus, we need consider only the cold plasma case. This situation has been treated by Fejer and Lee [1967], and the reader is referred to that paper for more details. Fejer and Lee also consider the effect of inhomogeneities, and show that unless there are strong perpendicular density gradients, geometrical guiding by the magnetic field is maintained and $k_\perp B = $ constant. Without dispersion the Io created perturbation would thus remain confined to the IFT.

The perturbation created by Io has a rather complicated structure perpendicular to B_0 [Goertz, 1980a] and it is difficult to evaluate the dispersive spreading. Goertz describes the electric field perturbation as consisting of a uniform field $E_\perp$ inside the Io flux tube and a "dipolar" component outside (as seen from Jupiter). The boundary across which $E_\perp$ changes rapidly from dipolar to uniform is assumed to have a thickness d_0 of roughly one (Io) ionospheric scale height (~ 80 km) (see Chap. 3). It can be shown that as the packet propagates along **B**, dispersion will cause this transition region to broaden to an equivalent width w, where

$$w^2 \simeq d^2(1 + 4z^2\omega^2 c^4/V_A^2 \omega_{pe}^4 \pi d^4) = d^2(1 + z^2/z_0^2) \tag{9.22}$$

where $d = d_0 B_0/B$ is the thickness one would obtain without dispersion, z is distance along the field line, ω is a characteristic frequency of the wave packet related to the duration T that a field line is in contact with Io by $\omega \simeq \pi/T$. For a weak perturbation, T is the convection time past Io (about 60 s). However, in Chapter 10 it is argued that near Io the electric field is reduced by at least a factor of three and the duration is hence increased to at least 180 s. In the center of the torus the "dispersion length" z_0 is larger than 10^6 R_J and broadening of the transition due to dispersion should be entirely negligible in the torus. At high latitudes along the Io flux tube, dispersion can also be ignored. Therefore, in the following discussion, we assume that the transition from a dipolar electric field outside of the IFT to a uniform $E_\perp$ inside the IFT occurs over a width equal to d.

Alfvén waves generated by Io carry a field aligned current that flows in the transition region where $E_\perp$ changes. This current is carried by electrons that are accelerated by the field aligned electric field. The magnitude of this field can be estimated from either a guiding center Vlasov equation [Fejer and Kan, 1969] or a fluid model [Goertz and Boswell, 1979]. The electric field component $E_\parallel$ in the advancing wave front is related to the perpendicular electric field amplitude $E_\perp$ by

$$E_\parallel = \frac{m_e}{m_i} \frac{c}{\omega_{pi} \omega_{ci}} \frac{d}{dt} \nabla \cdot (\mathbf{E} \times \mathbf{B}/B) \tag{9.23}$$

Fig. 9.7. The emission geometry for the
Io-phase control of DAM. At the leading
and trailing edges of the Io-generated
Alfvén wave packet, parallel electric
fields accelerate and decelerate electrons.
The electrons then form beams between
the leading and trailing fronts. In this
model, these beams then generate
electrostatic waves that in turn couple
and up-convert to DAM as described in
Section 9.3.

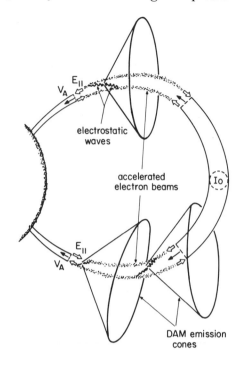

Electrons overtaken by the wave will be accelerated and gain parallel momentum
(recall that $\beta < m_e/m_i$ corresponds to $V_A^2 > k_B T_e/m_e$

$$\Delta p_\| = \int dt\, qE_\| = \frac{m_e}{m_i} \cdot \frac{c}{\omega_{pi}} \frac{q}{\omega_{ci}} \nabla \cdot (\mathbf{E} \times \mathbf{B}/B) \simeq \frac{m_e}{m_i} \frac{c}{\omega_{pi}} \frac{q}{\omega_{ci}} \frac{E_\perp}{d} \qquad (9.24)$$

An electron will then gain an amount of energy parallel to B_o equal to

$$\Delta W_{\|e} = \Delta p_\|^2/2m_e = \frac{c^2}{\omega_{pe}^2 d^2} \frac{m_i}{2} \left(c \frac{E_\perp}{B_0}\right)^2 \qquad (9.25)$$

Because of their larger mass, ions gain a much smaller amount of parallel energy.
They do, however, gain perpendicular energy as they are overtaken by the wave. In
Chapter 10 it is shown that $E_\perp$ near Io can be comparable to the corotational
field and hence $\Delta W_{\|e}$ can be of the order of several keV at high latitudes.

   These considerations suggest the following schematic model for the generation of
DAM which is illustrated in Figure 9.7. Between the leading and trailing edge of the
Alfvén wave packet, accelerated electrons exist with energies of several keV. The electrons form two beamlike distributions. The beams on the inner and outer sides of the
flux tube consist of electrons streaming toward and away from the ionosphere, respectively. As we have seen in Section 9.3, both beams will be unstable against the growth
of electrostatic waves propagating antiparallel to each other albeit on opposite sides of
the IFT. If these waves propagate across the flux tube, up-conversion may occur, but
in any event up-conversion can take place as the waves refract near the resonance
layers as described in Section 9.3. The efficiency of such a model and its relation of
DAM morphology needs to be investigated in greater detail.

Relation to DAM morphology

Recently there has been considerable interest in calculating what a distant observer will see if the emission pattern is indeed that of a conical sheet [Warwick et al., 1979b; Goldstein and Theiman, 1981; Pearce, 1981; Staelin, 1981; Warwick, 1981]. The problem is essentially that of computing the intersection of a cone with a plane. The axis of the cone is aligned with the direction of the local magnetic field. The plane contains the observer and is usually assumed coincident with, or parallel to, the equatorial plane of Jupiter. The intersection of a conical sheet with a plane is a hyperbola and the angle between the asymptotes of this hyperbola is 2μ where [Goldreich and Lynden-Bell, 1969]

$$\tan \mu = (\tan^2 \psi \cos^2 \iota - \sin^2 \iota)^{1/2} \tag{9.26}$$

In (9.26), ψ is the half angle of the cone, and ι measures the inclination of the axis of the cone to the plane. It is related to $\mathbf{B}(\mathbf{r})$ by

$$\iota = \arccos(B_c/B) \tag{9.27}$$

where B_c is the projection of $\mathbf{B}$ onto the Jovigraphic equatorial plane. The longitudes toward which the asymptotic lines point are $\lambda' = \phi \pm \xi$, where $\xi = \arcsin(B_\phi/B_c)$ and B_ϕ is the ϕ component of $\mathbf{B}$ at the apex of the cone (in a right-handed spherical coordinate system). The plus and minus signs refer to the southern hemisphere and northern hemisphere, respectively. The apparent source locations at the apex of the cone are then $\Lambda(\pm) = \lambda' \pm \mu$, where here the plus and minus refer to the two sides of the conical sheet.

If we now consider radiation coming from conical sheets aligned along only one magnetic flux tube, then at a particular time an observer may be receiving radiation from two or one (degenerate case) emission cones originating from different positions along the field line. Because the radiation is emitted close to the electron cyclotron frequency, two (or one) frequencies will be seen at that time on a frequency-time spectrogram. As Jupiter rotates, the observation geometry changes and a succession of active flux tubes will come into view. When the resulting pattern is computed using the formulas given above, the frequency-time spectrogram shows an arc structure. The shape of this arc will depend on where in the magnetosphere the active flux tube is located and on the magnitude of the emission cone angle ψ.

Pearce [1981] and Goldstein and Thieman [1981] have both tried to compare their calculations directly with dynamic spectra obtained from the Voyager PRA experiment. Pearce used a dipole approximation to the Jovian magnetic field and assumed that ψ remained constant at all frequencies along a given arc. He then attempted to find a source location and cone angle which would provide a good fit to selected observations. A typical result of such a calculation is shown in Figure 9.8, taken from Pearce [1981]. In the figure the cone is located on the $L = 5.5$ flux tube and ψ is $18°$. The calculation assumes that the observer is at $30°$ south latitude. During the time the observations were taken, the Voyager spacecraft was actually at $5°$ north latitude. Thus, Pearce's analysis seems to be internally inconsistent. In addition, the constant and small value of ψ implies that sources A and B are independent and could not represent opposite sides of the same conical emission pattern.

Goldstein and Thieman take a somewhat different approach. In contrast to Pearce, for whom the L shell of the source is a variable, Goldstein and Thieman assume that the active field lines are at $L = 6$, i.e., that the radiation is stimulated by the motion of Io. They also assume that the O4 octupole model provides a reasonably accurate representation of the magnetic field. In this case the computed "arcs" resemble the observed

Fig. 9.8. Arc model predictions (black
squares) superimposed on Voyager 1 data
from February 10, 1979. The model
calculation placed the source on the
$L = 5.5$ flux tube and assumed a half
angle for the conical emission pattern
of $\psi = 18°$. A dipole approximation
was used for the Jovian magnetic field.
[Taken from Pearce, 1981.]

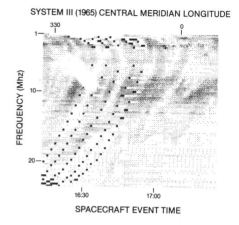

dynamic spectra only if the cone angle varies with frequency. They argue that refraction of the electromagnetic wave when it is close to ω_R could change ψ in the proper way. An example of the type of fit to the data that can be obtained if ψ varies with frequency while the observer is constrained to the equatorial plane is shown in Figure 9.9, taken from Goldstein and Thieman [1981]. This model requires somewhat different sets of input parameters to fit the main and early sources (A and B, respectively), and to differentiate between "greater" and "lesser" arcs. The model could be improved by incorporating growth rates and refraction predicted by a specific instability mechanism. However, fitting the longitude of the highest frequencies in source B may still be difficult. It is not clear whether the basic interpretive model is at fault, or whether the discrepancies would be removed if only a more detailed model of the magnetic field near the ionosphere were available.

Staelin [1981] and Hewitt, Melrose, and Rönmark [1981] have ascribed the variation of ψ with frequency to the way in which the relativistic resonance condition (9.4) and the variation of the maximum unstable wave number with frequency will change with both cyclotron frequency and the ratio ω_{pe}/ω_{ce}.

None of these arc models directly addresses the question of why the arcs appear discrete on the dynamic spectra. Boischot and Aubier [1981] have argued that the discreteness can arise from destructive interference between neighboring flux tubes, even if all field lines are emitting simultaneously. For this to work, the emission mechanism would have to produce coherent waves, which has not been a feature of the theories developed thus far. Nonetheless, by suitable variation of the relative phases of the interfering waves with frequency and time, Boischot and Aubier claim to be able to explain not only the discreteness of the arc pattern, but also the shapes of the arcs.

Gurnett and Goertz [1980a] attribute the discreteness to the nature of the interaction of the current carrying Alfvén wave with the ionosphere and torus. They point out that after the Alfvén wave is reflected from the ionosphere it is not expected to arrive back at the orbit of Io in time to close the dc circuit [Neubauer, 1980]. Therefore, many reflections of the wave could occur between hemispheres, thereby producing an extended standing wave current system downstream of Io. The three-dimensional geometry of this current system is illustrated in Figure 9.10. The apparent change in slope of the wave trajectory at the boundary of the torus is caused by the much lower plasma density outside of the torus. Because the reflected waves essentially produce a series of mirror images of the original Alfvén wave current system extending all around the Io L shell, this system of multiple reflections provides a simple explanation for the large number of decametric arcs observed in the radio emission spectrum

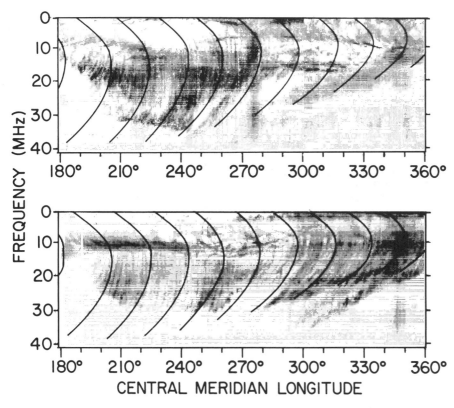

Fig. 9.9. An overlay of the modeled arcs on dynamic spectra observed by Voyager 2 on July 6–7, 1979 (upper panel), and July 3, 1979 (lower panel). The upper panel is an example of an Io- dependent source A event (the phase of Io is 240° when the central meridian longitude is 212°). The lower panel is an example of an Io-dependent source C event centered around 330° (Io phase is 240° at a longitude of 329°). The model calculation placed the source on the $L = 6$ flux tube and the half angle of the conical emission pattern varied as a function of frequency as described in the text. The O4 magnetic field model was used in these calculations [from Goldstein and Theiman, 1981].

(Chap. 7). In the figure, $\Delta\phi_0 = (360°)\cdot(2\ell M_A) / 2\pi(5.9R_J)$. Using 0.15 for M_A, the Alfvén mach number in the torus, and $2R_J$ for ℓ, the thickness of the torus, the longitudinal separation between successive reflections is $\Delta\phi_0 \simeq 5.8°$. There is also a minor periodicity $T_1 = (\Delta\phi_1/\Delta\phi_0)T_0$ introduced by the out-of-phase wave propagating in the opposite direction away from Io. Gurnett and Goertz [1980a] estimate that a large number (more than 10) of wave reflections could occur before the wave is damped by collisions in the ionosphere or collisionless processes in the magnetosphere, and that the temporal separation between successive arcs will range from 0 to 40 min.

Recently, Warwick [1981] has approached the problem of why the arcs are discrete by abandoning the hypothesis that the emission is directly related to the Jovian magnetic field as modeled by Pioneer 11 measurements. He attributes the sequence of arcs to a manifestation of widespread fine structure in the magnetic field. This fine structure is assumed to have dimensions similar in size and shape to the white ovals on the visible disk of Jupiter. Because this model is not related to the multipole models of the main Jovian magnetic field, it has the potential for accounting for the Io controlled source B location, which, as we have mentioned, has proved very difficult to explain in

Fig. 9.10. A schematic sketch of the current system that might arise from the interaction of Io with the Jovian ionosphere and plasma torus. The current is carried by an Alfvén wave that is reflected from the ionosphere. If the reflected wave does not close the dc circuit by returning to Io, the possibility arises that many reflections of the wave could occur between the northern and southern ionospheres and the torus, thereby providing a possible explanation for the discrete arc pattern that characterizes the decameter radiation [from Gurnett and Goertz, 1981].

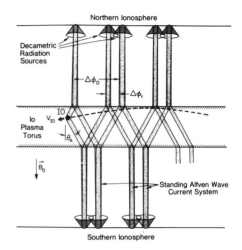

models which do attribute the source locations to properties of the global magnetic field. Several aspects of this model will have to be investigated in greater detail before its relationship to the observations is understood. For example, it is by no means obvious that the upper Jovian atmosphere can support islands of magnetic field signficantly more intense than in the surrounding medium; nor has it been conclusively demonstrated that the hypothesized field strengths could not in fact have been observed by Pioneer 11. Perhaps the most puzzling aspect of Warwick's model is that it divorces the observed decameter morphology from nearly all observable aspects of the magnetic field, in spite of the rather promising progress that has been made in our understanding of many phenomena (not withstanding the location of the Io-B source) utilizing the multipole models of the global magnetic field.

Implicit in the foregoing discussion of decameter arcs has been the assumption that a single large source region or a series of many small sources are radiating simultaneously to produce the large longitudinal extent of the decameter radiation. The individual sources are known from Very Long Baseline Interferometry (VLBI) to be quite small with projected linear dimensions along the line of sight less than 400 km [Dulk, 1970; Lynch et al., 1972]. This suggests that there must be multiple sources to account for the large longitudinal extent of the emission. That this is not necessarily the case has become evident from a rather novel interpretation of the arc pattern suggested by Lecacheux, Meyer-Vernet, and Daigne [1981]. In their model, the entire longitudinal extent of the emission can arise from a single small source located along a single field line.

The essential new feature in this theory is the inclusion of diffraction effects on the radiation as it propagates through the plasma torus near Io's orbit (see Chaps. 3, 6, and 11 for a description of properties of the torus). Lecacheux et al. argue that the plasma torus acts as a thin screen, changing the phase of the radio waves as in Fresnel diffraction. The resulting diffraction fringes, shown in Figure 9.11, resemble a pattern of arcs, either vertex early or vertex late, not dissimilar from the observed patterns. A careful comparison between the observations and this theory remains to be carried out, but it does seem clear that a pattern of repeating arcs can in principle be produced by a single localized source. Beaming of the radiation into a hollow conical sheet still appears to be required if sources A and B are to be interpreted as coming from the same physical location.

If the diffraction hypothesis is correct, then during the closest approach of Voyager 1, when the spacecraft was actually inside the Fresnel screen, the arc pattern

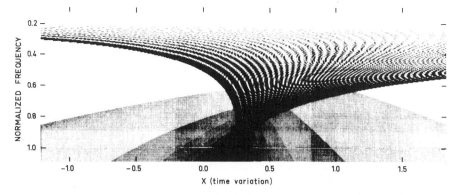

Fig. 9.11. Computed fringe patterns are shown for a ray tracing calculation in which the Io plasma torus is treated as a thin Fresnel screen. The frequency-time plot illustrates the intensity enhancements resulting from Fresnel diffraction of decameter waves [from Lecacheux et al., 1981].

should have disappeared. The observations do not suggest a clearcut resolution of this question, for although much of the familiar early and main source radiation is absent, some arcs are still present in the dynamic spectra [J. Alexander, private communication]. As far as we know, the close encounter data have not yet been analyzed in the context of any of the arc models.

9.5. Summary

In the previous sections we have concentrated on one question: What requirements can theory put on the particle distribution function necessary for the generation of DAM (or HOM and KOM for that matter)? The conclusion from our discussion must be that any anisotropic distribution of energetic particles can, in principle, be unstable against the emission of the fast X-mode, the most likely candidate for DAM. Direct mechanisms usually require a loss-cone type distribution whereas indirect mechanisms require a beamlike distribution for the generation of electrostatic waves that in turn produce the electromagnetic waves by one or another three-wave coupling mechanisms. This state of affairs is, of course, quite unsatisfactory, but it is very unlikely that we will ever know much more about the in situ plasma conditions including the existence or non-existence of particular electrostatic wave modes in the generation region of DAM. However, the similarity between DAM and TKR suggests that an understanding of TKR will also help to explain DAM. Obviously, such a view cannot be defended rigorously.

As yet there is no completely convincing theory of TKR (the problem being uniqueness not correctness) and so we cannot simply conclude that DAM is generated by exactly the same process, whatever that may be. Furthermore, there are significant differences between DAM and TKR, including the role of Io, spectral arcs, as well as the probability that although the source region of DAM is underdense in the sense $\omega_{pe}/\omega_{ce} \ll 1$, the ambient density still exceeds the beam density. Based on the discussions in Sections 9.3 and 9.4, we feel that the prominent Io-phase control of DAM is easier to understand if an indirect generation mechanism is operating. The argument is twofold. The highest DAM frequency of 39.5 MHz suggests generation in the northern hemisphere on field lines that connect Io to a region where the magnetic field is 14 G. Such a large field exists only in the northern hemisphere according to the presently known field models. If Io were only to enhance the pitch angle scattering of a trapped particle distribution into the loss-cone, the particles would precipitate into the southern hemisphere where the field is low [see Roederer, Acuña, and Ness, 1977; Dessler and Hill, 1979; Goldstein and Eviatar, 1979]. If, on the other hand, Io stimulates the generation of an electron beam, particles could reach the high magnetic fields in the northern hemisphere. One mechanism by which Io can generate beamlike distributions has been discussed in Section 9.4.

Clearly neither of these arguments are completely compelling. Even a precipitating beamlike distribution is capable of exciting loss-cone instabilities because the albedo distribution can resemble a loss-cone. Thus, even observation of electron beams would not necessarily provide a definitive resolution to this question. Furthermore, in neither of the four Jupiter fly-bys has there been clear indication of an Io injected electron beam, although the trajectories, viewing geometry and instrumentation have not been completely optimal. The same can be said about the existence or nonexistence of loss-cone features in the electron distribution.

We have briefly discussed some of the morphological aspects of DAM. Again we are faced with the problem of uniqueness. A case in point is the explanation of the decametric arcs in terms of at least three completely different models. However, more refined Earth based VLBI observations of DAM may distinguish between the models that require either one large or several small sources. It cannot be stressed enough that our ignorance about the source position and size is still the most serious obstacle to understanding DAM.

From our discussions and those in Chapter 7, it has become clear that the Io control consists essentially in the enhancement of a nearly continuous source. Either Io's influence extends over large distances away from Io as in the model of Gurnett and Goertz [1981], or as Io approaches this continuously active source, it only enhances the source intensity. Such an enhancement might be variable and related to the amount of plasma or neutrals Io injects into the torus region. Because the neutral and plasma density influences the intensity and spectral characteristics of the optical emission from the Io torus, there may be a correlation between the observed variability of the optical emission and DAM. If such a correlation exists, it would be an important clue to the understanding of DAM and the Io effect.

It is to be expected that neither the morphology nor the phenomenology of DAM will definitively distinguish between existing models. What is needed at this stage is, simply, a more detailed formulation of all the various theoretical models, more specific calculations to determine the range of parameters that each model can tolerate, more detailed estimates of the efficiency of each mechanism utilizing detailed global magnetic field and density models and the continuing effort to formulate observational tests of the models. Finally, we remark that the recent kilometric radiation from Saturn (SKR) by Kaiser et al. [1980] suggests that the emission of radio waves is a general feature of planetary magnetospheres. Perhaps by comparative study, we will be able to eliminate some presently viable theories. Although that will be difficult, the reward of understanding such an enigmatic phenomenon as planetary radio emission will be high.

APPENDIX

We first derive the probability of spontaneous wave emission by a single particle. Then, by integrating over a distribution of particles, emission from all particles can be computed. By use of the Einstein relations, and a similar integration over the particle distribution, we then calculate the coherent, or stimulated, emission.

We assume that the dielectric tensor $\epsilon_{ij}(\mathbf{k},\omega)$ for the background medium is Hermitian (no damping) and given by cold plasma theory [Stix, 1962]. When the Maxwell equations are Fourier transformed the results can be written in the form of an equation for the wave electric field $\mathbf{E}$

$$\Lambda_{ij}(\mathbf{k},\omega)E_j(\mathbf{k},\omega) = - \frac{4\pi i}{\omega} j_i(\mathbf{k},\omega) \tag{A.1}$$

where summation over repeated indices is assumed and $\mathbf{j}$ is a source current.

The tensor Λ_{ij} is related to the refractive index $n \equiv kc/\omega$ by

$$\Lambda_{ij}(\mathbf{k},\omega) = n^2(K_i K_j - \delta_{ij}) + \epsilon_{ij}(\mathbf{k},\omega) \tag{A.2}$$

$$\mathbf{K} = \mathbf{k}/k, k = |\mathbf{k}|$$

The current $\mathbf{j}$ could arise from either an uncorrelated random motion of charged particles, producing incoherent radiation; or an ordered motion producing coherent radiation. The average power radiated by $\mathbf{j}$ is given by

$$P = - \lim_{T \to \infty} \frac{1}{T} \int_{-T/2}^{T/2} dt \int d^3r \, \mathbf{j}(\mathbf{r},t) \cdot \mathbf{E}(\mathbf{r},t) \tag{A.3}$$

The radiation can thus be thought of as a dissipation of particle energy. The physical interpretation of Equations (A.1) and (A.3) is then transparent. A free current generates electric fields in a medium described by the dielectric tensor ϵ_{ij}. The power radiated is then equal to the energy loss $\mathbf{j} \cdot \mathbf{E}$. Expressing $\mathbf{j}$ and $\mathbf{E}$ in terms of their Fourier transforms and using the fact that negative and positive frequencies are physically equivalent we obtain

$$P = - \int d^3k \int_0^\infty d\omega [\mathbf{j}^* \cdot \mathbf{E} + \mathbf{j} \cdot \mathbf{E}^*]/(2\pi)^4 \tag{A.4}$$

The formal solution of (A.1) is

$$E_i(\mathbf{k},\omega) = - \frac{i}{\omega} D_{ij}(\mathbf{k},\omega) j_j(\mathbf{k},\omega) \tag{A.5}$$

where the dispersion tensor is

$$D_{ij} = 4\pi\lambda_{ij}(\mathbf{k},\omega) / \Lambda(\mathbf{k},\omega) \tag{A.6}$$

and $\lambda_{ij}(\mathbf{k},\omega)$ is the cofactor of Λ_{ij} defined by $\Lambda_{ij}\,\lambda_{jl} = \Lambda\delta_{il}$, $\Lambda(\mathbf{k},\omega)$ is the determinant of $\Lambda_{ij}(\mathbf{k},\omega)$. For normal modes $\Lambda = 0$ and D_{ij} is singular. The integration around this singularity is well known in plasma physics [see, e.g., Krall and Trivelpiece, 1973]. Because we are interested in $\mathbf{E}(t)$ due to the current $j(t_o)$ for $t_o < t$, we use retarded functions and integrate above the singularity $\omega = \omega(\mathbf{k})$ (or equivalently give ω a small positive imaginary part). Then

$$D_{ij} = \frac{4\pi\lambda_{ij}}{\partial\Lambda/\partial\omega} \left\{ P\frac{1}{\omega - \omega(\mathbf{k})} - i\pi\delta[\omega - \omega(\mathbf{k})] \right\} \tag{A.7}$$

The principal part does not contribute to the integral in Equation (A.4), and thus

$$P = \frac{1}{2\pi^2} \int d^3k \int (d\omega/\omega) \frac{|\lambda_{ss}| \, |\hat{e}_i \, j_i^*|^2 \, \delta[\omega - \omega(\mathbf{k})]}{\partial\Lambda/\partial\omega} \tag{A.8}$$

where $\hat{e} = \mathbf{E}/|\mathbf{E}|$ is the polarization vector of the wave, and λ_{ss} is the trace of λ_{ij}. This result holds for any current. We now let the current arise from a particle of charge q and mass m, that is,

$$\mathbf{j}(\mathbf{r},t) = q\mathbf{v}(t)\delta[\mathbf{r} - \mathbf{r}(t)] \tag{A.9}$$

where the particle's trajectory $\mathbf{r}(t)$ and the velocity $\mathbf{v}(t)$ satisfy the equation of motion in a uniform, unperturbed, magnetic field.

We choose a coordinate system in which $\mathbf{k} = (k_\perp,0,k_\parallel)$. Then the Fourier transform of $\mathbf{j}$ is [Melrose, 1968]

$$j_i(\mathbf{k},\omega) = \frac{\pi|q|c^2}{(m^2c^4 + p^2c^2)^{1/2}} \sum_{\nu=-\infty}^{\infty} \left(\frac{q}{|q|} \frac{k_\perp}{|k_\perp|} \right)^{\nu+1} \Gamma_i^\nu \delta(\omega - \nu\Omega - k_\parallel v_\parallel) \tag{A.10}$$

The quantities Γ_i^ν are related to the Bessel functions $J_\nu(z)$ and their derivatives $J_\nu'(z)$ via the following relationships

$$\Gamma_1^\nu = 2p_\perp(\nu/z)\,J_\nu(z) \tag{A.11}$$

$$\Gamma_2^\nu = -2ip_\perp(q/|q|)\,J_\nu'(z)$$

$$\Gamma_3^\nu = 2p_\parallel(k_\perp/|k_\perp|)\,J_\nu(z)$$

where $z = k_\perp v_\perp / \Omega = k_x p_x / m\omega_c$, $\omega_c = |q|B/mc$ is the cyclotron frequency of the species of interest, $\Omega = \omega_c/\gamma_r$, and $\gamma_r = (1 - v^2/c^2)^{1/2}$. One should note that unlike the situation in the Earth's magnetosphere, the relativistic treatment is sometimes necessary in the Jovian magnetosphere where particles can often be accelerated to high energies [see, e.g., Wu and Freund, 1977; Sentman and Goertz, 1978]. After inserting Equation (A.10) into (A.8) the power reduces to

$$P = \sum_{\nu=-\infty}^{\infty} \frac{q^2 c^2}{4\pi(m^2c^4 + p^2c^2)} \tag{A.12}$$
$$\cdot \int d^3k \int (d\omega/\omega)\,|\lambda_{ss}|\,|\hat{e}_i\Gamma_i^{\nu*}|^2\delta(\Lambda)\delta(\omega - \nu\Omega - k_\parallel v_\parallel)$$

In (A.12) the first delta function expresses the fact that we are calculating the power radiated into a particular cold plasma mode ($\Lambda = 0$), and the second delta function expresses a resonance condition for the parallel velocity.

The time average power radiated into a mode $\omega(\mathbf{k})$ can also be defined in terms of the following quantum-mechanical analog. Assume that a particle is in a quantum state s corresponding to a momentum $p_\parallel$ and $p_\perp$ where

$$p_\perp^2 = (qB/c)\hbar(2n + 1) \tag{A.13}$$

$$s = \{p_\parallel,n\}$$

Note that for all practical purposes the quantization step $\Delta p_\perp^2 = (qB/c)\hbar$ is negligibly small and the quantum states really form a continuum.

We define the probability of spontaneous emission $w_s^{s'}(\mathbf{k})$ as the probability that a particle in state s emits a photon $\hbar\omega$ with transition to s'. Then

$$P = \sum_{s'} \int \frac{d^3k}{(2\pi)^3} \hbar\omega(\mathbf{k})w_s^{s'}(\mathbf{k}) \tag{A.14}$$

where the sum is over all possible final states. If we use

$$s' = \{p_\| - \hbar k_\|, n - \nu)$$

then we can rewrite (A.14) as

$$P = \sum_{\nu=-\infty}^{\infty} \int \frac{d^3 k}{(2\pi)^3} \hbar\omega(\mathbf{k}) w(\mathbf{k})_{p_\|, n, \nu} \tag{A.15}$$

From (A.12), we obtain for the transition probability

$$\tag{A.16}$$

$$w(\mathbf{k})_{p_\|, n, \nu} = \frac{2\pi^2 q^2 c^4}{\hbar(m^2 c^4 + p^2 c^2)} \int \frac{d\omega}{\omega^2} |\lambda_{ss}| |\hat{e}_i \Gamma_i^{\nu*}|^2 \delta(\Lambda)\delta(\omega - \nu\Omega - k_\| v_\|)$$

In this quantum-mechanical description the distribution function $f(\mathbf{p})$ of the particles is replaced by $f(s)$, the occupation number of each momentum state s, so that

$$\frac{1}{V} \Sigma f(s) = \text{particle number density} \tag{A.17}$$

where V is the total volume of the system. Similarly, the photons have a distribution $N(\mathbf{k})$ and momentum $\hbar\mathbf{k}$ given by

$$\int \frac{d^3 k}{(2\pi)^3} N(\mathbf{k}) = \text{photon number density} \tag{A.18}$$

The rate of increase in the density of the waves due to spontaneous emission is then

$$\frac{\partial N^{se}}{\partial t}(\mathbf{k}) = \frac{1}{V} \sum_{s,s'} s_s^{s'}(\mathbf{k}) f(s) \tag{A.19}$$

The classical limit can be recaptured by using the replacements

$$\frac{\Sigma f(s)}{V} \to \int d^3 p f(\mathbf{p})$$

and

$$w_s^{s'}(\mathbf{k}) \to w(\mathbf{k})_{p_\|, n, \nu} \tag{A.20}$$

so that the spontaneously generated wavepower $[N\hbar\omega(\mathbf{k})]$ is

$$\frac{\partial P^{se}}{\partial t} \equiv \alpha(\mathbf{k}) = \sum_\nu \int d^3 p f(\mathbf{p}) \frac{2\pi^2 q^2 c^4}{(m^2 c^4 + p^2 c^2)} \tag{A.21}$$

$$\int \frac{d\omega}{\omega} |\lambda_{ss}| |\hat{e}_i \Gamma_i^{\nu*}|^2 \delta(\Lambda)\delta(\omega - \nu\Omega - k_\| v_\|)$$

For induced processes (coherent interactions), the change in $N(\mathbf{k})$ is given by the difference between the rate that photons with momentum $\hbar\mathbf{k}$ are emitted and the rate they are absorbed. The Einstein relations state that the probability of induced emission and absorption, $s \to s'$ and $s' \to s$, respectively, are both equal to $w_s^{s'}(\mathbf{k}) N^{ie}(\mathbf{k})$. Thus, for induced processes,

$$\frac{\partial N^{ie}}{\partial t}(\mathbf{k}) = \frac{1}{V} \sum_{s,s'} w_s^{s'}(\mathbf{k}) N^{ie}(\mathbf{k}) [\, f(s) - f(s')\,] \tag{A.22}$$

Again taking the classical limit, we obtain for the sum of coherent and incoherent emission, (see Melrose [1968] for details)

$$\frac{\partial P}{\partial t}(\mathbf{k}) = \sum_{\nu} \int d^3p\, [\,(1 + (1/2)\, D_\nu)\, w(\mathbf{k})_{p_\parallel, n, \nu}\, P(\mathbf{k})\,]\, D_\nu f(\mathbf{p}) \tag{A.23}$$

$$+ \frac{\partial P^{se}}{\partial t} = \gamma(\mathbf{k}) P(\mathbf{k}) + \alpha(\mathbf{k})$$

where

$$D_\nu = \hbar \left[\frac{\nu \omega_c\, m}{p_\perp} \frac{\partial}{\partial p_\perp} + k_\parallel \frac{\partial}{\partial p_\parallel} \right]$$

A Taylor expansion of $f(s')$ has been used in deriving (A.23). The expression for the growth rate $\gamma(\mathbf{k})$ can now be found from (A.23) for D_ν under the assumption that $f(\mathbf{p})$ is gyrotropic, the quantum correction $(1/2)\, D_\nu$ is negligible, and $w(\mathbf{k})$ is given by (A.16). The final result is Equation (2.1).

Under many conditions one can also show [Kennel, 1966; Melrose, 1968; Hasegawa, 1975]

$$\gamma(\mathbf{k}) = Im\,[\Lambda(\mathbf{k}, \omega)] \,/\, \{\partial Re\,[\Lambda(\mathbf{k}, \omega)\, /\, \partial \omega]\} \tag{A.24}$$

10
MAGNETOSPHERIC MODELS

T. W. Hill, A. J. Dessler, and C. K. Goertz

Theoretical ideas concerning Jovian magnetospheric phenomena are at least as diverse as the phenomena themselves, and there presently exists no single comprehensive model that encompasses all known phenomena within a unified theoretical framework. We identify here a number of important theoretical concepts, some subset of which (together with perhaps others yet unidentified) will ultimately provide the elements of such a comprehensive model. A number of ideas have been advanced to account for the copious plasma source associated with Io, but none of these has yet accounted satisfactorily for both the magnitude and the morphology of the inferred source. Nevertheless, given the observed fact that Io supplies the bulk of the magnetospheric plasma mass, and the corollary that the net plasma transport is predominantly outward, it follows that the rotational energy of Jupiter is an important if not dominant source of energy for magnetospheric phenomena. This rotational energy is expended in a variety of phenomena, including the electrodynamic Io–Jupiter interaction and associated radio and auroral emissions, the acceleration of charged particles to MeV energies, and the generation of a wide variety of spin-periodic phenomena as observed both remotely and in situ. The spin periodicities observed within the magnetosphere can be explained for the most part as resulting from the diurnal wobble of the magnetospheric current sheet caused by the offset between Jupiter's magnetic dipole axis and its spin axis. However, remotely observed spin periodicities (the "pulsar" phenomena) apparently require the existence of an intrinsic longitudinal asymmetry in the Jovian magnetosphere that corotates with Jupiter.

10.1. Introduction

Many celestial bodies have magnetospheres, that is, surrounding regions within which the motion of charged particles is influenced by the magnetic field of the central body. In addition to the terrestrial magnetosphere, which has been studied in increasing detail over the past three decades, the magnetospheres of Mercury, Jupiter, and Saturn have now been explored in situ. The Sun has a magnetosphere of sorts (the heliosphere), and there is evidence that pulsars, accreting binary star systems, and certain radio galaxies also have magnetospheres. Of the planetary magnetospheres that have been explored, Jupiter's is by far the largest in terms of both absolute size and size relative to the planetary body. It is also a magnetosphere largely dominated by rotational effects; as such, it offers unique insight to the study of inaccessible pulsar magnetospheres.

The present state of knowledge of Jupiter's magnetosphere is reminiscent of the state of knowledge of Earth's magnetosphere in the early 1960s – the observational data are sufficient to inspire a variety of theoretical developments but not sufficient in most cases to dictate a clear choice among them. Future observations will probably verify some of the theoretical concepts described below, and will probably disprove others. We shall concentrate here on those theoretical concepts that we believe to be consistent with available observations as described elsewhere in this book, and we shall explicitly point out areas of divergent views or inadequate study.

Our data base is relatively incomplete in coverage (local time, latitude, time, etc.), compared to the terrestrial magnetospheric data base, but there is a tendency to suppose that we can learn more now from a limited data base by drawing on several years'

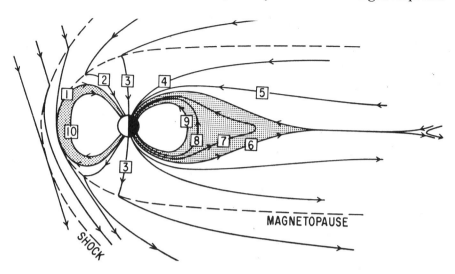

Fig. 10.1. The open model of Earth's magnetosphere [after Axford, 1969] in which, according to conventional wisdom, the solar-wind magnetic field becomes interconnected with the geomagnetic field on the dayside magnetopause (e.g., the field line labeled "1"), and the solar-wind motional electric field is transmitted to the polar cap to drive antisunward convection in the high-latitude magnetosphere (labels "2" through "5"). Conservation of total geomagnetic flux then requires a return sunward flow at lower latitudes (positions 6–10). Simple scaling arguments (see text) indicate that the Jovian polar cap formed in this manner would be small and the strength of the solar-wind-induced convection system would be correspondingly weak (in comparison with the rotational flow).

experience in studying Earth's magnetosphere. This is undoubtedly true in some important respects (for example, with respect to radiation-belt physics (Chaps. 4 and 5) or the generation of plasma waves (Chaps. 8 and 9). However, many Jovian magnetospheric phenomena have no apparent analogs in the terrestrial magnetosphere, and the fundamental differences can generally be attributed to the fact that the primary sources of plasma and energy are external in the case of Earth's magnetosphere but internal in the case of Jupiter's.

Perhaps the surest and most fundamental conclusions that have emerged from recent observational and theoretical work are that (1) the satellite Io is the primary source of Jovian magnetospheric plasma (Sec. 10.3), and that (2) the rotation of Jupiter is the primary source of energy for its magnetosphere (Sec. 10.4). These internal plasma and energy sources generate a wealth of phenomena that are subjects of active and continuing research, including the electrodynamic Io-Jupiter interaction (Sec. 10.5), the rapid acceleration of charged particles (Sec. 10.6), and the spin-periodic ("pulsar") behavior (Sec. 10.7), as well as plasma transport through the magnetosphere and the generation of the planetary wind (Chap. 11). These internally generated magnetospheric phenomena are of particular interest from a theoretical point of view inasmuch as they form a vital link between our relatively advanced understanding of the terrestrial magnetosphere (driven externally), and our relatively speculative understanding of astrophysical magnetospheres (believed to be driven internally). In order to set the context for our discussion of the internally driven phenomena, we first provide a sketch (Sec. 10.2) of what we mean by an externally driven Earthlike magnetosphere, and show why it is of limited applicability to Jupiter's magnetosphere.

10.2. An Earthlike model

In the case of Earth's magnetosphere, the solar wind drives a system of plasma convection across the polar caps in the antisolar direction with return sunward flow occuring at lower latitudes, and the Earth spins inside this Sun-oriented flow pattern. It is widely believed that this convection system is driven by the electric field induced by solar-wind flow and mapped into the polar cap along interconnected magnetic field lines (Fig. 10.1). The magnetospheric convection electric field is typically smaller than the unperturbed solar-wind electric field by a ratio $f \sim 0.1$. This ratio is frequently referred to as the "reconnection efficiency" because it represents the rate at which solar-wind magnetic flux becomes connected to the Earth, divided by the total available EMF (the solar-wind electric field times the diameter of the magnetospheric obstacle). Although this ratio is presently not predictable, its empirical value $f \sim 0.1$ is widely employed as a "universal constant" in scaling the Earth's magnetospheric convection system to other magnetospheres.

Given this "scaling law," and the other relevant parameters that are either measured directly or scaled straightforwardly, one can estimate the strength of solar-wind-induced convection at Jupiter. For example, Kennel and Coroniti [1977a] estimate a solar-wind-induced electric potential drop $\phi \sim f \times (5 \text{ MV}) \sim 0.5 \text{ MV}$ and an associated energy injection rate $K \sim f \times (4 \times 10^{14} \text{ W}) \sim 4 \times 10^{13} \text{ W}$. The available power is thus comparable to our estimate (in Sec. 10.4) of the rate at which energy is extracted from Jupiter's rotation to power magnetospheric phenomena. However, the available potential from the solar wind (~ 0.5 MV) is very small compared to the corotational potential $\Omega_J R_J^2 B_J \sim 376$ MV, which indicates that most of the power available from the solar wind is probably dissipated in a very small polar cap. Alternatively, we can compare the magnitudes of the solar-wind induced convection velocity and the corotation velocity. If we assume (for order-of-magnitude purposes) that the above potential (0.5 MV) is distributed uniformly across a magnetospheric diameter of 100 R_J, then the resulting convection velocity, within a dipole magnetic-field approximation, is much less than the corotation velocity throughout at least the dayside magnetosphere.

On the basis of such an argument, Brice and Ioannidis [1970] were the first to conclude that rotational effects may dominate solar-wind induced convection in determining the dynamics of Jupiter's magnetosphere. Their argument was generalized by Vasyliunas [1975], who removed the assumptions of uniform electric field and dipole magnetic field. Spacecraft observations tend to confirm this conclusion, and theoretical models have for the most part ignored solar-wind effects (although Coroniti and Kennel [1977] have attributed long-term variability [~ 1 week] to solar-wind-induced convection effects). The solar wind may, however, have significant influence on the Jovian magnetospheric tail and the magnetically connected polar caps, regions that have barely been explored by spacecraft. The solar wind is, in any case, responsible for establishing the day–night asymmetry of the magnetosphere, and this asymmetry has important theoretical implications. For example, the day–night asymmetry is the basis of the magnetic pumping process described in Section 10.6 below, and it is the coupling of the day–night asymmetry with an intrinsic longitudinal asymmetry that produces 10-hr variability (the "clock phenomena") within the magnetic anomaly model as described in Section 10.7.

10.3. The internal plasma source

Io, the innermost of the Galilean satellites, is the principal source of plasma for the Jovian magnetosphere. Secondary or negligible sources are the solar wind, the Jovian ionosphere, and the other Galilean satellites. As we will see in Section 10.4, considerations of the energetics of both the Io plasma torus and the more distant Jovian magnetosphere require an input of at least 6×10^{29} amu/s (approximately 1 ton/s) into the torus. This is the equivalent of about 3×10^{28} sulfur and oxygen ions/s; this requirement presents a considerable theoretical problem.

The Io source

A variety of approaches have been put forth to account for the Io source. They are quite different from one another and, in general, mutually exclusive. Furthermore, in their present form, none of them seem adequate to account for both the mass ejection rate and the other phenomena associated with the Io source. The source mechanisms depend on whether the atmosphere of Io is thick or thin and whether the material is ejected by direct escape, sputtering, magnetospheric pickup, ionospheric currents, and so forth. In this section, we discuss these various mechanisms and their difficulties.

Io is the most volcanically active body known. As a result of these volcanic emissions, its surface is covered principally with elemental sulfur and SO_2 frost [e.g., Fanale et al., 1979; Sagan, 1979]. The ejection velocity from the volcanic plumes is approximately a factor of three or more below the velocity of escape from Io [Smith et al., 1979a]. Material in the volcanic plumes rises along ballistic trajectories and falls back to the surface of Io, delivering kinetic energy at a rate of approximately 10^{11} W. The area where this material falls is heated slightly (less than 0.2 K) and the energy is radiated into space. Plume material that does not stick to the surface accommodates to the temperature of the surface. The scale height of an SO_2 atmosphere at the maximum daytime temperature of 135 K is only 10 km, and the mean thermal speed of the molecules is just over 100 m/s. The temperature at the volcanic vents is, of course, higher, but the vent exhaust speed is less than 1 km/s, which is still much less than the escape speed from Io (2.6 km/s). Thus, although one sees in the literature allusions to "impulsive injections into the Io torus by volcanic events," experimental and theoretical evidence is contrary to this concept. (However, see Mekler and Eviatar [1980] and Cheng [1980], for example, for an advocacy view.)

There appears to be a major problem in accounting for both the magnitude of the mass-injection rate from Io and the directionality of neutral atomic sodium escape from Io. Sodium is a relatively minor constituent of the torus, but it provides a useful tracer of the motion of neutral gas escaping Io, by virtue of its intense D-line radiation that is visible from Earth. Neutral sodium atoms are observed to stream predominantly from the side of Io that faces Jupiter (see review by Trafton [1981] and references therein). If the source were evaporation (Jeans escape), the sodium atoms would escape from Io more nearly isotropically; thus they do not appear to be evaporating from the top of a thick atmosphere.

Therefore, whereas a "thick" atmosphere of Io is apparently required to account for the large mass-injection rate, a "thin" atmosphere is apparently required to account for the directionality of the neutral sodium injection. A possible resolution of this dilemma would be at hand if it could be shown that there was a sudden and marked increase in the ionization rate in the torus just beyond the orbit of Io. One might then be able to allow an isotropic escape of sodium from Io, but the atoms departing from the side facing away from Jupiter would be quickly ionized and hence would not be detected optically.

In this context, a "thick" atmosphere is defined as one in which the base of the exosphere (the exobase) is several scale heights above the surface, so the surface is not subject to direct particle bombardment, and sputtering from the surface is suppressed.

A "thin" atmosphere is defined as one in which the exobase is near or below the surface (i.e., less than one scale height above the surface), and sputtering becomes a viable means of particle injection into the torus. The dividing line between thick and thin atmospheres occurs at a surface pressure of about 10^{-11} bar, which corresponds to a surface concentration of about 10^9 SO_2 molecules/cm^3 at a temperature of 135 K.

A useful upper limit to the surface atmospheric mass density is provided by the observed behavior of the volcanic plumes. The plumes appear to be formed by particles on ballistic (parabolic) trajectories. If the atmospheric mass density were sufficiently large, the atmospheric drag acting on dust particles in the plumes would produce noticeable distortions of the parabolic trajectories, and no such distortion is evident in the Voyager imagery, at least for the descending portion of the plumes [Cook, Shoemaker, and Smith, 1979]. For a rough estimate of this upper limit, we can set the drag force ($\rho v^2 A$) equal to the force of gravity (mg), where ρ is the atmospheric mass density, m, v, and A are the mass, velocity, and cross-sectional area of typical dust particles in the plumes, and g is the acceleration of Io's gravity. If we assume solid sulfur or SO_2 dust particles of radius 0.1 $\mu = 10^{-7}$ m [Collins, 1981] and velocity 1 km/s, we find an upper limit $\sim 10^{10}$/cm^3 for the number density of an SO_2 atmosphere. A similar but larger upper limit for the region immediately adjacent to an active plume is obtained if we reevaluate the results of Lee and Thomas [1980] using 0.1 μm particles rather than the 10 μm maximum diameter particles assumed in their analysis. This upper limit is a constraint on "thick" atmosphere models.

An entirely different limitation on the ionization rate in the immediate vicinity of Io is given by Shemansky [1980b], who derives an upper limit of 10^{27} ions/s of S^+ and O^+ ions on the grounds that any larger ionization rate would have created an ultraviolet glow around Io that would have been detected by Voyager. There are two ways around this limitation: (1) assume that the initial ionization near Io is of a molecular species (e.g., SO^+ or SO_2^+) that does not radiate at wavelengths that could be detected by Voyager, or (2) assume that material initially escapes from Io (or its atmosphere) in the form of neutral atoms, molecules, or dust, and is subsequently ionized some distance away (> 1 R_J) [Johnson, Morfill, and Grün, 1980; Brown and Ip, 1981].

Some material is stripped from the atmosphere of Io by a cometary type of interaction with the plasma in the Io torus, that is, atmospheric particles are ionized by electron impact and swept downstream in the corotational flow [Cloutier et al., 1978; Ip and Axford, 1980]. An upper limit of $S_n \sim 10^{26}$ ions/s for this source is given by Cloutier et al. [1978], who assume the atmosphere of Io is sufficiently thin to allow sputtering. A different approach is taken by Goertz [1980a] who assumes a relatively thick atmosphere with a concentration at the exobase $n_0 = 10^9$/cm^3, and a temperature $T = 1100$ K, and hence a scale height of 80 km. Assuming escape from the entire surface area of Io, he derives an escape flux of about 3×10^{28} ions/s. (One should reduce this estimate by a factor of two because the actual escape area is, at most, one-half the surface area of Io; ions formed on the Jupiter-facing side of Io are accelerated toward the surface by the corotation electric field as described below.) This estimate relies on the assumption that Io's atmosphere is thick (with respect to the mean-free path of incident ionizing electrons), because a thin neutral atmosphere (a) would have an "ionization efficiency" (number of ions produced per incident ionizing electron) less than unity, and (b) would maintain good thermal contact with the surface so that the temperature (and scale height) would be a factor of 10 less than assumed. Thus, for a thin atmosphere the source estimate would be reduced to $\leq 10^{27}$ ions/s (basically, the estimate of Cloutier et al. [1978] corrected with more recent observations of conditions in the Io torus).

As pointed out by Haff, Watson, and Yung [1981], the mass-injection rate can be made arbitrarily large for the case of a thick atmosphere, which can be heated significantly by particle bombardment at altitudes a few scale heights above the surface [Kumar, 1980]. The exobase for a thick, hot atmosphere can be far above the surface of Io where the gravitational attraction is diminished and the perturbing effect of Jupiter's gravity becomes important. Haff et al. [1981] give as one specific example an upper atmospheric temperature $T = 1500$ K and an exobase at 2.2 R_I. They propose atmospheric sputtering, and they show that sputtering efficiencies of 10^2 are expected. The required mass-injection rates can thus be realized.

The surface sputtering hypothesis, on the other hand, requires that the atmosphere of Io be thin enough (a surface pressure $\leq 10^{-11}$ bar) to allow both the impacting ions to reach Io's surface and the atoms sputtered from Io's surface to pass freely through the atmosphere into the torus [Johnson et al., 1976]. The observed unidirectional escape of sodium is accounted for in the sputtering hypothesis by noting that the Jupiter-facing side of Io is the one subject to ion bombardment caused by the $-\mathbf{v} \times \mathbf{B}$ electric field associated with the motion of Io relative to Jupiter and the plasma torus [Hill et al., 1979; Neubauer, 1980]. One possible source of this bombardment flux is the 10^6 A current inferred to flow toward the Jupiter-facing side of Io [Ness et al., 1979; Kivelson et al., 1979]. If we assume a distribution of ionization states such that, on the average, each ion carries a positive charge of 1.5 e, then the total ion flux delivered to Io by this current is $S_n \sim 4 \times 10^{24}$/s. The flux of energetic particles is much smaller [e.g., Krimigis et al., 1979]. In order to generate the required escape flux of at least 3×10^{28}/s the sputtering efficiency would have to be of the order of 10^4. This presents a problem because sputtering efficiencies this large are not expected [e.g., Brown et al., 1980].

Other bombardment fluxes that could cause the required sputtering are: (a) corotating ions in the plasma torus and (b) acceleration of ions created in the plumes or in Io's atmosphere. The maximum bombardment flux for source (a) is $S_n = nvA$ where n is the plasma-torus number density, $\mathbf{v}$ is the velocity of the torus ions relative to Io, and A is the cross-sectional area of Io. For $n = 10^3$ ion/cm^3, $v = 5.7 \times 10^4$ m/s, and $A = 1 \times 10^{13}$ m^2 we obtain $S_n = 6 \times 10^{26}$ ions/s, and the sputtering efficiency need be only 10^2. If these ions have an average atomic weight of 22 amu, their average streaming energy is 360 eV. The sputtering efficiency of such low energy ions is not known. The principal difficulty with this source is that the ions would strike Io on its trailing hemisphere (relative to its orbital motion), whereas the sodium observations indicate that the source is the hemisphere facing Jupiter.

Source (b) has the advantage of being in the correct hemisphere. Ions created near Io on its Jupiter facing side will be accelerated toward Io by the electric field associated with the motion of the plasma torus past Io [Cummings et al., 1980]. The potential across Io is about 400 kV, and the resulting external electric field is in the direction to accelerate ions toward Io if they are on its Jupiter-facing side. The bombarding flux is $S_n = SA$ where S is the column production rate of ions and A is either the cross-sectional area of Io or the area covered by volcanic plumes on Io's Jupiter facing side. Cloutier et al. [1978] estimate S to be 3.5×10^{12}/m^2-s, and A is either 1×10^{13} m^2 or 4×10^{11} m^2 (the area of the plumes). Thus for this source, S_n ranges from 3×10^{25}/s to 10^{23}/s, and the required sputtering efficiency ranges from 10^3 to 10^5. On the other hand, if we adopt a larger source rate, equal to the upper limit deduced by Shemansky [1980b] from the UV observations, diminished by a factor of 4 (the ratio of surface area to cross-sectional area), the source rate would be $S_n \sim 3 \times 10^{26}$/s and the required sputtering efficiency would be $\sim 10^2$. The energy of the impacting ions would

Table 10.1. *Injection of ions into the Jovian magnetosphere*

Source	S_m (kg/s)	S_m (amu/s)	Avg atomic wt. of ions
Io	$\geq 10^3$	$\geq 6 \times 10^{29}$	~ 22
Solar wind	$< 10^2$	$< 6 \times 10^{28}$	1
Jovian ionosphere	≥ 20	$\geq 10^{28}$	1
Other Galilean satellites	1	6×10^{26}	6

range from zero to the full corotational energy in Io's reference frame (540 eV for sulfur ions, 270 eV for oxygen ions).

Finally, the plasma-arc mechanism suggested by Gold [1980] merits consideration. Is it possible that the $10^6 A$ current flowing to and from Io could concentrate into arcs at the location of volcanic events and heat some portion of the plume sufficiently to cause direct escape into the torus? Or, could some of the dust particles in the plume become charged by this current and then be electrically accelerated to escape speed [Johnson, Morfill, and Grün, 1980]? For example, jets or "linear features" of un-ionized sodium extending outward from Io have been reported [Pilcher and Strobel, 1981] (see also Chap. 6). This is the sort of ejection that might be expected from a concentrated plasma-arc source. Also, Schaber [1980] reports that nearly all of the volcanic vents on Io are in the equatorial zone, again as expected from a current driven source [Gold, 1980]. The plasma-arc mechanism has not been pursued beyond Gold's original suggestion, except for one attempt to find the small, bright spots that might be expected at the feet of the plasma arcs. Instead of spots, more diffuse "electric-glow discharges" were observed by Cook et al. [1981], presumably at the tops of the plumes, which they refer to as auroras. A quantitative analysis of how the $10^6 A$ current flows into Io, whether diffuse or concentrated into spots, and whether the power associated with this current (as large as 4×10^{11} W, which is comparable to the 10^{11} W expended by the plumes) influences the structure and dynamics of the volcanic plumes, is yet to be developed.

All the above considerations and difficulties may be complicated considerably if, as has been suggested, Io has an intrinsic magnetic moment that creates a magnetosphere that would protect most of the atmosphere from charged particle bombardment [e.g., Neubauer, 1978; Southwood et al., 1980; Kivelson and Southwood, 1981]. The theoretical problem of explaining how at least 1 ton/s of Io material is delivered to the Io plasma torus awaits a solution supported by a full and careful analysis (and perhaps a mechanism or mechanisms yet to be proposed). However, that Io is the primary source of plasma for the Jovian magnetosphere is not in doubt (see Table 10.1).

The solar-wind source

It is commonly accepted that the solar wind is a primary source of plasma for Earth's magnetosphere (although recent measurements have drawn attention to the importance of Earth's ionosphere as an additional source [e.g., Johnson, 1979]). The mass flux of solar-wind plasma entering a planetary magnetosphere is $S_m = \rho_m A V_s \eta$, where ρ_m is the mass density of the solar wind, A is the cross-sectional area of the magnetosphere, V_s is the solar wind speed, and η is the fraction of incident solar-wind particles absorbed by the magnetosphere. For the Earth, η is inferred to be of the order of 10^{-3} [Hill, 1974], which yields $S_m = 5 \times 10^{-2}$ kg/s. If we assume the same value of η for Jupiter, and use a circle of 100 R_J radius for the magnetospheric cross section, we obtain $S_m = 20$ kg/s. Even if we increase η to 10^{-2}, S_m is only about 100 kg/s, which is the upper limit listed in Table 10.1. There is a theoretical expectation [e.g., Lemaire, 1977], and some evidence from measurements of the abundances of energetic particles, that solar-wind injection may be important near the front of the magnetosphere [Krimigis et al., 1979; Vogt et al., 1979]. However, this is a region in which there are relatively few particles from any source, so we conclude that the solar wind is negligible as an overall source of plasma for the Jovian magnetosphere.

The ionospheric source

It has been confidently expected, in analogy with Earth, that Jupiter's ionosphere is a source of plasma for the Jovian magnetosphere [Gledhill, 1967; Melrose, 1967; Ioannidis and Brice, 1971; Goertz, 1973; Michel and Sturrock, 1974; Hill et al., 1974; Carbary et al., 1974; Mendis and Axford, 1974; Swartz et al., 1975; Goertz, 1976b], and this expectation is supported by direct observation of energetic H_2^+ and H_3^+ ions [Hamilton et al., 1980].

One ionospheric source arises from development of a relatively hot photoelectron population (energies greater than about 10 eV) that pulls ions along as the photoelectrons escape from the ionosphere. Swartz et al. [1975] estimate that this flux amounts to about 10^{28}/s integrated over Jupiter's ionosphere. Processes such as precipitation of energetic particles [Dessler and Hill, 1975] or dissipation of energy from Birkeland currents [Dessler and Chamberlain, 1979] create additional hot electrons and enhance the escape flux from the ionosphere. Finally, the escape flux can be enhanced by heating of the ionosphere by precipitation of soft electrons [Hunten and Dessler, 1977; Dessler et al., 1981], or the heating of electrons by the precipitation of sulfur and oxygen ions with energies in the low MeV range [Thorne, 1981] (see also Chap. 12).

None of these energization mechanisms makes the mass flux from the ionosphere competitive with the Io source. Direct measurements show that the ionic composition within the magnetosphere is dominated by sulfur and oxygen, with hydrogen ions (the predominant ionospheric constituent) making up no more than 1% of the mass density or 20% of the number density in the inner magnetosphere [Sullivan and Bagenal, 1979; Krimigis et al., 1979]. However, this does not mean that the ionospheric source is unimportant in its effect on Jovian magnetospheric dynamics. One important function of ionospheric plasma is to carry the Birkeland currents that enforce (partial) corotation and link Jupiter's ionosphere with the Io torus; other important effects of ionospheric plasma are discussed in Section 10.7 and Chapter 12.

Other satellite sources

With Io being such an important source of plasma for the Jovian magnetosphere, it is natural to ask what the input might be from the other satellites, particularly the other three Galilean satellites. The significance of such plasma sources was investigated by Hill and Michel [1976] before it was recognized that Io was such an overwhelming source. The problem with getting much plasma from the other satellites (a problem that is avoided by Io) is that, once sputtering has removed the easy elements, the surface is then coated with a protective residue of material that is difficult to sputter. As an example of the severity of this problem, Lanzerotti et al. [1978] estimate that of the order of 1 km of exposed ice could be eroded from Europa over a period of 10^9 yr. (More recently, Johnson et al. [1981] have revised this erosion estimate downward by a factor of ten). It is reasonable to expect that some contaminant that is difficult to sputter, such as silicon or iron, will be concentrated on the surface as the ice is sputtered away; such a residue layer acts to prevent further sputtering. Io alone (from which about 500 meters of material would be sputtered in 10^9 yr at the presently inferred rate) avoids this limitation by the continuous resurfacing action of its volcanoes [Johnson et al., 1979].

Additional evidence that sputtering of water ice from the other satellites is not an important plasma source is provided by the ion composition measurements that show that H^+ is a negligible ionic constituent [Sullivan and Bagenal, 1979; Bagenal and Sullivan, 1981]. If a significant number of H_2O molecules were being sputtered into the Jovian magnetosphere and subsequently dissociated and ionized, the proton population would be more evident. Finally, the most recent proposal to account for the high concentrations of sulfur on the surfaces of Europa and Ganymede tends to make those satellites sinks rather than sources of magnetospheric particles [Eviatar et al., 1981b].

Time variations

Ground-based and spacecraft data show that both the temperature and the content of the Io plasma torus vary with time (see Chap. 6). These changes tend to be slow (time periods of months), and their cause has not yet been determined. The analysis of such time variations from Earth-based observations is complicated by the fact that the optical torus exhibits a gross longitudinal asymmetry [Trafton, 1980; Pilcher and Morgan, 1980; Trauger et al., 1980], being two to five times brighter in singly ionized sulfur emissions within the "active sector" ($\lambda_{III} \approx 175°$ to $320°$; see Sec. 10.7) than at other longitudes. Some analyses in which time variations have been reported have tacitly assumed that the torus was axially symmetric and hence have attributed all brightness variations to time variations rather than to the rotation of the bright portion of the torus into and out of the field of view [e.g., Mekler and Eviatar, 1980]. Similarly, Walker and Kivelson [1981] assume axial symmetry in a comparison of V 1 inbound and P 10 inbound torus data, and they attribute the inferred difference in torus ion concentration to time variations alone. What they have reported as time variations are due, at least in part, to spatial rather than temporal variations of the torus ion concentration.

Examination of the groundbased data of Trafton [1980] and of Voyager 1 and 2 spacecraft observations of Sandel et al. [1979], together with new analysis of the Pioneer 10 solar-wind detector data by Intriligator and Miller [1981] and L. A. Frank (private communication, 1981), indicate that the plasma torus may not deviate by much more than a factor of three from its average state. This is what one might expect from Io, a satellite whose surface is continually renewed by volcanic activity.

There is, however, a considerable body of opinion supporting the proposition that the torus content and geometry can change by more than a factor of 10 within a time scale of a year or two. The principal evidence to support this view is the Pioneer 10 UV spectrometer data, which were interpreted as showing only a thin torus, incomplete in longitude [Carlson and Judge, 1974, 1975]. This is in contrast to the dense, complete torus discovered by Voyager 1 and confirmed by Voyager 2 and ground observations (see Chap. 6). Does the Io plasma torus have a weak, incomplete, quiescent state as suggested by the Pioneer UV spectrometer? If so, magnetospheric theorists face yet another formidable challenge.

10.4. The internal energy source

Because of the vast size and rapid rotation of Jupiter's magnetosphere, it is widely believed that the rotation of Jupiter provides the dominant source of energy for magnetospheric phenomena (unlike terrestrial magnetospheric phenomena, for which the

Table 10.2. *Jupiter's energy budget*

Phenomenon	Power source (sink) (W)			
	Ionosphere	Io torus	Magnetosphere	External
Solar UV	(4×10^{12})			4×10^{12}
Particle precipitation	(10^{14})	$<<10^{14}$	10^{14}	
Secondary electron emission	$\lesssim 10^{14}$	$(\lesssim 10^{14})$		
Photoelectron emission	2×10^{9}	$(<<2 \times 10^{9})$	(2×10^{9})	
Enforcement of corotation:				
newly created ions	$\sim 10^{12}$	$(\sim 10^{12})$		
radially transported ions	$\gtrsim 5 \times 10^{13}$		$(\gtrsim 5 \times 10^{13})$	
UV aurora	10^{12}			(10^{12})
X-ray emission	8×10^{9}			(8×10^{9})
Io Alfvén waves	$(<10^{12})$	$<10^{12}$		
Torus UV emissions		2×10^{12}		(2×10^{12})
Particle acceleration (e.g., merging, magnetic pumping, etc.)	$\sim 10^{14}?$	$(?)$	$(\sim 10^{14})$	
Planetary wind			2×10^{13}	(2×10^{13})
Solar-wind-induced convection	$(\sim 2 \times 10^{13})$		$(\sim 2 \times 10^{13})$	$\sim 4 \times 10^{13} \ (?)$
Total	$\sim 1.3 \times 10^{14}$	$(\lesssim 10^{14})$	$(\gtrsim 5 \times 10^{13})$	$\sim 2 \times 10^{13}$

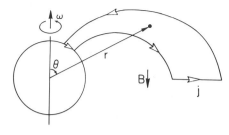

Fig. 10.2. The current system that transmits torque from Jupiter's ionosphere to plasma in the magnetosphere (e.g., in the Io torus) [Hill, 1979]. The ionospheric Pedersen current draws angular momentum from the corotating atmosphere, and the equatorial transverse current adds angular momentum to plasma as it is injected into the torus or transported outward from the torus. The ionospheric and equatorial transverse currents are linked by Birkeland (magnetic-field-aligned) currents.

dominant energy source is considered to be the solar-wind interaction). This rotational energy source is clearly responsible for enforcing (partial) corotation on plasma that is produced within the magnetosphere and transported outward. It probably also powers, directly or indirectly, various other phenomena such as auroral emissions (Chaps. 2 and 12), plasma heating (Chap. 12) and the resultant optical and UV line emissions (Chap. 6), the system of radial plasma transport in the magnetosphere (Chap. 11), and the emission of energetic electrons (Chap. 5) and radio noise (Chaps. 7 and 9). Our present estimates of the energy budget of Jupiter's magnetosphere are summarized in Table 10.2. (Altogether, these phenomena expend less than 10^{15} W, which represents an insignificant drain on Jupiter's 6×10^{34} J of rotational kinetic energy.)

The most obvious mechanism for tapping this rotational energy source consists of the Birkeland (magnetic-field-aligned) current system (Fig. 10.2) that extracts rotational energy from the Pedersen-conducting layer of Jupiter's atmosphere and invests that energy in producing and maintaining the (partial) corotation of plasma that is injected and transported outward in the magnetosphere. The local corotation energy at distance r for a particle of mass m is $W_c = m\Omega_J^2 r^2/2$, but newly injected ions generally attain an initial energy of $2W_c$ near the equatorial plane, regardless of where they are produced along the guiding field line. If the ions originate in Jupiter's ionosphere and are drawn away by hot photoelectrons as described in Section 10.3, the centrifugal slinging effect [Hill et al., 1974] produces a field-aligned streaming energy $W_\| = W_c - mM_JG/r \approx W_c$ in addition to the corotational drift energy ($= W_c$). If the ions are instead produced near the equatorial plane (e.g., by ionization of torus molecules as described in Sec. 10.3), the corotation electric field imparts a cyclotron energy $W_\perp = W_c - mM_JG/r \approx W_c$ in addition to the corotational drift energy. For intermediate injection points the energy will be distributed in an intermediate way between parallel and perpendicular degrees of freedom [see, e.g., Cummings et al., 1980] but the equatorial energy will generally be approximately $2W_c$. If the injection rate is S_n (ions/s), the corresponding rate of energy extraction is $2S_nW_c$.

A considerably larger investment of rotational energy is required to maintain the state of (partial) corotation as the plasma moves outward from its source [e.g., Dessler, 1980; Eviatar and Siscoe, 1980; Hill, 1981; Hill, Dessler, and Maher, 1981]. The primary source of magnetospheric plasma is the Io plasma torus at $L \approx 6$ (Sec. 10.3), and, as this plasma moves outward (probably in a rotation-driven convection system as described in Sec. 10.7 and Chap. 11), it falls through a net outward centrifugal-gravitational force field

$$g = \omega^2 r - M_J G/r^2 \tag{10.1}$$

$$\approx \omega^2 r \; (r \gtrsim 6 \, R_J)$$

where $\omega(r) < \Omega_J$ is the angular frequency of partial corotation (see Chap. 11). If plasma mass is transported outward through this effective gravity field at a rate S_m (kg/s), the rate of extraction of rotational energy can be estimated as

$$K_c = (S_m)\left[\frac{1}{2}\,\Omega_J^2(6R_J)^2 - M_J G/(6R_J) + \int_{6R_J}^{\infty}\omega^2\,r\,dr\right] \tag{10.2}$$

The first two terms represent the work required to accelerate the ions to corotation at Io's orbit, and the third term is the work expended in maintaining the state of (partial) corotation during the subsequent outward transport. It has been shown [Hill, 1979] that corotation is approximately enforced ($\omega \approx \Omega_J$) out to distances $r \sim R_c$ where

$$(R_c/R_J)^4 = \pi\Sigma B_J^2 R_J^2/S_m \tag{10.3}$$

where Σ is the height-integrated Pedersen conductivity of Jupiter's atmosphere. (See the detailed discussion in Chap. 11.) Nominal values $\Sigma \sim 0.1$ mho, $S_m \sim 1.7 \times 10^3$ kg/s yield the estimate $R_c \approx 20\ R_J$ [Hill, 1980], consistent with the Voyager observations described in Chapter 3. For $r \gtrsim R_c$, $\omega \propto r^{-2}$ and the above integral can thus be approximated by

$$K_c \approx \frac{1}{2}\,S_m\,\Omega_J^2\,R_c^2 \tag{10.4}$$

$$= (\pi/4)^{1/2}\,(\Sigma S_m)^{1/2}\,B_J\,R_J^3\,\Omega_J^2$$

$$\approx (5 \times 10^{13}\ \text{W})\,(\Sigma_{-1}\,S_{m\,30})^{1/2}$$

where $\Sigma_{-1} = \Sigma/(0.1$ mho) and $S_{m30} = S_m/(10^{30}$ amu/s). We have used "nominal" estimates for Σ and S_m, the actual values of which are not well determined, although their ratio is approximately determined by the Voyager observations (within the above model). Estimates of Σ range from <0.1 mho to ~ 10 mho, depending on the rate of energetic electron precipitation (Chap. 2). If, for example, Σ and S_m were both increased by a factor of 50, as suggested by Hill, Dessler, and Maher [1981], the above estimate of the energy deposition rate K_c would be increased by the same factor, while the estimate of R_c (which is constrained by the observations) would remain unchanged. The estimated energy deposition rate is at least comparable to that required to produce the observed Jovian aurora (Chap. 6). The same Birkeland current system dissipates a comparable amount of energy by Joule heating in Jupiter's atmosphere (Chap. 11), which represents an important high-latitude atmospheric heat source (Chap. 2).

The rotational energy source is evidently responsible for a variety of important effects on the structure and dynamics of Jupiter's outer magnetosphere. The centrifugal stress of the (partially) corotating plasma is partially responsible for the outward distortion of the magnetic field in a disclike structure, and hence for the overall inflation of the magnetosphere. Various theoretical models of the resulting plasma/field configuration have been developed [e.g., Goertz, 1976b; Carbary and Hill, 1978; Hill and Carbary, 1978; Vickers, 1978], and are discussed in Chapter 11.

The plasma injected by Io (and other sources interior to the magnetosphere) must ultimately escape the magnetosphere at the same average rate, and this escape is generally thought to occur in the form of a super-Alfvénic outflow called a planetary wind or a magnetospheric wind. (Other loss processes such as recombination or precipitation into Jupiter's atmosphere are probably unimportant beyond $L = 6$.) This planetary wind is sometimes decribed by analogy with axially symmetric stellar-wind models [e.g., Kennel and Coroniti, 1977a] which, however, do not really apply to Jupiter for at least two reasons: they assume a steady-state outflow and they lack the day–night asymmetry imposed by the solar wind. Because of this day–night asymmetry, Hill et al. [1974] have proposed that the planetary wind develops only on the night side of Jupiter, where the restraining solar-wind pressure is absent. Voyager data (see Chap. 4) confirm this basic geometry. A self-consistent quantitative model of a centrifugally driven planetary wind has not yet been developed (see Chap. 11).

If the rotational energy source is coupled with a corotating longitudinal asymmetry of the plasma mass distribution, as required by the magnetic-anomaly model described in Section 10.7, the result is a convective flow pattern that corotates with Jupiter [Vasyliunas, 1978; Hill, 1981; Hill, Dessler, and Maher, 1981]. This corotating convection system may provide an important radial transport mechanism (see Chap. 11). Other important rotational effects include the Io–Jupiter interaction, the acceleration of energetic particles, and the spin-periodic variations in the outer magnetosphere and interplanetary space. The remainder of this chapter is devoted to a discussion of these three effects.

10.5. The Io-Jupiter interaction

As Io moves through the corotating plasma torus, it disturbs the electric and magnetic fields as well as the particle distributions in its vicinity (see Chaps. 1, 3, 4, 5, and 6). The nature of this disturbance depends on the conductivity of Io or its ionosphere, the strength and orientation of its intrinsic magnetic field (if any), the parameters of the surrounding plasma, and the boundary conditions imposed by the Jovian ionosphere. The disturbance can be described in terms of waves ranging from zero frequency in Io's frame (i.e., a pattern carried around by Io) to very high frequencies (e.g., whistler-mode waves). Four types of models have been considered: low-frequency wave models, steady-state unipolar inductor models, acceleration models, and particle sweeping models. None of these models can be regarded as independent of the others; they all represent different aspects of the same interaction. For example, the amount of particle sweeping (Chap. 5) depends on the electric field in Io's vicinity, which in turn is related to the transverse current flowing at Io as discussed below. The transverse current is connected to Birkeland (magnetic-field aligned) currents that may be carried by low-frequency Alfvén waves. If these Birkeland currents are sufficiently large, parallel (magnetic field aligned) electric fields may occur, resulting in the direct acceleration of particles along the field lines. Although Voyager 1 passed very close to the Io flux tube, we still have insufficient data to decide which type of interaction is dominant. It may even be that the nature of the interaction changes with time in response to time-varying plasma parameters at Io's orbit.

Alfvén wave model

Independent of the exact nature of the perturbation in Io's vicinity, energy will be radiated away in the form of Alfvén waves. This is true in general for large conducting bodies moving through space [Drell et al., 1963]. This idea was originally applied to Io by Warwick [1961], Schmahl [1970], and Goertz and Deift [1973], and more recently by Neubauer [1980], Southwood et al. [1980], and Goertz [1980a].

Fig. 10.3. Sketch of the magnetic-field
distortion caused by Alfvén waves that
are generated by the relative motion **V**
between Io and the corotating
magnetospheric plasma. The waves
propagate along the background
magnetic field **B**$_0$ in the plasma reference
frame; in the Io reference frame (shown
here), the waves propagate along
characteristics **s** whose angle θ_A with **B**$_0$
is determined by the Alfvén mach
number of the relative flow. Reflection
from the Jovian ionosphere is neglected
in this picture.

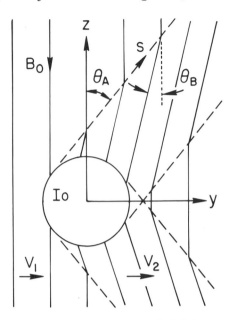

Suppose that near Io (e.g., in its ionosphere) the plasma velocity as seen in Io's frame
is v_2, and that upstream from Io the plasma velocity v_1 is the local corotation velocity
transformed to Io's frame: $v_1 = v_c - v_I$. The deceleration of the torus plasma from v_1 to
v_2 is accomplished by the **J** × **B** force in the extended ionosphere of Io. The associated
magnetic-field perturbations communicate the stress between Io and the torus plasma
flow. These perturbations propagate along the field lines by means of Alfvén waves,
while being carried downstream with the plasma. Thus, the current is confined
between "Alfvén wings" which are parallel to the characteristics **s** inclined relative to
the background field **B**$_0$ by the Mach angle $\theta_A = \tan^{-1} v_1/V_A$ where $V_A = B_0/(\mu_0 nm)^{1/2}$ is
the Alfvén speed (Fig. 10.3). For $v_1 = 57$ km/s and $V_A \sim 400$ km/s (corresponding to
$nm \sim 10^4$ amu/cm³), $\theta_A \sim 8°$. Between the wings the field is tilted by an angle $\theta_B < \theta_A$
which depends on the amplitude of the perturbation.

   Here we discuss only a simplified two-dimensional problem as illustrated in Figure
10.3, and we explicitly assume that Io has no significant magnetic field of internal
origin. The Birkeland current is given by [Goertz and Boswell, 1979; Neubauer,
1980; Southwood et al., 1980; Goertz, 1980a]

$$j' = \int j_B dx = \Sigma_A B_0 (v_1 - v_2) \tag{10.5}$$

where

$$\Sigma_A = 1/(\mu_0 V_A)$$

This current is supplied by Io, whose effective conductivity Σ_I includes Pedersen, Hall,
and "pick-up" conductivities as described below. The current through Io or its
atmosphere is

$$j'_{Io} = \Sigma_I B_0 v_2 \tag{10.6}$$

Combining (10.5) and (10.6) gives

$$\frac{v_2}{v_1} = \frac{\Sigma_A}{\Sigma_I + \Sigma_A} \tag{10.7}$$

$$j' = B_o v_1 \frac{\Sigma_I \Sigma_A}{\Sigma_I + \Sigma_A} \tag{10.8}$$

A three-dimensional analysis is required to find the flow pattern around Io [see, e.g., Neubauer, 1980; Goertz, 1980a]. However, the simple relations above illustrate the salient features of this local model for the interaction. The plasma velocity v_2 and hence the electric field in Io's frame depend on the ratio Σ_I/Σ_A. For small Σ_I/Σ_A the torus plasma impinges freely onto Io, and Io's low-energy plasma absorption cross section is equal to its geometrical cross section. (If Io has an intrinsic magnetic field its cross section may even be larger [Ip, 1981].) As Σ_I/Σ_A increases, the flow lines tend to avoid Io and the cross section for absorbing plasma becomes smaller. Because the amount of sweeping is proportional to this cross section, sweeping models cannot be evaluated independently of the Alfvén-wave model.

Steady-state unipolar model

Several authors [e.g., Piddington and Drake, 1968; Goldreich and Lynden-Bell, 1969; Gurnett, 1972; Shawhan et al., 1973, 1974; Shawhan, 1976; Dessler and Hill, 1979] have argued that the strength of the Io current is actually determined by the conductance Σ_J of the Jovian ionosphere, that is,

$$j' = B_o v_1 \frac{\Sigma_I \Sigma_J}{\Sigma_I + \Sigma_J} \tag{10.9}$$

Note that Σ_J instead of Σ_A determines j' in these models (cf. (10.8) and (10.9). These nonlocal models can be viewed as extensions of the Alfvén wave model. Thus far we have treated the field lines as infinitely extended, whereas they actually terminate (insofar as Alfvén wave propagation is concerned) in the conducting ionosphere of Jupiter. When the Alfvén wave reaches the ionosphere, currents there are driven by the electric field E of the wave. The current in the ionosphere, however, is not simply the Ohmic current $\Sigma_J E$, unless that current happens to match the polarization current of the Alfvén wave (i.e., unless $\Sigma_A = \Sigma_J$). In all other cases the wave electric field will be partially reflected, and the magnitude of the reflected wave depends on the ratio Σ_J/Σ_A. (Although Σ_A is rather well determined from in situ (Voyage 1) measurements, Σ_J is very uncertain as discussed in Chap. 2.) The reflection coefficient for Alfvén waves [Scholer, 1970] is

$$R = \frac{1 - \Sigma_J/\Sigma_A}{1 + \Sigma_J/\Sigma_A} \tag{10.10}$$

The reflected wave propagates back toward Io along the distorted magnetic field and is carried along with the plasma (flowing at the reduced velocity v_2). Thus between the wings of the Alfvén waves the return wave propagates along the characteristic s' as indicated in Figure 10.4. Regions 1 and 2 are as above; in region 3 the velocity is $v_3 = v_2(1 + R) - v_1 R$. If the reflected wave returns to Io (as illustrated in Fig. 10.4), it will be reflected back toward the Jovian ionosphere, resulting in a further deceleration of the plasma in region 4, and so forth. The distance δ that the plasma moves downstream during one wave bounce period is given by the integrals along the characteristics s and s':

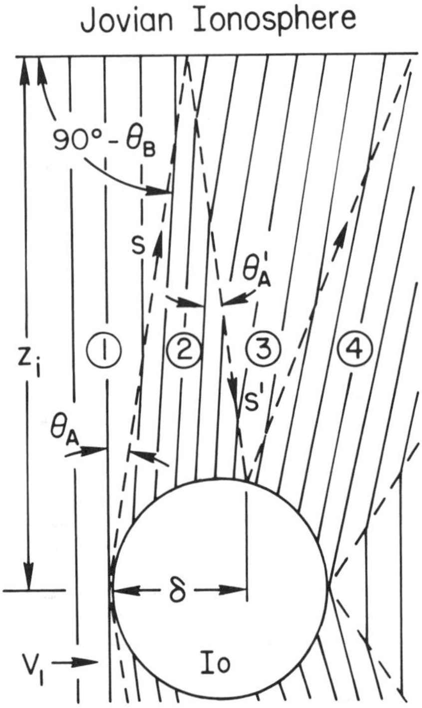

Fig. 10.4. The modification to Figure
10.3 caused by a single reflection of the
Alfvén wave from Jupiter's ionosphere.
The flow velocity (in Io's frame) is
reduced in region 2 compared to its
upstream (region 1) value, and the
characteristic wavefront inclination θ'_A of
the reflected wave is therefore different
from that of the initial wave (θ_A). The
wave is convected downstream a
distance δ before returning to the
equatorial plane after reflection from the
ionosphere; strong Io–Jupiter coupling
requires $\delta \lesssim 2 R_I$.

Jovian Ionosphere

$$\delta = \int_0^{z_i} \tan\theta_A \, dz - \int_0^{z_i} \tan(\theta_B - \theta'_A) dz \qquad (10.11)$$

The steady-state unipolar inductor model corresponds to the extreme case $\Sigma_I >> \Sigma_A$,
$\Sigma_J >> \Sigma_A$, for which $v_2 \approx 0$, $R \approx -1$, $\tan\theta_B \approx \tan\theta_A = v_1/V_A$, $\tan\theta_A' \approx 0$, and
$\delta \approx 0$, that is, the wave returns exactly to Io and after one round trip the current is
twice the original Alfvén wave current. The process is then repeated until the current
through Io is equal to $\Sigma B(v_1 - v_2)$ and a steady state is reached, which is described by
Equation (10.9) [Goertz and Deift, 1973]. The condition for such a steady state is
$\delta << 2 R_I$, or

$$\int \frac{ds}{V_A} << R_I/v_2 \qquad (10.12)$$

That is, in order for the return wave to hit Io, the round-trip Alfvén wave travel time
must be much less than the time required for the plasma to convect past Io $(2R_I/v_2)$.
Before the Voyager 1 encounter, the Alfvén-wave travel time was estimated to be less
than $2R_I/v_1 \sim 60$ s [Goldreich and Lynden-Bell, 1969; Goertz and Deift, 1973] and this
condition was believed to be fulfilled. However, the Alfvén speed is reduced within the
plasma torus, and the wave travel time is now known to be much larger (generally of
the order of 1000 s) [Neubauer, 1980]. It appears that condition (10.12) is not generally
satisfied, and that the currents driven by Io and the associated perturbations of electric
field and plasma flow are determined largely by local plasma conditions rather than by
the Jovian conductivity. (Note, however, that the time required for the plasma to convect past Io may also be increased if $\Sigma_I \gtrsim \Sigma_A$.) In this case the maximum current that
Io can drive in the magnetosphere is

$$j'_{max} = \frac{1}{2} B v_1 \Sigma_A \sim 10^6 \; A/R_I \tag{10.13}$$

Acuña (Chap. 1) has shown that the magnetic-field perturbation observed in the vicinity of Io's flux tube is compatible with this value. Likewise, the perturbations of the plasma flow near Io's flux tube are in apparent agreement with the Alfvén wave picture (Chap. 3).

Note that the Alfvén-wave travel time quoted above (~ 1000s) applies to torus parameters established by Voyager 1 inbound, and to those times when Io is deeply imbedded within the torus. When the Jovian magnetic dipole points toward or away from Io (i.e., when Io is near $\lambda_{III} = 200°$ or $20°$, respectively), the path length for Alfvén-wave propagation within the torus is less because Io is near the edge (northern or southern, respectively) of the torus. At these longitudes the wave travel time to the Jovian ionosphere (northern or southern, respectively) would be reduced to something like the pre-Voyager estimates, and a direct (unipolar-inductor) interaction might be anticipated at these longitudes if $\Sigma_I \gtrsim \Sigma_A$ (i.e., if the plasma flow past Io is significantly impeded; see Sec. 10.4).

Parallel electric fields

Several models have considered the consequences of parallel (magnetic-field-aligned) electric fields within the Io flux tube, usually within a steady state formulation (although parallel electric fields may also be expected to occur in a nonsteady interaction). Let us denote the Jovian height integrated Pedersen current by $j' = \Sigma_p E'$ where E' is the field in the ionosphere as seen in the planet's frame, the parallel current density by j_z, the parallel potential drop between the equatorial plane and the ionosphere along a field line by ϕ, and the electric field caused by the motion of Io relative to the magnetosphere, suitably mapped to the planet's surface, by E'_s. Then the current continuity equation reads

$$\partial j'_y/\partial y = -j_z \tag{10.14}$$

with the coordinate system as defined in Figure 10.5. The condition of steady state ($\nabla \times E = 0$) implies

$$\int (\nabla \times E)_x \, dz = \frac{j'_y}{\Sigma_p} - E'_{sy} - \frac{\partial \phi}{\partial y} = 0 \tag{10.15}$$

Combining Equations (10.14) and (10.15) and assuming that the field E'_s is constant inside the Io flux tube, we obtain

$$\frac{\partial^2 \phi}{\partial y^2} + \frac{j_z(y)}{\Sigma_p} = 0 \tag{10.16}$$

with the boundary condition

$$\partial \phi/\partial y = -E_{sy} \tag{10.17}$$

at the edge of the flux tube. Once a constituting relation between j_z and ϕ is formulated one can solve for $\phi(y)$. We see that if $j_z \neq 0$ and $E_s \neq 0$, a potential drop must occur.

Fig. 10.5. Illustration of equipotential
contours (heavy lines) associated with
a field-aligned potential drop, and the
coordinate system used in the text
(Sec. 10.5) to relate j' to ϕ.

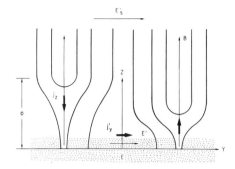

Various relations between ϕ and j_z have been proposed. Goldreich and Lynden-Bell [1969] used the approximation

$$j_z = -ne \left(\frac{2e|\phi|}{m} \right)^{1/2} \phi/|\phi| \tag{10.18}$$

which is valid if $e\phi \ll kT_e$, the electron thermal energy. Gurnett [1972] used the approximation

$$j_z = -j_e \, \phi/|\phi| \tag{10.19}$$

where J_e is the electron thermal current, which is valid in the opposite limit. Smith and Goertz [1978] give a similar but somewhat more complicated form. If the density of upward current at the inner edge of the Io flux tube exceeds the thermal current of Jovian ionospheric ions, then the current must be supplied by precipitating magneto-spheric electrons, in which case the magnetic mirror force becomes significant in the determination of $j_z(\phi)$. For this case Knight [1973] has derived a complicated expression $j_z(\phi)$ that may be approximated by $j_z \approx -Mj_e(e\phi/kT_e)$ for a wide range of $e\phi/kT_e$, where M is the magnetic mirror ratio.

In any case, the parallel potential drop ϕ can be easily obtained as a function of y once the relation $j_z(\phi)$ is given. Smith and Goertz [1978] show that in all the above cases the maximum potential drop occurs at the edge of the flux tube and that under special conditions the parallel potential drop on one side of the flux tube may approach the total potential across the satellite, which in the case of Io exceeds 400 kV. Such a potential could produce large numbers of 400 keV charged particles [Gurnett, 1972], as have apparently been observed [see, e.g., Fillius, 1976].

The conductivity of Io

The character of the Io interaction depends on the conductivity of the satellite, either its bulk and surface conductivity or that of its ionosphere. Pioneer 10 data [Kliore et al., 1974, 1975] indicate that Io possesses an ionosphere. If the atmosphere is sufficiently thick, the currents may flow through the ionosphere of Io rather than through the surface and body of the satellite. If Io's atmosphere and ionosphere are maintained directly by volcanic outgassing [Kumar, 1979], then one might assume that the ionosphere is as variable as the volcanic activity. There is evidence for a variation of volcanic activity (compare, e.g., Voyager 1 and Voyager 2 images), and the role of the ionosphere as the conducting medium might thus be variable (see the discussion below). On the other hand, if a thick atmosphere is maintained by surface sublimation [Kumar and Hunten, 1981], then the ionospheric conductivity would be less variable. If the atmosphere is thin enough to allow sputtering (see Sec. 10.3), then the atmosphere is not an important conductor in any case.

Cloutier et al. [1978] have argued that an Earth-like gravitationally-bound ionosphere would be blown away by the ram force of the corotating Jovian magnetospheric plasma (see also Ip and Axford [1980]). That conclusion was strengthened by the Voyager discovery of the dense plasma torus (Cloutier et al. calculated the ram force on the basis of an assumed plasma density of 50 protons/cm³; for a sulfur ion concentration of 1000/cm³ the ram force is 640 times larger). Note, however, that if the atmosphere proves to be thick, then the above analysis is invalid; a sufficiently dense atmosphere can resist the corotating torus plasma, as pointed out by Herbert and Lichtenstein [1980].

Cloutier et al. suggest that the ionosphere of Io resembles that of a comet [see also Goertz, 1980a]. The corotating torus plasma causes ionization by electron impact. The ions thus created are swept past Io and the structure of the ionosphere is determined by a balance between electron-impact ionization and the loss of ionization by convection past Io. The loss rate scales with the plasma velocity relative to Io, which in turn is determined by the electric field in Io's ionosphere. Cloutier et al. show that the observed structure of the ionosphere can be explained satisfactorily if the velocity is reduced to approximately one-fifth of the corotational value. They invoke the critical velocity phenomenon proposed by Alfvén [1954] to explain this reduction. Alfvén proposed that a tenuous but un-ionized cloud of gas cannot pass through a magnetized plasma if the kinetic energy per atom of the neutral gas relative to the plasma is greater than the ionization potential of the un-ionized gas. The physics of this phenomenon is not well understood, although it has been demonstrated in laboratory plasmas [Daniellsson, 1970] and in space plasmas [Lindeman et al., 1974].

Goertz [1980a] does not invoke the critical-velocity phenomenon but points out that after the new ions are created (at the rate S/m³s), they are accelerated by the electric field in the atmosphere and displaced by approximately one "cyclotron radius" R_i in the direction of the electric field. This displacement represents a motion of charge and hence a "pickup" current

$$\mathbf{j}_{pu} = q_i S \mathbf{R}_i \tag{10.20}$$

where the effective cyclotron radius $\mathbf{R}_i$ is given by [Goertz, 1980a; Cummings et al., 1980]

$$\mathbf{R}_i = \frac{m_i}{q_i B^2} \mathbf{E} \tag{10.21}$$

Combining (10.20) and (10.21) yields

$$\mathbf{j}_{pu} = \frac{m_i S}{B^2} \mathbf{E} \equiv \sigma_{pu} \mathbf{E} \tag{10.22}$$

The pick-up conductance $\Sigma_{pu} = \int \sigma_{pu} \, dz$ can be calculated once the rate of ionization is known. Electron impact ionization by the torus electrons is faster than photoionization, and we will consider only electron impact ionization. Scudder et al. [1981] have demonstrated that the torus electron distribution is not in thermal equilibrium but combines a cold and hot component; they report values for the concentrations and temperatures of these components not at Io's orbit but at 7.8 R_J and 5.5 R_J. Apparently, the hot component in the warm torus has a concentration of about 1–3/cm^3 and a temperature of 1 keV. The cold component has a maximum concentration of 2 × 10^3/cm^3 and a temperature of about 6 eV. Further analysis of the PLS data is needed to establish these numbers more precisely at Io's orbit. Using relations provided by Book [1980], we calculate the ionization rate for a combination of cold electrons ($T_c = 6$ eV, $n_c = 2 \times 10^3$/cm^3) and hot electrions ($T_h = 1$ keV, $n_h = 2$/cm^3) to be

$$S_{O^+} = 3 \times 10^{-6} n_O [\text{cm}^{-3} \text{ s}^{-1}]$$

$$S_{S^+} = 8 \times 10^{-6} n_S [\text{cm}^{-3} \text{ s}^{-1}]$$

where n_O and n_S are the concentrations of neutral oxygen and sulfur, respectively.

As noted in Section 10.3, the neutral concentration of Io's atmosphere is not well known. If, for example, we assume a thick neutral atmosphere with an exospheric temperature $T \sim 1100$ K and exobase concentration $n_{SO_2} \approx 10^9$/cm^3 [Ip and Axford, 1980; Kumar, 1980; Goertz, 1980a], then the scale height is $H \sim 80$ km, the pick-up conductivity is $\sigma_{pu} \sim 1.6 \times 10^{-4}$ mho/m, and the total pick-up conductance $\Sigma_{pu} \sim 12$ mho. This is only an order-of-magnitude estimate but it is larger than the usual Pedersen conductance estimated on the basis of ion-neutral collisions for the same atmospheric model. On the other hand, if the atmosphere is thin and therefore in thermal contact with the surface (see Sec. 10.3), the temperature is no greater than 135 K, the scale height is reduced by an order of magnitude, and the ionization rate is reduced by a factor depending on the thinness of the atmosphere. In this case the estimate of Σ_{pu} is reduced by at least an order of magnitude. (The Pedersen conductivity is also reduced by a similar factor.)

It can be shown [see, e.g., Goertz, 1980a; Southwood et al., 1980] that the electric field in the vicinity of Io is reduced (because of the shorting effect of Io's ionosphere) from its corotational value E_0 to the value

$$E = E_0 \frac{\Sigma_A}{\Sigma_I + \Sigma_A} \tag{10.23}$$

where Σ_I is the effective conductivity of Io's atmosphere (including both Pedersen and pick-up conductivities). For the thick atmosphere model with $\Sigma_I = \Sigma_{pu} \sim 12$ mho and $\Sigma_A \sim 4$ mho we find that the electric field in Io's ionosphere is reduced by a factor of 4, which is close to the value suggested by Cloutier et al. [1978]. On the other hand, the thin atmosphere model with $\Sigma_{pu} \lesssim 1$ mho produces a more modest ($\sim 20\%$) reduction of the electric field. It should be noted that this reduction of the electric field owing to the pick-up effect could be avoided altogether if the corotating plasma were magnetically deflected around Io, that is, if Io had its own intrinsic magnetic field as suggested by Neubauer [1978], Kivelson et al. [1979], and Southwood et al. [1980].

For the thick-atmosphere model, the total ionization rate is 2 × 10^{28} O$^+$ ions and 3 × 10^{28} S$^+$ ions per second; for a thin atmosphere the total ionization rate would be $\lesssim 10^{27}$/s. It is not presently known whether Io's atmosphere is thick or thin (see Sec. 10.3), so we cannot presently assess the importance of Io's atmosphere in the Io–Jupiter interacton.

10.6. Particle acceleration

We have discussed in Section 10.4 how the injection of ions into Jupiter's magnetosphere results in their initial acceleration up to twice the local corotation energy (namely, $1/2\ m\Omega_j^2 r^2$ in rotational energy and another $1/2\ m\Omega_j^2 r^2$ either in field-aligned streaming energy, for ions slung out from Jupiter's ionosphere, or in cyclotron energy, for ions injected near the equatorial plane). Because these centrifugal accelerations are mass-independent (producing a common velocity independent of mass), the resulting energies are significant for ions but not for electrons. The energies produced are of the order of $(60\ eV)(r/6\ R_J)^2\ (m/m_p)$. We now turn our attention to mechanisms for the acceleration of particles to energies much larger than the local corotation energy.

We shall classify the various acceleration mechanisms according to the degree to which they violate the three adiabatic invariants of particle motion in a dipolelike field: the first (magnetic moment) invariant, the second (bounce integral) invariant, and the third (magnetic flux shell) invariant (for definitions of these invariants see, for example, Northrop [1963]). We shall use the term *adiabatic* to describe processes that involve inward transport through violation of the third invariant to produce adiabatic compression through conservation of the first and second invariants. We shall describe as *quasiadiabatic* those processes that involve repeating cycles of adiabatic compression alternating with nonadiabatic scattering. We shall call *nonadiabatic* those processes in which the first or second invariants are violated as a necessary condition of the acceleration process itself. (These rather specialized definitions are introduced as a convenient shorthand; they should not be confused with the conventional thermodynamic definition of adiabaticity.)

Adiabatic processes

The classical mechanism of magnetospheric particle acceleration is adiabatic compression wherein particles are transported radially inward by violation of the third invariant. Conservation of the first invariant implies a betatron acceleration ($w_\perp \propto B$) and conservation of the second invariant implies a Fermi acceleration ($w_\parallel \propto \ell^{-2}$ where ℓ is the distance between mirror points measured along the guiding field line). Thus, in a dipole field, $w_\perp \propto r^{-3}$ while $w_\parallel \propto r^{-2}$, and the inward-moving particles tend toward a pancake pitch-angle distribution with $w_\perp > w_\parallel$ on the average. On the other hand, if pitch-angle scattering is sufficiently rapid to maintain isotropy during the inward motion, the total particle energy then increases as $w \propto r^{-8/3} \propto V^{-2/3}$ where V is the flux-tube volume, and we have a precise analogy with the adiabatic compression of an ideal monatomic gas.

The inward transport may be provided either by a systematic convection system of the type described in Section 10.7, or by "radial diffusion," which, in this context, means a stochastic system of radial convective motions (see Chap. 11). The timescale for this radial transport is not accurately known, but it is inferred to be much greater than the 10-hr Jovian rotation period (see Chap. 5).

Before the in situ observations by Pioneer 10, it was widely held that adiabatic compression through inward diffusion of solar-wind particles was the principal means of powering the Jovian radiation belts (whose properties were inferred from the synchrotron radio emissions) [e.g., Coroniti, 1974]. We now know that simple adiabatic compression of solar-wind particles is inadequate to produce the observed radiation belts (compare Fig. 10.6 with the results in Chaps. 4 and 5). We also now know that a principal source, if not the dominant source, of energetic particles lies in the inner magnetosphere (at Io's orbit), and these particles are instead subject to adiabatic expan-

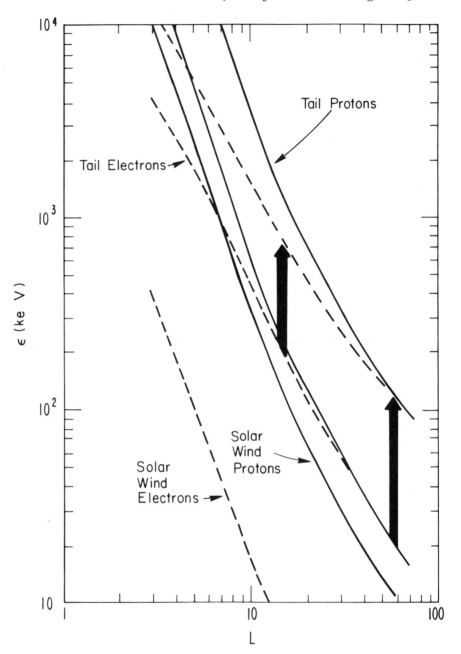

Fig. 10.6. Particle energies attainable in the Jovian magnetosphere as the result of betatron acceleration of solar-wind particles, and of particles from the magnetospheric tail having initial magnetic-moment invariants larger than solar-wind values as a result of magnetic merging in the tail [Carbary, Hill, and Dessler, 1976].

sion, and hence cooling, as they are transported outward in the magnetosphere. More-over, the radial transport is an intrinsically slow process and cannot account for the rapid timescale (~ 10 hr) associated with certain acceleration processes that are inferred from the observations [Chap. 5; Fillius and Knickerbocker, 1979]. Thus we

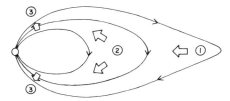

Fig. 10.7. Multiple-diffusion process proposed by Nishida [1976]. Inward radial diffusion from 1 to 2 increases $w_\perp$ by first-invariant conservation; pitch-angle scattering at 2 moves some particles to low altitudes (3) where they may diffuse across field lines (owing to atmospheric winds) without betatron deceleration, thus returning to point 1 with large field-aligned velocities.

require a faster and more powerful acceleration mechanism, although adiabatic compression remains important in determining the overall structure and energetics of the radiation belt, and particularly of the innermost synchrotron radiating region ($L \lesssim 3$).

Quasiadiabatic processes

The effect of adiabatic compression is limited by its reversibility – a particle in a dipole field, for example, gains a factor of 10^3 in perpendicular energy while moving from $L = 50$ to $L = 5$, but loses the same factor in moving back out to $L = 50$. This limitation can, however, be surmounted by alternating cycles of adiabatic compression and nonadiabatic relaxation.

A model of this type, proposed by Nishida [1976], is illustrated in Figure 10.7. Step 1, inward radial diffusion conserving the first and second invariants, is as described above; this produces a pancake pitch-angle distribution (peaked at 90°) at point 2. At this point pitch-angle scattering becomes important, as the result of cyclotron-mode waves that are amplified by the anisotropic distribution itself (see Chap. 12). The pitch-angle scattering allows some particles to mirror at low altitudes (3) where they are subject to meridional diffusion as a result of upper-atmospheric dynamo winds. This meridional diffusion displaces some particles toward the pole such that they return along the field line to the outer magnetosphere (back to point 1). If most of the poleward displacement occurs at low altitudes, the particles do not suffer the betatron deceleration that would have accompanied the equivalent radial displacement in the equatorial plane. The net result is that particles return to point 1 with field-aligned energies comparable to the cyclotron energy they gained in moving from point 1 to point 2. (Because the process is diffusive in nature, the opposite result is also possible: some particles will be displaced from 1 to 2 without benefit of any betatron acceleration.)

Sentman, Van Allen, and Goertz [1975] have added an additional pitch-angle scattering process at point 1 so that the field-aligned energy there is partially converted to cyclotron energy. (A plausible mechanism for such scattering has been discussed by Barbosa [1981].) Particles can then, in principle, repeat the cycle an arbitrary number of times and hence gain arbitrarily high energies. Sentman et al. recognized, however, that this recirculation process is inefficient because of its diffusive nature – during each cycle a particle is as likely to lose energy as gain energy, so that arbitrarily high energies are attainable only for an arbitrarily small fraction of the initial population. The process is also slow as it consists of multiple cycles of inward radial diffusion, itself a slow process. The process may, however, provide an explanation for the "dumbbell" (field-aligned) pitch-angle distributions of energetic particles observed by Pioneers 10 and 11 within the inner magnetosphere [cf., Sentman, Van Allen, and Goertz, 1974, 1978].

Another quasiadiabatic acceleration process was proposed by Goertz [1978], utilizing the day-night asymmetry of the magnetospheric field. Because the night-side magnetic field is weaker at a given distance than the day-side field, a corotating particle is subject to a small degree of betatron acceleration in drifting from midnight to noon, and a corresponding degree of betatron deceleration in drifting from noon to midnight. Without scattering the process is reversible and results in no net energy gain. However, if pitch-angle scattering occurs preferentially at the extreme points of the drift orbit (noon and midnight), then some of the perpendicular energy gained in the midnight-to-noon half of the orbit can be "stored" in the parallel component at noon and thus escape the corresponding betatron deceleration during the noon-to-midnight half, then returned to the perpendicular component at midnight, resulting in a net increase of the first invariant and of the total energy. (The parallel-energy component is also subject to adiabatic gain and loss through second-invariant conservation, but to a lesser degree than the perpendicular component – see above.) In the absence of scattering, the pitch-angle distribution would tend toward pancake anisotropy at noon and dumbbell anisotropy at midnight – it is the relaxation of these anisotropies through scattering that allows a systematic energy gain after a full rotation. Possible mechanisms for such scattering have been discussed by Northrop and Schardt [1980].

This mechanism is a close analog of the general magnetic pumping process first proposed by Alfvén [1949] to account for the acceleration of galactic cosmic rays. Because it systematically increases the average particle energy, it is more efficient than the recirculation process described above. It is also faster, requiring n rotations for a factor of 2^n increase in energy according to the estimate of Goertz [1978]. Thus, magnetic pumping may account for the bulk acceleration of particles ejected into interplanetary space – some observations of interplanetary electrons of Jovian origin apparently require an energy e-folding time in the magnetosphere of about 10 hr (Chap. 5). Even so, the magnetic pumping process is too slow to account for particle acceleration and loss on timescales ≤ 10 hr as seems to be required by some observations [e.g., Simpson et al., 1974b (Fig. 3); Fillius and Knickerbocker, 1979; Chap. 5 (Figs. 5.7, 5.12, 5.13, 5.17)].

Nonadiabatic processes

Three non-adiabatic processes – magnetic merging, parallel (magnetic-field-aligned) electric fields, and plasma wave heating – are possible candidates for producing the transient acceleration events (timescales ≤ 10 hr) in the outer magnetosphere.

Magnetic merging (or "field annihilation," or "reconnection") is a general process for converting magnetic-field energy into particle energy (bulk flow energy and/or thermal energy) in a region of magnetic-field reversal. An extensive theoretical literature has been reviewed by Vasyliunas [1975] – see also Hill [1975]. Likely regions of merging in Jupiter's magnetosphere include the magnetodisc current sheet (especially its extension into the magnetospheric tail) and perhaps the magnetosphere of Io, should one exist [e.g., Southwood et al., 1980].

In steady state models, the merging process generally accelerates particles (mostly ions) to an average energy corresponding to the magnetic energy density just outside the current sheet divided by the particle density within the current sheet, that is, $B^2/(\mu_0 n)$ [see Alfvén, 1968; Vasyliunas, 1975; Hill, 1975]. (Such an average energy is required to balance the pressure of the opposing magnetic fields on either side of the field reversal.) If we take $B \sim 10$ nT and $n \sim 0.1/\text{cm}^3$ as representative values (see Chaps. 1, 3, and 4), the average energy per particle is about 5 keV. Individual particles may gain energy increments larger than $B^2/(\mu_0 n)$ (limited ultimately by the total EMF across the system), but only insofar as their initial energy exceeds $B^2/(\mu_0 n)$ [Hill, 1975].

Thus, steady state merging by itself does not produce MeV particles in the outer magnetosphere, although it constitutes an important input to the adiabatic compression cycle described above, providing a factor-of-ten enhancement of the initial value of the first invariant $w_\perp/B$ compared to simple injection of solar-wind particles [Coroniti, 1974; Carbary et al., 1976].

The total average EMF associated with magnetic merging cannot presently be estimated because it depends on the (unknown) strength of the magnetospheric convection/planetary wind system. Whatever its average value may be, the EMF may be dramatically enhanced by "sporadic" merging associated with rapid changes in the magnetic-field configuration, either the relaxation of a highly stressed field toward a more dipolelike configuration or the formation of magnetic "bubbles" (closed loops) in the disc current sheet. This latter case may be appropriate to the production of energetic ion fluxes that have been observed streaming away from the equatorial current sheet (Chap. 5). The sporadic merging phenomenon has not been satisfactorily modeled theoretically, but observations in the Earth's magnetospheric tail [e.g., Sarris, Krimigis, and Armstrong, 1976] suggest that particle energies far in excess of that corresponding to the average EMF can be expected to accompany rapid reconfigurations of the disc magnetic field.

Carbary, Hill, and Dessler [1976] have proposed that rapid merging in the magnetospheric tail occurs with a 10-hr periodicity, peaking after each rotation through the tail of the active sector of the magnetic-anomaly model (see Sec. 10.7). Thus, they propose to account simultaneously for the 10-hr periodicities in both particle flux and energy spectral index observed outside and inside the magnetosphere.

Particle acceleration by parallel electric fields is evidently important in the auroral regions of the Earth's magnetosphere, and, by analogy, it is expected to be important wherever sufficiently intense Birkeland currents flow. In particular, parallel electric fields have been discussed in connection with the Io–Jupiter interaction (see Sec. 10.5). The field-aligned potential drop may be largely confined to small "double-layer" regions imbedded in the Io flux tube [see Smith and Goertz, 1978].

The observation of enhanced fluxes of electrons of several hundred keV energy near the Io flux tube (Chap. 5) tends to confirm the importance of field-aligned electric fields in the Io–Jupiter interaction. The total particle energy is, however, limited by the total EMF of the Io–Jupiter circuit, which is the potential across Io generated by the flow of torus plasma past Io:

$$\phi = (6\Omega_J R_J - V_I) B_I (2R_J) \sim 400 \text{ kV} \tag{10.24}$$

Particles accelerated near Io are subject to betatron deceleration (adiabatic expansion) as they move radially outward, and thus they do not contribute directly to the energetic particle population of the outer magnetosphere.

A potentially more powerful source of field-aligned electric fields would be the differential rotation of different parts of the same magnetic flux tube in the outer magnetosphere, as might occur if there were insufficient plasma to carry the Birkeland currents needed to enforce corotation; this phenomenon might be important if and when the (partial) corotation becomes super-Alfvénic. This possibility was mentioned by Fillius and McIlwain [1974] but has not yet been explored theoretically. The electric potential associated with corotation in the equatorial plane is $\phi = \Omega_J B_J R_J^2/L \sim (376 \text{ MV})/L$, and if any significant fraction of this potential were to appear along the field lines, a powerful linear accelerator would be produced. The direction of the corotation electric field is such that if the equatorial portion of a flux tube were to rotate more slowly than the near-Jupiter part (the expected sense), the field aligned electric field

Table 10.3. *Jupiter's pulsar behavior*

Phenomenon or property	Jupiter	Pulsar
1. Time variation of electromagnetic emission	Tied to planetary spin period	Neutron-star spin period
2. Release of energetic particles	(a) Jupiter is source of heliospheric cosmic-ray electrons with $E < 30$ MeV. (b) Outflow of relativistic electrons is modulated at planetary spin period.	(a) Crab pulsar is source of relativistic electrons in Crab Nebula. (b) Not known.
3. Source of energy	Kinetic energy of rotation of planet, i.e., solar wind unimportant.	Kinetic energy of rotation of neutron star, i.e., external sources unimportant.
4. Source of magnetospheric plasma	Internal to magnetosphere (principally Io and ionosphere), i.e., solar wind unimportant.	Internal to magnetosphere, i.e., external sources unimportant.
5. Magnetic moment	1.5×10^{20} T-m^3 (for equatorial surface field 4.2×10^{-4} T).	1.0×10^{20} T-m^3 (for surface field 10^8 T).
6. Fit to empirical power-loss relationship for pulsars $P \propto M^2 \Omega^n$.	Extrapolates to Crab pulsar if $n = 3.1$	$n = 3.5 \pm 0.5$

would be directed toward Jupiter so as to accelerate electrons toward the equatorial plane. This hypothetical differential rotation should not be confused with the partial corotation discussed in Section 10.4 and in Chapter 11, in which entire flux tubes rotate at a common angular velocity which is, however, less than that of the neutral atmosphere.

Particle acceleration by plasma waves (especially by cyclotron resonant interactions) is another nonadiabatic mechanism of potential importance which has not been adequately explored either observationally or theoretically (See Chap. 12). Ion cyclotron waves, for example, are thought to be responsible for accelerating the upward beams of O^+ ions observed over the Earth's auroral zones [e.g., Ashour-Abdalla et al., 1981], and such waves may also be important in the acceleration of heavy ions in Jupiter's magnetosphere, especially in the vicinity of the Io torus where the wave propagation speed, and hence the minimum resonant particle energy, are drastically reduced. (Plasma waves, of course, do not provide an energy source, but rather a mechanism for tapping a given source of free energy to provide plasma heating.)

In summary, the three nonadiabatic processes mentioned here (and possibly others) are not well developed theoretically, but they presently hold the greatest promise for explaining the rapid acceleration that seems to occur at times in the outer magnetospheric radiation belt ($\gtrsim 25\ R_J$) on a timescale less than one Jovian rotation period.

10.7. Spin periodicity

In almost every phenomenon with which it is associated, Jupiter exhibits evidence of modulation at its spin period. No other planet shows such a variety of spin-dependent behavior, although Saturn shows an interesting spin modulation of its radio emissions [e.g., Warwick et al., 1981]. In this respect, Jupiter is more like a pulsar than a planet (see Table 10.3), a suggestion made when pulsars were first discovered (e.g., Dowden [1968]). The only certain knowledge we have about pulsars is that they emit electromagnetic radiation with a well-regulated, time-varying intensity. The remainder of pulsar theory, for example, the hypothesis that the pulsation period is tied to the spin period of a rapidly rotating central object (commonly supposed to be a neutron star), is founded solely on logical inference and theoretical modeling. The Pioneer and Voyager encounters with Jupiter have enabled us to make in situ measurements of a body that exhibits weak but genuine pulsar behavior. Application of our understanding of the physics of the Jovian magnetosphere to pulsars is already forthcoming [e.g., Michel and Dessler, 1981; Michel, 1982]. However, the reader should be aware that there is, as usual, a contrary view; for example, Douglas-Hamilton [1968] and Kennel and Coroniti [1975] have considered the question of whether Jupiter's magnetosphere is like a pulsar's, and they have concluded it is not. Nevertheless, we concentrate on the spin periodic phenomena because they offer the most definitive challenge to magnetospheric modelers. One could feel some confidence in a model that could quantitatively describe the various spin-periodic phenomena that have been observed. The few magnetospheric phenomena that show no dependence on either Jupiter's spin phase angle or Io's orbital position should be fully contained within the time-dependent models.

Spin-periodic phenomena

The strongest variation of any observed magnetospheric quantity such as plasma density, temperature, particle flux, magnetic field strength, optical emission, synchrotron radiation, and so forth, is its variation with latitude. For example, the Io torus density and associated optical emissions vary by two orders of magnitude over a latitude

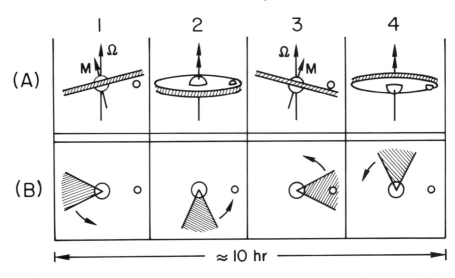

Fig. 10.8. Sketch illustrating the primary manifestations of the two basic magnetospheric models that have been put forth to account for certain of Jupiter's spin periodicities [after Hill, 1981]. Strip (*a*) shows the tilted dipole or disc model, and strip (*b*) shows the corotating active-sector or magnetic-anomaly model. The observer is at 0, and the time sequence is from left to right. Steps 1 through 4 represent one rotation of Jupiter, which requires approximately 10 hr. The most striking difference between the two models is that, for an observer near the spin equatorial plane, there are generally two events per planetary rotation for the disc model (*a*), but only one event per rotation for the magnetic-anomaly model (*b*).

range of 10° but only by a factor of, at most, 5 with longitude. This latitudinal confinement varies with distance and local time, but is observed throughout the explored portion of the magnetosphere. A smaller, but nevertheless quite important effect is the longitudinal modulation of these parameters. The effects of latitudinal confinement and longitudinal modulation, taken separately, are illustrated in Figure 10.8. It has become customary to associate the latitudinal confinement effects with the "disc model" and the effects of longitudinal variation with the "magnetic anomaly model," although many observed spin-periodic phenomena require both latitudinal and longitudinal effects for a complete explanation.

The models are described in the next subsection. The various phemonena known as of this writing are listed in Table 10.4 and are described briefly below. These are among the phenomena that a comprehensive theoretical model of the Jovian magnetosphere would be expected to explain.

a. Decametric radio emission. Before the discovery of the Io plasma torus, it was thought that the decametric radio emission that was independent of the position of Io in its orbit around Jupiter was somehow a separate phenomenon from the radio emission that occurred only when the angle between Io, Jupiter, and the observer was in one of two specific ranges (see Table 7.4 and the attendant discussion in Chap. 7). However, now, particularly with the information obtained with Voyager, it appears that the Io-dependent and Io-independent emissions are basically the same phenomenon, but that the emission is simply enhanced when Io has the preferred phase angle relative to the observer. Nearly all of the decametric radio emissions appear to come from a fixed sector of Jovian longitude with most of the emission coming from the northern hemisphere, but with specific important contributions originating in the southern hemi-

Table 10.4. *Spin-periodic phenomena*

Phenomenon	Comments

a. Decametric radio emission
  (i) Io-unrelated
  (ii) Io-related
  (Chap. 7)
	a(i) and a(ii) appear to be the same phenomenon except presence of Io at proper Jovian longitude and phase angle relative to observer causes emission intensity to increase. Primarily a longitudinal effect.

b. Modulated injection of relativistic electrons into interplanetary space (Chap. 5)	Seen primarily in modulation of spectral index, which is a minimum when active sector faces tail. Primarily a longitudinal effect.
c. Longitudinal asymmetry of Io plasma torus (Chaps. 6 and 11)	Seen only in emission from singly ionized sulfur. Torus brightest in active sector. Primarily a longitudinal effect.
d. Hydrogen bulge (Chap. 2)	Corotating bulge of atomic hydrogen north of spin equator and opposite active sector. Primarily a longitudinal effect.
e. Longitudinal asymmetry in radial extent of torus plasma	Outward motion of singly ionized sulfur from torus in active sector. Primarily a longitudinal effect.
f. Longitudinal asymmetry of plasma temperature in outer magnetosphere.	Higher proton temperature at active-sector current-sheet crossing. Primarily a longitudinal effect.
g. Modulation of magnetospheric plasma, energetic particles, and magnetic fields. (Chaps. 1, 3, and 5)	Inconsistent with extreme version of magnetic-anomaly model. Firm evidence for existence of thin disc in predawn quadrant to 60–80 R_J. Primarily a latitudinal effect.
h. Asymmetry of longitudinal shifts in leading and trailing particle intensity maxima as seen by Voyager (Chap. 5)	Ambiguous interpretation. Possible magnetic-anomaly effect within disc, effect of magnetotail, or wave motion. A combination of latidudinal and longitudinal effects.

sphere (see Fig. 10 of Alexander et al. [1981] for a graphical representation of the single northern-hemisphere source and Chap. 7 for a general discussion). The radio emission arises from the region where downward Birkeland currents, that is, upward moving electrons, are expected to flow because of a longitudinal asymmetry in the Io plasma torus that is discussed below (item c) [Dessler, 1980a].

b. Modulated injection of relativistic electrons into interplanetary space. Jupiter is a powerful source of interplanetary electrons; essentially all heliospheric cosmic-ray electrons with energies less than about 30 MeV originate at Jupiter (see Chap. 5). From the standpoint of magnetospheric models, the most important feature of these electrons is a "clock" modulation of their spectral index. Specifically, if the energy spectrum is represented by a power law $E^{-\gamma}$ with γ the spectral index, Chenette, Conlon, and Simpson [1974] discovered that the spectral index of Jovian electrons in interplanetary space varies by about a factor of two, and the variation is synchronized with Jupiter's spin period. The spin-periodicity in spectral index is observed to occur both in interplanetary space and within the outer part of the magnetosphere [see Fig. 9 of Chenette et al., 1974; and Fig. 3 of Simpson et al., 1975]. On the outbound pass, the clock phenomenon commences as soon as the spacecraft leaves the region of disclike modulation and enters the planetary or magnetospheric-wind region [Schardt, McDonald, and Trainor, 1981].

Vasyliunas [1975] noted that the spectral index is a minimum (or the escape of the most energetic electrons into interplanetary space is a maximum) when Jupiter is oriented so that the subsolar longitude is near $\lambda_{\mathrm{III}} = 60°$ (or when $\lambda_{\mathrm{III}} = 240°$ faces the tail). This discovery, which was made using data from the Pioneer 10 and 11 spacecraft, has been confirmed by data from the Voyager spacecraft [Schardt, McDonald, and Trainor, 1981]. This finding was labeled by its discoverers as a magnetospheric "clock" in that the effect is seen essentially everywhere in interplanetary space (if the spacecraft is in a flux tube that passes near Jupiter) whenever a given Jovian System III longitude faces the magnetospheric tail (see Chap. 5).

From the standpoint of conceptual magnetospheric modeling, this is a significant discovery; it shows that there is some longitudinal asymmetry fixed in System III coordinates (i.e., corotating with Jupiter's internal magnetic field) that perhaps influences the magnetosphere/solar-wind interaction and certainly influences the escape of relativistic electrons into interplanetary space. The importance of this particular spin dependence must be emphasized. Something at or near the surface of Jupiter modulates the escape of relativistic electrons from the distant magnetosphere to produce a 10-hr periodicity in interplanetary space. It was the discovery of this 10-hr modulation of relativistic electrons in interplanetary space that inspired the development of the magnetic-anomaly model [Dessler and Hill, 1975] (see Sec. 10.7 and Chap. 5).

The phase of the modulation (a minimum in the spectral index when $\lambda_{\mathrm{III}} \approx 240°$ faces away from the Sun) is not affected by interplanetary conditions, such as the orientation of the interplanetary magnetic field, sector boundary crossings, or strength of the solar wind. A compelling analysis of the connection between the surface field of Jupiter (the magnetic anomaly to be discussed in Sec. 10.7) and the System III longitude dependence of the phase of the modulation is provided by Schardt, McDonald, and Trainor [1981] (see their Fig. 18 and their attendant discussion).

The clock effect cannot be caused by the rotation of a tilted dipole because that would produce a 5-hr variation (Fig. 10.8); some sort of interconnection between the Jovian and interplanetary magnetic fields might produce a 10-hr variation, but it would also produce a shift in the phase of the modulation as the orientation of the interplanetary magnetic field changed (e.g., the phase would be expected to shift by $\approx 180°$ when a sector boundary swept past Jupiter, as illustrated in Fig. 10.13). Neither the 5-hr variation nor the 180° phase shift is observed.

c. Longitudinal asymmetry of the Io plasma torus. A series of independent, ground-based observations shows that the optical torus has a pronounced longitudinal asymmetry in the brightness of singly ionized sulfur emissions [Pilcher and Morgan, 1980; Trafton, 1980; Trauger et al., 1980]. These various observations, which extend over a period of four years, show that the maximum torus brightness usually lies near $\lambda_{\mathrm{III}} \approx 250°$, although both the location and amplitude can vary with time. The reported

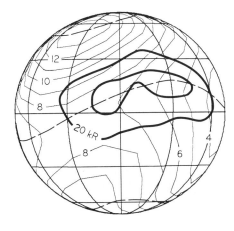

Fig. 10.9. The hydrogen bulge (a mountain of atomic hydrogen in Jupiter's upper atmosphere) as seen by Voyager 1 in resonantly scattered solar Lyman-alpha radiation [Dessler, Sandel, and Atreya, 1981]. The heavy closed curves show the 20 and the 22 kilo-Rayleigh isophotes that delineate the bulge. The view is from above the spin equator and the $\lambda_{III} = 110°$ meridian. The dashed line passing through the hydrogen bulge is the charged-particle drift equator. The lighter curves are the surface magnetic field contours in Gauss (10^{-4} T). The coincidence of the bulge with the drift equator and the corotation of the bulge with the planetary magnetic field show that the hydrogen bulge is created by the effect of charged particles on Jupiter's upper atmosphere.

range of longitudes of the maximum brightness is between 180° and 260°, and the reported asymmetry ratio of maximum to minimum brightness varies from 5 to 2. This longitudinal variation in brightness is attributed to a longitudinal variation in plasma concentration [Pilcher and Morgan, 1980; Trafton, 1980].

This asymmetry has so far shown itself only in the [S II] emission from singly ionized sulfur. No explanation has yet been offered as to why a similar asymmetry is not seen in the emission from the more highly ionized components in the outer torus, for example, [S III] or [S IV], but data obtained with the Voyager ultraviolet spectrometer instrument, which was sensitive to only these more highly ionized states, show the torus to be axially symmetric to within about 10% [Sandel et al., 1979; Sandel and Broadfoot, 1982]. Although the symmetry of the [S III] and [S IV] emissions is a puzzle, the asymmetry in the [S II] emission is not; it is fully expected in the context of the magnetic-anomaly model [e.g., Dessler and Vasyliunas, 1979; see their Prediction 1].

d. The hydrogen bulge. Sandel, Broadfoot, and Strobel [1980], on the basis of data obtained with the Voyager Ultraviolet Spectrometer (UVS), discovered another feature that is, thus far, unique to Jupiter – the hydrogen bulge. This feature is a mountain of atomic hydrogen in Jupiter's upper atmosphere that remains fixed in System III coordinates. The hydrogen bulge is located at longitude $\lambda_{III} \approx 90°$ and a latitude 10° to 15° north of the spin equator. Its longitude is approximately 180° removed from the longitude of the maximum plasma concentration in the Io torus, and its latitude is coincident with that of the particle drift equator. The contours of the hydrogen bulge as seen in resonantly scattered solar H Lyman Alpha (Lyα) are shown in Figure 10.9. The particle drift equator is the locus of the minimum in the magnetic field strength at ionospheric levels, that is, it is the locus of energetic particles at ionospheric height having 90° equatorial pitch angles. The coincidence of the hydrogen bulge and the particle drift equator, and the fact that the neutral atomic hydrogen enhancement that constitutes the hydrogen bulge corotates with the magnetic field (System III) rather than with the clouds and underlying neutral atmosphere (System II), make it obvious that the hydrogen bulge is created by some sort of interaction between the upper atmosphere and the energetic charged particles trapped in the Jovian magnetic field.

The existence and the stability of the hydrogen bulge are established by independent observations by Clarke et al. [1980] and Clarke, Moos, and Feldman [1981]. The observations, which were obtained over a period of nearly two years, suggest that the longitude of the bulge varies by up to about 30°, and the amplitude of the bulge (the excess column content of atomic hydrogen) varies by about a factor of two.

The hydrogen bulge is explained by the magnetic-anomaly model in terms of a two-cell convection pattern that corotates with Jupiter [Dessler, Sandel, and Atreya, 1981]. The convection is driven by the asymmetrical mass-loading of the plasma torus described in the following subsection (see also Chap. 11). An alternative explanation in terms of centrifugally driven flow of atomic hydrogen from the northern auroral zone is offered by Clarke, Moos, and Feldman [1981].

e. Longitudinal asymmetry in radial extent of torus plasma. Ground-based observations of singly ionized sulfur by Pilcher et al. [1981] show transient extensions of this plasma out to 7 or 8 R_J in the same System III longitude range as the maximum in torus density described previously. In this longitude range, the transient [S II] extensions are always outward from the torus, never toward Jupiter. At longitudes removed roughly 180° from these outward extensions, the motion of the sulfur is never away from Jupiter [Pilcher et al., 1981]. These transient extensions of the torus plasma from its average position can be explained as another manifestation of corotating convection within the Jovian magnetosphere. The sporadic nature of these plasma motions, and the variability in the amplitude and location of the hydrogen bulge described in the preceding subsection, would indicate that the convection is not a steady state phenomenon.

f. Longitudinal temperature asymmetry. There is evidence that the temperature of plasma within the magnetosphere shows a longitudinal asymmetry. Specifically, the results from the Low-Energy Charged Particle Experiment, described in Chapter 4, indicate that the temperature of protons is systematically higher in one of the two equatorial crossings than the other. This experimental result is shown in Figure 4.17. The higher temperature peak is marked A, in reference to the active sector of the magnetic anomaly model. (See the following discussion for a description of this model.) Evidence of a higher temperature in the active sector is also observed closer to Jupiter. The decimetric emission, which is generated by synchrotron radiation from relativistic electrons at about 2 R_J, is observed by de Pater [1980b] to be brightest at $\lambda_{III} \approx 270°$. The proton temperature maximum in Figure 4.17. appears only at $\lambda_{III} \approx 300°$ because of limitations imposed by the flyby trajectory. The actual maximum might be at $\lambda_{III} \approx 270°$, but the spacecraft did not cross the equator at this longitude.

The relative relationships in longitude of the phenomena described in the preceeding subsections a through f are illustrated in Figure 10.10.

g. Modulation of magnetospheric plasma, energetic particles, and magnetic fields. The most striking feature of the Jovian magnetosphere seen during the outbound trajectories of P 10, V 1, and V 2 is the confinement of magnetospheric plasma and energetic particles to a thin disc. By disc we mean a configuration that is thin in latitude compared to its extent in longitude. Although the current sheet extends inward to about 5R_J, the field remains principally dipolar in character with the disc geometry starting at about 20 R_J and extending to at least 60 to 80 R_J in the predawn quadrant (see Chap. 1,

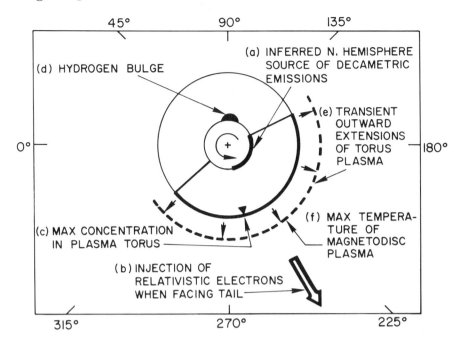

45°   90°   135°

0°   180°

315°   270°   225°

(d) HYDROGEN BULGE

(a) INFERRED N. HEMISPHERE
SOURCE OF DECAMETRIC
EMISSIONS

(e) TRANSIENT
OUTWARD
EXTENSIONS
OF TORUS
PLASMA

(c) MAX CONCENTRATION
IN PLASMA TORUS

(f) MAX TEMPERA-
TURE OF
MAGNETODISC
PLASMA

(b) INJECTION OF
RELATIVISTIC ELECTRONS
WHEN FACING TAIL

Fig. 10.10. Illustrative sketch showing various phenomena that can be explained in terms of the magnetic-anomaly model. The angular extent of each phenomenon is shown approximately to scale, but the radial distances are not.

particularly Figs. 1.20 and 1.23). Beyond that distance, with but a few exceptions, only one contact was made with the plasma and energetic particles during each planetary rotation. This behavior is discussed in some detail in Chapter 5, see particularly Figure 5.6, and by Fillius and Knickerbocker [1979]. The effect of a magnetodisc confinement is to produce, for an observer near the spin equator, two passages through the disc during each planetary rotation, as illustrated schematically in Figure 10.8a.

h. *Asymmetric spiraling of magnetodisc.* Pioneer 10 and Voyagers 1 and 2 traversed the predawn magnetosphere at sufficiently low latitude to encounter the magnetodisc, once each planetary rotation for P 10 and twice each rotation for V 1 and V 2. It was noted, first using P 10 data, that the System III longitude at which the disc crossed the extended dipole magnetic equator increased with increasing radial distance [Northrop, Goertz, and Thomsen, 1974; Eviatar and Ershkovich, 1976; Kivelson et al., 1978; Goertz, 1981]. The most common explanation for this effect is in terms of the finite speed of propagation of the position of the rotating tilted dipole to the distant magneto-disc. If the spiraling is represented as a longitudinal shift between the expected equatorial crossing of the disc and its observed position, a curious effect was noted by the Voyager experimenters [Barbosa et al., 1979; Bridge et al., 1979a,b; Krimigis et al., 1979a,b; Vogt et al., 1979a,b]: of each pair of consecutive crossings, the two were systematically but oppositely displaced relative to the position predicted from the formula derived earlier by Kivelson et al. [1978] with the assumption of a constant radial propagation speed. That is, in the course of each planetary rotation, one of the two crossings always occurred earlier than expected, and the other always occurred later; the tightness of the spiraling appears to be a function of longitude.

Two distinct explanations have been offered to account for this effect. Vogt et al. [1979a,b] suggest that the speed of radial propagation from the rotating, tilted dipole to the disc is a function of longitude, the propagation speed being a minimum in the active sector as defined by the magnetic-anomaly model (see Secs. 5.4 and 10.7). This explanation has been further developed by Vasyliunas and Dessler [1981] who show that, with no additional assumptions, a single set of parameters can be selected that fit the P 10 as well as the V 1 and V 2 outbound observations. Adopting an alternative approach, Bridge et al. [1979b], Carbary [1980], Behannon, Burlaga, and Ness [1981], and Goertz [1981] have developed similar quantitative models in which the disc, in addition to spiraling, is hinged at about 40 R_J in the tailward sector. That is, the disc is bent away from the magnetic equator toward the spin equator, and this bending starts at about 40 R_J. In order to fit both P 10 data, which do not show evidence for hinging, and V 1 and V 2 data, which do, the hinging action is attributed to the action of the magnetospheric tail [Goertz, 1981], which is assumed to be pulled to within about 3° of the spin equator by the solar wind. There seems to be no clear test with presently available data to differentiate between these alternative explanations.

Models advanced to account for spin periodicities

There are two basic models that pertain to various of the observed spin periodicities: (a) the tilted dipole or disc model and (b) the corotating active-sector or magnetic-anomaly model. These are illustrated schematically in Figure 10.8.

Disc and magnetic-anomaly models – phenomenological distinction. With the disc model, an observer O, at or near the spin equator, sees two similar phenomena each planetary rotation, either at Frame (a)1 and (a)3 of Figure 10.8 when respectively the south and north polar caps are tipped toward the observer or at Frames (a)2 and (a)4 when the observer passes through the plane of the magnetic equator or through the plane of the disc. There are variations on this model that include spiraled, bent, warped, and wavy discs, but all have the feature that an observer near the spin equator sees two events per planetary rotation.

With the extreme version of the magnetic-anomaly model, there is only one event per rotation, either as in Frame (b)3 when the active sector points toward the observer, or, for certain phenomena, as in Frame (b)1 when the active sector points away from the observer. The magnetic-anomaly model also includes "clock" behavior wherein some phenomenon that can be propagated omnidirectionally occurs each time the active sector points in some particular direction relative to the solar wind, for example, either toward the Sun or toward the Jovian magnetospheric tail. Thus, if the solar wind were coming from the left in Figure 10.8, a particular clock-type phenomenon would occur at either Frame (b)1 or (b)3. As with the disc model, there are variations to this elementary form of the magnetic-anomaly model that involve propagation and convective effects.

Before the Voyager flybys of Jupiter, some proponents of the magnetic-anomaly model posed an "either or" case for the magnetic-anomaly model vs. the disc model [see, for example, Dessler and Vasyliunas, 1979]. It is now clear that Jupiter's magnetosphere is complex enough to involve, in one way or another, nearly all ideas that have been developed thus far. In particular, the idea of a thin disc in the predawn quadrant of the magnetosphere is fully confirmed by the Voyager flybys, but it also appears that the disc involves magnetic-anomaly effects.

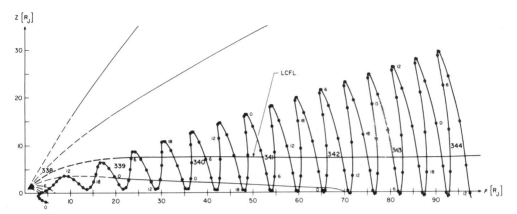

Fig. 10.11. Magnetic field lines superposed on the "wiggle-plot" trajectory of Pioneer 10 outbound, which is projected on a ρ, z plot where ρ is the axial distance from the tilted magnetic dipole and $z = 0$ is the spiraled magnetic equator. The line marked LCFL is the last closed field line above the $z = 0$ plane; the magnetodisc is presumed to be contained within that line and its mirror image below the $z = 0$ plane [after Goertz et al., 1976].

Magnetodisc models. One of the most striking features of the Jovian magnetosphere, particularly in the predawn quadrant, is the confinement of plasma and energetic particles to a narrow latitudinal range with relatively small asymmetries in longitude. That is, the latitudinal gradient in plasma concentration and energetic particle flux is large while, for many purposes, the longitudinal gradient can be neglected. The relative thinness of the disc is illustrated in Figure 10.11 (although its absolute thickness, as great as 4 to 5 R_J, is such that it could encompass the entire terrestrial magnetosphere). Liu [1982] has concluded that the thickness of the disc may be a function of both radial distance and longitude, but, for many purposes, such variations are unimportant. The day side region between about 0900 and noon local time (LT) (P 10 in, P 11 in and out, V 1 in, and V 2 in) does not exhibit the sharp latitudinal confinement that is characteristic of the night side region thus far observed between about 0300 and 0500 LT (P 10 out, V 1 out, and V 2 out). (See Fig. 5.1 for spacecraft trajectories.) However, there is enough concentration of plasma and energetic particles toward the magnetic equator to allow one to carry the concept of the magnetodisc to the day side [Jones et al., 1981], even though the term "disc" may not be strictly applicable.

Fillius and Knickerbocker [1979] have pointed out that the term "disc" may not be applicable to the day side because the energetic particles are not confined to a narrow range of latitude as they are on the night side. Specifically, Pioneer 11, which was at least 30° north of the magnetic equator during its daytime outbound passage, saw much the same time variation as other spacecraft that passed through the equator on the dayside. Moreover, the magnetic signature of the disc is less pronounced on the day side than in the predawn quadrant. Thus, although there is some concentration of plasma and energetic particles toward the magnetic equator, there is other evidence against the existence of a simple disc on the day side. It is assumed (sometimes tacitly) by proponents of the disc model that the thin disc is broadened or modified on the day side by solar-wind effects, but that disc confinement of plasma and energetic particles prevails through most of the night side magnetosphere.

There are, at present, neither direct measurements nor even qualitative theory to tell us how the plasma and energetic particle populations are confined through the rest

Fig. 10.12. Magnetodisc models [after
Carbary, 1980]. The rigid disc (*a*) is
formed by a hot plasma, the bent or
hinged disc (*b*) occurs either for a plasma
sufficiently cool that centrifugal force
dominates the magnetic mirror force or
when external stress such as that
produced by a magnetotail becomes
important. A wavy disc (*c*), which
corresponds to (*a*), or (*d*), which
corresponds to (*b*), results if propagation
delays are important. All of these disc
models spiral out of the meridional plane.
Although the illustrations imply some
degree of axial or reflection symmetry,
the plasma and energetic particles on the
dayside region investigated thus far do
not organize themselves into such a
symmetrical thin disc.

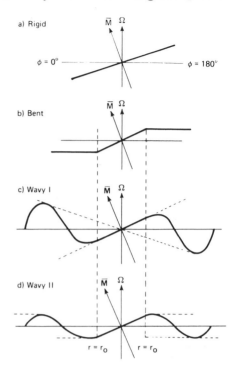

of the magnetosphere. That is, measurements within a 2-hr local-time interval in the
night side magnetosphere show a well-formed disc; on the day side there is a
3-hr local-time interval in which the sharp disc appears to be broadened in latitude.
Extrapolation of these results to the remaining 19 hr of local time remains uncertain
pending theoretical development and future observations.

   The plasma that determines the configuration of the magnetodisc is hotter than its
principal source, namely the plasma in the Io torus (Chaps. 3, 4, 11, and 12). Outward
transport should cool the torus plasma by simple adiabatic expansion, but this cooling
tendency is overpowered by a heating mechanism yet to be identified. (A possible
ionospheric heating source is discussed in Chap. 12; see also Barbosa [1981].)

   Goertz [1976b] developed a quantitative model of a Jovian hot plasma disc, and
showed that "hot" for the case of a magnetodisc means that the average ion cyclotron
speed is significantly greater than the corotation speed at a given radial distance [see
also Hill et al., 1974]. A hot plasma disc would be rigidly aligned with the magnetic
equatorial plane whereas a cold disc would "hinge" toward the rotational equatorial
plane because of centrifugal stress [Hill et al., 1974; Goertz, 1979]; the rigid disc model
appears to be consistent with data from Pioneer 10 outbound [Goertz, 1976b]. The
Pioneer 10 data alone are not unambiguous, however. Specifically, Jones et al. [1980]
were able to get an acceptable fit with a hinged, spiraled disc, although their model
requires the somewhat ad hoc assumption of a longitudinally confined corrugation of
the disc. The difference between a rigid and a hinged or bent disc is shown schematically in Figures 10.12*a* and 10.12*b*. Analysis of detailed relationships between magnetic field magnitude and energetic proton flux led Walker et al. [1978] to conclude
that "low energy (≤ 5 keV) plasma contributes less than 3% to the current-sheet
energy density." These conclusions are confirmed by direct measurements by Krimigis
et al. [1979a, 1979b, and Chap. 4] and indirectly by Barbosa et al. [1979], so the weight

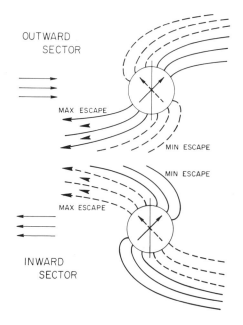

OUTWARD
SECTOR

MAX ESCAPE

MIN ESCAPE

MIN ESCAPE

MAX ESCAPE

INWARD
SECTOR

Fig. 10.13. Sketch illustrating how the phase of the clock governing the escape of energetic electrons from the Jovian magnetosphere would shift by five hours after a sector boundary crossing if the rate of escape were controlled by magnetic merging between the Jovian and interplanetary magnetic fields. The solid arrow shows the orientation of the dipole at some arbitrary time, and the dashed arrow shows the dipole five hours later. The circle represents the magnetosphere, and the dashed and solid lines external to the circle represent the interplanetary magnetic field. The phase shift expected by this model is not observed.

of the evidence indicates that the plasma is hot so that centrifugal force is not important in determining the morphology of the portion of the magnetodisc thus far observed.

Although the Pioneer 10 outbound pass at about 0500 LT did not show any appreciable deflection or "hinging" of the magnetodisc from the dipole equator of Jupiter's magnetic field, the Voyager 1 and 2 outbound measurements indicated what has been interpreted as evidence for hinging, even though the plasma in the magnetodisc fulfills the criterion for a hot plasma, which should form a rigid disc. Centrifugal effects should not be important for a hot plasma, but solar-wind stress, transmitted through Jupiter's magnetotail, might be. Thus, it has been suggested by Ness et al. [1979b] that hinging has been observed because of an antisunward stress on the disc produced by the Jovian magnetic tail (see Chaps. 1 and 5). This conclusion has been challenged by Vasyliunas and Dessler [1981] and defended by Goertz [1981] and Thomsen and Goertz [1981a]. The matter is discussed further in Chapter 11. It is possible that this issue can be resolved by more detailed, quantitative analysis of existing data.

Magnetic-anomaly models. The development of the magnetic-anomaly model was initially put forth to explain the discovery by Chenette, Conlon, and Simpson [1974] of the clock modulation of the spectral index of relativistic electrons in interplanetary space (see Sec. 10.7b). To understand this phenomenon, it is necessary to find some means of relating the orientation of the internal magnetic field of Jupiter relative to the solar wind with a modulation in the escape of relativistic electrons from the Jovian magnetosphere into interplanetary space. A simple interaction between a magnetodisc and the solar wind or the magnetopause will not work because a 5-hr variation would be produced as suggested schematically in Figure 10.8a, and the observed modulation has a 10-hr periodicity. Nor will a simple connection between the Jovian magnetic field and the interplanetary magnetic field suffice. Figure 10.13 illustrates how such a scheme would lead to a 180°, or 5-hr, shift in phase whenever the direction of the interplanetary field switched by 180°, as occurs at sector boundary crossings, for example.

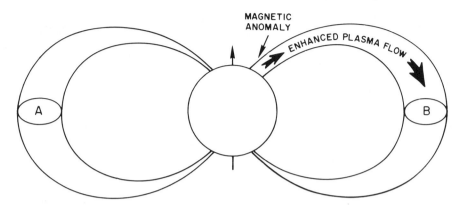

Fig. 10.14. Schematic illustration of how a magnetic anomaly enhances the flow of ionospheric plasma into the Io torus and the magnetosphere. Flux tubes A and B have the same cross-sectional area at the equator. However, flux tube B, which enters a negative magnetic anomaly (a weak-field region) in the northern hemisphere has a larger foot than either foot of flux tube A. The effect of a larger foot is that, for a given ionospheric escape flux, more ionospheric plasma will be transported to the equatorial plane by flux tube B.

Because such phase shifts are not observed, some other internal linkage must be sought. The only other idea that has thus far been developed is the magnetic-anomaly model wherein an extensive area of weak surface magnetic field is reflected to large radial distances as an enhancement in plasma concentration [Dessler and Hill, 1975].

The magnetic-anomaly model is simple in concept. The basic idea is that high-order multipoles in Jupiter's internal magnetic field produce, through one or more plasma processes, a gross longitudinal asymmetry within the magnetosphere at unexpectedly large Jovicentric distances. As can be seen by comparing the total surface field intensity with the total field intensity at $2\,R_J$ in Figure 1.1, the direct contribution of the quadrupole and octopole moments are very small even at a distance of $2\,R_J$; at $6\,R_J$, the Jovicentric distance of the Io plasma torus, the magnetic field is completely dominated by the dipole component.

In planetary magnetism, a magnetic anomaly is defined as a deviation of the surface magnetic field from that expected from the best-fit displaced dipole. Thus, Figure 1.5 is a magnetic-anomaly map, that is, it shows the surface field minus the displaced dipole field. The magnetic anomaly of interest is the depressed field region in the northern hemisphere centered near $\lambda_{III} = 260°$. There may also be some magnetic anomaly influence from the smaller area of depressed field in the southern hemisphere around $\lambda_{III} = 45°$.

There are at least two ways a magnetic anomaly might affect the distribution of plasma within the Jovian magnetosphere so as to produce a longitudinal asymmetry. The two suggestions advanced thus far are (1) a longitudinal variation in the escape flux of plasma from the Jovian ionosphere [Dessler and Hill, 1975] and (2) a longitudinal variation of the height-integrated conductivity of the Jovian ionosphere [Dessler and Hill, 1979]. The most obvious observed manifestation of a plasma asymmetry within the magnetosphere is the longitudinal asymmetry of the Io plasma torus described above. Much of the controversy surrounding the magnetic-anomaly model centers on how much weight should be given to this asymmetry, which is seen only in the cold torus and not in the hot torus. Needless to say, proponents of the magnetic-anomaly model regard the observed cold-torus asymmetry with some reverence.

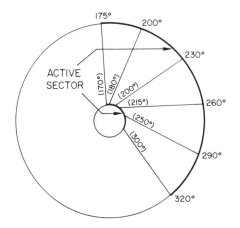

Fig. 10.15. Schematic view from above the north pole showing the longitudinal extent of the active sector. The outer circle represents the Io plasma torus and the longitude extremes of $\lambda_{III} = 175°$ and 320° are the approximate limits of the active sector within the torus. The System III longitudes of intermediate positions are marked as well as the longitudes of these positions when followed down magnetic field lines to the northern hemisphere ionosphere. These ionospheric longitudes, which are shown in parentheses, extend from $\lambda_{III} = 170°$ to 300°.

The way in which a longitudinal asymmetry in magnetospheric plasma loading from the ionosphere can be caused by a surface magnetic anomaly is illustrated in Figure 10.14. Given two magnetic flux tubes with equal equatorial cross-sectional areas and at the same equatorial distance from the displaced dipole, it is obvious that the flux tube with its foot in the magnetic anomaly has a larger cross-sectional area at iono-spheric heights, and therefore it conducts a larger escape flux of ionospheric plasma to the equatorial plane.

The second mechanism is related to an enhancement of ionospheric conductivity within the magnetic-anomaly region. The conductivity enhancement is a twofold effect of the reduced magnetic field within the magnetic anomaly. First, the conductiv-ity is increased because the Pedersen conductivity in the ionosphere is approximately inversely proportional to the magnetic field strength. Second, the ionization rate by particle bombardment within the magnetic-anomaly region should be greater because of the reduced mirror altitude and correspondingly larger loss cone for trapped parti-cles where the surface magnetic field is weaker. Dessler and Hill [1979] estimate that the height-integrated Pedersen conductivity within the magnetic-anomaly region is increased by an order of magnitude or more over the conductivity at other longitudes.

Given these two direct consequences of a magnetic anomaly, enhanced plasma loading of the magnetosphere in the longitude range of the magnetic anomaly can be accomplished by one of a number of processes [Dessler and Hill, 1975, 1979; Dessler, Sandel, and Atreya, 1981]. The most common theme is that un-ionized gas escapes from Io, goes into Keplerian orbit around Jupiter, and is subsequently ionized, with the ionization occurring most rapidly in the magnetic-anomaly region. Some such process would account for the observed longitudinal asymmetry in the cold torus, although the question remains as to why the asymmetry is not mimicked in the hot outer torus.

The magnetic-anomaly region and the magnetically connected region of the Io torus is the active sector defined by Vasyliunas [1975b] as the System III longitude range that is associated with non-Io-related radio noise when facing the Sun and the release of relativistic electrons into interplanetary space when facing the tail (see also Sakurai [1976]). Because of the contribution of higher-order multipoles to Jupiter's magnetic field, connections between the torus and the ionosphere are not along meridional planes, as is illustrated in Figure 10.15; the longitudes of magnetically connected torus and ionospheric elements can differ by as much as 45°. The longitude range of the active sector is presently uncertain to within perhaps ±25°.

A longitudinally asymmetric loading of the magnetosphere leads immediately to a longitudinal asymmetry in the radial transport of magnetospheric plasma; specifically,

Fig. 10.16. The denser portion of the Io plasma torus moves outward because of the action of centrifugal force. This motion establishes an electric-field pattern that causes the rest of the magnetospheric plasma to move in a corotating convection pattern roughly as shown. The magnetic anomaly region corresponds to flux tube B in Figure 10.14. The convection is probably sporadic.

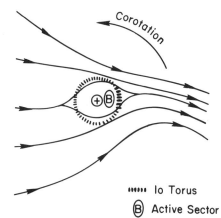

it has been proposed that the transport occurs in the form of a two-cell convection pattern that corotates with Jupiter [Vasyliunas, 1978; Dessler, Sandel, and Atreya, 1981; Hill, Dessler, and Maher, 1981]. The convection is outward from the active sector and inward in the conjugate sector (180° removed in longitude), as illustrated in Figure 10.16. A two-celled convection pattern is thus established that corotates with the planet.

The convection is likely to be a time-varying, if not impulsive, phenomenon. There are two reasons for this conclusion: (1) The convection period is calculated by Hill, Dessler, and Maher [1981] to be as short as 15 hr, with 30 hr being more probable. The escape of material from Io to supply the plasma torus appears to be too slow to allow such rapid convection to proceed continuously. The introduction of the concept of a duty cycle into the corotating convection model reduces the required rate of supply to an acceptable level. (2) Phenomena that are proposed to be related to or controlled by convection show an erratic, time-varying behavior. For example, the amplitude and longitude of the hydrogen bulge vary, the decametric radio emissions from Jupiter show evidence of nulling and timing jitter (in pulsar parlance), the amplitude and longitude of the maximum in the mass density in the cold torus can change appreciably on a time scale of one Jovian rotation period, and the escape of relativistic electrons into interplanetary space, while a reproducible phenomenon, is not a continuous one. The time and longitude variations in these phenomena can be understood on the basis of time variation in the pattern of corotating magnetospheric convection.

An explanation of the first five phenomena listed in Table 10.4, and graphically illustrated in Figure 10.10, follows naturally as a consequence of the corotating convection that results from the magnetic anomaly model. For example:

a. Decametric radio emission is generated principally in the active sector. The ionospheric conductivity has its maximum value there, and the most intense Birkeland currents should flow between the active sector and the density maximum in the torus. These intense Birkeland currents presumably lead to the generation of radio noise by current driven instabilities [Dessler and Hill, 1979; Dessler, 1980; see Chap. 9]. When Io is near the active sector, additional plasma injection into the torus causes the Birkeland currents to be enhanced and the radio emissions to brighten correspondingly.

b. Modulation of the energy spectrum of electrons released into interplanetary space occurs because magnetospheric plasma and energetic particles escape primarily through the Jovian magnetic tail; this flow is enhanced once each planetary rotation when the corotating convective outflow is directed tailward [Hill, Carbary, and Dessler, 1974; Hill and Dessler, 1975].

c. The longitudinal asymmetry of the Io plasma torus probably arises from an ionizing interaction between neutral particles escaping Io and the active sector. This asymmetry is more correctly regarded as a cause than a consequence of corotating convection.

d. The hydrogen bulge, presently the strongest evidence for the existence of corotating convection, is caused by inflow of hot magnetospheric plasma which dissociates CH_4 and H_2 in Jupiter's upper atmosphere [Dessler, Sandel, and Atreya, 1981].

e. The longitudinal asymmetry of the radial extent of plasma from the Io torus, as reported by Pilcher et al. [1981], can be regarded as visual evidence of the plasma flow expected from a corotating convection system.

The longitudinal asymmetry of plasma temperature (item f) might also be related to the corotating convection pattern, but the nature of this connection is not clear at present because the nature of the plasma acceleration mechanism itself is not clear. Whatever this acceleration mechanism may be, an important clue may be provided by the fact that it appears to be more effective in the active sector.

10.8. Conclusion

We have described a number of theoretical concepts that have been applied more-or-less successfully to the explanation of observed Jovian magnetospheric phenomena. The emerging picture of Jovian magnetospheric dynamics differs radically from that of the terrestrial magnetosphere, although our progress in understanding the Jovian magnetosphere is undoubtedly aided by the experience gained in several years' detailed study of the terrestrial magnetosphere. The important differences apparently result from the internal sources of plasma and energy within Jupiter's magnetosphere, giving rise to a variety of spin-periodic phenomena. These phenomena are generally attributed to the diurnal precession of Jupiter's tilted magnetic dipole (the disc model) and/or the corotation of a longitudinally asymmetric magnetospheric plasma feature (the magnetic anomaly model). Each type of model has had some success at explaining (as opposed to merely describing) some of the phenomena, although neither type of model, as presently formulated, can account for all of the observed phenomena. Moreover, the Saturnian magnetosphere exhibits spin-periodic control of low-frequency radio emissions reminiscent of that in the Jovian magnetosphere [Warwick et al., 1981], even though Saturn's intrinsic magnetic field does not exhibit the significant tilt required for modulation by a disc model. The Voyager trajectories were not adequate to detect a modest nondipole component of the type required for a magnetic anomaly model.

Further theoretical work (and perhaps acquisition of new data) are required before we will be in a position to set forth a comprehensive Jovian magnetosphere model that encompasses the known phenomenology of Jupiter's magnetosphere and elucidates the connection (or lack of connection) between spin-periodic phenomena in the Jovian, Saturnian, and pulsar magnetospheres. Such a model must address (at least) the following fundamental unanswered questions:

What is the nature of the electrodynamic Io–Jupiter interaction, and why does the strength of this interaction appear to depend on Jovian longitude?

How is material removed from Io and ionized to form the plasma torus? (This question is discussed further in Chap. 6.)

How is the Io torus plasma transported outward to form the magnetospheric current sheet and planetary wind? (This question is discussed in detail in Chap. 11.) In particular, why does a persistent longitudinal asymmetry appear in the inner, cooler torus but not in the outer, warmer, UV-emitting torus?

By what mechanism is torus plasma heated (rather than adiabatically cooled) as it moves outward in the magnetosphere?

How are ions and electrons from the torus accelerated to form the extensive radiation belt?

How do relativistic electrons escape from Jupiter's magnetosphere with evidence of modulation at the Jovian spin period?

A comprehensive model of the Jovian magnetosphere should be expected not only to address these questions individually but also to expose the physical connections among the answers to these and many subsidiary questions. We have identified a number of conceptual ingredients that may be important in formulating such a model, and a number of problem areas where additional theoretical work is called for in the context of present observational knowledge. The development of a comprehensive model will, however, probably require a much broader data base than is available now. Only then will we be in a position to apply our understanding of Jupiter's magnetosphere with confidence to other magnetospheres outside the solar system.

ACKNOWLEDGMENTS

We are grateful to V. M. Vasyliunas, A. W. Schardt, F. Bagenal, D. F. Strobel, J. K. Alexander, R. A. Brown, F. C. Michel, D. L. Matson, J. F. Carbary, J. E. P. Conner-ney, J. A. Simpson, and R. E. Johnson for helpful comments. The work of TWH and AJD was supported in part by the National Science Foundation under grant ATM 80-19425 and the National Aeronautics and Space Administration under grants NAGW-166 and NAGW-168; the work of CKG was supported by the National Science Foundation under grant ATM76-82739 and the National Aeronautics and Space Administration under Contract NAS2-6553.

11

PLASMA DISTRIBUTION AND FLOW

Vytenis M. Vasyliunas

The highly extended magnetic field-line configuration of the Jovian magnetosphere with a near-equatorial current sheet and associated plasma sheet arises from mechanical stresses in the rotating plasma balanced by magnetic stresses. The relative geometrical thinness of the current sheet permits the use of several approximations in the description of the stress balance, each with a specified regime of validity in terms of taillike vs. dipolar field and hot vs. cold plasma; these include pressure balance, a simplified tangential stress balance, and an estimate of current-sheet thickness. A number of simple but quantitative models of the magnetic field are now available, including both theoretical models based on various assumptions about the distribution and degree of corotation of the plasma and empirical models intended to represent the observations. From the empirical models, values of plasma parameters required to maintain stress balance can be estimated. To obtain agreement between the estimated and the observed mass density values, it is necessary to assume that the azimuthal velocity of the plasma decreases significantly below rigid corotation in the outer magnetosphere. The uncertainties in the magnetic field component normal to the current sheet lead to sizable discrepancies among various estimates of the density or of the current sheet thickness. Azimuthal magnetic fields over the midnight-to-dawn quadrant are nearly independent of local time, in contrast to the situation in the terrestrial magnetosphere; they imply radial currents whose closure through the ionosphere is related to partial corotation. Generalization of Parker-spiral arguments to include a finite ionospheric conductivity provides a quantitative model for the azimuthal field. To keep the angular acceleration of the plasma within the required bounds, a mass flow of at least some 10^{30} to 10^{31} amu/s must be assumed but it is not yet clear whether this is a net outflow or primarily a circulation. Dipole tilt effects on the current sheet can be quantitatively modeled within the rigid corotation region; at larger distances, only qualitative propagating-wave descriptions plus empirical fits are available. Plasma flow models so far are mostly qualitative except for the description of partial corotation. The inability of magnetic stresses to maintain centripetal acceleration of a given flux tube content of plasma beyond a limiting distance is expected to result in a radial outflow (the planetary wind) and associated changes of magnetic field topology; the implied flow pattern is very similar qualitatively to the observed magnetospheric wind.

11.1. Introduction

The role of magnetospheric plasma in determining the configuration and dynamics of the magnetosphere is considerably more important at Jupiter than it is at Earth. If we imagine a sphere centered on the planet and require that the effects of the plasma inside the sphere on the magnetic field represent no more than a fractional perturbation of the total field, then the radius of the sphere could be allowed to reach almost the distance to the subsolar magnetopause at Earth but hardly a tenth of that at Jupiter. Consistency between the magnetic field configuration and the plasma distribution and flow is thus a major constraint on the physical description of almost the entire Jovian magnetosphere, whereas in the terrestrial case it is for most purposes a significant requirement only in the outermost boundary regions and the magnetotail. The main concern of this chapter is the extent to which various aspects of Jovian magnetospheric configuration and dynamics can be understood in a unified and self-consistent fashion on the basis of present knowledge of the plasma population (Chaps. 3 and 4) and of the magnetic field (Chap. 1) together with appropriate theoretical models (Chap. 10). Most emphasis is placed here on the middle and outer magnetosphere, beyond the Io torus (the physics of the torus itself is discussed in Chaps. 3 and 6).

11.2. Plasma configuration in the middle and outer magnetosphere

1. Formulation of the problem

Beyond a distance of some 20 R_J from the planet, the Jovian magnetosphere contains highly distended magnetic field lines, with a near-equatorial current sheet and associated plasma sheet; the observational evidence for this configuration (which is a clear permanent feature of the observed predawn sector and is perhaps somewhat less clear on the dayside) is summarized in Chapters 1, 3, 4, 5, and 8. The physical description of such a stretched-out configuration requires considering stress balance both in a local and in a global context. Locally, there exists a magnetic force density that can be viewed either as arising from the Maxwell stress of the stretched-out field lines or, equivalently, as the $\mathbf{j} \times \mathbf{B}$ force of the magnetic field acting on the current sheet, balanced by mechanical plasma stresses to be identified and described. Globally, the large-scale pattern of the currents and the source of the mechanical stresses need to be considered.

A useful analog is provided by the tail of the terrestrial magnetosphere, which also contains a stretched-out field configuration with a current sheet and an associated plasma sheet. Locally, stress balance is maintained by counteracting the magnetic tension on the current sheet by a plasma pressure gradient in the direction along the magnetotail axis, by a similar gradient of plasma flow kinetic energy, by the tension resulting from a plasma pressure anisotropy, or by some combination of all three [see, e.g., Rich, Vasyliunas, and Wolf, 1972, for a detailed discussion]. The requirement of local stress balance strongly constrains the spatial variation of plasma and magnetic field properties; for example, in the two- dimensional approximation, detailed quantitative models of the plasma and field configuration can be constructed from suitable boundary conditions and the assumption that a plasma pressure gradient alone balances the magnetic stress [see, e.g., Schindler, 1975, 1979]. Globally, the stretched-out magnetotail configuration is the result of solar wind flow and specifically of the tangential stress exerted by the solar wind in its interaction with the magnetosphere. The power extracted by this tangential stress acting against solar wind flow represents the principal source of energy dissipated in the terrestrial magnetosphere [see, e.g., Siscoe and Cummings, 1970; Siscoe and Crooker, 1974; Akasofu, 1981].

The physical description of the middle and outer Jovian magnetosphere is in many ways similar to that of the terrestrial magnetotail, but with the difference that Jupiter's rotation rather than solar wind flow is generally thought to play the dominant role. Locally, the centripetal acceleration associated with any corotational motion of the plasma must be added to the list of possible contributors to stress balance. Globally, the stretched-out ('magnetodisc') configuration is considered to be the result of outward stresses ultimately derivable in some way from the planetary rotation. Unlike the case of a magnetotail formed by solar wind interaction, these stresses are predominantly perpendicular to the associated (corotational) flow and thus do not by themselves extract energy from the rotation; dissipation of rotational energy requires the additional presence of outward flow or azimuthal stress. Finally, one does expect that at sufficiently large distances the importance of any rotational effects decreases and the solar wind becomes dominant in shaping the field-line configuration; it is, however, still a matter of debate where this transition occurs.

The quantitative formulation of local stress balance is contained in the momentum equation for the plasma

$$\rho(d\mathbf{V}/dt - \mathbf{g}) + \nabla \cdot \mathsf{P} = \mathbf{j} \times \mathbf{B} \tag{11.1}$$

together with two of Maxwell's equations

$$\nabla \times \mathbf{B} = \mu_0 \mathbf{j} \tag{11.2}$$

$$\nabla \cdot \mathbf{B} = 0 \tag{11.3}$$

In Equation (11.1), ρ is the mass density, $\mathbf{g}$ is the gravitational field (negligible for most purposes in the middle and outer magnetosphere),

$$d\mathbf{V}/dt = \partial\mathbf{V}/\partial t + \mathbf{V} \cdot \nabla\mathbf{V} \tag{11.4a}$$

is the acceleration of plasma bulk flow with respect to inertial space, and the rest of the symbols have their usual meaning. (In the rest of this chapter, the gravitational term will be omitted; it can always be reintroduced, if necessary, by the simple substitution $d\mathbf{V}/dt \rightarrow d\mathbf{V}/dt - \mathbf{g}$.) If the velocity $\mathbf{V}$ is referred not to inertial space but to a system of coordinates rotating with angular velocity ω, Equation (11.4a) is replaced by

$$\begin{aligned} d\mathbf{V}/dt = {} & \partial\mathbf{V}/\partial t + \mathbf{V} \cdot \nabla\mathbf{V} + \omega \times (\omega \times \mathbf{r}) \\ & + 2\omega \times \mathbf{V} - \mathbf{r} \times (\partial/\partial t + \mathbf{V} \cdot \nabla)\omega \end{aligned} \tag{11.4b}$$

where the added terms represent centrifugal, Coriolis, and (possible) differential rotation effects, respectively. The pressure tensor P should be well approximated by the gyrotropic form

$$\mathsf{P} = P_\perp \mathsf{I} + (P_\parallel - P_\perp) \mathbf{bb} \tag{11.5}$$

because the length scales of practically all magnetospheric structures of interest are large compared to typical particle cyclotron radii. It often proves convenient to eliminate $\mathbf{j}$ from Equation (11.1) by using (11.2) and (11.3) to write $\mathbf{j} \times \mathbf{B}$ in either of two equivalent forms:

$$\mu_0 \mathbf{j} \times \mathbf{B} = \mathbf{B} \cdot \nabla\mathbf{B} - 1/2 \, \nabla B^2 \tag{11.6a}$$

$$= \nabla \cdot (\mathbf{BB} - 1/2 \, B^2 \mathsf{I}) \tag{11.6b}$$

Equations (11.1)–(11.6) express the connection between the distribution of the plasma and the configuration of the magnetic field; at least in principle, they allow either one to be calculated from the other.

In applying these equations to construct models that might represent stress balance in the Jovian magnetosphere, several simplifications become possible. First, the observed magnetic field configuration indicates a well-defined and relatively thin current sheet; the length scales for spatial variations normal and transverse to the current sheet may be assumed to be of the order of, respectively, the sheet thickness h and the radial distance r, and hence the geometrical thin-sheet approximation $h \ll r$ may be consistently used. Second, most of the quantitative models developed to date are based on the physical assumption that the direction of the plasma bulk flow velocity is azimuthal, that is, corotational; this assumption, whose basis is discussed in Section 11.4.2, does not mean necessarily that other flow components are absent but merely that they are sufficiently small to have negligible direct dynamical effects. Third, axial symmetry and time independence are often imposed as simplifying assumptions, neglecting any effects of dipole tilt relative to the rotation axis as well as any magnetic-anomaly effects; this makes the models more tractable mathematically but obviously restricts them to describing those aspects of the magnetosphere where tilt and anomaly effects are not important.

2. Implications of a thin current sheet

The thin-current-sheet approximation seems to be rather generally valid, and it is useful to consider its consequences for the stress balance equations before proceeding to more detailed (and more restricted) models. The approximation enables one (a) to derive an equation that describes directly the tangential stress on the current sheet, (b) to establish a pressure balance relation between the plasma and the magnetic field, and (c) to estimate the thickness of the current sheet and hence to check the validity of the assumed thin-sheet approximation.

For the quantitative description of local stress balance in the neighborhood of a point on the current sheet, introduce Cartesian coordinates with the z axis along the normal to the current sheet at that point, and let the subscript t denote (vector) quantities tangential to the current sheet; thus, for example, $\mathbf{B} = B_z\,\hat{z} + \mathbf{B}_t$ and $\nabla = \hat{z}\,\partial/\partial z + \nabla_t$. Maxwell's Equations (11.2) and (11.3) become

$$\mu_0\,\mathbf{j} = \hat{z} \times (\partial\mathbf{B}_t/\partial t - \nabla_t\,B_z) + \nabla_t \times \mathbf{B}_t \qquad (11.7)$$

$$0 = \partial B_z/\partial z + \nabla_t \cdot \mathbf{B}_t \qquad (11.8)$$

The thin-sheet approximation is then expressed by the ordering

$$\partial/\partial z \sim O(1)/h \qquad \nabla_t \sim O(\epsilon)/h \qquad (11.9)$$

where $\epsilon << 1$ is the ratio of current sheet thickness h to a gradient length scale in the tangential direction. It is convenient to define

$$\mathbf{B}_p \equiv \mu_0 \int dz\,\mathbf{j} \times \hat{z} \qquad (11.10)$$

which is equivalent to the relation

$$\mu_0\,\mathbf{j}_t = \hat{z} \times \partial\mathbf{B}_p/\partial z \qquad (11.11)$$

$\mathbf{B}_p$ may be thought of as the "planar" magnetic field associated with the local sheet current [cf., Mead and Beard, 1964] but it may also be viewed as simply a convenient representation of the current per unit length obtained by integrating the current density across either part of or the entire width of the current sheet. The advantage of introducing $\mathbf{B}_p$ is that it represents the only part of the magnetic field that varies on a spatial scale of current sheet thickness: with the magnetic field written as $\mathbf{B} = \mathbf{B}_p + (\mathbf{B} - \mathbf{B}_p)$, it is readily shown from (11.7), (11.8), and (11.11) that

$$(\partial/\partial z)\,(\mathbf{B} - \mathbf{B}_p) = \nabla_t\,B_z - \hat{z}\,\nabla_t \cdot \mathbf{B}_t \qquad (11.12)$$

If the current sheet is sufficiently thin, the quantity $\mathbf{B} - \mathbf{B}_p$ may be treated as constant, to within terms of $O(\epsilon)$, over its width. Note that

$$\mathbf{B}_t - \mathbf{B}_p = \int dz\,\nabla_t\,B_z \qquad (11.13)$$

and thus $\mathbf{B}_p$ may be equated to $\mathbf{B}_t$, the tangential component of the entire field, again to $O(\epsilon)$ for a sufficiently thin sheet. The Lorentz force density may be written as

$$\mu_0\,\mathbf{j} \times \mathbf{B} = -\hat{z}\,\mathbf{B}_p \cdot \partial\mathbf{B}_p/\partial z + \mu_0\,\mathbf{j}_t \times (\mathbf{B} - \mathbf{B}_p) + (\nabla_t \times \mathbf{B}_t) \times \mathbf{B}_t \qquad (11.14)$$

This expression is exact; the thin-sheet approximation consists in neglecting the last term (associated with an obviously small z component of $\mathbf{j}$ due, for example, to an azimuthally varying width of the current sheet) and in treating $\mathbf{B} - \mathbf{B}_p$ in the second term as independent of z.

Tangential stress balance. The components of the momentum Equation (11.1) tangential to the current sheet may now be written as

$$(11.15)$$

$$\rho(d\mathbf{V}/dt)_t + \nabla_t P_\perp + \nabla_t \cdot (P_\| - P_\perp)\, \mathbf{b}_t \mathbf{b}_t + (\partial/\partial z)(P_\| - P_\perp)\, b_z \mathbf{b}_t = (1/\mu_0)\, B_z\, \partial\mathbf{B}_p/\partial z$$

where Equations (11.5), (11.14) with the previously mentioned thin-sheet approximations, (11.11), and the fact that $\hat{z} \cdot (\mathbf{B} - \mathbf{B}_p) = B_z$ have been used; $(d\mathbf{V}/dt)_t$ is of course the tangential part of the acceleration (to be distinguished from the time derivative of the tangential velocity!). Equation (11.15) balances the tangential components of inertial stress and pressure gradient (first two terms on the LH side) against the tension of magnetic field lines crossing the current sheet (RH side), with both the pressure gradient and the magnetic tension modified by additional terms dependent on pressure anisotropy (last two terms on the LH side). If the plasma flow is predominantly azimuthal,

$$\mathbf{V} = \omega \times \mathbf{r} \tag{11.16}$$

where ω may or may not be equal to Ω_J (see Sec. 11.D.2), then the acceleration is

$$d\mathbf{V}/dt = \omega \times (\omega \times \mathbf{r}) = -\omega^2\,(\mathbf{r} - \hat{\omega}\mathbf{r} \cdot \hat{\omega}) \equiv -\omega^2\,\mathbf{R} \tag{11.17}$$

and points inward toward the rotation axis; the pressure normally decreases with increasing radial distance in the outer magnetosphere and hence the pressure gradient also points inward. The magnetic tension is thus required to point inward, which it does as long as the signs of $\mathbf{B}_p$ (or $\mathbf{B}_t$) and B_z are those appropriate to stretched-out but closed field lines from the planet.

Two variants of Equation (11.15) are of interest for further developments. On the field reversal surface, defined by the condition $\mathbf{B}_t = 0$, (11.15) can be cast into the rather simple form

$$\rho(d\mathbf{V}/dt)_t + \nabla_t P_\perp + (P_\| - P_\perp)\,\nabla_t B_z /B_z = (B_z\,\xi/\mu_0)\,\partial\mathbf{B}_p/\partial z \tag{11.18}$$

where $\xi \equiv 1 - \mu_0(P_\| - P_\perp)/B^2$ (note that $B = B_z$ on the field reversal surface). Equation (11.18) is actually exact (for the case $\mathbf{B}_t = 0$), with no thin-sheet approximations. One may also integrate (11.15) across the full width of the current sheet to obtain the equation for the tangential force per unit area:

$$\sigma(d\mathbf{V}/dt)_t + \nabla_t \int dz\, P_\perp + \nabla_t \cdot \int dz\,(P_\| - P_\perp)\,\mathbf{b}_t \mathbf{b}_t = (1/\mu_0) B_z\, [\mathbf{B}_p] \tag{11.19}$$

Here $\sigma \equiv \int dz\,\rho$ is the mass density per unit area, and $[\mathbf{B}_p]$ is the vector difference of $\mathbf{B}_p$ above and below the current sheet; from Equation (11.10),

$$[\mathbf{B}_p] = \mu_0\,\mathbf{j}' \times \hat{z} \tag{11.20}$$

with $\mathbf{j}'$ the sheet current density per unit length. To obtain Equation (11.19) several thin-sheet approximations have been used, in particular constancy of B_z and of $(d\mathbf{V}/dt)_t$ over the sheet thickness and neglect of $P_\| - P_\perp$ compared to B^2/μ_0 outside the sheet. Furthermore, if the current sheet is sufficiently thin in the sense that its thickness is allowed to become small while σ and P are held fixed, the pressure terms in (11.19) may be dropped.

Pressure balance. The component of the momentum Equation (11.1) normal to the current sheet may be written, after some manipulation of (11.14), as

$$\rho(d\mathbf{V}/dt) \cdot \hat{\mathbf{z}} + \partial P_{zz}/\partial z + \nabla_t \cdot (P_{\parallel} - P_{\perp})\mathbf{b}_t b_z \tag{11.21}$$

$$= - (1/\mu_0)\left[\frac{\partial}{\partial z}\left(\frac{1}{2} B_p^2 \right) + (\mathbf{B} - \mathbf{B}_p) \cdot \partial\mathbf{B}_p/\partial z \right]$$

The third term on the LH side is quite generally $O(\epsilon)$ compared to the second and vanishes entirely if the plasma pressure is isotropic. On the RH side, the thin-sheet approximation of $\mathbf{B} - \mathbf{B}_p$ independent of z may be invoked. Equation (11.21) may then be rewritten

$$\frac{\partial}{\partial z}\left[P_{zz} + B_p^2/2\mu_0 + \mathbf{B}_p \cdot (\mathbf{B} - \mathbf{B}_p)/\mu_0 + \int^z dz' \, \rho(d\mathbf{V}/dt) \cdot \hat{\mathbf{z}} \right] = 0 \tag{11.22}$$

which, when integrated with respect to z, states that the quantity in brackets remains constant across the width of the current sheet [more precisely, undergoes a fractional variation of no more than $O(\epsilon)$]. The magnetic terms in (11.22), which may be rewritten as

$$(1/2)\, B_p^2 + \mathbf{B}_p \cdot (\mathbf{B} - \mathbf{B}_p) = (1/2)\, B_t^2 - (1/2)\, |\mathbf{B}_t - \mathbf{B}_p|^2 \tag{11.23}$$

represent the pressure of the tangential magnetic field (minus the pressure of its curl-free part $\mathbf{B}_t - \mathbf{B}_p$ if it is not negligible). Thus, Equation (11.22) is the familiar statement of pressure balance between the plasma and the magnetic field modified by an added term representing the inertial stress of acceleration normal to the sheet. If the plasma pressure is isotropic, P_{zz} is simply the pressure P; otherwise, from Equation (11.5),

$$P_{zz} = P_{\parallel} b_z^2 + P_{\perp}(1 - b_z^2). \tag{11.24}$$

Note in particular that $P_{zz} = P_{\parallel}$ at the field reversal surface where $\mathbf{B}_t = 0$. (Note also that the thermal energy density of the plasma is always $(1/2)\,(P_{xx} + P_{yy} + P_{zz}) = (1/2)\, P_{\parallel} + P_{\perp}$, which is not equal to P_{zz} except in the singular case when $2P_{\perp}/P_{\parallel} = 1 - (B_t/B_z)^2$; the balancing of plasma and magnetic energy densities instead of pressures, as in Walker, Kivelson, and Schardt [1978] or Lanzerotti et al. [1980], is incorrect, being in error by a factor of 3/2.)

Thickness of the current sheet. We may now obtain an order-of-magnitude estimate for h, starting with the approximate definition

$$\frac{1}{h} \simeq \frac{1}{B_p^*}\left| \frac{\partial}{\partial z} \mathbf{B}_p \right| \tag{11.25}$$

where $\partial\mathbf{B}_p/\partial z$ is to be evaluated at the field reversal surface from the tangential stress balance relation (11.18) and $2B_p^* = |[\mathbf{B}_p]|$, the magnitude of the vector change of $\mathbf{B}_p$ across the entire width of the current sheet (the factor 2 makes $B_p^* = |\mathbf{B}_p|$ on either side of the current sheet for the symmetric case and means that h is actually the halfwidth). It is apparent from Equations (11.11) and (11.10) or (11.20) that this estimate gives the thickness as the ratio of current per unit length to the current density at a representative location within the current sheet. A specific model for the mechanical stresses on the LH side of (11.18) must be adopted. Following common practice, we assume the acceleration to be that associated with azimuthal flow [Eqs. (11.16) and

(11.17)] with speed $V_\phi = \omega r$ and the gradients of the pressure and the magnetic field to be radial; we write

$$\nu \equiv -\partial \log P_\perp / \partial \log r \qquad \nu_z \equiv -(1/B_z)\, \partial B_z / \partial \log r \qquad (11.26)$$

and define the thermal speeds $w_\perp$, $w_\parallel$ by

$$P_\perp \equiv \rho w_\perp^2 \qquad P_\parallel \equiv \rho w_\parallel^2 \qquad (11.27)$$

Combining (11.25)–(11.27) with (11.18) yields an estimate for h

$$\frac{h}{r} \approx \left| \frac{B_p{}^*}{B_z} \right| \frac{(B_z^2/\mu_0 \rho) + w_\perp^2 - w_\parallel^2}{V_\phi^2 + \nu w_\perp^2 + \nu_z(w_\parallel^2 - w_\perp^2)} \qquad (11.28)$$

where all the quantities on the RH side (with the exception of $B_p{}^*$) are to be evaluated at the field reversal surface. It is convenient to rewrite (11.28) in the form

$$\frac{h}{r} \approx \left| \frac{B_p{}^*}{B_z} \right| \frac{[B_z^2/\mu_0 \rho(V_\phi^2 + \nu w_\perp^2)]\,(1 - \chi)}{1 + \nu_z \chi[B_z^2/\mu_0 \rho(V_\phi + \nu w_\perp^2)]} \qquad (11.29)$$

where

$$\chi \equiv \mu_0 \rho(w_\parallel^2 - w_\perp^2)/B_z^2 \qquad (11.30)$$

On physical grounds one expects $|\chi| < 1$ (if $|\chi| \geq 1$, the plasma will be subject to either the firehose or the mirror instability, depending on the sign of χ), and Equation (11.29) then makes it apparent that the order of magnitude of h/r is not greatly affected by plasma pressure anisotropies as long as these are no larger than what is plausible.

The main question about the thickness is not whether it is small in the merely geometrical sense $h \ll r$ but whether it is small enough to ensure the validity of the various results derived from the thin-sheet approximation, which are principally the following:

(1) Pressure balance, Equation (11.22). The crucial step in the derivation was the neglect of the term $\mathbf{B}_p \cdot (\partial/\partial z)\,(\mathbf{B} - \mathbf{B}_p)$ compared to $(\partial/\partial z)\, B_p^2/2$; with $(\partial/\partial z)\,(\mathbf{B} - \mathbf{B}_p)$ given by (11.12) and $\partial B_p/\partial z$ approximated by $B_p{}^*/h$, the criterion for its validity is readily shown to be

$$\nu_z \left| \frac{B_z}{B_p{}^*} \right| \frac{h}{r} \ll 1 \qquad (11.31)$$

(2) The near equality $\mathbf{B}_p \approx \mathbf{B}_t$. From Equation (11.13) we obtain

$$\mathbf{B}_t - \mathbf{B}_p \sim h\,\nabla_t B_z \sim -\nu_z B_z\, h/r \qquad (11.32)$$

so that the criterion for $|\mathbf{B}_t - \mathbf{B}_p| \ll B_p{}^*$ is the same as (11.31).

(3) Constancy of B_z across the thickness of the current sheet, used in deriving the tangential force Equation (11.19). The change of B_z is given by the integral of (11.8) as

$$\delta B_z = -\int dz\, \nabla_t \cdot \mathbf{B}_t \sim \nu_r B_t\, h/r \qquad (11.33)$$

where ν_r, analogously to ν_z, is a measure of the radial gradient of B_t. Taking into account the relation between $\mathbf{B}_t$ and $\mathbf{B}_p$ given by (11.32), with due regard to signs, we obtain as the criterion for $|\delta B_z/B_z| \ll 1$

$$\nu_r \left[\frac{h}{r} \left| \frac{B_p{}^*}{B_z} \right| + \nu_z \left(\frac{h}{r} \right)^2 \right] \ll 1 \qquad (11.34)$$

(4) Localization of field lines crossing the current sheet. The radius vector to a point on a magnetic field line undergoes, across a half-thickness of the current sheet, a tangential change given by

$$\delta r_t = \int dz\, B_t/B_z \sim hB_t/B_z \tag{11.35}$$

where the integration in this case follows the field line; however, if $|\delta r_t| \ll r$, integrals along the field line may be equated to simple integrals across the width of the current sheet at a fixed location. Comparison of (11.35) with (11.33) shows that the criterion for $|\delta r_t/r| \ll 1$ is the same as that for $|\delta B_z/B_z| \ll 1$ and hence given by (11.34), except for absence of the factor ν_t.

There are thus two distinct thin-sheet approximations, embodied in conditions (11.31) and (11.34), respectively. The first implies pressure balance and the identification $B_p \sim B_r$; then B_p* can be determined either from observations of the actual magnetic field just outside the current sheet or by applying the pressure balance relation (11.22) between the field reversal surface and the outside with the result

$$B_p*^2 \simeq 2\mu_0 \rho w_\|^2 \tag{11.36}$$

(the inertial terms have here been neglected, as their normal component is usually relatively small). To satisfy the inequality in (11.31) it is sufficient and (if we ignore the improbable singular case $\chi \simeq 1$) necessary to have

$$B_z^2/\mu_0 \rho (V_\phi^2 + \nu w_\perp^2) \ll 1 \tag{11.37}$$

which can be rewritten, since (11.36) holds if the inequality does, as

$$(B_z/B_p*)^2 [2w_\|^2/(V_\phi^2 + \nu w_\perp^2)] \ll 1 \tag{11.38}$$

With a limit on the plasma pressure anisotropy set by

$$\frac{w_\|^2 - w_\perp^2}{w_\|^2} = \chi \frac{B_z^2}{\mu_0 \rho w_\|^2} = 2\chi \left(\frac{B_z}{B_p*}\right)^2 \tag{11.39}$$

with $|\chi| < 1$, it is readily shown that in the case $|B_z| \ll |B_p*|$ (which we call a *taillike* field) the inequality (11.38) and hence (11.31) is automatically satisfied without any additional assumptions [a result that can also be seen directly from (11.31) if we recall that h/r should in no case be larger than $O(1)$], whereas in the case of a nontaillike or nearly dipolar field $|B_z| \gtrsim |B_p*|$ or $|B_z| \gg |B_p*|$ the inequality requires $V_\phi^2 \gg w_\perp^2$ and $V_\phi^2 \gg w_\|^2$ as well as $B_z^2/\mu_0 \rho V_\phi^2 \ll 1$. Pressure balance, therefore: (a) always holds, whether the plasma be cold or hot, in the case of a highly stretched-out, taillike field configuration, such as is found in the outer regions of the Jovian magnetosphere; (b) in the nearly dipolar field of the inner regions, holds only if the plasma is sufficiently cold, that is, the azimuthal flow speed V_ϕ exceeds the thermal speed, and if furthermore V_ϕ appreciably exceeds the Alfvén speed in the current sheet; (c) does not hold in a strongly dipolar field defined by the condition $B_z^2 \gtrsim \mu_0 \rho V_\phi^2$. Whenever pressure balance does hold, it is convenient for most purposes to express the thickness of the current sheet as

$$\frac{h}{r} = \left|\frac{B_z}{B_p*}\right| \frac{2w_\|^2(1 - \chi)}{V_\phi^2 + \nu w_\perp^2} \tag{11.40}$$

obtained from (11.29) by using (11.36) to eliminate ρ and (11.37) to drop small terms. Equation (11.40) is the basis for most observational models of current sheet thickness to be discussed in Section 11.3.3.

Table 11.1. *Validity of thin-sheet approximations*

	Plasma:	Pressure balance, $\mathbf{B}_p = \mathbf{B}_t$		B_z = constant, $(\delta \mathbf{r}_t)_{\text{field line}} \ll r$	
		Hot	Cold	Hot	Cold
Magnetic field:					
$B_p^* \gg B_z$	Tail-like	Yes	Yes	No	Yes
$B_p^* \lesssim B_z$ or $B_p^* \ll B_z$ $B_z^2 \ll \mu_0 \rho V_\phi^2$	Nearly dipolar	No	Yes	No	Yes
$B_z^2 \gtrsim \mu_0 \rho V_\phi^2$	Strongly dipolar	No	No	No	Yes

The other thin-sheet approximation, given by inequality (11.34), implies constancy of B_z and negligible change of tangential location of a field line across the thickness of the current sheet; an important corollary is that the plasma content of a flux tube per unit magnetic flux, defined by an integral along the field line as

$$\eta \equiv \int d\ell \, \frac{\rho}{B} \tag{11.41}$$

can then be simply approximated as

$$\eta = \int d\ell \, \frac{\rho}{B} = \int dz \, \frac{\rho}{B_z} \simeq \frac{\sigma}{B_z} \tag{11.42}$$

The condition $h/r \ll 1$ is evidently necessary for (11.34) to hold; in the case $|B_z| \gtrsim |B_p^*|$ or $|B_z| \gg |B_p^*|$ it is also sufficient, but in this same case $h/r \ll 1$ is implied by pressure balance [cf., Eq. (11.29) and inequality (11.37)]. Thus, for a nearly dipolar field, the constant B_z approximation holds under the same conditions when pressure balance does, that is, when the plasma is cold. It will shortly be shown that for a strong dipolar field the constant B_z approximation also holds when the plasma is cold. Finally, for a taillike field, substituting h/r from (11.40) and using the fact that [Eq. (11.39)] $w_\perp^2 \simeq w_\parallel^2 = w^2$ to within $O(B_z/B_p^*)^2$ we reduce (11.34) to the approximate form

$$\frac{2\nu_r w^2}{V_\phi^2 + \nu w^2} \ll 1 \tag{11.43}$$

in which the inequality is satisfied if and only if $V_\phi^2 \gg w^2$. Thus, quite generally, the constant B_z approximation holds only when the plasma is cold; the presure gradient terms in the tangential force Equation (11.19) should be neglected whenever constancy of B_z has been invoked. (Statements by Gleeson and Axford [1976] and by Goertz [1976b] that B_z = constant follows merely from $h/r \ll 1$ are thus not correct, although the calculation of Gleeson and Axford is self-consistent inasmuch as they drop the tangential pressure gradients, whereas Goertz does not.)

Table 11.1 summarizes the results on the validity of both the thin-sheet approximations under various magnetic field and plasma conditions.

Although the method described so far for estimating the thickness of the current sheet is in principle valid generally, in practice it does not provide useful estimates of h/r when pressure balance does not hold, since there is then no ready way of evaluating B_p^* in Equation (11.28) or (11.29): one can neither calculate B_p^* from the pressure balance relation nor equate it to the measured tangential component of the field. However, an alternative approach to estimating the thickness may be taken. The component of the momentum Equation (11.1) parallel to the magnetic field may be written, with a gyrotropic pressure tensor (11.5) and after some manipulation,

$$\rho \left(\frac{d\mathbf{V}}{dt} \right) + \frac{\partial}{\partial \ell} P_\| - (P_\| - P_\perp) \frac{\partial}{\partial \ell} \log B = 0 \tag{11.44}$$

where $\partial/\partial \ell \equiv \hat{\mathbf{b}} \cdot \nabla$ and the RH side is of course zero since $\mathbf{j} \times \mathbf{B}$ has no parallel component. The parallel component of the acceleration, with the usual assumption of azimuthal flow, may be written

$$\left(\frac{d\mathbf{V}}{dt} \right)_\| = -\omega^2 \mathbf{R} \cdot \hat{\mathbf{b}} = -\frac{\partial}{\partial \ell} \left(\frac{1}{2} \omega^2 R^2 \right) \tag{11.45}$$

where $\mathbf{R}$ is the cylindrical radial distance from the rotation axis [cf., Eq. (11.17)] and the constancy of ω along a field line, implied by Ferraro's isorotation theorem, has been used. Equation (11.44) may be considered an equation for the variation of $P_\|$ along a field line:

$$\frac{\partial}{\partial \ell} \log P_\| = \frac{\rho}{P_\|} \frac{\partial}{\partial \ell} \left(\frac{1}{2} \omega^2 R^2 \right) + \left(1 - \frac{P_\perp}{P_\|} \right) \frac{\partial}{\partial \ell} \log B \tag{11.46}$$

Given the necessary assumptions about the pressure-density relation and the magnetic field geometry, Equation (11.46) can be integrated to obtain the profile of ρ along the field line and hence the thickness h; note that the thickness now refers, in the first instance, to the plasma sheet (as distinct from the current sheet). This approach is in practice useful primarily in regions of strong dipolar field where $\mathbf{B}_p$ is a small perturbation of the total field $\mathbf{B}$, so that the field geometry is approximately known independently of the current sheet (whereas in regions of taillike field the geometry is strongly influenced by the current sheet itself and cannot be specified without already knowing the sheet thickness); its range of applicability thus neatly complements that of the previously described approach based on stress balance.

An important special case is that of isotropic pressure, when (11.46) reduces to

$$\frac{\partial}{\partial \ell} \log P = \frac{\rho}{P} \frac{\partial}{\partial \ell} \left(\frac{1}{2} \omega^2 R^2 \right) \tag{11.47}$$

(Note that Eq. (11.47) implies, with no additional assumptions, that the maximum pressure on any given field line is found at the point where the field line reaches its greatest distance R_0 from the rotation axis, or equivalently where $\mathbf{B} \cdot \mathbf{R} = 0$, a fact of significance for the discussion of dipole tilt effects in Sec. 11.3.5.) With the further assumption of isothermal behavior, $P/\rho = w^2$ constant along a given field line, Equation (11.47) can be explicitly integrated to yield

$$\rho = \rho_0 \exp \left\{ -\frac{1}{2} \frac{\omega^2}{w^2} (R_0^2 - R^2) \right\} \tag{11.48}$$

To relate $R_0^2 - R^2$ to the distance z from the center of the sheet a model for the magnetic field lines must be adopted. For a dipolar field one obtains

$$R_0^2 - R^2 = 3z^2 \tag{11.49}$$

to lowest order in $(z/R_0)^2$; the factor 3 is the ratio of field-line radius of curvature to radial distance near the equator [cf., e.g., Roederer, 1970, pp. 53, 58]. Equation (11.48) then represents a Gaussian profile

$$\rho = \rho_0 \exp\left\{-\frac{3}{2}\frac{\omega^2 z^2}{w^2}\right\} \tag{11.50}$$

and the plasma sheet thickness is given by the width of the Gaussian as

$$\frac{h}{r} \simeq \left(\frac{2}{3}\right)^{1/2}\frac{w}{\omega r} = \left(\frac{2}{3}\right)^{1/2}\frac{w}{V_\phi} \tag{11.51}$$

whence follows the previously stated conclusion that $h/r \ll 1$ in a strongly dipolar field if the plasma is cold. The results (11.50) and (11.51) are well known in the literature, for example, Hill and Michel [1976], Siscoe [1977]; h given by (11.51) is sometimes called the centrifugal scale height.

Explicit density profiles can be derived by integration of (11.46) in a few other cases. If adiabatic behavior is assumed, $P/\rho^\gamma = $ constant, one obtains

$$\rho = \rho_0\left\{1 - \frac{\gamma - 1}{2\gamma}\frac{\omega^2}{w_0^2}(R_0^2 - R^2)\right\}^{\frac{1}{\gamma - 1}} \tag{11.52}$$

where w_0 is the thermal speed at the center of the sheet; this result, which of course reduces to (11.48) as $\gamma \to 1$, was discussed by Mendis and Axford [1974] in the somewhat different context of an ionospheric plasma source. Assumption of the Chew-Goldberger-Low (CGL) double adiabatic relations

$$P_\perp/\rho B = \text{constant} \quad P_\parallel B^2/\rho^3 = \text{constant} \tag{11.53}$$

[e.g., Rossi and Olbert, 1970, p. 358] can be shown to give

$$\rho = \rho_0\frac{B}{B_0}\left\{1 - \frac{\omega^2}{3w_\parallel^2}(R_0^2 - R^2) - \frac{2}{3}\left(\frac{w_\perp}{w_\parallel}\right)^2\left(\frac{B}{B_0} - 1\right)\right\}^{1/2} \tag{11.54}$$

where $w_\parallel$ and $w_\perp$ represent values at the center of the sheet, although the subscript 0 has been omitted for ease of reading; with the use of (11.49) and a corresponding near-equatorial approximation for B/B_0, (11.54) becomes, to lowest order in $(z/R_0)^2$,

$$\rho \simeq \rho_0\left\{1 - \frac{z^2}{r^2}\frac{V_\phi^2 + 3w_\perp^2}{w_\parallel^2}\right\}^{1/2} \tag{11.55}$$

However, the results (11.52) and (11.54) are mainly of academic interest because the assumed equation of state — either adiabatic or CGL — holds, if at all, along a plasma flow line; there is no reason to expect either one to hold along a magnetic field line, unless the plasma flow were indeed predominantly field aligned.

Fig. 11.1. Dipole magnetic latitude as a function of System III longitude for a spacecraft at a fixed angle θ above the rotational equator [from Vasyliunas and Dessler, 1981]. The bulk of available observations in the outer Jovian magnetosphere lie between $\theta = 3°$ and $\theta = 5°$ (Voyager 1 and 2) and near $\theta \approx 9°$ (Pioneer 10), plus the high-latitude pass ($\theta \approx 30°$) of Pioneer 11 outbound near the local noon. Note the absence of coverage near the magnetic equator for the longitude range 120°–300°, which includes the active sector.

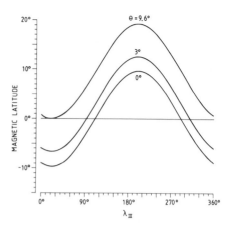

11.3. Models of the magnetic field and stress balance

After the preceding survey of general principles and approximations, we turn to a discussion of specific models for the magnetic field in the middle and outer Jovian magnetosphere. Chapter 1 discusses the purely magnetic aspects of many of the models, their relation to observations of the field, and the complex and longitudinally asymmetric nondipolar internal magnetic field that provides the basis for the magnetic anomaly model (see Chap. 10) but otherwise has no significant direct effects at distances beyond a few R_J; in this chapter, we are concerned with the relation of the magnetic field to stress balance, the implications for the distribution of plasma, and the agreement (or otherwise) with observed plasma properties.

The models may be grouped into two classes, conveniently labeled theoretical and empirical, respectively, although a more fundamental distinction is whether the magnetic field is calculated from the plasma distribution or vice versa. Theoretical models as a rule start with assumptions about the plasma and thence derive the configuration of the magnetic field; empirical models are to a greater or lesser extent constructed as representations of the observed magnetic field and serve as a basis for inferences about properties of the plasma.

Axial symmetry plays a role, explicitly or implicitly, in most existing models. Theoretical models generally assume no dependence either on local time or on longitude (thereby necessarily treating the planetary magnetic dipole moment as aligned with the rotation axis). Empirical models are often derived from observations over a limited range of local times and thus nominally require no global assumptions about local time dependence (although derivatives with respect to local time over that limited range are as a rule set to zero or ignored), but neglect of any dependence on Jovian longitude other than the geometrical effects of dipole tilt (see Sec. 11.3.5) has been their essential even if often unstated assumption. Simplicity is not the sole motive for assuming axial symmetry. The orbit of any spacecraft imposes an inevitable latitude–longitude correlation upon the observations, discussed by Vasyliunas and Dessler [1981] and illustrated in Figure 11.1, as a result of which the available observations, either of the plasma or of the magnetic field, are seriously incomplete in longitudinal coverage at any fixed magnetic latitude and do not provide an adequate basis for a fully longitude-dependent model.

Another assumption widely invoked in constructing theoretical and applying empirical models is that the plasma pressure is isotropic, $P_\perp = P_\parallel = P$. Again, the motive is in part simplicity and in part lack of any clear guidance in the available limited observations toward possible alternatives. An exception are models that assume the plasma in the magnetosphere to be predominantly of ionospheric origin, in which case one expects $P_\parallel \simeq \rho \, \Omega_J^2 \, r^2 \gg P_\perp$ [e.g., Carbary and Hill, 1978]; however, such models are now of little interest inasmuch as it is well established that the major source of plasma in the Jovian magnetosphere is the Io torus and not the ionosphere (see Chapts. 2, 3, and 10).

1. Theoretical models

The geometrical thinness of the plasma sheet in the Jovian magnetosphere, which is an observationally established property (see Chap. 3) as well as a theoretically expected consequence of the equatorially confined Io source [Hill and Michel, 1976; Siscoe, 1977], permits the construction of a model to be carried out in two steps. First, the large-scale structure of the magnetic field can be calculated from the integrated tangential stress balance Equation (11.19), which with the usual assumption of centripetal acceleration and the idealization of the system as axially symmetric (hence with the current sheet in the common magnetic-rotational equator) may be written in the form

$$-\sigma\omega^2 r = B_z j_\phi' \tag{11.56}$$

where j_ϕ' is the current in the azimuthal direction per unit radial length and r is the cylindrical radial coordinate with respect to the rotation axis. (The pressure terms have been dropped, as discussed in Sec. 11.2.2.) The quantity $\sigma/B_z \equiv \eta$ represents, as pointed out already, the plasma content of a flux tube per unit magnetic flux. Specifying $\eta\omega^2$ as a function of r suffices to determine j_ϕ' everywhere, and the magnetic field can then be calculated by treating the current sheet thickness as infinitesimal and applying standard boundary value methods [e.g., Jackson, 1962, especially his Sec. 5.5 and Problem 5.4]. As the second step, the thickness of the current sheet and the properties of plasma within it can be calculated from the normal component of the momentum equation. The calculation is particularly simple when pressure balance holds: the plasma pressure at the center of the current sheet then is given by

$$P = B_r^2/2\mu_0 \tag{11.57}$$

(where $B_r \simeq B_p^*$ is the field just outside the sheet), and the general order-of-magnitude relation $\sigma \simeq \rho h$ may be written as

$$\rho h \omega^2 = B_z \, [\sigma\omega^2/B_z] \tag{11.58}$$

giving two equations whose RH sides contain only quantities already available from the first step, either as its input ($\sigma\omega^2/B_z$) or as its result (B_r, B_z). Evidently, from Equations (11.58) and (11.59) the density and the thermal speed can be calculated if $\omega^2 h$ is independently known as a function of r, or alternatively $\omega^2 h$ and ρ can be calculated if a pressure-density relation is known (this includes, as an important special case, specifying the thermal speed as a function of r).

In this section we discuss only the first step, models of the large-scale magnetic field derived from the radial distribution of the single quantity $\eta\omega^2$; current sheet thickness and related questions are deferred to Section 11.3.3. Furthermore, the toroidal magnetic field component B_ϕ is relatively small except in the outermost regions of the Jovian magnetosphere and is neglected in most models (it is discussed in Sec. 11.3.4). With the assumption of axial symmetry and neglect of B_ϕ, **B** is most conveniently derived from the vector potential **A** which now has only a ϕ-component $A_\phi(r, z)$. A representation of the field by the Euler potential functions f, g (see, e.g., Chap. 1, Sec. 1.3) is then given by

$$f = rA_\phi(r, z) \qquad g = \phi \tag{11.59}$$

The curves defined by $rA_\phi(r, z) = $ constant are magnetic field lines on a meridian plane, and the value of $2\pi rA_\phi$ at any point represents the amount of magnetic flux between the magnetic shell surface through that point and the field line, defined by $rA_\phi = 0$, emanating from the poles of the dipole.

A family of models was obtained by Gleeson and Axford [1976], who assumed for j_ϕ' a family of analytic expressions

$$j_\phi'(r) = \eta_n \, \Omega_J^2 \, r \, / \, [1 + (r/a_n)^2]^{n + (1/2)} \tag{11.60}$$

which allows **B** to be calculated in closed form. Here n is a positive integer, with $n = 1$, 2 being the cases treated in detail, and η_n and a_n are constants (our notation differs from that of Gleeson and Axford). Comparison of (11.60) with (11.56) implies that the flux tube content times ω^2 varies in the models as

$$\eta \omega^2 = \eta_n \, \Omega_J^2 / [1 + (r/a_n)^2]^{n + (1/2)} \tag{11.61}$$

that is, constant for $r \ll a_n$ and decreasing as an inverse power of r for $r \gg a_n$ (it makes no difference, as far as the large-scale structure of the model field is concerned, whether the decrease at large r is associated with a decrease of η or ω^2 or both). The vector potential is given by

$$A_\phi(r, z) = \frac{Mr}{(r^2 + z^2)^{3/2}} + \frac{MK_1}{a_1 r} \left[1 - \frac{Z_1}{(Z_1^2 + r^2)^{1/2}} \right] \tag{11.62}$$

$$+ \frac{MK_2}{3} \frac{r}{(Z_2^2 + r^2)^{3/2}}$$

where

$$Z_n \equiv |z| + a_n, \qquad K_n \equiv \mu_0 \, a_n^4 \, \eta_n \, \Omega_J^2 / 2M$$

The first term in (11.62) represents the Jovian dipole field with moment M and the terms multiplied by the constants K_1 and K_2 arise from j_ϕ' of Equation (11.60) with $n = 1$ and $n = 2$, respectively. (Note a misprint in the corresponding equation of Gleeson and Axford, their Equation (20): in the numerator a_2^2 should be a_2^3.) The magnetic field component normal to the current sheet, derived from (11.62), is

$$B_z(r, z=0) = - \frac{M}{r^3} \left\{ 1 - \frac{K_1 \, r^3}{(a_1^2 + r^2)^{3/2}} - \frac{K_2 \, r^3 \, (2a_2^2 - r^2)}{3(a_2^2 + r^2)^{5/2}} \right\} \tag{11.63}$$

From (11.63) and (11.61) one can at once calculate the variation of $\sigma \omega^2$ with r.

It is convenient to deal with two separate models, $n = 1$ and $n = 2$, obtained by setting K_2 and K_1 to zero in turn. In both models, the magnetic-field-line configuration is stretched out in comparison to a pure dipole, the degree of stretching increasing with increasing value of K. This is graphically illustrated in a series of figures presented by Gleeson and Axford, but it can also be qualitatively deduced from Equation (11.62) by noting that the value of rA_ϕ at any given point is always larger than the dipole value, implying that the field line through that point emanates from a lower dipole latitude (for an ideal dipole field at a radial distance of 1 R_J, the Euler potential f has the value

$$f = rA_\phi = M \sin^2 \theta / R_J \qquad \text{(at } r^2 + z^2 = R_J^2 \text{)} \qquad (11.64)$$

where θ is the colatitude). However, in the case $n = 2$, the stretching-out becomes insignificant at large distances: when $r^2 + z^2 \gg a_2^2$, the field line configuration approaches that of a dipole with moment $M[1 + (K_2/3)]$, and $rA_\phi \to 0$ as $r \to \infty$ with the result that the model contains no open field lines — all the field lines emanating from the dipole eventually close across the equatorial plane (unless, of course, departures from the model at large distances are introduced, caused, for example, by magnetotail currents). On the contrary, in the case $n = 1$ the field line configuration remains non-dipolar out to arbitrarily large distances and $rA_\phi \to MK_1/a_1$ as $r \to \infty$ at fixed z; thus, in this model, all field lines emanating poleward of the dipole colatitude defined by

$$\sin^2 \theta = K_1 R_J / a_1 \qquad (11.65)$$

are truly open and do not cross the equatorial plane at any distance, no matter how large. These open field lines are present for any value of $K_1 > 0$, even though for $K_1 < 1$ the model does not contain any magnetic singular lines, of $\times$ type or other, either at a finite distance or in the limit $r \to \infty$.

As expected from the stretched-out configuration, the normal component B_z given by (11.63) is reduced in magnitude below the dipole value, everywhere for $n = 1$ and out to a distance $r = (2)^{1/2} a_2$ for $n = 2$. Eventually, as K_1 or K_2 is increased to a critical value, B_z is reduced to zero at some point, indicating the incipient appearance of magnetic singular lines. For $n = 1$, the critical value of K_1 is $K_{1c} = 1$ and the singular line appears at infinity; if $K_1 > 1$, a magnetic $\times$ line is found at a finite distance. For $n = 2$, $K_{2c} = 3/2 \, (5/2)^{5/2} \approx 14.8$ and the singular line appears at $r_c = a_2 \, (2/3)^{1/2}$; for $K_2 > K_{2c}$, it splits into an $\times - \bigcirc$ line pair with associated closed loops of magnetic field lines. (Note: Gleeson and Axford give incorrect numerical expressions for K_{2c} and r_c, as well as for rA_ϕ at r_c which should be $(3/2)^{5/2} M/a_2$.) When the critical value of K is exceeded, B_z given by the model has a reversed sign over a portion of the current sheet, but this is unphysical because Equation (11.56) together with (11.60) then implies a negative mass density σ. Hence follows the important conclusion that, within the Gleeson–Axford family of models, corotating plasma can exist in stress balance only if $K \leq K_c$, which is effectively a restriction on the distance of corotation for a given level of flux tube content, as can be seen by using the definition of K_n to write the restriction in the form

$$a_n^4 \leq 2K_c \, M / \mu_0 \, \eta_n \, \Omega_J^2 \qquad (11.66)$$

The distance $r \approx a_n$ represents in a qualitative sense the limit of corotating plasma, because for $r \gg a_n$ we have $\eta_n \omega^2 \ll \eta_n \Omega_J^2$, that is, either the plasma moves much more slowly than corotation or else there is much less of it (or both). The constant η_n may be equated to the flux tube content at the location of the plasma source, that is, $r = r_s = 6 \, R_J$ for an Io source, provided $a_n \gg r_s$:

$$\eta_n = \eta_s \equiv \rho_s h_s / B_s \qquad (11.67)$$

where ρ_s is the plasma mass density and h_s the plasma sheet thickness at the source location (averaged over all longitudes because the model has axial symmetry). By writing $B_s = M/r_s^3$, $M = B_J R_J^3$, using (11.67) and expressing all distances in units of R_J we may cast (11.66) into a more familiar form

$$\left(\frac{a_n}{R_J}\right)^4 \le \frac{2K_c B_J^2}{\mu_0 \rho_s h_s \Omega_J^2 R_s L_s^3} \tag{11.68}$$

where $L_s \equiv r_s/R_J$.

The model of Hill and Carbary [1978] makes the assumption that rigid corotation and constancy of flux tube content are maintained throughout the closed field line region of the magnetosphere; thus $\eta \omega^2 = \eta_s \Omega_J^2$ as long as $B_z > 0$, out to a distance $r = r_0$ to be determined, and j_ϕ' is given by

$$j_\phi'(r) = \eta_s \Omega_J^2 r \qquad r \le r_0 \tag{11.69}$$

For $r > r_0$ the plasma can no longer be corotating in stress balance and may be presumed instead to form a planetary/magnetospheric wind as discussed in Section 11.4.3, but for the construction of the model it suffices to assume merely that

$$B_z(r, z = 0) \qquad r \ge r_0 \tag{11.70}$$

Conditions (11.69) and (11.70) together with continuity of j_ϕ' across $r = r_0$ and a dipole at the origin define a boundary value problem with a unique solution (including determination of the value of r_0) which must be obtained numerically. Hill and Carbary found that condition (11.70) could be satisfied to an adequate accuracy by setting

$$j_\phi'(r) = \eta_s \Omega_J^2 r_0 [1.5 (r_0/r)^2 - 0.5 (r_0/r)^3] \qquad r \ge r_0 \tag{11.71}$$

[note that j_ϕ' here is *not* connected with $\eta \omega^2$ by Equation (11.56)]; with j_ϕ' thus specified everywhere, its contribution to **B** can be calculated by numerical integration and expressed as a function of r/r_0 multiplied by the scale factor $\eta_s \Omega_J^2 r_0$. In particular, the part of $B_z(r=r_0, z=0)$ contributed by the current sheet alone, which must be equal and opposite to the dipole field to give $B_z = 0$, may be written as equal to the quantity $B_r/\alpha = \mu_0 \eta_s \Omega_J^2 r_0/2\alpha$, where $\alpha \approx 2$ is a constant given by the numerical calculation; thus r_0 is obtained by setting

$$\mu_0 \eta_s \Omega_J^2 r_0/2\alpha = M/r_0^3 \tag{11.72}$$

which yields

$$r_0^4 = 2\alpha M/\mu_0 \eta_s \Omega_J^2 \tag{11.73}$$

or equivalently

$$\left(\frac{r_0}{R_J}\right)^4 = \frac{2\alpha B_J^2}{\mu_0 \rho_s h_s \Omega_J^2 R_J L_s^3} \tag{11.74}$$

identical to the equalities in (11.66) or (11.68) except for the replacement of K_c by α. The limiting colatitude of open field lines is found to be

$$\sin^2 \theta = 1.2 R_J/r_0 \tag{11.75}$$

to be compared with the corresponding result (11.65) for the $n = 1$ Gleeson–Axford model.

Evidently the model of Hill and Carbary has several points of similarity with the Gleeson–Axford models, particularly with their $n = 1$ case (the $n = 2$ case is of relatively little interest because it has no open field lines). The most highly stretched Gleeson–Axford model has $K_1 = K_c = 1$; therefore the ratio $r_0/a_1 = 2^{1/4} \approx 1.2$, which leads to the remarkable result that the amount of open flux in that model is the same as in that of Hill and Carbary. Both models have the same asymptotic form $j_\phi' \sim r^{-2}$ as $r \to \infty$; the total current diverges logarithmically if integrated to infinity (contrary to the impression given in the paper of Gleeson and Axford) but all field and plasma quantities remain finite everywhere. The main difference is that Hill and Carbary maintain full corotation and source-imposed flux tube plasma content out to a limiting distance r_0 beyond which stress balance breaks down altogether as field lines no longer cross the equator, whereas Gleeson and Axford maintain stress balance and field line closure at all distances (although a significant amount of magnetic flux remains unclosed even as $r \to \infty$) but allow the flux tube plasma content and/or degree of corotation to fall off gradually at and beyond distances comparable with r_0. In some ways, the Gleeson–Axford model resembles a smoothed-out version of the Hill–Carbary model. A comparison of the two and of other models is presented later in Section 11.3.2.

Vickers [1978] attempted to extend the Gleeson-Axford models by including the effects of tangential pressure gradients. However, these effects are of the same order of magnitude as those due to $\partial B_z/\partial z \neq 0$ within the current sheet, as discussed in Section 11.2.2, and it is shown in Appendix A that this inconsistency invalidates his model. Sozzou [1978] developed a similarity model of a specialized mathematical form with singularities at the origin, which illustrates some effects of rotation and pressure gradients.

2. Empirical models and their implications

Several quantitative models of the magnetic field in the Jovian magnetosphere have been constructed on the basis of empirical fits to some subset of the available in situ magnetic field observations; the reader is referred to Chapter 1, Section 1.3 for a detailed review. Stress balance plays no direct role in the construction of such a model but it does impose significant constraints on the plasma properties if these are to be compatible with the model; in some cases the plasma properties required for compatibility may be physically unreasonable, indicating that some revision of the model is called for. The source of the constraints is the momentum Equation (11.1): with the magnetic field $\mathbf{B}(\mathbf{r})$ given by the model and $\mathbf{j}(\mathbf{r})$ obtained from $\nabla \times \mathbf{B}$, the RH side $\mathbf{j} \times \mathbf{B}$ of (11.1) is a known function of space, and the spatial variation of pressure and density must be such as to yield the same function for the LH side. In the particular case of isotropic pressure and axially symmetric configuration with corotational flow, Equation (11.1) suffices to determine $\rho(r, z)$ and $P(r, z)$, as pointed out by Goldstein [1977].

Written out explicitly in cylindrical coordinates, with the assumption of isotropic pressure and azimuthal flow, Equation (11.1) becomes

$$(\partial P/\partial z)_r = (\mathbf{j} \times \mathbf{B}) \cdot \hat{\mathbf{z}} \tag{11.76}$$

$$-\rho \omega^2 r + (\partial P/\partial r)_z = (\mathbf{j} \times \mathbf{B}) \cdot \hat{\mathbf{r}} \tag{11.77}$$

where the fact that P is a function of r and z is explicitly indicated by showing as subscripts the variables held constant during partial differentiation (it will soon prove useful to introduce another set of variables). With $\mathbf{j} \times \mathbf{B}$ known from a field model, $P(r, z)$ can be calculated by direct integration of (11.76) with a suitable boundary condition, for example, $P \to 0$ as $z \to \infty$; differentiation then yields $(\partial P/\partial r)_z$, and $\rho(r, z)$ can be calculated directly from (11.77). If axial symmetry is assumed, there is no dependence on ϕ so that the spatial variation of P and ρ is known completely. Furthermore, with $\mathbf{j} = j(r, z) \hat{\phi}$ and $B_\phi = 0$ the components of $\mathbf{j} \times \mathbf{B}$ assume the simple form

$$(\mathbf{j} \times \mathbf{B}) \cdot \hat{\mathbf{z}} = -jB_r \tag{11.78}$$

$$(\mathbf{j} \times \mathbf{B}) \cdot \hat{\mathbf{r}} = jB_z \tag{11.79}$$

Equations (11.76)–(11.79) provide a complete and readily usable prescription for determining the plasma distribution needed to maintain a given magnetic field configuration in stress balance. They can be cast into a form that is more elegant (although not necessarily more useful for calculation) by introducing the Euler potential or magnetic flux function $f = rA_\phi$ and treating P and ρ as functions of r and f rather than r and z. The partial derivatives are readily transformed by writing

$$dP = (\partial P/\partial r)_z \, dr + (\partial P/\partial z)_r \, dz \tag{11.80}$$

$$df = (\partial f/\partial r)_z \, dr + (\partial f/\partial z)_r \, dz = rB_z \, dr - rB_r \, dz \tag{11.81}$$

and forming the ratios dP/df with $dr = 0$, and so forth; the resulting expressions allow the reduction of (11.76)–(11.79) to

$$(\partial P/\partial f)_r = j/r \tag{11.82}$$

$$(\partial P/\partial r)_f = \rho\omega^2 r \tag{11.83}$$

Equations (11.82)–(11.83) were derived by Goldstein [1977] on the basis of a restrictive assumption that the plasma electron and ion velocity distribution functions are special solutions of the Vlasov equation depending on the two explicitly known constants of motion, energy, and canonical angular momentum about the symmetry axis (whereas the general solution for a time-independent axially symmetric system depends on four constants of motion). The derivation given here makes it apparent that Goldstein's results are more general, presupposing only the momentum equation with isotropic pressure and axial symmetry.

An even simpler form of the equations results if an effective potential U for the acceleration is introduced,

$$U \equiv (1/2)\omega^2 r^2 \tag{11.84}$$

(this form of U presupposes that ω depends at most only on f, that is, ω is constant along a field line, in accordance with Ferraro's isorotation theorem): with P and ρ now treated as functions of U and f, it is readily shown that

$$(\partial P/\partial f)_U = j/r \tag{11.85}$$

$$(\partial P/\partial U)_f = \rho \tag{11.86}$$

It is now also evident that the equations can be generalized to the case when the effective acceleration of the plasma does not have the simple centripetal form, by merely changing U appropriately; for example, if the gravitational term is nonnegligible, (11.84) is replaced by

$$U \equiv (1/2)\omega^2 r^2 - M_J G/(r^2 + z^2)^{1/2} \tag{11.87}$$

In whichever form the equations are used, it is possible to calculate both the plasma mass density and the pressure everywhere, given *any* field model; there are no general restrictions on the model magnetic field other than effective axial symmetry (but also there is no a priori assurance that the calculated ρ and P will be plausible or even physically realizable). A contrasting situation arises when the inertial terms in the momentum equation are negligibly small, as for the terrestrial magnetosphere in models of the symmetric ring current [e.g., Carovillano and Siscoe, 1973; and references therein] and in two-dimensional models (with a planar rather than cylindrical geometry) of the plasma sheet [e.g., Schindler, 1975, 1979]: in these cases only the pressure distribution can be calculated, no information on the density being obtainable from the magnetic field, and the field model itself is required to satisfy the rather restrictive constraint that j/r (or j, for a planar geometry) must be constant along field lines, a restriction that follows from Equation (11.82) together with the constancy of P along field lines implied by Equation (11.83) when $\rho\omega^2 r$ is negligible. Rotational stresses in the Jovian magnetosphere are thus indispensable if inferences about the plasma distribution are to be drawn from empirical models of the magnetic field, constructed for the most part without imposing any particular constraints.

Empirical models developed to a sufficient extent to enable the derivation of plasma parameters by the method described above include (see Chap. 1 for a detailed description) the model of Goertz et al. [1976] (slightly modified by Jones, Melville, and Blake [1980]), applicable at distances between 20 and 80 R_J in the predawn local time sector, and the model of Connerney, Acuña, and Ness [1981], applicable out to a distance of about 30 R_J at all local times. There is also the model of Barish and Smith [1975], commonly thought to represent the magnetic field in the dayside magnetosphere; however, as shown in Appendix A, this model in its main features differs considerably from what is either observed or theoretically expected.

For the model of Goertz et al., Goldstein [1977] has calculated the implied distribution of plasma mass density and pressure in full detail, using Equations (11.82) and (11.83). An alternative simpler treatment has been presented by Goertz et al. [1979]. For many purposes, the quantities of prime interest are the pressure and mass density at the center of the current sheet, the rest of the spatial distribution being adequately described by the value of current sheet thickness and the statement that P and ρ are negligibly small outside the current sheet. The magnetic field configuration, within the distance range where the model is applicable, is sufficiently taillike so that integration of Equation (11.76) to obtain P leads to a simple pressure balance relation, as in Section 11.2. Hence the plasma pressure at the center is well approximated by the magnetic pressure outside the current sheet or

$$P(r, z = 0) = b_0^2/2\mu_0 D^2 r^{2a+2} \tag{11.88}$$

in terms of the usual parameters of the model: $b_0 = 9 \times 10^3\, nT R_J^{2+a}$ (note the units of b_0, which are not properly given either in Goertz et al. [1976] or in Chap. 1), $D = 1\, R_J$, the half thickness of the current sheet, and $a = 0.7$. [Equation (11.88) differs slightly but not significantly from the results of Goertz et al. [1979], because of an ambiguity in the model at large values of $|z|$, discussed in Appendix A.] From Equation (11.77) and the radial derivative of Equation (11.88) one obtains the mass density $\rho(r, z = 0)$ with the result given, to within $O(D^2/r^2)$, by

$$\rho\omega^2 = (b_0 M/\mu_0 D^2 r^{a+5})(1 - c_1 r^{1-a}) \tag{11.89}$$

where M is the Jovian dipole moment,

$$c_1 \equiv b_0(aC + a + 1)/M \tag{11.90}$$

and the value of the model parameter C is given as $C = 10$ by Goertz et al. [1976] and $C = 15$ by Jones, Melville, and Blake [1980].

For the model of Connerney, Acuña, and Ness [1981], no estimates of the implied plasma parameters have yet appeared in the literature. They can, however, be calculated fairly easily, owing to the simple form of the azimuthal current density in the model:

$\mu_0 j_\phi = B_0/r \qquad$ for $|z| < D$

$$r_0 < r < r_1 \tag{11.91}$$

$\qquad = 0 \qquad$ elsewhere

where the adopted parameter values are $D = 2.5\,R_J$, $r_0 = 5\,R_J$, $r_1 = 50\,R_J$, and $B_0 = 450\,nT$ or $300\,nT$ if the model is fitted to Voyager 1 or Voyager 2 observations, respectively (see Chap. 1, Sec. 1.3 for a more detailed presentation). The model's intended range of applicability extends deep into the inner magnetosphere where the magnetic field is nearly dipolar and the pressure balance approximation is *not* valid. The pressure must thus be calculated by an explicit integration of Equation (11.76), but with j given by (11.91) and with the model field represented by the analytical approximations given by Connerney, Acuña, and Ness [1981] (reproduced here in Chap. 1, Appendix A) the integration can be carried out in closed form. With the boundary condition $P = 0$ at $z = \pm D$, required by the fact that $\mathbf{j} \times \mathbf{B} = 0$ for $|z| > D$ in the model, the calculated pressure is given by

$$P(r, z) = (MB_0/\mu_0)\,\{(r^2 + z^2)^{-3/2} - (r^2 + D^2)^{-3/2}\} \tag{11.92}$$

$$+ (B_0^2/2\mu_0)\,\{(D^2 - z^2)/r^2 + G_1 + G_2 + (r_0^2/4r^2)G_3\}$$

where

$$\tag{11.93}$$

$$2r^2 G_1 \equiv (z + D)\,[(z + D)^2 + r^2]^{1/2} - (z - D)\,[(z - D)^2 + r^2]^{1/2}$$

$$- 2D[4D^2 + r^2]^{1/2}$$

$$2G_2 \equiv \sinh^{-1}[(z + D)/r] - \sinh^{-1}[(z - D)/r] - \sinh^{-1}[2D/r]$$

$$G_3 \equiv 2D[4D^2 + r^2]^{-1/2} + (z - D)\,[(z - D)^2 + r^2]^{-1/2}$$

$$- (z + D)\,[(z + D)^2 + r^2]^{-1/2}.$$

Equation (11.92) holds for $r > r_0$, $|z| < D$; the model implies zero pressure for $r < r_0$ and for $|z| > D$, whereas for $r \to r_1$ the model becomes inapplicable and we shall use the model results only out to about $r = 0.6\,r_1 = 30\,R_J$. The terms in the first set of curly brackets in (11.92) arise from the current sheet's interaction with the dipole field alone and the rest from its interaction with its own field; as r increases, the former become negligibly small and the latter approach the expected pressure-balance value

$$P(r, z) \approx (B_0^2/2\mu_0)\,(D^2 - z^2)/r^2 \tag{11.94}$$

The plasma mass density profile $\rho(r, z)$ implied by the model, obtained from Equation (11.77) with the radial pressure gradient from (11.92), is given by

$$\rho\omega^2 r^2 = (MB_0/\mu_0)\,\{3r^2(r^2 + D^2)^{-5/2} - 2(r^2 + z^2)^{-3/2}\} \tag{11.95}$$

$$- (B_0^2/2\mu_0)\,\{2(D^2 - z^2)/r^2 + 2D[(z^2 + r^2)^{-1/2} - (z^2 + r_1^2)^{-1/2}]$$

$$+ G_1 + (r_0^2/2r^2)\,G_3\} + r_0 P_0\,\delta(r - r_0)$$

where G_1 and G_3 are given by (11.93) and P_0 is the value of $P(r, z)$ given by (11.92) as $r \to r_0^+$. The density is zero for $r < r_0$ and for $|z| > D$. The delta function in density at

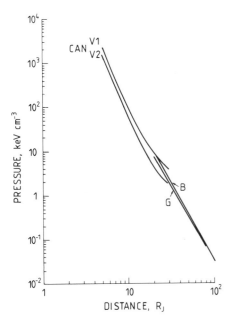

Fig. 11.2. Plasma pressure at the center of the current sheet calculated from the magnetic field model of Connerney, Acuña, and Ness [1981] (CAN) and the magnetic field model of Goertz et al. [1979] (G), and calculated from pressure balance with the observed field by Barbosa et al. [1979] (B).

the inner edge of the current sheet results from the assumed abrupt termination of j; a gradual fall-off over a distance range Δr would lead to a finite density spike of radial width Δr.

Figure 11.2 shows the calculated pressure at the center of the current sheet ($z = 0$) for both the Connerney, Acuña, and Ness model and the Goertz et al. model; in addition, an empirical determination from magnetic field observations and pressure balance published by Barbosa et al. [1979] is shown for comparison. The various curves are in fair agreement with each other, and the general radial profile is smooth and shows no unexpected or questionable features. The numerical values are roughly consistent with the direct observational determinations of the plasma pressure, available (albeit with sizable uncertainties) at distances beyond 10 R_J [Lanzerotti et al., 1980; Krimigis et al., 1981].

The situation with the calculated mass density, shown in Figure 11.3, is far less satisfactory. Three problem areas are at once evident from the figure: (1) negative density in some regions, (2) disagreement between the two models where they overlap, (3) discrepancy, beyond 20 R_J, between the density radial gradient derived from the models and estimated from plasma wave observations [Barbosa et al., 1979].

(1) The mass density implied by either the V 1 or V 2 version of the Connerney, Acuña, and Ness model is negative in the equatorial plane from about 6.2 R_J inward, until the (positive) δ-function spike at 5 R_J. This occurs because the inward-directed radial pressure gradient exceeds the (likewise inward-directed) radial component of the $j \times B$ force density, and the requirement of stress balance then forces the centripetal acceleration term $-\rho\omega^2 r$ to be directed outward. In turn, the excessively large pressure is traceable to a too large thickness D. At these close distances, the current sheet interacts primarily with the dipole field (hence, incidentally, any inaccuracies in the analytical approximations to the model field are of no consequence); P at $z = 0$ varies nearly as $B_0 D^2$ [see Eq. (11.92)], one factor D coming from the length of the integration in $\int_0^D dz \, jB_r$ and the other from the z dependence of the dipolar B_r, and the radial component of $j \times B$ varies as B_0. According to Connerney, Acuña, and Ness [1981], the quantity $B_0 D$ is well determined by the observations, whereas D is rather uncertain.

Fig. 11.3. Plasma mass density times $(\omega/\Omega_J)^2$ at the center of the current sheet, calculated from the same models as in Figure 11.2; for the magnetic field model of Goertz et al., parameter values given by Goertz et al. [1976] (G) and by Jones, Melville, and Blake [1980] (JMB) were used. The mass density values inferred from Barbosa et al. [1979] (B) assume $\omega = \Omega_J$ and mean atomic mass number either $A = 1$ or $A = 16$.

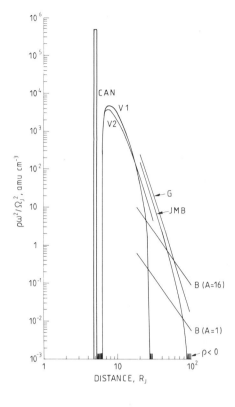

Fig. 11.4. Plasma mass density times $(\omega/\Omega_J)^2$ and pressure at the center of the current sheet at $r = 6R_J$, calculated from the V 1 model of Connerney, Acuña, and Ness [1981], as a function of assumed current sheet thickness D (for a fixed value of B_0D).

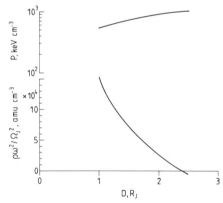

Hence, with fixed B_0D, $\partial P/\partial r \sim D$ and $(\mathbf{j} \times \mathbf{B}) \cdot \hat{\mathbf{r}} \sim 1/D$, and a negative mass density is obtained if D is chosen too large. This is illustrated in Figure 11.4, where the equatorial mass density and pressure at $r = 6\ R_J$ are calculated from the Connerney, Acuña, and Ness model for various values of D. As expected, ρ falls sharply with increasing D and becomes negative for $D \geq 2.4\ R_J$; to match the density of $\sim 3 \times 10^4$ amu/cm^3 at $r = 6\ R_J$ inferred from plasma observations (see Chap. 3, Figs. 3.8, 3.11, 3.12), one needs $D \approx 1.9\ R_J$.

It is thus possible to eliminate the inner negative density region (and with it the sharp density decrease at 7–8 R_J, produced by the same cause) merely by choosing a smaller thickness for the model current sheet; the associated reduction of the pressure, also shown in Figure 11.4, is not very large and does not significantly affect the pressure

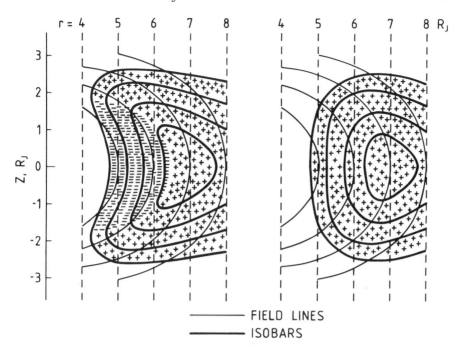

Fig. 11.5. Qualitative configuration of constant pressure curves and positive and negative current density regions near the inner edge of the current sheet, deduced from Equation (11.85) (sign of j_ϕ depends on whether the pressure decreases or increases along a centrifugal equipotential) and Equation (11.86) (positive density requires pressure to decrease with decreasing centrifugal potential along a field line). Left: isobars at the inner edge nearly aligned with field lines, giving rise to positive and negative current regions. Right: current everywhere positive, requiring isobars to be curved away from the centrifugal equipotentials.

profile in Figure 11.2. However, the model then still contains an unphysical feature, the infinite density spike at 4–5 R_J; the infinity can of course be removed by assigning a finite radial width Δr to the inner edge of the current sheet, but even with $\Delta r = 0.5\ R_J$, assumed in Figure 11.3 (and close to the width of observed density features, e.g., Fig. 3.7), the maximum value of ρ exceeds by an order of magnitude anything that has been observed. This large (positive) mass density enhancement is required in order for the centripetal acceleration to balance the *outward* radial pressure gradient that exists at the inner edge of the current sheet, it being assumed in the model that $\mathbf{j} \times \mathbf{B}$ remains inwardly pointing throughout. In reality, however, if this huge density enhancement is not present, the system will achieve stress balance by developing a reversed current at the sheet's inner edge, so that $\mathbf{j} \times \mathbf{B}$ there points outward (a possibility not considered when the model was constructed and fitted to observations). The expected qualitative configuration, easily deduced with the help of Equations (11.85) and (11.86), is sketched in Figure 11.5: if the isobars in the meridian plane are curved in the same sense as the magnetic field lines, there is a crescent-shaped region of reversed current at the inner edge where the pressure decreases with decreasing radius. (In order for j to have the same sign everywhere, the isobars in the region of decreasing pressure must curve in the opposite sense from the field lines, away from the rotation axis.) What effect the inclusion of a current sheet inner edge of finite width and with reversed current would have on the model, in particular on estimates of the sheet thickness and calculation of the implied mass density, is at present unknown. Clearly, the Connerney–Acuña–Ness model at distances less than about 8 R_J is in some need of revision.

Other regions of negative mass density occur beyond about 27 R_J (near the end of the model's range of applicability) for the Voyager 1 version only of the Connerney–Acuña–Ness model and beyond 87.5 R_J for the model of Goertz et al. with the parameters of Jones, Melville, and Blake ($C = 15$) only. The physical reason for the negative density is the same as before; however, at these distances pressure balance holds and the radial pressure gradient is fairly reliably determined by the magnetic field observations, hence the fault must lie in the estimate of jB_z. Either j in the models is too small (presumably because of a too large assigned current sheet thickness, again) or B_z has been underestimated; unlike the case of the inner region where B_z is reliably given by the dipole field, B_z here is sensitive to the model assumptions. For the model of Goertz et al., the occurrence of negative ρ is analytically predicted from Equation (11.89) for distances $r > r_d$ defined by

$$r_d^{1-a} = M/b_0(aC + a + 1) \tag{11.96}$$

For $C = 15$, $r_d = 87.5$ R_J as previously stated; for $C = 10$, $r_d = 270$ R_J lies outside the model's range of applicability. The distance r_d is less than the distance r_{nl} of the neutral line where B_z changes sign, given by

$$r_{nl}^{1-a} = M/b_0 \, aC \tag{11.97}$$

with values 144 R_J and 558 R_J for $C = 15$ and 10, respectively, in both cases outside the model's range of applicability. Note that the mass density implied by the empirical models can be negative even in regions where B_z is not reversed in sign compared to the dipole field; in such a case, $\mathbf{j} \times \mathbf{B}$ still has the appropriate inward orientation and is merely too small in magnitude to balance the radial pressure gradient. Only for theoretical models (such as those of Gleeson and Axford) that assume negligible current sheet thickness is the appearance of negative mass density tantamount to a reversal of B_z.

(2) In the distance range 20–30 R_J both models should be applicable and (with the exception of the V 1 model of Connerney, Acuña, and Ness) neither contains evident unphysical features, yet the calculated mass density values for the Goertz et al. model are systematically higher than for the Connerney–Acuña–Ness V 2 model, by a factor of about 4.5 (if $C = 10$) or 2.7 (if $C = 15$). The current sheet thickness in the Goertz et al. model, however, is smaller by a factor of 2.5, so that the mass density integrated across the thickness has rather comparable values in the two models. This suggests that the density difference is related to the as yet unresolved discrepancy in the current sheet thickness. Comparison of Figure 11.3 with the observed equatorial values shown in Figures 3.12 and 3.14 tends to favor the Connerney–Acuña–Ness model.

(3) At larger distances, the mass density implied by the model of Goertz et al. may be compared with the density radial profile between 20 and 100 R_J reported by Barbosa et al. [1979]. This profile, derived from plasma wave observations, refers to the electron concentration and must be multiplied by the average ion mass-to-charge ratio A/Z to obtain the mass density (Fig. 11.3 shows the profile for $A/Z = 1$ and for $A/Z = 16$); furthermore, it is of course independent of any assumptions about corotation, whereas the model predicts not ρ but $\rho(\omega/\Omega_J)^2$. It is at once apparent that the two differ markedly in the radial dependence of the density: n_e inferred from observations decreases with increasing distance as $1/r^{2.75}$, and $\rho(\omega/\Omega_J)^2$ decreases somewhat faster than $1/r^{5.7}$ [see Eq.

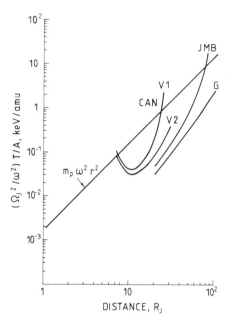

Fig. 11.6. Plasma temperature per unit mass times $(\Omega_J/\omega)^2$ at the center of the current sheet calculated from various magnetic field models (same nomenclature as in Fig. 11.3). The straight line corresponds to thermal speed equal to the actual azimuthal speed (for any value of A and Ω_J/ω).

(11.89)] — a discrepancy by a factor r^3. In principle, there are several possible explanations. A change in the ion composition from mostly sulfur and oxygen at close distances to mostly hydrogen farther out would make ρ decrease faster than n_e, but a change in A/Z by a factor of about $(100\ R_J/20\ R_J)^3 = 125$ is surely out of the question. If the current sheet thickness actually decreases with increasing distance, the correct mass density will decrease more slowly than that inferred from the model with assumed constant thickness, but the required variation of the thickness is $\sim 1/r^3$, that is, a decrease from, say, $4\ R_J$ at $r = 20\ R_J$ to $0.03\ R_J$ at $r = 100\ R_J$, which seems rather farfetched and implausible. The most likely explanation is that $(\omega/\Omega_J)^2$ decreases with increasing distance; there is both theoretical and observational evidence for such a partial corotation of magnetospheric plasma, discussed in Section 11.4.4. Quantitatively, to reconcile the two profiles one needs to assume that, approximately, $\omega \sim 1/r^{3/2}$ or equivalently $V_\phi \sim 1/r^{1/2}$.

Having calculated P and $\rho(\omega/\Omega_J)^2$ implied by the various models, one may form their ratio and obtain a temperature per unit mass divided by $(\omega/\Omega_J)^2$, shown in Figure 11.6 ($P/\rho \equiv kT/A \equiv w^2$ where w is the thermal speed); note that, with A taken to be the average mass of the *ions*, kT is the *sum* of electron and ion temperatures, and the average ion kinetic energy of thermal motion is equal to $(3/2)Aw^2/(1 + T_e/T_i)$. The differences between the temperatures implied by the various models reflect, of course, merely the differences in the densities (Fig. 11.3) which have already been discussed. The minimum at $r \sim 10\ R_J$ for the model of Connerney, Acuña, and Ness is probably an artifact of too low density estimates as the negative density region is approached. The model of Goertz et al. predicts, if $\omega = \Omega_J$ is assumed, a pronounced increase of temperature with increasing radius, as noted by Goldstein [1977] and Goertz et al. [1979]; if, on the other hand, $\omega \sim 1/r^{3/2}$ as just discussed, the temperature decreases nearly as $1/r$ [see also Barbosa et al., 1979]. If the model quantities in Figure 11.6 are divided by twice the corotational energy per unit mass, $\Omega_J^2\ r^2$ (also shown as a line in the figure), the result is $(w/V_\phi)^2$, the square of the ratio of the thermal speed to the actual (not the rigidly corotating) azimuthal speed, whose value is given independently of any assumptions about A or ω/Ω_J; according to both models, this ratio lies below unity, although

Fig. 11.7. Current density integrated
across the half-thickness of the current
sheet and expressed as an equivalent
planar magnetic field, for various
theoretical models [Gleeson and Axford,
1976, GA; Hill and Carbary, 1978, HC]
and empirical models (same
nomenclature as in Fig. 11.2).

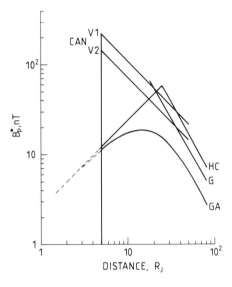

not by more than a factor of 10, nearly everywhere (the few exceptions occur close to
regions of apparent negative density and are therefore suspect).

Finally, a comparison between the empirical models and the theoretical models discussed in Section 11.3.1 is presented in Figures 11.7 and 11.8. Suitable quantities for
the comparison are the integrated current density, expressed as the equivalent planar
field B_p^* (Fig. 11.7), which measures the tangential magnetic stress per unit flux, and
the normal component of the magnetic field at the current sheet B_z (Fig. 11.8), which
describes the magnetic flux crossing the current sheet; both of these quantities can be
specified without reference to the sheet thickness, which is neglected in the theoretical
models and and uncertain in the empirical ones. To fix the numerical values of the
theoretical model parameters, we write the plasma content per unit magnetic flux η_s at
the source distance $r_s = 6\,R_J$ as

$$\eta_s = \rho_s h_s/B_s = [\mu_0 \rho_s/B_s^2]\,(M/r_s^3)\,h_s/\mu_0 \tag{11.98}$$

and take $h_s = 1\,R_J$ and (from Fig. 3.11) $B_s/(\mu_0 \rho_s)^{1/2} = 250$ km/s. The Hill–Carbary
model is then completely specified; the distance to the singular line is $r_0 = 24.2\,R_J$. The
Gleeson–Axford model has an additional parameter K_1 (or equivalently the distance a_1)
which has been set to the critical value $K_1 = 1$ (corresponding, with the above value of
η_s, to $a_1 = 20.4\,R_J$) at which a singular line first appears at infinity; even if the singular
line is allowed to move in to, say, 120 R_J, the model remains practically unchanged,
with $K_1 = 1.044$ and $a_1 = 20.6\,R_J$.

The most striking feature of Figure 11.7 is the discrepancy, reaching an order of
magnitude, between the empirical and the theoretical models in the inner magnetosphere. At distances from 6 to 8 R_J, the theoretical calculation of the integrated
current implied by the centripetal acceleration of the observed plasma is subject to relatively few uncertainties. The much larger current deduced from the empirical models
indicates, therefore, that the dominant tangential stress in the inner magnetosphere
arises from plasma pressure gradients rather than rotational stresses. (The importance
of plasma pressure here has been argued also on direct observational grounds, e.g.,
Krimigis et al. [1981] and Chap. 4.) In the outer magnetosphere, beyond 20 R_J, the
models exhibit considerably more similarity in the integrated current and it is at least

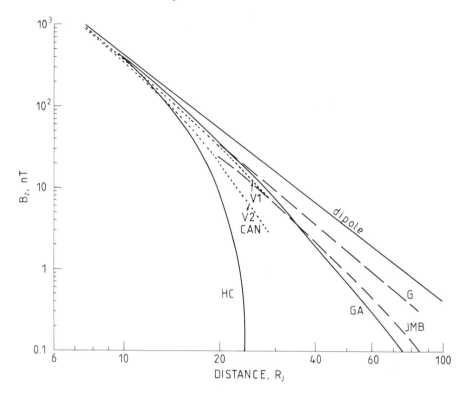

Fig. 11.8. Magnetic field component normal to the current sheet, for various theoretical and empirical models (same nomenclature as in Fig. 11.3 and 11.7).

plausible that rotational effects constitute a major part of the total stress. As to the normal component shown in Figure 11.8, there is a great diversity of model representations, with few common trends other than the fact that B_z is everywhere reduced below its dipole value. Evidently, the distribution of magnetic flux (and hence of the plasma content per unit flux) in the outer Jovian magnetosphere is still poorly understood, from either a theoretical or an empirical viewpoint.

3. Radial variation of current-sheet thickness

To determine the thickness of the current sheet or of the associated plasma sheet from direct observations of the magnetic field or of particle intensities presents a difficult problem because the observational data are functions of time along the spacecraft trajectory and a specific model of current-sheet shape and motion is needed to relate them to distance from the center of the sheet. As an example of how model-dependent such determinations are, Thomsen and Goertz [1981a] show that a particular set of particle observations yields a plasma sheet thickness that is either nearly independent of radial distance or increases in proportion to it, according to whether one assumes a bent-disc model or a magnetic-anomaly-based model for the shape of the sheet [see Vasyliunas and Dessler, 1981; Chap 10, Section 7). Any theoretical constraints on the thickness are therefore of considerable interest. Stress-balance arguments leading to an estimate of the current-sheet thickness have been discussed in Section 11.2.2; for the outer magnetosphere (the region of prime interest for this purpose) the basic result is given in Equation (11.40) which, specialized to the case of isotropic pressure, can be rewritten as

$$h = [(2r/\nu) |B_z/B_p^*|] [1/(\mu + 1)] \tag{11.99}$$

where ν is given in (11.26) and

$$\mu \equiv V_\phi^2/\nu w^2 = \omega^2 r^2/\nu w^2 \tag{11.100}$$

Equation (11.99) for h is conveniently viewed as a product of two factors. The quantity within the first set of brackets depends only on the magnetic field, specifically on the integrated current and the normal component as functions of r; the quantity within the second set of brackets depends on the ratio of azimuthal flow speed to thermal speed and does not explicitly depend on the magnetic field (except insofar as the field fixes ν through pressure balance). The first quantity also represents the maximum thickness h_{max}, reached in the limit $\mu \to 0$, that is, azimuthal flow speed negligible compared to thermal speed.

As discussed already in Section 11.3.1, calculation of h by this method is a logical second step in the case of theoretical models of the field, which supply the first factor and then require further assumptions on the plasma temperature and flow for the second. In the case of empirical models, on the contrary, it must be clearly understood that this approach is not complementary but alternative to the one described in Section 11.3.2. To calculate the plasma parameters implied by an empirical magnetic field model, the thickness must have already been specified as part of the model; if the calculated plasma parameters are inserted along with the model field into Equation (11.99), one obtains no new information but merely recovers the previously assumed model thickness. It is possible, on the other hand, to use only the magnetic field from an empirical model and to evaluate μ on the basis of some other assumptions about the plasma. One obtains then an independent model for h as a function of distance, which in general will not be consistent with the thickness parameters of the magnetic field model.

Two calculations of the current sheet thickness by means of the above semiempirical method have appeared in the literature, Goertz [1976b] and Liu [1982]. Both make use of the empirical field model of Goertz et al., and the difference between the approaches of the two illustrate some of the ambiguities of the method. According to its usual parametrization, the model of Goertz et al. has

$$B_p^* = b_0/Dr^{1+a}, \quad B_z = (M/r^3) - (ab_0 C/r^{2+a}) \tag{11.101}$$

hence, with $\nu = 2(1 + a)$ implied by pressure balance,

$$h = [aCD/(1 + a)(1 + \mu)] [(M/ab_0 Cr^{1-a}) - 1] \tag{11.102}$$

$$= [aCD/(1 + a)(1 + \mu)] [(r_{nl}/r)^{1-a} - 1]$$

where r_{nl} is the distance to the neutral line given by Equation (11.97). Liu [1982] assumed that the power-law dependences $B_r \sim 1/r^{1+a}$ and $B_z - B_{dipole} \sim 1/r^{2+a}$ are the primary feature of the Goertz et al. model, to be preserved in the calculation, which demands that D in Equations (11.101) and (11.102) be considered a constant. Specification of h as a function of r is then completed by inserting into (11.102) the numerical values of the constants b_0, a, C, D, M and obtaining μ by an independent argument (to be discussed later).

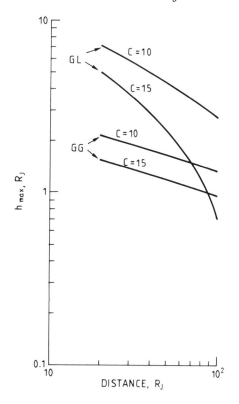

Fig. 11.9. Maximum thickness of the current sheet, reached in the limit $w^2 \gg V_\phi^2$, calculated from the model of Goertz et al. [1976] as interpreted by Goertz [1976b] (GG) and by Liu [1982] (GL).

Goertz [1976b] assumed, on the other hand, that the model parameter D is to be identified with the actual and variable current sheet thickness h; to preserve the observationally well-established dependence $B_r \sim 1/r^{1+a}$, he parameterized the model as

$$B_p^* = b_0'/r^{1+a}, \; B_z = (M/r^3) - (ab_0' \, CD/r^{2+a}) \tag{11.103}$$

where $b_0' \equiv b_0/D = 9 \times 10^3 \, nT \, R_J^{1+a}$ (note the units for b_0') is to be kept constant as D varies. (Note: in the paper of Goertz [1976b], b_0' is designated b_0 and given the units of Gauss, or $10^5 \, nT$; other than the formulas, there is no hint anywhere that the model parametrization differs from that of Goertz et al. [1976].) Instead of (11.102) one then has

$$h = [aC/(1 + a) (1 + \mu)] [(M/ab_0' \, Cr^{1-a}) - D] \tag{11.104}$$

and it is assumed that $D = h$, whereupon the equation can be solved for h:

$$h = M/\{b_0' \, r^{1-a} [aC + (1 + a) (1 + \mu)]\} \tag{11.105}$$

With this value for $h = D$, $B_z - B_{\text{dipole}}$ no longer varies as $1/r^{2+a}$; combining (11.105) with (11.103) we have

$$B_z = (M/r^3) \{1 - aC/[aC + (1 + a) (1 + \mu)]\} \tag{11.106}$$

that is, B_z is now simply the dipole field reduced by a constant factor. In this version of the Goertz et al. model $B_z \neq 0$ at all finite distances; there is no neutral line and hence no last closed field line.

The maximum thickness h_{max} predicted by both models in the hot plasma limit $\mu \to 0$ is shown in Figure 11.9. Equation (11.102) predicts an h_{max} that is relatively large at

$r = 20\ R_J$ and decreases markedly with increasing distance, though nowhere near as fast as the $1/r^3$ dependence looked for in Section 11.3.2; h_{max} approaches zero, of course, at $r = r_{nl}$ where $B_z \rightarrow 0$. Equation (11.105) predicts a simple power-law dependence $h_{max} \sim 1/r^{1-a} = 1/r^{0.3}$, a rather slow decrease cited by Goertz [1976b] and Thomsen and Goertz [1981a] as being close to a constant thickness originally assumed in the model; that advantage, as we have seen, has been bought at the price of significantly modifying B_z and eliminating the identification of open field lines that was given much emphasis in the original model [Goertz et al., 1976; see also Fig. 10.11 and its caption]. (It may be pointed out, though, that the last closed field line in the original model is identified by its equatorial crossing at $r = r_{nl}$, the location of the neutral line, and since r_{nl} lies well outside the model's range of applicability the significance of the identification is perhaps doubtful in any case.)

To calculate actual values of h and not just h_{max}, assumptions about the quantity μ must be made. Goertz [1976b] assumed that μ was a constant and set $V_\phi^2/w^2 = 2$; his h is simply scaled down from h_{max} by a constant factor. Liu [1982] assumed that $V_\phi = \omega r$ with $\omega \sim 1/r^2$, derived from a model of partial corotation (see Sec. 11.4.4), and that $w^2 \equiv P/\rho$ is obtainable from a relation $P \sim \rho^\gamma$ with P given by pressure balance with the magnetic field. He treated the adiabatic ($\gamma = 5/3$) and isothermal ($\gamma = 1$) cases with several different initial ratios V_ϕ^2/w^2 specified at the reference distance $r = 20\ R_J$, obtaining a number of h vs. r profiles which all lie under the h_{max} vs. r curve of Figure 11.9, of course, and generally decrease with increasing r, although for the isothermal case with $V_\phi^2/w^2 > 1.72$ he finds that h initially increases with increasing r and reaches a maximum before decreasing.

Theoretical estimation of the thickness of the current sheet and its dependence on the distance is thus at present rather uncertain. There is no unique model, in large part because of still unresolved uncertainties about the magnetic field, particularly its B_z component, and inability to decide among several still viable hypotheses about the ratio of plasma thermal speed to azimuthal flow speed.

4. Azimuthal magnetic fields and angular momentum transfer

So far in this chapter, the azimuthal (or nonmeridional or toroidal) magnetic field component B_ϕ has been neglected, partly on the grounds that it is relatively small and partly because it plays no direct role in the stress balance along the radial and normal directions. However, the existence of well-defined large-scale azimuthal magnetic fields which, in the outermost regions of the magnetosphere, can become comparable in magnitude to the other field components, is a prominent feature of the observations. The interpretation of these fields has become a matter of considerable controversy, involving some fundamental questions about the physics of the Jovian magnetosphere, in particular about the relative roles of rotation, plasma outflow, and solar wind effects. Chapter 1, Section 1.3 contains a brief account, with references, of the observations and models (presented, however, entirely from the point of view of one side in the controversy).

With $\hat{\phi}$ defined as positive in the direction of corotation, the observed B_ϕ is negative in regions above the current sheet and positive below; within the current sheet B_ϕ is greatly reduced in magnitude, consistent with its required reversal. Looked at from above the pole, the magnetic field lines thus have a spiral appearance of lagging behind the planetary rotation. Such a configuration is readily interpreted as a rotation-associated effect, a "wrapping up" of the field lines as the plasma in the outer magnetosphere is slowed down in its azimuthal motion, either by a tangential drag between it and the solar wind or by its own inertia if it is moving radially outward. Theoretical suggestions of the lagging spiral configuration and of both these possible explanations were made [Piddington, 1969] long before there were any in situ observations. Implicit in this interpretation is the assumption that the azimuthal field pattern has the same character all around the planet, that is, the sign of B_ϕ is, for the most part, independent of local time, although a local-time dependence of the magnitude of B_ϕ is not precluded. The dominant mechanical effect of the stresses in the magnetic field is a *torque* opposed to the planetary rotation and extracting energy from it.

An alternative interpretation, championed principally by Ness and his coworkers [Ness et al., 1979a,b,c; Behannon, Burlaga, and Ness, 1981; Chap. 1], considers the azimuthal fields to be primarily associated not with Jupiter's rotation but with its interaction with the solar wind; the observed B_ϕ is ascribed to the bending of magnetic field lines away from the Sun by the action of a solar-wind-aligned magnetospheric tail, similar to what is observed in the terrestrial magnetosphere [e.g., Fairfield, 1968], and the resemblance to the rotationally lagging spiral is viewed as merely the result of the coincidental fact that all the available observations have been made in the postmidnight–dawn–prenoon side of the Jovian magnetosphere. Implicit in this interpretation is the assumption that the azimuthal magnetic fields in the yet unobserved dusk side also point away from the Sun, that is, B_ϕ reverses sign at or near the noon–midnight meridian. To first approximation, the magnetic torques from the dawn and dusk sides cancel; the dominant mechanical effect of the stresses in the magnetic field is now not a torque but a *force* in the direction of solar wind flow (see discussions of the terrestrial analog by Siscoe [1966], Siscoe and Cummings [1969], and Carovillano and Siscoe [1973]).

It is obvious that observations during even a single traversal of the outer magnetosphere in the premidnight or dusk local time sector should settle the controversy conclusively because the two interpretations firmly predict opposite signs for B_ϕ there. However, spacecraft missions with the required trajectories have not been flown yet, and the interpretation of B_ϕ (only observed in regions where both models predict the same sign) remains in dispute. On one side, many details of magnetic field observations have been fitted into the framework of the solar-wind-aligned magnetotail interpretation, as summarized in Chapter 1, Sections 1.3 and 1.4 (although the impression given there, that the controversy is as good as settled in favor of this model, is the personal opinion of the chapter's authors and does not represent a consensus). On the other side, Vasyliunas and Dessler [1981] argue that an Earthlike magnetotail can have significant effects only in regions where the magnetic pressure is at least an order of magnitude lower than the solar wind ram pressure, hence at distances no closer than 80–100 R_J in the nightside Jovian magnetosphere (see also Sec. 11.3.5). Recently, Vasyliunas [1982a] has reexamined the magnetic field evidence adduced for a Jovian magnetotail and concluded that it is equally consistent with the rotational interpretation (the "wrapped-around magnetotail" in the phrasing of Piddington [1969]); he also finds that the terrestrial analog predicts a local-time variation of B_ϕ not seen at Jupiter.

In the midnight-to-dawn sector, the region traversed by the outbound passes of Pioneer 10, Voyager 1, and Voyager 2, the quantitative description of the observed B_ϕ is relatively simple. Goertz et al. [1976] found that Pioneer 10 observations of B_ϕ could be fitted by the form (in cylindrical coordinates)

$$B_\phi/rB_r = (1/r_f) F(r) \tag{11.107}$$

where $r_f = 160\ R_J$ is a constant and $F(r)$ is a slowly varying function nearly equal to unity; Goertz et al. obtained as the best fit

$$F(r) = \exp(r/500\ R_J) \tag{11.108}$$

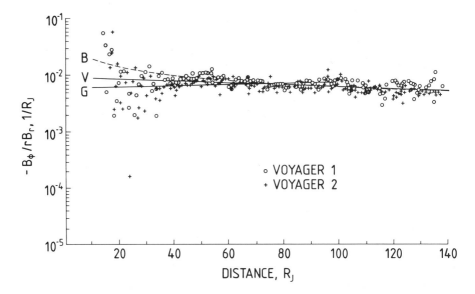

Fig. 11.10. The ratio B_ϕ/rB_r observed during the Voyager 1 and 2 outbound passes [from Figs. 11 and 12 of Behannon, Burlaga, and Ness, 1981, with omission of measurements within the current sheet and the depressed field region]. The lines are various empirical fits: G represents Equation (11.108) (drawn over the radial range of Pioneer 10 observations only), B is Equation (11.109) and V is Equation (11.110).

which varies from 1.04 to 1.17 over the distance range 20 to 80 R_J of the observations. Behannon, Burlaga, and Ness [1981] found a very similar result from Voyager 1 and 2 observations: B_ϕ given by Equation (11.107) but with a somewhat different $F(r)$; with the choice $r_f = 160 \, R_J$ again, the fit of Behannon, Burlaga, and Ness is

$$F(r) = (100 \, R_J/r)^{1/2} \qquad (11.109)$$

which decreases from 2.24 at 20 R_J to 0.845 at 140 R_J. In both cases, the result applies only outside the current sheet region; inside it, B_ϕ may be significantly reduced in magnitude from the value given by (11.107). The fact that B_ϕ/rB_r appears to be nearly constant or only slowly decreasing has played a major role in the interpretation of the observed azimuthal field, as discussed further on.

Figure 11.10 shows B_ϕ/rB_r outside the current sheet as a function of radial distance: the combined Voyager 1 and 2 observations and the fitted curve representing the Pioneer 10 observations, together with other suggested fits, Equation (11.109) and my own suggestion

$$F(r) = 1.44 \exp - (r/260 \, R_J) \qquad (11.110)$$

Except at distances $r < 40 \, R_J$, where Voyager observations show considerable scatter, the observed values do vary only slowly with r, consistent with Equation (11.107). There is also good agreement among data from the three spacecraft despite their different positions in local time; only for $r \geq 120 \, R_J$ is there some indication of a significant systematic difference, Voyager 1 values tending to become higher than those from Voyager 2. For the distance range $40 \, R_J \leq r \leq 80 \, R_J$ (only for $r \leq 80 \, R_J$ are all three spacecraft well inside the magnetosphere), Figure 11.11 shows B_ϕ/rB_r, expressed as observed range within each 10 R_J interval, as a function of local time; no significant local-time dependence is apparent.

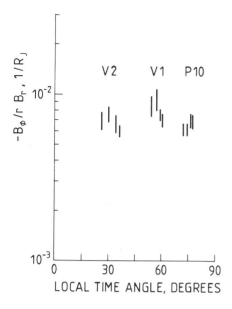

Fig. 11.11. The ratio B_ϕ/rB_r observed during the Pioneer 10, Voyager 1, and Voyager 2 outbound passes, as a function of local time angle measured from midnight. The bars represent the range that contains 50% of all observations within the radial distance intervals (from left to right for each pass) 40-50, 50-60, 60-70, and 70-80 R_J, taken from Figure 11.10 and from Figure 3 of Goertz et al. [1976]; the choice of 50% range eliminates extreme values associated with entry into or near the current sheet [from Vasyliunas, 1982a].

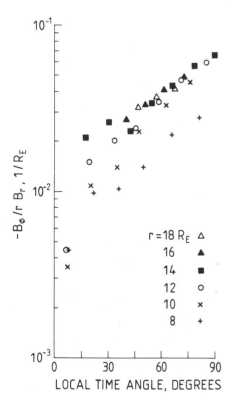

Fig. 11.12. The ratio B_ϕ/rB_r observed in the terrestrial magnetosphere as a function of local time, scaled from Figure 3 of Fairfield [1968] [from Vasyliunas, 1982a].

For comparison, Figure 11.12 shows B_ϕ/rB_r observed in the terrestrial magnetosphere [from Fairfield, 1968] as a function of local time in the midnight-to-dawn quadrant. In this case there is an obvious local-time variation, B_ϕ/rB_r decreasing toward midnight (as it should if it is to become zero there and to reverse sign on the dusk side); the variation is present over a wide range of radial distances, including 12 to 18 R_E where B_ϕ/rB_r is nearly independent of r. The fact that at Earth the magnetic field lines on the nightside are azimuthally bent by an amount that varies with local time, as expected from the action of a solar-wind-aligned magnetospheric tail, while at Jupiter the azimuthal bending of the field lines exhibits no corresponding variation and is essentially independent of local time (as expected from a rotation-related effect) within the observed regions of the nightside magnetosphere (except possibly in its very distant part, beyond 120 R_J) would seem to be a strong argument against assuming that all azimuthal fields in the magnetosphere arise from a common physical mechanism at both planets.

Local-time asymmetries do exist in the Jovian magnetosphere but they are observed primarily as day–night differences and not as variations within the midnight-to-dawn quadrant. On the dayside, the azimuthal magnetic field is generally smaller than on the nightside, although it still often exhibits a spiraling trend, evidenced by a qualitative tendency for B_ϕ and B_r to have correlated magnitude variations and opposite signs [Jones, Thomas, and Melville, 1981]; a quantitative systematic radial variation analogous to Equation (11.107) has not been reported yet, however. A particularly weak and irregular B_ϕ was observed during the Pioneer 11 outbound pass, but whether this is a local time or a latitude effect – Pioneer 11 outbound went both closer to local noon and up to higher latitudes than any other spacecraft at Jupiter to date – is not yet entirely clear [see Smith, Davis, and Jones, 1976; Jones, Thomas, and Melville, 1981].

Modeling and theoretical studies of the azimuthal field have concentrated for the most part on the nightside magnetosphere, where Equation (11.107) may be applied and furthermore the usual assumption of axial symmetry has some direct observational support as noted previously. Among the topics of interest are the following: (1) relation of B_ϕ to the previously discussed models of the magnetic field, (2) global models of current systems associated with B_ϕ, (3) quantitative models for spiraling of the magnetic field, (4) azimuthal stress balance and angular momentum transfer.

(1) With separation of the magnetic field into meridional and azimuthal components, $\mathbf{B} = \mathbf{B}_m + B_\phi \hat{\phi}$, and assumption of (at least local) axial symmetry, $(\partial/\partial\phi) = 0$, the Lorentz force density may be cast into the form

$$\mu_0 \mathbf{j} \times \mathbf{B} = (\nabla \times \mathbf{B}_m) \times \mathbf{B}_m - (1/2r^2) \nabla r^2 B_\phi^2 + \hat{\phi} (B/r) \cdot \nabla rB_\phi \qquad (11.111)$$

A nonzero B_ϕ results in the appearance of an azimuthal stress component (third term) as well as modification of the magnetic stress in the meridional plane (second term). This modification, however, is slight as long as B_ϕ is relatively small compared to $|B_m|$. A model magnetic field calculated by neglecting B_ϕ will be changed, if B_ϕ is included, from $\mathbf{B}_m$ to $\mathbf{B}_m + \delta\mathbf{B}_m$ where

$$(\nabla \times \delta\mathbf{B}_m) \times (\mathbf{B}_m + \delta\mathbf{B}_m) + (\nabla \times \mathbf{B}_m) \times \delta\mathbf{B}_m = (1/2r^2) \nabla r^2 B_\phi^2 \qquad (11.112)$$

and it is apparent that, in order of magnitude,

$$\delta B_m/B_m \sim O(B_\phi^2/B_m^2) \sim O(r^2/r_f^2) \qquad (11.113)$$

where r_f is defined in Equation (11.107). Hence, stress balance in the radial and vertical directions is little affected by the observed B_ϕ and most of the considerations and models of Sections 11.3.1, 11.3.2, and 11.3.3 remain applicable, until distances of the order of, or in excess of, some 80–100 R_J are reached. Stress balance in the azimuthal direction is to be dealt with separately later.

(2) Associated with an azimuthal magnetic field there are in general electric current components in the radial and/or vertical directions, in addition to the azimuthal

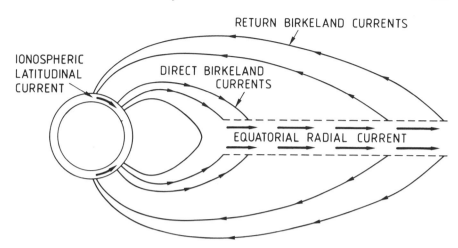

Fig. 11.13. Sketch of the current system associated with the azimuthal magnetic field, projected on a meridional surface. The size of Jupiter and the thickness of the ionosphere are greatly exaggerated for clarity [from Vasyliunas, 1982b].

currents implied by the stretched-out configuration of the meridional magnetic field. The guiding principle in attempting to infer the large-scale current system associated with the observed B_ϕ is that strong currents perpendicular to **B** are possible only within the thin current sheet in the equatorial magnetosphere and within the Jovian ionosphere where the plasma stresses can be large enough to balance the $\mathbf{j} \times \mathbf{B}$ forces. Elsewhere, in the low-density region between the current sheet and the ionosphere, $\mathbf{j} \times \mathbf{B}$ must be very small and the magnetic field very nearly force-free, with currents flowing as Birkeland currents (i.e., along the magnetic field lines). Given these constraints, the qualitative configuration of the current system can be sketched immediately and is shown in Figure 11.13. The equatorial radial current density, integrated over (half) the thickness of the current sheet, is directly related to B_ϕ just outside the sheet,

$$\mu_0 j_r' = -B_\phi \tag{11.114}$$

and can thus be inferred from observations. The various proposed models of the current system differ primarily in the assumed location of the Birkeland currents.

Parish, Goertz, and Thomsen [1980] considered two alternative models, with the direct Birkeland currents either (1) distributed over the entire radial extent of the current sheet, with uniform current density $j_\parallel$ near the ionosphere, or (2) concentrated at the inner edge of the current sheet, assumed to lie at $r = 10\ R_J$; they compared both models with Pioneer 10 observations of B_ϕ and concluded that either model gave an adequate fit, within the observational uncertainties. Connerney [1981b; see also Thomsen and Goertz, 1981b] adopted their second model but gave a much simpler mathematical treatment and placed the inner edge with its Birkeland currents at $r = 5\ R_J$. The return Birkeland currents were emphasized by Connerney and mentioned by Parish, Goertz, and Thomsen (who placed them near the magnetopause) but not treated in detail in either paper.

With axial symmetry, the force-free assumption for the region above the current sheet implies, from Equation (11.111), that rB_ϕ is constant along a field line; with the help of Ampère's law it then follows that

$$rj_r' = R_J j_\theta' \sin \theta \tag{11.115}$$

where j_i' is the same as in Equation (11.114) and j_θ' is the height-integrated current in the ionosphere at the colatitude θ joined by a field line to the radial distance r in the current sheet. The Birkeland current density $j_\parallel$ is given by current continuity together with Equation (11.114) as

$$\mu_0 j_\parallel B_z/B = (1/r)\,(\partial/\partial r)\,rB_\phi \tag{11.116}$$

just above the current sheet; within the force-free region $j_\parallel/B$ is constant along a field line. It is evident that if the Birkeland currents are concentrated at the inner and outer radial edges, B_ϕ above most of the current sheet must vary as $1/r$, as pointed out by Connerney [1981b], whereas distributed Birkeland currents imply a correspondingly different radial variation of B_ϕ. (Note: the radial dependence of the ratio j_i'/j_θ' has no relevance to any of this, contrary to a remark by Jones, Melville, and Blake [1980].) Now Equation (11.107) with $F = 1$ and B_z from the model of Goertz et al. yields $rB_\phi \sim r^{0.3}$, or a change between $r = 20\,R_J$ and $r = 80\,R_J$ by a factor 1.52 [reduced to 1.20 if F given by Equation (11.110) is assumed]. Such changes hardly constitute a significant departure from constancy, given the uncertainties and fluctuations of the observed B_ϕ values; it is thus not surprising that it has not proved possible to extract from the available observations a unique model for the Birkeland current distribution.

If the dependence $rB_\phi \sim r^{0.3}$ is nevertheless taken at face value, Equation (11.116) implies, with B_z from the model of Goertz et al.,

$$j_\parallel/B \sim r^{1.3}/[1 - (r/r_{nl})^{0.3}] \tag{11.117}$$

that is, $j_\parallel/B$ increases with increasing r, by a factor 8.7 if $C = 10$ or 16.8 if $C = 15$ between $r = 20\,R_J$ and $r = 80\,R_J$. By contrast, the two models of Parish, Goertz, and Thomsen assume $j_\parallel/B = $ constant and $j_\parallel/B \sim \delta(r - 10\,R_J)$, respectively, the second assumption implying $rB_\phi = $ constant. Widely different assumptions about the distribution of Birkeland currents can thus lead to rather similar radial profiles of B_ϕ.

(3) The configuration of the azimuthal magnetic field described by Equation (11.107) is identical in form to the spiral pattern of magnetic field lines in interplanetary space, which results from the interaction of solar rotation with the radially outward solar wind flow [e.g., Parker, 1963; Hundhausen, 1972]. Hence, it has been widely assumed that B_ϕ within the nightside Jovian magnetosphere may be similarly attributed to the interaction of the planet's rotation with a radial outflow of magnetospheric plasma, such as the planetary/magnetospheric wind discussed in Section 11.4.3. The quantitative description of this interaction is based on the MHD relation between the electric field $\mathbf{E}$ and the plasma bulk flow velocity $\mathbf{V}$:

$$\mathbf{E} + \mathbf{V} \times \mathbf{B} = 0 \tag{11.118}$$

With the use of * to designate quantities referred to a frame of reference rigidly corotating with Jupiter, Equation (11.118) may also be written as

$$\mathbf{E}^* + \mathbf{V}^* \times \mathbf{B} \equiv [\mathbf{E} + (\Omega_J \times \mathbf{r}) \times \mathbf{B}] + [\mathbf{V} - (\Omega_J \times \mathbf{r}) \times \mathbf{B} = 0 \tag{11.119}$$

Now if the ionospheric conductivity is sufficiently high, $\mathbf{E}^*$ becomes vanishingly small in regions of space connected to the ionosphere by magnetic field lines (see Sec. 11.4.2 for a more detailed argument) and Equation (11.119) then implies that $\mathbf{V}^*$ and $\mathbf{B}$ are aligned, whence in particular (in cylindrical coordinates)

$$B_\phi/B_r = (V_\phi - \Omega_J r)/V_r \tag{11.120}$$

and the further assumptions $V_\phi \ll \Omega_J r$, $V_r \sim$ constant yield an equation of the form

(11.107). To account for the observed spiraling of the field in the nightside Jovian magnetosphere, a value of the radial outflow speed

$$V_r \sim \Omega_J r_f = 2 \times 10^3 \text{ km/s}$$

is required at distances from about 20–40 R_J to 100–120 R_J. This value is too large to be consistent with planetary wind theory, which predicts that the assumptions $V_\phi \ll \Omega_J r$ and $V_r \sim$ constant will hold only at distances where $V_r \gg \Omega_J r$ (see Sec. 11.4.3). Furthermore, plasma observations do not indicate any strong radial outflow at distances smaller than 130–150 R_J (Chaps. 3 and 4 and references therein). Thus a straightforward analogy with the Parker spiral in the solar wind does not provide an adequate model for B_ϕ in the Jovian magnetosphere.

 An obvious weak point in the derivation of Equation (11.120) is the assumption that the Jovian ionospheric conductivity is sufficiently high to enforce corotation. As shown by Hill [1979] and discussed here in Section 11.4.4, plausible values of the ionospheric conductivity in fact allow large departures from corotational motion. Each constant-latitude strip of the ionosphere may then be characterized by the angular frequency Ω of its (longitude-averaged) actual azimuthal motion, with $\Omega \neq \Omega_J$ in general. This partial corotation will extend, by virtue of Equation (11.118), to the entire shell of magnetic field lines emanating from the fixed latitude, and the same derivation may be carried through as before to obtain Equation (11.120) but with Ω_J replaced by Ω everywhere. We then have

$$B_\phi/rB_r = (\omega - \Omega)/V_r \tag{11.121}$$

where $\omega \equiv V_\phi/r$ is the angular frequency of the azimuthal motion of plasma in the magnetosphere, to be distinguished both from Ω (which refers to the foot of the field line, in the ionosphere) and from Ω_J. A latitudinal current in the ionosphere is driven by the electric field associated with the difference between Ω_J and Ω:

$$j_\theta' = 2\Sigma \, B_J \, R_J \, (\Omega_J - \Omega) \sin \theta \, [(1 + 3 \cos^2 \theta)/4 \cos \theta] \tag{11.122}$$

where Σ is the height-integrated Pedersen conductivity of the ionosphere and the factor in brackets, which comes from the field inclination angle and its effect on the horizontal conductivity, may be set equal to one for all practical purposes (its value at a latitude as low as the $L = 6$ Io shell is 0.959). The ionospheric current is in turn related to the equatorial radial current and thence to B_ϕ by Equations (11.115) and (11.114), yielding

$$j_\theta' = -rB_\phi/\mu_0 R_J \sin \theta \tag{11.123}$$

Combining Equations (11.121), (11.122), and (11.123) and solving for B_ϕ and for Ω one obtains

$$B_\phi/rB_r = (\omega - \Omega_J)/[V_r + (1/\mu_0 \Sigma \chi)] \tag{11.124}$$

$$\Omega = \omega + (\Omega_J - \omega)V_r/[V_r + (1/\mu_0 \Sigma \chi)] \tag{11.125}$$

where

$$\chi \equiv 2B_J(R_J \sin \theta)^2/r^2 B_r \tag{11.126}$$

The foregoing derivation has been given by Vasyliunas [1982b].

 The dimensionless quantity χ, effectively a mapping factor between the equatorial plane and the ionosphere, depends only on the geometry of the magnetic field. From

Fig. 11.14. Theoretical values of B_ϕ/rB_r (dashed lines) calculated from the theory described in the text with an assumed $\Sigma = 0.4$ mho and the field model of Goertz et al. [1976] with parameter $C = 10$ or 15, compared with various empirical fits (solid lines; see Fig. 11.10 for the nomenclature and note that the curve B is not a good fit to the observations for $r < 40\ R_J$) [from Vasyliunas, 1982b].

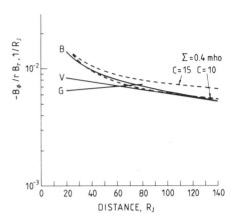

equality of magnetic flux through corresponding segments of the ionosphere and the equatorial plane it can be shown that

$$B_J \left(R_J \sin \theta\right)^2 = M/r + \int_{R_J}^{r} r\,dr\,(B_{\text{dipole}} - B_z) \tag{11.127}$$

and thus χ can be evaluated from a field model that specifies B_r and B_z in the equatorial plane. Vasyliunas [1982b] calculated χ by using the model of Gleeson and Axford [1976] to carry out the integration in (11.127) from $r = 1\ R_J$ out to $r = 20\ R_J$ (in this range the model approximates well to the V 2 model of Connerney, Acuña, and Ness, see Fig. 11.8) and the model of Goertz et al. [1976] at larger distances; he found χ to be a slowly decreasing function of r, with values near 2 at $r = 25\ R_J$ and near 1 at $r = 100\ R_J$.

Two limiting cases are of interest. In the high-conductivity limit,

$$\mu_0 \Sigma \chi V_r = \chi(\Sigma/1\ \text{mho})\ (V_r/796\ \text{km s}^{-1}) \gg 1 \tag{11.128}$$

Equation (11.124) reduces to (11.120) and (11.125) reduces to $\Omega = \Omega_J$; we recover the expected full corotation of the ionosphere and the Parker spiral for the magnetic field. More relevant to the Jovian magnetosphere is the opposite, low-conductivity limit. With the inequality in (11.128) reversed, Equations (11.124)–(11.125) reduce to

$$B_\phi/rB_r = (\omega - \Omega_J)\ \mu_0 \Sigma \chi \tag{11.129}$$

$$\Omega = \omega \tag{11.130}$$

The ionosphere now corotates with the magnetospheric plasma but not necessarily with the planet (the value of ω, governed by azimuthal stress balance, is not determined by the model in any limit); the spiraling angle of the magnetic field is now proportional to the ionospheric conductivity and does not depend on V_r.

Numerical values of B_ϕ/rB_r predicted by Equation (11.129) for the case $\omega \ll \Omega_J$, with $\Sigma = 0.4$ mho and χ calculated as described previously, are shown in Figure 11.14 together with the several fits to observed values (see Fig. 11.10 and associated discussion). The agreement between theory and observation is quite good (except at close distances, where the assumption $\omega \ll \Omega_J$ may not hold) for the model with $C = 10$ with the chosen conductivity; an equally good agreement for the model with $C = 15$ can be obtained by choosing $\Sigma = 0.33$ mho. These conductivity estimates are to be regarded as lower limits; if either ω or V_r are nonnegligible, higher values of Σ are required to maintain the fit to the observations. (V_r is insignificant unless it approaches or exceeds 10^3 km/s.)

The observed azimuthal magnetic fields in the nightside Jovian magnetosphere can thus be quantitatively accounted for as a simple consequence of finite ionospheric conductivity and partial corotation, without the necessity of postulating unrealistically large radial outflow speeds. In simplest physical terms, the model states that the ionospheric continuation of the equatorial radial current associated with B_ϕ (see Fig. 11.13) is simply the current driven by the meridional electric field that is applied to the conducting ionosphere as a result of partial corotation. There is a maximum possible value of this electric field, corresponding to near-zero corotational motion ($\Omega \to 0$); thus from observations of B_ϕ, which together with a magnetic field model for the mapping factor χ determine a definite amount of current to be driven through the ionosphere, a minimum required ionospheric conductivity of some 0.3–0.4 mho can be inferred.

(4) The azimuthal component of the tangential stress balance Equation (11.18), can be written in the form

$$\sigma (d/dt) \omega r^2 = 2r B_\phi B_z/\mu_0 \tag{11.131}$$

where

$$(d/dt) \equiv \partial/\partial t + V_r \partial/\partial r + \omega \, \partial/\partial\phi \tag{11.132}$$

and $2B_\phi$ is the (vector) difference of the oppositely directed azimuthal field components above and below the current sheet. The pressure gradient term has been neglected, both on the general grounds discussed in Section 11.2.2 and because $\partial P/\partial\phi$ is likely to be small in any case and vanishes for axial symmetry. Equation (11.131) states that the torque exerted by the azimuthal magnetic field acts to change the angular momentum of the plasma. It is convenient to introduce the quantity

$$S \equiv r \int d\phi \, \sigma V_r \tag{11.133}$$

(with the integration carried out over an angular sector of width $\Delta\phi$) which represents the integrated net mass flux across the cylinder (or cylinder segment if $\Delta\phi \neq 2\pi$) at a radial distance r. If $\partial/\partial t$ and $\partial/\partial\phi$ in (11.132) are either assumed negligible or averaged over, Equation (11.131) may be integrated to give

$$S(d/dr) \langle \omega \rangle r^2 = 2r^2 \int d\phi \, B_\phi B_z/\mu_0 = 2r^2 B_\phi B_z \, \Delta\phi/\mu_0 \tag{11.134}$$

where the second form assumes axial symmetry of the magnetic field. The average local rotation frequency $\langle \omega \rangle$ is defined by

$$\langle \omega \rangle \equiv r \int d\phi \sigma V_r \omega/S \tag{11.135}$$

If, furthermore, inflow of plasma into the current sheet from above and below (e.g., from an ionospheric source) is neglected, then, by conservation of mass, S is independent of r (except within the Io torus or other plasma source regions).

Note that $\langle \omega \rangle$ is an average weighted by the radial mass flux σV_r. If the outward mass transport occurs predominantly by simple radial outflow so that $\sigma V_r > 0$ at all ϕ, then $\langle \omega \rangle$ approximates to a straightforward average over ϕ. In this case the observed fact that B_ϕ has the same sign at all r implies, together with Equation (11.129), that $\langle \omega \rangle \leq \Omega_J$; that is, there can be on the average no supercorotation of the plasma as that would imply an opposite sense of B_ϕ to what is observed. The angular momentum added to the plasma by the magnetic torque is in this case transported outward by simple advection, the flowing plasma carrying angular momentum with it. If, on the other hand, the predominant plasma motion is one of circulation, radial flow occurring both inward and outward and a net outward mass transport resulting only from a larger mass flux in the outflow regions compared to the inflow (as in models of corotating

Fig. 11.15. The value of B_ϕ/rB_r needed to maintain full corotation of the plasma for various values of the mass flux S (solid lines), calculated with use of the Goertz et al. [1976] model with the two values of C, and the various empirical fits (dashed lines, see Fig. 11.10).

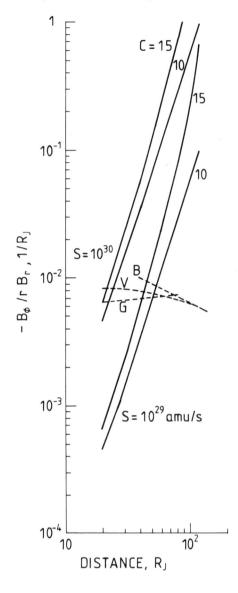

convection or of radial diffusion, described in Sections 11.4.6 and 11.4.7), then $\langle\omega\rangle$ may be written as

$$\langle\omega\rangle = (S_{out}\,\omega_{out} - S_{in}\,\omega_{in})\,/\,(S_{out} - S_{in}) \tag{11.136}$$

where S_{out} is the mass flux integrated over the part of the cylindrical surface where $\rho V_r > 0$, ω_{out} the associated average ω, and S_{in} with ω_{in} defined similarly where $\rho V_r < 0$; since $S \equiv S_{out} - S_{in}$, (11.136) may be rewritten as

$$\langle\omega\rangle = \omega_{out} + (S_{in}/S)\,(\omega_{out} - \omega_{in}) \tag{11.137}$$

It is thus possible now for $\langle\omega\rangle$ to become much larger than Ω_J, even when $\omega_{out} \leq \Omega_J$ and $\omega_{in} \leq \Omega_J$, provided S_{in}/S is sufficiently larger than one, that is, the net outward mass flux is much smaller than the total mass flux in the circulating motion. In this case the angular momentum can be transported outward by means of an effective eddy viscosity: outflowing plasma carries more angular momentum than inflowing plasma ($\omega_{out} > \omega_{in}$) with the result of a net outward angular momentum flux even if the net outward mass flux is small or negligible (hence $\langle\omega\rangle$, defined as the ratio of the two fluxes, is allowed to become very large).

Given the (constant) value of S, Equation (11.134) constitutes a relation between $\langle\omega\rangle$ and B_ϕ, from which either can be calculated given the other. In the following calculations we take $\Delta\phi = 90°$, corresponding to the midnight-to-dawn quadrant, approximately within which a well-defined, nearly axially symmetric structure of B_ϕ has been observed; values of S refer then to this quandrant, but they can be easily scaled for other assumptions, for example, doubled if outflow over the entire 180° of the nightside is assumed. Values of B_z (and of B_r where needed) are taken from the model of Goertz et al. [1976].

Setting $\langle\omega\rangle = \Omega_J$ in Equation (11.134) determines the B_ϕ needed to maintain the plasma within the current sheet in rigid corotation; the calculated B_ϕ scales in proportion to S and is shown in Figure 11.15 for two values of S (actually the quantity B_ϕ/rB_r is plotted, for convenience in comparison with the observations). In contrast to the nearly constant observed values, B_ϕ/rB_r required to maintain rigid corotation increases sharply with increasing r, a consequence of the fact that the angular momentum per unit mass of rigidly corotating plasma increases as r^2 and hence must be supplied, if the plasma is transported outward from a source in the inner magnetosphere, by an ever-increasing magnetic torque. It is also apparent from the figure that if $S = 10^{30}$ amu/s, generally considered a reasonable estimate (see Chap. 10 and references therein), then the observed B_ϕ is far too small to enforce corotation at distances beyond about 20–25 R_J. If $S = 10^{29}$ amu/s, the observed B_ϕ is at first far larger than what is needed but then also becomes too small at distances beyond about 45–55 R_J.

It is also possible to set B_ϕ in Equation (11.134) equal to the observed values and, by integrating the equation, to obtain the implied $\langle\omega\rangle$ as a function of r. The result may be written as

$$\langle\omega\rangle = (T/r^2 S) + \omega_1 (r_1/r)^2 \tag{11.138}$$

where

$$T = 2\Delta\phi \int_{r_1}^{r} dr' (r')^2 B_\phi B_z/\mu_0 \tag{11.139}$$

is the integrated magnetic torque on a segment of the current sheet from r_1 to r and ω_1 is the boundary value of $\langle\omega\rangle$ at $r = r_1$. Numerical results for $\omega_1 = \Omega_J$, at $r_1 = 20$ R_J, with B_ϕ taken from Equation (11.107) with $F = 1$, are shown in Figure 11.16a for various values of S. Consistent with Figure 11.15 and associated discussion, for $S = 10^{31}$ amu/s $\langle\omega\rangle$ decreases sharply with increasing r and is not much above the $\omega_1 (r_1/r)^2$ value for negligible magnetic torque per unit mass ($S \to \infty$), while for $S = 10^{29}$ amu/s the initially large magnetic torque raises $\langle\omega\rangle$ to values much above Ω_J, with an increase of angular momentum that then suffices to maintain $\langle\omega\rangle > \Omega_J$ even to $r = 100$ R_J where B_ϕ has become much smaller. For $S = 10^{30}$ amu/s, with the observed B_ϕ somewhat above the corotational value just for a short interval beyond 20 R_J, $\langle\omega\rangle$ initially rises slightly above Ω_J and then decreases sharply. (The critical value of S, above which there is no initial rise above Ω_J at $r_1 = 20$ R_J, can be shown to be $S = 4.0 \times 10^{30}$ amu/s for the $C = 10$ model and $S = 2.8 \times 10^{30}$ amu/s for $C = 15$.) Similar results are obtained if the plasma is assumed to be only partially corotating at $r_1 = 20$ R_J already, as shown in Figure 11.16b for $\omega_1 = 0.5$ Ω_J.

Fig. 11.16. (*a*) Average angular speed of the plasma calculated from an empirical fit to the observed B_ϕ and the two versions of the Goertz et al. [1976] model, with assumed full corotation at $r = 20\ R_J$, for various values of the mass flux S. (*b*) The same but with $\omega = 0.5\ \Omega_J$ at $r = 20\ R_J$ assumed as the boundary condition.

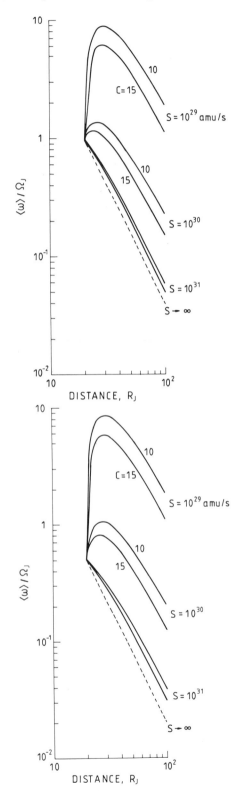

As mentioned previously, values of $\langle \omega \rangle > \Omega_J$ are acceptable only if $S \ll S_{in}$; otherwise, in a simple radial outflow, B_ϕ would decrease to zero as $\omega \to \Omega_J$ and thus no angular acceleration to $\omega > \Omega_J$ could occur. The following conclusion can therefore be drawn: if the net mass outflow rate S is less than about 10^{30} amu/s, the angular momentum must be transported outward by means of an effective eddy viscosity mechanism and an amount of mass much larger than S must be circulating in and out through the outer magnetosphere; if, on the other hand, the angular momentum is simply advected by the outward-flowing plasma, S must be significantly larger than 10^{30} amu/s. In any case, a mass flow of something like 10^{30}–10^{31} amu/s from the inner to the outer magnetosphere seems to be almost unavoidable, given the torques implied by the observed azimuthal magnetic field, and the only question is whether a major fraction of this mass returns to the inner magnetosphere (forming a vast circulation/convection system) or whether most of it flows out of the magnetosphere (in the form of a planetary/magnetospheric wind).

The total magnetic torque implied by the observed B_ϕ or equivalently the angular momentum flux in the magnetic field can of course be calculated independently of any assumptions about the mass flux. The magnetic torque on the current sheet integrated from $r = 20$ to $100\ R_J$ over a $\Delta\phi = 90°$ segment is, from evaluation of Equation (11.139),

$$T_{cs} = 1.4 \times 10^{18}\ \text{Nm (if } C = 10)$$
$$= 8.1 \times 10^{17}\ \text{Nm (if } C = 15)$$

(11.140)

This is the angular momentum per unit time transferred from the magnetic field to the plasma within the current sheet. In addition, the magnetic field above and below the current sheet at $r = 100\ R_J$ implies a magnetic angular momentum flux, through a cylindrical surface, given by

$$T_B = 2r^3 \sin\theta\ \Delta\phi\ B_r\ B_\phi/\mu_0 = 2.4 \times 10^{18}\ \text{Nm}$$

(11.141)

where the latitudinal extent θ of the region with B_ϕ equal to the observed values has been set to 9.6° (the dipole tilt angle, roughly equal to the maximum magnetic latitude sampled by Pioneer 10 outbound) and $\Delta\phi = 90°$ again; this represents a torque exerted on plasma lying beyond $100\ R_J$. For comparison, the angular momentum flux associated with a mass flux S corotating at $r = 20\ R_J$ is

$$T_m = S\,\Omega_J\,r^2 = (S/10^{30}\ \text{amu/s})\ 6.0 \times 10^{17}\ \text{Nm}$$

(11.142)

which presumably equals the total magnetic torque at distances $r < 20\ R_J$ that acts to maintain the plasma in corotation.

All three quantities – T_{CS}, T_B, and T_m – also represent torques acting against the rotation of Jupiter and extracting energy from it, at a rate given by Ω_J times torque. The implied power is, for each of the three torques,

$$P_{CS} = 4.9 \times 10^{14}\ \text{W (if } C = 10)$$
$$= 2.9 \times 10^{14}\ \text{W (if } C = 15)$$

(11.143)

$$P_B \approx 8.5 \times 10^{14}\ \text{W}$$

$$P_m = (S/10^{30}\ \text{amu/s})\ 1.1 \times 10^{14}\ \text{W}$$

(For the estimates of P_{CS} and P_B, the torques given by (11.140) and (11.141) were multiplied by a factor of 2, to allow for an equal contribution from the dusk–midnight quadrant.) These numbers represent the total power supplied by loss of rotational energy.

Not all of it, however, may be available for "interesting" phenomena like particle energization or auroral radiation. Dessler [1980b] and Eviatar and Siscoe [1980] have pointed out that one half of P_m is required to supply the corotational kinetic energy of the outflowing plasma. Joule dissipation in the ionosphere also claims a share of the total power. By noting that the torque can be calculated from the Lorentz force on the ionospheric current j_θ' given by Equation (11.122) while the Joule dissipation is obtained from $(j_\theta')^2/\Sigma$, it is readily shown that a rough estimate for the fraction of the total power that must go into Joule dissipation is $(\Omega_J - \Omega)/\Omega_J$; hence if $\Omega \approx \omega \ll \Omega_J$ in the outer magnetosphere, much if not most of P_{CS} and P_B may be expended in Joule heating of the ionosphere.

The foregoing discussion has taken for granted the fundamentally rotational origin of azimuthal magnetic fields in the Jovian magnetosphere and hence the expected similarity of dawn and dusk sides, with the same sign of B_ϕ (for this reason the magnetic torques were doubled to obtain the power). If instead a solar-wind-aligned magnetospheric tail origin is assumed, so that the dawn and dusk sides are again similar but with reversed signs of B_ϕ, then the magnetic torques from the two sides cancel and P_{CS} and P_B vanish; P_m remains as the sole power supplied by the planetary rotation. The discussion given of magnetic field spiraling and azimuthal stress balance would remain, on the whole, valid but applicable to the dawnside only. On the duskside, a similar discussion but with reversed B_ϕ presents many problems that have hardly been considered yet by either the proponents or the opponents of the solar wind interpretation. Fundamentally, the spiraling and the azimuthal stress are opposed: on the dawnside, as we have seen, the spiraling requires that the plasma lag behind rigid corotation, while the azimuthal magnetic stress tends to bring it back to corotation, and a sufficiently large rate of mass through-flow is needed to prevent the magnetic stress from destroying the spiraling (and itself with it). A nonmagnetic stress opposed to corotation, required to set up the plasma lag in the first place, is provided by the inertia of outflowing plasma (and could in principle also be provided by solar wind drag, as originally suggested by Piddington). If now B_ϕ on the duskside is reversed in sign, the spiraling will require that the plasma rotate faster than rigid corotation, in opposition to the azimuthal magnetic stress, again demanding a sufficiently large rate of mass through-flow to maintain the configuration and a nonmagnetic stress in the direction of corotation to set it up. Such a nonmagnetic stress could be provided by the inertia of *inflowing* plasma (with sufficient initial angular momentum), as discussed recently in another context by Hill, Goertz, and Thomsen [1982]; solar wind drag would be effective for this purpose only if the flow speed of solar wind plasma near the magnetopause exceeds the speed of corotation.

5. Dipole tilt and solar wind effects

When the relatively small but nonzero tilt angle $\alpha \approx 9.6°$ between Jupiter's rotation axis and its magnetic dipole moment is no longer neglected, it becomes necessary to consider explicitly where the current sheet is located because now there is no plane perpendicular to both the rotation axis and the dipole moment. The momentum Equation (11.1) resolved into components normal and tangential to the current sheet at a given point may be written

$$\hat{n}\,[\rho(d\mathbf{V}/dt) \cdot \hat{n} + (\partial/\partial n)\,(P + B_p^2/2\mu_0)] = \hat{n}\,\hat{n} \cdot \mathbf{j}_t \times (\mathbf{B} - \mathbf{B}_p) \qquad (11.144)$$

$$\rho(d\mathbf{V}/dt)_t + \nabla_t P = \mathbf{j}_t \times \hat{n}\,(\mathbf{B} - \mathbf{B}_p) \cdot \hat{n} \qquad (11.145)$$

(for the definition of $\mathbf{B}_p$, see Equation (11.10) and associated discussion). Equation (11.144) is obtained from (11.11) and (11.21) (the symbol $\hat{z}$ has been replaced by $\hat{n}$ to emphasize that we are dealing with directions normal and tangential to the actual current sheet and not with an independently prescribed coordinate system); pressure anisotropy and the small normal component of $\mathbf{j}$ (given by $\nabla_t \times \mathbf{B}_t$) have been neglected. These equations imply that $\mathbf{B} - \mathbf{B}_p$ is perpendicular to the vector formed by their LH sides. We now define the center of the current sheet as the point where $P + B_p^2/2\mu_0$ reaches its maximum value and hence $(\partial/\partial n)(P + B_p^2/2\mu_0) = 0$. It follows that

$$(\mathbf{B} - \mathbf{B}_p) \cdot (\rho \, dV/dt + \nabla_t P) = 0 \tag{11.146}$$

at the center of the current sheet. An equivalent result involving quantities integrated across the thickness of the current sheet can be derived by noting that the integral of $(\partial/\partial n)(P + B_p^2/2\mu_0)$ yields zero because B_p^2 by definition has the same value on both sides of the current sheet and P is negligibly small outside it.

Equation (11.146) or its integral counterpart serves to determine the location of the current sheet. It states that the external magnetic field at the current sheet (i.e., the field minus the local planar field of the current sheet itself) must be perpendicular to the vector formed by the inertial stress plus the tangential part of the pressure gradient – the normal pressure gradient and the Lorentz force of the planar field balance each other. In the cold-plasma limit, with $\nabla_t P$ negligible in comparison to the inertial term, $\mathbf{B} - \mathbf{B}_p$ must be perpendicular to the plasma acceleration. In particular, with corotating plasma and $\mathbf{B} - \mathbf{B}_p$ approximated by the dipole field $\mathbf{B}_d$, the center of the current sheet lies in the surface defined by

$$\mathbf{B}_d \cdot (\mathbf{r} - \hat{\Omega}\,\hat{\Omega} \cdot \mathbf{r}) = 0 \tag{11.147}$$

that is, the locus of points where each field line reaches its maximum distance from the rotation axis, a result familiar in the literature [e.g., Gledhill, 1967; Hill, Dessler, and Michel, 1974]. This surface, known as the centrifugal symmetry surface or the centrifugal equator, is easily shown to be given by the equation (in cylindrical coordinates about the *rotation* axis)

$$z/r \cos \phi = -(4/3) \tan \alpha/\{1 + [1 + (8/9) \cos^2 \phi \tan^2 \alpha]^{1/2}\} \tag{11.148}$$

where ϕ is the longitude angle measured from the prime meridian plane containing $\hat{\Omega}$ and the dipole moment $\mathbf{M}$ (assumed tilted by an angle α toward the positive x axis). Because of the factor $\cos^2 \phi$ in the denominator, this surface is not exactly a plane (contrary to a statement by Hill, Dessler, and Michel) but the deviations from planarity amount to no more than a fractional change in z/x by a factor $\leq 1 + (2/9) \tan^2 \alpha = 1.006$, far smaller than the neglected effects of nondipolar field components or of gravitational acceleration. The surface lies between the rotational and the magnetic equator and forms an angle β with the magnetic equator given by $\tan(\beta - \alpha) = $ RH side of (11.148) with $\cos \phi = \pm 1$, which can be shown to yield

$$\tan \beta = (2/3) \tan \alpha/\{1 + [1 + (8/9) \tan^2 \alpha]^{1/2}\} \tag{11.149}$$

a formula given by Hill, Dessler, and Michel. With $\alpha = 9.6°$, Equation (11.149) gives $\beta = 3.2°$.

In the opposite limit of a hot plasma, with the pressure gradient dominant over the inertial term, $\mathbf{B} - \mathbf{B}_p$ must be perpendicular to $\nabla_t P$ and hence normal to the current sheet; if $\mathbf{B} - \mathbf{B}_p \approx \mathbf{B}_d$ the current sheet lies in the magnetic equator. In the general case, then, the center of the current sheet, defined as above by the maximum of

$P + B_p^2/2\mu_0$, is expected to lie between the magnetic equator and the centrifugal symmetry surface, approaching the one or the other as the ratio of thermal to corotational speed becomes large or small. At the same time, the maximum of the plasma pressure P alone always occurs at the centrifugal symmetry surface, for any value of the temperature, as shown in Section 11.2, Equation (11.47). It is thus obvious that, with a nonnegligible dipole tilt, the current sheet is necessarily asymmetric about its center and its width is large enough to encompass the centrifugal symmetry surface, reaching a value $h \gtrsim r \tan \beta$ in the hot-plasma limit.

Quantitative modeling of the current sheet location on the basis of Equation (11.146) with neither term neglected has so far proved feasible only in the prime meridian plane, where, by symmetry, $\nabla_r P$ is coplanar with $d\mathbf{V}/dt$. The following description is based on the approach of Goertz [1976b], with some corrections and generalizations. Let r, z be cylindrical coordinates with respect to the *dipole* axis, so that in the prime meridian

$$\hat{\Omega} = -\hat{r} \sin \alpha + \hat{z} \cos \alpha \tag{11.150}$$

Let $z = Z(r)$ be the location of the center of the current sheet; then the direction normal to the current sheet is

$$\hat{n} = [\hat{r}(dZ/dr) + \hat{z}] / [1 + (dZ/dr)^2]^{1/2} \tag{11.151}$$

It may be assumed that $\alpha \ll 1$, $|Z/r| \ll 1$, and $|dZ/dr| \ll 1$, with the calculation carried out to lowest order in these quantities. With this approximation, the components of the centripetal acceleration and of the dipole field normal and tangential to the current sheet, evaluated at its center, are

$$(d\mathbf{V}/dt) \cdot \hat{n} = -\Omega_J^2 r [\sin \alpha + dZ/dr] \tag{11.152}$$

$$(d\mathbf{V}/dt) \cdot \hat{t} = -\Omega_J^2 r$$

$$\mathbf{B}_d \cdot \hat{t} = -M/r^3$$

$$\mathbf{B}_d \cdot \hat{t} = (M/r^3) [3Z/r + dZ/dr]$$

(Note: Goertz [1976b] mistakenly omits the terms dZ/dr, resolving his vectors along $\hat{z}$, $\hat{r}$ rather than $\hat{n}$, $\hat{t}$.) Furthermore, we set

$$\mathbf{B} - \mathbf{B}_p = \mathbf{B}_d + b_n \hat{n} + b_t \hat{t} \tag{11.153}$$

where $\mathbf{b}$ represents any external fields other than the planetary dipole, and parameterize the pressure gradient by the expressions (11.26) and (11.100). Inserting these together with (11.152) and (11.153) into (11.146) and rearranging yields a differential equation for $Z(r)$

$$[1 + (\mu r^3 b_n/M)] \, dZ/dr + 3(1 + \mu) \, Z/r - [1 - (r^3 b_n/M)] \, \mu \sin \alpha \tag{11.154}$$
$$- (1 + \mu) \, r^3 b_t/M = 0$$

to be solved with the boundary condition $Z \to 0$ as $r \to 0$.

Assume first that $b_n = b_t = 0$, that is, only the dipole field is present in addition to the planar field of the current sheet itself. The solution of Equation (11.154) for any specified radial dependence of μ is given by

$$g(r) \, Z(r) = \sin \alpha \int dr \, g(r') \mu(r') \tag{11.155}$$

where

$$g(r) \equiv \exp \{3 \int [1 + \mu(r)] \, dr/r\} \tag{11.156}$$

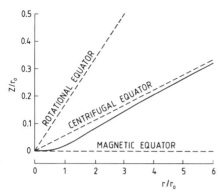

If μ is independent of r (the thermal speed increases in proportion to the corotational speed),

$$Z = r \sin \alpha / [3 + (4/\mu)] \tag{11.157}$$

a straight line whose slope varies from zero (current sheet in the magnetic equator) in the hot-plasma limit $\mu = 0$ to $(1/3) \sin \alpha$ (current sheet in the centrifugal symmetry surface, with $\sin \alpha \approx \tan \alpha$ within the approximations used) in the cold-plasma limit $\mu \to \infty$, as expected from the previous qualitative discussion; when $\mu = 1$, that is, the pressure gradient and the centrifugal force density are equal, $Z/r = (1/7) \sin \alpha$ and the current sheet is approximately half way between the two limits. If the thermal speed is independent of r so that μ varies as r^2, the solution is

$$(3/2)^{1/2} Z = (r_0/3) \sin \alpha \; \{s^4 - 2s^2 + 2[1 - \exp(-s^2)]\}/s^3 \tag{11.158}$$

where $s \equiv (3/2)^{1/2} r/r_0$ and r_0 is the distance at which $\mu = 1$; for small and large values of r,

$$Z = (r_0/6) \sin \alpha \; (r/r_0)^3 \; [1 - (3/2) (r/r_0)^2 + \ldots] \qquad r \ll r_0 \tag{11.159}$$

$$Z = (r/3) \sin \alpha \; [1 - (4/3) (r_0/r)^2 + \ldots] \qquad r \gg r_0$$

and the complete solution is shown in Figure 11.17. The current sheet in this case is appreciably curved, and the expected transition, from the magnetic equator at $r \ll r_0$, where the corotational speed is small compared to the thermal speed, toward the centrifugal equator at $r \gg r_0$ where the corotational speed is large, is clearly apparent.

When $\mathbf{b}$ is not neglected, the solution is a straight line if and only if the quantities μ, $r^3 b_n$ and $r^3 b_t$ are all independent of r. Goertz [1976b] set $r^3 b_n / M$ to the constant value implied by his field model with variable current sheet thickness (see Sec. 11.3.3) and given here by Equation (11.106); he further implicitly assumed that $b_t = 0$ and $\mu = $ constant. With these assumptions, the solution of Equation (11.154) is

$$Z = r \sin \alpha / [3 + (4/\mu) + 4a \, C/\mu \, (1 + a)] \tag{11.160}$$

or in terms of the parameter $k \equiv 1/\mu \, (1 + a)$ used by Goertz

$$Z = r \sin \alpha / [3 + 4k(1 + a + aC)] \tag{11.160'}$$

(the corresponding equation of Goertz has a coefficient $3k$ instead of $4k$, a result of his neglecting dZ/dr). As the hot-plasma limit is approached, Z now decreases toward the magnetic equator $Z = 0$ much more rapidly than in the case when $b_n = 0$, because of the added term proportional to aC in the denominator. However, the validity of this result may be questionable because it depends on the assumption $b_t = 0$ and even relatively small values of b_t (such as may be expected from the previously noted necessary asymmetry of the current sheet, whose center defined by the condition $\partial/\partial n$ $(P + B^2/2\mu_0) = 0$ need not coincide with the surface defined by either $\mathbf{B}_t = 0$ or $\mathbf{B}_p = 0$) may have a significant effect; as a simple example, if we set $b_t = \epsilon b_n \sin \alpha$ ($b_t = 0$ of course when $\alpha = 0$), Equation (11.160') is replaced by

$$Z = r \sin \alpha \, [1 + \epsilon kaC] / [3 + 4k(1 + a + aC)] \tag{11.161}$$

so that a term proportional to aC now appears in the numerator as well.

By definition, $\mathbf{b}$ includes magnetic fields from all sources other than the planetary dipole and the local sheet current. In general, the direction of $\mathbf{b}$ has no simple relation to the local normal to the current sheet; thus b_n and b_t may depend linearly on dZ/dr, with a consequent change in the coefficients of the differential Equation (11.154). It has been suggested [Hill, Dessler, and Michel, 1974; Smith et al., 1974a] that at large distances the current sheet should bend away from the centrifugal symmetry surface and become parallel to the rotational equator. Setting $\mathbf{b}$ equal to the magnetic field of the entire current sheet minus the field of the local sheet current (the "curvature" field as distinct from the "planar" field, in the terminology of Mead and Beard [1964]) should provide a quantitative description of such an effect, but no specific models have yet been developed.

The geometrical configuration of the current sheet is expected to be stationary in the rigidly corotating frame of reference where the planetary dipole is fixed (except in the outermost regions of the magnetosphere where local-time variations are significant). It has been tacitly assumed throughout the foregoing discussion that the plasma is effectively at rest in this corotating frame. When the plasma motion deviates significantly from rigid corotation, the plasma becomes subject to vertical accelerations as the tilted and possibly curved current sheet configuration rotates with respect to it. The term $(d\mathbf{V}/dt) \cdot \hat{\mathbf{n}}$ in Equation (11.144) is then no longer equal to the centripetal acceleration and in general cannot be specified a priori. It becomes more appropriate to view Equation (11.144) as a statement that an imbalance of applied normal stresses determines the instantaneous vertical acceleration of the current sheet and hence the time history of its location. In effect, the current sheet now becomes a wave, generated by a rotating tilted dipole and propagating radially outward. Discussions of the current sheet from this point of view (mostly qualitative or based on simple analogies to Alfvén waves) have been given by, among others, Northrop, Goertz, and Thomsen [1974], Prakash and Brice [1975], Eviatar and Ershkovich [1976], Kivelson et al. [1978], Carbary [1979], and Goertz [1981]. The various empirical representations of the current sheet and their relation to these as well as to alternate models are reviewed in Chapters 1, 5, and 10. A longitude-averaged effective radial propagation speed for the shape of the current sheet of 40–45 R_J/hr, independent of radial distance, provides a good fit to the Pioneer 10 outbound observations [Kivelson et al., 1978] as well as to Voyager 1 and 2 [Bridge et al., 1976b; Carbary, 1979; Goertz, 1981; Vasyliunas and Dessler, 1981].

The extent of solar-wind influence on the magnetic field configuration within the Jovian magnetosphere is a matter of considerable controversy, some aspects of which have already been discussed in Section 11.3.4 as well as in Chapters 1 and 10. Sufficiently far downstream of Jupiter, a magnetotail analogous to that observed at Earth should develop, with magnetic field lines and the current sheet oriented approximately parallel to the solar wind flow and with the current flowing in a θ-shaped pattern across the magnetotail from one side to another and returning along the magnetopause above and below. What is in dispute is not the existence of such a structure but the distance

Table 11.2. *Magnetotail parameters derived from magnetopause model*

		Jupiter	
	Earth	V1	V2
Distance to subsolar point, x_s	$10.9\ R_E$	$57\ R_J$	$68\ R_J$
$\cos^2 \psi$			
at $x = -x_s$	0.20	0.27	0.16
$x = -1.5x_s$	0.17	0.23	0.13
$-\nu_T \equiv \partial \log B_T / \partial \log \|x\|$			
at $x = -x_s$	0.20	0.18	0.23
$x = -1.5x_s$	0.25	0.21	0.26

where it begins. In a terrestrial-type magnetotail, the pressure of the magnetic field B_T in the lobes above and below the current sheet must equal the pressure P_{MS} of the solar wind plasma in the adjacent magnetosheath, which is given fairly adequately by the Newtonian approximation

$$B_T^2/\mu_0 = P_{MS} = 0.9\ P_{SW}\ (\cos^2 \psi + \epsilon) \tag{11.162}$$

where ψ is the angle between the upstream solar wind flow direction and the normal to the magnetopause, P_{SW} is the solar wind dynamic pressure, and $\epsilon \ll 1$ is the ratio of the thermal to the dynamic pressure in the upstream solar wind [see, e.g., Spreiter, Alksne, and Summers, 1968]. The magnetic field gradient with distance $-x$ down the magnetotail is then given by

$$-\nu_T \equiv \partial \log B_T/\partial \log \|x\| = (1/2)\ (\partial \cos^2 \psi/\partial \log \|x\|)/(\cos^2 \psi + \epsilon) \tag{11.163}$$

and can be calculated from $\cos^2 \psi$ given by a model for the magnetopause surface. Table 11.2 shows some values of $\cos^2 \psi$ and ν_T calculated from the parabolic models of the V 1 and V 2 magnetopause crossings given by Lepping, Burlaga, and Klein [1981] (also in Chap. 1, Sec. 1.4), together with values from a similar model for the terrestrial magnetopause given by Fairfield [1971]. (It is obvious that ϵ is small enough to be neglected in comparison to $\cos^2 \psi$.) The predicted values are very similar for both the terrestrial and the Jovian magnetospheres, but whereas the observed $\nu_T \approx 0.3$ at Earth [Behannon, 1968] is slightly larger than but close to the predicted values, at Jupiter the observed field magnitude gradient [Behannon, Burlaga, and Ness, 1981; also Chap. 1, Sec. 1.4] corresponds to values of ν_T ranging from 1.4 to 1.7, larger than the predicted values by a factor between 5 and 8; furthermore, at Jupiter the observed field magnitude depends on the radial distance r rather than the distance $-x$ along the solar wind flow direction. From the values of $\cos^2 \psi$ one expects the observed magnetic pressure to be a fraction 1/4 to 1/8 of the solar wind dynamic pressure, but the two are nearly comparable over the 40–80 R_J segments of the V 1 and V 2 outbound passes, as pointed out by Vasyliunas and Dessler [1981]. The observed field magnitude and especially its gradient thus seem to be in quantitative disagreement with the hypothesis that either Voyager 1 or Voyager 2 reached a fully developed terrestrial-type magnetotail.

Fig. 11.18. Magnetic field magnitude observed during the Pioneer 10 and Voyager 1 and 2 outbound passes [after Behannon, Burlaga, and Ness, 1981, with omission of P 10 values beyond the magnetopause at 98 R_J). Solid curve: B calculated from the Goertz et al. [1976] model for B_r together with the dipole field and with B_ϕ given by Equations (11.107) and (11.110). Dotted curve: $B \sim 1/r^{0.3}$.

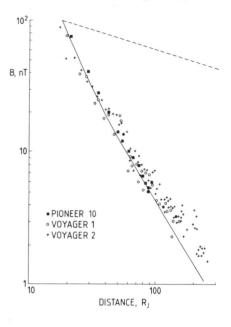

The observed field magnitude as a function of radial distance for the Pioneer 10 and Voyager 1 and 2 outbound pass is shown in Figure 11.18 (the same data as in Figure 1.30 but plotted on a common scale and with P 10 values beyond the magnetopause at 98 R_J omitted). To be noted is the good agreement of all three spacecraft over their common distance range $r < 98\ R_J$, indicating that the field magnitude in the midnight-to-dawn quadrant is nearly independent of local time (and thus in particular is a function of r and not of x, as already mentioned), but also a slight but definite decrease of slope for $r > 100\ R_J$, although not to a value anywhere near $\nu_r \approx 0.3$ yet. (This decrease together with the increasing radial range of the observations may account at least in part for the progressive flattening of the power-law fits from P 10 to V 1 to V 2 reported by Behannon, Burlaga, and Ness [1981].) The flattening is larger than that expected from $(B_\phi^2 + B_r^2)^{1/2}$ with B_ϕ given by Equation (11.110) and $B_r \sim 1/r^{1.7}$ and may very well signify that V 1 and V 2 were at least approaching the expected terrestrial-type magnetotail with its slowly varying $(B_r \sim 1/|x|^{0.3})$ field.

The fact that the solar wind flow direction is very nearly perpendicular to Jupiter's rotation axis may make it difficult to distinguish between rotational and solar-wind-related effects by purely geometrical arguments: a vector parallel to the solar wind flow is also parallel to the rotational equator. The most reliable geometrical indicator of a solar wind influence is a dependence on local time. The magnetosphere within the midnight-to-dawn sector, out to a distance of at least 100 R_J, appears to be remarkably independent of local time [Vasyliunas, 1982a]: the ratio B_ϕ/B_r, the magnetic field magnitude, and the effective radial propagation speed for the shape of the current sheet all show little or no significant difference between the outbound passes of Pioneer 10 and Voyager 1 and 2. By contrast, there are clear local-time variations observed within the dawn-to-noon sector; between P 10 outbound near dawn and P 11 outbound near noon, the azimuthal sheet current density as indicated by B_r decreases by about a factor of 3 [Jones, Thomas, and Melville, 1981]. Detailed models of such effects and their physical interpretation remain to be developed. As discussed already in Section 11.3.4, a crucial test for rotational vs. solar-wind effects should be provided by observations (if and when they are obtained) within the duskside of the magnetosphere.

11.4. Plasma flow models

1. General principles

The rest of this chapter presents a brief summary of models for the plasma flow in the magnetosphere of Jupiter. The brevity is imposed primarily by limitations of chapter length, although it is also true that both observation and theory are far less extensively developed for the flow than for the magnetic field and plasma distribution. Available observations of the flow are reviewed in Chapters 3 and 4.

The formulation of the basic physical principles governing the flow is contained in the theory of magnetosphere–ionosphere interaction, extensively developed and applied in the case of the terrestrial magnetosphere [see, e.g., Vasyliunas, 1970, 1972a, 1975c; Boström, 1974; Wolf, 1974, 1975, and references therein]. Within the magnetosphere, the electric field $\mathbf{E}$ and the plasma bulk flow velocity $\mathbf{V}$ are connected by the MHD relation, Equation (11.118) (see, e.g., Vasyliunas [1975a] for a discussion of its applicability and limitations). Within the ionosphere, the height-integrated current density $\mathbf{j}'$ is given by

$$\mathbf{j}' = \Sigma(\mathbf{E} + \mathbf{V}_n \times \mathbf{B}) \tag{11.164}$$

where Σ is the height-integrated conductivity and $\mathbf{V}_n$ the velocity of the neutral atmosphere at ionospheric altitudes. Because of the thinness of the ionosphere compared to $1\ R_J$, the horizontal electric field is essentially independent of height up to and including the lowest regions of the magnetosphere just above the ionosphere; hence, comparing Equations (11.118) and (11.164) we have

$$\mathbf{j}' = \Sigma(\mathbf{V}_n - \mathbf{V}) \times \mathbf{B} \tag{11.165}$$

where $\mathbf{V}$ is the plasma velocity just above the ionosphere. Nonzero horizontal ionospheric currents exist if and only if the plasma velocity just above the ionosphere differs from the velocity of the neutral atmosphere within the ionosphere. A nonzero divergence of $\mathbf{j}'$ gives rise to Birkeland (magnetic-field-aligned) currents between the ionosphere and the magnetosphere; these currents then close through $\mathbf{j}$ perpendicular to $\mathbf{B}$ in the equatorial regions, subject to the momentum Equation (11.1) and the stress balance considerations discussed in Section 11.2.

Well-known consequences of Equation (11.118) are that the magnetic flux through any loop moving with the plasma remains constant and that plasma elements initially on a magnetic field line remain on a field line [see, e.g., Stern, 1966; Vasyliunas, 1972b]. Hence, for a given magnetic field configuration, the plasma velocity $\mathbf{V}$ just above the ionosphere, at the foot of a magnetic field line, determines $\mathbf{V}$ perpendicular to $\mathbf{B}$ everywhere along the field line. The component of $\mathbf{V}$ parallel to $\mathbf{B}$ is governed by continuity and stress balance along $\mathbf{B}$ and is less easily determined. The large size of the Jovian magnetosphere implies that the timescale to establish equilibrium along $\mathbf{B}$ may be comparatively long, and some flow observations have been interpreted as parallel flows related to this effect [Belcher and McNutt, 1980; Chap. 3].

2. Corotation

If the ionospheric conductivity Σ is very large, Equation (11.165) implies that $\mathbf{V} = \mathbf{V}_n$, that is, the magnetospheric plasma moves with the neutral atmosphere. This is the motion that is meant when one speaks of the corotation of magnetospheric plasma, since $\mathbf{V}_n \approx \Omega_J \times \mathbf{r}$ to an accuracy adequate for most purposes. Note that: (1) corotation refers to motion of the plasma and not of the magnetic field pattern as such – dipole tilt or another longitudinal asymmetry of the field is neither necessary nor sufficient for corotation; (2) the plasma moves with the upper neutral atmosphere and not with the planetary rotation except insofar as it is imposed on the atmosphere. (Strictly speaking, the corotation frequency is that of System I or II, depending on latitude, and not of System III, although the differences between them are far smaller than the actual differences between **V** and **V**$_n$.)

If the plasma flow in the magnetosphere is to be at least approximately azimuthal with the angular velocity of planetary rotation, four conditions must be met: (1) sufficiently effective vertical transport of momentum in the atmosphere to maintain corotation of the neutral atmosphere at ionospheric altitudes (planet-atmosphere coupling); (2) high ionospheric conductivity (atmosphere-ionosphere coupling); (3) validity of the MHD approximation (11.118) (ionosphere-magnetosphere coupling); and (4) adequate magnetic stresses to balance the centripetal acceleration of the plasma.

(1) That condition (1) is satisfied is generally taken for granted by magnetospheric physicists, although Kennel and Coroniti [1975] have expressed some doubts.

(2) The value of Σ needed for $\mathbf{V} \approx \mathbf{V}_n$ increases in direct proportion to mechanical stresses in the magnetosphere (or at its boundary) that tend to oppose or modify the corotational motion of the plasma, since these stresses must balance the $\mathbf{j} \times \mathbf{B}$ force associated with the magnetospheric closure of currents proportional to $\Sigma(\mathbf{V}_n - \mathbf{V})$ [see Eq. (11.165)]. Specific examples are discussed in Sections 11.4.4, 11.4.5, and 11.4.6.

(3) As a rule, the MHD approximation is valid for structures with length scales much larger than all microscopic lengths (gyroradii, etc.) of the plasma [see, e.g., Vasyliunas, 1975a]. Within the Earth's magnetosphere, the MHD approximation provides a good description of nearly all global and intermediate-scale features [see, e.g., Vasyliunas, 1976], a conclusion that holds a fortiori for the Jovian magnetosphere with its even larger size.

These three conditions suffice to make the components of the plasma flow perpendicular to **B** correspond to corotational motion. It is, however, still possible for **V** in the magnetosphere to deviate significantly from azimuthal motion as a result of flow parallel to **B** or of time variations of **B**. To preclude this, condition (4) is necessary. Its specific requirements have been discussed in detail in Sections 11.2.2, 11.3.1, and 11.3.2.

The defining characteristic of corotation is that plasma at the foot of each magnetic flux tube circles about the rotation axis (not the dipole axis); the perpendicular flow throughout the rest of the magnetosphere is then determined by tracing the magnetic field lines attached to the moving plasma. If there are significant distortions of the magnetic field lines that depend on local time (or, more generally, are not time-independent when viewed from a corotating frame of reference), the resulting motion in the equatorial plane may deviate considerably from simple azimuthal flow. In particular, if field lines are extended on the nightside and compressed by the solar wind on the dayside, as expected for the outermost regions of the magnetosphere, corotational motion has not only azimuthal but also radial components in the equatorial plane, inward on the dawnside and outward on the duskside. Such a flow is crucial for the acceleration mechanism proposed by Goertz [1978] (see also Chap. 10, Sec. 6).

3. Radial outflow and planetary/magnetospheric wind

There is a limit to the distance at which a flux tube of given plasma content can be maintained in rigid corotation by an inward magnetic stress. It is often stated that at the limiting distance the Alfvén speed equals the corotation speed, but in fact that gives only an order-of-magnitude estimate; the actual value of the limiting distance is set by the tangential stress balance (Secs. 11.2 and 11.3). For the Gleeson–Axford and the Hill–Carbary models discussed in Section 11.3.1, the limiting distance is given by the (virtually identical) Equations (11.68) and (11.74). (It may be noted that, with the assumptions of plasma sheet thickness $h \approx r$ and mass density $\rho \sim 1/r^4$, appropriate to an ionospheric plasma source in a dipolar field assumed by Hill, Dessler, and Michel [1974] and Michel and Sturrock [1974], these equations predict a limiting distance larger by only a factor of 1.2 than the distance given by the simple $V_A = \Omega r$ estimate.) Corotation out to a considerably larger distance is seemingly allowed by the magnetic field model of Goertz et al. [1976], but this is possible only because either the flux tube content or the corotation speed implied by the model decreases with increasing distance, as shown in Section 11.3.2.

If, then, plasma on a magnetic flux tube is transported outward beyond the limiting distance and neither the flux tube content nor the corotation frequency of the ionosphere are allowed to decrease, the plasma will begin to move in a path approaching a straight line, unless deflected by stresses other than the now too small magnetic tension. On the dayside, the constraint of remaining within the magnetosphere may force the plasma to continue moving in a curved path similar to corotation, with solar wind pressure on the magnetopause providing the necessary inward stress, but on the nightside the plasma may move nearly freely outward into the low-pressure, low-density magnetotail; the magnetic field lines, attached to the corotating ionosphere at one end and to the outward-flowing plasma at the other, become highly extended and eventually break open. This is the concept of rotationally driven plasma radial outflow described by Hill, Dessler, and Michel [1974] and Michel and Sturrock [1974], named by them the planetary wind. Figure 11.19 shows a sketch of the expected flow pattern in the equatorial plane and the associated topological changes of the magnetic field. The extended magnetic field lines become detached through the formation of singular lines of $\times$ and $\bigcirc$ type (see Vasyliunas [1976] and references therein, for a discussion of analogous topological changes in the terrestrial magnetotail). At the $\times$ line, a magnetic-merging process [e.g., Vasyliunas, 1975a] is expected to occur, with consequent flow pattern as shown in Figure 11.19, plasma heating, and particle acceleration [Carbary, Hill, and Dessler, 1976; Chap. 10, Sec. 6]. The precise expected location of the $\times$ and $\bigcirc$ lines is at present unknown (although an alignment of the $\bigcirc$ line with the plasma flow is implied by the results of Vasyliunas [1980]), and the locations shown in Figure 11.19 are a free sketch.

There is a strong qualitative similarity between the flow pattern of Figure 11.19, expected on the basis of the planetary wind model, and the flow pattern inferred from Voyager charged particle observations [Krimigis et al., 1979b, their Fig. 9; Chap. 4, Fig. 4.7]. The radial outflow observed by Krimigis et al. and named by them the magnetospheric wind may thus be plausibly identified with the planetary wind, specifically with the expected strong tailward flow beyond the $\times$ line near the dawn magnetopause.

The planetary-wind model as such makes no predictions about either the source or the temperature of the plasma – these must be provided as inputs, and the assumption of a cold ionospheric plasma in Hill, Dessler, and Michel [1974] and Michel and Sturrock [1974] is not part of the model but simply the pre-Voyager prevailing view. The limiting distance, beyond which radial outflow should begin, given by Equation (11.74) with the use of Voyager determinations of the plasma mass density ρ and the plasma sheet thickness h, is 24 R_J (the value of about 10 R_J given in Chap. 4 results from using ρ but ignoring h and tacitly assuming that the plasma sheet extends over the entire flux tube), but this estimate presupposes rigid corotation of the ionosphere at all latitudes, which is almost certainly not the case (see Secs. 11.3.4 and 11.4.4); allowance

EQUATORIAL PLANE   MERIDIAN SURFACE

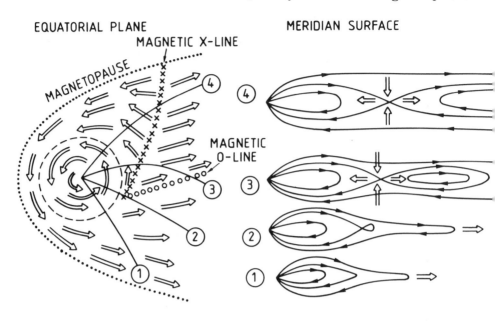

Fig. 11.19. Qualitative sketch of plasma flow in the equatorial plane (left) and of the associated magnetic field and plasma flow in a sequence of meridian surfaces (right) expected from the planetary wind model. The locations of the × and ○ lines are speculative but have *not* been adjusted to take account of any observations (they are identical to the pre-Voyager locations drawn by Vasyliunas in 1978 for a conference presentation but not published; thus this figure and Fig. 9 of Krimigis et al. [1979b] are independent).

of partial corotation will, of course, increase the limiting distance. Furthermore, it is apparent from Figure 11.19 that strong radial outflow in the predawn sector is expected only beyond the × line, as a result of superimposed flows from magnetic merging.

If the solar wind pressure is reduced and the magnetopause recedes to great distances, it is possible for radial outflow and the associated formation of magnetic loops to occur at all local times. Models with such an axially symmetric planetary wind as well as possible transitions between symmetric and nightside-only outflow in response to solar wind pressure changes have been considered by Kennel and Coroniti [1977a] and Coroniti and Kennel [1977] but have not received much observational support.

The distinction between the rotationally driven stellar wind model of Mestel [1968] and the planetary wind model is not always clearly made [Kennel and Coroniti, 1977a; Goertz, 1979]. The Mestel model describes radial outflow along open magnetic field lines, with a plasma source at the inner end (i.e., the ionosphere) and vanishing pressure at large distances; by contrast, the planetary wind model has field lines that for the most part are either closed or form detached loops, radial outflow perpendicular to the magnetic field in the equatorial plane, and plasma that is concentrated toward the equatorial regions. The Mestel model has a critical Alfvén radius r_A, with near-corotation of plasma for $r \ll r_A$ and super-Alfvénic outflow for $r \gg r_A$, but this is a transition within a single (open) flux tube, not a limiting distance in the equatorial plane; furthermore, r_A is defined by the condition $B_r^2/\mu_0 \rho = V_r^2$, where V_r is the radial outflow and not the corotational speed, and thus differs from both the precise and the order-of-magnitude expressions for the limiting distance of the planetary wind model. It is possible that a Mestel-type outflow may exist, on the open field lines within the Jovian magnetosphere well above the current sheet, as a rotationally driven species of polar wind [cf. Fig. 2 of Hill, Dessler, and Michel, 1974] distinct from the planetary/magnetospheric wind.

4. Partial corotation

With outward transport of plasma from a source within the inner magnetosphere, corotation requires a magnetic torque exerted by radial currents which close through the ionosphere (see Sec. 11.3.4). Hill [1979] pointed out that the difference between the plasma and the neutral atmosphere velocities required to drive these currents [see Eq. (11.165)] leads to a reduction of the azimuthal plasma flow speed in the ionosphere and therefore also in the magnetosphere, that is, a partial corotation of the plasma, with angular frequency $\omega < \Omega_J$. The theory of partial corotation combines two aspects already considered in detail separately in Section 11.3.4: the relations of B_ϕ to the ionospheric current, leading to Equation (11.129), on the one hand, and to the torque on the plasma, Equation (11.134), on the other. Elimination of B_ϕ between these two equations yields a differential equation for ω as a function of r, which was derived by Hill without explicit consideration of B_ϕ. With the meridional magnetic field approximated by a dipole, Hill showed that ω decreases gradually, falling appreciably below Ω_J at a characteristic distance r_c given by

$$(r_c/R_J)^4 = \pi \Sigma B_J^2 R_J^2 / S \tag{11.166}$$

where S is the net outward mass flux; for $r \gg r_c$,

$$\omega \approx \pi^{1/2} \Omega_J (r_c/r)^2 \tag{11.167}$$

corresponding to negligible torque on the plasma. Numerically,

$$r_c = [(\Sigma/1\ \text{mho}) (10^{30}\ \text{amu/s})/S]^{1/4} 36\ R_J \tag{11.168}$$

From a comparison with Voyager 1 observations, Hill [1980] estimated $r_c \approx 20\ R_J$. (The comparison requires some care because ω given by the theory is actually the radial-flux-weighted average $\langle \omega \rangle$ defined in Equation (11.135).) The model of B_ϕ discussed by Vasyliunas [1982b] and in Section 11.3.4 identifies the radial currents implied by the observed B_ϕ with those responsible for partial corotation in Hill's model.

The characteristic distance r_c, beyond which there is no rigid corotation because condition (2) of Section 11.4.2 (atmosphere–ionosphere coupling) is not satisfied, is entirely distinct from the limiting distance r_0, given in Equation (11.74) and discussed in Section 11.4.3, beyond which condition (4) (centrifugal stress balance) is not satisfied. [Equation (11.74) is applicable only if $r_0 \leq r_c$, because rigid corotation of the ionosphere was assumed in deriving it.] The ratio r_c/r_0 is given by

$$(r_c/r_0)^4 = \pi \mu_0 \Sigma \rho_s h_s \Omega_J^2 L_s^3 / 4S \tag{11.169}$$

which can be written as

$$(r_c/r_0)^4 = \mu_0 \Sigma (\Omega_J r_s)^2 / 8 \langle V_r \rangle \tag{11.170}$$

where $\langle V_r \rangle$ is the average speed of radial transport out of the Io torus defined by

$$S = 2\pi r_s \rho_s h_s \langle V_r \rangle. \tag{11.171}$$

Empirical estimates give a ratio r_c/r_0 close to one. Whether this is merely a remarkable coincidence or whether the outward transport process adjusts $\langle V_r \rangle$ to give this value is an interesting unanswered question.

The angular frequency of corotation, whether rigid or partial, of the ionospheric and magnetospheric plasmas has no direct connection with the periodic modulation of Jovian radio emissions (see Chapters 7 and 9). The modulation implies that the spatial configuration of the emission sources has some azimuthal asymmetry, and it is the effective rotation of this asymmetric spatial pattern that determines the modulation period. For example, the modulation is at Jupiter's rotation period if the asymmetry is associated with dipole tilt or magnetic anomaly effects and at Io's orbital period if it is associated with the Io flux tube, regardless of what the plasma motions are. If the Jovian magnetosphere were axially symmetric in all respects, there could be no periodic modulation of the radio emissions but the physics of corotation and its limits would not be changed in any substantial way.

Effects analogous to partial corotation, associated with azimuthal stresses produced by inward and outward motions that result from day–night asymmetries or from compressions and expansions of the magnetosphere, have been considered by Nishida and Watanabe [1981].

5. Solar-wind-driven magnetospheric convection

Plasma flow analogous to magnetospheric convection at Earth, induced by the flow of the solar wind past the magnetosphere, is discussed in Chapter 10, Section 2 and shown to be of little importance at Jupiter.

6. Corotating magnetospheric convection

One of the main tenets of the magnetic anomaly model is an enhancement of the plasma content within the Io torus over the longitude range of the active sector (see Chap. 10 for a detailed discussion). The corotation of the plasma implies a corresponding enhancement of the centrifugal stress and hence of the balancing $\mathbf{j} \times \mathbf{B}$ force. Because the enhanced azimuthal current is confined to the active sector, continuity requires it to couple to Birkeland currents at the edges of the active sector, with current closure through the ionosphere [see, e.g., Dessler, 1980a]. Vasyliunas [1978] pointed out that the electric field (or, equivalently, the difference between the plasma and the neutral atmosphere velocities) required to drive the closing ionospheric currents implies a two-cell pattern of circulating plasma flow, with outward flow over the active sector and inward flow at other longitudes. The flow pattern is stationary when viewed from a frame of reference rotating with the planet, rather like an atmospheric cyclonic feature (the ionospheric projection of the flow has the appearance of four huge eddies, two in each hemisphere, with their centers at approximately 90° longitude to each side of the active sector); it thus differs from solar-wind-driven magnetospheric convection, which is also basically a two-cell circulating flow but fixed in local time rather than in System III longitude. The mean speed of the flow or equivalently the characteristic circulation frequency scales as the ratio of mass content asymmetry in the Io torus to the ionospheric conductivity. Limited quantitative models have been developed by Hill, Dessler, and Maher [1981] and by Summers and Siscoe [1982], and possible applications to various observed phenomena are discussed by Hill, Dessler, and Maher [1981] and by Dessler, Sandel, and Atreya [1981] (see also Chap. 10, Section 7).

7. Radial diffusion

There are several mechanisms by which plasma circulatory motions may arise that are somewhat similar to magnetospheric convection (either corotating or solar-wind driven) but are transient and variable in time and/or longitude rather than stationary; they may also have much smaller scales, with many small cells instead of two large ones. One mechanism is the centrifugal instability of the Io torus, which is in effect an unsteady succession of corotating magnetospheric convection patterns, produced not by a fixed asymmetry but by random enhancements of the plasma content (which grow as a consequence of the flow). Another is the coupling of ionospheric plasma flow to circulatory motions of the neutral atmosphere, with $\mathbf{j}' = 0$ and $\mathbf{V} = \mathbf{V}_n$ in Equation (11.165). Radial diffusion is the transport of plasma and energetic particles that results from the averaging over such stochastic circulatory motions. Historically, atmospheric winds were the first to be suggested as the mechanism of radial diffusion in the Jovian magnetosphere [Brice and McDonough, 1973]. For a detailed discussion of radial diffusion driven by centrifugal effects of the Io torus, see Siscoe and Summers [1981] and references therein. The mathematical description of the transport by radial diffusion, largely independent of the specific mechanism, is discussed in Chapter 5. Its application to the plasma distribution is considered by Siscoe [1978b]; for application to radiation belt particles, see, for example, Schulz [1979] and Chapter 5.

11.5. Conclusion

The dominant structure of the Jovian magnetosphere, the extended near-equatorial sheet of plasma and electric current, is fairly well understood at least qualitatively in physical terms of stress balance between the plasma and the magnetic field, but a number of aspects need further quantitative development and others remain with some basic physical questions unanswered. There is for the most part an order-of-magnitude agreement between the observed plasma parameters and those implied by stress balance with the observed magnetic field, but few detailed comparisons have been carried out as yet. The plasma density implied by stress balance depends rather sensitively on the assumed field model; purely empirical fits to the observed magnetic field without taking into account any self-consistency requirements are liable to yield models that imply some unphysical properties for the plasma. There is continuing controversy on the relative importance of rotational vs. solar wind effects in the outermost regions of the magnetosphere, in particular for the interpretation of local time variations or lack of them; the range of viable alternatives would be greatly narrowed by even limited observations from the so far unexplored duskside. Both direct observations and inferences from field-plasma self-consistency indicate that the azimuthal flow of the plasma decreases well below rigid corotation at large distances at least on the nightside, an effect that is well understood theoretically and can be quantitatively related to the observed azimuthal magnetic field. Understanding of plasma flow other than partial corotation remains fragmentary, and the nature of the outward radial transport mechanism – convection or diffusion, and the ratio of net outflow to circulating mass flux – has not been firmly established.

APPENDIX: COMMENTS ON SOME MAGNETIC FIELD MODELS

(1) The model of Vickers [1978] is intended as a generalization of the Gleeson–Axford model to include a radial pressure gradient. In the radial stress balance equation

$$-\rho\omega^2 r + \partial P/\partial r = B_z(\partial B_r/\partial z)/\mu_0 \tag{11.172}$$

[equivalent to (11.15) with isotropic pressure, centripetal acceleration, and $B_p = B_r$], one may set (in our notation) $\rho = P/w^2$ and $P = (B_0^2 - B_r^2)/2\mu_0$ given by pressure balance ($B_0 \equiv B_p^*$ is the field outside the current sheet), and solve for B_z to obtain

$$B_z = [(\partial/\partial r) (B_0^2 - B_r^2) - (\omega^2 r/w^2) (B_0^2 - B_r^2)]/2 (\partial B_r/\partial z) \tag{11.173}$$

If we assume that B_r has the form

$$B_r = B_0(r) \tanh (z/h) \tag{11.174}$$

then the RH side of (11.173) is independent of z and

$$B_z = [dB_0/dr - \omega^2 rB_0/2w] h \tag{11.175}$$

provided *either* h is independent of r or the radial pressure gradient term (and hence dB_0/dr) is neglected entirely. Gleeson and Axford [1976] choose the second alternative; then h is in general a function of r, and Equation (11.175) with dB_0/dr omitted becomes simply an expression for the current sheet thickness in terms of field and thermal quantities, identical in fact to Equation (11.99) with $\mu \gg 1$. Vickers, however, retains the pressure gradient term and argues that B_r given by (11.174) with constant h is a general solution because B_z given by (11.175) gives $\partial B_z/\partial z = 0$ in agreement with $\nabla \cdot \mathbf{B} = 0$ for a thin sheet; with both B_z and B_0 related to the current density $j_\phi' (r)$, an integrodifferential equation for j_ϕ' is obtained which has a unique solution, in contrast to the freedom of Gleeson and Axford to choose j_ϕ' arbitrarily.

The apparent uniqueness of the solution is simply a consequence of the assumed constant thickness h and otherwise has no direct connection with inclusion of the radial pressure gradient. It would be possible to assume $h = $ constant and use Equation (11.175) without dB_0/dr to obtain a unique j_ϕ' for the Gleeson–Axford model. In fact, there is no basis for assuming a strictly constant h, and the argument given by Vickers is incorrect: from $\nabla \cdot \mathbf{B} = 0$ and B_r given by (11.174), it is readily shown that

$$B_z(r, z) = B_z(r, 0) - h(dB_0/dr + B_0/r) \log \operatorname{sech} (z/h) \tag{11.176}$$

so that Equation (11.175) is *not* satisfied unless $dB_0/dr = 0$ and $h/r \to 0$ because one side depends on z and the other does not. Equation (11.174) thus does not give a solution except in the Gleeson–Axford limit of negligible radial gradient and $h \ll r$ an unspecified function.

(2) The model of Barish and Smith [1975] is intended to illustrate how the stretched-out magnetic field lines can nevertheless close within the compressed dayside magnetosphere and to represent the Pioneer 10 inbound magnetic field observations. Actually, it fulfills neither purpose. As to the comparison with observations, the model has nearly radial field lines in the outer magnetosphere which reverse across the current sheet; hence, it predicts a strong 10-hr modulation to be observed by Pioneer 10 as the current sheet sweeps up and down past the spacecraft. In fact, the striking feature of the Pioneer 10 observations between 95 R_J and 55 R_J inbound is that the magnetic field was more nearly normal to the equatorial plane and showed little evidence of 10-hr modulation [Smith et al., 1976, especially their Fig. 5].

As to the intended closure of field lines, the magnetic flux function of the model in the equatorial plane may be written as $f = M\alpha$, where M is the Jovian dipole moment and α is given by

$$\alpha^5(\alpha - \alpha_d) - \epsilon \log (R/r) = 0 \tag{11.177}$$

where $\alpha_d = 1/r$ describes a dipole field, $\epsilon = 1.623 \times 10^{-10} \, R_J^{-6}$, and $R = 100 \, R_J$ (see also Chap. 1, Sec. 1.3). It is easily shown that for $r \leq R$ Equation (11.177) has two real solutions α_1 and α_2, with $\alpha_1 = 0$ and $\alpha_2 = \alpha_d$ at $r = R$. From Equation (11.177) it is then obvious that $\alpha_1 \leq 0$ and $\alpha_2 \geq \alpha_d$ for all $r \leq R$, and it can further be shown that $d\alpha_1/dr > 0$, whereas $d\alpha_d/dr < 0$; thus α_1 must be rejected because it represents a magnetic field with B_z opposite in sign throughout the magnetosphere to the field of the assumed planetary dipole (note that α is intended to represent not just the perturbation but the total field, including the dipole). The other solution does represent a suitable field with extended field lines, but the magnetic flux is not contained within $r = R$, the intended distance of the magnetopause, where $\alpha_2 = \alpha_d$ and thus the unclosed flux is equal to the dipole flux. It can be shown that the two solution branches come together at a distance $r = R_2 = 100.04 \, R_J$, where $\alpha_1 = \alpha_2 = (5/6)\alpha_d$ and $|d\alpha/dr| \to \infty$; for $r > R_2$, Equation (11.177) has no real solutions. A fraction 1/6 of the unclosed flux at $r = R$ thus closes between $r = 100.00$ and $r = 100.04$ and the rest closes through a singular layer of zero thickness and infinite field. It may be noted that the model gives $|B_z| = 1.42 \, nT$ at $r = 90 \, R_J$, $2.15 \, nT$ at $r = 95 \, R_J$ and $68.6 \, nT$ at $r = 100 \, R_J$.

(3) The azimuthal current density in the model of Goertz et al. [1976] (obtained by taking the curl of the full model field as a function of r and z) has two components: one varying as $\mathrm{sech}^2 (z/D)$ and another extending over a wide latitude range and depending only on z/r when $|z| \gg D$. The first evidently represents the actual current sheet and the second is usually ignored; it may be an artifact resulting from specifying both B_z and B_r at the current sheet (assumption of $\nabla \times \mathbf{B} = 0$ outside the current sheet would give a boundary value problem whose solution determines B_z given B_r or vice versa). In computing the pressure from vertical stress balance in the model, there is thus some ambiguity as to where P should be set to zero, since $j_\phi \neq 0$ at all finite distances. In Section 11.3.2, only the field associated with the $\mathrm{sech}^2 z/D$ component was included in the pressure balance. Goertz et al. [1979], on the other hand, set $P = 0$ at the last closed field line, and Goldstein [1977] treats the pressure in the equatorial plane as a boundary condition to be specified by an independent assumption.

ACKNOWLEDGMENTS

I am grateful to A. J. Dessler, T. W. Hill, W. I. Axford, C. K. Goertz, H. Goldstein, G. Burgess, and J. E. P. Connerney for useful comments on the manuscript and to J. E. P. Connerney for providing a table of calculated field values from the Connerney-Acuña-Ness model.

12

MICROSCOPIC PLASMA PROCESSES IN THE JOVIAN MAGNETOSPHERE

Richard Mansergh Thorne

Plasma waves observed near Jupiter exhibit characteristics similar to waves in the terrestrial magnetosphere and a quantitative analysis suggests a universality of the generation mechanisms. Electromagnetic whistler-mode waves are enhanced within the high-density plasma torus surrounding the orbit of Io and evidence for electron stable trapping is found in the inner torus. The observed intensity of low-frequency hiss is sufficient to scatter resonant ($\gtrsim$ 100 keV) electrons and cause continuous precipitation loss to the atmosphere at a rate comparable to 5% of the limit imposed by strong pitch-angle diffusion. The energy flux into the Jovian atmosphere is estimated to be approximately 3 mW/m^2 over a broad invariant latitude range $65° \lesssim \Lambda \lesssim 70°$ mapping from the Io torus. Total power dissipation may therefore approach 10^{13} W. But because the energetic electron deposition occurs deep in the Jovian atmosphere, little of the concomitant auroral emission should be detectable by Voyager. The intermittent bursts of electromagnetic chorus and the equatorially confined electrostatic $(n + 1/2) f_{c,e}$ waves can at times scatter lower energy (keV) electrons on strong diffusion. The net energy deposition, however, is typically less than 0.5 mW/m^2 and the total power dissipation over the entire auroral zone should be $\lesssim 10^{12}$ W. Energetic ions exhibit evidence for rapid precipitation loss throughout the plasma torus. Although the waves responsible for such scattering have not yet been identified, ion precipitation loss near the strong diffusion limit apparently offers the only viable mechanism to excite the observed auroral emissions. Over the reported energy range ($E_p \gtrsim$ 500 keV) the anticipated power dissipation can exceed 2×10^{13} W, but an extrapolation down to ion energies comparable to 100 keV could increase this to 10^{14} W, which is comparable to the auroral requirements.

It is unlikely that the solar wind plays a significant role in the energetics of the Jovian magnetosphere. The most viable power supply for the intense auroral dissipation is the rotational energy of Jupiter. If the mechanism for tapping this energy is mass loading in the torus, the observed auroral emissions require an injection rate of heavy thermal ions from Io in excess of 10^{29}/s. The anticipated precipitation of energetic ions into the Jovian atmosphere can also produce a significant flux of suprathermal ($>$ 20 eV) secondary electrons. If the precipitation is dominated by energetic ($\gtrsim$ 100 keV) heavy ions, the energy flux of escaping secondary electrons could provide a heat source ($\gtrsim 10^{12}$ W) to the torus comparable to the observed radiative cooling and the ambipolar outflow could inject cold hydrogen ions into the magnetosphere at a rate comparable to heavy ion injection from Io.

12.1. Introduction

Despite the acknowledged success of the macroscopic (MHD) description of planetary magnetospheres, there are many fundamental plasma phenomena that can be understood only through a detailed treatment of the microscopic plasma processes. Resonant wave-particle interactions, during which energy can be efficiently transferred between the energetic particle population and plasma waves, have been demonstrated to be of paramount importance in the Earth's magnetosphere [e.g., Andronov and Trakhtengerts, 1964; Kennel and Petschek, 1966; Cornwall, 1966; Cornwall, Coroniti, and Thorne, 1970; Lyons, Thorne, and Kennel, 1972; Thorne, 1976] and for a number of years it has been conjectured that similar processes should occur at Jupiter [Thorne and Coroniti, 1972; Kennel, 1972; Coroniti, 1974; Coroniti, Kennel, and Thorne, 1974; Baker and Goertz, 1976; Barbosa and Coroniti, 1976; Fillius et al., 1976; Scarf, 1976; Scarf and Sanders, 1976; Sentman and Goertz, 1978]. Although the adiabatic invariants provide a useful kinematical framework, a strictly adiabatic description of the particle dynamics is inappropriate because the particle distribution function can be radically modified during the interaction with plasma waves. When the waves are reasonably intense, the concomitant wave-particle scattering can be far more effective than collisional processes in causing the injection, acceleration, nonadiabatic transport and ultimate loss of particles from the radiation belts. The in situ particle and field data obtained from Voyager 1 and 2 permit the first quantitative assessment to be made of the importance of such wave-particle interactions in the Jovian magnetosphere.

A detailed review of the plasma waves observed in the Jovian magnetosphere [Scarf, Gurnett, and Kurth, 1979] is given in Chapter 8 together with a brief description of potential mechanisms for wave generation. The remarkable similarity to waves observed in the terrestrial magnetosphere suggests that the knowledge gained from the extensive study of wave-particle processes in the Earth's magnetosphere can be directly applied to Jupiter. It is therefore appropriate to first briefly review the concepts and implications of quasilinear scattering of trapped particles during resonant interactions with various plasma waves with emphasis placed on understanding the mechanisms for wave origin and establishing the rate of particle removal due to pitch-angle scattering loss to the atmosphere. The finite geometrical size of the atmospheric loss-cone places an absolute upper limit on the rate of particle removal. This is attained under strong pitch-angle diffusion [e.g., Kennel, 1969] when particles are scattered across the loss-cone within a bounce period. Estimates are presented for the amplitude of plasma waves required for strong diffusion and these are subsequently compared with Voyager wave observations.

Although the Voyager instrumentation was unable to resolve particle fluxes within the loss-cone, estimates of the precipitation flux can be obtained from the theoretically predicted scattering lifetimes and observations of the trapped particle populations. The particle lifetimes presented here admittedly reflect the author's personal bias and in certain instances have a weak observational basis. Nevertheless, a direct comparison of the theoretically predicted precipitation flux with that required to excite the intense H and H_2 auroral emissions observed in the Jovian polar regions [Broadfoot et al., 1979; Sandel et al., 1979] provides a stringent test on our current ability to adequately model the microscopic plasma processes.

The rapid particle loss from the Jovian magnetosphere places stringent limits on the transport mechanisms for energetic plasma and the ultimate energy source available for intense auroral dissipation. One can also anticipate a strong coupling, both in terms of electrodynamics and particle sources, between the Jovian upper atmosphere and the magnetosphere. Our discussion of such processes, although speculative, is intended as an initial attempt to provide an holistic description of the Jovian plasma physics.

12.2. Wave–particle interactions in the terrestrial magnetosphere

Under adiabatic conditions, the kinematics of energetic radiation belt particles can be uniquely described by the magnetic field topology and the conservation of three adiabatic invariants (μ, J, Φ) associated with basic periodicities in the particle motion [e.g., Northrop, 1963]. Violation of the invariants, however, can occur when the charged particles are subject to a force field fluctuating on a timescale comparable to or faster than that of the respective periodic motion. For the first two invariants this requires a fluctuating field with characteristic frequencies near $\omega \gtrsim \omega_c$; τ_b^{-1} respectively, where ω_c is the particle cyclotron frequency and τ_b the bounce time for travel between magnetic mirror points. Violation of either invariant leads to pitch-angle scattering and an ultimate loss of trapped particles to the atmosphere. In contrast, violation of the third invariant, which involves fluctuations on a timescale less than the azimuthal drift time τ_d results in nonadiabatic radial transport, which can act either as a source or sink for radiation belt particles.

In the Earth's magnetosphere, fluctuating plasma waves occur spanning the entire range of frequencies required for violation of each invariant, and numerous estimates have been made of the effect of specific wave modes on the trapped particle population. The waves can be grouped into two broad categories, electromagnetic or electrostatic, depending on whether or not there is a measurable fluctuating magnetic field. A detailed review of the principal wave modes observed in the Earth's magnetosphere has recently been provided by Shawhan [1979] and the properties of similar waves detected by Voyager in the Jovian magnetosphere are discussed by Gurnett and Scarf in Chapter 8 of this volume.

In general, the temporal evolution of the radiation-belt environment can be described in terms of a phase-averaged distribution function $f_o(\mu,J,\Phi,t)$ in which nonadiabatic processes are treated as a diffusion with respect to the three adiabatic invariants. The basic form of the multi-dimensional diffusion equation can be written as [e.g., Schulz and Lanzerotti, 1974]

$$\frac{\partial f_o}{\partial t} = \sum_{i,j} \frac{\partial}{\partial J_i} \left(D_{ij} \frac{\partial f_o}{\partial J_j} \right) + S - L \tag{12.1}$$

where J_i are the action integrals associated with the adiabatic periodicities of the motion, D_{ij} are diffusion coefficients and S and L are source and loss terms that are introduced to account for processes that inject or remove particles. Although, in general, all three invariants can be subject to simultaneous violation, it is conceptually convenient to distinguish between velocity space diffusion (in which μ, J are not conserved) and radial diffusion (which violates Φ). Because of the rapid timescales involved in μ, J violation, it is often appropriate to assume that Φ is ostensibly constant and simply treat the diffusion in particle pitch-angle and energy. This approach is taken here to evaluate the role of wave-particle interactions in the scattering loss for trapped particles. The complementary approach of explicitly treating violation of Φ through the radial diffusion equation

$$\frac{\partial f_o}{\partial t} = L^2 \frac{\partial}{\partial L} \frac{D_{LL}}{L^2} \frac{\partial f_o}{\partial L} + S - L \tag{12.2}$$

in which μ or J violation is parameterized by an effective loss term $L \approx f_o/\tau_L$, has been discussed in Chapters 10 and 11 of this volume. We shall also utilize (12.2) in Section 12.3 to establish whether radial diffusion can act as a viable source for energetic particles capable of exciting high-frequency instabilities in the Jovian magnetosphere.

Quasilinear scattering by plasma waves

The scattering of geomagnetically trapped particles by plasma waves is dominated by resonant interactions at harmonics of the particle bounce and cyclotron frequency. However, because emphasis here is placed on processes that lead to a precipitation of particles into the atmosphere, bounce resonance is not specifically treated because it is usually important only for particles mirroring near the geomagnetic equator [e.g., Roberts and Schulz, 1968]. Scattering into the atmospheric loss cone is predominantly controlled by plasma waves that are Doppler shifted in frequency to some integral multiple ℓ of the particle cyclotron frequency

$$\omega - k_\| v_\| = \frac{\ell \omega_c}{\gamma} \tag{12.3}$$

where $k_\|$, $v_\|$ are components of the wave propagation vector and particle velocity along the ambient magnetic field direction and $\gamma = (1 - v^2/c^2)^{-1/2}$. During the interaction, resonant particles are constrained to diffuse in velocity space along surfaces [Kennel and Englemann, 1966] such that

$$(v_\| - \omega/k_\|)^2 + v_\perp^2 = \text{constant} \tag{12.4}$$

The Landau ($\ell = 0$) resonance thus simply involves energy transfer between the wave and the particle velocity parallel to $\bar{B}$. For the cyclotron ($\ell \neq 0$) resonances, diffusion occurs predominantly in pitch-angle ($v \approx$ constant) for high-energy resonant particles but significant energy diffusion also occurs when the particle velocity becomes comparable to the wave phase speed. However, near the edge of the loss-cone ($v_\perp/v_\| \ll 1$), cyclotron resonant interactions result in essentially pure pitch-angle diffusion for all resonant particles.

The rate of diffusion in velocity space depends sensitively on the resonant wavemode and the distribution of the fluctuating power spectral density $P(\omega, \bar{k})$ as a function of frequency and wave normal. For the idealized case of pure pitch-angle diffusion, (12.1) can be rewritten in the form

$$\frac{\partial f_o}{\partial t} = \frac{1}{\sin \alpha} \frac{\partial}{\partial \alpha} \sin \alpha \sum_\ell \left(D_{\alpha\alpha}^{(\ell)} \frac{\partial f_o}{\partial \alpha} \right) + S - L \tag{12.5}$$

where $\alpha = \arctan(v_\perp/v_\|)$ is the instantaneous pitch-angle and $D_{\alpha\alpha}^{(\ell)} = \langle \Delta\alpha \rangle^2/2\Delta t$ is the pitch-angle diffusion coefficient associated with each cyclotron harmonic resonance for a prescribed band of waves. Landau resonant diffusion, although strictly confined to diffusion in $v_\|$, can also be included within the above formalism by defining an effective rate of pitch-angle scattering [Roberts and Schulz, 1968; Lyons, Thorne, and Kennel, 1972].

Equation (12.5) prescribes the effect of diffusion at one point in space. To evaluate the time-averaged evolution of the particle distribution function it is first necessary to identify all particles by their equatorial pitch-angle α_o and then average (12.5) over the bounce trajectory between magnetic mirror points to obtain [Lyons, Thorne, and Kennel, 1972]

$$\frac{\partial f_o}{\partial t} = \frac{1}{T(\alpha_o) \sin 2\alpha_o} \frac{\partial}{\partial \alpha_o} \mathscr{D}_\alpha(\alpha_o) T(\alpha_o) \sin 2\alpha_o \frac{\partial f_o}{\partial \alpha_o} \tag{12.6}$$

$$+ \langle S \rangle - \langle L \rangle$$

where $\mathscr{D}_\alpha(\alpha_o)$ is the net bounce averaged effective diffusion coefficient, $\langle S \rangle$ and $\langle L \rangle$ are average source and loss functions, and $T(\alpha_o) \approx 1.38 - 0.64 \sin^{3/4} \alpha_o$ represents the dependence of the particle bounce period on equatorial pitch-angle [Davidson, 1976].

Under conditions when the particle source is negligible ($\langle S \rangle \to 0$) and losses are due to collisions with the atmosphere, $L(\alpha_o) \sim f_o(\alpha_o)/\tau_L(\alpha_o)$, one can solve (12.6) simultaneously [Spjeldvik and Thorne, 1975] for the precipitation lifetime

$$\frac{1}{\tau_P(E,L)} = \left\langle \frac{1}{\tau_L(E,L,\alpha_o)} \right\rangle \tag{12.7}$$

$$= \frac{\displaystyle\int_o^{\pi/2} \frac{1}{\tau_L(E,L,\alpha_o)} \, f_o(E,L,\alpha_o) T(\alpha_o) \sin 2\alpha_o \, d\alpha_o}{\displaystyle\int_o^{\pi/2} f_o(E,L,\alpha_o) T(\alpha_o) \sin 2\alpha_o \, d\alpha_o}$$

and the equilibrium pitch-angle distribution

$$f_o(E,L,\alpha_o) = f_o\left(E,L, \frac{\pi}{2} \right) - \int_o^{\pi/2} \frac{d\alpha_o'}{\mathcal{D}_\alpha(E,L,\alpha_o')T(\alpha_o')\sin 2\alpha_o'} \tag{12.8}$$

$$\times \int_{\alpha_o'}^{\pi/2} \left(\frac{1}{\tau_P(E,L)} - \frac{1}{\tau_L(E,L,\alpha_o'')} \right) f_o(E,L,\alpha_o'')T(\alpha_o'') \sin 2\alpha_o'' \, d\alpha_o''$$

Alternatively, if the source strength is prescribed, a steady state solution ($\partial f_o/\partial t \to 0$) to (12.6) can be obtained by equating the rate of injection to the diffusion flux into the loss cone [e.g., Kennel and Petschek, 1966].

In the limit of weak pitch-angle diffusion, $\mathcal{D}_\alpha(\alpha_L) << \alpha_L^2/\tau_b$, when particles are unable to diffuse significantly across the loss-cone α_L within a bounce time τ_b the precipitation lifetime $\tau_P \approx 1/\mathcal{D}_\alpha(\alpha_L)$ and the particle flux exhibits a steep gradient at the edge of the loss-cone. At the opposite extreme under strong diffusion (when $\mathcal{D}_\alpha(\alpha_L) >> \alpha_L^2/\tau_b$) particles can readily diffuse across the dimensions of the loss-cone within a bounce period; the particle flux approaches isotropy, and the bounce averaged precipitation time [Lyons, 1973] approaches an asymptotic limiting value, controlled only by the particle bounce time and the geometric size of the loss cone $\alpha_L^2 = B/B_A \approx 1/2\,L^3$ (for $L >> 1$),

$$\tau_{SD} = \tau_b/3\alpha_L^2 \approx 3.6\,R_P L^4/v \tag{12.9}$$

where R_P is the planetary radius and B_A is the magnetic field strength at the top of the atmosphere.

Wave amplitudes required for strong diffusion

The strong diffusion lifetime (12.9) places an absolute upper limit on the rate of removal of radiation belt particles; this is valid irrespective of the intensity of the scattering waves. To obtain an estimate of the minimum fluctuating wave intensity required for strong diffusion scattering, one must first evaluate the effective rate of pitch-angle scattering for any particular resonant wave mode. In general, this involves a complex integration over wave frequency and $\bar{k}$-space together with a summation over each harmonic resonance [e.g., Lyons, Thorne, and Kennel, 1971]. For the case of electromagnetic waves this can be approximated by

$$\mathcal{D}_\alpha \approx \frac{\omega_c}{\gamma} \left(\frac{B'}{B} \right)^2 \tag{12.10}$$

where B' is the total wide-band amplitude of the resonant fluctuating magnetic field [Kennel and Petschek, 1966]. For electrostatic waves polarized with $k_\perp/k_\parallel \gg 1$, the effective pitch-angle scattering coefficient is

$$\mathcal{D}_\alpha \approx \frac{\omega_c}{\gamma} \left(\frac{\mathcal{E}'}{\beta B} \right)^2 \tag{12.11}$$

where $\beta = v/c$ and $\mathcal{E}'$ is the wide-band fluctuating electric field amplitude. In either case, the required minimum wave amplitudes can then be obtained by equating the bounce averaged diffusion coefficient to τ_{SD}^{-1}. This yields

$$B'_{SD} \approx (\gamma^2 - 1)^{1/4} (cB^2/3.6 \, \delta \, R_P L^4 \omega_c)^{1/2} \tag{12.12}$$

and

$$\mathcal{E}'_{SD} \approx \beta \, B'_{SD} \tag{12.13}$$

where δ represents the fraction of the particle bounce orbit spent in resonance with the waves. For a dipole field with $\delta \approx 0.2$, one obtains numerical values

$$B'_{SD,e} \approx 10^2 \, (\gamma_e^2 - 1)^{1/4} L^{-7/2} \, nT \tag{12.14}$$

for relativistic electrons and

$$B'_{SD,p} \approx 3 \times 10^2 \, E_p^{1/4} \, (keV) \, L^{-7/2} \, nT \tag{12.15}$$

for nonrelativistic protons with kinetic energy E_p. The corresponding strong diffusion electrostatic wave amplitudes are

$$\mathcal{E}'_{SD,e} \approx 3 \times 10^4 (\gamma_e^2 - 1)^{3/4} \, \gamma_e^{-1} \, L^{-7/2} \, mV/m \tag{12.16}$$

for relativistic electrons and

$$\mathcal{E}_{SD,p} \approx 10^2 \, E_p^{3/4} \, (keV) \, L^{-7/2} \, mV/m \tag{12.17}$$

for nonrelativistic protons.

A comparison of the wave amplitudes required to scatter electrons and protons on strong diffusion at $L = 6$ is shown in Figure 12.1. Fortuitously, because the ratio between the surface magnetic field strength and the planetary radius is essentially the same for the Earth and Jupiter, this graph applies equally well to either magnetosphere. It is immediately clear that electrostatic waves have a lower threshold for causing strong diffusion scattering particularly in the case of nonrelativistic ions. For all nonrelativistic particles $B'_{SD} \sim m^{1/4} E^{1/4}$ whereas $\mathcal{E}'_{SD} \sim m^{-1/4} E^{1/4}$. For heavy ion scattering, the values plotted in Figure 12.1 would have to be scaled up or down by the fourth root of the ion atomic mass. As the particle energy increases, the conditions for strong diffusion scattering become more severe and the required power spectral density in electrostatic waves approaches that for electromagnetic waves. In either case, the required amplitudes scale as $L^{-7/2}$ so that strong-diffusion scattering is more likely to occur at large radial distance from the planet.

Possible waves for scattering and resonant particle energies

Within the terrestrial magnetosphere, four distinct classes of plasma waves have been identified that are capable of cyclotron resonant interaction with trapped radiation belt particles. These are the electromagnetic electron cyclotron (sometimes referred to as

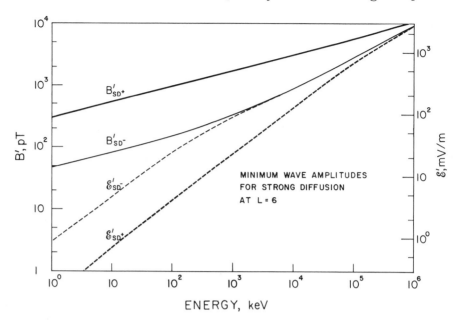

Fig. 12.1. Minimum fluctuating electric or magnetic wave amplitudes required to scatter protons (+) or electrons (−) or strong diffusion at $L = 6$. At other locations the required amplitudes scale as $L^{-7/2}$. The curves apply equally well to Earth or Jupiter.

whistler-mode) and ion-cyclotron waves, which are restricted to frequencies below the respective cyclotron frequency, and the electrostatic ion and electron cyclotron waves which typically occur in bands centered near odd half harmonics of the appropriate cyclotron frequencies $(n + 1/2)\omega_c$. For each class of wave, the observed spectral properties and typical resonant electron and ion energies [obtained from (12.3)] are presented in Table 12.1. Resonant energies with electromagnetic waves are controlled by the magnetic energy per particle $E_M = B^2/8\pi n$, while those for electrostatic waves are controlled by the thermal energy E_{th} of the plasma. The assumption has been made here that resonant ions are nonrelativistic, $E_{P,res} \ll E_{o,p} = m_p c^2$. Resonant electrons, however, must in general be treated as relativistic with $E_{e,res} \rightarrow E_{o,e} (\gamma_{e,res}^2 - 1)/2$ in the nonrelativistic limit as $\gamma_{e,res} \rightarrow 1$.

Electromagnetic ion cyclotron waves are generally observed in the dusk sector or dayside region of the Earth's magnetosphere [Bossen, McPherron, and Russell, 1976]. This is consistent with theoretical expectations which favor most rapid instability in regions of enhanced plasma density [e.g., Cornwall, Coroniti, and Thorne, 1970; Thorne, 1972; Cuperman, Gomberoff, and Stemlieb, 1975] where the resonant ion flux should maximize. In such preferred regions, E_M is typically 1–10 keV. Electrons resonant with such waves are therefore relativistic ($\gamma_e \gg 1$), while the resonant ion energies span the thermal population of the magnetosphere ring current. Because the observed broadband wave amplitudes are several nanotesla, the resonant particles can be subject to strong diffusion scattering (Fig. 12.1) leading to intense precipitation of relativistic electrons [Thorne and Kennel, 1971; Thorne, 1974] and ring current ions [Cornwall, Coroniti, and Thorne, 1970, 1971; Thorne, 1972].

Electromagnetic whistler-mode waves on the other hand occur throughout the terrestrial magnetosphere. A broadband hiss emission predominates at low L within the high density plasmasphere. This emission is thought to be the principal source of

Table 12.1. *Wave–particle resonant interactions*

	Electromagnetic		Electrostatic	
	Ion cyclotron	Whistler-mode	Ion cyclotron	Electron cyclotron
Waves class	Pc1, IPDP micropulsations	Sferics or whistlers plasmaspheric hiss outer zone chorus auroral hiss	Broad-band electrostatic noise	$(n + 1/2)\omega_{c,e}$ harmonic bands ω_{UH} upper hybrid waves
Spectral properties	$\left(\dfrac{k_\| c}{\omega}\right)^2 \approx \dfrac{\omega_p^2}{\omega_{c,e}\omega_{c,p}}\;\dfrac{1}{(1 - \omega/\omega_{c,p})}$	$\left(\dfrac{k_\| c}{\omega}\right)^2 \approx \dfrac{\omega_p^2}{\omega\,\omega_{c,e}}\;\dfrac{1}{(1 - \omega/\omega_{c,e})}$	$k_\| \sim \dfrac{\omega_{c,p}}{v_{th,p}}$; $\omega \approx \left(n + \dfrac{1}{2}\right)\omega_{c,p}$	$k_\| \sim \dfrac{\omega_{c,e}}{v_{th,e}}$; $\omega \approx \left(n + \dfrac{1}{2}\right)\omega_{c,e}$
Resonant energies $(\gamma_{res,e}^2 - 1)\,E_{0,e}/2$	$E_M\left(\dfrac{m_p}{m_e}\right)\left(\dfrac{\omega_{c,p}}{\omega}\right)^2\left(1 - \dfrac{\omega}{\omega_{c,p}}\right)$	$E_M\left(\dfrac{\omega_{c,e}}{\omega}\right)(1 - \omega/\omega_{c,e})^3$	$E_{th,p}\left(\dfrac{m_p}{m_e}\right)\left(\dfrac{\ell}{k_\| \rho_p}\right)^2$	$E_{th,e}\left(\dfrac{n - \ell + \frac{1}{2}}{k_\| \rho_e}\right)^2$
$E_{res,p} = \dfrac{m_p v_{res,p}^2}{2}$	$E_M\left(\dfrac{\omega_{c,p}}{\omega}\right)^2\left(1 - \dfrac{\omega}{\omega_{c,p}}\right)^3$	$E_M\left(\dfrac{m_p}{m_e}\right)\left(\dfrac{\omega}{\omega_{c,e}}\right)\left(1 - \dfrac{\omega}{\omega_{c,e}}\right)$	$E_{th,p}\left(\dfrac{n - \ell + \frac{1}{2}}{k_\| \rho_p}\right)^2$	$\left(\dfrac{m_p}{m_e}\right)E_{th,e}\left(\dfrac{n + \frac{1}{2}}{k_\| \rho_e}\right)^2$

weak-diffusion scattering [Lyons, Thorne, and Kennel, 1972] for energetic electrons
($\gtrsim$ 50 keV) leading to the formation of the slot between the inner and outer radiation
belts [Lyons and Thorne, 1973]. Intense bursts of "chorus" emissions occur in the outer
zone during substorm periods [Tsurutani and Smith, 1974, 1977]. Observed wave
amplitudes are sufficient to drive resonant electrons onto strong diffusion, leading to
intense precipitation into the atmosphere. For either emission, the resonant ions are
typically in the energy range above an MeV and the rate of pitch-angle scattering is
relatively weak.

Both classes of observed electrostatic waves are polarized with $k_\perp \gg k_\parallel$. For such
oblique waves, higher-order cyclotron resonances become important, thus broadening
the resonant energy range. This is in sharp contrast to the case of nearly parallel propagating electromagnetic waves where first order resonance predominates. Terrestrial
electron electrostatic waves occur predominantly near the equatorial plane [Kennel
et al., 1970; Christiansen et al., 1978; Kurth et al., 1979b] in the outer radiation zone.
They are considered to be important in scattering plasma sheet electrons leading to
diffuse auroral precipitation [Lyons, 1974]. Analogous waves at frequencies just
above the proton cyclotron frequency are characteristically present on auroral field
lines [Gurnett and Frank, 1977]. Precise frequency resolution into the anticipated
harmonic band structure is not possible using satellite-borne instruments because of
the large Doppler shift caused by the satellite motion. The observed waves therefore
appear as broadband electrostatic noise, but the measured amplitudes appear to be sufficient to account for the observed diffuse auroral ion precipitation [Ashour-Abdalla
and Thorne, 1977, 1978].

12.3. Plasma instability and quasilinear scattering in the Jovian magnetosphere

The spectral properties of plasma waves observed in the Jovian magnetosphere
(Chapter 8 of this volume), together with simultaneous information on the trapped particle population (Chapters 4 and 5) can be utilized to both identify the potential mechanisms for wave instability and evaluate the rate of scattering loss of the resonant
particles. The linear growth or damping rate of any particular wave mode can be
expressed in terms of velocity-space gradients in the resonant particle distribution function [e.g., Kennel and Wong, 1967; Melrose, 1968]. The interested reader is referred to
Chapter 9 (Sec. 9.2) for a summary of the basic formalism. Because the Landau ($\ell = 0$)
resonant growth rate is directly proportional to $\partial f_o/\partial v_\parallel$, it will in general lead to wave
attenuation. Landau amplification, however, can occur if the plasma carries a net cur-
rent (i.e., an electron drift relative to the ions) or if there is a high-energy particle beam
moving relative to the thermal population. Cyclotron ($\ell \neq 0$) resonant growth, on the
other hand, can occur when the plasma exhibits a pitch-angle anisotropy (e.g., due to
the mirror magnetic field geometry) or positive gradients in perpendicular velocity
space $\partial f_o/\partial v_\perp > 0$. In our subsequent discussion, the observed distribution of various
plasma waves is compared with theoretical expectations, and the wave intensities are
employed to compute the rate of velocity space scattering.

However, before analyzing the specifics of any particular wave-particle resonance, it
is useful to review certain general properties of energetic particle injection and removal
from the Jovian radiation belts. In the absence of a local acceleration source, a steady-
state solution to the radial diffusion equation (12.2) requires a balance between the
cross-L transport and precipitation loss:

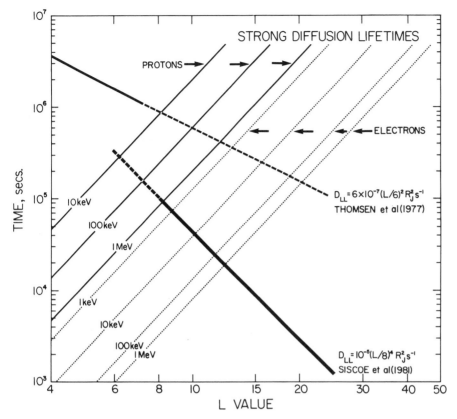

Fig. 12.2. Strong-diffusion lifetimes for energetic electrons and protons as a function of radial location in the Jovian magnetosphere. Two estimates of the time-scale for radial diffusion, $\tau_D \sim D_{LL}^{-1}$, are shown for comparison. The results are strictly valid only over the solid regions of the curves. The empirically determined results of Thomsen et al. [1977] represent an upper limit for diffusion driven by ionospheric winds. The results of Siscoe et al. [1981] are deduced from the distribution of heavy ions in the Io torus and have been interpreted in terms of centrifugally driven interchange instability.

$$L^2 \frac{\partial}{\partial L} \left[\frac{D_{LL}}{L^2} \frac{\partial f_o}{\partial L} \right] = \frac{f_o}{\tau_L} \qquad (12.18)$$

Representative (empirically determined) values for the time-scale of radial diffusion $\tau_D \sim D_{LL}^{-1}$ caused either by centrifugal driven interchange instability of heavy ions in the outer region of the Io plasma torus [Froidevaux, 1980; Siscoe and Summers, 1981; Siscoe et al., 1981] or by neutral winds in the Jovian ionosphere [e.g., Brice and McDonough, 1973; Thomsen et al., 1977] are illustrated in Figure 12.2. [One should note that, in contrast to Jupiter, both interchange instability and neutral atmospheric winds provide relatively ineffective radial diffusion in the terrestrial magnetosphere.] Also shown for comparison are the minimum lifetimes (12.9) of electrons and protons subject to strong diffusion pitch-angle scattering. In the outer magnetosphere, where the rate of radial diffusion exceeds the rate of precipitation loss ($D_{LL} \tau_L \gg 1$), rapid cross-L transport leads to a uniform plasma distribution function ($f_o(L) \approx$ constant). For a given magnetic moment, $\mu = p_\perp^2/2mB$, the differential flux $j(E) = p^2 f$ thus scales in direct proportion to the ambient field strength

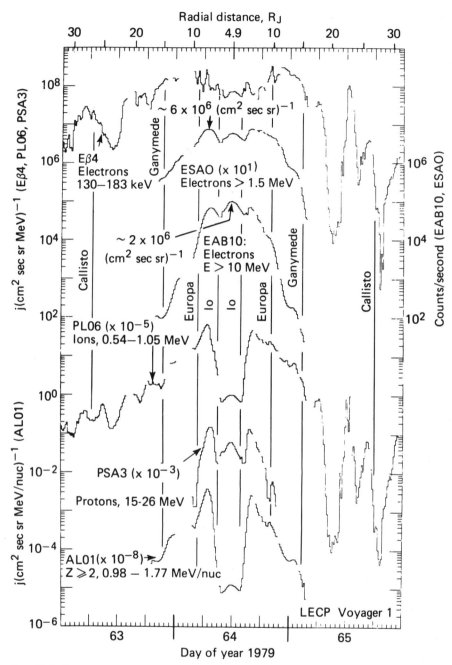

Fig. 12.3. Energetic charged particles measured in the inner Jovian magnetosphere by the LECP instrument on Voyager 1 [Krimigis et al., 1979a].

$j(\mu = \text{constant}) \sim L^{-3}$. The corresponding integral flux $J(> E)$ scales as $L^{-4.5}$ or L^{-6} depending on whether or not the particle energies are relativistic [e.g., Thorne and Coroniti, 1972]. However, once the rate of removal exceeds the rate of injection (namely at lower L), the particle distribution function develops a steep radial gradient to maintain a balance in (12.18) between injection and loss. As long as other transport processes (e.g., convection; Chapters 10 and 11) are relatively weak, one expects the energetic particle flux to peak near the region where $D_{LL} \tau_L \approx 1$.

The energetic particle fluxes observed on Voyager 1 [Krimigis et al., 1979a] exhibit clear examples of an inward diffusion induced precipitation boundary in the inner magnetosphere (Fig. 12.3). There is little evidence for pronounced longitudinal asymmetry that would be expected if large scale convection were the dominant transport mechanism. A comparison of the relevant time scales in Figure 12.2 suggests that ions with $E_p \sim 1$ MeV must be scattered near the strong diffusion rate to account for the pronounced flux depletion inside $L \approx 8$. The modest drop in energetic (> 1.5 MeV) electron flux in approximately the same location (upper panels of Fig. 12.3) requires only weak pitch-angle scattering. Mechanisms responsible for the ion and electron loss are discussed in more detail below. We simply note here that the combined ion and electron observations support the concept that inward diffusion driven in response to interchange instability of heavy torus ions provides the dominant source for energetic particles in the middle Jovian magnetosphere exterior to the orbit of Io. This conclusion does not rule out the possible importance of large scale convection as a viable transport mechanism for the thermal plasma [e.g., Hill, Dessler, and Maher, 1981] but, in contrast to the Earth's magnetosphere, convection does not appear to be a major source for the Jovian ring current plasma.

Inward diffusion also tends to enhance the pitch-angle anisotropy of the energetic particles (because the equatorial perpendicular and parallel momenta scale as $p_\perp \sim L^{-3/2}$ and $p_\parallel \sim L^{-1}$ for constant μ and J) and thus also provides a source of free energy capable of exciting plasma instabilities. Even when pitch-angle scattering becomes strong, inward diffusion can maintain a reasonable level of anisotropy in the region where $D_{LL} \tau_L \gg 1$. The necessary conditions for instability of various plasma waves are discussed below, and a comparison is made with Voyager observations.

Whistler-mode waves

Figure 12.4 contains an overview of plasma waves observed on Voyager 1 near closest approach to Jupiter [Scarf et al., 1979a]. Several distinct classes of waves can be identified; we concentrate first on waves below the electron cyclotron frequency. As already discussed in Chapter 8 of this volume, definitive identification of the wave mode is impossible because the Voyager plasma wave instrument only measures fluctuating electric fields. Nevertheless, the spectral characteristics of the waves observed well above the proton gyrofrequency but below the electron cyclotron frequency suggest that the dominant emissions are analogous to terrestrial electromagnetic whistler-mode hiss and chorus. Auroral hiss [Gurnett, Kurth, and Scarf, 1979b] and discrete lightning generated whistlers [Gurnett et al., 1979b] are not considered here because their amplitudes are relatively weak (see Chapter 8), and the waves consequently have little effect on the trapped particles.

Although the spectrum channel data provide a useful overview of the wave spectral properties, the distinction between chorus and hiss is more apparent in the wide-band analog data. Examples of these high-resolution data have been shown in Figures 14 and 15 of Chapter 8. The most persistent and intense wave emission in the inner Jovian magnetosphere is a band of hiss with a peak power spectral intensity near a few hundred hertz, namely below the lower-hybrid resonance frequency. As illustrated in Figure 12.4 the intensity of this broadband emission is dramatically enhanced within the high-density torus surrounding the orbit of Io. Chorus-like emissions, on the other hand, occur intermittently. The wave frequency is typically centered near $\omega_{c,e}/2$ and the waves are confined to the magnetic equatorial region.

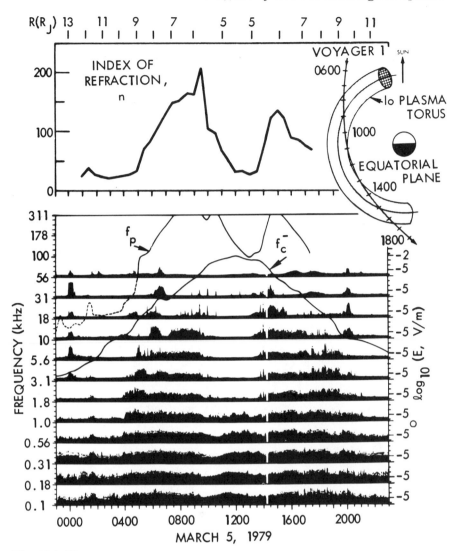

Fig. 12.4. Voyager 1 plasma wave observations near closest approach [Scarf et al., 1979a]. Preliminary values for the plasma frequency and electron gyrofrequency are shown for comparison. The index of refraction (top panel) can be employed to infer the fluctuating magnetic field intensity, which is dominant for whistler-mode signals.

A potential candidate for the generation of either electromagnetic emission is cyclotron resonant instability, driven by pitch-angle anisotropy in the electron distribution. For relativistic electrons, computation of the wave growth (or damping) rate involves a complicated integration along the resonance contour (12.3) which is hyperbolic in momentum space [Liehmohn, 1967; Schulz and Vampola, 1975; Barbosa and Coroniti, 1976]. A much simpler and more physically understandable result is obtained in the nonrelativistic limit [e.g., Kennel and Petschek, 1966]; the temporal growth rate for parallel propagating whistler-mode waves can then be expressed analytically as

$$\gamma_{cyc} \approx \pi\omega_{c.e}\left(1 - \frac{\omega}{\omega_{c.e}}\right)^2 \eta_e(v_{res})\,(A_e(v_{res}) - A_c) \tag{12.19}$$

where

$$\eta_e(v_{res}) = 2\pi\, v_{res} \int_0^\infty v_\perp\, f_{o,e}(v_\perp, v_\parallel = v_{res})\, dv_\perp \tag{12.20}$$

is the fractional number of resonant electrons,

$$A_e(v_{res}) = \left. \frac{\displaystyle\int_0^\infty v_\perp \tan\alpha\, \frac{\partial f_{o,e}}{\partial\alpha}\, dv_\perp}{\displaystyle 2\int_0^\infty v_\perp\, f_{o,e}\, dv_\perp} \right|_{v_\parallel = v_{res}} \tag{12.21}$$

is the pitch-angle anisotropy of resonant electrons and

$$A_c = \left(\frac{\omega_{c,e}}{\omega} - 1 \right)^{-1} \tag{12.22}$$

is the critical anisotropy, required for instability. This nonrelativistic cool plasma approximation (12.19) will generally be adequate for our discussion of instability in the high density region of the Io plasma torus.

Because the resonant electron energy (Table 12.1) is proportional to $B^2/8\pi n$, the high density plasma torus is a region of lower resonant energy and (for an electron energy spectrum decreasing with E) correspondingly higher resonant flux. Characteristic values for $B^2/8\pi n$ are sketched in Figure 12.5; they range from a few keV inside the equatorial torus to several hundred keV at higher latitudes (or low $L < 5$). This variation effectively confines instability to a limited region near the equatorial plane. The net gain G in wave power during field aligned transit across the unstable equatorial region (of length ℓ^*) is given by $\ln G = 2\gamma_{cvc}\,\ell^*/v_g$, where v_g is the average wave group speed. To ensure significant wave amplification, the resonant integral electron flux must exceed a critical value

$$J_e^*(E_e > E_{res}) \approx \frac{B}{A_e - A_c} \cdot \frac{\ln G}{\ell^*} \cdot \frac{c}{2\pi^2 e} \tag{12.23}$$

which is equivalent to the maximum stably trapped flux defined earlier by Kennel and Petschek [1966]. For $\ln G \approx 3$ (≈ 20 fold increase in wave power), $\ell^* \sim 2R_J$ and $A_e - A_c \sim 0.2$ the critical flux for wave instability in the Io plasma torus is approximately

$$J_e^* \approx 10^{14}\, L^{-3}/\mathrm{m}^2\,\mathrm{s} \tag{12.24}$$

The variation in J_e^* with L is slightly different from the dependence in the terrestrial magnetosphere [Kennel and Petschek, 1966] owing to the assumption here that instability in the torus is restricted to the limited equatorial region controlled by centrifugal confinement of ions injected from Io. A more precise relativistic treatment of the wave growth [Barbosa and Coroniti, 1976] yields a result almost identical to (12.24) and this value will therefore be used in our subsequent discussion.

From (12.19) and (12.22), it is clear that cyclotron damping provides a natural upper limit to the band of unstable waves at a normalized frequency

$$(\omega/\omega_{c,e})_{max} \approx A_e/(1 + A_e) \tag{12.25}$$

Fig. 12.5. The scaling energy $B^2/8\pi n$
for electromagnetic wave-particle
resonance as a function of location in the
vicinity of the Io plasma torus. This
energy exceeds an MeV at high latitude
and attains minimum values below 10
keV in the high density equatorial torus.

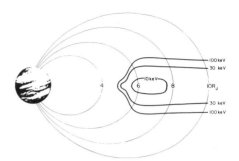

Furthermore, because the resonant electron energy (Table 12.1) increases as $\omega/\omega_{c,e}$ decreases, one can expect fewer resonant electrons and correspondingly weaker amplification at low frequency. From the resonant electron energies given in Table 12.1, one can therefore estimate a lower frequency limit to the unstable wave spectrum,

$$(\omega/\omega_{c,e})_{\min} \approx \frac{(\omega_{c,e}/\omega_p)^2}{(\gamma_{res,e}^2 - 1)_{\max}}$$ (12.26a)

or in the nonrelativistic limit

$$(\omega/\omega_{c,e})_{\min} \approx E_M/E_{e,\max}$$ (12.26b)

where $E_{e,\max}$ is the maximum resonant electron energy for which $J(E_e > E_{e,\max}) \gtrsim J_e^*$.

A comparison between the critical flux required for instability (12.24) and electron fluxes observed at three locations on Voyager 1 is shown in Figure 12.6. The plotted integral flux has been computed from direct measurements of the integral spectrum above 1.5 and 10 MeV reported by Krimigis et al. [1979a], differential fluxes in the energy range 130–183 keV [Krimigis et al., 1979a] and values of the low energy ($E \leq 4$ keV) electron distribution function $f_{o,e}(v)$ [Scudder et al., 1981], which can be employed to obtain the differential flux $j_e(v) \sim v^2 f_{o,e}(v)/\gamma m_e$. At $L = 18$, the entire electron population below ≈ 1 MeV lies above the critical flux required to excite whistler-mode emissions. Intense instability can therefore be anticipated over a broad range of frequency. In the absence of a strong particle source, one would expect pitch-angle scattering to eventually reduce the flux to the critical level J_e^* [Kennel and Petschek, 1966]. However, at $L \approx 18$, interchange driven radial diffusion is considerably more rapid than precipitation loss, even under strong pitch-angle diffusion (see Fig. 12.2). The observed electron flux profile should therefore be controlled by inward radial transport from the outer Jovian magnetosphere. However, we have already established that the integral electron flux in the diffusion dominated regime should scale as $J(> E) \sim L^{-4.5}$ or L^{-6} depending on whether or not the electrons are relativistic. The relative increase in the inward diffusing flux is therefore larger than the increase in $J_e^*(L)$, and values exceeding J_e^* consequently occur in the inner magnetosphere. This concept of a radial diffusion dominated region in the outer Jovian magnetosphere was discussed earlier by Thorne and Coroniti [1972], Coroniti [1974] and Coroniti et al., [1974]. The only difference here is our use of the more rapid rate of radial transport by interchange instability [Siscoe and Summers, 1981] rather than ionospheric winds or solar-wind induced electric fields.

At $L \approx 7.7$, the integral flux of electrons at energies below ~ 600 keV is also above the critical level; instability can again be anticipated over a broad range of frequency. However, the extent by which J_e exceeds J_e^* is less dramatic and a comparison between the observed fluxes at 7.7 and those anticipated on the basis of adiabatic transport from $L = 18$ (the dashed line) indicates a significant departure from simple loss-free inward

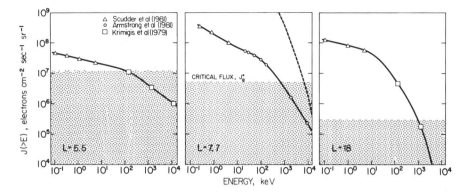

Fig. 12.6. The integral electron flux observed on Voyager at three locations in the middle Jovian magnetosphere is compared to the anticipated critical flux for onset of electron-cyclotron resonant instability with electromagnetic whistler-mode waves. For $L \gtrsim 7.7$, intense instability is predicted over a broad range of resonant frequency. The observed flux at $L = 7.7$ is consequently smaller than the value predicted (dashed line) on the basis of loss-free inward diffusion from $L = 18$. In the region of relatively slow inward diffusion inside the orbit of Io, rapid scattering loss has depleted the flux at $L = 5.5$ to values comparable to the stably trapped level.

injection. But because the timescale for strong diffusion loss is much smaller than the radial diffusion time at $L \approx 7.7$ (Fig. 12.2), it is clear that energetic electrons in the torus are subject only to weak diffusion scattering. In fact, to reach this location the electron precipitation lifetimes must typically exceed $\sim 10^5$ s. From (12.10) an upper limit can be placed on the allowed power spectral intensity of the resonant whistler-mode waves, which, as shown later, is generally consistent with the average properties of observed waves in this region of the torus.

Compelling evidence for the reduction in trapped electron flux to the critical stably trapped level J_e^* is apparent at $L = 5.5$. The entire electron population between 1 and 100 keV lies within a factor of three of the predicted level. This reduction is also consistent with the marked decrease in the observed wave intensities inside this location (Fig. 12.4). Radial diffusion inside the orbit of Io is apparently no longer able to provide an effective source to combat even weak diffusion losses in the inner torus. A dramatic change in the dominant mechanism for radial transport is anticipated inside $L \approx 5.9$ [Richardson et al., 1980] because centrifugally driven interchange instability can only carry Io injected ions to larger L. Radial diffusion inside the orbit of Io is probably driven by neutral winds in the Jovian ionosphere but even the most optimistic estimates for such transport (see Fig. 12.2) yields cross-L diffusion times in excess of 10^6 seconds.

The integral electron fluxes observed on Voyager [Krimigis et al., 1979a] have been employed by Thorne and Tsurutani [1979] to predict the lower frequency cut-off (12.26) of unstable whistler-mode waves in the vicinity of the Io plasma torus. Their results are reproduced here in Figure 12.7. The rapid increase in f_{min} inside $L \approx 5.5$ is mainly caused by the dramatic increase in the electromagnetic resonant scaling energy $B^2/8\pi n$ owing to lower plasma density inside the orbit of Io. The observed waves (Fig. 12.4) generally lie within the theoretically predicted unstable band. The only notable discrepancy is the existence of very low frequency ($\lesssim 1$ kHz) waves inside $L \sim 5.5$. This feature, however, can be understood [Scarf et al., 1979] in terms of waves originating within the torus and subsequently propagating to lower L across the

Fig. 12.7. The predicted band (shaded) of unstable cyclotron resonant whistler-mode waves in the inner Jovian magnetosphere compared to the equatorial cyclotron $f_{c,e}$ and plasma frequency f_p. The upper frequency cut-off is controlled by the available pitch-angle anisotropy (here assumed to be $A_e \approx 1$) of resonant (≈ 1 keV) electrons. The lower frequency cut-off is determined by the maximum energy of electrons with flux J_{res} exceeding the threshold for significant wave growth. The steep increase in f_{min} inside $L = 6$ is mainly due to the dramatic increase in resonant electron energy at low L. The predicted unstable band is generally consistent with Voyager observations [Scarf et al., 1979a].

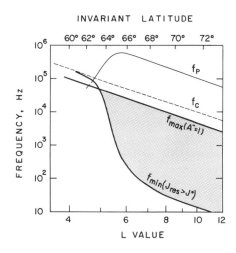

magnetic field lines, which is possible only at frequencies below the lower hybrid resonance frequency.

Although the general properties of the observed whistler-mode waves are consistent with a cyclotron-resonant instability, the detailed spectral properties have yet to be understood. The occurrence of chorus at frequencies above $f_{c,e}/2$ requires the presence of resonant ~ keV electrons with substantial pitch-angle anisotropy, $A_e > 1$ (see 12.25). Furthermore, the intermittent bursty nature of chorus suggests rapid temporal variations in either the flux or the anisotropy of low-energy resonant electrons. But it has already been shown in Figure 12.6 that the flux of ~ keV electrons at $L \approx 7.7$ and 18 is approximately two orders of magnitude above the level required for instability. The intermittent nature of chorus must therefore be related to variability in the electron anisotropy. The absence of intense chorus emissions over most of the Voyager trajectory through the torus can realistically be explained only by requiring that the pitch-angle anisotropy be small ($A_e << 1$) for the low-energy electron population. This could be maintained by processes such as Coulomb collisions or by pitch-angle scattering by other plasma waves. As an example, the characteristic time for Coulomb isotropization (i.e., the time to scatter one radian) of electrons is $\tau_c \sim 3 \times 10^8 E_e^{3/2}$ (keV)/n seconds. Coulomb scattering should therefore be effective at maintaining an isotropic low energy ($E_e < 1$ keV) distribution in the high density ($n \sim 10^3$ cm^{-3}) torus where τ_c is less than the timescale for inward radial diffusion. Unless some additional rapid source for injecting $\approx$ keV electrons with flat pitch-angles ($\alpha \sim \pi/2$) is available, high-frequency ($\omega/\omega_{c,e} \gtrsim 1/2$) whistler-mode waves could not be excited in the high density torus, and this is consistent with the absence of chorus at low L. However, in the outer torus ($L \gtrsim 8$) where $N \lesssim 10^2$ cm^{-3}, the characteristic Coulomb deflection time for $E_e \sim 1$ keV exceeds the interchange radial diffusion time (Fig. 12.2) and, inward diffusion could therefore establish significant electron anisotropy. Stochastic acceleration by intense electrostatic waves provides an alternative source for anisotropic low energy electrons (see the following section). The confinement of such waves to the equator might also explain why the observed chorus emissions are located near the equatorial plane since the perpendicular adiabatic acceleration is most effective there.

When chorus is observed the emission is particularly intense. The example reported by Coroniti et al., [1980] (also see Fig. 8.15), which occurred near $L \sim 8$ at 0620:45, had a wide band amplitude of approximately 10 picotesla. By scaling the required strong diffusion amplitudes given in Figure 12.1 to $L \approx 8$, it is clear that such waves are sufficient to drive resonant ($\approx$ keV) electrons into strong pitch-angle diffusion. The concomitant rapid pitch-angle scattering should lead to a relaxation of the anisotropy and thus provides a natural explanation for the bursty nature of the emission.

Why broadband hiss should be most intense at frequencies well below the electron cyclotron frequency is also something of a puzzle. The peak growth rate (12.19) should typically occur at frequencies just below the upper cut-off (12.25); unless the electron anisotropy is exceedingly small, one would normally expect the most intense waves near a few tenths of the electron cyclotron frequency. This is clearly not the case, which suggests that propagation effects might play an important role in the origin of Jovian hiss as it does for similar waves in the Earth's plasmasphere [Thorne et al., 1980], but this conjecture also remains to be tested.

On the assumption that the emissions observed below the electron cyclotron frequency are indeed propagating in the whistler-mode, one can use the measured fluctuating electric fields to obtain estimates for the rate of resonant particle pitch-angle scattering. If the waves are polarized with $\bar{k}$ parallel to $\bar{B}$, the fluctuating magnetic field, $B'(f) = n_{\parallel} \, \mathcal{E}'(f)$. This is also a reasonably good approximation for oblique propagation. Because the whistler-mode refractive index is high ($n_{\parallel} \approx 10$–100) within the torus (see Fig. 12.4), the fluctuating power spectral density is dominated by the wave magnetic field. One can therefore use (12.10) to evaluate the resonant pitch-angle diffusion coefficient. Independent estimates for the rate of scattering by the observed broadband hiss [Scarf et al., 1979a; Thorne and Tsurutani, 1979] indicate that the diffusion is most rapid for electrons above 100 keV. Figure 12.8 (from Scarf et al., [1979a]) provides a summary of the average scattering properties at different locations near the inner edge of the plasma torus. The top panels show the average fluctuating electric fields observed by Voyager 1 during four 96 s intervals. The lower panels display the computed wave refractive index, the fluctuating magnetic field, the resonant electron energy, and the anticipated timescale for local pitch-angle scattering ($\tau_s \sim 1/D_{\alpha\alpha}$) near the orbit of Voyager. Because the wave-particle interactions should be strongly confined to the equatorial torus region, electrons spend a small fraction ($\delta \sim 0.1$) of their orbit in resonance. The timescale for precipitation loss to the atmosphere is consequently an order of magnitude longer than the scattering times shown in Figure 12.8. Nevertheless, the expected rate of removal of relativistic electrons in resonance with $f \lesssim 1$ kHz waves can become very rapid. For example, at 1530 (near $L \approx 6$) the electron lifetime for $E_e \gtrsim 500$ keV is typically between 10^4 and 10^5 s. Although this is longer than the lifetime expected under strong pitch-angle diffusion it is considerably shorter than the most optimistic estimates for radial diffusion times in the vicinity of Io, as exhibited in Figure 12.2.

In the absence of a strong source capable of accelerating electrons locally to relativistic energies, the energetic ($\gtrsim$ MeV) electron flux would be rapidly depleted by such intense scattering. Furthermore, because the flux of relativistic electrons is generally below J_c^* (see Fig. 12.6) the pitch-angle diffusion is parasitic in the sense that the scattering waves are produced by lower energy electrons. Similar parasitic scattering occurs in the Earth's plasmasphere [Lyons et al., 1972] leading to an almost total removal of energetic electrons between the inner and outer radiation belts [Lyons and Thorne, 1973]. Because the removal of relativistic ($>$ MeV) electrons is not expected to be limited by the stably trapped flux, rapid precipitation loss poses a serious question on whether inward radial diffusion can provide a viable source of synchrotron radiating electrons in the inner Jovian magnetosphere as discussed by Coroniti [1974]. It also appears unlikely that corotating convection [Hill, Dessler, and Maher, 1981] provides a more rapid transport mechanism because the anticipated longitudinal asymmetries

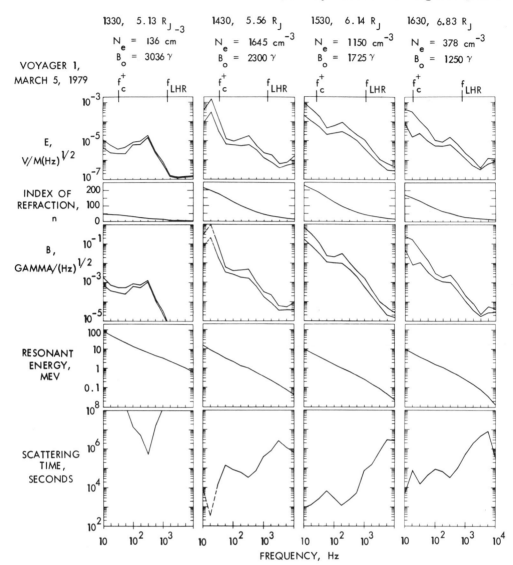

Fig. 12.8. Detailed analyses of the wave particle interactions for four 96 s intervals as Voyager 1 traveled from the inner edge of the plasma torus (1330) to the outer torus region (1630) [Scarf et al., 1979b]. The plasma density and magnetic field strength for the four intervals are: (1330) $n_e = 136$ cm^{-3}, $B_o = 3036$ gammas; (1430) $n_e = 1645$ cm^{-3}, $B_o = 2300$ gammas; (1530) $n_e = 1150$ cm^{-3}, $B_o = 1725$ gammas; (1630) $n_e = 378$ cm^{-3}, $B_o = 1250$ gammas. At the inner edge of the torus, the long local scattering times indicate that the electron distribution is relatively stable to precipitation losses. Within the torus, electrons with $E_e > 0.1$ MeV undergo strong local scattering and have precipitation lifetimes that may be comparable to radial diffusion source times.

have not been observed. There is, however, some uncertainty in the identification of the low frequency (≤ 60 Hz) signals as exclusively whistler-mode waves [Scarf et al., 1979a]. The observed wave power spectral density is highly variable and often exhibits an enhancement associated with the proton gyrofrequency (see Fig. 12.8). One cannot exclude the possibility that electrostatic ion-mode waves contribute to the observed low frequency emission. If so, the relativistic electron ($<$ few MeV) scattering times would be much longer than the values exhibited in Figure 12.8.

Even at frequencies above 100 Hz, where the wave identification is more secure, the observed scattering wave amplitudes are highly variable (Fig. 12.8). Average precipitation lifetimes in the torus ($L \gtrsim 5.5$) probably exceed 10^5 seconds at energies near 1 MeV and rise to values above 10^6 s near 100 keV. Inward radial diffusion driven in response to interchange instability of the torus heavy ions should therefore provide an effective source of electrons, even at relativistic energies throughout most of the Io torus. Although the rate of inward diffusion should decrease dramatically inside the orbit of Io, the observed intensity of the scattering waves is also considerably lower. Furthermore, the scaling energy for resonance with electromagnetic waves, $B^2/8\pi n$, rise to values above 100 keV for $L \lesssim 5$ (Fig. 12.5), and even MeV electrons will have too low an energy to resonate with the residual low frequency ($\omega < \omega_{LH}$) waves which presumably originate in the torus (see panel 1 in Fig. 12.8). The relativistic ($\sim$ MeV) electrons observed at $L \approx 5.5$ (Fig. 12.6) should therefore experience relatively little depletion during their subsequent slow inward diffusion into the synchrotron radiating region ($L \lesssim 3$) of the inner Jovian magnetosphere. The initial Voyager 1 observations reported by Krimigis et al. [1979a], while limited to the region $L \gtrsim 4.9$, support these theoretical expectations. As shown in Figure 12.3, the energetic (> 1.5 MeV) electron flux exhibits significant depletion in the inner torus ($5.5 \lesssim L \lesssim 8$) but then increases with decreasing L in the inner magnetosphere ($L < 5.5$). In the absence of further scattering loss, the electron distribution function $f_o(\mu = $ constant) should be conserved and by scaling the observed flux at $L = 5.5$ (Fig. 12.6) one can anticipate a relativistic ($E_e > 1$ MeV) integral spectrum $J_e(> E) \approx 2 \times 10^{12}/E_e$(MeV) electrons/m²-s-sr at $L \approx 3$ near the outer portion of the synchrotron radiating region. It is also of interest to note that the lower energy portion ($E_e \approx$ MeV) of the synchrotron spectrum between ($2 \lesssim L \lesssim 3$) originates from electron fluxes ($E_e \approx$ keV) limited by whistler-mode stable trapping in the inner torus (Fig. 12.6); a concept initially suggested by Coroniti [1974]. However, the more energetic electrons ($E_e \gtrsim 10$ MeV) in the synchrotron region should be subject to parasitic scattering loss ($J_e \ll J^*$) during transit through the torus. Their flux could thus exhibit significant long-term variability, which should be reflected in the power of higher frequency decimetric radiation.

Electrostatic emissions above $\omega_{g,e}$

In addition to the broad-band whistler-mode emissions, Figure 12.3 shows several examples of waves above the electron cyclotron frequency that, in analogy with terrestrial observations, have been identified as electrostatic upper hybrid or electron cyclotron harmonic waves [Kurth et al., 1980b; also see Chapter 8, Sec. 8.4]. Both classes of wave occur in narrow frequency bands centered between multiples of the electron cyclotron frequency. Strong cyclotron damping by the thermal plasma is expected at exact multiples of $\omega_{c,e}$. Although highly variable, the wave intensities exhibit a systematic increase with decreasing radial distance and reach peak amplitudes of a few mV/m in the outer torus. The observed emissions are also very tightly confined (within a few degrees) to the magnetic equator; this limits the average rate of scattering of any resonant electrons.

Barbosa and Kurth [1980] have developed a theoretical model for the generation of the Jovian electrostatic waves using free plasma energy associated with a loss-cone distribution (for which $\partial f_o/\partial v_\perp > 0$) of suprathermal electrons superimposed on a dense thermal background plasma. This approach has been successfully used to explain

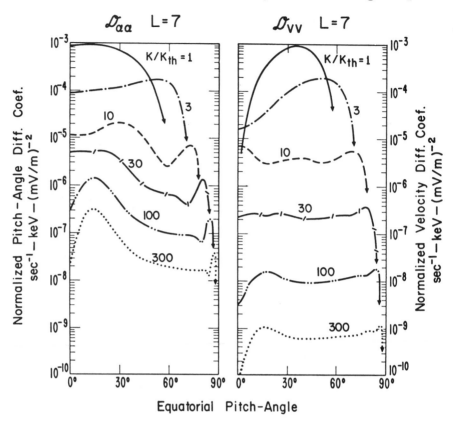

Fig. 12.9. Normalized coefficients for diffusion in pitch-angle ($D_{\alpha\alpha}$) and for diffusion in speed D_{vv} plotted as a function of equatorial pitch-angle at $L = 7$ for representative values of the electron energy normalized to the thermal energy of the warm electrons [Lyons, 1974]. The actual bounce-orbit averaged diffusion coefficients can be obtained by multiplying $D_{\alpha\alpha}$ and D_{vv} by $|\mathscr{E}'|^2/K_{th}$ with $|\mathscr{E}'|$ expressed in mV/m and K_{th} expressed in keV.

electron harmonic waves in the terrestrial magnetosphere [Young, Cullen, and McCune, 1973; Ashour-Abdalla and Kennel, 1978a; Hubbard and Birmingham, 1978; Curtis and Wu, 1979]. Barbosa and Kurth [1980] have also computed wave ray paths to evaluate the net convective growth of unstable waves. Their theoretical simulation shows that waves originating only a few degrees away from the equator experience a rapid change in resonant velocity (12.3), which significantly reduces the total convective gain. Furthermore, waves amplified at the equator are effectively trapped (by their subsequent ray paths) leading to the observed equatorial confinement. Barbosa and Kurth [1980] have also made estimates for the critical electron flux required to produce significant convective wave amplification. Their "best" estimate requires an integral resonant electron flux $J_e \, (> E_{res})$ between 2×10^{11} and 6×10^{11}/m²-s-sr to excite the first harmonic $(3f_{c,e}/2)$ band over the radial range $20 \leq L \leq 8$, respectively. Proportionately larger flux is required for higher $(n + 1/2) f_{c,e}$ harmonic excitation. Even for the first harmonic band the required electron flux is considerably larger than J_e^* for whistler-mode instability (12.24), but a direct comparison with observed fluxes (see Fig. 12.6) indicates that anisotropic electrons in the energy range below ≈ 30 keV should nevertheless be able to excite the electrostatic waves in the outer torus. However, in the inner torus at $L \approx 5.5$, the marked reduction in trapped electron flux to the whistler stably trapped level could quench the instability of electrostatic electron-cyclotron harmonic waves.

Provided that resonance is possible, the reported peak wave amplitudes (a few millivolts per meter) would appear to be marginally sufficient to cause strong diffusion scattering of low energy ($\lesssim$ 10 keV) electrons (see Fig. 12.1). However, because the observed waves are confined within a few degrees of the equator, resonant electrons spend only a few percent of their bounce orbit in the interaction region. Electric field amplitudes required for strong diffusion scattering consequently are about three times larger than shown in Figure 12.1. Even so, resonant $\approx$ keV electrons should still be strongly scattered by the most intense waves ($\mathscr{E}' \approx$ few mV/m) in the outer torus.

A detailed evaluation of the bounce averaged rate of pitch-angle scattering $D_{\alpha\alpha}$ and energy diffusion D_{rr} has previously been performed by Lyons [1974] for electrons resonant with narrow band first harmonic ($3 f_{c.e}/2$) waves in the Earth's magnetosphere. His generalized results, reproduced here in Figure 12.9, should apply equally well to Jupiter. It is clear that both energy diffusion and pitch-angle scattering are most effective for electrons near the thermal energy of the warm plasma. In the middle Jovian magnetosphere ($10 \leq L \leq 20$), characteristic electron thermal energies are probably near a few keV [Scudder et al., 1981]. The most rapid scattering by the narrow-band cyclotron harmonic waves should therefore occur for electrons below approximately 10 keV. At higher energies the relative rate of pitch-angle scattering scales as $D_{\alpha\alpha} \sim (E_e/E_{th})^{-3/2}$ and the energy diffusion rate $D_{rr} \sim (E_e/E_{th})^{-5/2}$.

Without detailed information on the electrostatic fluctuating electric field amplitudes as a function of L, only approximate estimates can be made for the resonant electron lifetimes. But we have already established that the peak wave amplitudes observed in the inner magnetosphere (a few mV/m) should be marginally adequate to scatter the thermal keV electrons on strong diffusion. To proceed, we assume that this is true throughout the middle magnetosphere; namely, the lifetime for $\sim$ keV electrons is given by the strong diffusion lifetime (12.9) plotted in Figure 12.2. However, as exhibited in Figure 12.9, the rate of scattering decreases with increasing electron energy. Energetic resonant electrons therefore are subject only to weak diffusion and the corresponding precipitation lifetime $\tau_L \approx \mathscr{D}_{\alpha\alpha}^{-1} \sim (E_e/E_{th})^{+3/2}$, which can be compared to the strong diffusion lifetime $\tau_{SD} \sim E_e^{-1/2}$. Provided that thermal electrons are never subject to pitch-angle diffusion at a rate exceeding the strong-diffusion level (which appears to be true for the keV electrons in resonance with electrostatic waves in the Jovian magnetosphere) the precipitation lifetime for energetic electrons can be simply expressed in terms of the appropriate strong diffusion lifetime (12.9) as

$$\tau_L(E_e)/\tau_{SD}(E_e) \approx (E_e/E_{th,e})^2 \left[\tau_L(E_{th,e})/\tau_{SD}(E_{th,e})\right] \tag{12.27}$$

When the thermal ($\sim$ keV) electrons are indeed subject to strong diffusion, the energetic electron lifetime can be directly computed using Figure 12.2. For example at $L = 10$ the electrostatic scattering lifetime for 100 keV electrons would be (10^3 to 10^4)τ_{SD} or 10^7 to 10^8 s, which is much longer than the lifetimes established for scattering by whistler-mode hiss (see previous section). Because this estimate is based on the peak electrostatic wave amplitudes observed at Jupiter, it represents an absolute lower limit on the resonant electron lifetime. The nonadiabatic dynamics of energetic (> 100 keV) electrons is apparently controlled by whistler-mode waves throughout the middle magnetosphere.

Electrostatic waves were observed only when the Voyager trajectory lay within a few degrees of the magnetic equator, and even then the intensities were highly variable. It is therefore difficult to make precise estimates on their overall scattering properties. However, near $L \sim 8$ on the inbound trajectory, the plasma wave instrument

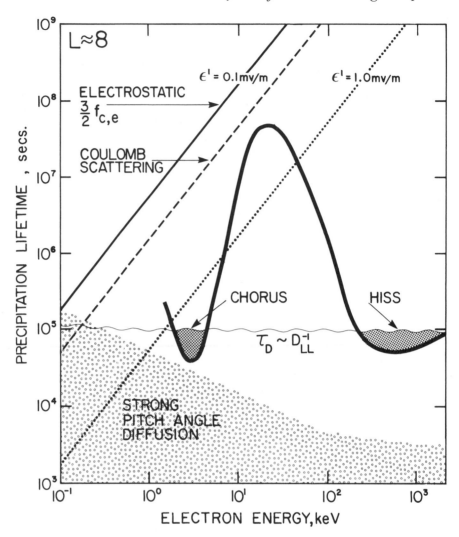

Fig. 12.10. Precipitation lifetimes for electrons resonant with electrostatic $3\,f_{ce}/2$ waves and electromagnetic whistler-mode chorus and hiss observed in the Io torus near $L \approx 8$.

detected electromagnetic hiss and chorus together with electrostatic cyclotron harmonic waves. It is therefore possible to compare scattering lifetimes directly in this region of the magnetosphere. The relevant electron pitch-angle scattering times due to resonance with the electromagnetic emissions have been evaluated by Thorne and Tsurutani [1979] and also shown in Fig. 8.15. The effective electron precipitation time has been plotted in Figure 12.10 by scaling these scattering times upward by a factor of 10 to account for the fraction of the electron orbit spent in resonance with the waves. As discussed in the previous section, the intense chorus emissions cause rapid removal of 3 keV electrons near the limit imposed by strong diffusion, while the scattering by hiss is most effective for electrons above 100 keV. At this location, both electromagnetic emissions appear to be able to remove resonant electrons at a rate comparable to interchange driven inward diffusion. The trapped electron flux consequently exhibits a marked depletion in phase space density at lower L; this is consistent with the high energy flux profiles reported by Krimigis et al. [1979a] (also see Fig. 12.3).

The maximum fluctuating electric field amplitude in the first $(3 f_{ce}/2)$ harmonic band was approximately 10^{-1} mV/m near $L \approx 8$ (see Fig. 12.4); this should be compared to the amplitude $\mathscr{E}'_{SD,e} \approx 1$ mV/m required for strong diffusion scattering of thermal keV) electrons. The corresponding electron precipitation lifetimes are therefore approximately $\tau_L(E_{th}) \sim 100 \, \tau_{SD}(E_{th})$ for thermal electrons and considerably longer (12.27) at higher energy. As illustrated in Figure 12.10, the expected precipitation lifetime due to resonance with the observed electrostatic waves exceeds the scattering time for electromagnetic resonant interactions over the entire energy range. The electrostatic scattering lifetimes are even longer than the angular deflection time due to Coulomb scattering in the torus assuming $n \sim 200$ [Bagenal and Sullivan, 1981]. It might therefore appear that the electrostatic waves are relatively unimportant at this location even for lower energy electrons. However, should the wave amplitudes reach values near 1 mV/m (which is entirely conceivable owing to the limited equatorial coverage on Voyager) the scattering by electrostatic waves would become significant at all energies below ~ 50 keV. This could be particularly important for lower energy electron scattering in locations where chorus is not present.

A second potentially important role played by electrostatic plasma waves is their ability to provide stochastic acceleration. As exhibited in Figure 12.9, the rate of energy diffusion is comparable to that for pitch-angle diffusion for resonant electrons near the thermal energy, but it becomes progressively less important at higher energy. The net result of any energy diffusion is an acceleration of particles to higher energy in an attempt to flatten out the velocity space gradient. If the scattering waves are polarized with $k_\perp \gg k_\parallel$ the acceleration occurs predominantly in the perpendicular direction. This tends to enhance the pitch-angle anisotropy of the thermal plasma and thereby provides an additional source of free energy to excite other plasma waves such as electromagnetic whistler-mode chorus (see previous section). For example, if the electrostatic $3 f_{ce}/2$ waves near $L \approx 8$ attain amplitudes approaching 1 mV/m, localized energy diffusion could provide a source of thermal electrons ($E_e \lesssim$ few eV) at a rate comparable to or faster than cross-L transport (see Fig. 12.10). Furthermore, the predominantly perpendicular acceleration enhances the low-energy electron anisotropy at a rate significantly faster than that of Coulomb isotropization, thereby providing a viable mechanism for high frequency chorus excitation at this location. Because the timescale for energy diffusion is proportional to $(E_e/E_{th})^{5/2}$, perpendicular electrostatic acceleration becomes negligible above 10 keV. The band of unstable chorus excited by this acceleration should therefore be limited to frequencies comparable to or above $f_{ce}/2$.

Possible waves for ion scattering

The radial profiles of energetic ions observed on Voyager 1 [Krimigis et al., 1979a; Armstrong et al., 1981; Lanzerotti et al., 1981] exhibit evidence for rapid precipitation loss in the inner ($L \lesssim 8$) torus (see Fig. 12.3). This depletion cannot be due to collisional processes such as Coulomb scattering or charge exchange because the relevant removal rates are much smaller than the anticipated rate of inward radial diffusion. Scattering by plasma turbulence offers the only viable loss mechanism and several possible scattering wave-modes have been considered.

Conditions for excitation of electromagnetic ion-cyclotron waves in the Jovian magnetosphere and their potential influence on trapped ions were initially discussed by Kennel [1972] and Thorne and Coroniti [1972] and more recently applied by Goertz [1980b] to describe the ion losses observed on Voyager. The instability is directly analogous to the whistler-mode interaction with energetic electrons as outlined in an earlier section. The temporal growth rate for parallel propagating waves [e.g., Kennel and Petschek, 1966] is

$$\gamma_{cyc} \approx \frac{\pi}{2}\,\omega_{c,p}\left(\frac{\omega_{c,p}}{\omega}\right)\frac{(1-\omega/\omega_{c,p})^2}{(1-\omega/2\omega_{c,p})}\,\eta_p(v_{res})\,(A_p(v_{res})-A_c) \qquad (12.28)$$

where $\eta_p(v_{res})$ and $A_p(v_{res})$ are the fraction number and anisotropy of the resonant ions and $A_c=(\omega_{c,p}/\omega-1)^{-1}$ is the critical ion anisotropy for instability. When the convective properties of the ion-cyclotron waves are applied to the region of the Io plasma torus, one can define a critical ion flux for instability that is essentially identical to the electron-whistler value (12.24).

If the ion energy spectrum and pitch-angle anisotropy were known, predictions (analogous to Fig. 12.7) could be made for the unstable band of ion-cyclotron turbulence. The energetic ion spectrum in the middle magnetosphere, however, has been reported only at energies above 0.5 MeV (see Fig. 12.3). Nevertheless, the fluxes reported by Armstrong et al. [1981] near $L \approx 7.7$ are best modeled by an integral spectrum $J(>E_p) \approx 8 \times 10^{10}/E_p$ (MeV)/m^2-s-sr, so that all ions below $E_p \approx 1$ MeV would appear to have sufficient flux (12.24) to excite ion-cyclotron waves. This energy is substantially larger than $B^2/8\pi n \approx 10$ keV; one might therefore (in analogy with Fig. 12.7) anticipate a broadband of unstable waves resonant with 10 keV to 1 MeV ions. Unfortunately at $L = 7.7$ the predicted unstable band (1–5 Hz) occurs in a frequency range just below the lower limit of the plasma- wave instrument. But intense low-frequency electromagnetic waves could also be identified by the magnetometer on Voyager 1. To assess whether this is indeed feasible, we recall that limits on the wave amplitude can be obtained from the rate of pitch-angle scattering. The rapid depletion of the energetic ion flux inside $L \approx 8$ (Fig. 12.3) when compared to the expected rate of injection by interchange driven radial diffusion requires pitch-angle scattering near the strong diffusion rate (Fig. 12.2). If the scattering were indeed due to electromagnetic ion-cyclotron turbulence, it would require wave amplitudes $B' \gtrsim 1\ nT$ (Fig. 12.1). This should be within the resolution of the Voyager magnetometer [Acuña, private communication, 1980] but the anticipated scattering waves have not yet been reported.

Rapid ion scattering apparently occurs throughout the inner torus and by $L \approx 6$ the flux above 0.5 MeV has dropped by more than two orders of magnitude (Fig. 12.3) below the critical level J_p^* for ion-cyclotron instability. The persistent scattering loss of high-energy ions could, nevertheless, be maintained parasitically [as apparently also occurs for electrons with $E_e > 1$ MeV (see Fig. 12.6)] provided that lower energy ions (10–100 keV) remain available with adequate flux ($J_p > J_p^*$) to excite the plasma waves. Although this cannot be substantiated, it is entirely reasonable because the strong diffusion timescale for 10 keV ion removal is an order of magnitude longer than at 1 MeV (Fig. 12.2).

Low-frequency electrostatic waves offer an alternative mechanism for rapid ion removal from the inner torus. One potential wave mode is the electrostatic loss-cone instability [Post and Rosenbluth, 1966], which can lead to a broad band of unstable waves (centered just below the ion plasma frequency) in the high β magnetospheric environment [Coroniti, Fredericks, and White, 1972]. Plasma waves with the appropriate spectral properties have been detected beyond the terrestrial plasmapause [Anderson and Gurnett, 1973] and Scarf, Gurnett, and Kurth [1981] have reported broadband electrostatic noise bursts, which suggest that a modified version of this instability could be operative in the outer region of the Io plasma torus. However, wave amplitudes near 10 mV/m are required to provide strong diffusion scattering of ions with $E_p \gtrsim 1$ MeV. Although this requirement appears to be larger than the observed wave intensities [Scarf, private communication, 1980], one cannot exclude such waves being the ion scattering mechanism owing to the limited spectral range of the plasma wave detector on Voyager.

The Voyager plasma-wave instrument also detected intense low frequency electric field noise near each crossing of the outer boundary of the Jovian plasma sheet [Barbosa et al., 1981]. As reported in Chapter 8, this emission is similar to the broadband electrostatic noise that is continuously present on auroral field lines in the terrestrial magnetosphere [Gurnett and Frank, 1977]. Potential candidates for this emission are electrostatic ion-cyclotron waves, excited either by the loss-cone instability or field-aligned currents [Ashour-Abdalla and Thorne, 1978] or lower-hybrid drift waves driven by perpendicular currents [Huba, Gladd, and Papadopoulos, 1978]. In either case, a significant Doppler shift would be required to explain the observed broadband spectrum. Analogous waves in the terrestrial environment adequately account for the observed strong diffusion scattering of ions ($E_p \lesssim 100$ keV) on auroral field lines and are thus thought to be the major mechanism for the diffuse proton aurora [Ashour-Abdalla and Thorne, 1978]. However, the wave amplitudes in the Jovian magnetosphere are significantly smaller (typically $\mathcal{E}' \sim 0.1$ mV/m; Fig. 8.20) and at best provide only weak diffusion. Such waves are certainly not responsible for the dramatic ion depletion inside $L \approx 8$.

Even though energetic ions in the middle Jovian magnetosphere may be subject to strong pitch-angle scattering, a comparison of the timescales for injection and removal (Fig. 12.2) indicates that the ion flux distribution $J(L)$ in the outer torus ($L \geq 8$) is controlled by inward radial diffusive injection from the outer magnetosphere. Significant depletion by pitch-angle scattering and ultimate precipitation loss to the atmosphere occur only in the inner torus ($L \leq 8$). Within this loss dominated region, one can anticipate a reduction in the trapped flux to the critical level $J_p{}^*$ appropriate for the excitation of either electromagnetic ion-cyclotron waves [Thorne and Coroniti, 1972] or the electrostatic loss-cone instability [Coroniti, Kennel, and Thorne, 1974]. Furthermore, high-energy ions may be scattered parasitically and thereby be reduced to flux levels well below $J_p{}^*$.

In an attempt to simulate the observed radial depletion in the phase-space density of energetic ions [Armstrong et al., 1981], one can solve the steady-state radial diffusion equation (12.18) under the assumption that losses approach the strong diffusion limit (12.9). The reported observations were for ions with a fixed magnetic moment, $\mu \approx 70$ MeV/G, for which the minimum lifetime is $\tau_{SD,p} = \tau_o L^{11/2}$ with $\tau_o \approx 1$ s for energetic protons. If the radial diffusion coefficient is modeled empirically by $D_{LL} = D_0 L^m$, the diffusion equation (12.18) has an analytical solution [e.g., Irving and Mullineux, 1959 (see their Eq. 6.21)]

$$f_o(L_*) \sim L_*^{(m-3)/2} K_\nu(z) \tag{12.29}$$

where $K_\nu(z)$ is a modified Bessel function of the second kind of order $\nu = (2m-6)/(2m+7)$ and argument $z = L_*^{-(2m+7)/4}$ and

$$L_* = \left\{ \left(\frac{2m+7}{4} \right)^2 D_o \tau_o \right\}^{2/2m+4} L \tag{12.30}$$

is a renormalized radial coordinate.

Fig. 12.11. The phase-space density of
energetic (70 MeV/G) ions observed by
Voyager 1 as a function of radial
location. The dramatic decrease inside
$L \approx 8$ has been modeled by balancing
the rate of inward radial diffusion with
strong diffusion precipitation loss. The
best fit requires a change in the rate of
diffusion inside $L = 8$ consistent with
the models of centrifugally driven
interchange instability recently proposed
by Siscoe et al. [1981].

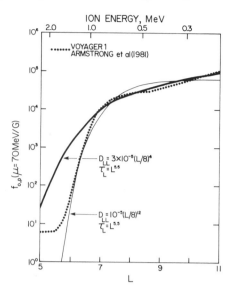

A comparison between the observed distribution function [Armstrong et al., 1981]
and two theoretical solutions for different values of the radial diffusion index m is
shown in Figure 12.11. Each theoretical solution is arbitrarily normalized, and the
scaling coefficient D_o has been optimized to best simulate the radial gradient $\partial f/\partial L$ in
the torus. As discussed by Siscoe, et al. [1981], radial diffusion with an index m = 4 is
appropriate for centrifugally driven interchange instability in the outer ($L \geq 8$) region
of the torus. This scaling was deduced from the observed radial gradient of outward
diffusing thermal (40 eV) heavy ions. This same scaling with a similar magnitude D_o for
the radial diffusion coefficient also provides an excellent fit (in the region $L \geq 7$) to the
radial profile for inward diffusing energetic ions subject to strong diffusion scattering
loss. Inside $L \approx 8$, however, the strong radial gradient in the energy density (or pressure) of energetic ions can seriously inhibit the outward rate of diffusion of warm torus
ions. The observed steep density gradient in the thermal ion distribution [Bagenal and
Sullivan, 1980] near the outer edge ($L \approx 7.5$) of the UV emitting torus has been
explained by Siscoe, et al. [1981] in terms of such a marked reduction in the rate of
radial diffusion. They further suggest that a radial diffusion coefficient with an index
$m = 12$ is appropriate for this region of the torus. This scaling has been adopted for the
second theoretical solution shown in Figure 12.11, and it gives an excellent fit to
the observed energetic ion distribution between $6 \leq L \leq 8$.

This result lends further support to the concept that radial diffusion driven in
response to interchange instability of thermal torus ions is the major source for energetic plasma throughout the middle Jovian magnetosphere. It also provides compelling
evidence that the energetic ions are indeed lost by precipitation into the Jovian atmosphere at a rate comparable to the limit imposed by strong pitch-angle diffusion.
Although the precise mechanism for such scattering cannot be identified using available plasma-wave information from Voyager, the apparent absence of effective scattering loss inside $L \approx 5.5$ (Fig. 12.11) points towards the electromagnetic ion-cyclotron
mode, which should be more easily excited in the high density torus. In the terrestrial
magnetosphere such waves attain peak amplification well away from the equator
[Joselyn and Lyons, 1976]. If this is also true for Jupiter, it might account for
the absence of intense waves (with sufficient intensity to provide strong diffusion
scattering) along the Voyager trajectory.

12.4. Precipitation fluxes and the Jovian aurora

Even before Voyager 1 entered the Jovian environment the extreme ultraviolet spectrometer made two important discoveries. The first were heavy ion (S^+, O^+, etc.) emission lines from a warm ($T_e \approx 10$ eV) dense ($n \approx 10^3/cm^3$) torus surrounding the orbit of Io, and the second were intense auroral HL$y\alpha$ and H_2 Lyman and Werner band emissions originating from the Jovian polar regions apparently mapping along field lines from the torus [Broadfoot et al., 1979]. The auroral emissions were particularly intense with a total radiated power of approximately 2×10^{12} W in HL$y\alpha$ and 3×10^{12} W in H_2 band emissions. Relevant cross-sections for auroral excitation by electron impact on H_2 have recently been measured by Yung et al. [1982] and the results have been employed to model the entire spectrum of auroral emissions from the Jovian atmosphere. They reported significant discrepancy with the excitation cross-sections published earlier by Stone and Zipf [1972], and, at the time of writing, these differences have not been reconciled. The best estimate to account for the total observed radiated power would require $(0.3$ to $1.2) \times 10^{14}$ W to be dissipated by energetic electron precipitation. The quoted range reflects remaining uncertainties in the excitation cross-section and the problem of radiation extinction in the atmosphere. This required energy deposition is somewhat larger than the most recent value reported by Broadfoot et al. [1981], but it is comparable to the approximate estimates by Thorne [1981] based on the Lyα excitation cross-section computed by Miles, Thompson, and Green [1972].

From a detailed comparison of the color ratio in various wavelength ranges, Yung et al. [1982] also suggest a factor of two extinction in the emitted short wavelength radiation. This would place the required auroral energy input above 6×10^{13} W for electron impact with a typical precipitating electron energy near 10 keV. On the other hand, if ion precipitation were dominant, the auroral excitation would be due primarily to secondary electrons with a mean energy near 30 eV [Kuyatt and Jorgensen, 1963]. Because the efficiency for auroral excitation is considerably smaller for such electrons [Miles, Thompson, and Green, 1972] the required ion precipitation flux would exceed 10^{14} W [Thorne, 1981] and energies near 100 keV would be needed to account for the apparent atmospheric extinction at short wavelengths.

Although the energetic particle detectors on Voyager could not resolve fluxes within the atmospheric loss cone, an estimate for the average unidirectional precipitation flux inside the loss cone

$$j_P(E) = \frac{2}{\alpha_L^2} \int_0^{\alpha_L} j(E,\alpha) \sin \alpha \, d\alpha \tag{12.31}$$

can be obtained from the average directional trapped flux

$$j_T(E) = \int_{\alpha_L}^{\pi/2} j(E,\alpha) \sin \alpha \, d\alpha \tag{12.32}$$

and knowledge of the pitch-angle scattering lifetime [Coroniti and Kennel, 1970],

$$j_P(E)/j_T(E) \simeq \tau_{SD}(E)/\tau_L(E) \tag{12.33}$$

The precipitation flux is therefore always less than the averaged trapped flux, but the two become comparable under strong pitch-angle diffusion. Estimates for the particle lifetime obtained in Section 12.3 can thus be combined with information on the observed trapped fluxes to compute the net energy deposition into the Jovian atmosphere. Such estimates are given for each class of observed scattering wave turbulence, and comparisons are made with the energy input required to account for the Jovian auroral luminosity.

The intense band of low frequency electromagnetic hiss observed throughout the Io plasma torus typically resonates with electrons above 100 keV. Near $L \approx 8$, the trapped energetic ($E_e \geq 100$ keV) differential spectrum reported by Armstrong et al. [1981] can be modeled by the functional form

$$j \left(E_e, \alpha \approx \frac{\pi}{2} \right) \approx 3 \times 10^{13}/E_e^2 (\text{keV})/\text{m}^2\text{-s-sr-keV} \tag{12.34}$$

For a pitch-angle distribution $j(E, \alpha) \sim \sin^2 \alpha$ the average trapped flux (12.32) is $j_T(E_e) \sim 2 \times 10^{13}/E_e^2/\text{m}^2\text{-s-sr-keV}$. The observed wave amplitude at this location provides scattering at a rate comparable to 5% of the strong diffusion limit (Fig. 12.10) which from (12.33) yields an average precipitation flux $j_p(E_e) \sim 10^{12}/E_e^2/\text{m}^2\text{-s-sr-keV}$. The net energy flux deposited into the Jovian atmosphere is therefore

$$\epsilon_P \approx \int E_e j_p(E_e) \, dE_e \, d\Omega \approx 2\pi 10^8 \, \ell\text{n} \, (E_{max}/E_{min}) \text{ keV/cm}^2 \text{ s} \tag{12.35}$$

For resonant electrons in the range $150 \text{ keV} \leq E_e \leq 3 \text{ MeV}$ (Fig. 12.10) the total precipitation energy flux due to scattering by electromagnetic hiss is $\epsilon_P \approx 3 \text{ mW/m}^2$. Because the trapped electron flux exhibits relatively minor (less than a factor of 3) variability throughout the torus $6 \leq L \leq 10$ (Fig. 12.3) and because the rate of pitch-angle scattering at $L \approx 8$ is reasonably representative of average values throughout this region (see Fig. 12.8) one can anticipate an average energy deposition comparable to 3 mW/m^2 over an extended invariant latitude range $65° \leq \Lambda \leq 70°$ mapping from the torus. The total area of this extended diffuse auroral zone amounts to 2×10^{15} m^2 in both hemispheres. Total power dissipated by the energetic (≥ 150 keV) electron precipitation is therefore approximately 6×10^{12} W, as initially established by Thorne and Tsurutani [1979]. Although such intense precipitation constitutes a major energy source for the Jovian ionosphere, it is substantially less than the input required to explain the observed auroral emissions. Furthermore, even if the energetic electrons were scattered near the strong diffusion rate, the energy deposition would occur deep in the atmosphere where short wavelength extinction is catastrophic [Yung et al., 1982]. One must therefore look for an alternative source for the Jovian aurora.

The role played by electromagnetic chorus emissions, which predominantly resonate with electrons near 1 keV was initially considered by Coroniti et al. [1980]. Observational data on the resonant electron flux were unavailable at the time of their analysis, and the computed ($E_e \approx$ keV) energy deposition was therefore based purely on theoretical considerations. The distribution function of electrons below 6 keV has now been reported by Scudder, Sittler, and Bridge [1981a], and one can therefore make a more realistic assessment of the low energy precipitation flux. Over the relevant energy range, $300 \text{ eV} \leq E_e \leq 4 \text{ keV}$, the electron differential flux near $L \approx 8$ is best fit by the spectrum

$$j \left(E_e, \frac{\pi}{2} \right) \approx 10^{12}/E_e (\text{keV})/\text{m}^2\text{-s-sr-keV} \tag{12.36}$$

To excite chorus at frequencies above $f_{ce}/2$, the resonant electrons must be highly anisotropic ($A_e \geq 1$). Because observational information is unavailable, we optimistically assume a pitch-angle distribution $j(E_e,\alpha) \sim \sin^4 \alpha$ that yields an average trapped flux $j_T(E_e) \sim 5 \times 10^{11}/E_e/\text{m}^2\text{-s-sr-keV}$. The intensity of chorus emissions observed near

$L \approx 8$ provides scattering of 2–4 keV electrons within a factor of two of the strong diffusion rate (Fig. 12.10). The average precipitation flux is therefore $j_P(E_e) \approx 2.5 \times 10^{11}/E_e$ /m²-s-sr-keV, and the net energy deposition into the atmosphere is

$$\epsilon_P(2 \text{ keV} \le E_e \le 4 \text{ keV}) \approx 0.5 \text{ mW/m}^2 \tag{12.37}$$

This is substantially smaller than the total precipitation energy flux of higher energy electrons scattered by hiss. Also, even if chorus were continually present throughout the torus (which apparently is not the case), the total power dissipation by the resonant low-energy electrons would be less that 10^{12} W. Scattering by chorus therefore plays a relatively minor role in Jovian auroral energetics.

The importance of the third class of wave capable of causing electron precipitation, namely the electrostatic $(n + 1/2)$ cyclotron harmonic waves, has not previously been considered in the auroral context. The waves are continuously present near the equatorial plane in the outer torus and predominantly resonate with lower energy electrons. Estimates for the precipitation lifetime near $L \approx 8$ are illustrated in Figure 12.10. Observed wide-band wave amplitudes near $L \approx 8$ ($\mathscr{E}' \sim 0.1$ mV/m) can provide strong diffusion scattering at $E_e \approx 100$ eV, but the scattering becomes progressively weaker at higher energy. Using (12.27) for the resonant electron lifetime and (12.36) for the trapped electron flux, one obtains an average precipitation flux

$$j_P(E_e) \approx 5 \times 10^9/E_e^3 \text{ (keV)/m}^2\text{-s-sr-keV} \tag{12.38}$$

and a net energy deposition $\epsilon_P(E_e > 300 \text{ eV}) \approx 2 \times 10^{-2}$ mW/m². This is substantially smaller than the contribution by either chorus or hiss. Thus, even if the electrostatic wave amplitudes attain their peak value near 1 mV/m [Kurth et al., 1980b] the net energy deposition flux would only approach $\epsilon_P(E_e > 1 \text{ keV}) \sim 0.5$ mW/m² and the total power dissipation over the entire latitude range mapping from the torus would amount to less than 10^{12} W. Electrostatic waves are therefore unable to significantly contribute to the auroral deposition mechanism.

Because the most liberal estimates for overall electron energy deposition into the Jovian atmosphere ($\lesssim 10^{13}$ W) fall short of the required auroral input, one must turn to energetic ion precipitation as an alternative mechanism. Unfortunately, our ability to quantitatively model the ion input has a weaker observational basis. Although there is compelling evidence for strong diffusion scattering loss of energetic ions in the torus, the scattering mechanism cannot be established, and the trapped ion flux has only been reported for energies above 500 keV. Nevertheless, in the central region of the torus ($L \sim 7.7$) the trapped ion flux between 500 keV and 2 MeV [Armstrong et al., 1981], can be represented by the spectrum

$$j\left(E_p, \frac{\pi}{2}\right) \approx 7 \times 10^{13}/E_p^2 \text{(keV)/m}^2\text{-s-sr-keV} \tag{12.39}$$

The ion pitch-angle distribution varies as $j(E_p, \alpha) \sim \sin^2 \alpha$ [Lanzerotti et al., 1981], which yields an average trapped flux $j_T(E_p) \sim 5 \times 10^{13}/E_p^2$/m²-s-sr-keV. In the absence of any information on the scattering waves, one can simply resort to the excellent fit to the radial diffusion profile (Fig. 12.11) and infer that ions are indeed scattered at a rate comparable to strong pitch-angle diffusion. Conservatively, we may assume scattering at 30% of the strong-diffusion limit and take an average precipitation flux $j_P(E_p) \sim 0.3 j_T(E_p)$. This provides an energy deposition flux

$$\epsilon_P(E_{min} \le E_p \le E_{max}) \approx 10^{14} \, \ell\text{n} \, (E_{max}/E_{min}) \text{ keV/m}^2\text{-s} \tag{12.40}$$

Because the ions are subject to such rapid scattering loss, intense precipitation is limited to a relatively narrow range of invariant latitude $67° \leq \Lambda \leq 70°$ mapping from the middle torus $7 \leq L \leq 9$ (see Fig. 12.11); namely over an area $\approx 10^{15}$ m^2 in both hemispheres. The total power dissipation is therefore $\approx 1.5 \times 10^{13}$ ln (E_{max}/E_{min}) W. Over the observed energy range (0.5 MeV $\leq E_p \leq$ 2.0 MeV), the anticipated power dissipation should be approximately 2×10^{13} W in rough agreement with the earlier estimate by Goertz [1980b]. Although this is still below the input required to excite the observed aurora, an extrapolation of the measured trapped ion spectrum (12.39) to energies below 100 keV enhances the total power dissipation to values comparable to the anticipated auroral requirements $\approx 10^{14}$ W. Furthermore, in contrast to the energetic electron precipitation, ions will deposit their energy in the higher altitude region of the Jovian atmosphere where extinction of auroral emissions is relatively unimportant. Precipitating ions in the energy range near 100 keV should therefore be considered the leading candidate for excitation of the diffuse Jovian aurora.

12.5. Energy transfer processes

Although the energy flux of plasma waves in the Jovian magnetosphere is miniscule in comparison to the observed energy flux in trapped radiation belt particles (see Table 12.2), the ability to efficiently scatter resonant particles in velocity space can cause rapid particle loss and significant energy deposition into the Jovian atmosphere. The most intense particle precipitation is anticipated over a relatively broad range of invariant latitudes mapping from the high density plasma torus surrounding the orbit of Io. The Jovian aurora should therefore be predominantly diffuse in nature, in contrast to the Earth where the most intense energy deposition occurs in discrete auroral arcs. A second major distinction between the Jovian and terrestrial aurora is that energetic ions probably provide the major energy input at Jupiter. Estimates based on the observed trapped-particle flux and the intensity of plasma waves in the Jovian magnetosphere indicates that the total power dissipation by low energy ($\sim$ few keV) precipitating electrons should barely exceed 10^{12} W (Sec. 12.4). In analogy with the terrestrial aurora the energy deposition could be enhanced by invoking a field aligned potential difference between the magnetosphere and the atmosphere but this would have to exceed several tens of keV over an extended range of L to account for the observed auroral emissions. More energetic electrons ($>$ 100 keV) may dissipate up to 10^{13} W, but the energy deposition occurs so deep in the atmosphere that most of the resulting auroral emissions would be reabsorbed. It therefore appears that ion precipitation near the limit imposed by strong diffusion offers the only viable mechanism to account for the observed auroral emissions. Unfortunately, although there is compelling evidence for rapid loss of energetic ($>$ 500 keV) ions from the region of the torus, lower energy ions, which should contribute most of the energy deposition, were not detected in the inner magnetosphere by Voyager 1. Furthermore, the precise mechanism for strong diffusion scattering cannot be established owing to the absence of plasma-wave observations in the appropriate resonant frequency range. These deficiencies should be remedied by instrumentation carried on the future Galileo mission.

The enormous power that is continuously dissipated in the Jovian auroral zone places stringent limits on the mechanisms responsible for injecting energy into the magnetosphere. As an example, the total power available from the solar wind incident over the entire Jovian magnetosphere is approximately 10^{15} W. If this were to provide the ultimate source for auroral dissipation, the energy transfer mechanism would have to be far more efficient (namely 10%) than the analogous processes at Earth. It is

Table 12.2. *Wave-particle energetics in the Jovian magnetosphere*

Wave mode	Amplitude	E_{res}	Energy flux (mW/m^2)		
			Waves	$\epsilon_T(E_{res})$	$\epsilon_p(E_{res})$
Whistler-mode hiss	$B' \sim 30$ mγ	$E_e \gtrsim 100$ keV	2×10^{-6}	60	3
Chorus	$B' \sim 10$ mγ	$E_e \sim$ few keV	10^{-6}	1	0.5
Electrostatic electron cyclotron	$\mathscr{E}' \sim 1$ mV/m	$E_e \sim (1\text{-}10)$ keV	2×10^{-8}	3	0.5
Electromagnetic ion-cyclotron	$B' \sim 1\gamma$	$E_p \gtrsim 10$ keV	2×10^{-4}	200	70

therefore unlikely that the solar wind contributes significantly to the energetics of the Jovian magnetosphere. The rapid rotation of Jupiter, however, has the potential of providing significant power input to the magnetosphere [e.g., Gold, 1976; Carbary, Hill, and Dessler, 1976]. One specific mechanism for tapping this rotational energy is through the torque exerted by Jupiter on the heavy ions injected into the torus surrounding the orbit of Io [e.g., Dessler, 1980b]. The total power input is proportional to the rate of ion injection. Half of this power is consumed in driving bulk motion of the plasma and the remainder represents an upper limit on the power available for auroral dissipation [Eviatar and Siscoe, 1980]. The inferred aurora input of approximately 10^{14} W therefore places a lower bound on the rate of ion production in the torus $dM/dt \gtrsim 2 \times 10^{30}$ amu/s or roughly 10^{29} ions s^{-1} if we adopt $A_+ \approx 25$ for the mean atomic mass of heavy ions in the torus [Bagenal and Sullivan, 1981]. For the reported total torus population of 5×10^{34} ions [Bagenal and Sullivan, 1981] the mean residence time in the torus would be approximately 7 days, which is not inconsistent with the rapid rate of interchange driven radial diffusion computed by Siscoe et al. [1981] (also see Fig. 12.2) or the convection time estimated by Hill, Dessler, and Maher [1981].

It should be emphasized, however, that such rapid mass loading and short residence times in the torus are not universally accepted. On the assumption that most of the neutral atoms are ionized within 1 R_J of Io, Shemansky [1980b] concludes that the absence of detectable UV emissions near Io limits the ion production to values near 10^{27} ions s^{-1}. Such a low mass loading, however, would provide little more than 10^{12} W to the magnetosphere. This would barely be sufficient to balance the radiative cooling of the torus (the total power in UV emission is approximately 2.5×10^{12} W [Sandel et al., 1979]) and would be totally inadequate to drive the more intense and continuous auroral dissipation. One must therefore conclude that if mass loading is indeed the ultimate energy source for energizing the magnetosphere and subsequently exciting the aurora, the neutrals released from Io must either be ionized over a more extended region, or the initially produced ions species must be of a form that cannot be detected by the extreme ultraviolet spectrometer (e.g., SO_2^+).

A further consequence of the rapid mass loading is that the cold electrons ($T_- \approx 10^{-2}$ eV) and the warm ($T_+ \gtrsim 300$ eV) heavy ions initially produced in the corotating torus cannot equilibrate by direct Coulomb energy transfer which requires a time [Spitzer, 1962].

$$\tau_{\pm} \sim \frac{3 \times 10^5 \, A_+ A_e}{z_+^2 \, n_e} \left\{ \frac{T_+}{A_+} + \frac{T_e}{A_e} \right\}^{3/2} \qquad (12.41)$$

where $A_\pm$ are the ion and electron mass in amu, $T_\pm$ are the ion and electron temperatures in eV, and n_e is the electron number density. Adopting average properties for plasma observed in the torus $A_+ \sim 25$, $Z_+ \sim 1.5$, $n_e \sim 10^3/\text{cm}^3$, $T_e \sim 10 \, \text{eV}$ [Bagenal and Sullivan, 1981; Scudder, Sittler, and Bridge, 1981a] one obtains a Coulomb energy transfer time $\tau_c \approx 5 \times 10^6$ s, which is longer than any realistic residence time. Furthermore, the total power input to the electron gas due to Coulomb interaction with heavy ions, is $n_+ T_+ V/\tau_c$ where $V \approx 5 \times 10^{25} \, \text{m}^3$ is the effective volume of the radiating torus. For the observed mean ion energy, $T_+ \sim 40 \, \text{eV}$, this provides a total power supply $\approx 5 \times 10^{10}$ W. Because of the sensitivity of these estimates to the average electron density and temperature in the torus, neither of which are precisely known, Brown [1982] has suggested that the rate of energy transfer could be considerably larger. But even with the most optimistic choice of plasma parameters, the rate of electron heating is below the known radiative-cooling rate ($\sim 3 \times 10^{12}$ W) of the torus. One must therefore look for an alternative source to maintain the torus electrons above the temperature ($\geq 10 \, \text{eV}$) required to excite the observed UV emissions.

An important clue to the nature of this heat source has recently been provided by Scudder, Sittler, and Bridge [1981a] who reported a bimodal electron distribution in the outer torus. Near $L \approx 8$, the dense Maxwellian population, with $T_e \sim 20 \, \text{eV}$, is found in conjunction with a suprathermal tail (containing comparable energy flux) extending to energies above several keV. Because the time-scale for energy equilibration between hot and cold electrons is substantially shorter than between electrons and heavy ions, these suprathermal electrons could provide a viable heat source for the torus. For example $\leq 100 \, \text{eV}$ suprathermal electrons will equilibrate with the cold torus electrons within a few hours.

One possible source for these suprathermal electrons in the magnetosphere are secondary electrons produced during the intense ion precipitation into the Jovian atmosphere. The average energy of secondary electrons emitted during the energetic proton impact on H_2 is typically 20 to 40 eV [Kuyatt and Jorgensen, 1963; Rudd, Sautter, and Bailey, 1966] but a significant number can also be produced with energies up to several keV. Those produced in the upward hemisphere above an altitude where the mean free path exceeds the amospheric scale height, can escape into the magnetosphere with little energy loss. The differential flux of escaping secondaries with energy E has recently been computed by Thorne [1981],

$$j(E) \simeq [\tilde{\sigma}_s(E)/\sigma_{ion}(E)] \, J_P(> E_{+,min}) \qquad (12.42)$$

where $\tilde{\sigma}_s(E)$ is the average differential cross-section for secondary electron production, $\sigma_{ion}(E)$ is the electron ionization cross-section and J_P is the integral flux of precipitation ions above the minimum energy $E_{+,min} \approx (E + 15.6 \, \text{eV}) \, M_+/4m_-$ required for secondary electron production in H_2 where the ionization potential is 15.6 eV. Using the experimentally determined cross-sections and adopting an average precipitation ion flux $J_P(> E_p) \approx 10^{14}/E_p(\text{keV})/\text{m}^2$-s as deduced in Section 12.4, one can integrate (12.42) to obtain both the net integral flux in escaping secondary electrons and the heat flux into the magnetosphere. Note that to escape from the atmosphere the secondary electrons must carry with them an equal flux of cold ionosphere H^+ ions; this is possible because their energy exceeds the gravitational potential ($\approx 10 \, \text{eV}$) for ambipolar outflow from the Jovian atmosphere [Ioannidis and Brice, 1971].

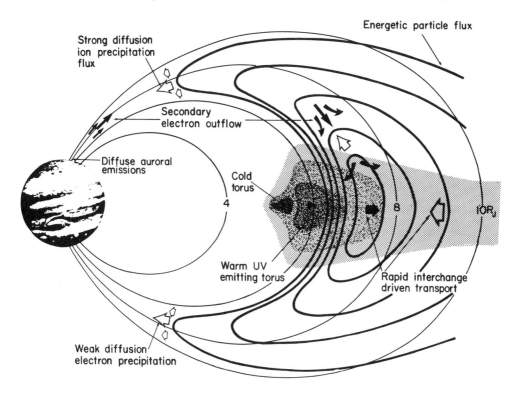

Fig. 12.12. A schematic view of physical processes associated with the heavy ion torus (shaded) surrounding Io. The flux of energetic ring current ions exhibits a steep inner edge near $L \approx 7$, presumably owing to strong diffusion precipitation. Energetic ion precipitation leads to the outflow of superthermal secondary electrons and cold H^+ from the Jovian auroral zone. The upflowing ionospheric particles are scattered onto magnetospherically trapped orbits during their initial transit through the outer torus and subsequently redistributed in radius by eddies associated with interchange instability of the heavy ions injected by Io.

If energetic proton precipitation predominates, the anticipated heat flux to the magnetosphere from the entire auroral zone should amount to approximately 3×10^{11} W and cold H^+ ions would be injected into the torus at a rate comparable to 10^{28} ions/s. However, if the auroral precipitation is dominated by heavy ions (S^+ or O^+), the energy dissipation occurs higher in the atmosphere, $\tilde{\sigma}_s$ will typically be larger by the ratio M_+/M_p, and a larger fraction of the secondary electrons produced in the atmosphere can escape. The heat flux to the magnetosphere could then exceed 10^{12} W and cold H^+ ions would be injected into the torus at a rate on the order of 2×10^{29} s which is comparable to the most optimistic estimates for heavy ion injection from Io. A schematic view of this coupling between the auroral zone and the torus is illustrated in Figure 12.12.

Eviatar, et al. [1982] have considered the implications of such a substantial ionospheric input. The combination of the direct electron heat flux and the more rapid Coulomb energy transfer from heavy ions due to the substantial H^+ concentrations can account for both the apparent cooling of the heavy ions to the observed temperatures near 40 eV and the maintenance of the electron temperature above 10 eV. Because of the ambipolar electric fields, the peak H^+ ion concentrations are expected well away from the equatorial plane in a region not explored by Voyager. This conclusion is supported by whistler dispersion observations [Gurnett et al., 1981b] which require substantially more columnar electron density than can be accounted for by heavy ions in the equatorial torus. Furthermore, although the plasma science instrument on the Voyager spacecraft could not directly detect thermal H^+ ions in the inner torus, significant H^+ concentrations ($\approx 30\%$) have been reported by McNutt, Belcher, and Bridge [1981] in the middle magnetosphere ($L \geq 10$), and Hamilton et al. [1980] have discovered energetic H_3^+ and H_2^+ ions in the outer Jovian magnetosphere. Both observations suggest that the ionosphere is an important plasma source.

The injected ionospheric plasma should be rapidly scattered onto trapped orbits in the magnetosphere during the initial transit of the high density plasma torus and subsequently redistributed in radial location by the eddies associated with interchange instability of the heavy ions injected from Io or by corotating magnetospheric convection [e.g., Hill, Dessler, and Maher, 1981]. However, the outward flow of thermal heavy ions exhibits no evidence for systematic adiabatic cooling. Some additional heat input is therefore required to maintain the ion temperatures near 100 eV. Plasma waves or secondary-electron heat flux from the extended Jovian aurora zone are potential candidates. An important implication of the high thermal ion temperatures in the outer magnetosphere is that subsequent inward transport leads to adiabatic heating to energies comparable to 100 keV in the region of the torus. Rapid inward radial transport can therefore provide a means of maintaining the energy content of the Jovian ring current plasma, which is the reservoir for auroral dissipation. Whether this transport is dominated by interchange eddies, corotating magnetospheric convection or other processes remains to be determined.

ACKNOWLEDGMENTS

The author wishes to thank F. V. Coroniti, A. J. Dessler, A. Eviatar, D. A. Gurnett, F. L. Scarf, M. Schulz, G. L. Siscoe, B. T. Tsurutani, and Y. L. Yung for advice and constructive criticism during the preparation of this report. S. R. Church computed the Bessel functions used in the ion radial diffusion solution. The work was supported in part by N.S.F. Grants ATM 81-10517 and ATM 81-19544.

APPENDIX A
SYMBOLS AND ACRONYMS

Symbols

A amps
   collecting area of telescope
   angstroms
   atomic mass number
    A_+ mean atomic number of ions

$\tilde{\mathbf{A}}$ magnetic vector potential

A_p geomagnetic index

A_0 spin averaged flux
A_1 first order anisotropy in particle flux
A_2 second order anisotropy in particle flux
A_n n^{th} order anisotropy in particle flux

A_c critical anisotropy of electrons required for instability
A_e pitch angle anisotropy of resonant e^-

B magnetic field
   B_r, B_θ, B_ϕ right-hand spherical components of **B**
   B_ρ, B_ϕ, B_z right-hand cylindrical components of **B**
   B_A magnetic field (strength) at the top of the atmosphere
   B_c projection of **B** onto Jovigraphic equatorial plane
   B_{eq} equatorial magnetic field
   B_I magnetic field at Io
   B_{max} maximum hourly average magnetic field
   B_T magnetic field in tail lobes
   $\mathbf{B}_0$ background magnetic field
   $\hat{\mathbf{B}}$ unit magnetic field vector
   $\hat{\mathbf{B}}\hat{\mathbf{B}}$ unit dyadic of magnetic field

b any magnetic field external to the current sheet, other than the planetary
   dipole and the planar field of the current sheet itself

$\hat{\mathbf{b}}$ unit vector parallel to **b**

c speed of light

D dispersion (units s-$\mathrm{Hz}^{1/2}$)

D_{LL} radial diffusion coefficient
D_o diffusion coefficient at $L = 1$ ($D_{LL} = D_o L^n$)
D_{ij} diffusion coefficient
D_{rr} energy diffusion coefficient
$D_{\alpha\alpha}$ pitch angle diffusion coefficient
$\mathcal{D}_\alpha$ pitch angle diffusion coefficient, bounce averaged

D length scale for thickness or depth of a region

D_E Jovicentric declination of the Earth (degrees)
           angular distance of the Earth from Jupiter's spin equator
D_S Jovicentric declination of the Sun

$\mathbf{E}$ electric field

$\hat{e}$ unit vector in the direction of the electric field

E energy
           E_a energy of ion species a
           E_e electron energy
           E_i ion energy
           E_p proton energy
           E_M magnetic energy per particle
           E_{th} thermal energy (of plasma)

e fundamental charge (charge of electron $= -e$)

f reconnection efficiency $= E_{magnetospheric \atop convection} / E_{solar \atop wind}$
           dynamic flattening of Jupiter
           Euler potential
           magnetic flux function
           frequency
           f_c cyclotron frequency
           f_{ce} electron cyclotron frequency
           f_{ci} ion cyclotron frequency
           f_p plasma frequency
           f_{pe} electron plasma frequency
           f_{pi} ion plasma frequency
           f_{LHR} lower hybrid resonance frequency
           f_{UHR} upper hybrid resonance frequency
           distribution function
           f_e electron distribution function
           f_o phase-averaged distribution function

G Newton's gravitational constant
           geometric factor

$\mathbf{g}$ gravitational acceleration

g	local acceleration of gravity at a planet or moon
	Euler potential
H	local horizontal field component
	scale height
h	distance variable
	Planck's constant
$\hbar$	Planck's constant$/2\pi$
I	Stokes parameter (total flux density)
	dip angle
	total current (amps)
J	second adiabatic invariant
	integral particle flux
J_i	action integral associated with i^{th} adiabatic periodicity of motion
$J_e{}^*$	resonant integral electron flux
$J_p{}^*$	critical proton integral flux for ion-cyclotron instability
$\tilde{\mathbf{j}}$	current density
	$j_\perp$ current perpendicular to **B**
	$j_\parallel$ current parallel to **B**, field-aligned (Birkeland) current
j	differential particle flux
	j_e differential particle flux for electrons
	j_T trapped differential flux
j_B	Birkeland (field-aligned) current density
$\tilde{\mathbf{j}}_t$	current tangential to the current sheet
j'	current density (amps/m)
	height-integrated current density
K	Kelvin degrees
	energy injection rate
	eddy diffusion coefficient
K_h	eddy diffusion coefficient at Jupiter's homopause
K_ν	modified Bessel function of the second kind of order ν
K_{th}	thermal energy

k	Boltzmann constant
	rate coefficient

$\overline{k}$ wave propagation vector

$\quad\quad k_\perp$ component of wave propagation vector perpendicular to B

$\quad\quad k_\parallel$ component of wave propagation vector parallel to B

L magnetic field line equatorial crossing distance at R_J

loss rate

$d\overline{L}$ path element

$\mathscr{L}$ loss rate

ℓ distance between points measured along a magnetic field line

M planetary magnetic dipole moment

M_J mass of Jupiter

M_A Alfvén Mach number

M_s sonic Mach number

m particle mass

$\quad\quad m_a$ mass of an ion of species "a"

$\quad\quad m_e$ mass of electron

$\quad\quad m_i$ mass of ion

$\quad\quad m_0$ particle (rest) mass

$\quad\quad m_p$ mass of proton

N total number of ions/unit L-shell

column number abundance of emitting species (per unit area)

total number – as opposed to n: number/volume

n refractive index

plasma number density or concentration (number/volume) – as distinct from N: total number or column number density (number/area)

$\quad\quad n_e$ e^- concentration (number/volume)

$\quad\quad n_i$ ion concentration (number/volume)

$\quad\quad n_p$ concentration of protons (amu/volume)

$\quad\quad n_H$ hydrogen concentration (number/volume)

P pressure

power (in watts)

P_c degree of circular polarization (sign indicates sense)

P_L degree of linear polarization

P_e	production rate/unit volume (electrons)
P_i	production rate/unit volume (ions)
$P_n^m (\cos \theta)$	associated Legendre polynomial
p	momentum vector
Q	Stokes parameter
q	particle's charge
R	count rate reflection coefficient
R_J	radius of Jupiter
R_I	radius of Io
R_p	planetary radius
r	axial ratio (sign indicates RH or LH) (characterizes polarization of radiation) Jovicentric distance
r_m	effective radius of one of Jupiter's moons
S	total flux density source term with units phase space density/time column production rate of ions
s	a quantum state corresponding to momentum $p_\parallel$ and $p_\perp$
T	temperature

$\quad\quad\quad\quad T_\perp \quad$ temperature characterizing thermal motion perpendicular to **B**
$\quad\quad\quad\quad T_\parallel \quad$ temperature characterizing thermal motion parallel to **B**
$\quad\quad\quad\quad T_e \quad$ electron temperature
$\quad\quad\quad\quad T_i \quad$ ion temperature

t	time
U	potential Stokes parameter
V	Stokes parameter (sign specifies polarization sense $[-\ \text{RH},\ +\ \text{LH}]$ magnetic potential
V	plasma bulk velocity

$\quad\quad\quad\quad V_A \quad$ Alfvén velocity
$\quad\quad\quad\quad V_I \quad$ orbital velocity of Io
$\quad\quad\quad\quad V_s \quad$ solar-wind velocity

v	velocity of particle
	$\quad v_\parallel \quad$ component of particle velocity parallel to **B**
	$\quad v_\perp \quad$ component of particle velocity perpendicular to **B**
	$\quad \tilde{\mathbf{v}}_n \quad$ velocity of the neutral atmosphere
	$\quad v_C \quad$ corotation velocity
	$\quad v_g \quad$ average wave group speed
	$\quad v_{ph} \quad$ wave phase velocity
w	total particle energy
	$\quad w_\parallel \quad$ energy due to particle motion parallel to **B**
	$\quad w_\perp \quad$ energy due to particle motion perpendicular to **B**
Y	magnetic flux shell density
Z	atomic charge in units of e
z	altitude
α	pitch angle
	angle between Jupiter's rotational and magnetic axes ($\sim 10°$)
	alpha particle (He nucleus)
	polarizability of an atom (in units of Bohr radii cubed) recombination rate
	$\quad \alpha_r \quad$ radiative recombination coefficient
β	plasma parameter = particle thermal pressure/magnetic pressure
	angle between the centrifugal symmetry surface and the magnetic equator of Jupiter
γ	$= 10^{-5}$ Gauss $= 10^{-9}$ Tesla
	index in power law spectrum $\sim E^{-\gamma}$
	adiabatic index
γ_I	Io phase angle
γ_r	relativistic contraction factor
θ	co-latitude in right-hand polar coordinates
Λ	invariant latitude
λ	wavelength
λ_D	Debye length
λ_{III}	system III longitude
μ	first adiabatic invariant (magnetic moment)
	reduced mass
μ_o	magnetic permeability of free space

ρ	mass density particle's cyclotron radius

Σ	conductivity
	    Σ_p height-integrated Pedersen conductivity
	    Σ_A Alfvén conductance

σ	cross section
τ	characteristic lifetime of particles
τ_b	bounce time for travel between magnetic mirror points
Φ	electric potential drop third adiabatic invariant
Φ_{sc}	spacecraft potential (volts)
χ	solar zenith angle
ψ_m	magnetic latitude
Ω_j	angular frequency of Jupiter

ω	frequency (angular)
	    ω_c cyclotron frequency of a given species
	    ω_{ce} electron cyclotron frequency
	    ω_{ci} ion cyclotron frequency
	    ω_{pi} ion plasma frequency
	    ω_{pe} electron plasma frequency
	    ω_{LHR} frequency of lower hybrid resonance
	    ω_{UHR} frequency of upper hybrid resonance
	    ω_R right-hand cutoff frequency
	    ω_L left-hand cutoff frequency

Acronyms

BP-HD	bent plane/hinged disc model of Jovian magnetodisc
BP/WP	bent plane/wave-propagating-outward model of Jovian magnetodisc
BS	bow shock
bKOM	broadband component of KOM
CA	closest approach
CIR	corotating interaction region
CML	central meridian longitude

DAM   Jupiter radiation component with spectral peak at wavelengths of
            decameters

DIM   Jupiter radiation component with spectral peak at wavelengths
            of decimeters

DOY   day of year

D_4 magnetic field model for Jupiter of Smith, Davis, and Jones [1976] – offset,
            tilted dipole

E-E   limited amplitude wave model of Eviatar and Ershkovich for Jovian
            magnetodisc

GSFC O_4 Goddard model of Jupiter's magnetic field

HG   heliographic

HOM   Jupiter radiation peaking in the hectometer wavelengths (between 100 m
            and 1000 m)

IFT   Io flux tube

IMF   interplanetary magnetic field

IRIS   infrared spectrometer

KOM   Jupiter radiation component with spectral peak at kilometer wavelengths
            (subdivided nKOM and bKOM)

LCFL   last closed field line

LECP   Low Energy Charged Particle detector

LEMPA   Low Energy Magnetospheric Particle Analyzer

LEPT   Low Energy Particle Telescope

LET-B   proton detector

LH   left-hand circularly polarized

LT   local time

LTE   local thermodynamic equilibrium

MS   magnetosheath

MP   magnetopause

nKOM   narrow band component of KOM

OTD   offset, tilted dipole

O_4	magnetic field model of Acuña and Ness [1976] (same as GSFC O_4)
P10	Pioneer 10
P11	Pioneer 11
P.A.	position angle of the plane of linear polarization
PLS	low-energy plasma instrument
PWS	plasma-wave instrument
RH	right-hand circularly polarized
RP/RD	rocking plane-rotating disc model of Jovian magnetodisc
RSS	radioscience instrument
S/C	spacecraft
SCET	spacecraft event time
SCM	spacecraft maneuver
SKR	Saturn kilometric radiation
SZA	solar zenith angle
TD	tangential discontinuity
TKR	terrestrial kilometric radiation
UT	universal time
UV	ultraviolet
UVS	ultraviolet spectrometer
V1	Voyager 1
V2	Voyager 2

APPENDIX B
COORDINATE SYSTEMS

A. J. Dessler

Jovian coordinate systems are not complicated or cabalistic, but they are different. The following is a description of these systems, as relevant to this book. I will also try to explain why things are as they are. There is logic behind the present system, even if some of the results seem curious or unfortunate.

B.1. Jovian longitude conventions

Latitude and longitude coordinates are usually established relative to some solid surface. Because Jupiter does not have a solid surface (at least none that is visible through the clouds), arbitrary, but convenient, coordinate grids have been prescribed. A spin equator is rather easily made out from observations of cloud motion, so the direction of the planetary spin axis is determined with relatively good accuracy. However, the determination of longitude is an entirely different matter.

Longitudes on a planet are fixed relative to an arbitrary, but well defined, prime- or zero-longitude meridian. For example, the Earth's prime meridian is the one that passes through the central cross-hair of the transit telescope at the Greenwich Royal Observatory. Its location is unique, and it stays put*. The selection of this meridian as the prime or zero-longitude meridian was initially arbitrary, but the selection, once made, fixes the longitude grid with precision. The problem immediately faced in establishing a Jupiter longitude system is that the mean rotation period of the clouds is a function of latitude. The equatorial region rotates faster than the temperate and polar regions, as is common in all planetary upper atmospheres. The difference is large enough that a cloud feature near the equator completely laps a cloud feature at higher latitude in about 120 Jupiter rotations, or 50 days. (Unless otherwise stated, "day" refers to a 24-hour terrestrial day.) Thus, a single longitude system cannot conveniently be used to keep track of motions of cloud features at different latitudes.

The solution selected was to define two separate longitude grids. System I applies to cloud features within about 10° of the equator. System II applies to higher latitudes. Rotation rates were established and a Central Meridian Longitude (i.e., sub-Earth longitude, commonly abbreviated CML) was selected for each of these systems. All of this was done some time ago, as evidenced by the starting time, Greenwich noon on July 14, 1897 that was used to define the locations of the prime meridians for Systems I and II.

The rotation of the longitude grid is defined in terms of a rotation rate (877.90°/day for System I and 870.27°/day for System II). The rotation period (which is usually quoted) is derived from these rotation rates. This explains the seemingly meaningless number of significant figures in the quoted rotation periods (9h 50m 30.0034s for System I and 9h 55m 40.6322s for System II). It is the rotation rate that is exact by

---

* It can be argued that continental drift does not move the land on which the Greenwich transit telescope rests. By definition it is the other land masses that move.

498

Fig. B.1. Jovian coordinate convention.
The view is from above the north pole.
The standard astronomical definition of
the eastern and western sides of Jupiter
as seen from the Earth are indicated by
an E and W, respectively. Unfortunately
the reverse (or the terrestrial convention),
with east on the right and west on the
left is also prevalent in the literature. The
reader should be wary. The prime
meridian rotates counterclockwise at a
constant rate Ω_J. Longitude is measured
clockwise from this prime meridian. The
System III longitude of the Earth, or of
any observer, is called the Central
Meridian Longitude (CML). This
illustration also shows the longitude of
some object (a satellite, spacecraft, or
phenomenon) marked X. Although this
figure shows λ_{III}, the same conventions
apply to λ_I and λ_{II} the only difference
being the numerical value of Ω_J, which is
different for each system, and which
causes the three prime meridians to move
relative to each other.

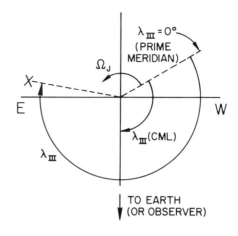

definition; the rotation period is only a numerical approximation. These rotation
periods have never been revised, and the 1897 convention is still the adopted one.

A third longitude system became necessary when, a half-century later, radio signals
were detected that gave evidence for a planetary magnetic field that rotates at a rate
intermediate between Systems I and II (although only about 0.3°/day faster than
System II). After about five years of observations, a rotation period of 9h 55m 29.37s
(not a rate as for Systems I and II) was selected, and a starting time of 00 UT on January 1, 1957 was picked. This system is called System III(1957.0). It is System III,
which describes the rotation of the Jovian magnetic field, that is of primary use to
magnetospheric physics.

Within less than a decade, it became apparent that the defined period for System
III(1957.0) was in error by less than 1 part in 10^5, the period being too short by about
1/3 s. The result was that a given radio phenomenon drifted steadily in longitude.
This drift amounts to only 3.4×10^{-3} degrees/Jovian rotation, but in a year it grows
to $3.4 \times 10^{-3} \times 365$ d/yr $\times 2.4$ rev/d $= 3°$/yr.

The direction of drift is determined by the sense of the longitude system, which is left
handed, as illustrated in Figure B.1. That is, looking down on Jupiter from above the
north pole, longitude increases in a clockwise direction. In a Mercator projection, longitude is usually shown increasing from left to right. However, for Jupiter, longitude is
usually shown increasing from right to left, as, for example, figure 1.1 or 1.3. The
advantage of a left-handed coordinate system is that, as viewed from the Earth (or from
any other distant or slowly moving observation point), the longitude directly beneath
the observer (the Central Meridian Longitude), increases with time. It should also be
noted that east and west on Jupiter are usually (but not always) defined in terms of
their direction on the Earth. Thus, for an observer in the Earth's northern hemisphere
looking at Jupiter in the southern sky, the western side of Jupiter is on the observer's

right (or terrestrial west), and the eastern side of Jupiter is on the observer's left (or terrestrial east). This is standard astronomical usage. However, the reader should be cautioned that the opposite convention is frequently used where east on Jupiter is the direction that an observer standing (or floating) on Jupiter would see the Sun rise, and west is the direction the observer would see the Sun set. The usage of east and west is inconsistent as applied to the planets (other than Earth); if the direction is important, be sure you know the convention being used in each individual paper.

The outstanding problem with a coordinate system that does not rotate at the same rate as the planet arises when one wishes to compare one data set with another. One must know the epoch of each data set, that is, the time when each were obtained, and make a correction for the drift (i.e., correct for the cumulative error in the inaccurately defined planetary spin period). For example, if one wished to compare Pioneer 10 flyby data (obtained in December 1973) with radio astronomy data obtained in, say, 1960, a correction of approximately 45° of longitude must be introduced. At best, this is an inconvenience (one must remember whether to add or subtract the correction), but, if the date (epoch) the data were obtained is not given, useful comparison is difficult or impossible.

In 1976, the International Astronomical Union (IAU) adopted a new longitude system, known as System III(1965). A more accurate rotation rate was selected so that the drift in longitude of magnetically related phenomena has been effectively stopped. Undoubtedly, some small error is still present in the selected value of the spin rate. For example, May, Carr, and Desch [1979] conclude the IAU period may be in error by about one part in 10^6 (consistent with the stated uncertainty in the IAU value). This could lead to a drift of 0.19°/year, or less than 2°/decade, which is a drift rate that can be safely ignored by magnetospheric physicists for decades to come.

It is often necessary to compare System III(1957.0) data with System III(1965) data, not only to be able to compare older radio data with more recent measurements, but to compare spacecraft data. Pioneer 10 and 11 data were reported in 1957.0 coordinates, whereas Voyager data were reported in 1965 coordinates, and a correction of about 30° is required if magnetic longitudes between these two missions are to be correlated.

The transformation from $\lambda_{III}(1957.0)$ to $\lambda_{III}(1965)$ on a given Julian date t is given by Riddle and Warwick [1976], and in a slightly more precise form by Seidelman and Devine [1977], as

$$\lambda_{III}(1965) = \lambda_{III}(1957.0) - 0.0083169 \, (t - 2438761.5) \tag{B.1}$$

The number inside the final parentheses on the right is the Julian date for January 1, 1957 at 00 UT. Equation (B.1) may be written more conveniently (although with slightly less accuracy) as

$$\lambda_{III}(1965) = \lambda_{III}(1957.0) - 3.04T \tag{B.2}$$

where T is the time in years and decimal fraction of a year since 00 UT on January 1, 1965. For example, Pioneer 10 made its closest approach to Jupiter on December 4, 1973. Thus, $T = 8.9$, and from Equation (B.2) we find we must subtract $3.04 \times 8.9 = 27°$ from longitudes related to the Pioneer 10 flyby to convert them to System III (1965). In cases where an approximate correction is adequate, the value of the correction can be scaled from Figure B.2.

Much of the early literature for the Pioneer 10 and 11 encounters contains reference to $\lambda_{III}(1973.9)$ or $\lambda_{III}(1974.9)$. This usage is in error. Whenever it appears, the reader would probably be safe in assuming that System III(1957.0) is what was used, and the

Fig. B.2. Difference between $\lambda_{III}(1957.0)$ and $\lambda_{III}(1965)$. The two longitude systems were in agreement at 00 UT on January 1, 1965. The 1957.0 coordinate system drifts slowly relative to Jupiter's magnetic field. To convert a $\lambda_{III}(1957.0)$ longitude to $\lambda_{III}(1965)$ simply subtract $\Delta\lambda_{III}$ from the 1957.0 value. In the equation, T is the time in years since 1965.0. The times for the Jupiter encounters for Pioneer 10 and 11 are shown on the 1957.0 line and for Voyager 1 and 2 on the 1965 line. (See also Eqs. (B.1) and (B.2).)

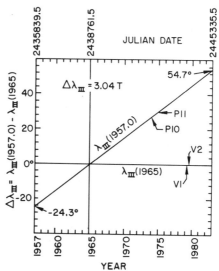

referenced observations were obtained in 1973.9, 1974.9, or whatever year is given in parentheses. By 1978, System III(1965) was in common usage. As far as I am aware, all of the 1979 Voyager encounter data (even the cloud imaging data) and the subsequent analyses are presented in System III(1965) coordinates. Papers published before 1977, with exceptions, use System III(1957.0).

Finally, it is sometimes necessary to convert a System II longitude to System III longitude. The conversion is, from Seidelman and Devine [1977],

$$\lambda_{III}(1965) = \lambda_{II} + 81.2 + 0.266\,(t - 2438761.5) \tag{B.3}$$

Readers interested in or requiring more detailed information on Jovian System III longitudes are referred to the explanatory papers by Mead [1974] (covers 1957 coordinates and the Pioneer flybys), Riddle and Warwick [1976], and particularly Seidelmann and Devine [1977] (contains the final IAU definitions for the 1965 system). Those wishing to calculate values of λ_{III} for satellites or for the Central Meridian will need to consult the *Astronomical Almanac*. Before 1981 these volumes were published as *The American Ephemeris and Nautical Almanac* and *The Nautical Alamanac and Astronomical Ephemeris*. These volumes are produced annually and are sold in the United States by the U.S. Government Printing Office and in the United Kingdom by Her Majesty's Stationery Office. There is also an *Explanatory Supplement to the Astronomical Ephemeris and the American Ephemeris and Nautical Alamanac*, which contains explanations and derivations relevant to the *Astronomical Almanac* and its predecessors.

The following is a specific example of how to use the *Almanac* to find the phase angle and System III longitude of a specific satellite (T. D. Carr, private communication). For this example we will find the position of Europa at 1720 UT on August 24, 1976. Page references are to the *American Ephemeris and Nautical Almanac, 1976*.

First we calculate Europa's orbital phase γ_E (see next section for definitions). From page 393 of the *Almanac* we see that Europa, *Satellite II* was a Superior Geocentric Conjunction ($\gamma_E = 0°$) at 0206 UT on August 23. Using Europa's period from page 390, we find that by 1720 on the 24th the satellite has moved

$$\gamma_E = 360° \times 1.6347/3.5541 = 166°$$

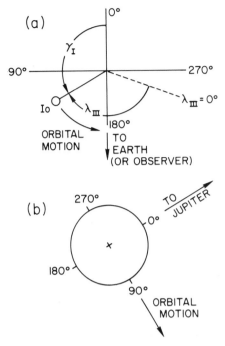

Fig. B.3. Satellite coordinate convention. The view is from above the north pole. (*a*) The orbital phase angle of the satellite (Io in this figure) is γ_1, which is measured counterclockwise starting from geocentric superior conjunction (i.e., the anti-Earth meridian). Note that the phase angle can be similarly defined for an observing point away from the Earth, for example, a spacecraft. $\gamma = 0°$ is always the anti-observer meridian. The value of λ_{III} for the satellite is determined as in Figure B.1. (*b*) The longitude on a satellite is measured from its prime meridian, which is the sub-Jupiter meridian. This meridian is fixed on the Jovian satellites because the same face of each always points toward Jupiter. This is a left-handed coordinate system (longitude increases in a clockwise direction as viewed from above the north pole) so that, as seen from the Earth, longitude increases with time.

Next we find the Central Meridian Longitude. From page 383, $\lambda_{II} = 219.5°$ at 00 UT on August 24. (Note that System III CML is not listed in the *Almanac* until 1981. The *Alamanac* is not quick to adopt the latest fads.) The Julian date of 00 UT August 24 is 2443014.5 (from page 17). We convert to λ_{III} using equation (B.3), which yields $\lambda_{III} = 1432.0°$ or $\lambda_{III}(1965) = 352°$. The elapsed time from 00 UT to our desired time is 0.7222 day. The change in λ_{III} is $\Delta\lambda_{III} = 870.536°/\text{day} \times 0.7222 \text{ day} = 629°$. Therefore at 1720 UT, $\lambda_{III}(1965)$ of Europa is $352 + 629 = 981°$ or $\lambda_{III}(1965) = 261°$.

B.2. Orbital phase angle and longitude conventions for satellites

The position of a satellite in its orbit around Jupiter is described by an orbital phase angle. In addition, the larger satellites have longitude systems of their own (these two coordinates are illustrated in Fig. B.3). Because of its popularity in magnetospheric circles, Io is the satellite in the Figure B.3(a), but the system is the same for all of Jupiter's satellites. The orbital phase angle γ is measured counterclockwise (as viewed from north of the ecliptic) from superior geocentric conjunction. This is a right-handed system with $\gamma = 0°$ when the satellite is directly behind Jupiter as seen from the Earth. The specific satellite is indicated by a subscript, such as γ_i for Io's phase angle. Like CML, γ is sometimes referenced to an extraterrestrial observer (for example, a spacecraft) instead of to Earth. In such a case, $\gamma = 0°$ at the orbital meridian that is 180° from the observer's meridian.

Longitude is measured clockwise around each satellite starting from the meridian that points toward Jupiter. This definition of a prime meridian is possible for the Jovian satellites because the same side of a given satellite always faces Jupiter, as does the Earth's Moon. (The actual definition of the prime meridian is more complex than indicated here because slight ellipticity of a satellite orbit causes some periodic (libration) motion of the sub-Jupiter meridian. However, the above definition, illustrated in Fig. B.3(b), should be adequate for magnetospheric study.)

Fig. B.4. Jovian dipole equators. The spin equator is a plane that passes through the center of the planet and is perpendicular to the spin axis, which is shown here having an angular velocity $\tilde{\Omega}$. The magnetic equator is the surface defined by the locus of points of minimum magnetic field strength along a magnetic line of force from dipole moment **M**. This point is located where r reaches its maximum value. The centrifugal equator is the surface defined by the locus of points that are the maximum distance (or where ρ is a maximum) from the spin axis for given lines of force. If Jupiter's magnetic dipole is tilted at an angle $\alpha = 10°$ toward λ_{III} $= 200°$, the centrifugal equator is tipped $7°$ toward this same longitude.

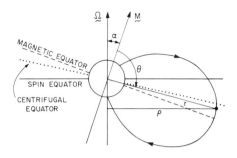

B.3. Latitude conventions

There are, as for the Earth, two latitude systems for Jupiter. There is the conventional Jovigraphic system with latitude measured positive northward from the spin equator. In addition, there is a magnetic latitude system defined by a centered tilted dipole. The northern end of this dipole is tilted $10°$ toward $\lambda_{III}(1965) = 200°$, and the southern end is, of course, tilted $10°$ toward $\lambda_{III} = 20°$ (values given for the tilt differ by about $\pm 1°$ and values of λ_{III} by about $\pm 3°$, see Tables 1.2 and 1.3). The magnetic equator is the plane that is perpendicular to the centered dipole and passes through Jupiter's center. Latitude is measured positive northward.

A possible source of confusion is that the symbol λ has been usurped to signify Jovian longitude, whereas λ is commonly used to designate latitude on the Earth. In this book we have selected ψ for the Jovigraphic latitude symbol, and ψ_m for magnetic latitude.

Magnetic and Jovigraphic latitudes agree where the two equators cross ($\lambda_{III} = 110°$ and 290°). Near the equator, the difference between magnetic and Jovigraphic latitude as a function of System III longitude is approximated by

$$\Delta\psi = \psi_m - \psi = 10° \sin (\lambda_{III} - 110°) \tag{B.4}$$

Because the Io plasma torus is important, and because it consists principally of relatively heavy ions with low energies, one other "equator" is necessary to fully describe the Jovian magnetosphere, and that is the "centrifugal equator." Energetic charged particles bounce (mirror) about the magnetic equator. An energetic particle with $90°$ pitch angle will drift along the magnetic equator. Because of the large quadrupole and octupole moments in Jupiter's magnetic field, particles close to Jupiter drift along a more complex path, the "particle drift equator," shown in Figure 1.12. Note that at distances of only 6 R_J, the drift equator (shown as the dotted line) is essentially that of a simple dipole.

Particles of sufficiently low energy are affected by centrifugal force in Jupiter's large, rapidly rotating magnetosphere. They do not follow the magnetic equator. Consider first a particle of zero magnetic moment (or perpendicular energy) in the corotating frame. It will slide along a magnetic field line and settle at the point (or oscillate

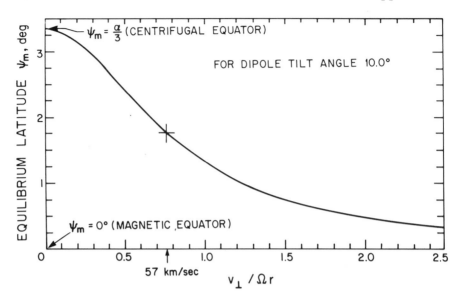

Fig. B.5. The equilibrium latitude of a charged particle as determined by centrifugal and magnetic-moment forces. The abscissa is the particle speed perpendicular to the magnetic field ($v_\perp$) in units of corotation speed (Ωr) where Ω is the corotation angular velocity and r is the Jovicentric distance to the particle. The $v_\perp$ of an ion freshly injected from Io is indicated (57 km/s). The plane of equilibrium latitude for such particles is tilted about 1.8° from the magnetic equator. Note that the equilibrium latitude depends only on particle speed, not charge or mass [after Cummings et al., 1980].

about the point) that is a maximum distance from the spin axis. The locus of such points defines the centrifugal equator. The centrifugal equator lies between the magnetic and spin equators, as illustrated in Figure B.4. It is the locus of points for which ρ is a maximum. If the magnetic field is that of a simple dipole, the centrifugal equator is tipped away from the magnetic equator and toward the spin equator by 1/3 of the dipole tilt angle α. If the field is not dipolar, the angle between the magnetic equator and the centrifugal equator is $\alpha r_c/L$ where r_c is the radius of curvature of the field line at the centrifugal equator, and L is the equatorial crossing distance of the field line.

A warm particle having a nonzero magnetic moment experiences a mirroring force that pushes it toward the magnetic equator. Such a particle has an equilibrium position somewhere between the centrifugal and magnetic equators, as illustrated in Figure B.4. The displacement of this equilibrium position from the magnetic equator as a function of the particle's cyclotron speed, normalized in terms of the corotation speed, is shown in Figure B.5. Note that, in this presentation, the equilibrium latitude is independent of particle mass or charge.

ACKNOWLEDGMENTS

I wish to thank J. K. Alexander, F. A. Bozyan, M. L. Kaiser, and particularly T. D. Carr for their kind, patient, and sometimes repeated explanations of the mysterious workings of System III longitudes.

APPENDIX C

JUPITER AND IO: SELECTED PHYSICAL PARAMETERS

Jupiter

Heliocentric distance	$5.20 \text{ AU} = 7.78 \times 10^8 \text{ km}$
Sidereal period	11.86 years
Synodic period	398.88 days $= 13.10$ months $= 1.092$ years
Mean orbital speed	13.06 km/sec $= 15.7 R_J$/day
Equatorial radius $(1 R_J)$	$7.14 \times 10^4 \text{ km} = 1 R_J$ (by definition)
Polar radius	$6.68 \times 10^4 \text{ km}$
Practical radius (nearly the mean and easily remembered)	$7 \times 10^7 \text{ m}$
Mass	$317.8 M_\oplus = 1.901 \times 10^{27} \text{ kg}$
Escape speed	61 km/sec
Gravitational acceleration	$25.9 \text{ m/sec}^2 = 2.64 g_\oplus$
Escape energy for Hydrogen atom	19.4 eV
System III (1965) sidereal spin period	9 hr 55 min 29.71sec (derived from angular velocity) $= 3.573 \times 10^4 \text{ sec} = 9.925$ hr
System III (1965) angular velocity	$1.76 \times 10^{-4} \text{ rad/sec} = 870.536°$/day (by definition)
System I (equatorial clouds) sidereal spin period	9 hr 50 min 30.00 sec $= 9.842$ hr
System II (polar clouds) sidereal spin period	9 hr 55 min 40.63 sec $= 9.928$ hr
Rotational (spin) kinetic energy	$3.6 \times 10^{34} \text{ J}$
Spin equator inclined to orbital plane	3° 5′
Main magnetic-dipole moment	$4.2 \times 10^{-4} \text{ } TR_J^3$
Magnetic dipole tilt	$9.8° \pm 0.3°$
Tilt toward	$\lambda_{III} = 200° \pm 2°$
Dipole displaced	$0.12 \pm 0.02 R_J$ toward $\lambda_{III} = 149° \pm 6°$

Io

Jovicentric distance	$5.91 R_J = 4.216 \times 10^5 \text{ km}$
Sidereal period	1.769 days $= 42.46$ hr
Mean orbital speed about Jupiter	17.34 km/sec
Angular velocity about Jupiter	$4.112 \times 10^{-5} \text{ rad/sec}$
Radius	$1.82 \times 10^3 \text{ km} = 2.55 \times 10^{-2} R_J$
Mass	$8.91 \times 10^{22} \text{ kg}$
Escape speed	2.56 km/sec
Gravitational acceleration	$1.80 \text{ m/sec}^2 = 0.184 g_\oplus$
Corotation speed at Io's orbit	74.2 km/sec
Angular velocity of corotation relative to Io	$1.35 \times 10^{-4} \text{ rad/sec}$
Speed of corotation relative to Io	56.8 km/sec
Corotation electric field at Io	0.113 V/m outward from Jupiter
Max potential across Io	411 kV

REFERENCES

Acuña, M. H., and N. F. Ness. Results from the GSFC fluxgate magnetometer on Pioneer 11. In *Jupiter*, ed. T. Gehrels, pp. 830-847. University of Arizona Press, Tucson, 1976a.

Acuña, M. H., and N. F. Ness. The main magnetic field of Jupiter. In *Magnetospheric Particles and Fields*, ed. B. M. McCormac, pp. 311-323. D. Reidel, Dordrecht, 1976b.

Acuña, M. H., and N. F. Ness. The main magnetic field of Jupiter. *J. Geophys. Res. 81*:2917-2922, 1976c.

Acuña, M. H., F. M. Neubauer, and N. F. Ness. Standing Alfvén wave current system at Io: Voyager 1 observations. *J. Geophys. Res. 86*:8513-8522, 1981.

Aitken, D. K., and M. F. Harrison. Measurement of the cross-sections for electron impact ionization of multi-electron ions. I. O^+ to O^{2+} to O^{3+}. *J. Phys. B 4*:1176-1188, 1971.

Akasofu, S.-I. Energy coupling between the solar wind and the magnetosphere. *Space Sci. Rev. 28*:121-190, 1981.

Alfvén, H.. On the solar origin of cosmic radiation. *Phys. Rev. 75*:1732-1735, 1949.

Alfvén, H. *On the Origin of the Solar System*, Chapter 2. Oxford at the Clarendon Press, London, 1954.

Alfvén, H. Some properties of magnetospheric neutral surfaces. *J. Geophys. Res. 73*:4379-4381, 1968.

Alexander, J. K. Note on the beaming of Jupiter's decameter-wave radiation and its effect on radio rotation period determinations. *Astrophys. J. 195*:227-233, 1975.

Alexander, J. K., T. D. Carr, J. R. Thieman, J. J. Schauble, and A. C. Riddle. Synoptic observations of Jupiter's radio emissions: Average statistical properties oberved by Voyager. *J. Geophys. Res. 86*:8529-8545, 1981.

Alexander, J. K., M. D. Desch, M. L. Kaiser, and J. R. Thieman. Latitudinal beaming of Jupiter's low-frequency radio emissions. *J. Geophys. Res. 84*:5167-5174, 1979.

Alexander, J. K., and M. L. Kaiser. Terrestrial kilometric radiation: 1. Spatial structure studies. *J. Geophys. Res. 81*:5948-5956, 1976.

Allcock, G. McK. A study of the audio-frequency radio phenomena known as "dawn chorus." *Austr. J. Phys. 10*:286-298, 1957.

Allcock, G. McK., and L. H. Martin. Simultaneous occurrence of "dawn chorus" at places 600 km apart. *Nature 178*:937-938, 1956.

Allen, C. W. *Astrophysical Quantities*. Athlone Press, London, 1973.

Anderson, R. R., and D. A. Gurnett. Plasma wave observations near the plasmapause with the S^3-A satellite. *J. Geophys. Res. 78*:4756-4764, 1973.

Anderson, R. R., G. K. Parks, T. E. Eastman, D. A. Gurnett, and L. A. Frank. Plasma waves associated with energetic particles streaming into the solar wind from the Earth's bow shock. *J. Geophys. Res. 86*:4493-4510, 1981.

Andronov, A. A., and V. Y. Trakhtengerts. Kinetic instability of the earth's outer radiation belt. *Geomagn. Aeron. 4*:181-188, 1964.

Angerami, J. J., and J. O. Thomas. Studies of planetary atmospheres. *J. Geophys. Res. 69*:4537-4560, 1964.

506

Armstrong, T. P., M. T. Paonessa, S. T. Brandon, S. M. Krimigis, and L. J. Lanzerotti. Low energy charged particle observations in the 5-20 R_J region of the Jovian magnetosphere. *J. Geophys. Res.* *86*:8343-8355, 1981.

Asbridge, J. R., S. J. Bame, W. C. Feldman, and M. D. Montgomery. Helium and hydrogen velocity differences in the solar wind. *J. Geophys. Res.* *81*:2719-2727, 1976.

Ashour-Abdalla, M., and C. F. Kennel. Nonconvective and convective electron cyclotron harmonic instabilities. *J. Geophys. Res.,* *83*:1531-1543, 1978a.

Ashour-Abdalla, M., and C. F. Kennel. Multi-harmonic electron cyclotron instabilities. *Geophys. Res. Lett.* *5*:711-714, 1978b.

Ashour-Abdalla, M., H. Okuda, and C. Z. Cheng. Acceleration of heavy ions on auroral field lines. *Geophys. Res. Lett.* *8*:795-798, 1981.

Ashour-Abdalla, M., and R. M. Thorne. The importance of electrostatic ion-cyclotron instability for quiet-time proton auroral precipitation. *Geophys. Res. Lett.* *4*:45-48, 1977.

Ashour-Abdalla, M., and R. M. Thorne. Towards a unified view of diffuse auroral precipitation. *J. Geophys. Res.* *83*:4755-4766, 1978.

Atreya, S. K., and T. M. Donahue. Model ionospheres of Jupiter. In *Jupiter*, ed. T. Gehrels, pp. 304-318. University of Arizona Press, Tucson, 1976.

Atreya, S. K., and T. M. Donahue. The atmosphere and ionosphere of Jupiter. In *Vistas in Astronomy*, ed. A. Beer, *25*:315-335. Pergamon Press, Oxford, 1982.

Atreya, S. K., T. M. Donahue, and M. C. Festou. Jupiter: Structure and composition of the upper atmosphere. *Astrophys. J.* *247*:L43-L47, 1981.

Atreya, S. K., T. M. Donahue, and M. B. McElroy. Jupiter's ionosphere: Prospects for Pioneer 10. *Science 184*:154-156, 1974.

Atreya, S. K., T. M. Donahue, B. R. Sandel, A. L. Broadfoot, and G. R. Smith. Jovian upper atmospheric temperature measurement by the Voyager 1 UV spectrometer. *Geophys. Res. Lett.* *6*:795-798, 1979.

Atreya, S. K., T. M. Donahue, and J. H. Waite, Jr. An interpretation of the Voyager measurement of Jovian electron density profiles. *Nature 280*:795-796, 1979.

Atreya, S. K., and J. H. Waite, Jr. Saturn ionosphere: Theoretical interpretation. *Nature (Lond.), 292*:682-683, 1981.

Atreya, S. K., Y. L. Yung, T. M. Donahue, and E. S. Barker. Search for Jovian auroral hot spots. *Astrophys. J. 218*:L83-L87, 1977.

Aumann, H. H., C. M. Gillespie, Jr., and F. J. Low. The internal power and effective temperature of Jupiter and Saturn. *Astrophys. J. Lett. 157*:L69-L72, 1969.

Axford, W. I. Magnetospheric convection. *Rev. Geophys.* *7*:421-459, 1969.

Bagenal, F. The inner magnetosphere of Jupiter and the Io plasma torus. Ph.D. thesis, Massachusetts Institute of Technology, Cambridge, Mass., June 1981.

Bagenal, F., and J. D. Sullivan. Direct plasma measurements in the Io torus and inner magnetosphere of Jupiter. *J. Geophys. Res.* *86*:8447-8466, 1981.

Bagenal, F., J. D. Sullivan, and G. L. Siscoe. Spatial distribution of plasma in the Io torus: *Geophys. Res. Lett.* *7*:41-44, 1980.

Baker, D. N., and C. K. Goertz. Radial diffusion in Jupiter's magnetosphere. *J. Geophys. Res. 81*: 5215-5219, 1976.

Baker, D. N., and J. A. Van Allen. Energetic electrons in the Jovian magnetosphere. *J. Geophys. Res. 8l*:617-632, 1976.

Baldwin, D. E., I. B. Bernstein, and M. P. H. Weenink. Kinetic theory of plasma waves in a magnetic field. *Adv. in Plasma Phys., Vol. 3*, ed. A. Simon and W. B. Thompson, pp. 1-124. Interscience, New York, 1969.

Baluja, K. L., R. G. Burke, and A. E. Kingston. Electron impact excitation of semi-forbidden transitions of O III. *J. Phys. B 13*:829-838, 1980.

Barber, D. The polarization, periodicity, and angular diameter of the radiation from
    Jupiter at 610 Mc. *Mon. Not. Roy. Astron. Soc. 133*:285-308, 1966.

Barbosa, D. D. Electrostatic mode coupling at $2\omega_{UH}$: A generation mechanism for auroral
    kilometric radiation. Ph.D. thesis, University of California, Los Angeles, 1976.

Barbosa, D. D. Fermi-Compton scattering due to magnetopause surface fluctuations in
    Jupiter's magnetospheric cavity. *Astrophys. J.*, *243*:1076-1087, 1981a.

Barbosa, D. D. On the injection and scattering of protons in Jupiter's magnetosphere.
    J. Geophys. Res. 86:8981-8990, 1981b.

Barbosa, D. D., and F. B. Coroniti. Relativistic electrons and whistlers in Jupiter's
    magnetosphere. *J. Geophys. Res. 81*:4531-4536, 1976.

Barbosa, D. D., D. A. Gurnett, W. S. Kurth, and F. L. Scarf. Structure and properties of
    Jupiter's magnetoplasmadisc, *Geophys. Res. Lett. 6*:785-788, 1979.

Barbosa, D. D., and W. S. Kurth. Suprathermal electrons and Bernstein waves in Jupiter's
    inner magnetosphere. *J. Geophys. Res. 85*:6729-6742, 1980.

Barbosa, D. D., F. L. Scarf, W. S. Kurth, and D. A. Gurnett. Broadband electrostatic
    noise and field-aligned currents in Jupiter's middle magnetosphere. *J. Geophys.
    Res. 86*:8357-8369, 1981.

Barish, F. D., and R. A. Smith. An analytical model of the Jovian magnetosphere.
    Geophys. Res. Lett. 2:269-272, 1975.

Bar-Nun, A. Thunderstorms on Jupiter. *Icarus 24*:86-94, 1975.

Barrow, C. H. Association of corotating magnetic sector structure with Jupiter's
    decameter-wave radio emission. *J. Geophys. Res. 84*:5366-5372, 1979.

Barrow, C. H., and J. K. Alexander. Maximum frequency of the decametric radiation
    from Jupiter. *Astron. Astrophys. 90*:L4-L6, 1980.

Barrow, C. H., and M. D. Desch. Non-Io decameter radiation from Jupiter at
    frequencies above 30 MHz. *Astron. Astrophys. 86*:339, 1980.

Barrow, C. H., and D. P. Morrow. The polarization of the Jupiter radiation at 18 Mc/s.
    Astrophys. J. 152:593-608, 1968.

Bartels, J. Eccentric dipole approximating the Earth's magnetic field. *Terrest. Magn.
    41*:225-250, 1936.

Bates, D. R. (ed.). *Atomic and Molecular Processes*. Academic Press, New York, 1962.

Bates, D. R., and A. Dalgarno. Electronic recombination. In *Atomic and Molecular
    Processes*, ed. D. R. Bates, p. 245. Academic Press, New York, 1962.

Beard, D. B., and D. J. Jackson. The Jovian magnetic field and magnetosphere shape.
    J. Geophys. Res. 81:3399-3400, 1976.

Beard, D. B., and J. L. Luthey. Analysis of the Jovian electron belts: 2. Observations of
    the decimeter radiation. *Astrophys. J. 183*:679-689, 1973.

Beck, A. J. Conclusion. In *Proceedings of the Jupiter Radiation Belt Workshop*, ed. A. J.
    Beck, pp. 473-485. NASA TM 33543, 1972.

Behannon, K. W. Mapping of the earth's bow shock and magnetic tail by Explorer 33.
    J. Geophys. Res. 73:907-930, 1968.

Behannon, K. W. Observations of the interplanetary magnetic field between 0.46 and 1
    AU by the Mariner 10 spacecraft. Ph.D. Thesis, Catholic University of America.
    NASA GSFC document X-692-76-2, January, 1976.

Behannon, K. W., L. F. Burlaga, and N. F. Ness. The Jovian magnetotail and its current
    sheet. *J. Geophys. Res. 86*:8385-8401, 1981.

Belcher, J. W., C. K. Goertz, and H. S. Bridge. The low energy plasma in the Jovian
    magnetosphere. *Geophys. Res. Lett. 7*:17-20, 1980.

Belcher, J., C. K. Goertz, J. D. Sullivan, and M. H. Acuña. Plasma observations of the
    Alfvén wave generated by Io. *J. Geophys. Res. 86*:8508-8512, 1981.

Belcher, J. W., and R. L. McNutt, Jr. The dynamic expansion and contraction of the
    Jovian plasma sheet. *Nature 287*:813-815, 1980.

Ben-Ari, M. The decametric radiation of Jupiter: A nonlinear mechanism. Ph.D. thesis, Tel-Aviv University, Tel Aviv, 1980.

Berge, G. L. An interferometric study of Jupiter at 10 and 21 cm. *Radio Sci. 69D*:1552-1556, 1965.

Berge, G. L. An interferometric study of Jupiter's decimetric radio emission. *Astrophys. J. 146*:767-798, 1966.

Berge, G. L. The position and Stokes parameters of the integrated 21 cm radio emission of Jupiter and their variation with epoch and central meridian longitude. *Astrophys. J. 191*:775-784, 1974.

Berge, G. L., and S. Gulkis. Earth-based radio observations of Jupiter: Millimeter to meter wavelengths. In *Jupiter*, ed. T. Gehrels, pp. 621-692. University of Arizona Press, Tucson, 1976.

Bergstralh, J. T., D. L. Matson, and T. V. Johnson. Sodium D-line emission from Io: Synoptic observations from Table Mountain Observatory. *Astrophys. J. Lett. 195*:L131-L135, 1975.

Bergstralh, J. T., J. W. Young, D. L. Matson, and T. V. Johnson. Sodium D-line emission from Io: A second year of synoptic observation from Table Mountain Observatory. *Astrophys. J. 211*:L51-L55, 1975.

Bhadra, K., and R. J. W. Henry. Oscillator strengths and collision strengths for S IV. *Astrophys. J. 240*:368-373, 1980.

Bigg, E. K. Influence of the satellite Io on Jupiter's decametric emission. *Nature 203*:1008-1010, 1964.

Bigg, E. K. Periodicities in Jupiter's decametric radiation. *Planet. Space Sci. 14*:741-758, 1966.

Biraud, F., J. C. Ribes, J. D. Murry, and J. A. Roberts. Circular polarization of Jupiter's 1.4 GHz radio emission. *Astron. Astrophys. 58*:433-435, 1977.

Birmingham, T. J. Jovian magnetic models and the polarization angle of Jovian decimetric radiation. *Astrophys. J. 245*:736-742, 1981.

Birmingham, T. J., J. K. Alexander, M. D. Desch, R. F. Hubbard, and B. M. Pedersen. Observations of electron gyroharmonic waves and the structure of the Io torus. *J. Geophys. Res. 86*:8497-8507, 1981.

Birmingham, T., W. Hess, T. Northrop, R. Baxter, and M. Lojko. The electron diffusion coefficient in Jupiter's magnetosphere. *J. Geophys. Res. 79*:87-97, 1974.

Birmingham, T. J., and T. G. Northrop. Theory of flux anisotropies in a guiding center plasma. *J. Geophys. Res. 84*:41-45, 1979.

Boischot, A., and M. Aubier. The Jovian decametric arcs as an interference pattern. *J. Geophys. Res., 86*:8561-8563, 1981.

Boischot, A., A. Lecacheux, M. L. Kaiser, M. D. Desch, J. K. Alexander et al. Radio Jupiter after Voyager: An overview of the Planetary Radio Astronomy observations. *J. Geophys. Res. 86*:8213-8226, 1981.

Book, D. L. *NRL Plasma Formulary* (NRL Memorandum Report No. 3332), 1980.

Bossen, M., R. L. McPherron, and C. T. Russell. A statistical study of Pc 1 micropulsations at synchronous orbit. *J. Geophys. Res. 81*: 6083-6091, 1976.

Boström, R. Ionosphere-magnetosphere coupling. In *Magnetospheric Physics*, ed. B. M. McCormac, pp. 45-59. D. Reidel, Dordrecht, 1974.

Bozyan, F. A., and J. N. Douglas. Directivity and stimulation in Jovian decametric radiation. *J. Geophys. Res. 81*:3387-3392, 1976.

Branson, N. J. B. A. High resolution radio observations of the planet Jupiter. *Mon. Not. Roy. Astron. Soc. 139*:155-162, 1968.

Brice, N. Fundamentals of very low frequency emission generation mechanisms. *J. Geophys. Res. 69*:4515-4522, 1964.

Brice, N. M., and G. A. Ioannidis. The magnetospheres of Jupiter and Earth. *Icarus 13*:173-183, 1970.

Brice, N. M., and T. R. McDonough. Jupiter's radiation belts. *Icarus 18*:206-219, 1973.

Bridge, H. S., J. W. Belcher, R. J. Butler, A. J. Lazarus, A. M. Mavretic et al. The plasma experiment on the 1977 Voyager mission. *Space Sci. Rev. 21*:259-287, 1977.

Bridge, H. S., J. W. Belcher, A. J. Lazarus, J. D. Sullivan, R. L. McNutt et al. Plasma observations near Jupiter: Initial results from Voyager 1. *Science 204*:987-991, 1979a.

Bridge, H. S., J. W. Belcher, A. J. Lazarus, J. D. Sullivan, F. Bagenal et al. Plasma observations near Jupiter: Initial results from Voyager 2. *Science 206*:972-976, 1979b.

Broadfoot, A. L., M. J. S. Belton, P. Z. Takacs, B. R. Sandel, D. E. Schemansky et al. Extreme ultraviolet observations from Voyager 1 encounter with Jupiter. *Science 204*:979-982, 1979.

Broadfoot, A. L., B. R. Sandel, D. E. Shemansky, J. C. McConnell, G. R. Smith et al. Overview of the Voyager ultraviolet spectrometry results through Jupiter encounter. *J. Geophys. Res. 86*:8259-8284, 1981.

Brook, E., M. F. A. Harrison, and A. C. H. Smith. Electron ionization of He, C, N, and O atoms. *J. Phys. B 11*:3115-3132, 1978.

Brown, L. W. Spectral behavior of Jupiter near 1 MHz. *Astrophys. J. 194*:L159-L162, 1974.

Brown, R. A. Optical line emission from Io. In *Exploration of the Planetary System*, ed. A. Woszczyk and C. Iwaniszewska, pp. 527-531. D. Reidel, Hingham, Mass., 1974.

Brown, R. A. A model of Jupiter's sulfur nebula. *Astrophys. J. 206*:L179-L183, 1976.

Brown, R. A. Measurements of S II optical emission from the thermal plasma of Jupiter. *Astrophys. J. Lett. 224*:L97-L98, 1978.

Brown, R. A. The Jupiter hot plasma torus: observed electron temperature and energy flows. *Astrophys. J. 244*:1072-1080, 1981a.

Brown, R. A. Heavy ions in Jupiter's environment. In *Planetary Aeronomy and Astronomy*, ed. S. K. Atreya and J. J. Caldwell. *Adv. Space Res. 1*:75-82, 1981b.

Brown, R. A. The probability distribution of S^+ gyrospeeds in the Io torus. *J. Geophys. Res. 87*:230-234, 1982.

Brown, R. A., R. M. Goody, F. J. Murcray, and F. H. Chaffee. Further studies of line emission from Io. *Astrophys. J. 200*:L49-L53, 1975.

Brown, R. A., and W.-H. Ip. Atomic clouds as distributed sources for the Io plasma torus. *Science 213*:1493-1495, 1981.

Brown, R. A., and N. M. Schneider. Sodium remote from Io. *Icarus, 48*:519-535, 1981.

Brown, R. A., and D. E. Shemansky. On the nature of SII emission from the Io plasma torus. *Atrophys. J.,* in press, 1982.

Brown, R. A., D. E. Shemansky, and R. E. Johnson. Observed deficiency of O III in the Io torus. *Astrophys. J.,* in press, 1983.

Brown, R. A., and Y. L. Yung. Io, its atmosphere and optical emissions. In *Jupiter*, ed. T. Gehrels, pp. 1102-1145. University of Arizona Press, Tucson, 1976.

Brown, W. L., W. M. Augustyniak, L. J. Lanzerotti, R. E. Johnson, and R. Evatt. Linear and nonlinear processes in the erosion of H_2O ice by fast light ions. *Phys. Rev. Lett. 45*:1632-1635, 1980.

Burke, B. F., and K. L. Franklin. Observations of a variable radio source associated with the planet Jupiter. *J. Geophys. Res. 60*:213-217, 1955.

Burlaga, L. F., J. W. Belcher, and N. F. Ness. Disturbances observed near Ganymede by Voyager 2. *Geophys. Res. Lett. 7*:21-24, 1980.

Burtis, W. J., and R. A. Helliwell. Magnetospheric chorus: Occurrence patterns and normalized frequency. *Planet. Space Sci. 24*:1007-1024, 1976.

Burton, E. T., and E. M. Boardman. Audio-frequency atmospherics. *Proc. IRE 21*:1476-1494, 1933.

Busse, F. Theory of planetary dynamos. In *Solar System Plasma Physics, Vol II*, ed. C. F. Kennel, L. J. Lanzerotti, and E. N. Parker, pp. 293-317. North Holland, Amsterdam, 1979.

Capone, L. A., J. Dubach, R. C. Whitten, and S. S. Prasad. Cosmic ray ionization of the Jovian atmosphere. *Icarus 39*:433-449, 1979.

Capone, L. A., R. C. Whitten, J. Dubach, S. S. Prasad, and W. T. Huntress, Jr. The lower ionosphere of Titan. *Icarus 28*:367-378, 1976.

Capone, L. A., R. C. Whitten, S. S. Prasad, and J. Dubach. The ionospheres of Saturn, Uranus, and Neptune. *Astrophys. J. 215*:977-983, 1977.

Carbary, J. F. Periodicities in the Jovian magnetosphere: Magnetodisc models after Voyager. *Geophys. Res. Lett. 7*:29-32, 1980.

Carbary, J. F., and T. W. Hill. A self-consistent model of a corotating Jovian magnetosphere. *J. Geophys. Res. 83*:2603-2608, 1978.

Carbary, J. F., T. W. Hill, and A. J. Dessler. Planetary spin period acceleration of particles in the Jovian magnetosphere. *J. Geophys. Res. 81*:5189-5195, 1976.

Carbary, J. F., and S. M. Krimigis. Energetic electrons ($E_e \geq 20$ keV) at Jupiter, (Abstract). *Trans. AGU 61*:1090, 1980.

Carbary, J. F., S. M. Krimigis, E. P. Keath, G. Gloeckler, W. I. Axford, and T. P. Armstrong. Ion anisotropies in the outer Jovian magnetosphere. *J. Geophys. Res. 86*:8285-8299, 1981.

Carbary, J. F., S. M. Krimigis, and E. C. Roelof. Ion anisotropies in the Jovian magnetodisc. In preparation, 1982.

Carlson, R. W., and D. L. Judge. Pioneer 10 ultraviolet photometer observations at Jupiter encounter. *J. Geophys. Res. 79*:3623-3633, 1974.

Carlson, R. W., and D. L. Judge. Pioneer 10 ultraviolet photometer observations of the Jovian hydrogen torus: The angular distribution. *Icarus 24*:395-399, 1975.

Carlson, R. W., D. L. Matson, and T. V. Johnson. Electron impact ionization of Io's sodium emission cloud. *Geophys. Res. Lett. 2*:469-472, 1975.

Carlson, R. W., D. L. Matson, T. V. Johnson, and J. T. Bergstralh. Sodium D-line emission from Io: Comparison of observed and theoretical line profiles. *Astrophys. J. 223*:1082-1086, 1978.

Carovillano, R. L., and G. L. Siscoe. Energy and momentum theorems in magnetospheric processes. *Rev. Geophys. Space Phys. 11*:289-353, 1973.

Carr, T. D. *Proceedings of the Jupiter Radiation Belts Workshop*, NASA-JPL Report 616-48, 1972a.

Carr, T. D. Jupiter's decametric rotation period and the Source A emission beam. *Phys. Earth Planet. Int. 6*:21-28, 1972b.

Carr, T. D., G. W. Brown, A. G. Smith, C. S. Higgins, H. Bollhagen et al. Spectral distribution of the decametric radiation from Jupiter in 1961. *Astrophys. J. 140*:778-795, 1964.

Carr, T. D., and M. D. Desch. Recent decametric and hectometric observations of Jupiter. In *Jupiter*, ed. T. Gehrels, pp. 693-735. University of Arizona Press, Tucson, 1976.

Carr, T. D., M. D. Desch, and J. K. Alexander. Phenomenology of magnetospheric radio emission. This volume, Chapter 7.

Carr, T. D., and S. Gulkis. The magnetosphere of Jupiter. *Ann. Rev. Astron. Astrophys. 7*:577-618, 1969.

Carr, T. D., A. G. Smith, F. F. Donovan, and H. I. Register. The twelve-year periodicities of the decametric radiation of Jupiter. *Radio Sci. 5*:495-503, 1970.

Chamberlain, J. W. *Physics of the Aurora and Airglow*. Academic Press, New York, 1961.

Chang, D. B. Amplified whistlers as the source of Jupiter's sporadic decametric radio emission. *Astrophys. J. 138*:1231-1241, 1963.

Chang, D. B., and L. Davis, Jr. Synchrotron radiation as the source of Jupiter's polarized decimeter radiation. *Astrophys. J. 136*:567-581, 1962.

Chapman, S., and J. Bartels. *Geomagnetism.* Oxford University Press, London, 1940.

Chen, R. H. Studies of Jupiter's lower ionospheric layers. *J. Geophys. Res., 86*:7792-7794, 1981.

Chenette, D. L. The propagation of Jovian electrons to Earth. *J. Geophys. Res. 85*:2243-2256, 1980.

Chenette, D. L., T. F. Conlon, and J. A. Simpson. Bursts of relativistic electrons from Jupiter observed in interplanetary space with the time variation of the planetary rotation period. *J. Geophys. Res. 79*:3551-3558, 1974.

Cheng, A. F. Effects of Io's volcanoes on the plasma torus and Jupiter's magnetosphere. *Astrophys. J. 242*:812-827, 1980.

Christiansen, P., P. Gough, G. Martelli, J. J. Block, N. Cornilleau et al. Geos 1: Identification of natural magnetospheric emissions. *Nature 272*:682-686, 1978.

Chu, K. R., and J. L. Hirshfield. Comparative study of the axial and azimuthal bunching mechanisms in electromagnetic cyclotron instabilities. *Phys. Fluids 21*:461-466, 1978.

Clarke, J. N. A synchrotron model for the decimetric radiation of Jupiter. *Radio Sci. 5*:529-533, 1970.

Clarke, J. T., H. W. Moos, and P. D. Feldman. IUE monitoring of the spatial distribution of the H Lyα emission from Jupiter. *Astrophys. J. Lett. 245*:127-129, 1981.

Clarke, J. T., H. A. Weaver, P. D. Feldman, H. W. Moos, W. G. Fastie, and C. B. Opal. Spatial imaging of Hydrogen Lyman A emission from Jupiter. *Astrophys. J. 240*:696-701, 1980.

Cloutier, P. A., R. E. Daniell, Jr., A. J. Dessler, and T. W. Hill. A cometary ionosphere model for Io. *Astrophys. Space Sci. 55*:93-112, 1978.

Conlon, T. F. The interplanetary modulation and transport of Jovian electrons. *J. Geophys. Res. 83*:541-552, 1978.

Connerney, J. E. P. The magnetic field of Jupiter: A generalized inverse approach. *J. Geophys. Res. 86*:7679-7693, 1981a.

Connerney, J. E. P. Comment on 'Azimuthal magnetic field at Jupiter' by J. L. Parish, C. K. Goertz, and M. F. Thomsen. *J. Geophys. Res. 86*:7796-7797, 1981b.

Connerney, J. E. P., M. H. Acuña, and N. F. Ness. Modeling the Jovian current sheet and inner magnetosphere. *J. Geophys. Res. 86*:8370-8384, 1981.

Connerney, J. E. P., M. H. Acuña, and N. F. Ness. Voyager 1 assessment of Jupiter's planetary magnetic field, *J. Geophys. Res., 86*:3623-3627, 1982.

Cook, A. F., E. M. Shoemaker, and B. A. Smith. Dynamics of volcanic plumes on Io. *Nature 280*:743-746, 1979.

Cook, A. F., E. M. Shoemaker, B. A. Smith, G. E. Danielson, T. V. Johnson, and S. P. Synnott. Volcanic origin of the eruptive plumes on Io. *Science 211*:1419-1422, 1981.

Cornwall, J. M. Micropulsations and the outer radiation zone. *J. Geophys. Res. 71*:2185-2199, 1966.

Cornwall, J. M., F. V. Coroniti, and R. M. Thorne. Turbulent loss of ring current protons. *J. Geophys. Res. 75*:4699-4709, 1970.

Cornwall, J. M., and Schulz, M. Electromagnetic ion-cyclotron instabilities in multicomponent magnetospheric plasmas. *J. Geophys. Res. 76*:7791-7796, 1971.

Coroniti, F. V. Energetic electrons in Jupiter's magnetosphere. *Astrophys. J. Suppl. 27*:261-281, 1974.

Coroniti, F. V. Denouement of Jovian radiation belt theory. In *The Magnetospheres of the Earth and Jupiter*, ed. V. Formisano, pp. 391-410. D. Reidel, Dordrecht, 1975.

Coroniti, F. V., F. L. Scarf, C. F. Kennel, W. S. Kurth, and D. A. Gurnett. Detection of Jovian whistler mode chorus: Implications for the Io torus aurora. *Geophys. Res. Lett. 7*:45-48, 1980a.

Coroniti, F. V., L. A. Frank, D. J. Williams, R. P. Lepping, F. L. Scarf et al. Variability of plasma sheet dynamics. *J. Geophys. Res. 85*:2957-2977, 1980b.

Coroniti, F. V., R. W. Fredricks, and R. White. Instability of ring current protons beyond the plasmapause during injection events. *J. Geophys. Res. 77*:6243-6248, 1972.

Coroniti, F.V., and C. F. Kennel. Electron precipitation pulsations. *J. Geophys. Res. 75*:1279-1289, 1970.

Coroniti, F. V., and C. F. Kennel. Possible origins of time variability in Jupiter's outer magnetosphere, 1, Variations in solar wind dynamic pressure. *Geophys. Res. Lett. 4*:211-214, 1977.

Coroniti, F. V., C. F. Kennel, and R. M. Thorne. Stably trapped proton fluxes in the Jovian magnetosphere. *Astrophys. J. 189*:383-388, 1974.

Cravens, T. E., G. A. Victor, and A. Dalgarno. The absorption of energetic electrons by molecular hydrogen gas. *Planet. Space Sci. 23*:1059-1070, 1975.

Cummings, W. D., A. J. Dessler, and T. W. Hill. Latitudinal oscillations of plasma within the Io torus. *J. Geophys. Res. 85*:2108-2114, 1980.

Cupermann, S., L. Gomberoff, and A. Stemlieb. Absolute maximum growth rates and enhancement of unstable electromagnetic ion cyclotron waves in mixed warm-cold plasmas. *J. Plasma Phys. 12*:259-272, 1975.

Curtis, S. A. A theory for chorus generation by energetic electrons during substorms. *J. Geophys. Res. 83*:3841-3848, 1978.

Curtis, S. A., and C. S. Wu. Electrostatic and electromagnetic cygroharmonic emissions due to energetic electrons in magnetospheric plasmas. *J. Geophys. Res. 84*:2057-2075, 1979.

Dalgarno, A. Diffusion and mobilities. In *Atomic and Molecular Processes*, ed. D. R. Bates, pp. 643-661. Academic Press, New York, 1962.

Dalgarno, A. Application in aeronomy. In *Physics of Electronic and Atomic Collisions*, Vol. VII, ICPEAC, ed. T. T. Grover and F. J. de Heer, pp. 381-398. North-Holland, Amsterdam, 1971.

Dalgarno, A., M. R. C. McDowell, and A. Williams. The mobilities of ions in unlike gases. *Phil. Trans. A250*:411, 1958.

Danielsson, L. Experiment on the interaction between a plasma and a neutral gas. *Phys. Fluids 13*:2288-2294, 1970.

Davidson, G. T. An improved empirical description of the bounce motion of trapped particles. *J. Geophys. Res. 81*:4029-4030, 1976.

Davidson, R. C. *Methods in Nonlinear Plasma Theory*, Chap. 6. Academic Press, New York, 1972.

Davis, L., Jr., D. E. Jones, and E. J. Smith. The magnetic field of Jupiter. Presented at AGU Fall Meeting, San Francisco, December, 1975.

Davis, L., Jr., and E. J. Smith. The Jovian magnetosphere and magnetopause. In *Magnetospheric Particles and Fields*, ed. B. M. McCormac, pp. 301-310. Reidel, Dordrecht, 1976.

Degioanni, J. J. C. The radiation belts of Jupiter. *Icarus 23*:66-88, 1974.

Degioanni, J. J. C., and J. R. Dickel. Jupiter's radiation belts and upper atmosphere. In *Exploration of the Planetary System*, ed. A. Woszczyk and C. Iwaniszewska, pp. 375-384. Reidel, Dordrecht, 1974.

de Pater, I. 21 cm maps of Jupiter's radiation belts from all rotational aspects. *Astron. Astrophys. 88*:175-183, 1980a.

de Pater, I. Observations and models of the decimetric radio emission from Jupiter. Ph.D. Thesis, State University of Leiden, 1980b.

de Pater, I. A comparison of the radio data and model calculations of Jupiter's synchrotron radiation: I. The high energy electron distribution in Jupiter's inner magnetosphere. *J. Geophys. Res. 86*:3397-3422, 1981a.

de Pater, I. A comparison of the radio data and model calculations of Jupiter's synchrotron radiation: II. E-W asymmetry in the radiation belts as a function of Jovian longitude. *J. Geophys. Res. 86*:3423-3429, 1981b.

de Pater, I. Radio maps of Jupiter's radiation belts and planetary disk at $\lambda = 6$ cm. *Astron. Astrophys. 93*:370-381, 1981c.

de Pater, I., and H. A. C. Dames. Jupiter's radiation belts and atmosphere. *Astron. Astrophys. 72*:148-160, 1979.

Desch, M. D. Groundbased and spacecraft studies of Jupiter at decameter and hectometer wavelengths. Ph.D. thesis, University of Florida, 1976.

Desch, M. D. Io phase motion and Jovian decameter source locations. *Nature 272*:339-340, 1978.

Desch, M. D. Io control of Jovian radio emission. *Nature 287*:815-816, 1980.

Desch, M. D., and T. D. Carr. Decametric and hectometric observation of Jupiter from the RAE-1 satellite. *Astrophys. J. 194*:L57-L59, 1974.

Desch, M. D., and T. D. Carr. Modulation of the Jovian emission below 8 MHz. *Astron. J. 83*:828-837, 1978.

Desch, M. D., T. D. Carr, and J. Levy. Observations of Jupiter at 26.3 MHz using a large array. *Icarus 25*:12-17, 1975.

Desch, M. D., R. S. Flagg, and J. May. Jovian S-burst observations at 32 MHz. *Nature 272*:38-40, 1978.

Desch, M. D., and M. L. Kaiser. The occurrence rate, polarization character and intensity of broadband Jovian kilometric radiation. *J. Geophys. Res. 85*:4248-5256, 1980.

Dessler, A. J. The Jovian boundary layer as formed by magnetic-anamoly effects. *Proc. of Magnetospheric Boundary Layers Conference, ESASP-148*:225-226, 1979.

Dessler, A. J. Corotating Birkeland currents in Jupiter's magnetosphere: an Io plasma-torus source. *Planet. Space Sci. 28*:781-788, 1980a.

Dessler, A. J. Mass-injection rate from Io into the Io plasma torus. *Icarus, 44*:291-295, 1980b.

Dessler, A. J., and J. W. Chamberlain. Jovian longitudinal asymmetry in Io-related and Europa-related auroral hot spots. *Astrophys. J. 230*:974-981, 1979.

Dessler, A. J., and T. W. Hill. High-order magnetic multipoles as a source of gross asymmetry in the distant Jovian magnetosphere. *Geophys. Res. Lett. 2*:567-570, 1975.

Dessler, A. J., and T. W. Hill. Jovian longitudinal control of Io-related radio emissions. *Astrophys. J. 227*: 664-675, 1979.

Dessler, A. J., B. R. Sandel, and S. K. Atreya. The Jovian hydrogen bulge: evidence for co-rotating magnetospheric convection. *Planet. Space Sci. 29*:215-224, 1981.

Dessler, A. J., and V. M. Vasyliunas. The magnetic anomaly model of the Jovian magnetosphere: Predictions for Voyager. *Geophys. Res. Lett. 6*:37-40, 1979.

Dickel, J. R. 6-cm observations of Jupiter. *Astrophys. J. 148*:535-540, 1967.

Dickel, J. R., J. J. Degioanni, and G. C. Goodman. The microwave spectrum of Jupiter. *Radio Sci. 5*:517-527, 1970.

Divine, N. Jupiter charged-particle environment for Jupiter Orbiter Probe 1981/1982 mission. *J.P.L. Publ. 66624* (revised version), Jet Propulsion Lab., Pasadena, August 1976.

Douglas, J. N. Decametric radiation from Jupiter. *IEEE Trans. on Mil. Elec., MIL-8*:173-187, 1964.

Douglas, J. N., and H. J. Smith. Interplanetary scintillation in Jovian decametric radiation. *Astrophys. J. 148*:885-903, 1967.

Douglas-Hamilton, D. H. Pulsar Io effect? *Nature 218*:1035, 1968.

Dowden, R. L. Polarization measurements of Jupiter radio bursts at 10.1 Mc/s. *Austr. J. Phys. 16*:398-410, 1963.

Dowden, R. L. A Jupiter model of pulsars. *Proc. Astron. Soc. Austr. 1*:159, 1968.

Drake, F. D., and H. Hvatum. Non-thermal microwave radiation from Jupiter. *Astron. J. 64*:329-330, 1959.

Drell, S. D., H. M. Foley, and M. A. Ruderman. Drag and propulsion of large satellites in the ionosphere: An Alfvén propulsion engine in space. *J. Geophys. Res. 70*:3131-3145, 1965.

Dufton, P. L., and A. E. Kingston. Electron impact excitation of the $^2P^o$-$^4P^e$ transition in S IV. *J. Phys. B: Atom. Mol. Phys. 13*:4277-4284, 1980.

Dulk, G. A. Io-related radio emission from Jupiter. Ph.D. thesis, University of Colorado, Boulder, 1965.

Dulk, G. A. Apparent changes in the rotation rate of of Jupiter. *Icarus 7*:173-182, 1967.

Dulk, G. A. Characteristics of Jupiter's decametric radio source measured with arc-second resolution. *Astrophys. J. 159*:671-684, 1970.

Duncan, R. A. Jupiter's rotation. *Planet. Space Sci. 19*:391-398, 1971.

Eckersley, T. L. Musical atmospherics. *Nature 135*:104-105, 1935.

Ellis, G. R. A. The decametric radio emission of Jupiter. *Radio Sci. 69D*:1513-1530, 1965.

Ellis, G. R. A. The Jupiter radio bursts. *Proc. Astron. Soc. Austr. 2*:1-8, 1974.

Ellis, G. R. A. Spectra of the Jupiter radio bursts. *Nature 253*:415-417, 1975.

Ellis, G. R. A. An atlas of selected spectra of the Jovian S bursts. University of Tasmania, Physics Dept., Hobart, Tasmania, 1979.

Ellis, G. R. A., and P. M. McCulloch. The decametric radio emissions of Jupiter. *Austr. J. Phys. 16*:380-397, 1963.

Engle, I. M., and D. B. Beard. Idealized Jovian magnetosphere shape and field. *J. Geophys. Res. 85*:579-592, 1980.

Eshleman, V. R., G. L. Tyler, G. E. Wood, G. F. Lindal, J. D. Anderson et al. Radio science with Voyager 1 at Jupiter: Preliminary profiles of the atmosphere and ionosphere. *Science 204*:976-978, 1979a.

Eshleman, V. R., G. L. Tyler, G. E. Wood, G. F. Lindal, J. D. Anderson et al. Radio science with Voyager at Jupiter: Initial Voyager 2 results and a Voyager 1 measure of the Io torus. *Science 206*:959-962, 1979b.

Eviatar, A., and A. I. Ershkovich. Plasma density in the outer Jovian magnetosphere. *J. Geophys. Res. 81*:4027-4028, 1976.

Eviatar, A., C. F. Kennel, and M. Neugebauer. Possible origins of time variability in Jupiter's outer magnetosphere, 3. Variations in the heavy ion plasma. *Geophys. Res. Lett. 5*:287-290, 1978.

Eviatar, A., Y. Mekler, N. Brosch, and T. Mazah. Ground-based observations of the Io torus during Voyager 1 encounter: Indications of enhanced injection and transport. *Geophys. Res. Lett. 8*:249-252, 1981a.

Eviatar, A., and G. L. Siscoe. Limit on rotational energy available to excite Jovian aurora. *Geophys. Res. Lett. 7*:1085-1088, 1980.

Eviatar, A., G. L. Siscoe, T. V. Johnson, and D. L. Matson. Effects of Io ejecta on Europa. *Icarus 47*:75-83, 1981b.

Eviatar, A., G. L. Siscoe, and Yu. Mekler. Temperature anisotropy of the Jovian sulfur nebula. *Icarus 39*:450-458, 1979.

Eviatar, A., G. L. Siscoe, and R. M. Thorne. Thermodynamics of the Io torus ultraviolet radiation, (in preparation) 1982.

Fairfield, D. H. Average magnetic field configuration of the outer magnetosphere. *J. Geophys. Res. 73*:7329-7338, 1968.

Fairfield, D. H. Average and unusual locations of the earth's magnetopause and bow shock. *J. Geophys. Res. 76*:6700-6716, 1971.

Fanale, F. P., R. H. Brown, D. P. Cruikshank, and R. N. Clarke. Significance of absorption features on Io's IR reflectance spectrum. *Nature 280*:761-763, 1979.

Fanale, F. P., T. V. Johnson, and D. L. Matson. Io: A surface evaporite deposit? *Science 186*:922-925, 1974.

Fejer, J. A., and J. R. Kan. A guiding center Vlasov equation and its application to Alfvén waves. *J. Plasma Phys. 3*:331-351, 1969.

Fejer, J. A., and K. F. Lee. Guided propagation of Alfvén waves in the magnetosphere. *J. Plasma Phys. 1*:387-406, 1967.

Festou, M. C., S. K. Atreya, T. M. Donahue, D. E. Shemansky, B. R. Sandel, and A. L. Broadfoot. Composition and thermal profiles of the Jovian upper atmosphere determined by the Voyager ultraviolet stellar occultation experiment. *J. Geophys. Res., 86*:5715-5725, 1981.

Field, G. B. The source of radiation from Jupiter at decimeter wavelengths. *J. Geophys. Res. 64*:1169-1177, 1959.

Filbert, P. C., and P. J. Kellogg. Electrostatic noise at the plasma frequency beyond the bow shock. *J. Geophys. Res. 84*:1369-1381, 1979.

Fillius, R. W. The trapped radiation belts of Jupiter. In *Jupiter*, ed. T. Gehrels, pp. 896-927. University of Arizona Press, Tucson, 1976.

Fillius, W., W. H. Ip, and P. Knickerbocker. Interplanetary electrons: What is the strength of the Jupiter source. *Proc. Cosmic Ray Conf. 5*:232, 1977.

Fillius, R. W., and P. Knickerbocker. The phase of the ten-hour modulation in the Jovian magnetosphere (Pioneers 10 and 11). *J. Geophys. Res. 84*:5763-5772, 1979.

Fillius, R. W., and C. E. McIlwain. Measurements of the Jovian radiation belts. *J. Geophys. Res. 79*:3589-3599, 1974a.

Fillius, R. W., and C. E. McIlwain. Radiation belts of Jupiter. *Science 183*:314-315, 1974b.

Fillius, W., C. McIlwain, A. Mogro-Campero, and G. Steinberg. Evidence that pitch-angle scattering is an important loss mechanism in the inner radiation belt of Jupiter. *Geophys. Res. Lett. 3*:33-36, 1976.

Fillius, R. W., A. Mogro-Campero, and C. McIlwain. Radiation belts of Jupiter: A second look. *Science 188*:465-467, 1975.

Fjeldbo, G., A. Kliore, B. Seidel, D. Sweetnam, and D. Cain. The Pioneer 10 radio occultation measurements of the ionosphere of Jupiter. *Astron. Astrophys. 39*:91-96, 1975.

Fjeldbo, G., A. Kliore, B. Seidel, D. Sweetnam, and P. Woiceshyn. The Pioneer 11 radio occultation measurements of the Jovian ionosphere. In *Jupiter*, ed. T. Gehrels, pp. 239-246. University of Arizona Press, Tucson, 1976.

Flagg, R. S., and M. D. Desch. Simultaneous multifrequency observations of Jovian S bursts. *J. Geophys. Res. 84*:4238-4244, 1979.

Flagg, R. S., D. S. Krausche, and G. R. Lebo. High-resolution spectral analysis of the Jovian decametric radiation, II. The band-like emission. *Icarus 29*:477-482, 1976.

Frank, L. A., K. A. Ackerson, and R. P. Lepping. On hot tenuous plasmas, fireballs, and boundary layers in the earth's magnetotail. *J. Geophys. Res. 81*:5859-5881, 1976.

Frank, L. A., K. L. Ackerson, J. H. Wolfe, and J. D. Mihalov. Observations of plasmas in the Jovian magnetosphere. *J. Geophys. Res. 81*:457-468, 1976.

Frankel, M. S. LF radio noise from the Earth's magnetosphere. *Radio Sci. 8*:991-1005, 1973.

Franklin, K. L., and B. F. Burke. Radio observations of Jupiter. *Astron. J. 61*:177, 1956.

Fredricks, R. W. Plasma instability at $(n + 1/2)f_c$ and its relationship to some satellite observations. *J. Geophys. Res. 76*:5344-5348, 1971.

Fredricks, R. W., C. F. Kennel, F. L. Scarf, G. M. Crook, and I. M. Green. Detection of electric-field turbulence in the Earth's bow shock. *Phys. Rev. Lett. 21*:1761-1764, 1968.

Fredricks, R. W., and F. L. Scarf. Recent studies of magnetospheric electric field emissions above the electron gyrofrequency. *J. Geophys. Res. 78*:310-314, 1973.

French, R. G., and P. J. Gierasch. Waves in the Jovian upper atmosphere. *J. Atmos. Sci. 31*:1707-1712, 1974.

Froidevaux, L. Radial diffusion in Io's torus: Some implications from Voyager 1. *Geophys. Res. Lett. 7*:33-35, 1980.

Fung, P. C. W. Excitation of backward Doppler-shifted cyclotron radiation in a magnetoactive plasma by an electron stream. *Planet. Space Sci. 14*:335-346, 1966a.

Fung, P. C. W. Excitation of cyclotron radiation in the forward subluminous mode and its application to Jupiter's decametric emissions. *Planet. Space Sci. 14*:469-481, 1966b.

Fung, P. C. W. Excitation of cyclotron electromagnetic waves in a magnetoactive plasma by a stream of charged particles, including temperature effects of the stream. *Austr. J. Phys. 19*:489-499, 1966c.

Gallet, R. M. Radio observations of Jupiter. In *Planets and Satellites*, ed. G. P. Kuiper, p. 500. University of Chicago Press, Chicago, 1961.

Gardner, F. F., and C. A. Shain. Further observations of radio emission from the planet Jupiter. *Austr. J. Phys. 11*:55-69, 1958.

Gardner, F. F., and J. B. Whiteoak. Linear polarization observations of Jupiter at 6, 11, and 21 cm wavelengths. *Astron. Astrophys. 60*:369-375, 1977.

Gary, B. An investigation of Jupiter's 1400 MHz radiation. *Astron. J. 68*:568-572, 1963.

Gary, P. S. Ion-acoustic-like instabilities in the solar wind. *J. Geophys. Res. 83*:2504-2510, 1978.

Gauss, C. F. Allgemeine theorie des Erdmagnetismus. Resultate aus den Beobachtungen des magnetischen Vereins im Jahre 1838. Leipzig, 1839 (reprinted in *Werke 5*:119-193. Königliche Gesellschaft der Wissenschaften, Göttingen, 1877).

Gehrels, T., (ed.). *Jupiter*, pp. 1254. University of Arizona Press, Tucson, 1976.

Genova, F., and Y. Leblanc. Interplanetary scintillation and Jovian DAM emission. Submitted to *Astron. Astrophys.*, 1981.

Gerard, E. Long term variations of the decimeter radiation of Jupiter. *Radio Sci. 5*:513-516, 1970.

Gerard, E. Long term variations of the decimetric radio emission of Jupiter (and Saturn?). In *The Magnetospheres of the Earth and Jupiter*, ed. V. Formissano, pp. 237-239. Reidel, Dordrecht, Holland, 1975.

Gerard, E. Variation of the radio emission of Jupiter at 21.3 and 6.2 cm wavelength. *Astron. Astrophys. 50*:353-360, 1976.

Ginzburg, V. L. *The Propagation of Electromagnetic Waves in Plasmas*, 2nd edition, pp. 428f. Pergamon, Oxford, 1970.

Gledhill, J. A. Magnetosphere of Jupiter. *Nature 214*:155-156, 1967.

Gleeson, L. J., and W. I. Axford. The Compton-Getting effect. *Astrophys. Space Sci. 2*:431, 1968.

Gleeson, L. J., and W. I. Axford. An analytical model illustrating the effects of rotation on a magnetosphere containing low energy plasma. *J. Geophys. Res. 81*:3403-3406, 1976.

Gleeson, L. J., M. P. C. Legg, and K. C. Westfold. On the radio frequency spectrum of Jupiter. *Proc. Astron. Soc. Austr. 1*:320-322, 1970.

Gloeckler, G., D. Hovestadt, and L. A. Fisk. Observed distribution functions of H, He, C, O, and Fe in corotating energetic particle streams: Implications for interplanetary acceleration and propagation. *Astrophys. J. 230*:L191-L195, 1979.

Goertz, C. K. Jupiter's magnetosphere: Particles and fields. In *Jupiter*, ed. T. Gehrels, pp. 23-58. University of Arizona Press, Tucson, 1976a.

Goertz, C. K. The current sheet in Jupiter's magnetosphere. *J. Geophys. Res. 81*:3368-3372, 1976b.

Goertz, C. K. Comment on 'Longitudinal asymmetry of the Jovian magnetosphere and the periodic escape of energetic particles' by T. W. Hill and A. J. Dessler. *J. Geophys. Res. 81*:5601, 1976c.

Goertz, C. K. Energization of charged particles in Jupiter's outer magnetosphere. *J. Geophys. Res. 83*:3145-3150, 1978.

Goertz, C. K. The Jovian magnetodisc. *Space Sci. Rev. 23*:319-343, 1979.

Goertz, C. K. Io's interaction with the plasma torus. *J. Geophys. Res. 85*:2949-2956, 1980a.

Goertz, C. K. Proton aurora on Jupiter's nightside. *Geophys. Res. Lett. 7*:365-388, 1980b.

Goertz, C. K. The orientation and motion of the predawn current sheet and Jupiter's magnetotail. *J. Geophys. Res. 86*:8429-8434, 1981.

Goertz, C. K., and R. W. Boswell. Magnetosphere-ionosphere coupling. *J. Geophys. Res. 84*:7239-7246, 1979.

Goertz, C. K., and P. A. Deift. Io's interaction with the magnetosphere. *Planet. Space Sci. 21*:1399-1415, 1973.

Goertz, C. K., D. E. Jones, B. A. Randall, E. J. Smith, and M. F. Thomsen. Evidence for open field lines in Jupiter's magnetosphere. *J. Geophys. Res. 81*:3393-3398, 1976.

Goertz, C. K., A. W. Schardt, J. A. Van Allen, and J. L. Parish. Plasma in the Jovian current sheet. *Geophys. Res. Lett. 6*:495-498, 1979.

Goertz, C. K., and M. Thomsen. The dynamics of the Jovian magnetosphere. *Rev. Geophys. Space Phys. 17*:731-743, 1979a.

Goertz, C. K., and M. F. Thomsen. Radial diffusion of Io-injected plasma. *J. Geophys. Res. 84*:1499-1504, 1979b.

Gold, T. The magnetosphere of Jupiter. *J. Geophys. Res. 81*:3401-3402, 1976.

Gold, T. Electrical origin of the outbursts on Io. *Science 206*:1071-1073, 1979.

Goldberg, B. A., R. W. Carlson, D. L. Matson, and T. V. Johnson. A new asymmetry in Io's sodium cloud. *Bull. Amer. Astron. Soc. 10*:579, 1978.

Goldberg, B. A., Y. Mekler, R. W. Carlson, T. V. Johnson, and D. L.Matson. Io's sodium emission cloud and the Voyager 1 encounter. *Icarus 44*:305-317, 1981.

Goldreich, P., and D. Lynden-Bell. Io, a Jovian unipolar inductor. *Astrophys. J. 156*:59-78, 1969.

Goldstein, H. Theory of the plasma sheet in the Jovian magnetosphere. *Planet. Space Sci. 25*:673-679, 1977.

Goldstein, M. L., and A. Eviatar. The plasma physics of the Jovian decameter radiation. *Astrophys. J. 175*:275-283, 1972.

Goldstein, M. L., and A. Eviatar. An emission mechanism for the Io-independent Jovian decameter radiation. *Astrophys. J. 230*:261-273, 1979.

Goldstein, M. L., A. Eviatar, and J. R. Thieman. A beaming model of the Io-independent Jovian decameter radiation based on multipole models of the Jovian magnetic field. *Astrophys. J. 229*:1186-1197, 1979.

Goldstein, M. L., and J. R. Thieman. The formation of arcs in the dynamic spectra of Jovian decameter bursts. *J. Geophys. Res.*, *86*:8569-8578, 1982.

Goldstein, M. L., R. R. Sharma, M. Ben-Ari, A. Eviatar, and K. Papadopoulos. A theory of Jovian decameter radiation. *J. Geophys. Res.*, in press, 1982.

Goodrich, C. C., J. D. Sullivan, and H. S. Bridge. Voyager predictions of the standoff distance of the Jovian bow shock. *MIT Report CSR-TR-80-4*, June 1980.

Goody, R. M., and J. Apt. Observations of the sodium emission from Jupiter, Region C. *Planet. Space Sci. 25*:603-604, 1977.

Gosling, J. T., J. R. Asbridge, S. J. Bame, W. C. Feldman, R. D. Zwickl et al. Interplanetary ions during an energetic storm particle event: The distribution function from solar wind thermal energies to 1.6 MeV. *J. Geophys. Res. 86*:547-554, 1981.

Gower, J. F. R. The flux density of Jupiter at 81.5 Mc/s. *Observatory 88*:264-267, 1968.

Grabbe, C. L. Auroral kilometric radiation: A theoretical review. *Rev. Geophys. Space Sci., 19*:627-633, 1981a.

Grabbe, C. L., and P. J. Palmadesso. A coherent nonlinear theory of auroral kilometric radiation, II. Dynamical evolution. *J. Geophys. Res.*, submitted, 1981b.

Grabbe, C. L., K. Papadopoulos, and P. J. Palmadesso. A coherent nonlinear theory of auroral kilometric radiation, I. Steady state model. *J. Geophys. Res. 85*:3337-3346, 1980.

Grard, R. J. L., S. E. DeForest, and E. C. Whipple, Jr. Comment on low energy electron measurements in the Jovian magnetosphere. *Geophys. Res. Lett. 4*:247-248, 1977.

Green, J. L., and D. A. Gurnett. Ray tracing of Jovian kilometric radiation. *Geophys. Res. Lett. 7*:65-68, 1980.

Greenstadt, E. W., and R. W. Fredricks. Shock systems in collisionless space plasmas. In *Solar System Plasma Physics*, Vol. III, ed. L. Lanzerotti, C. F. Kennel, and R. W. Parker, pp. 3-43. North-Holland, Amsterdam, 1979.

Greenstadt, E. W., R. W. Fredricks, C. T. Russell, F. L. Scarf, R. R. Anderson, and D. A. Gurnett. Whistler-mode wave propagation in the solar wind near the bow shock. *J. Geophys. Res., 86*:4511-4516, 1981.

Gross, S. H., and S. I. Rasool. The upper atmosphere of Jupiter. *Icarus 3*:311-322, 1964.

Gubbins, D. Theories of the geomagnetic and solar dynamos. *Rev. Geophys. Space Phys. 12*:137-154, 1974.

Gulkis, S., and T. D. Carr. Radio rotation period of Jupiter. *Science 154*:257-259, 1966.

Gulkis, S., B. Gary, M. Klein, and C. Stelzreid. Observations of Jupiter at 13 cm wavelength during 1969 and 1971. *Icarus 18*:181-191, 1973.

Gurevich, A. V., A. L. Krylov, and E. N. Fedorov. Inductive interaction of conducting bodies with a magnetized plasma. *Sov. Phys. JETP 48*:1074-1078, 1978.

Gurnett, D. A. A satellite study of VLF hiss. *J. Geophys. Res. 71*:5599-5615, 1966.

Gurnett, D. A. Sheath effects and related charge-particle acceleration by Jupiter's satellite Io. *Astrophys. J. 175*:525-533, 1972.

Gurnett, D. A. The Earth as a radio source: Terrestrial kilometric radiation. *J. Geophys. Res. 79*:4227-4238, 1974.

Gurnett, D. A. The Earth as a radio source: The nonthermal continuum. *J. Geophys. Res. 80*:2751-2763, 1975.

Gurnett, D. A., and R. R. Anderson. Plasma wave electric fields in the solar wind: Initial results from Helios 1. *J. Geophys. Res. 82*:632-650, 1977.

Gurnett, D. A., R. R. Anderson, B. T. Tsurutani, E. J. Smith, G. Paschmann et al. Plasma wave turbulence at the magnetopause: Observations from ISEE 1 and 2. *J. Geophys. Res. 84*:7043-7058, 1979a.

Gurnett, D. A., and L. A. Frank. VLF hiss and related plasma observations in the polar magnetosphere. *J. Geophys. Res. 77*:172-190, 1972.

Gurnett, D. A., and L. A. Frank. Electron plasma oscillations associated with type III radio emissions and solar electrons. *Solar Phys. 45*:477-493, 1975.

Gurnett, D. A., and L. A. Frank. Continuum radiation associated with low energy electrons in the outer radiation zone. *J. Geophys. Res. 81*:3875-3885, 1976.

Gurnett, D. A., and L. A. Frank. A region of intense plasma wave turbulence on auroral field lines. *J. Geophys. Res. 82*:1031-1050, 1977.

Gurnett, D. A., and L. A. Frank. Ion acoustic waves in the solar wind. *J. Geophys. Res.* 83:58-74, 1978a.

Gurnett, D. A., and L. A. Frank. Plasma waves in the polar cusp: Observations from Hawkeye 1. *J. Geophys. Res. 83*:1447-1462, 1978b.

Gurnett, D. A., L. A. Frank, and R. P. Lepping. Plasma waves in the distant magnetotail. *J. Geophys. Res. 81*:6059-6071, 1976.

Gurnett, D. A., and C. K. Goertz. Multiple Alfvén wave reflections excited by Io: Origin of the Jovian decametric arcs. *J. Geophys. Res. 86*:717-722, 1981.

Gurnett, D. A., W. S. Kurth, and F. L. Scarf. Plasma wave observations near Jupiter: Initial results from Voyager 2. *Science 206*:987-991, 1979a.

Gurnett, D. A., W. S. Kurth, and F. L. Scarf. Auroral hiss observed near the Io plasma torus. *Nature 280*:767-770, 1979b.

Gurnett, D. A., W. S. Kurth, and F. L. Scarf. The structure of the Jovian magnetotail from plasma wave observations. *Geophys. Res. Lett. 7*:53-56, 1980.

Gurnett, D. A., J. E. Maggs, D. L. Gallagher, W. S. Kurth, and F. L. Scarf. Parametric interaction and spatial collapse of beam-driven Langmuir waves in the solar wind. *J. Geophys. Res., 86*:8833-8841, 1981a.

Gurnett, D. A., F. L. Scarf, W. S. Kurth, R. R. Shaw, and R. L. Poynter. Determination of Jupiter's electron density profile from plasma wave observations. *J. Geophys. Res. 86*:8199-8212, 1981b.

Gurnett, D. A., and R. R. Shaw. Electromagnetic radiation trapped in the magnetosphere above the plasma frequency. *J. Geophys. Res. 78*:8136-8149, 1973.

Gurnett, D. A., R. R. Shaw, R. R. Anderson, W. S. Kurth, and F. L. Scarf. Whistlers observed by Voyager 1: Detection of lightning on Jupiter. *Geophys. Res. Lett.* 6:511-516, 1979b.

Haff, P. K., and C. C. Watson. The erosion of planetary and satellite atmospheres by energetic atomic particles. *J. Geophys. Res. 84*:8436-8442, 1979.

Haff, P. K., C. C. Watson, and Yuk L. Yung. Sputter ejection of matter from Io. *J. Geophys. Res. 86*:6933-6938, 1981.

Hamilton, D. C., G. Gloeckler, S. M. Krimigis, C. O. Bostrom, T. P. Armstrong et al. Detection of energetic hydrogen molecules in Jupiter's magnetosphere by Voyager 2: Evidence for an ionospheric plasma source. *Geophys. Res. Lett. 7*:813-816, 1980.

Hamilton, D. C., G. Gloeckler, S. M. Krimigis, and L. J. Lanzerotti. Composition of non-thermal ions in the Jovian magnetosphere. *J. Geophys. Res. 86*:8301-8318, 1981.

Hanel, R., B. Conrath, M. Flasar, V. Kunde, P. Lowman et al. Infrared observations of the Jovian system from Voyager 2. *Science 206*:952-956, 1979.

Harris, E. G. Unstable plasma oscillations in a magnetic field. *Phys. Rev. Lett. 2*:34-36, 1959.

Harris, E. G. Plasma instabilities. In *Physics of Hot Plasmas*, ed. B. J. Rye and J. C. Taylor, pp. 145-201. Oliver and Boyd, Edinburgh, 1968.

Harris, E. G. Classical plasma phenomena from a quantum mechanical viewpoint. In *Adv. in Plasma Phys., vol. 3*, ed. A. Simon and W. B. Thompson, pp. 157-241. Interscience, New York, 1969.

Hasegawa, A. *Plasma Instabilities and Nonlinear Effects*, p. 22. Springer-Verlag, New York, 1975.

Hasted, J. B. Charge transfer and collisional detachment. In *Atomic and Molecular Processes*, ed. D. R. Bates, pp. 696-721. Academic Press, New York, 1962.

Helliwell, R. A. *Whistlers and Related Ionospheric Phenomena*. Stanford University Press, Palo Alto, 1965.

Helliwell, R. A. A theory of discrete VLF emissions from the magnetosphere. *J. Geophys. Res. 72*:4773-4790, 1967.

Helliwell, R. A., and T. R. Crystal. A feedback model of cyclotron interaction between whistler-mode waves and energetic electrons in the magnetosphere. *J. Geophys. Res.* 78:7357-7371, 1973.

Henry, R. J. W., and M. B. McElroy. The absorption of extreme ultraviolet solar radiation by Jupiter's upper atmosphere. *J. Atmos. Sci.* 26:912-917, 1969.

Herbert, F., and B. R. Lichtenstein. Joule heating of Io's ionosphere by unipolar induction currents. *Icarus* 44:296-304, 1980.

Hewitt, R. G., D. B. Melrose, and K. G. Rönnmark. A cyclotron theory for the beaming pattern of Jupiter's decametric radio emission. Proc. Astron. Soc. Australia, submitted, 1981.

Hide, R., and D. Stannard. Jupiter's magnetism: Observations and theory. In *Jupiter*, ed. T. Gehrels, pp. 767-787. University of Arizona Press, Tucson, 1976.

Hill, T. W. Origin of the plasma sheet. *Rev. Geophys. Space Phy.* 12:379-388, 1974.

Hill, T. W. Magnetic merging in a collisionless plasma. *J. Geophys. Res.* 80:4689-4699, 1975.

Hill, T. W. Interchange stability of a rapidly rotating magnetosphere. *Planet. Space Sci.* 24:1151-1154, 1976.

Hill, T. W. Inertial limit on corotation. *J. Geophys. Res.* 84:6554-6558, 1979.

Hill, T. W. Corotation lag in Jupiter's magnetosphere: A comparison of observation and theory. *Science* 207:301-302, 1980.

Hill, T. W. The Jovian magnetosphere: A post-Voyager view. *EOS* 62:25-27, 1981.

Hill, T. W., and J. F. Carbary. Centrifugal distortion of the Jovian magnetosphere by an equatorially confined current sheet. *J. Geophys. Res.* 83:5745-5749, 1978.

Hill, T. W., J. F. Carbary, and A. J. Dessler. Periodic escape of relativistic electrons from the Jovian magnetosphere. *Geophys. Res. Lett.* 1:333-336, 1974.

Hill, T. W. and A. J. Dessler. Longitudinal asymmetry of the Jovian magnetosphere and the periodic escape of energetic particles. *J. Geophys. Res.* 81, 3383-3386, 1976.

Hill, T. W., A. J. Dessler, and F. P. Fanale. Localized deposition and sputtering of Jovian ionospheric sodium on Io. *Planet. Space Sci.* 27:419-424, 1979.

Hill, T. W., A. J. Dessler, and L. J. Maher. Corotating magnetospheric convection. *J. Geophys. Res.* 86:9020-9028, 1981.

Hill, T. W., A. J. Dessler, and F. C. Michel. Configuration of the Jovian magnetosphere. *Geophys. Res. Lett.* 1:3-6, 1974.

Hill, T. W., C. K. Goertz, and M. F. Thomsen. Some consequences of corotating magnetospheric convection. *J. Geophys. Res.*, in press, 1982.

Hill, T. W., and F. C. Michel. Planetary magnetospheres. Rev. Geophys. Space Phys. 13:967-974, 1975.

Hill, T. W., and F. C. Michel. Heavy ions from the Galilean satellites and the centrifugal distortion of the Jovian magnetosphere. *J. Geophys. Res.* 81:4561-4565, 1976.

Hirshfield, J. F., and G. Bekefi. Decameter radiation from Jupiter. *Nature* 198:20-22, 1963.

Hones, E. W.. The magnetotail: Its generation and dissipation. In *Physics of Solar Planetary Environments*, ed. D. J. Williams, p. 558. AGU, Washington, D. C., 1976.

Huba, J. D., N. T. Gladd, and K. Papadopoulos. Lower-hybrid drift wave turbulence in the distant magnetotail. *J. Geophys. Res.* 83:5217-5226, 1978.

Hubbard, R. F., and T. J. Birmingham. Electrostatic emissions between electron gyroharmonics in the outer magnetosphere. *J. Geophys. Res.* 83:4837-4850, 1978.

Hundhausen, A. J. *Coronal Expansion and Solar Wind*. Springer-Verlag, New York, 1972.

Hunten, D. M., and A. J. Dessler. Soft electrons as a possible heat source for Jupiter's thermosphere. *Planet. Space Sci.* 25:817-821, 1977.

Huntress, W. T., Jr. A review of Jovian ionospheric chemistry. *Advan. At. Mol. Phys.* *10*:295-340, 1974.

Huntress, W. T., Jr. Laboratory studies of bimolecular reactions of positive ions in interstellar clouds, in comets, and in planetary atmospheres of reducing composition. *Astrophys. J. Suppl. 33*:495-514, 1977.

Intriligator, D. S., and W. D. Miller. Detection of the Io plasma torus by Pioneer 10. *Geophys. Res. Lett. 8*:409-412, 1981.

Intriligator, D. S., and J. H. Wolfe. Plasma electron measurements in the outer Jovian magnetosphere. *Geophys. Res. Lett. 4*:249-250, 1977.

Ioannidis, G. A., and N. M. Brice. Plasma densities in the Jovian magnetosphere: Plasma slingshot or Maxwell demon? *Icarus 14*:360-373, 1971.

Ip, W.-H. On the drift motions of magnetospheric charged particles in the vicinity of a planetary satellite with dipole field. *J. Geophys. Res. 86*:1596-1600, 1981.

Ip, W.-H., and W. I. Axford. A weak interaction model for Io and the Jovian magnetosphere. *Nature 283*:180-183, 1980.

Ipavich, F. M. The Compton-Getting effect for low energy particles. *Geophys. Res. Letts. 1*:149-152, 1974.

Irving, J., and N. Mullineux. *Mathematics in Physics and Engineering.* Academic Press, New York, 1959.

Jackson, J. D. *Classical Electrodynamics.* Wiley, New York, 1975.

Jacobs, V. L., J. Davis, and J. E. Rogerson. Ionization equilibrium and radiative energy loss rates for C, N, and O ions in low density plasmas. *J. Quant. Radiat. Trans. 19*:591-598, 1978.

Jacobs, V. L., J. Davis, J. E. Rogerson, and M. Blaha. Dielectronic recombination rates, ionization equilibrium, and radiative energy-loss rates for neon, magnesium, sulfur ions in lower-density plasmas. *Astrophys. J. 230*:627-638, 1979.

Johnsen, R., and M. A. Biondi. Measurement of positive ion conversion and removal reactions relating to the Jovian ionosphere. *Icarus 23*:139-143, 1974.

Johnson, R. E., L. J. Lanzerotti, W. L. Brown, and T. P. Armstrong. Erosion of Galilean satellite surfaces by Jovian magnetospheric particles. *Science 212*:1027-1030, 1981.

Johnson, R. G. Energetic ion composition in the Earth's magnetosphere. *Rev. Geophys. Space Phys. 17*:696-705, 1979.

Johnson, T. V., A. F. Cook, II, C. Sagan, and L. A. Soderblom. Volcanic resurfacing rates and implications for volatiles on Io. *Nature 280*:746-750, 1979.

Johnson, T. V., D. L. Matson, and R. W. Carlson. Io's atmosphere and ionosphere: New limits on surface pressure from plasma models. *Geophys. Res. Lett. 3*:293-296, 1976.

Johnson, T. V., G. E. Morfill, and E. Grün. Dust in Jupiter's magnetosphere: An Io source. *Geophys. Res. Lett. 7*:305-308, 1980.

Jones, D. Source of terrestrial nonthermal radiation. *Nature 260*:686-689, 1976.

Jones, D. Latitudinal beaming of planetary radio emissions. *Nature 288*:225-229, 1980.

Jones, D. E., J. G. Melville, II, and M. L. Blake. Modeling Jupiter's current disc: Pioneer 10 outbound. *J. Geophys. Res. 85*:3329-3336, 1980.

Jones, D. E., B. T. Thomas, and J. G. Melville, II. Equatorial disc and dawn dusk currents in the frontside magnetosphere of Jupiter: Pioneer 10 and 11. *J. Geophys. Res. 86*:1601-1605, 1981.

Joselyn, J. A., and L. R. Lyons. Ion-cyclotron wave growth calculated from observations of the proton ring current during storm recovery. *J. Geophys. Res. 81*:2275-2282, 1976.

Judge, D. L., and R. W. Carlson. 1974. Pioneer 10 observations of the ultraviolet glow in the vicinity of Jupiter. *Science 183*:317-318.

Judge, D. L., R. W. Carlson, F. M. Wu, and U. G. Hartmann. Pioneer 10 and 11 ultraviolet photometer observations of the Jovian satellites. In *Jupiter*, ed. T. Gehrels, pp. 1068-1101. University of Arizona Press, Tucson, 1981.

Kaiser, M. L. A low-frequency radio survey of the planets with RAE 2. *J. Geophys. Res. 82*:1256-1260, 1977.

Kaiser, M. L., and J. K. Alexander. The Jovian decametric rotation period. *Astrophys. Lett. 12*:215-217, 1972.

Kaiser, M. L., and J. K. Alexander. Periodicities in the Jovian decametric emission. *Astrophys. Lett. 14*:55-58, 1973.

Kaiser, M. L., and J. K. Alexander. Terrestrial kilometric radiation. 3. Average spectral properties. *J. Geophys. Res. 82*:3273-3280, 1977.

Kaiser, M. L., and M. D. Desch. Narrow-band Jovian kilometric radiation: A new radio component. *J. Geophys. Res. 7*:389-392, 1980.

Kaiser, M. L., M. D. Desch, A. C. Riddle, A. Lecacheux, J. B. Pearce et al. Voyager spacecraft radio observations of Jupiter: Initial cruise results. *Geophys. Res. Lett. 6*:507-510, 1979.

Kaiser, M. L., M. D. Desch, J. W. Warwick, and J. B. Pearce. Voyager detection of nonthermal radio emission from Saturn. *Science 209*:1238-1240, 1980.

Keath, E. P., E. C. Roelof, C. O. Bostrom, and D. J. Williams. Fluxes of $\geq$ 50 keV protons and $\geq$ 30 keV electrons at $\sim$ 35 R_E, 2. Morphology and flow patterns in the magnetotail. *J. Geophys. Res. 81*:2315-2326, 1976.

Kennedy, D. Polarization of the decametric radiation from Jupiter. Ph.D. thesis, University of Florida, Gainesville, Fla., 1969.

Kennel, C. F. Low-frequency whistler mode. Phys. Fluids 9:2190-2202, 1966.

Kennel, C. F. Stably trapped proton limits for Jupiter. In *Proc. of the Jupiter Radiation Belt Workshop*, ed. A. J. Beck, pp. 347-361. NASA Tech. Memo 33-543, Jet Propulsion Laboratory, Pasadena, California, 1972.

Kennel, C. F. Magnetospheres of the planets. *Space Sci. Rev. 14*:511-533, 1973.

Kennel, C. F., and F. V. Coroniti. Is Jupiter's magnetosphere like a pulsar's or Earth's? In *The Magnetospheres of the Earth and Jupiter*, ed. V. Formisano, pp. 451-477. D. Reidel, Dordrecht, 1975.

Kennel, C. F., and F. V. Coroniti. Jupiter's magnetosphere. *Ann. Rev. Astron. Astrophys. 15*:389-486, 1977a.

Kennel, C. F., and F. V. Coroniti. Possible origins of time variability in Jupiter's outer magnetosphere: 2. Variations in solar wind magnetic field. *Geophys. Res. Lett. 4*:215-218, 1977b.

Kennel, C. F., and F. V. Coroniti. Jupiter's magnetosphere and radiation belts. In *Solar System Plasma Physics, Vol. II*, eds. C. F. Kennel, L. J. Lanzerotti, E. N. Parker, pp. 106-181. North Holland, Amsterdam, 1979.

Kennel, C. F., and F. Engelmann. Velocity space diffusion from weak plasma turbulence in a magnetic field. Phys. Fluids. 9:2377-2388, 1966.

Kennel, C. F., and H. E. Petschek. Limit on stably trapped particle fluxes. *J. Geophys. Res. 71*:1-28, 1966.

Kennel, C. F., F. L. Scarf, R. W. Fredricks, J. H. McGehee, and F. V. Coroniti. VLF electric field observations in the magnetosphere. *J. Geophys. Res. 75*:6136-6152, 1970.

Kennel, C. F., and H. V. Wong. Resonant particle instabilities in a uniform magnetic field. *J. Plasma Phys. 1*:75-80, 1967.

Kindel, J. M., and C. F. Kennel. Topside current instabilities. *J. Geophys. Res. 76*:3055-3078, 1971.

Kintner, P. M., M. C. Kelley, and F. S. Mozer. Electrostatic hydrogen cyclotron waves near one Earth radius altitude in the polar magnetosphere. *Geophys. Res. Lett. 5*:139-142, 1978.

Kirsch, E., S. M. Krimigis, J. W. Kohl, and E. P. Keath. Upper limit for X-ray and energetic neutral particle emission from Jupiter: Voyager-1 results. *Geophys. Res. Lett. 8*:169-172, 1981.

Kivelson, M. G. Jupiter's distant environment. In *Physics of Solar Planetary Environments*, Vol. 2, ed. D. J. Williams, pp. 836-853. AGU, Washington, D.C., 1976.

Kivelson, M. G., P. J. Coleman, Jr., L. Froidevaux, and R. Rosenberg. A time-dependent model of the Jovian current sheet. *J. Geophys. Res. 83*:4823-4829, 1978.

Kivelson, M. G., J. A. Slavin, and D. J. Southwood. Magnetospheres of the Galilean satellites. *Science 205*:491-493, 1979.

Kivelson, M. G., and D. J. Southwood. Plasma near Io: Estimates of some physical parameters. *J. Geophys. Res. 86*:10122-10126, 1981.

Kivelson, M. G., and G. R. Winge. Field-aligned currents in the Jovian magnetosphere: Pioneer 10 and 11. *J. Geophys. Res. 81*:5853-5858, 1976.

Klein, M. J. The variability of the total flux density and polarization of Jupiter's decimetric radio emission. *J. Geophys. Res. 81*:3380-3382, 1976.

Klein, M. J., S. Gulkis, and C. T. Stelzreid. Jupiter: New evidence of long term variations of its decimetric flux density. *Astrophys. J. 176*:L85-L88, 1972.

Kliore, A., D. L. Cain, G. Fjeldbo, B. L. Seidel, and S. I. Rasool. Preliminary results on the atmospheres of Io and Jupiter from the Pioneer 10 S-band occultation experiment. *Science 183*:323-324, 1974.

Kliore, A., G. Fjeldbo, B. L. Seidel, D. N. Sweetnam, T. T. Sesplaukis, and P. M. Woiceshyn. Atmosphere of Io from Pioneer 10 radio occultation measurements. *Icarus 24*:407-410, 1975.

Kliore, A. J., G. F. Lindal, I. R. Patel, D. N. Sweetnam, and H. B. Hotz. Vertical structure of the ionosphere and upper neutral atmosphere of Saturn from Pioneer 11 Saturn radio occultation. *Science 207*:446-450, 1980.

Knight, S. Parallel electric fields. *Planet. Space Sci. 21*:741-750, 1973.

Komesaroff, M. M., and P. M. McCulloch. The radio rotation period of Jupiter. *Astrophys. Lett. 1*:39-41, 1967.

Komesaroff, M. M., and P. M. McCulloch. Asymmetries in Jupiter's magnetosphere. *Mon. Not. Roy. Astron. Soc. 172*:91-95, 1975.

Komesaroff, M. M., and P. M. McCulloch. The position angle of Jupiter's linearly polarized synchrotron emission and the Jovian magnetic field configuration. *Mon. Not. Roy. Astron. Soc. 195*:775-785, 1981.

Komesaroff, M. M., P. M. McCulloch, G. L. Berge, and M. J. Klein. The position angle of Jupiter's linearly polarized synchrotron emission – Observations extending over 16 years. *Mon. Not. Roy. Astron. Soc.*, *193*:745-759, 1980.

Komesaroff, M. M., D. Morris, and J. A. Roberts. Circular polarization of Jupiter's decimetric emission and the Jovian magnetic field strength. *Astrophys. Lett. 7*:31-36, 1970.

Korchak, A. A. On the polarization of synchrotron radiation in a dipole magnetic field. *Geomagn. Aeron. 3*:394-396, 1962.

Krall, N., and A. Trivelpiece. *Principles of Plasma Physics*. McGraw-Hill, New York, 1973.

Kraus, J. D. *Radio Astronomy*. McGraw-Hill, New York, 1966.

Krausche, D. S., R. S. Flagg, G. R. Lebo, and A. G. Smith. High-resolution spectral analysis of the Jovian decametric radiation, I. Burst morphology and drift rates. *Icarus 29*:463-475, 1976.

Krimigis S. M. Planetary magnetospheres: The in situ astrophysical laboratories. *Proc. 17th International Cosmic Ray Conference, 10*:229-272, Paris, 1981.

Krimigis, S. M., T. P. Armstrong, W. I. Axford, C. O. Bostrom, C. Y. Fan et al. The low energy charged particle (LECP) experiment on the Voyager spacecraft. *Space Sci. Rev. 21*:329-354, 1977.

Krimigis, S. M., T. P. Armstrong, W. I. Axford, C. O. Bostrom, C. Y. Fan et al. Low-energy charged particle environment at Jupiter: A first look. *Science 204*:998-1003, 1979a.

Krimigis, S. M., T. P. Armstrong, W. I. Axford, C. O. Bostrom, C. Y. Fan et al. Hot plasma environment at Jupiter: Voyager 2 results. *Science 206*:977-984, 1979b.

Krimigis, S. M., T. P. Armstrong, W. I. Axford, C. O. Bostrom, C. Y. Fan et al. Energetic (~ 100 keV) tailward-directed ion beam outside the Jovian plasma boundary. *Geophys. Res. Lett. 7*:13-16, 1980.

Krimigis, S. M., J. F. Carbary, E. P. Keath, C. O. Bostrom, W. I. Axford et al. Characteristics of hot plasma in the Jovian magnetosphere: Results from the Voyager spacecraft. *J. Geophys. Res. 86*:8227-8257, 1981.

Krimigis, S. M., E. T. Sarris, and T. P. Armstrong. Observations of Jovian electron events in the vicinity of Earth. *Geophys. Res. Lett. 2*:561-564, 1975.

Kumar, S. The stability of an SO_2 atmosphere on Io. *Nature 280*:758-760, 1979.

Kumar, S. A model of the SO_2 atmosphere and ionosphere of Io. *Geophys. Res. Lett. 7*:9-12, 1980.

Kumar, S., and D. M. Hunten. The atmospheres of Io and other satellites. In *The Satellites of Jupiter*, ed. D. Morrison, University of Arizona Press, Tucson, 1981.

Kunc, J. A., and D. L. Judge. Electron-capture cross sections for some astrophysical processes. *Geophys. Res. Lett. 8*:177-178, 1981.

Kupo, I., Y. Mekler, and A. Eviatar. Detection of ionized sulfur in the Jovian magnetosphere. *Astrophys. J. Lett. 205*:L51-L53, 1976.

Kurth, W. S., D. D. Barbosa, F. L. Scarf, D. A. Gurnett, and R. L. Poynter. Low frequency radio emissions from Jupiter: Jovian kilometric radiation. *Geophys. Res. Lett. 6*:747-750, 1979a.

Kurth, W. S., J. D. Craven, L. A. Frank, and D. A. Gurnett. Intense electrostatic waves near the upper hybrid resonance frequency. *J. Geophys. Res. 84*:4145-4164, 1979b.

Kurth, W. S., D. A. Gurnett, and F. L. Scarf. High-resolution spectrograms of ion-acoustic waves in the solar wind. *J. Geophys. Res. 84*:3413-3419, 1979.

Kurth, W. S., L. A. Frank, M. Ashour-Abdalla, D. A. Gurnett, and B. G. Burek. Observations of a free-energy source for intense electrostatic waves. *Geophys. Res. Lett. 7*:293-296, 1980a.

Kurth, W. S., D. D. Barbosa, D. A. Gurnett, and F. L. Scarf. Electrostatic waves in the Jovian magnetosphere. *Geophys. Res. Lett. 7*:57-60, 1980b.

Kurth, W. S., D. A. Gurnett, and F. L. Scarf. Spatial and temporal studies of Jovian kilometric radiation. *Geophys. Res. Lett. 7*:61-64, 1980.

Kurth, W. S., D. A. Gurnett, and R. R. Anderson. Escaping nonthermal continuum radiation: A terrestrial analog to Jovian narrowband kilometric radiation. *J. Geophys. Res. 86*:5519-5531, 1981.

Kutcher, G. J., M. G. Heaps, and A. E. S. Green. Electron energy deposition and electron escape in the Jovian ionosphere. *EOS Trans. 56*:406, 1975.

Kuyatt, C. E., and T. Jørgensen, Jr. Energy and angular dependence of the differential cross-section for production of electrons by 50-100 keV protons in hydrogen gas. *Phys. Rev. 130*:1444-1455, 1963.

Lanzerotti, L. J., W. L. Brown, J. M. Poate, and W. M. Augustyniak. On the contribution of water products from Galilean satellites to the Jovian magnetosphere. *Geophys. Res. Lett. 5*:155-158, 1978.

Lanzerotti, L. J., C. G. Maclennan, T. P. Armstrong, S. M. Krimigis, R. P. Lepping, and N. F. Ness. Ion and electron angular distributions in the Io torus region of the Jovian magnetosphere. *J. Geophys. Res. 86*:8491-8496, 1981.

Lanzerotti, L. J., C. G. Maclennan, S. M. Krimigis, T. P. Armstrong, K. Behannon, and N. F. Ness. Statics of the nightside Jovian plasma sheet. *Geophys. Res. Lett.* *7*:817-820, 1980.

Larson, J., and L. Stenflo. Parametric instabilities of waves in magnetized plasmas. *Beitrage aus der Plasma Phys. 16*:79-85, 1976.

Leblanc, Y. On the arc structure of the DAM Jupiter emission. *J. Geophys. Res. 86*:8564-8568, 1981.

Leblanc, Y., M. G. Aubier, C. Rosolen, F. Genova, and J. de la Noe. The Jovian S bursts. II. Frequency drift measurements at different frequencies through several storms. *Astron. Astrophys. 86*:349-354, 1980.

Leblanc, Y., and F. Genova. The Jovian S-burst sources. *J. Geophys. Res. 86*:8546-8560, 1981.

Leblanc, Y., F. Genova, and J. de la Noe. The Jovian S bursts. I. Occurrence with L-bursts and frequency limit. *Astron. Astrophys. 86*:342-348, 1980.

Lecacheux, A. Periodic variations of the position of Jovian sources in longitude (System III) and phase of Io. *Astron. Astrophys. 37*:301-304, 1974.

Lecacheux, A., N. Meyer-Vernet, and G. Daigne. Jupiter's decametric radio emission: A nice problem of optics. *Astron. Astrophys., 94*:L9-L12, 1981.

Lee, L. C., J. R. Kan, and C. S. Wu. Generation of auroral kilometric radiation and the structure of the auroral acceleration region. *Planet. Space Sci. 28*:703-711, 1980.

Lee, L. C., and C. S. Wu. Amplification of radiation near cyclotron frequency due to electron population inversion. *Phys. Fluids 23*:1348-1354, 1980.

Lee, S. W., and P. C. Thomas. Near-surface flow of volcanic gases on Io. *Icarus 44*:280-290, 1980.

Legg, M. P., and K. C. Westfold. Ellipitical polarization of synchrotron radiation. *Astrophys. J. 154*:499-514, 1968.

Lemaire, J. Impulsive penetration of filamentary plasma elements into the magnetospheres of the Earth and Jupiter. *Planet. Space Sci. 25*:887-890, 1977.

Lepping, R. P., and K. W. Behannon. Magnetic field directional discontinuities: 1. Minimum variance errors. *J. Geophys. Res. 85*:4695-4703, 1980.

Lepping, R. P., L. F. Burlaga, and L. W. Klein. Jupiter's magnetopause, bow shock, and 10-hour modulated magnetosheath: Voyagers 1 and 2. *Geophys. Res. Lett. 8*:99-102, 1981.

Lepping, R. P., L.F. Burlaga, L. W. Klein, J. Jessen, and C. C. Goodrich. Observations of the magnetic field and plasma flow in Jupiter's magnetosheath. *J. Geophys. Res. 86*:8141-8156, 1981.

Leu, M. T., M. A. Biondi, and R. Johnsen. Measurements of the recombination of electrons with H_3O^+ $(H_2O)_n$ series ions. *Phys. Rev. A 7*:292-298, 1973.

Levitskii, L. S., and B. M. Vladimirski. The effect of interplanetary magnetic field sector structure on the decametric-wave emission of Jupiter. *Izvest. Krysmk. Astrofiz. Observ. 59*:104-109, 1979.

L'Heureux, J., and P. Meyer. Quiet-time increases of low-energy electrons: The Jovian origin. *Astrophys. J. 209*:955-960, 1976.

Liehmohn, H. B. Cyclotron – resonance amplification of VLF and ULF whistlers. *J. Geophys. Res. 72*:39-55, 1967.

Lin, R. P., C.-I. Meng, and K. A. Anderson. 30 to 100 keV protons upstream from the earth's bow shock. *J. Geophys. Res. 79*:489-498, 1974.

Lindhard, J., and M. Scharff. Energy dissipation by ions in the keV region. *Phys. Rev. 124*:128-130, 1961.

Liu, Z. X. Modified disc model of Jupiter's magnetosphere. *J. Geophys. Res., 87*:1691-1694, 1982.

Lotz, W. Electron-impact ionization cross-sections and ionization rate coefficients for atoms and ions. *Astrophys. J. Suppl. 14*:207-238, 1967.

Lynch, M. A., T. D. Carr, and J. May. VLBI measurements of Jovian S bursts. *Astrophys. J. 207*:325-328, 1976.

Lynch, M. A., T. D. Carr, J. May, W. F. Block, V. M. Robinson, and N. F. Six. Long-baseline analysis of a Jovian decametric L burst. *Astrophys. Lett. 10*:153-158, 1972.

Lyons, L. R. Comments on pitch-angle diffusion in the radiation belts. *J. Geophys. Res. 78*:6793-6797, 1973.

Lyons, L. R. Electron diffusion driven by magnetospheric electrostatic waves. *J. Geophys. Res. 79*:575-580, 1974.

Lyons, L. R., and R. M. Thorne. Parasitic pitch-angle diffusion of radiation belt particles by ion-cyclotron waves. *J. Geophys. Res. 77*:5608-5616, 1972.

Lyons, L. R., and R. M. Thorne. Equilibrium structure of radiation belt electrons. *J. Geophys. Res. 78*:2124-2142, 1973.

Lyons, L. R., R. M. Thorne, and C. F. Kennel. Electron pitch-angle diffusion driven by oblique whistler-mode turbulence. *J. Plasma Phys. 6*:589-606, 1971.

Lyons, L. R., R. M. Thorne, and C. F. Kennel. Pitch-angle diffusion of radiation belt electrons within the plasmasphere. *J. Geophys. Res. 77*:3455-3474, 1972.

Lysak, R. L., M. K. Hudson, and M. Temerin. Ion heating by strong electrostatic turbulence. *J. Geophys. Res. 85*:678-686, 1980.

Macy, W., and L. M. Trafton. Io's sodium emission clouds. *Icarus 25*:432-438, 1975a.

Macy, W., and L. M. Trafton. A model for Io's atmosphere and sodium cloud. *Astrophys. J. 200*:510-519, 1975b.

Maeda, K., P. H. Smith, and R. R. Anderson. VLF emission from ring current electrons. *Nature 263*:37-41, 1976.

Maggs, J. E. Coherent generation of VLF hiss. *J. Geophys. Res. 81*:1707-1724, 1976.

Maggs, J. E. Electrostatic noise generated by the auroral electron beam. *J. Geophys. Res. 83*:3173-3199, 1978.

Maier, H. N., and R. W. Fessenden. Electron-ion recombination rate constants for some compounds of moderate complexity. *J. Chem. Phys. 162*:4790-4795, 1975.

Marsch, E., H. Rosenbauer, W. Pilipp, K. H. Mulhauser, and R. Schwenn. Solar wind three dimensional alpha distribution in the solar wind observed by Helios between 0.3 AU and 1 AU. In *Solar Wind*, p. 4, ed. H. R. Rosenbauer. Springer-Verlag, New York, 1979.

Masursky, H., G. G. Schaber, L. A. Soderblom, and R. G. Strom. Preliminary geological mapping of Io. *Nature 280*:725-729, 1979.

Matson, D. L., B. A. Goldberg, T. V. Johnson, and R. W. Carlson. Images of Io's sodium cloud. *Science 199*:531-533, 1978.

Matson, D. L., T. V. Johnson, and F. P. Fanale. Sodium D-line emission from Io: Sputtering and resonant scattering hypotheses. *Astrophys. J. 192*:L43-L46, 1974.

May, J., T. D. Carr, and M. D. Desch. Decametric radio measurements of Jupiter's rotation period. *Icarus 40*:87-93, 1979.

McAdam, W. B. The extent of the emission region on Jupiter at 408 Mc/s. *Planet. Space Sci. 14*:1041-1046, 1966.

McClain, E. F., and R. M. Sloanaker. Preliminary observations at 10 cm wavelength using the NRL 84-foot radio telescope. In *Proceedings IAU Symposium No. 9 URSI Symposium No. 1*, ed. R. Bracewell, pp. 61-68. Stanford University Press, Stanford, 1959.

McConnell, J. C., B. R. Sandel, and A. L. Broadfoot. Airglow from Jupiter's nightside and crescent: Ultraviolet spectrometer observations from Voyager 2. *Icarus 43*:128-142, 1980.

McConnell, J. C., B. R. Sandel, and A. L. Broadfoot. Voyager UV spectrometer observations of He 584 A dayglow at Jupiter. *Planet. Space Sci., 29*:283-292, 1981.

McCulloch, P. M. Long term variation in Jupiter's 11 cm radio emission. *Proc. Astron. Soc. Austr. 2*:340-342, 1975.

McDonald, F. B., A. W. Schardt, and J. H. Trainor. Energetic protons in the Jovian magnetosphere. *J. Geophys. Res. 84*:2579-2596, 1979.

McDonough, T. R., and N. M. Brice. New kind of ring around Saturn. *Nature 242*:513, 1973.

McElroy, M. B.. The ionospheres of the major planets. *Space Sci. Rev. 14*:460-473, 1973.

McElroy, M. B., and Y. L. Yung. The atmosphere and ionosphere of Io. *Astrophys. J. 196*:227-250, 1975.

McEwen, D. J., and R. E. Barrington. Some characteristics of the lower hybrid resonance noise bands oberved by the Alouette 1 satellite. *Can. J. Phys. 45*:13-19, 1967.

McIlwain, C. E., and R. W. Fillius. Differential spectra and phase space densities of trapped electrons at Jupiter. *J. Geophys. Res. 80*:1341-1345, 1975.

McKibben, R. B., and J. A. Simpson. Evidence from charged particle studies for the distortion of the Jovian magnetosphere. *J. Geophys. Res. 79*:3545-3549, 1974.

McLaughlin, W. I. Prediscovery evidence of planetary rings. *J. Brit. Interplanet. Soc. 33*:287-294, 1980.

McNutt, R. L., J. W. Belcher, and H. S. Bridge. Positive ion observations in the middle magnetosphere of Jupiter. *J. Geophys. Res. 86*:8319-8342, 1981.

McNutt, R. L., Jr., J. W. Belcher, J. D. Sullivan, F. Bagenal, and H. S. Bridge. Departure from rigid corotation of plasma in Jupiter's dayside magnetosphere. *Nature 280*:803, 1979.

Mead, G. D. Magnetic coordinates for the Pioneer 10 Jupiter encounter. *J. Geophys. Res. 79*:3514-3521, 1974.

Mead, G. D., and D. B. Beard. Shape of the geomagnetic field solar wind boundary. *J. Geophys. Res. 69*:1169-1179, 1964.

Mead, G. D., and W. N. Ness. Jupiter's radiation belts and the sweeping effect of its satellites. *J. Geophys. Res. 78*:2793-2811, 1973.

Mekler, Yu., and A. Eviatar. Spectroscopic observations of Io. *Astrophys. J. Lett. 193*:L151-L152, 1974.

Mekler, Yu., and A. Eviatar. Thermal electron density in the Jovian magnetosphere. *J. Geophys. Res. 83*:5679-5684, 1978.

Mekler, Yu., and A. Eviatar. Time analysis of volcanic activity on Io by means of plasma observations. *J. Geophys. Res. 85*:1307-1310, 1980.

Mekler, Yu., A. Eviatar, and F. V. Coroniti. Sodium in the Jovian magnetosphere. *Astrophys. Space Sci. 40*:63-72, 1976.

Melrose, D. B. Rotational effects on the distribution of thermal plasma in the magnetosphere of Jupiter. *Planet. Space Sci. 15*:381-393, 1967.

Melrose, D. B. The emission and absorption of waves by charged particles in magnetized plasmas. *Astrophys. Space Sci. 2*:171-235, 1968.

Melrose, D. B. An interpretation of Jupiter's decametric radiation and the terrestrial kilometric radiation as direct amplified gyro-emission. *Astrophys. J. 207*:651-662, 1976.

Melrose, D. B. A theory for the nonthermal radio continua in the terrestrial and Jovian magnetosphere. *J. Geophys. Res., 86*:30-36, 1981.

Mendis, D. A., and W. I. Axford. Satellites and magnetospheres of the outer planets. *Ann. Rev. of Earth and Planet. Sci. 2*:419-474, 1974.

Menietti, J. D., and D. A. Gurnett. Whistler propagation in the Jovian magnetosphere. *Geophys. Res. Lett. 7*:49-52, 1980.

Mestel, L. Magnetic braking by a stellar wind, I. *Mon. Not. Roy. Astron. Soc. 138*:359, 1968.

Mewaldt, R. A., E. C. Stone, and R. E. Vogt. Observations of Jovian electrons at 1 AU. *J. Geophys. Res. 81*:2397-2400, 1976.

Michel, F. C. The astrophysics of Jupiter. *Space Sci. Rev. 24*:381-406, 1979.

Michel, F. C.. Theory of pulsar magnetospheres. *Rev. Mod. Phys. 54*:1-66, 1982.

Michel, F. C., and A. J. Dessler. Pulsar disk systems. *Astrophys. J. 251*:654-664, 1981.

Michel, F. C., and P. A. Sturrock. Centrifugal instability of the Jovian magnetosphere and its interaction with the solar wind. *Planet. Space Sci. 22*:1501-1510, 1974.

Mihalov, J. D. Result of least-squares fitting of Jupiter's inner trapped proton fluxes. NASA Technical Memorandum *TM-73*:260, June, 1977.

Mihalov, J. D., D. S. Colburn, R. G. Currie, and C. P. Sonett. Configuration and reconnection of the geomagnetic tail. *J. Geophys. Res. 73*:943-959, 1968.

Mihalov, J. D., H. R. Collard, D. D. McKibben, J. H. Wolfe, and D. S. Intriligator. Pioneer 11 encounter: Preliminary results from the Ames Research Center plasma analyzer experiment. *Science 188*:448-451, 1975.

Miles, W. T., R. Tompson, and A. E. S. Green. Electron-impact cross sections and energy deposition in molecular hydrogen. *J. Appl. Phys. 43*:678-686, 1972.

Mitchell, D. G., and E. C. Roelof. Thermal iron ions in high speed solar wind streams: Detection by the IMP 7/8 energetic particle experiments. *Geophys. Res. Lett. 7*:661-664, 1980.

Mitchell, D. G., E. C. Roelof, W. C. Feldman, S. J. Bame, and D. J. Williams. Thermal iron ions in high speed solar wind streams, 2. Temperatures and bulk velocities. *Geophys. Res. Lett. 8*:827-830, 1981.

Mogro-Campero, A. Absorption of radiation belt particles by the inner satellites of Jupiter. In *Jupiter*, ed. T. Gehrels, pp. 1190-1214. University of Arizona Press, Tucson, 1976.

Mogro-Campero, A., and R. W. Fillius. The absorption of trapped particles by the inner satellites of Jupiter and the radial diffusion coefficient of particle transport. *J. Geophys. Res. 81*:1289-1295, 1976.

Montgomery, D. C., and D. A. Tidman. *Plasma Kinetic Theory*, pp. 129-153. McGraw Hill, New York, 1964.

Moore, C. E. *Atomic Energy Tables*. NSRDS-NBS 35, Vol.1, 1971.

Moos, H. W., and J. T. Clarke. Ultraviolet observations of the Io torus from earth orbit using the IUE Observatory. *Astrophys. J. 247*:354-361, 1981.

Morabito, L. A., S. P. Synnott, P. M. Kupferman, and S. A. Collins. Discovery of currently-active extraterrestrial volcanism. *Science 204*:972, 1979.

Morgan, J. S., and C. B. Pilcher. Plasma characteristics of the Io torus. *Astrophys. J. 253*:406-421, 1982, and *254*:420, 1982.

Morris, D., and G. L. Berge. Measurements of the polarization and angular extent of the decimeter radiation of Jupiter. *Astrophys. J. 136*:276-282, 1962.

Morris, D., J. B. Whiteoak, and F. Tonking. The linear polarization of radiation from Jupiter at 6 cm wavelength. *Austr. J. Phys. 21*:337-340, 1968.

Münch, G. and J. T. Bergstralh. Io: Morphology of its sodium emission region. *Pub. Astron. Soc. Pac. 89*:232-237, 1977.

Münch, G., J. Trauger, and F. Roesler. Interferometric studies of the emissions associated with Io. *Bull. Am. Astron. Soc. 8*:468, 1976.

Munson, M. S. B., and F. H. Field. Reactions of gaseous ions, XVII, Methane + unsaturated hydrocarbons. *J. Amer. Chem. Soc. 91*:3413-3418, 1969.

Murcray, F. J., and R. M. Goody. Pictures of the Io sodium cloud. *Astrophys. J. 226*:327-335, 1978.

Nagy, A. F., W. L. Chameides, R. H. Chen, and S. K. Atreya. Electron temperatures in the Jovian ionosphere. *J. Geophys. Res. 81*:5567-5569, 1976.

Nash, D. B. Jupiter sulfur plasma ring. *EOS 60*:307, 1979.

Nash, D. B., and R. M. Nelson. Spectral evidence for sublimates and adsorbates on Io. *Nature 280*:763-766, 1979.

Neidhofer, J., R. S. Booth, D. Morris, W. Wilson, F. Biraud et al. New measurements of the Stokes parameters of Jupiter's 11 cm radiation. *Astron. Astrophys. 61*:321-328, 1977.

Neidhofer, J., D. Morris, and W. Wilson. Observations of the Jovian radiation at 11 and 18 cm wavelength. *Astron. Astrophys. 83*:297-302, 1980.

Ness, N. F. The Earth's magnetic tail. *J. Geophys. Res. 70*:2989-3005, 1965.

Ness, N. F., M. H. Acuña, R. P. Lepping, L. F. Burlaga, K. W. Behannon, and F. M. Neubauer. Magnetic field studies at Jupiter by Voyager 1: Preliminary results. *Science 204*:982-987, 1979a.

Ness, N. F., M. H. Acuña, R. P. Lepping, K. W. Behannon, L. F. Burlaga, and F. M. Neubauer. Jupiter's magnetic tail. *Nature 280*:799-802, 1979b.

Ness, N. F., M. H. Acuña, R. P. Lepping, L. F. Burlaga, K. W. Behannon and F. M. Neubauer. Magnetic field studies at Jupiter by Voyager-2: Preliminary results. *Science 206*:966-971, 1979c.

Neubauer, F. M. Possible strengths of dynamo magnetic fields of the Galilean satellites and of Titan. *Geophys. Res. Lett. 5*:905-908, 1978.

Neubauer, F. M. Nonlinear standing Alfvén wave current system at Io: Theory. *J. Geophys. Res. 85*:1171-1178, 1980.

Neugebauer, M., and A. Eviatar. An alternative interpretation of Jupiter's "plasmapause." *Geophys. Res. Lett. 3*:708-710, 1976.

Neugebauer, M., and C. W. Snyder. The mission of Mariner II: Preliminary observations. *Science 138*:1095-1096, 1962.

Newburn, R. L., Jr., and S. Gulkis. A survey of the outer planets: Jupiter, Saturn, Uranus, Neptune, Pluto, and their satellites. *Space Sci. Rev. 3*:179-271, 1973.

Nicholson, D. R., M. V. Goldman, P. Hoyng, and J. C. Weatherall. Nonlinear Langmuir waves during type III solar radio bursts. *Astrophys. J. 223*:605-619, 1978.

Nishida, A. Outward diffusion of energetic particles from the Jovian radiation belt. *J. Geophys. Res. 81*:1771-1773, 1976.

Nishida, A., and Y. Watanabe. Joule heating of the Jovian ionosphere by corotation enforcement currents. *J. Geophys. Res. 86*:9945-9952, 1981.

Northrop, T. G. *The Adiabatic Motion of Charged Particles*, p. 41. Interscience, New York, 1963.

Northrop, T. G. Residence lifetimes of 1.79-2.15 MeV protons in the equatorial region of the Jovian magnetosphere. *J. Geophys. Res. 84*:5813-5816, 1979.

Northrop, T. G., T. J. Birmingham, and A. W. Schardt. Anisotropies in the fluxes of Pioneer 10 protons. *J. Geophys. Res. 84*:47-55, 1979.

Northrop, T. G., C. K. Goertz, and M. F. Thomsen. The magnetosphere of Jupiter as observed with Pioneer 10: 2. Nonrigid rotation of the magnetodisc. *J. Geophys. Res. 79*:3579-3582, 1974.

Northrop, T. G., and A. W. Schardt. Instability of equatorial protons in Jupiter's mid-magnetosphere. *J. Geophys. Res. 85*:25-32, 1980.

Ortwein, N. R., D. B. Chang, and L. Davis. Synchrotron radiation in a dipole field. *Astrophys. J. Suppl. 12*:323-389, 1966.

Osterbrock, D. E. *Astrophysics of Gaseous Nebulae*. Freeman, San Francisco, 1974.

Oya, H. Origin of Jovian decameter wave emissions – conversion from the electron cyclotron plasma wave to the ordinary mode electromagnetic wave. *Planet. Space Sci. 22*:687-708, 1974.

Oya, H. A, T. Kondo, and A. Morioka. The energy sources of Jovian decametric radio waves. In *Highlights of the Japanese IMS Program*, Instit. Space Aeron. Sci., University of Tokyo, 1980.

Oya, H., A. Morioka, and T. Kondo. Locations of Jovian decametric radiation sources. *Planet. Space Sci.* *27*:963-972, 1979.

Pacholczyk, A. G. *Radio Astrophysics: Non-Thermal Processes in Galactic and Extragalactic Sources.* Freeman, San Francisco, 1970.

Palmadesso, P., T. R. Coffey, S. L. Ossakow, and K. Papadopoulos. Generation of terrestrial kilometric radiation by a beam-driven electromagnetic instability. *J. Geophys. Res. 81*:1762-1770, 1976.

Papadopoulos, K., M. L. Goldstein, and R. A. Smith. Stabilization of electron streams in type III solar radio bursts. *Astrophys. J. 190*:175-185, 1974.

Parish, J. L., C. K. Goertz, and M. F. Thomsen. Azimuthal magnetic field at Jupiter. *J. Geophys. Res. 85*:4152-4156, 1980.

Parker, E. N. *Interplanetary Dynamical Processes.* Wiley-Interscience, New York, 1963.

Parker, G. D., G. A. Dulk, and J. W. Warwick. *Astrophys. J. 157*:439-448, 1969.

Peach, G. Ionization of neutral atoms with outer 2p and 3p electrons by electron and proton impact. *J. Phys. B 1*:1088-1188, 1968.

Peach, G. Ionization of atoms and positive ions by electron and proton impact. *J. Phys. B 4*:1670-1677, 1971.

Peale, S. J., P. Cassen, and R. T. Reynolds. Melting of Io by tidal dissipation. *Science 203*:892, 1979.

Pearce, J. B. A heuristic model for Jovian decametric arcs. *J. Geophys. Res.*, *86*:8579-8580, 1981.

Pearl, J., R. Hanel, V. Kunde, W. Maguire, K. Fox et al. Identification of gaseous SO_2 and new upper limits for other gases on Io. *Nature 280*:757-758, 1979.

Penrose, O. Electrostatic instabilities of a uniform non-Maxwellian plasma. *Phys. Fluids 3*:258-265, 1960.

Pesses, M. E., and C. K. Goertz. Jupiter's magnetotail as the source of interplanetary Jovian MeV electrons observed at Earth. *Geophys. Res. Lett. 3*:228-230, 1976.

Petterson, J., and I. Martinson, in preparation, 1982.

Piddington, J. H. *Cosmic Electrodynamics*, Chapter 10. Wiley-Interscience, New York, 1969.

Piddington, J. H., and J. F. Drake. Electrodynamic effects of Jupiter's satellite Io. *Nature 217*:935-937, 1968.

Pilcher, C. B. Images of Jupiter's sulfur ring. *Science 207*:181-183, 1980a.

Pilcher, C. B. As reported in *Science 208*:384-386, 1980b.

Pilcher, C. B. Transient sodium ejection from Io. *Bull. Amer. Astron. Soc. 12*:675, 1980c.

Pilcher, C. B., and J. S. Morgan. Detection of singly ionized oxygen around Jupiter. *Science 205*:297-298, 1979.

Pilcher, C. B., and J. S. Morgan. The distribution of [S II] emission around Jupiter. *Astrophys. J. 238*:375-380, 1980.

Pilcher, C. B., J. S. Morgan, J. H. Fertel, and C. C. Avis. A movie of the Io plasma torus. *Bull. Amer. Astron. Soc. 13*:731, 1981.

Pilcher, C. B., and W. V. Schempp. Jovian sodium emission from Region C_2. *Icarus 28*:1-11, 1979.

Pilcher, C. B., and D. F. Strobel. Emissions from neutrals and ions in the Jovian magnetosphere. In *Satellites of Jupiter*, ed. D. Morrison, University of Arizona Press, Tucson, 1982.

Poquerusse, M., and A. Lecacheux. First direct measurement of the beaming of Jupiter's decametric radiation. *Nature 275*:111-113, 1978.

Post, R. F., and M. N. Rosenbluth. Electrostatic instabilities in finite mirror-confined plasmas. *Phys. Fluids. 9*:730-749, 1966.

Prakash, A., and N. Brice. Magnetospheres of Earth and Jupiter after Pioneer 10. In *The Magnetospheres of the Earth and Jupiter*, ed. V. Formisano, pp. 411-423. D. Reidel, Dordrecht, 1975.

Prasad, S. S., and A. Tan. The Jovian ionosphere. *Geophys. Res. Lett. 1*:337-340, 1974.

Preece, W. H. Earth currents. *Nature 49*:554, 1894.

Pustovalov, V. V., and V. P. Silin. Nonlinear theory of the interaction of waves in a plasma. In *Proceedings of the P. N. Lebedev Phys. Inst. No. 61, Theory of Plasmas*, ed. D. V. Skobel'tsyn, pp. 37ff. Consultants Bureau, New York, 1975.

Pyle, K. R., and J. A. Simpson. The Jovian relativistic electron distribution in interplanetary space from 1 to 11 AU: Evidence for a continuously emitting "Point" source. *Astrophys. J. 215*:L89-L93, 1977.

Radhakrishnan, V., and J. A. Roberts. Polarization and angular extent of the 960 Mc/s radiation from Jupiter. *Phys. Rev. Lett. 4*:493-494, 1960.

Ratner, M. I. Very-long-baseline observations of Jupiter's millisecond radio bursts. Ph.D. Thesis, University of Colorado, Boulder, 1976.

Rebbert, R. E., S. G. Lias, and P. Ausloos. Pulse radiolysis of methane. *J. Res. Nat. Bur. Stand., Sect. A 77*:249-257, 1973.

Reyes, F., and J. May. Beaming of Jupiter's decametric radiation. *Nature 291*:171, 1981.

Rich, F. J., V. M. Vasyliunas, and R. A. Wolf. On the balance of stresses in the plasma sheet. *J. Geophys. Res. 77*:4670-4676, 1972.

Richardson, J. D., and G. L. Siscoe. Factors governing the ratio of inward to outward diffusion of flux of satellite ions. *J. Geophys. Res. 86*:8485-8490, 1981.

Richardson, J. D., G. L. Siscoe, F. Bagenal, and J. D. Sullivan. Time dependent plasma injection by Io. *Geophys. Res. Lett. 7*:37-40, 1980.

Riddle, A. C., and J. W. Warwick. Redefinition of System III longitude. *Icarus 27*:457-459, 1976.

Riihimaa, J. J. Structured events in the dynamic spectra of Jupiter's decametric radio emission. *Astron. J. 73*:265-270, 1968.

Riihimaa, J. J. Radio spectra of Jupiter. *Report S22*. Dept. Elec. Eng., University of Oulu, Oulu, Finland, 1971.

Riihimaa, J. J. Modulation lanes in the dynamic spectra of Jupiter's decametric radio emission. *Ann. Acad. Sci. Fenn. A VI*:1-38, 1974.

Riihimaa, J. J. S bursts in Jupiter's decametric radio spectra. *Astrophys. Space Sci. 51*:363-383, 1977.

Riihimaa, J. J. Drift rates of Jupiter's S bursts. *Nature 279*:783-785, 1979.

Riihimaa, J. J., and T. D. Carr. Interactions of S- and L-bursts in Jupiter's decametric radio spectra. *The Moon and the Planets, 25*:373-387, 1981.

Riihimaa, J. J., G. A. Dulk, and J. W. Warwick. Morphology of the fine structure in the dynamic spectra of Jupiter's decametric radiation. *Astrophys. J. Suppl. 172 19*:175-192, 1970.

Roberts, C. S., and M. Schulz. Bounce resonance scattering of particles trapped in the Earth's magnetic field. *J. Geophys. Res. 73*:7361-7376, 1968.

Roberts, J. A. The pitch angles of electrons in Jupiter's radiation belt. *Proc. Astron. Soc. Austr. 3*:53-55, 1976.

Roberts, J. A., and R. D. Ekers. The position of Jupiter's Van Allen belt. *Icarus 5*:149-153, 1968.

Roberts, J. A., and M. M. Komesaroff. Observations of Jupiter's radio spectrum and polarization in the range from 6 cm to 100 cm. *Icarus 4*:127-156, 1965.

Roberts, J. A., and M. M. Komesaroff. Circular polarization of the 1.4 GHz emission from Jupiter's radiation belts. *Icarus 29*:455-461, 1976.

Roberts, J. A., and G. J. Stanley. Radio emission from Jupiter at a wavelength of 31 centimeters. *Publ. Astron. Soc. Pacific. 71*:485-496, 1959.

Roble, R. G., R. E. Dickinson, E. C. Ridley, and Y. Kamide. Thermospheric response to the November 8-9, 1969, magnetic disturbances. *J. Geophys. Res. 84*:4207-4216, 1979.

Rodriguez, P. Magnetosheath electrostatic turbulence. *J. Geophys. Res. 84*:917-930, 1979.

Rodriguez, P., and D. A. Gurnett. Electrostatic and electromagnetic turbulence associated with the Earth's bow shock. *J. Geophys. Res. 80*:19-31, 1975.

Roederer, J. G. *Dynamics of Geomagnetically Trapped Radiation.* Springer-Verlag, New York, 1970.

Roederer, J. G., M. H. Acuña, and N. F. Ness. Jupiter's internal magnetic field geometry relevant to particle trapping. *J. Geophys. Res. 82*:5187-5194, 1977.

Roelof, E. C. Scatter-free collimated convection and cosmic-ray transport at 1 AU. *Proc. 14th International Cosmic Ray Conference, Munich 5*:1716, 1975.

Roelof, E. C., E. P. Keath, C. O. Bostrom and D. J. Williams. Fluxes of $\geq$ 50 keV protons and $\geq$ 30 keV electrons at 35 R_E, 1. Velocity anisotropies and plasma flow in the magnetotail. *J. Geophys. Res. 81*:2304-2314, 1976.

Rönnmark, K., H. Borg, P. S. Christiansen, M. P. Gough, and D. Jones. Banded electron cyclotron harmonic instability – a first comparison of theory and experiment. *Space Sci. Rev. 22*:401-417, 1978.

Rossi, B., and S. Olbert. *Introduction to the Physics of Space.* McGraw-Hill, New York, 1970.

Roux, A., and R. Pellat. Coherent generation of the auroral kilometric radiation by nonlinear beatings between electrostatic waves. *J. Geophys. Res. 84*:5189-5198, 1979.

Rowland, H., P. J. Palmadesso, and K. Papadopoulos. Anomalous resistivity on auroral field lines. *Geophys. Res. Lett., 8*:1257-1260, 1981.

Rudd, M. E., C. A. Sautter, and C. L. Bailey. Energy and angular distributions of electrons ejected from hydrogen and helium by 100 to 300 keV protons. *Phys. Rev. 151*:20-27, 1966.

Russell, C. T., R. E. Holzer, and E. J. Smith. Observations of ELF noise in the magnetosphere: 1. Spatial extent and frequency of occurrence. *J. Geophys. Res. 74*:755-777, 1969.

Sagan, C. Sulfur flows on Io. *Nature 280*:750-753, 1979.

Sagan, C. E., E. R. Lippincoltt, M. O. Dayhoff, and R. V. Eck. Organic molecules and the coloration of Jupiter. *Nature 213*:273-274, 1967.

Sakurai, K. Non-Io-related radio emissions and the modulation of relativistic electrons in the Jovian magnetosphere. *Planet. Space Sci. 24*:657-659, 1976a.

Sakurai, K. A theoretical model of the bow shock and the magnetosphere of Jupiter. *Planet. Space Sci. 24*:661-664, 1976b.

Sakurai, K. A possible method of accelerating relativistic electrons in the neutral sheet in the Jovian outer magnetosphere. *Planet. Space Sci. 24*:1207-1208, 1976c.

Sandel, B. R. Longitudinal asymmetrics in Jupiter's aurora and Io's plasma torus: Voyager observations. Presentation at 'Physics of the Jovian Magnetosphere" Conference, Rice University, Houston, February 1980.

Sandel, B. R., and A. L. Broadfoot. Io's hot plasma torus – A synoptic view from Voyager. *J. Geophys. Res. 87*:212-218, 1982a.

Sandel, B. R., and A. L. Broadfoot. Discovery of an Io-correlated energy source for Io's hot plasma torus. *J. Geophys. Res., 87*:2231-2240, 1982b.

Sandel, B. R., A. L. Broadfoot, and D. F. Strobel. Discovery of a longitudinal asymmetry in the hydrogen Lyman-alpha brightness of Jupiter. *Geophys. Res. Lett. 7*:5-8, 1980.

Sandel, B. R., D. E. Shemansky, A. L. Broadfoot, J. L. Bertaux, J. E. Blamont et al. Extreme ultraviolet observations from Voyager 2 encounter with Jupiter. *Science 206*:962-966, 1979.

Sarris, E. T., S. M. Krimigis, and T. P. Armstrong. Observations of magnetospheric bursts of high-energy protons and electrons at ~ 35 R_E with IMP 7. *J. Geophys. Res.* *81*:2341-2355, 1976.

Sarris, E. T., S. M. Krimigis, C. O. Bostrom, and T. P. Armstrong. Simultaneous multispacecraft observations of energetic proton bursts inside and outside the magnetosphere. *J. Geophys. Res.* *83*:4289-4305, 1978.

Scarf, F. L. Plasma physics and wave-particle interaction at Jupiter. In *Jupiter*, p. 870-895, edited by T. Gehrels. University of Arizona Press, Tuscon, 1976.

Scarf, F. L., and N. L. Sanders. Some comments on the whistler-mode instability at Jupiter. *J. Geophys. Res.* *81*:1787-1790, 1976.

Scarf, F. L., and D. A. Gurnett. A plasma wave investigation for the Voyager mission. *Space Sci. Rev.* *21*:289-308, 1977.

Scarf, F. L., R. W. Fredricks, L. A. Frank, C. T. Russell, P. J. Coleman, Jr., and M. Neugebauer. Direct correlations of large amplitude waves with suprathermal protons in the upstream solar wind. *J. Geophys. Res.* *75*:7316-7322, 1970.

Scarf, F. L., R. W. Fredricks, L. A. Frank, and M. Neugebauer. Nonthermal electrons and high-frequency waves in the upstream solar wind: 1. Observations. *J. Geophys. Res.* *76*:5162-5171, 1971.

Scarf, F. L., R. W. Fredricks, C. F. Kennel, and F. V. Coroniti. Satellite studies of magnetospheric storms on August 15, 1968. *J. Geophys. Res.* *78*:3119-3130, 1973.

Scarf, F. L., D. A. Gurnett, and W. S. Kurth. Jupiter plasma wave observations: An initial Voyager 1 overview. *Science 204*:991-995, 1979.

Scarf, F. L., F. V. Coroniti, D. A. Gurnett, and W. S. Kurth. Pitch-angle diffusion by whistler-mode waves near the Io plasma torus. *Geophys. Res. Lett.* *6*:653-656, 1979a.

Scarf, F. L., D. A. Gurnett, W. S. Kurth, and R. L. Poynter. Plasma wave turbulence at Jupiter's bow shock. *Nature 280*:796-797, 1979b.

Scarf, F. L., D. A. Gurnett, and W. S. Kurth. Measurements of plasma wave spectra in Jupiter's magnetosphere. *J. Geophys. Res.* *86*:8181-8198, 1981.

Scarf, F. L., W. W. L. Taylor, C. T. Russel, and R. C. Elphic. Pioneer-Venus plasma wave observations: The solar wind Venus interaction. *J. Geophys. Res.*, *85*:7599-7612, 1980.

Schaber, G. G. The surface of Io: Geologic units, morphology, and tectonics. *Icarus 43*:302-333, 1980.

Schardt, A. W., and T. J. Birmingham. Discrepancy in proton flux extrapolation along field lines in the middle magnetosphere. *J. Geophys. Res.* *84*:56-62, 1979.

Schardt, A. W., F. B. McDonald, and J. H. Trainor. Acceleration of protons at 32 Jovian radii in the outer magnetosphere of Jupiter. *J. Geophys. Res.* *83*:ll04-ll08, 1978.

Schardt, A. W., F. B. McDonald, and J. H. Trainor. Energetic particles in the pre-dawn magnetotail of Jupiter. *J. Geophys. Res.* *86*:8413-8428, 1981.

Schatten, K. H. A great red spot dependence on solar activity? *Geophys. Res. Lett.* *6*:593-596, 1979.

Schatten, K. H., and N. F. Ness. The magnetic-field geometry of Jupiter and its relation to Io-modulated Jovian decametric radio emission. *Astrophys. J.* *165*:621-631, 1971.

Scheuer, P. A. G. Synchrotron radiation formulae. *Astrophys. J. Lett.* *151*:L139-L142, 1968.

Schindler, K. Plasma and field in the magnetospheric tail. *Space Sci. Rev.* *17*:589-614, 1975.

Schindler, K. Theories of tail structures. *Space Sci. Rev.* *23*:365-374, 1979.

Schmahl, E. J. *Io, an Alfvén Wave Generator*. Ph.D. thesis, University of Colorado, Boulder, 1970.

Scholer, M. On the motion of artificial ion clouds in the magnetosphere. *Planet. Space Sci. 18*:977-1004, 1970.

Scholer, M., G. Gloeckler, F. M. Ipavich, D. Hovestadt, and B. Kleckler. Upstream particle events close to the bow shock and 200 R$_E$ upstream: ISEE-1 and ISEE-2 observations. *Geophys. Res. Lett. 7*:73-76, 1980.

Schulz, M. Jupiter's radiation belts. *Space Sci. Rev. 23*:277-318, 1979.

Schulz, M., and A. Eviatar. Charged-particle absorption by Io. *Astrophys. J. Lett. 211*:L149-L154, 1977.

Schulz, M., and L. J. Lanzerotti. *Particle Diffusion in the Radiation Belts.* Springer Verlag, New York, 1974.

Schulz, M., and A. L. Vampola. Magnetospheric stability criteria for artificial electron belts. *Aerospace Corp. Report SPL-75*, 6200-6220, 1975.

Scudder, J. D., E. C. Sittler, Jr., and H. S. Bridge. A survey of the plasma electron environment of Jupiter: A view from Voyager. *J. Geophys. Res. 86*:8157-8179, 1981.

Seaquist, E. R. Circular polarization of Jupiter at 9.16 cm. *Nature 224*:1011-1012, 1969.

Seidelman, P. K., and N. Divine. Evaluation of Jupiter longitudes in System III (1965). *Geophys. Res. Lett. 4*:65-68, 1977.

Sentman, D. D., and C. K. Goertz. Whistler mode noise in Jupiter's inner magnetosphere. *J. Geophys. Res. 83*:3151-3165, 1978.

Sentman, D. D., and J. A. Van Allen. Angular distributions of electrons of energy $E_e > 0.06$ MeV in the Jovian magnetosphere. *J. Geophys. Res. 81*:1350-1360, 1976.

Sentman, D. D., J. A. Van Allen, and C. K. Goertz. Recirculation of energetic particles in Jupiter's magnetosphere. *Geophys. Res. Lett. 2*:465-468, 1975.

Sentman, D. D., J. A. Van Allen, and C. K. Goertz. Correction to 'Recirculation of energetic particles in Jupiter's magnetosphere.' *Geophys. Res. Lett. 5*:621-622, 1978.

Sharp, R. D., E. G. Shelley, R. G. Johnson, and A. G. Ghielmetti. Counterstreaming electron beams at altitudes of ~ 1 R$_E$ over the auroral zone (abstract). *EOS Trans. AGU 59*:1158, 1978.

Shaw, R. R., and D. A. Gurnett. Electrostatic noise bands associated with the electron gyrofrequency and plasma frequency in the outer magnetosphere. *J. Geophys. Res. 80*:4259-4271, 1975.

Shaw, R. R., B. D. Strayer, D. A. Gurnett, W. S. Kurth, and F. L. Scarf. Observations of whistlers from Jupiter. *J. Geophys. Res.*, in preparation, 1981.

Shawhan, S. D. Io sheath-accelerated electrons and ions. *J. Geophys. Res. 81*:3373-3379, 1976.

Shawhan, S. D. Magnetospheric plasma waves. In *Solar System Plasma Physics*, Vol III, ed. L. J. Lanzerotti, G. F. Kennel, and E. Parker, pp. 211-270. North-Holland, Amsterdam, 1979.

Shawhan, S. D., C. K. Goertz, R. F. Hubbard, D. A. Gurnett, and G. Joyce. Io-accelerated electrons and ions. In *The Magnetospheres of Earth and Jupiter*, ed. V. Formisano, pp. 375-389. D. Reidel, Dordrecht, 1975.

Shawhan, S. D., M. W. Hodges, and S. R. Spangler. The 1975.9 Jovian decimetric spectrum. *J. Geophys. Res. 82*:1901-1905, 1977.

Shawhan, S. D., R. F. Hubbard, G. Joyce, and D. A. Gurnett. Sheath acceleration of photoelectrons by Jupiter's moon Io. In *Photon and Particle Interactions with Surfaces in Space*, ed. R. Grard, pp. 405-413. D. Reidel, Dordrecht, 1973.

Shemansky, D. E. Radiative cooling efficiencies and predicted spectra of species of the Io plasma torus. *Astrophys. J. 236*:1043-1054, 1980a.

Shemansky, D. E. Mass-loading and diffusion-loss rates of the Io plasma torus. *Astrophys. J. 242*:1266-1277, 1980b.

Shemansky, D. E., and B. R. Sandel. The injection of energy into the Io plasma torus. *J. Geophys. Res.*, *87*:219-229, 1982.

Shemansky, D. E., and G. R. Smith. The Voyager EUV spectrum of the Io plasma torus. *J. Geophys. Res. 86*:9179-9192, 1981.

Sherril, W. M. Polarization measurements of the decameter emission from Jupiter. *Astrophys. J. 142*:1171-1185, 1965.

Simpson, J. A., D. C. Hamilton, R. B. McKibben, A. Mogro-Campero, K. R. Pyle et al. The protons and electrons trapped in the Jovian dipole magnetic field region and their interaction with Io. *J. Geophys. Res. 79*:3522-3544, 1974a.

Simpson, J. A., D. Hamilton, G. A. Lentz, R. B. McKibben, A. Mogro-Campero et al. Protons and electrons in Jupiter's magnetic field: Results from the University of Chicago experiment on Pioneer 10. *Science 183*:306-309, 1974b.

Simpson, J. A., D. C. Hamilton, G. A. Lentz, R. B. McKibben, M. Perkins et al. Jupiter revisited: First results from the University of Chicago charged particle experiment on Pioneer 11. *Science 188*:455-459, 1975.

Simpson, J. A., and R. B. McKibben. Dynamics of the Jovian magnetosphere and energetic particle radiation. In *Jupiter*, ed. T. Gehrels, pp. 738-766. University of Arizona Press, Tucson, 1976.

Siscoe, G. L. On the equatorial confinement and velocity space distribution of satellite ions in Jupiter's magnetosphere. *J. Geophys. Res. 82*:1641-1645, 1977.

Siscoe, G. L. Towards a comparative theory of magnetosphere. In *Solar System Plasma Physics*, ed. C. F. Kennel, L. J. Lanzerotti, and E. N. Parker. North-Holland, Amsterdam, 1978a.

Siscoe, G. L. Jovian plasmaspheres. *J. Geophys. Res. 83*:2118-2126, 1978b.

Siscoe, G. L., and C. K. Chen. Io: A source for Jupiter's inner plasmasphere. *Icarus 31*:1-10, 1977.

Siscoe, G. L., and N. Crooker. A theoretical relation between Dst and the solar wind merging electric field. *Geophys. Res. Lett. 1*:17-19, 1974.

Siscoe, G. L., and W. D. Cummings. On the cause of geomagnetic bays. *Planet. Space Sci. 17*:1795-1802, 1969.

Siscoe, G. L., A. Eviatar, R. M. Thorne, J. D. Richardson, F. Bagenal, and J. D. Sullivan. Ring current impoundment of the Io plasma torus. *J. Geophys. Res. 86*:8480-8484, 1981.

Siscoe, G. L., and D. Summers. Centrifugally driven diffusion of Iogenic plasma. *J. Geophys. Res. 86*:8471-8479, 1981.

Slee, O. B., and G. A. Dulk. 80 MHz measurements of Jupiter's synchrotron emission. *Austr. J. Phys. 25*:103-105, 1972.

Sloanaker, R. M. Apparent temperature of Jupiter at a wavelength of 10 cm. *Astron. J. 64*:346, 1959.

Smith, B. A., and S. A. Smith. Upper limits for an atmosphere on Io. *Icarus 17*:218-222, 1972.

Smith, B. A., E. M. Shoemaker, S. W. Kieffer, and A. F. Cook, II. The role of SO_2 in volcanism on Io. *Nature 280*:738-743, 1979a.

Smith, B. A., L. A. Soderblom, T. V. Johnson, A. P. Ingersoll, S. A. Collins et al. The Jupiter system through the eyes of Voyager 1. *Science 204*:13-31, 1979b.

Smith, E. J., L. Davis, Jr., and D. E. Jones. Jupiter's magnetic field and magnetosphere. In *Jupiter*, ed. T. Gehrels, pp. 788-829. University of Arizona Press, Tucson, 1976.

Smith, E. J., L. Davis, Jr., D. E. Jones, P. J. Coleman, D. S. Colburn et al. The planetary magnetic field and magnetosphere of Jupiter: Pioneer 10. *J. Geophys. Res. 79*:3501-3513, 1974.

Smith, E. J., L. Davis, D. E. Jones, P. J. Coleman, P. Dyal et al. Jupiter's magnetic field, magnetosphere, and interaction with the solar wind: Pioneer 11. *Science 188*:451-454, 1975.

Smith, E. J., R. W. Fillius, and J. H. Wolf. Compression of Jupiter's magnetosphere by the solar wind. *J. Geophys. Res. 83*:4733-4742, 1978.

Smith, E. J., and S. Gulkis. The magnetic field of Jupiter: A comparison of radio astronomy and spacecraft observations. *Ann. Rev. Earth Planet. Sci. 7*:385-415, 1979.

Smith, R. A. Models of Jovian decametric radiation. In *Jupiter*, ed. T. Gehrels, pp. 1146-1189. University of Arizona Press, Tucson, 1976a.

Smith, R. A. Stop zones in the Jovian plasmasphere. *Astrophys. Lett. 17*:167-172, 1976b.

Smith, R. A., and C. K. Goertz. On the modulation of the Jovian decametric radiation by Io. 1. Acceleration of charged particles. *J. Geophys. Res. 83*:2617-2627, 1978.

Smoluchowski, R. Jupiter's molecular hydrogen layer and the magnetic field. *Astrophys. J. 200*:L119-121, 1975.

Smyth, W. H. Io's sodium cloud: Explanation of the east-west asymmetries. *Astrophys. J. 234*:1148-1153, 1979.

Smyth, W. H., and M. B. McElroy. The sodium and hydrogen gas clouds of Io. *Planet. Space Sci. 25*:415-431, 1977.

Smyth, W. H., and H. B. McElroy. Io's sodium cloud: Comparison of models and two-dimensional images. *Astrophys. J. 226*:336-346, 1978.

Smythe, W. D., R. M. Nelson, and D. B. Nash. Spectral evidence for SO_2 frost or adsorbate on Io's surface. *Nature 280*:766, 1979.

Sonnerup, B. U. Ö., and L. J. Cahill. Magnetopause structure and attitude from Explorer 12 observations. *J. Geophys. Res. 72*:171-183, 1967.

Southwood, D. J., M. G. Kivelson, R. J. Walker, and J. A. Slavin. Io and its plasma environment. *J. Geophys. Res. 85*:5959-5968, 1980.

Sozzou, C. A similarity model illustrating the effect of rotation on an inflated magnetosphere. *Planet. Space Sci. 26*:311-317, 1978.

Spitzer, L., Jr. *Physics of Fully Ionized Gases*. Interscience, John Wiley, New York, 1962.

Spjeldvik, W. N., and R. M. Thorne. The cause of storm after effects in the middle latitude D-region. *J. Atmos. Terrest. Phys. 37*:777-795, 1975.

Spreiter, J. R., A. Y. Alksne, and A. L. Summers. External aerodynamics of the magnetosphere. In *Physics of the Magnetosphere*, ed. R. L. Carovillano, J. F. McClay, and H. R. Radoski, pp. 301-375. D. Reidel, Dordrecht, Netherlands, 1968.

Staelin, D. H. Character of the Jovian decameter arcs. *J. Geophys. Res. 86*:8581-8584, 1981.

Stannard, D., and R. G. Conway. Recent observations of the decimetric radio emission from Jupiter. *Icarus 27*:447-452, 1976.

Stansberry, K.G., and R. S. White. Jupiter's radiation belt. *J. Geophys. Res. 79*:2331-2342, 1974.

Stebbings, R. F., A. C. Smith, and H. Ehrhardt. In *Atomic Collision Processes*, ed. M. R. C. McDowell, p. 814. North-Holland, Amsterdam, 1964.

Stern, D. P. The motion of magnetic field lines. *Space Sci. Rev. 6*:147-173, 1966.

Stern, D. P. Euler potentials. *Amer. J. Phys. 38*:494-501, 1970.

Stern, D. P. Representation of magnetic fields in space. *Rev. Geophys. Space Phys. 14*:199-214, 1976.

Stevenson, D. J. Planetary magnetism. *Icarus 22*:403-415, 1974.

Stix, T. *The Theory of Plasma Waves*. McGraw-Hill, New York, 1962.

Stone, E. C., R. E. Vogt, F. B. McDonald, B. J. Teegarden, J. H. Trainor et al. Cosmic ray investigation for the Voyager mission; energetic particle studies in the outer heliosphere–and beyond. *Space Sci. Rev. 21*:355-376, 1977.

Stone, E. J., and E. C. Zipf. Excitation of the Werner bands of H_2 by electron impact. *J. Chem. Phys. 56*:4646-4650, 1972.

Strobel, D. F. The F2-layer at middle latitudes. *Planet. Space Sci. 18*:1181-1202, 1970.

Strobel, D. F. Aeronomy of the major planets: Photochemistry of ammonia and hydrocarbons. *Rev. Geophys. Space Phys. 13*:372-382, 1975.

Strobel, D. F. The ionosphere of the major planets. *Rev. Geophys. Space Phys. 17*:1913-1922, 1979.

Strobel, D. F., and J. Davis. Properties of the Io plasma torus inferred from Voyager EUV data. *Astrophys. J. 238*:L49-L52, 1980.

Strobel, D. F., and G. R. Smith. On the temperature of the Jovian thermosphere. *J. Atmos. Sci. 30*:718-725, 1973.

Sullivan, J. D., and F. Bagenal. In situ identification of various ionic species in Jupiter's magnetosphere. *Nature 280*:798-799, 1979.

Summers, D., and G. L. Siscoe. Solutions to the equations for corotating magnetospheric convection. *Astrophys. J.*, in press, 1982.

Swartz, W. E., R. W. Reed, and T. R. McDonough. Photoelectron escape from the ionosphere of Jupiter. *J. Geophys. Res. 80*:495-501, 1975.

Teegarden, B. J., F. B. McDonald, J. H. Trainor, W. R. Webber, and E. C. Roelof. Interplanetary MeV electrons of Jovian origin. *J. Geophys. Res. 79*:3615-3622, 1974.

Terasawa, T., K. Maezawa, and S. Machida. Solar wind effect on Jupiter's non-Io-related radio emission. *Nature 273*:131-132, 1978.

Thieman, J. R. Higher resolution studies of Jupiter's decametric radio emissions. Ph.D. thesis, University of Florida, Gainesville, 1977.

Thieman, J. R., A. G. Smith, and J. May. Motion of Jupiter's decametric sources in Io phase. *Astrophys. Lett. 16*:83-86, 1975.

Thomsen, M. F. Jovian magnetosphere-satellite interactions: Aspects of energetic charged particle loss. *Rev. Geophys. Space Phys. 17*:369-387, 1979.

Thomsen, M. F., and C. K. Goertz. Further observational support for the limited-latitude magnetodisc model of the outer Jovian magnetosphere. *J. Geophys. Res. 86*:7519-7526, 1981a.

Thomsen, M. F., and C. K. Goertz. Reply. *J. Geophys. Res. 86*:7798, 1981b.

Thomsen, M. F., C. K. Goertz, and J. A. Van Allen. On determining magnetospheric diffusion coefficients from the observed effects of Jupiter's satellite Io. *J. Geophys. Res. 82*:5541-5550, 1977a.

Thomsen, M. F., C. K. Goertz, and J. A. Van Allen. A determination of the L dependence of the radial diffusion coefficient for protons in Jupiter's inner magnetosphere. *J. Geophys. Res. 82*:3655-3658, 1977b.

Thomsen, M. F., T. G. Northrop, A. W. Schardt, and J. A. Van Allen. Corotation of Saturn's magnetosphere: Evidence from energetic proton anisotropies. *J. Geophys. Res. 85*:5725-5730, 1980.

Thomsen, M. F., and D. D. Sentman. Precipitation fluxes of energetic electrons at Jupiter: An estimated upper limit. *J. Geophys. Res. 84*:1409-1418, 1979.

Thorne, K. S. The theory of synchrotron radiation from stars with dipole magnetic field. *Astrophys. J. Suppl. 8*:1-30, 1963.

Thorne, R. M. Stormtime instabilities of the ring current. In *Magnetosphere Ionosphere Interactions*, ed. K. Folkestad, pp. 185-202. University Press, Oslo, 1972.

Thorne, R. M. A possible cause of dayside relativistic electron precipitation events. *J. Atmos. Terrest. Phys. 36*:635-645, 1974.

Thorne, R. M. The structure and stability of radiation belt electrons as controlled by wave-particle interactions. In *Magnetospheric Particles and Fields*, p. 157-180, ed. B. M. McCormac. Reidel, Dordrecht, 1976.

Thorne, R. M. Jovian auroral secondary electrons and their influence on the Io plasma torus. *Geophys. Res. Lett. 8*:509-512, 1981.

Thorne, R. M., S. R. Church, and D. J. Gorney. On the origin of plasmaspheric hiss: The importance of wave propagation and the plasmapause. *J. Geophys. Res.* *84*:5241-5247, 1979.

Thorne, R. M., and F. V. Coroniti. A self consistent model for Jupiter's radiation belts. In *Proc. of the Jupiter Radiation Belt Workshop*, p. 363-380, ed. A. J. Beck. NASA JPL TM 33-543, Jet Propulsion Laboratory, Pasadena, California, 1972.

Thorne, R. M., and C. F. Kennel. Relativistic electron precipitation during magnetic storm main phase. *J. Geophys. Res.* *76*:4446-4453, 1971.

Thorne, R. M., E. J. Smith, R. K. Burton, and R. E. Holzer. Plasmaspheric hiss. *J. Geophys. Res.* *78*:1581-1596, 1973.

Thorne, R. M., and B. T. Tsurutani. Diffuse Jovian aurora influenced by plasma injection from Io. *Geophys. Res. Lett.* *6*:649-652, 1979.

Tidman, D. A., and N. A. Krall. *Shock Waves in Collisionless Plasmas.* p. 118. Interscience, New York, 1971.

Tokar, R. L., D. A. Gurnett, F. Bagenal, and R. R. Shaw. Light ion concentrations in Jupiter's inner magnetosphere. *J. Geophys. Res.*, in press, 1982.

Torrens, I. M. *Interatomic Potentials.* Academic Press, New York, 1972.

Trafton, L. High-resolution spectra of Io's sodium emission. *Astrophys. J. Lett.* *202*:L107-L112, 1975a.

Trafton, L. Detection of a potassium cloud near Io. *Nature* *258*:690-692, 1975b.

Trafton, L. Periodic variations in Io's sodium and potassium clouds. *Astrophys. J.* *215*:960-970, 1977.

Trafton, L. A survey of Io's potassium cloud. *Bull. Amer. Astron. Soc.* *11*:591-592, 1979.

Trafton, L. Jovian S II torus: Its longitudinal asymmetry. *Icarus* *42*:111-124, 1980.

Trafton, L. The atmospheres of the outer planets and satellites. *Rev. Geophys. Space Phys.* *19*:43-89, 1981.

Trafton, L., and W. Macy. An oscillating asymmetry to Io's sodium emission cloud. *Astrophys. J.* *202*:L155-L158, 1975.

Trafton, L., and W. Macy, Jr. Io's sodium emission profiles: Variations due to Io's phase and magnetic latitude. *Astrophys. J.* *215*:971-976, 1977.

Trafton, L., and W. Macy, Jr. On the distribution of sodium in the vicinity of Io. *Icarus* *33*:322-335, 1978.

Trafton, L., T. Parkinson, and W. Macy, Jr. The spatial extent of sodium emission around Io. *Astrophys. J. Lett.* *190*:L85-L89, 1974.

Trainor, J. H., F. B. McDonald, B. J. Teegarden, W. R. Webber, and E. C. Roelof. Energetic particles in the Jovian magnetosphere. *J. Geophys. Res.* *79*:3600-3613, 1974.

Trauger, J. T., G. Münch, and F. L. Roesler. A study of the Jovian [S II] and [S III] nebulae at high spectral resolution. *Bull. Amer. Astron. Soc.* *11*:591-592, 1979.

Trauger, J. T., G. Münch, and F. L. Rosler. A study of the Jovian [S II] nebula at high spectral resolution. *Astrophys. J.* *236*:1035-1042, 1980.

Tsurutani, B. T., and E. J. Smith. Postmidnight chorus: A substorm phenomena. *J. Geophys. Res.* *79*:118-127, 1974.

Tsurutani, B. T., and E. J. Smith. Two types of magnetospheric ELF chorus and their substorm dependences. *J. Geophys. Res.* *82*:5112-5128, 1977.

Tsytovich, V. N. *Nonlinear Effects in Plasmas.* Plenum Press, New York, 1970.

Twiss, R. Q. Radiation transfer and the possibility of negative absorption in radio astronomy. *Austr. J. Phys.* *11*:564-579, 1958.

Twiss, R. Q., and J. A. Roberts. Electromagnetic radiation from electrons rotating in an ionized medium under the action of a uniform magnetic field. *Austr. J. Phys.* *11*:424-444, 1958.

Tyler, G. L., V. R. Eshleman, J. D. Anderson, G. S. Levy, G. F. Lindal et al. Radio science investigations of the Saturn system with Voyager 1: Preliminary results. *Science 212*:201-206, 1981.

Van Allen, J. A. High energy particles in the Jovian magnetosphere. In *Jupiter*, ed. T. Gehrels, pp. 928-960. University of Arizona Press, Tucson, 1976.

Van Allen, J. A. Energetic electrons in Jupiter's dawn magnetodisc. *Geophys. Res. Lett.* 6:309-312, 1979.

Van Allen, J. A., D. N. Baker, B. A. Randall, and D. D. Sentman. The magnetosphere of Jupiter as observed with Pioneer 10: 1. Instrument and principal findings. *J. Geophys. Res. 79*:3559-3577, 1974a.

Van Allen, J. A., D. N. Baker, B. A. Randall, M. F. Thomsen, D. D. Sentman et al. Energetic electrons in the magnetosphere of Jupiter. *Science 183*:309-311, 1974b.

Van Allen, J. A., G. H. Ludwig, E. C. Ray, and C. E. McIlwain. Observations of high intensity radiation by satellites 1958 alpha and gamma. *Jet Propulsion 28*:588-592, 1958.

Van Allen, J. A., B. A. Randall, D. N. Baker, C. K. Goertz, D. D. Sentman et al. Pioneer 11 observations of energetic particles in the Jovian magnetosphere. *Science 188*:459-462, 1975.

Van Allen, J. A., B. A. Randall, and M. F. Thomsen. Sources and sinks of energetic electrons and protons in Saturn's magnetosphere. *J. Geophys. Res. 85*:5679-5694, 1980.

Vasil'ev, V. P., V. D. Volovik, and I. I. Zalyubovskii. Extensive air showers on Jupiter and its sporadic decameter radio emission. *Sov. Astron. — AJ 16*:337-341, 1972.

Vasyliunas, V. M. Mathematical models of magnetospheric convection and its coupling to the ionosphere. In *Particles and Fields in the Magnetosphere*, ed. B. M. McCormac, pp. 60-71. D. Reidel, Dordrecht, Netherlands, 1970.

Vasyliunas, V. M. The interrelationship of magnetospheric processes. In *The Earth's Magnetospheric Processes*, ed. B. M. McCormac, pp. 29-38. D. Reidel, Dordrecht, Netherlands, 1972a.

Vasyliunas, V. M. Nonuniqueness of magnetic field line motion. *J. Geophys. Res. 77*:6271-6274, 1972b.

Vasyliunas, V. M. Theoretical models of magnetic field-line merging, 1. *Rev. Geophys. Space Phys. 13*:303-336, 1975a.

Vasyliunas, V. M. Modulation of Jovian interplanetary electrons and the longitude variation of decametric emissions. *Geophys. Res. Lett. 2*:87-88, 1975b.

Vasyliunas, V. M. Concepts of magnetospheric convection. In *The Magnetospheres of the Earth and Jupiter*, ed. V. Formisano, pp. 179-188. D. Reidel, Dordrecht, 1975c.

Vasyliunas, V. M. An overview of magnetospheric dynamics. In *Magnetospheric Particles and Fields*, ed. B. M. McCormac, pp. 99-110. D. Reidel, Dordrecht, 1976.

Vasyliunas, V. M. A mechanism for plasma convection in the inner Jovian magnetosphere. Cospar Program/Abstracts, p. 66, Innsbruck, Austria. 29 May-10 June, 1978.

Vasyliunas, V. M. Local time independence of the nightside Jovian magnetosphere, in preparation, 1982a.

Vasyliunas, V. M. A model for the origin of azimuthal magnetic fields in the Jovian magnetosphere, in preparation, 1982b.

Vasyliunas, V. M., and A. J. Dessler. The magnetic-anomaly model of the Jovian magnetosphere: A post-Voyager assessment. *J. Geophys. Res. 86*:8435-8446, 1981.

Veverka, J., L. H. Wasserman, J. Elliot, C. Sagan, and W. Liller. 1974. The occultation of B Scorpii by Jupiter, I, The structure of the Jovian upper atmosphere. *Astron. J. 79*:73-84, 1974.

Vickers, G. T. A self-consistent model of the Jovian plasma sheet. *Planet. Space Sci. 26*:381-385, 1978.

Vogt, R. E., W. R. Cook, A. C. Cummings, T. L. Garrard, N. Gehrels et al. Voyager 1: Energetic ions and electrons in the Jovian magnetosphere. *Science 204*:1003-1007, 1979a.

Vogt, R. E., A. C. Cummings, T. L. Garrard, N. Gehrels, E. C. Stone et al. Voyager 2: Energetic ions and electrons in the Jovian magnetosphere. *Science 206*:984-987, 1979b.

Völk, H. J. Cosmic ray propagation in interplanetary space. *Rev. Geophys. Space Phys. 13*:547-566, 1975.

Walker, R., and M. Kivelson. Multiply reflected standing Alfvén waves in the Io torus: Pioneer 10 observations. *Geophys. Res. Lett. 8*:1281-1284, 1981.

Walker, R. J., M. G. Kivelson, and A. W. Schardt. High-β plasma in the dynamic Jovian current sheet. *Geophys. Res. Lett. 5*:799-802, 1978.

Warwick, J. W. Theory of Jupiter's decametric radio emission. *Ann. N. Y. Acad. Sci. 95*:39-60, 1961.

Warwick, J. W. Dynamic spectra of Jupiter's decametric emission, 1961. *Astrophys. J. 137*:41-60, 1963a.

Warwick, J. W. The position and sign of Jupiter's magnetic moment. *Astrophys. J. 137*:1317-1318, 1963b.

Warwick, J. W. Radio emission from Jupiter. *Ann. Rev. Astron. Astrophys. 2*:1-22, 1964.

Warwick, J. W. Radiophysics of Jupiter. *Space Sci. Rev. 6*:841-891, 1967.

Warwick, J. W. Particles and fields near Jupiter. NASA *CR-1685*, 1970.

Warwick, J. W. Models for Jupiter's decametric arcs. *J. Geophys. Res.*, 86:8585-8592, 1981.

Warwick, J. W., G. A. Dulk, and A. C. Riddle. Jupiter radio emission: January 1960 - March 1975. University of Colorado, Radio Astronomy Observatory Report, 1975.

Warwick, J. W., J. B. Pearce, R. G. Peltzer, and A. Riddle. Radiophysics of Jupiter. *Space Sci. Rev. 21*:309, 1977.

Warwick, J. W., J. B. Pearce, A. C. Riddle, J. K. Alexander, M. D. Desch et al. Voyager 1 planetary radio astronomy observations near Jupiter. *Science 204*:995-998, 1979a.

Warwick, J. W., J. B. Pearce, A. C. Riddle, J. K. Alexander, M. D. Desch et al. Planetary radio astronomy observations from Voyager 2 near Jupiter. *Science 206*:991-995, 1979b.

Warwick, J. W., J. B. Pearce, D. R. Evans, T. D. Carr, J. J. Schauble et al. Planetary radio astronomy observations from Voyager 1 near Saturn. *Science 212*:239-243, 1981.

Weibel, E. S. Spontaneously growing transverse waves in a plasma due to an anisotropic velocity distribution. *Phys. Rev. Lett. 2*:83-84, 1959.

Werner, M. G., G. Neugebauer, J. R. Houck, and M. G. Hauser. One-millimeter brightness temperatures of the planets. NASA TM-74989, 1978.

Whiteoak, J. B., F. f. Gardner, and D. Morris. Jovian linear polarization at 6 cm wavelength. *Astrophys. Lett. 3*:81-84, 1969.

Williams, D. J. The ion-electron magnetic separation and solid state detector detection system flown on IMP-7 and 8; $E_p \geq 50$ keV, $E_e \geq 30$ keV. *NOAA Tech. Rep. ERL-393, SEL-40*, 1977.

Williams, D. J. Ring current composition and sources. In *Dynamics of the Magnetosphere*, ed. S.-I. Akasofu, pp. 407-424. D. Reidel, Dordrecht, 1979.

Wolf, R. A. Calculations of magnetospheric electric fields. In *Magnetospheric Physics*, ed. B. M. McCormac, pp. 167-177. D. Reidel, Dordrecht, Netherlands, 1974.

Wolf, R. A. Ionosphere-magnetosphere coupling. *Space Sci. Rev. 17*:537-562, 1975.

Wolfe, J. H., H. R. Collard, J. D. Mihalov, and D. S. Intriligator. Preliminary Pioneer 10 encounter results from the Ames Research Center plasma analyzer experiment. *Science 183*:303-305, 1974.

Wu, C. S. Unified quasilinear theory of weakly turbulent plasmas. *Phys. Fluids 11*:1733-1744, 1968.

Wu, C. S., and H. P. Freund. Induced emission of Jupiter's decametric radiation by Io-accelerated electrons. *Astrophys. J. 213*:575-587, 1977.

Wu, C. S., and L. C. Lee. A theory of the terrestrial kilometric radiation. *Astrophys. J. 230*:621-626, 1979.

Young, T. S. T., J. D. Cullen, and J. E. McCune. High frequency electrostatic waves in the magnetosphere. *J. Geophys. Res. 78*:1082-1099, 1973.

Yung, Y. L., and D. F. Strobel. Hydrocarbon photochemistry and Lyman-alpha albedo of Jupiter. *Astrophys. J. 239*:395-402, 1980.

Yung, Y. L., G. R. Gladstone, K. M. Chang, J. M. Ajello, and S. K. Srivastara. H_2 fluorescence spectrum from 1200 to 1700 A by electron impact: Laboratory study and application to Jovian aurora. *Astrophys. J. Lett., 254*:L65-L69, 1982.

Zabriskie, F. R. Low-frequency radio emission from Jupiter. *Astron. J. 75*:1045-1051, 1970.

Zheleznyakov, V. V. The origin of Jovian radio emission. *Sov. Astron. — AJ 9*:617-625, 1966.

Zwickl, R. D., S. M. Krimigis, T. P. Armstrong, and L. J. Lanzerotti. Interplanetary Jovian ion events observed by Voyagers 1 and 2. *Geophys. Res. Lett. 7*:453-456, 1980.

Zwickl, R. D., S. M. Krimigis, J. F. Carbary, E. P. Keath, T. P. Armstrong et al. Energetic particles ($\geq$ 30 keV) of Jovian origin observed by Voyager 1 and 2 in interplanetary space. *J. Geophys. Res. 86*:8125-8140, 1981.

Zwickl, R. D., E. C. Roelof, R. E. Gold, S. M. Krimigis, and T. P. Armstrong. Z-rich solar particle event characteristics 1972-1976. *Astrophys. J. 225*:281-303, 1978.

INDEX

DATE DUE